Design and Construction of Modern Steel Railway Bridges

Design and Construction of Modern Steel Railway Bridges

Second Edition

By

John F. Unsworth

CRC Press
Taylor & Francis Group
Boca Raton London New York

CRC Press is an imprint of the
Taylor & Francis Group, an **informa** business

CRC Press
Taylor & Francis Group
6000 Broken Sound Parkway NW, Suite 300
Boca Raton, FL 33487–2742

Printed on acid-free paper

International Standard Book Number-13: 978-1-4987-3410-3 (Hardback)

Library of Congress Cataloging-in-Publication Data

Names: Unsworth, John F.
Title: Design and construction of modern steel railway bridges, / John F. Unsworth.
Description: Second edition. | Boca Raton : Taylor & Francis, CRC Press, 2017. | Includes bibliographical references and index.
Identifiers: LCCN 2017006499| ISBN 9781498734103 (hardback) | ISBN 9781315120775 (ebk)
Subjects: LCSH: Railroad bridges--Design and construction. | Iron and steel bridges.
Classification: LCC TG445 .U57 2017 | DDC 624.2-- dc23
LC record available at https://lccn.loc.gov/2017006499

Visit the Taylor & Francis Web site at
http://www.taylorandfrancis.com

and the CRC Press Web site at
http://www.crcpress.com

Dedication

To my extraordinary wife, Elizabeth, whose steadfast support and patience made this book and most other good things in my life possible.

Contents

Acknowledgment xv
Preface to the Second Edition xvii
Author xix

Chapter 1 History and Development of Steel Railway Bridges 1

1.1 Introduction 1
1.2 Iron Railway Bridges 2
1.2.1 Cast Iron Construction 2
1.2.2 Wrought Iron Construction 8
1.3 Steel Railway Bridges 23
1.4 The Development of Railway Bridge Engineering 31
1.4.1 Strength of Materials and Structural Mechanics 31
1.4.2 Railway Bridge Design Specifications 32
1.4.3 Modern Steel Railway Bridge Design 34
Bibliography 36

Chapter 2 Steel for Modern Railway Bridges 39

2.1 Introduction 39
2.2 Manufacture of Structural Steel 39
2.3 Engineering Properties of Steel 42
2.3.1 Strength 42
2.3.1.1 Elastic Yield Strength of Steel 42
2.3.1.2 Fatigue Strength of Steel 44
2.3.2 Ductility 45
2.3.3 Fracture Resistance 45
2.3.4 Weldability 46
2.3.5 Corrosion Resistance 46
2.4 Types of Structural Steel 47
2.4.1 Carbon Steels 47
2.4.2 High-Strength Low-Alloy Steels 47
2.4.3 Heat-Treated Low-Alloy Steels 48
2.4.4 High-Performance Steels 48
2.5 Structural Steel for Railway Superstructures 49
2.5.1 Material Properties 49
2.5.2 Structural Steel for Modern North American Railway Superstructures 50
References 54

Chapter 3 Planning and Preliminary Design of Modern Steel Railway Bridges 55

3.1 Introduction 55
3.2 Planning of Railway Bridges 56
3.2.1 Bridge Crossing Economics 56
3.2.2 Railroad Operating Requirements 57

3.2.3 Site Conditions (Public and Technical Requirements of Bridge Crossings) ... 58
3.2.3.1 Regulatory Requirements ... 58
3.2.3.2 Hydrology and Hydraulics of the Bridge Crossing ... 58
3.2.3.3 Highway, Railway, and Marine Clearances ... 69
3.2.3.4 Geotechnical Conditions ... 69
3.2.4 Geometry of the Track and Bridge ... 70
3.2.4.1 Horizontal Geometry of the Bridge ... 71
3.2.4.2 Vertical Geometry of the Bridge ... 82
3.3 Preliminary Design of Steel Railway Bridges ... 82
3.3.1 Bridge A Esthetics ... 82
3.3.2 Steel Railway Bridge Superstructures ... 83
3.3.2.1 Bridge Decks for Steel Railway Bridges ... 84
3.3.2.2 Bridge Framing Details ... 87
3.3.2.3 Bridge Bearings ... 88
3.3.3 Bridge Stability ... 90
3.3.4 Pedestrian Walkways ... 90
3.3.5 General Design Criteria ... 91
3.3.5.1 Structural Analysis for Modern Steel Superstructure Design ... 91
3.3.5.2 Structural Design for Modern Steel Superstructure Fabrication ... 92
3.3.6 Fabrication Considerations ... 100
3.3.7 Erection Considerations ... 102
3.3.8 Detailed Design of the Superstructure ... 102
References ... 102

Chapter 4 Loads and Forces on Steel Railway Bridges ... 105

4.1 Introduction ... 105
4.2 Dead Loads ... 105
4.3 Railway Live Loads ... 106
4.3.1 Static Freight Train Live Load ... 106
4.3.1.1 Cooper's Design Live Load for Projected Railway Equipment ... 115
4.3.1.2 Fatigue Design Live Load for Railway Equipment ... 118
4.3.2 Dynamic Freight Train Live Load ... 125
4.3.2.1 Rocking and Vertical Dynamic Forces ... 125
4.3.2.2 Design Impact Load ... 144
4.3.2.3 Longitudinal Forces due to Traction and Braking ... 145
4.3.2.4 Centrifugal Forces ... 155
4.3.2.5 Lateral Forces from Moving Freight Equipment ... 159
4.3.3 Distribution of Live Load ... 159
4.3.3.1 Distribution of Live Load for Open Deck Steel Bridges ... 159
4.3.3.2 Distribution of Live Load for Ballasted Deck Steel Bridges ... 160
4.3.3.3 Distribution of Live Load for Direct Fixation Deck Steel Bridges ... 162
4.4 Environmental and Other Steel Railway Bridge Design Forces ... 164
4.4.1 Wind Forces on Steel Railway Bridges ... 164

4.4.2 Thermal Forces from Continuous Welded Rail on Steel Railway Bridges 170
4.4.2.1 Safe Rail Separation Criteria 172
4.4.2.2 Safe Stress in the CWR to Preclude Buckling 173
4.4.2.3 Acceptable Relative Displacement between Rail-to-Deck and Deck-to-Span 175
4.4.2.4 Design for CWR on Steel Railway Bridges 186
4.4.3 Seismic Forces on Steel Railway Bridges 187
4.4.3.1 Equivalent Static Lateral Force 187
4.4.3.2 Response Spectrum Analysis of Steel Railway Superstructures 188
4.4.4 Loads Relating to Overall Stability of the Superstructure 190
4.4.4.1 Derailment Load 190
4.4.4.2 Other Loads for Overall Lateral Stability 192
4.4.5 Pedestrian Loads 192
4.5 Load and Force Combinations for Design of Steel Railway Superstructures 193
References 194

Chapter 5 Structural Analysis and Design of Steel Railway Bridges 197

5.1 Introduction 197
5.2 Structural Analysis of Steel Railway Superstructures 197
5.2.1 Live Load Analysis of Steel Railway Superstructures 197
5.2.1.1 Maximum Shear Force and Bending Moment due to Moving Concentrated Loads on Simply Supported Spans 199
5.2.1.2 Influence Lines for Maximum Effects of Moving Loads on Superstructures 213
5.2.1.3 Equivalent Uniform Loads for Maximum Shear Force and Bending Moment in Simply Supported Spans 234
5.2.1.4 Maximum Shear Force and Bending Moment in Simply Supported Spans from Equations and Tables 244
5.2.1.5 Modern Structural Analysis 245
5.2.2 Lateral Load Analysis of Steel Railway Superstructures 246
5.2.2.1 Lateral Bracing Systems 246
5.3 Structural Design of Steel Railway Superstructures 260
5.3.1 Failure Modes of Steel Railway Superstructures 261
5.3.2 Steel Railway Superstructure Design 262
5.3.2.1 Strength Design 262
5.3.2.2 Serviceability Design 264
5.3.2.3 Other Design Criteria for Steel Railway Bridges 273
References 274

Chapter 6 Design of Axial Force Steel Members 277

6.1 Introduction 277
6.2 Axial Tension Members 277
6.2.1 Strength of Axial Tension Members 277
6.2.1.1 Net Area, A_n, of Tension Members 278
6.2.1.2 Effective Net Area, A_e, of Tension Members 279
6.2.2 Fatigue Strength of Axial Tension Members 282
6.2.3 Serviceability of Axial Tension Members 284

6.2.4 Design of Axial Tension Members for Steel Railway Bridges......... 289
6.3 Axial Compression Members 291
6.3.1 Strength of Axial Compression Members 291
6.3.1.1 Elastic Compression Members 291
6.3.1.2 Inelastic Compression Members 296
6.3.1.3 Yielding of Compression Members 301
6.3.1.4 Compression Member Design for Steel Railway Superstructures 302
6.3.2 Serviceability of Axial Compression Members 302
6.3.3 Axial Compression Members in Steel Railway Superstructures 304
6.3.3.1 Buckling Strength of Built-Up Compression Members 304
References 325

Chapter 7 Design of Flexural Steel Members 327

7.1 Introduction 327
7.2 Strength Design of Noncomposite Flexural Members 327
7.2.1 Bending of Laterally Supported Beams and Girders 327
7.2.2 Bending of Laterally Unsupported Beams and Girders 329
7.2.3 Shearing of Beams and Girders 333
7.2.3.1 Shearing of Rectangular Beams 333
7.2.3.2 Shearing of I-Shaped Sections 335
7.2.3.3 Design for Shearing of Shapes and Plate Girders 336
7.2.4 Biaxial Bending of Beams and Girders 336
7.2.5 Preliminary Design of Beams and Girders 337
7.2.6 Plate Girder Design 339
7.2.6.1 Main Girder Elements 339
7.2.6.2 Secondary Girder Elements 356
7.2.7 Box Girder Design 360
7.2.7.1 Steel Box Girders 360
7.2.7.2 Steel–Concrete Composite Box Girders 360
7.3 Serviceability Design of Noncomposite Flexural Members 360
7.4 Strength Design of Steel and Concrete Composite Flexural Members 374
7.4.1 Flexure in Composite Steel and Concrete Spans 376
7.4.2 Shearing of Composite Beams and Girders 378
7.4.2.1 Web Plate Shear 378
7.4.2.2 Shear Connection between Steel and Concrete 379
7.5 Serviceability Design of Composite Flexural Members 381
References 397

Chapter 8 Design of Steel Members for Combined Forces 399

8.1 Introduction 399
8.2 Biaxial Bending 399
8.3 Unsymmetrical Bending (Combined Bending and Torsion) 400
8.4 Combined Axial Forces and Bending of Members 413
8.4.1 Axial Tension and Uniaxial Bending 413
8.4.2 Axial Compression and Uniaxial Bending 414
8.4.2.1 Differential Equation for Axial Compression and Bending in a Simply Supported Beam 415
8.4.2.2 Interaction Equations for Axial Compression and Uniaxial Bending 420

8.4.3 Axial Compression and Biaxial Bending ... 423
8.4.4 AREMA Recommendations for Combined Axial Compression and Biaxial Bending ... 423
8.5 Combined Bending and Shear of Plates ... 424
References ... 424

Chapter 9 Design of Connections for Steel Members ... 425
9.1 Introduction ... 425
9.2 Welded Connections ... 425
9.2.1 Welding Processes for Steel Railway Bridges ... 427
9.2.1.1 Shielded Metal Arc Welding ... 427
9.2.1.2 Submerged Arc Welding ... 427
9.2.1.3 Flux Cored Arc Welding ... 427
9.2.1.4 Stud Welding ... 427
9.2.1.5 Welding Electrodes ... 427
9.2.2 Weld Types ... 428
9.2.2.1 Groove Welds ... 428
9.2.2.2 Fillet Welds ... 429
9.2.3 Joint Types ... 429
9.2.4 Welded Joint Design ... 430
9.2.4.1 Allowable Weld Stresses ... 430
9.2.4.2 Fatigue Strength of Welds ... 431
9.2.4.3 Weld Line Properties ... 431
9.2.4.4 Direct Axial Loads on Welded Connections ... 432
9.2.4.5 Eccentrically Loaded Welded Connections ... 436
9.2.4.6 Girder Flange to Web "T" Joints ... 445
9.3 Bolted Connections ... 446
9.3.1 Bolting Processes for Steel Railway Superstructures ... 446
9.3.1.1 Snug-Tight Bolt Installation ... 446
9.3.1.2 Pretensioned Bolt Installation ... 446
9.3.1.3 Slip-Critical Bolt Installation ... 446
9.3.2 Bolt Types ... 447
9.3.2.1 Common Steel Bolts ... 447
9.3.2.2 High-Strength Steel Bolts ... 448
9.3.3 Joint Types ... 448
9.3.4 Bolted Joint Design ... 448
9.3.4.1 Allowable Bolt Stresses ... 448
9.3.4.2 Axially Loaded Members with Bolts in Shear ... 459
9.3.4.3 Eccentrically Loaded Connections with Bolts in Shear and Tension ... 471
9.3.4.4 Axially Loaded Connections with Bolts in Direct Tension ... 480
9.3.4.5 Axial Member Splices ... 482
9.3.4.6 Beam and Girder Splices ... 483
References ... 495

Chapter 10 Construction of Steel Railway Bridges: Superstructure Fabrication ... 497
10.1 Introduction ... 497
10.2 Fabrication Planning ... 498
10.2.1 Project Cost Estimating ... 498

10.2.2 Shop Drawings for Steel Fabrication ... 498
10.2.3 Fabrication Shop Production Scheduling and Detailed Cost Estimating ... 499
10.2.4 Material Procurement for Fabrication ... 500
10.3 Steel Fabrication Processes ... 502
10.3.1 Material Preparation ... 502
10.3.1.1 Layout and Marking of Plates and Shapes ... 502
10.3.1.2 Cutting of Plates and Shapes ... 502
10.3.1.3 Straightening, Bending, Curving, and Cambering of Plates and Shapes ... 505
10.3.1.4 Surface Preparation ... 505
10.3.1.5 Heat Treatment ... 506
10.3.2 Punching and Drilling of Plates and Shapes ... 508
10.3.2.1 Hole Quality ... 508
10.3.2.2 Punching and Drilling Accuracy for Shop and Field Fasteners ... 509
10.3.3 Shop Assembly for Fit-Up of Steel Plates and Shapes ... 509
10.3.3.1 Fabrication of Cambered Superstructure Assemblies ... 509
10.3.3.2 Shop Assembly of Longitudinal Beams, Girders, and Trusses ... 511
10.3.3.3 Progressive Shop Assembly of Longitudinal Beams, Girders, and Trusses ... 512
10.3.3.4 Shop Assembly of Bolted Splices and Connections ... 512
10.3.3.5 Fit-Up for Shop Welded Splices and Connections ... 514
10.3.3.6 Fabrication and Erection Tolerances ... 514
10.4. Bolting of Plates and Shapes ... 518
10.5 Welding of Plates and Shapes ... 520
10.5.1 Shop Welding Processes ... 520
10.5.2 Shop Welding Procedures ... 523
10.5.3 Effects of Welding on Plates and Shapes ... 524
10.5.3.1 Welding Flaws ... 524
10.5.3.2 Welding-Induced Cracking ... 525
10.5.3.3 Welding-Induced Distortion ... 526
10.5.3.4 Welding-Induced Residual Stresses ... 526
10.5.3.5 Welding-Induced Lamellar Tearing ... 528
10.6 Coating of Steel Plates and Shapes for Railway Superstructures ... 529
10.7 QC and QA of Fabrication ... 530
10.7.1 QC Inspection of Fabrication ... 530
10.7.2 QA Inspection of Fabrication ... 532
10.7.2.1 Shop or Detail Drawing Review ... 532
10.7.2.2 Inspection of Raw Materials ... 532
10.7.2.3 Inspection of Fabricated Members ... 532
10.7.2.4 Assembly Inspection ... 533
10.7.2.5 Bolting Inspection ... 534
10.7.2.6 Welding Inspection ... 534
10.7.2.7 Coatings Inspection ... 535
10.7.2.8 Final Inspection for Shipment ... 536
10.7.3 NDT for QC and QA Inspection of Welded Fabrication ... 536
10.7.3.1 Dye-Penetrant Testing ... 536
10.7.3.2 Magnetic Particle Testing (Figure 10.25) ... 536
10.7.3.3 Ultrasonic Testing (Figure 10.26) ... 536

10.7.3.4 Phased Array Ultrasonic Testing ... 537
10.7.3.5 Radiographic Testing (Figure 10.27) ... 537
Bibliography ... 537

Chapter 11 Construction of Steel Railway Bridges: Superstructure Erection ... 539

11.1 Introduction ... 539
11.2 Erection Planning ... 540
11.2.1 Erection Methods and Procedures Planning ... 541
11.2.2 Erection Methods and Equipment Planning ... 544
11.2.2.1 Erection with Cranes and Derricks ... 544
11.2.2.2 Erection on Falsework and Lateral Skidding of Superstructures ... 551
11.2.2.3 Erection by Flotation with Barges ... 554
11.2.2.4 Erection with Stationary and Movable Frames ... 556
11.2.2.5 Other Erection Methods ... 559
11.3 Erection Engineering ... 566
11.3.1 Erection Engineering for Member Strength and Stability ... 567
11.3.2 Erection Engineering for Cranes and Derricks ... 575
11.3.2.1 Stationary Derricks ... 575
11.3.2.2 Mobile Cranes ... 575
11.3.3 Erection Engineering for Falsework ... 580
11.3.4 Erection Engineering for Cranes, Derricks, and Falsework on Barges ... 582
11.3.5 Erection Engineering for Stationary and Movable Frames ... 584
11.3.6 Engineering for Other Erection Methods ... 585
11.3.6.1 Erection Engineering for Launching ... 585
11.3.6.2 Erection Engineering for Cantilever Construction ... 586
11.3.6.3 Engineering for Tower and Cable, and Catenary High-Line Erection ... 586
11.3.6.4 Engineering for SPMT Erection ... 586
11.4 Erection Execution ... 587
11.4.1 Erection by Mobile Cranes ... 588
11.4.2 Falsework Construction ... 588
11.4.3 Erection Fit-Up ... 589
11.4.4 Erection of Field Splices and Connections ... 590
11.4.4.1 Welded Field Splices and Connections ... 590
11.4.4.2 Bolted Field Splices and Connections ... 590
11.4.5 Field Erection Completion ... 592
Bibliography ... 593

Appendix A: Design of a Ballasted through Plate Girder (BTPG) Superstructure ... 595

Appendix B: Design of a Ballasted Deck Plate Girder (BDPG) Superstructure ... 635

Appendix C: Units of Measurement ... 669

Index ... 673

Acknowledgment

I respectfully acknowledge my father, who provided opportunities for an early interest in science and engineering, and my mother, whose unconditional support in all matters has been acutely treasured. My wife, Elizabeth, deserves singular recognition for everything she does, and the kind and thoughtful way in which she does it. She, my daughters, Tiffany and Genevieve, and granddaughter, Johanna, are perpetual sources of joy and inspiration. In addition, the professional guidance and friendship provided by many esteemed colleagues, in particular Dr. R. A. P. Sweeney, W. G. Byers, and W. B. Conway, must be gratefully acknowledged.

Preface to the Second Edition

The first edition of this book provided nine chapters with a focus on the design of new steel superstructures for modern railway bridges referencing the recommended practices of Chapter 15—Steel Structures in the 2008 edition of the American Railway Engineering and Maintenance-of-way Association (AREMA) Manual for Railway Engineering (MRE). This second edition updates the first edition by including changes precipitated by subsequent revisions to Chapter 15 of the MRE and the many valuable comments received from steel railway bridge design engineers, fabricators, students, academics, and researchers. Grateful appreciation is due to all those who offered comments and suggestions and, in particular, to the many members of AREMA Committee 15—Steel Structures, whose experience and expertise, helped improve the technical content of this second edition.

A notable amendment to this second edition is the use of Système Internationale (SI) units in addition to US Customary or Imperial units throughout* the book. Moreover, in this second edition, attention has been expanded to include two new chapters on the construction (fabrication and erection) of new steel superstructures for modern railway bridges.

This second edition is divided into eleven chapters. The first three chapters deliver introductory and general information as a foundation for the subsequent six chapters examining the detailed analysis and design, which precede two chapters concerning fabrication and erection, of modern steel railway superstructures.

Chapter 1 is retained as a brief history of iron and steel railway bridges. The chapter concludes with the evolution and advancement of structural mechanics and design practice precipitated by steel railway bridge development.

A discussion regarding the manufacture of structural steel (steel making) has been included in Chapter 2 as a prelude to the material concerning the engineering properties and types of structural steel used in modern railway superstructure design and fabrication.

The information in Chapter 3 concerning the planning of steel railway bridges is enhanced with additional material regarding bridge scour investigation in accordance with AREMA (2015). Added to the discussion about preliminary design is a brief introduction to probabilistic structural design in terms of modern steel railway superstructure design issues.

The next two chapters concerning the development of loads and structural analysis of modern steel railway bridge superstructures have been substantially updated.

The discussion of railway live loads in Chapter 4 is enhanced with a discussion of the historical development of modern freight train design live loads. Material concerning the fatigue design load, which was included in Chapter 5 of the first edition, is now more appropriately included in Chapter 4 as it specifically relates to the modern freight train design live load. The discussion of the freight train live load in Chapter 4 has been extensively revised using an approach originating with modern vehicle–bridge interaction (VBI) dynamics concepts. The VBI models are reduced to dynamic moving sprung mass, mass and force problems to examine the theoretical foundations of railway live load impact. The load combination table at the end of Chapter 4 has been updated based on thoughtful review by many members of AREMA Committee 15. Chapter 5 now also includes material concerning the lateral deflection of steel superstructures based on recent revisions to AREMA Chapter 15 that provide for better control of track geometry.† The material concerning fatigue strength or resistance remains in Chapter 5.

* In a very few cases only, US Customary or Imperial units are used. This typically occurs when equations or formulas are developed empirically in US Customary or Imperial units.

† As a safety measure, as train speeds increase, track geometry tolerances decrease (become more stringent).

Chapters 6–9 remain to outline the design of members and connections in accordance with AREMA (2015). The postbuckling shear strength of plate girder web plates is not included in AREMA (2015) due to intolerance of such behavior in railway superstructures. Nevertheless, a brief introduction to plate girder web plate postbuckling strength is included in Chapter 7 of the second edition as information concerning the ultimate behavior of plate girder superstructures. Chapter 9 includes updated information regarding the design shear strength of slip-critical connections. Other information in these four chapters has also been updated in accordance with the applicable revisions to AREMA Chapter 15 since the first edition of this book.

Chapters 10 and 11 are new to the second edition. Much of the subject matter considered in Chapters 2, 5, 6, 7, and 9 is affected by fabrication. Consequently, and because steel superstructure design ultimately culminates in fabrication, Chapter 10 concerning the planning, processes, execution, and inspection of fabricated members and assemblies has been incorporated. Since, steel superstructure erection logically trails fabrication and concludes the project, Chapter 11 outlining some typical practices of steel railway superstructure erection planning, equipment, engineering, and execution, follows Chapter 10 to conclude the book.

Appendices outlining the design of a ballasted through plate girder (BTPG) and a ballasted deck plate girder (BDPG) superstructure are included in the second edition to complement the material presented in the book. An appendix has also been included as a précis of the common engineering unit conversions used in the book. Conversions between SI and US Customary or Imperial units and vice versa are presented.

This second edition remains as only one constituent of the information essential for the design and construction of safe and reliable modern steel railway superstructures. Other sources of technical information are also necessary and, again, it is anticipated that, where such material is referenced in this book, proper attribution has been appropriately expressed.

John F. Unsworth
Cochrane, Alberta, Canada

Author

John F. Unsworth is a professional engineer (P Eng). Since his completion of a bachelor of engineering degree in civil engineering and a master of engineering degree in structural engineering, he has held professional engineering and management positions concerning track, bridge, and structures maintenance, design, and construction at the Canadian Pacific Railway.

He is a former president of the American Railway Engineering and Maintenance-of-way Association (AREMA) and has served as Chairman of AREMA Committee 15—Steel Structures. He currently serves as an emeritus member of AREMA Committee 15. In addition, he is the current Chair of the Association of American Railroads (AAR) Bridge Research Advisory Group and is a former member of the National Academy of Sciences Transportation Research Board (TRB) Steel Bridges Committee. He is also a member of the Canadian Society for Civil Engineering (CSCE), the American Society of Civil Engineers (ASCE), the American Institute of Steel Construction (AISC), and the International Association of Bridge and Structural Engineers (IABSE). He is a licensed professional engineer in six Canadian provinces.

He has written papers and presented them at AREMA Annual Technical Conferences, the International Conference on Arch Bridges, TRB Annual Meetings, CSCE Bridge Conferences, the ASCE Structures Congress, and the International Bridge Conference (IBC). He has also contributed to the fifth edition of the *Structural Steel Designer's Handbook* and the *International Heavy Haul Association (IHHA) Best Practices* books.

1 History and Development of Steel Railway Bridges

1.1 INTRODUCTION

The need for reliable transportation systems evolved with the industrial revolution. By the early 19th century, it was necessary to transport materials, finished goods, and people over greater distances in shorter times. These societal requirements, in conjunction with the development of steam power,[*] heralded the birth of the railroad. The steam locomotive with a trailing train of passenger or freight cars on iron rails became the principal means of transportation. Accordingly, as transportation improvements were required, the railroad industry became the primary catalyst in the evolution of materials and engineering mechanics in the latter half of the 19th century.

The railroad revolutionized the 19th century. Railroad transportation commenced in the UK on the Stockton to Darlington Railway in 1823 and on the Liverpool and Manchester Railway in 1830. The first commercial railroad in the United States was the Baltimore and Ohio (B&O) Railroad, which was chartered in 1827.

Construction of the associated railroad infrastructure required that a great many wood, masonry, and metal bridges be built. Bridges were required for live loads that had not been previously encountered by bridge builders.[†] The first railroad bridge in the United States was a wooden arch-stiffened truss built by the B&O in 1830. Rapid railroad expansion[‡] and increasing locomotive weights, particularly in the United States following the Civil War, provoked a strong demand for longer and stronger railway bridges. In response, many metal girder, arch, truss, and suspension bridges were built to accommodate railroad expansion, which was occurring simultaneously in the United States and the UK following the British industrial revolution.

In the United States, there was intense competition among emerging railroad companies to expand west. Nevertheless, crossing the Mississippi River was the greatest challenge to planned railroad growth. The first railway bridge across the Mississippi River was completed in 1856 by the Chicago, Rock Island, and Pacific Railroad.[§] The efforts of the B&O Railroad company to expand its business and to cross the Mississippi River at St. Louis, MO, commencing in 1839[¶] and finally realized in 1874, proved to be a milestone in steel railway bridge design and construction. Although the St. Louis Bridge[**] never served the volume of the railway traffic anticipated in 1869 at the start of construction, its engineering involved many innovations that provided the foundation for long-span railway bridge design for many years following its completion in 1874.

---

[*] Nicolas Cugnot is credited with production of the first steam-powered vehicle in 1769. Small steam-powered industrial carts and trams were manufactured in the UK in the early years of the 19th century and George Stephenson built the first steam locomotive, the "Rocket," for use on the Liverpool and Manchester Railway in 1829.

[†] Before steam locomotives, bridges carried primarily pedestrian, equestrian, and light cart traffic. Railroad locomotive axle loads were about 50 kN (11,000 lbs) on the Baltimore and Ohio Railroad in 1835.

[‡] For example, in the 1840s charters to hundreds of railway companies were issued by the British government.

[§] The bridge was constructed by the Rock Island Bridge Company after US railroad companies received approval to construct bridges across navigable waterways. The landmark Supreme Court case that enabled construction of this bridge across the Mississippi River also provided national exposure to the Rock Island Bridge Company solicitor, Abraham Lincoln.

[¶] In 1849, Charles Ellet, who designed the ill-fated suspension bridge at Wheeling, WV, was the first engineer to develop preliminary plans for a railway suspension bridge to cross the Mississippi River at St. Louis, MO. Costs were considered prohibitive, as were subsequent suspension bridge proposals by J. A. Roebling, and the project was never commenced.

[**] Now known as the Eads Bridge in honor of its builder, James Eads.

The need for longer and stronger railway bridges precipitated an evolution of materials from wood and masonry to cast and wrought iron, and eventually to steel. Many advances and innovations in engineering mechanics and construction technology can also be attributed to the development of the railroads and their need for more robust bridges of greater span.

1.2 IRON RAILWAY BRIDGES

1.2.1 Cast Iron Construction

A large demand for railway bridges was generated as railroads in the UK and the United States prospered and expanded. Masonry and timber were the principal materials of early railway bridge construction, but new materials were required to span the greater distances and carry the heavier loads associated with railroad expansion. Cast iron had been used in 1779 for the construction of the first metal bridge, a 30.5 m (100 ft) arch span over the Severn River at Coalbrookdale, UK. The first bridge to use cast iron in the United States was the 24.5 m (80 ft) arch, built in 1839, at Brownsville, PA. Cast iron arches* were some of the first metal railway bridges constructed, and their use expanded with the rapidly developing railroad industry. Table 1.1 indicates some notable cast iron arch railway bridges constructed between 1847 and 1861.

The oldest cast iron railway bridge in existence is the 14 m (47 ft) trough girder at Merthyr Tydfil in South Wales, which was built in 1793 to carry an industrial rail tram. The first iron railway bridge for use by the general public on a chartered railroad was built in 1823 by George Stephenson on the Stockton to Darlington Railway (Figure 1.1). The bridge consisted of 3.8 m (12.5 ft) long lenticular spans† in a trestle arrangement. This early trestle was a precursor to the many trestles that would be constructed by railroads to enable almost level crossings of wide and/or deep valleys. Table 1.2 summarizes some notable cast iron railway trestles constructed between 1823 and 1860.

George Stephenson's son, Robert, and Isambard Kingdom Brunel were British railway engineers, who understood cast iron material behavior and the effects of moving railway loads on arches. They successfully built cast iron arch bridges that were designed to act in compression. However, the

FIGURE 1.1 Gaunless River Bridge of the Stockton and Darlington Railway built in 1825 at West Auckland, UK. (From Chris Lloyd, The Northern Echo, Darlington.)

---

* Cast iron bridge connections were made with bolts because the brittle cast iron would crack under pressures exerted by rivets as they shrank from cooling.

† Also referred to as Pauli spans.

TABLE 1.1
Notable Iron and Steel Arch Railway Bridges Constructed during 1847–1916

Location	Railroad	Engineer	Year	Material	Hinges	Span m (ft)
Thirsk, UK	Leeds & Thirsk	—	1847	Cast iron	0	—
Newcastle, UK	Northeastern	R. Stephenson	1849	Cast iron	0	38 (125)
Oltwn, Switzerland	Swiss Central	Etzel & Riggenbach	1853	Wrought iron	0	31 (103)
Paris, France	Paris-Aire	—	1854	Wrought iron	2	45 (148)
Victoria, Bewdley, UK	—	J. Fowler	1861	Cast iron	—	
Albert, UK	—	J. Fowler	1861	Cast iron	—	
Coblenz, Germany	—	—	1864	Wrought iron	2	—
Albert, Glasgow, Scotland	—	Bell & Miller	1870	Wrought iron	—	—
St. Louis, MO	Various	T. Cooper & J. Eads	1874	Cast steel	0	159 (520)
Garabit, France		G. Eiffel	1884	Wrought iron	2	165 (540)
Paderno, Italy	—	—	1889	Iron	—	150 (492)
Stony Creek, BC	Canadian Pacific	H. E. Vautelet	1893	Steel	3	102 (336)
Keefers, Salmon River, BC	Canadian Pacific	H. E. Vautelet	1893	Steel	3	82 (270)
Surprise Creek, BC	Canadian Pacific	H. E. Vautelet	1897	Steel	3	88 (290)
Grunenthal, Germany	—	—	1892	Steel	2	156 (513)
Levensau, Germany	—		1894	Steel	2	163 (536)
Mungsten, Prussia	—	A. Rieppel	1896	Steel	0	170 (558)
Niagara Gorge[a], NY	—	L. L. Buck	1897	Steel	2	168 (550)
Viaur Viaduct, France	—	—	1898	Steel	0	220 (721)
Worms, Germany	—	Schneider & Frintzen	1899	Steel	—	66 (217)
Yukon, Canada	Whitepass & Yukon	—	—	Steel	0	73 (240)
Passy Viaduct, France	Western Railway of Paris	—	—	Steel	—	86 (281)

(Continued)

TABLE 1.1 (*Continued*)
Notable Iron and Steel Arch Railway Bridges Constructed during 1847–1916

Location	Railroad	Engineer	Year	Material	Hinges	Span m (ft)
Rio Grande, Costa Rica	Narrow gage	—	1902	Steel	2	137 (448)
Birmingham, AL	Cleveland & Southwestern Traction	—	1902	Steel	—	—
Kingsford, WI	Chicago, Milwaukee & St. Paul	—	1902	Steel	3	63 (207)
Mainz, Germany	—	—	1904	Steel	—	—
Estacada, OR	Oregon, Water, Power and Railway	G. Brown	1904	Steel	—	61 (200)
Paris, France	Metropolitan	—	1905	Steel	—	140 (460)
Song-Ma, China	Indo-China	—	—	Steel	3	162 (532)
Iron Mountain, MI	Iron ore	—	—	Steel	3	—
Zambesi, Rhodesia	—	G. A. Hobson	1905	Steel	—	152 (500)
Thermopylae, Greece	—	P. Bodin	1906	Steel	3	80 (262)
Nami-Ti Gorge, China	Yunnan	—	1909	Steel	3	55 (180)
Bend, OR	Spokane, Portland & Seattle	R. Modjeski	1911	Steel	2	107 (350)
Stillwater, MN	Wisconsin Central	C. Turner	1911	Steel	3	107 (350)
Lytton, BC	Canadian Northern	J. A. L. Waddell	1915	Steel	2	137 (450)
Hell Gate, NY	New England Connecting	G. Lindenthal	1916	Steel	2	298 (978)

[a] Indicates the second bridge constructed at the site.

TABLE 1.2
Notable Iron and Steel Viaduct Railway Bridges Constructed during 1823–1909

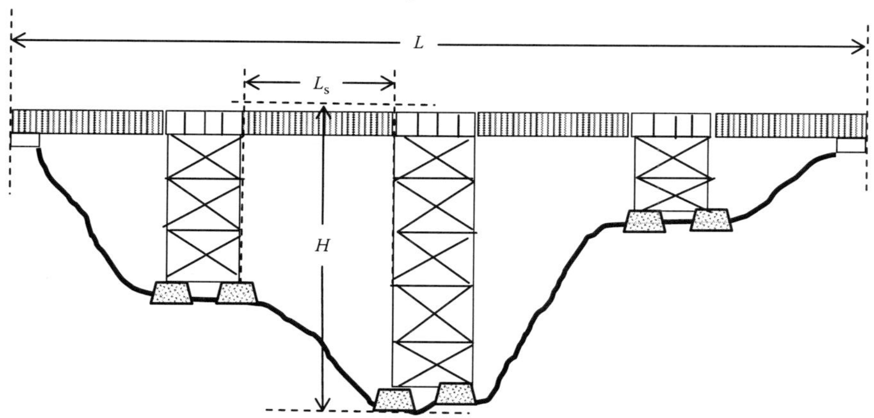

Viaduct	Railroad	Engineer	Year	Material	L_s m (ft)	L m (ft)	H m (ft)
Gauntless, UK	Stockton to Darlington	G. Stephenson	1823	Cast iron	3.8 (12.5)	15 (50)	~4.5 (~15)
Newcastle, UK	Northwestern	I. K. Brunel	1849	Cast and wrought iron	38 (125)	229 (750)	25(83)
Tray Run	Baltimore & Ohio	A. Fink	1853	Cast iron	—	136 (445)	18 (58)
Buckeye	Baltimore & Ohio	A. Fink	1853	Cast iron	—	107 (350)	14 (46)
Crumlin, UK	Newport & Hereford	Liddell & Gordon	1857	Wrought iron	46 (150)	549 (1800)	64 (210)
Guth, PA, Jordan Creek	Catasauqua & Fogelsville	F. C. Lowthorp	1857	Cast and wrought iron	30.5 (100), 33.5 (110)	342 (1122)	27 (89)
Belah, UK	—	Sir T. Bouch	1860	Cast and wrought iron	13.5 (45)	293 (960)	55 (180)
Weston, ON	Grand Trunk		1860	Iron	22 (72)	198 (650)	21 (70)
Fribourg, Switzerland	—	Mathieu	1863	Iron	48 (158)	396 (1300)	76 (250)
Creuse, Busseau, France	—	Nordling	1865	Iron	—	287 (940)	48 (158)
La Cere, France	Orleans	Nordling	1866	Iron	—	236 (775)	53 (175)
Assenheim, Germany	—	—	~1866	Iron	—	—	—
Angelroda, Germany	—	—	~1866	Iron	30.5 (100)	92 (300)	—

(Continued)

TABLE 1.2 (*Continued*)
Notable Iron and Steel Viaduct Railway Bridges Constructed during 1823–1909

Viaduct	Railroad	Engineer	Year	Material	L_S m (ft)	L m (ft)	H m (ft)
Bullock Pen	Cincinnati & Louisville	F. H. Smith	1868	Iron		143 (470)	18 (60)
Lyon Brook, NY	New York, Oswego & Midland	—	1869	Wrought iron	9 (30)	250 (820)	49 (162)
Rapallo Viaduct	New Haven, Middletown & Willimantic	—	1869	Iron	9 (30)	421 (1380)	18 (60)
St. Charles Bridge over Mississippi	—	—	1871	—	—	—	—
La Bouble, France	Commentary-Gannat	Nordling	1871	Wrought iron	49 (160)	396 (1300)	66 (216)
Bellon Viaduct, France	Commentary-Gannat	Nordling	1871	Steel	40 (131)	—	49 (160)
Verragus, Peru	Lima & Oroya	C. H. Latrobe	1872	Wrought iron	33.5 (110), 38 (125)	175 (575)	78 (256)
Olter, France	Commentary-Gannat	Nordling	1873	Steel	—	—	—
St. Gall, France	Commentary-Gannat	Nordling	1873	Steel	—	—	—
Horse Shoe Run	Cincinnati Southern	G. Bouscaren	~1873	Wrought iron	—	274 (900)	27 (89)
Cumberland	Cincinnati Southern	G. Bouscaren	~1873	Wrought iron	—	—	30.5 (100)
Tray Run[b]	Baltimore & Ohio	—	1875	Steel			18 (58)
Fishing Creek	Cincinnati Southern	G. Bouscaren	1876	Wrought iron	—	—	24 (79)
McKees Branch	Cincinnati Southern	G. Bouscaren	1878	Wrought iron	—	—	39 (128)
Portage, NY	Erie	G. S. Morison	1875	Iron	15.3 (50), 30.5 (100)	249 (818)	62 (203)
Staithes, UK	Whitby & Loftus	J. Dixon	1880	—	—	210 (690)	45 (150)
Oak Orchard, Rochester, NY	Rome, Watertown and Western	—	~1881	Steel	9 (30)	210 (690)	24 (80)
Kinzua[a], PA	New York, Lake Erie and Western	Clarke, Reeves & Co.	1882	Wrought iron	—	626 (2053)	92 (302)
Rosedale, Toronto, ON	Ontario & Quebec	—	1882	—	9 (30),18 (60)	—	—
Dowery Dell, UK	Midland	Sir T. Bouch	~1882				
Marent Gulch, MT	Northern Pacific	—	1884	Steel	35 (116)	244 (800)	61 (200)
Loa, Bolivia	Antofagasta	—	1885–1890	—	—	244 (800)	102 (336)
Malleco, Chile	—	A. Lasterria	1885–1890	—	—	366 (1200)	95 (310)
Souleuvre, France	—	—	1885–1890	—	—	366 (1200)	75 (247)
Moldeau, Germany	—	—	1885–1890	—	—	270 (886)	65 (214)

(*Continued*)

TABLE 1.2 (*Continued*)
Notable Iron and Steel Viaduct Railway Bridges Constructed during 1823–1909

Viaduct	Railroad	Engineer	Year	Material	L_S m (ft)	L m (ft)	H m (ft)
Schwarzenburg, Germany	—	—	1889	Steel	—	—	—
Panther Creek, PA	Wilkes-Barre & Eastern	—	1893	Steel	—	503 (1650)	47 (154)
Pecos, CA	—	—	1894	Steel	—	665 (2180)	98 (320)
Grasshopper Creek	Chicago & Eastern Illinois	—	1899	Steel	—	—	—
Lyon Brook[b], NY	New York, Ontario & Western	—	1894	Steel	30	250 (820)	49 (162)
Kinzua[b], PA	New York, Lake Erie and Western	C. R. Grimm	1900	Steel	—	626 (2052)	92 (302)
Gokteik, Burma	Burma	Sir A. Rendel	1900	Steel	—	690 (2260)	100 (320)
Boone, IA	Chicago & Northwestern	G. S. Morison	1901	Steel	13.5 (45), 23 (75), 92 (300)	819 (2685)	56 (185)
Portage, NY[b]	Erie	—	1903	Steel	15.2 (50), 30.5 (100)	249 (818)	62 (203)
Richland Creek, IN	—	—	1906	Steel	12 (40), 23 (75)	—	48 (158)
Moodna Creek	Erie	—	1907	Steel	12 (40), 24.5 (80)	976 (3200)	56 (182)
Colfax, CA	—	—	1908	Steel	—	247 (810)	58 (190)
Makatote, New Zealand	—	—	1908	Steel	—	262 (860)	92 (300)
Cap Rouge, QC	Transcontinental	—	1908	Steel	12 (40), 18.3 (60)	—	53 (173)
Battle River, AB	Grand Trunk Pacific	—	1909	Steel	—	~823 (~2700)	56 (184)
Lethbridge, AB	Canadian Pacific	Monsarrat & Schneider	1909	Steel	20 (67), 30.5 (100)	1625 (5328)	96 (314)

[a] Indicates the first bridge constructed at the site.
[b] Indicates the second bridge constructed at the site.

relatively level grades required for train operations (due to the limited tractive effort available to early locomotives) and use of heavier locomotives also provided motivation for the extensive use of cast iron girder and truss spans for railway bridges.

Commencing about 1830, Robert Stephenson built both cast iron arch and girder railway bridges in the UK. Cast iron plate girders were also built in the United States by the B&O Railroad in 1846, the Pennsylvania Railroad in 1853, and the Boston and Albany Railroad in 1860. The B&O Railroad constructed the first cast iron girder trestles in the United States in 1853. One of the first cast iron railway viaducts in Europe was constructed in 1857 for the Newport to Hereford Railway line at Crumlin, UK. Nevertheless, while many cast iron arches and girders were built in the UK and the United States, American railroads favored the use of composite trusses of wood and iron.

American railroad trusses built after 1840 were often constructed using cast iron, wrought iron, and timber members. In particular, Howe trusses with wood and cast iron compression members and wrought iron tension members were widely used in early American railroad bridge construction.

The failure of a cast iron girder railway bridge in 1847* stimulated an interest in wrought iron construction among British railway engineers.† British engineers were concerned with the effect of railway locomotive impact on cast iron railway bridges, and they were beginning to understand that, while strong, cast iron was brittle and prone to sudden failure. Concurrently, American engineers were becoming alarmed by cast iron railway bridge failures, and some even promoted the exclusive use of masonry or timber for railway bridge construction. For example, following the collapse of an iron truss bridge in 1850 on the Erie Railroad, some American railroads dismantled their iron trusses and replaced them with wood trusses. However, the practice of constructing railway bridges using iron was never discontinued on the B&O Railroad.

European and American engineers realized that a more ductile material was required to resist the tensile forces developed by heavy railroad locomotive loads. Wrought iron‡ provided this increase in material ductility, and it was integrated into the construction of many railway bridges after 1850. The use of cast iron for railway bridge construction in Europe ceased in about 1867.§ One of the last major railway bridges in Europe to be constructed using cast iron was Gustave Eiffel's 488 m (1600 ft) long Garonne River Bridge built in 1860. However, cast iron continued to be used in the United States (primarily in compression members), even in some long-span bridges for more than a decade after its demise in Europe.

1.2.2 Wrought Iron Construction

Early short- and medium-span railway bridges in the United States were usually constructed from girders or propriety trusses (e.g., the Bollman, Whipple, Howe, Pratt, and Warren trusses shown in Figure 1.2). An example of a Whipple truss is also shown in Figure 1.3. US patents were granted for small- and medium-span iron railway trusses after 1840, and they became widely used by American railroads. The trusses typically had cast iron or wood compression members and wrought iron tension members.¶

---

* This was Stephenson's cast iron girder bridge over the River Dee on the London-Chester-Holyhead Railroad. In fact, Stephenson had recognized the brittle nature of cast iron before many of his peers and reinforced his cast iron railway bridge girders with wrought iron rods. Nevertheless, failures ensued with increasing railway loads.

† Hodgekinson, Fairbairn, and Stephenson had also performed experiments with cast and wrought iron bridge elements between 1840 and 1846. The results of those experiments led to a general acceptance of wrought iron for railway bridge construction among British engineers.

‡ Wrought iron has much lower carbon content than cast iron and is typically worked into a fibrous material with elongated strands of slag inclusions.

§ In the United States, J. H. Linville was a proponent of all wrought iron truss construction in the early 1860s.

¶ Wrought iron bridge construction provided the opportunity for using riveted connections instead of bolts. The riveted connections were stronger due to the clamping forces induced by the cooling rivets.

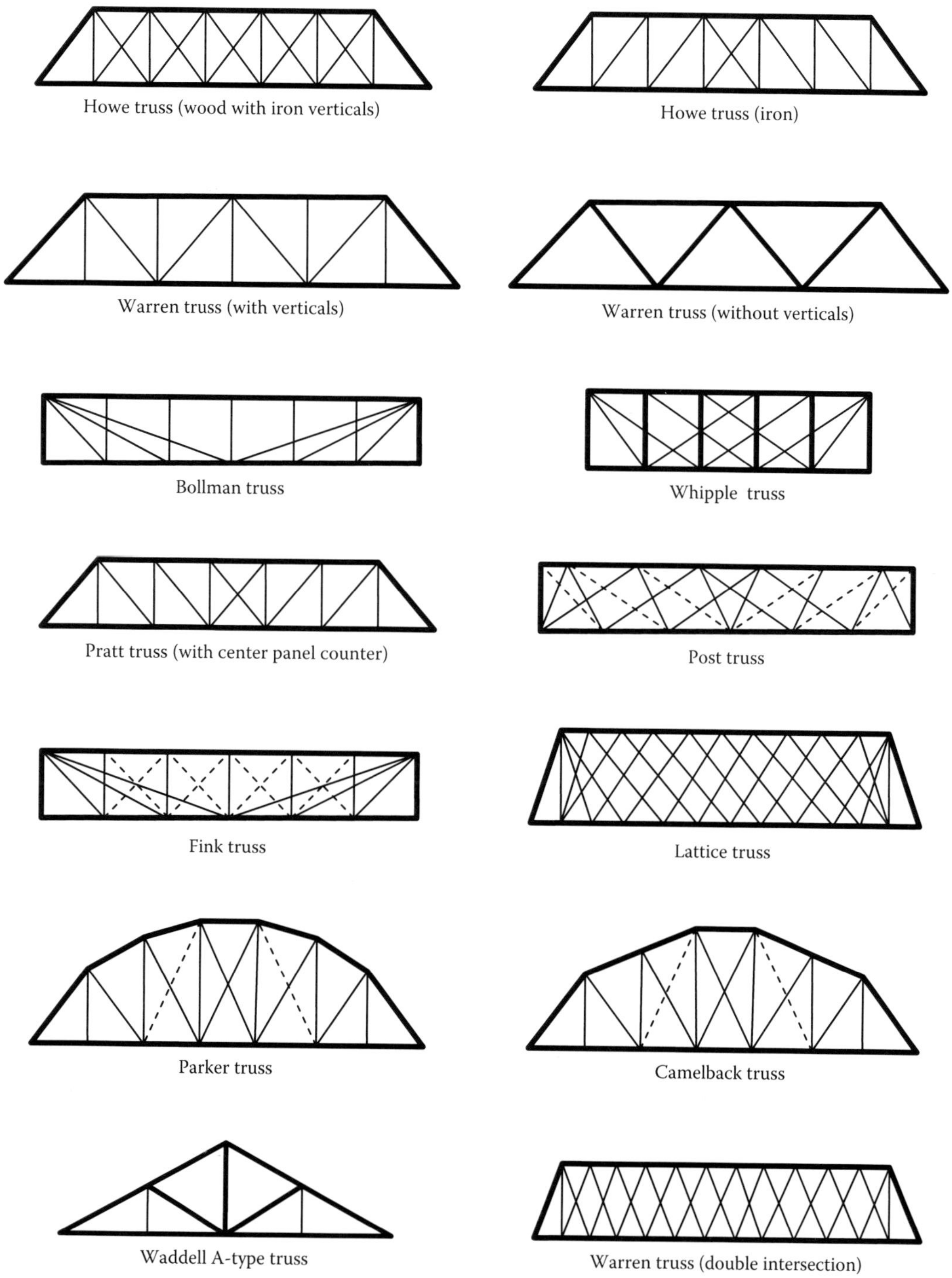

FIGURE 1.2 Truss forms used by railroads in the United States.

FIGURE 1.3 Whipple truss span. (Courtesy of the author, Canadian Pacific Engineering.)

The wooden Howe truss with wrought iron vertical members (patented in 1840) was popular on American railroads up to the 1860s.[*] The principal attraction of the Howe truss was the use of wrought iron rods that did not permit the truss joints to come apart when diagonal members were in tension from railway loading. However, the Howe truss form is statically indeterminate, and, therefore, many were built on early American railroads without the benefit of applied scientific analysis.[†]

The first railway bridge in the United States constructed entirely of iron was a Howe truss with cast iron compression and wrought iron tension members built by the Philadelphia and Reading Railroad in 1845 at Manayunk, PA. Iron Howe trusses were also constructed on the Boston and Albany Railroad in 1847 near Pittsfield, MA, and on the Harlem and Erie Railroad in 1850. Following this, iron truss bridges became increasingly common as American railroads continued their rapid expansion. Early examples of Pratt truss use were the Pennsylvania Railroad's cast and wrought iron arch-stiffened Pratt truss bridges of the 1850s. An iron railway bowstring truss, also utilizing cast iron compression and wrought iron tension members, was designed by Squire Whipple[‡] for the Rensselaer & Saratoga Railway in 1852. Fink and Bollman, both engineers employed by the B&O Railroad, used their own patented cast and wrought iron trusses extensively between 1840 and 1875.[§] Noteworthily, iron trusses were also built by the North Pennsylvania Railroad in 1856 (a Whipple truss) and the Catasauqua and Fogelsville Railroad in 1857. The Erie Railroad pioneered the use of iron post truss bridges in 1865, and they remained a standard of construction on the B&O Railroad for the next 15 years.

However, due to the bridge failures (predominantly of cast iron members) in the 30 years after 1840, the use of cast iron ceased, and wrought iron was used exclusively for railway girders and

* During construction of the railroad between St. Petersburg and Moscow, Russia (c. 1842), American Howe truss design drawings were used for many bridges. Timber Howe trusses were also used for the rapid construction of temporary bridges on the Canadian Pacific Railroad in the 1880s.

† Scientific analysis by engineers was becoming prevalent to ensure safety following the many railway bridge failures that occurred in the middle of the 19th century.

‡ In 1847, Whipple published *A Treatise on Bridge Building*, the first book on the scientific or mathematical analysis of trusses.

§ The first all-iron trusses on the B&O were designed by Fink in 1853.

trusses. Isambard Kingdom Brunel used thin-walled wrought iron plate girders in his designs for short and medium railway spans on the Great Western Railway in the UK during the 1850s. Between 1855 and 1859, Brunel also designed and constructed many noteworthy wrought iron lattice girder, arch, and suspension bridges for British railways. In particular, the Royal Albert Railway Bridge across the Tamar River, completed at Saltash in 1859, is a significant example of a Brunel wrought iron railway bridge using large lenticular trusses (Figure 1.4). Other important railway bridges built by Brunel on the Great Western Railway were the Wharncliffe Viaduct, Maidenhead, and Box Tunnel bridges. Table 1.3 lists some notable wrought iron truss railway bridges constructed between 1845 and 1877.

The English engineer William Fairbairn constructed a tubular wrought iron through-girder bridge on the Blackburn and Bolton Railway in 1846. Later, in partnership with Fairbairn, Robert Stephenson designed and built the innovative and famous wrought iron tubular railway bridges for the London–Chester–Holyhead Railroad at Conwy in 1848 and Menai Straits (the Britannia Bridge) in 1850. The Conwy bridge (Figure 1.5) is a simple tubular girder span of 125.6 m (412 ft) and the Britannia bridge consists of four continuous tubular girder spans of 70.1 m (230 ft), 140.2 m (460 ft), 140.2 m (460 ft), and 70.1 m (230 ft) (Figure 1.6).* Spans of 140.2 m (460 ft) were mandated for navigation purposes, making this the largest wrought iron bridge constructed. It was also one of the first uses of span continuity to reduce dead load bending moments in a bridge. Arch bridges were also proposed for the Menai Straits crossing by Stephenson† and Brunel.‡ However, arch bridges were rejected due to concerns about interference with navigation, and the four wrought iron tubular girder continuous spans were built to obtain the stiffness required to resist wind and train loadings.

FIGURE 1.4 The Royal Albert Bridge built in 1859 over Tamar River at Saltash, UK. (From Owen Dunn, http://en.wikipedia, June 2005.)

* The Britannia Bridge was destroyed by fire in 1970 and only the Conwy Bridge remains as an example of Stephenson's tubular railway bridges. Following the fire, the Britannia Bridge was rebuilt as a steel truss arch bridge, carrying both road vehicle and railway traffic.

† Stephenson had studied the operating issues associated with some suspension railway bridges, notably the railway suspension bridge built at Tees in 1830, and decided that suspension bridges were not appropriate for railway loadings. He proposed an arch bridge.

‡ To avoid the use of falsework in the channel, Brunel outlined the first use of the cantilever construction method in conjunction with his proposal for a railway arch bridge across Menai Straits.

TABLE 1.3
Notable Iron and Steel Simple Truss Span Railway Bridges Constructed during 1823–1907

Location	Railroad	Engineer	Year Completed	Type	Material	*L* m (ft)
West Auckland, UK	Stockton to Darlington	G. Stephenson	1823	Lenticular	Cast iron	3.8 (12.5)
Ireland	Dublin & Drogheda	G. Smart	1824	Lattice	Cast Iron	25.5 (84)
Manayunk, PA	Philadelphia and Reading	R. Osborne	1845	Howe	Cast and wrought iron	10.5 (34)
Pittsfield, MA	Boston & Albany	—	1847	Howe	Cast and wrought iron	9 (30)
Windsor, UK	Great Western	I. K. Brunel	1849	Bowstring	Iron	57 (187)
Newcastle, UK	Northwestern	I. K. Brunel	1849	Bowstring	Cast and wrought iron	38 (125)
—	Harlem & Erie	—	1850	Howe	Iron	—
Various	Pennsylvania	H. Haupt	1850s	Pratt with cast iron arch	Iron	—
Harper's Ferry	Baltimore & Ohio	W. Bollman	1852	Bollman	Cast and wrought iron	38 (124)
Fairmont, West VA	Baltimore & Ohio	A Fink	1852	Fink	Cast and wrought iron	62.5 (205)
—	Rennselaer & Saratoga	S. Whipple	1852	Whipple	Iron	—
Newark Dyke, UK	Great Northern	C. Wild	1853	Warren	Cast and wrought iron	79 (259)
—	North Pennsylvania	—	1856	Whipple	Iron	—
Guth, PA, Jordan Creek	Catasauqua & Fogelsville	F. C. Lowthorp	1857	—	Cast and wrought iron	33.5 (110)
Phillipsburg, NJ	Lehigh Valley	J. W. Murphy	1859	Whipple (pin-connected)	Iron	50 (165)
Plymouth, UK	Cornish (Great Western)	I. K. Brunel	1859	Lenticular	Wrought iron	139 (455)
Frankfort, Germany	—	—	1859	Lenticular	Iron	105 (345)
Various	New York Central	H. Carroll	1859	Lattice	Wrought iron	27.5 (90)
Kehl River, Germany	Baden State	Keller	1860	Lattice	Iron	60 (197)
Schuylkill River	Pennsylvania	J. H. Linville	1861	Whipple	Cast and wrought iron	58.5 (192)
Steubenville, OH	Pennsylvania	J. H. Linville	1863	Murphy-Whipple	Cast and wrought iron	97.5 (320)
Mauch Chunk, PA.	Lehigh Valley	J. W. Murphy	1863	—	Wrought iron	—
Liverpool, UK	London & Northwestern	W. Baker	1863	—	Iron	93 (305)
Blackfriar's Bridge, UK	—	Kennard	1864	Lattice	Iron	—

(Continued)

TABLE 1.3 (*Continued*)
Notable Iron and Steel Simple Truss Span Railway Bridges Constructed during 1823–1907

Location	Railroad	Engineer	Year Completed	Type	Material	*L* m (ft)
Orival, France	Western	—	~1865	Lattice	Iron	51 (167)
Various	Baltimore & Ohio	S. S. Post	1865	Post	Iron	—
Lockport, IL	Chicago & Alton	S. S. Post	~1865	Post	Cast and wrought iron	—
Schuylkill River	Connecting Railway of Philadelphia	J. H. Linville	1865	Linville	Wrought iron	—
Burlington[a], IA	Chicago, Burlington & Quincy	M. Hjorstberg	1868	—	Iron	—
Dubuque, IA	Illinois Central	J. H. Linville	1868	Linville	Wrought iron	76 (250)
Quincy, IL	Chicago, Burlington & Quincy	T. C. Clarke	1868	—	Cast and wrought iron	76 (250)
Kansas City[a], MO	Chicago, Burlington & Quincy	J. H. Linville & O. Chanute	1869	—	Iron	71 (234)
Louisville, KY	Baltimore & Ohio	A. Fink	1869	Subdivided Warren & Fink	Wrought iron	119 (390)
Parkersburg & Benwood, WV	Baltimore & Ohio	J. H. Linville	1870	Bollman	Iron	106 (348)
Hannibal[a], MO.	—	—	1871	—	Iron	—
Atcheson	Various	—	1875	Whipple	Iron	79 (260)
Cincinnati, OH	Cincinnati Southern	J. H. Linville & G. L. F. Bouscaren	1876	Linville	Wrought iron	157 (515)
Tay River[a], Scotland		Sir T. Bouch	1877	Lattice	Wrought iron	
Glasgow, MO	Chicago & Alton		1879	Whipple	Steel	732 (2402)
Bismark, ND		G. S. Morison & C. C. Schneider	1882	Whipple	Steel	
Hannibal[b], MO			1886		Steel	
Tay River[b], Scotland			1887		Steel	
Sioux City, IA			1888	—	Steel	122 (400)
Cincinnati, OH		W. H. Burr	1888	—	Steel	168 (550)

(*Continued*)

TABLE 1.3 ***(Continued)***
Notable Iron and Steel Simple Truss Span Railway Bridges Constructed during 1823–1907

Location	Railroad	Engineer	Year Completed	Type	Material	L m (ft)
Benares, India			1888	Lattice	Steel	109 (356)
Hawkesbury, Australia	—	—	1889	—	Steel	127 (416)
Henderson Bridge	Louisville & Nashville	—	~1889	Subdivided Warren	Steel	160 (525)
Cairo, IL	Illinois Central		1889	—	Steel	—
Ceredo RR Bridge		Doane & Thomson	~1890	—	Steel	159 (521)
Merchant's Bridge, St. Louis		G. S. Morison	1890	Petit	Steel	158 (517)
Kansas City[b], MO		—	1891	—	Steel	
Burlington[b], IA	Chicago, Burlington & Quincy	G. S. Morison	1892	—	Steel	—
Louisville, KY	—		1893	Petit	Steel	168 (550)
Nebraska City, NB	—	G. S. Morison	1895	Whipple	Steel	122 (400)
Sioux City, IA			1896		Steel	149 (490)
Montreal, QC	Grand Trunk		1897		Steel	106 (348)
Kansas City, MO	Kansas City Southern	J. A. L. Waddell	1900	Pratt	Steel	—
Rumford, ON	Canadian Pacific	C. N. Monsarrat	1907	Subdivided Warren	Steel	126 (412)

[a] Indicates the first bridge constructed at the site.
[b] Indicates the second bridge constructed at this site.

FIGURE 1.5 The Conwy Bridges: Stephenson's Tubular Railway Bridge built in 1848 and Thomas Telford's Suspension Highway Bridge built in 1826 at Conwy Castle, Wales. (From Stephen J. Hill, Redwood City, CA. With permission.)

FIGURE 1.6 The Britannia Bridge built in 1850 across the Menai Straits, Wales. (Postcard from the private collection of Jochem Hollestelle.)

The construction of the Conwy and Britannia tubular iron plate bridges also provided the opportunity for further investigations into issues of plate stability, riveted joint construction, lateral wind pressure, and thermal effects. Fairbairn's empirical work on fatigue strength and plate stability during the design of the Conwy and Britannia bridges is particularly significant.*

A small 16.8 m (55 ft) long, simple span tubular wrought iron plate girder bridge was built in the United States by the B&O Railroad in 1847. However, the only large tubular railway bridge constructed in North America was the Victoria Bridge built in 1859 for the Grand Trunk Railway over the St. Lawrence River at Montreal† (Figure 1.7). The Victoria Bridge was the longest bridge in the world upon its completion.‡ The bridge was replaced with steel trusses in 1898 due to the rivet

* In 1864, Fairbairn studied iron plate and box girder bridge models under a cyclical loading representative of railway traffic. These investigations assisted with the widespread adoption of wrought iron, in lieu of cast iron, for railway bridge construction in the latter quarter of the 19th century.

† The Victoria Bridge over the St. Lawrence River at Montreal was also designed by Stephenson.

‡ The longest span in the Victoria Bridge was 100 m (330 ft).

FIGURE 1.7 The Victoria Bridge under construction (completed in 1859) across the St. Lawrence River, Montreal, Canada. (Courtesy of William Notman, Library and Archives Canada.)

failures associated with increasing locomotive weights and ventilation problems harmful to passengers traveling across the 2788 m (9144 ft) river crossing with almost 2000 m (6560 ft) of tubular girders. Table 1.4 indicates some notable continuous span railway bridges constructed after 1850.

These tubular bridges provided the stiffness desired by their designers but proved to be costly. Suspension bridges were more economical but many British engineers were hesitant to use flexible suspension bridges for long-span railroad crossings.* Sir Benjamin Baker's 1867 articles on long-span bridges also promoted the use of more rigid bridges for railway construction. Furthermore, Baker had earlier recommended cantilever trusses for long-span railway bridges.† Also in 1867, Heinrich Gerber constructed the first cantilever bridge at Hanover, Germany; and following this, cantilever arch and truss bridges were built in New England and New Brunswick‡ between 1867 and 1885.

Nevertheless, railway suspension bridges were built in the United States in the last quarter of the 19th century. Unlike the almost universal aversion to railway suspension bridge design and construction that was prevalent among British railway engineers, some American engineers were using iron suspension bridges for long spans carrying relatively heavy freight railroad traffic. Modern suspension bridge engineering essentially commenced with the construction of the 250 m (820 ft) span railway suspension bridge over the Niagara Gorge in 1854. European engineers and many American engineers had expressed concern over the scope of such a suspension bridge.§ Nevertheless, this bridge, designed and constructed by John A. Roebling, was used by the Great Western, New York Central, Grand Trunk Railway, and successor railroads for over 40 years. Roebling had realized the need for greater rigidity

* An effort to construct a suspension bridge for the Stockton and Darlington Railroad in the 1820s had been a failure. The first railway suspension bridge built over the Tees River in 1830 in the UK [with a 91.5 m (300 ft) span] had performed poorly by deflecting in a very flexible manner that even hindered the operation of trains. It was replaced by cast iron and steel girders, respectively, in 1842 and 1905. The Basse-Chaine suspension bridge in France collapsed in 1850 as the suspension bridge at Wheeling, WV in 1854, illustrating the susceptibility of flexible suspension bridges to failure under wind load conditions.

† Baker's 1862 book *Long-Span Railway Bridges* and A. Ritter's calculations of the same year outlined the benefits of cantilever bridge design.

‡ For example, the railway bridge built in 1885 (replaced in 1922) over the reversing falls of the Saint John River in New Brunswick, Canada.

§ Only four American engineers expressed support of the proposal by the Great Western Railroad to connect to the New York Central Railroad with a suspension bridge. These were Charles Ellet, John A. Roebling, Edward Serrell, and Samuel Keefer.

TABLE 1.4
Notable Continuous Span Railway Bridges Constructed during 1850–1929

Location	Railroad	Engineer	Year	Type	Longest Span m (ft)
Torksey, UK	—	J. Fowler	1850	3 Span continuous tubular girder	40 (130)
Britannia Bridge, Menai Straits, UK	London-Chester-Holyhead	R. Stephenson	1850	4 Span continuous tubular	140 (460)
Montreal, QC	Grand Trunk	R. Stephenson	1860	25 Span continuous tubular	100 (330)
Montreal, QC	Canadian Pacific	C. Shaler Smith	1886	4 Span continuous trusses	124 (408)
Sciotoville, OH	Chesapeake & Ohio	G. Lindenthal & D. B. Steinman	1917	2 Span continuous truss	236 (775)
Allegheny River	Bessemer & Lake Erie	—	1918	3 Span continuous truss	158.5 (520)
Nelson River	Bessemer & Lake Erie	—	1918	3 Span continuous truss	122 (400)
Cincinnati, OH	C.N.O. & T.P.	—	1922	3 Span continuous truss	157 (516)
Cincinnati, OH	Cincinnati & Ohio	—	1929	3 Span continuous truss	206 (675)

in suspension bridge design after the failure of the Wheeling* and other suspension bridges. As a consequence, his Niagara Gorge suspension bridge was the first to incorporate stiffening trusses into the design (Figure 1.8). Rehabilitation work was required in 1881 and 1887, and it was replaced with a steel spandrel braced hinged arch bridge, designed by Leffert L. Buck, in 1897, due to capacity requirements for heavier railway loads. The railway suspension bridge constructed in 1840 over the Saone River in France was replaced only 4 years after completion due to poor performance under live load.† The railway suspension bridge constructed in 1860 at Vienna, Austria, was also prematurely replaced with an iron arch bridge in 1884 after concerns over the flexibility of the suspended span. The early demise of these and other suspension bridges generated new concerns among some American engineers over the lack of rigidity of cable-supported bridges under steam locomotive and moving train loads.

The first all-wrought-iron bridge in the United States, a lattice truss, was completed in 1859 by the New York Central Railroad.‡ In the same year, the Lehigh Valley Railroad built the first pin-connected truss. In 1861, the Pennsylvania Railroad pioneered the use of forged eyebars in a pin-connected truss over the Schuylkill River. After this, many American railway bridges were constructed with pinned connections, while European practice still favored the use of riveted construction. Riveted construction was considered superior, but pin-connected construction enabled the economical and rapid erection of railway bridges in remote areas of the United States. The principal exception was the New York Central Railroad, which used riveted construction exclusively for its iron railway bridges.

In 1863, the Pennsylvania Railroad successfully crossed the Ohio River using a 98 m (320 ft) iron truss span. The railroad used the relatively rigid Whipple truss for such long spans. This bridge construction encouraged greater use of longer span iron trusses to carry heavy freight railroad traffic in the United States. Another notable wrought iron railway truss was the 119 m (390 ft) span built by the B&O Railroad at Louisville, KY, in 1869.

In the 1870s, the Pratt Truss (patented in 1844) became predominant for short- and medium-span railway bridges in the United States. Pratt trusses are statically determinate, and their form is well

* The 308 m (1010 ft) wire rope suspension bridge, designed by Charles Ellet, over the Ohio River at Wheeling, WV collapsed due to wind loads in 1854, just 5 years after completion of construction.

† The suspension bridge was replaced by a stone masonry bridge.

‡ The New York Central Railroad also initiated the use of iron stringers (instead of wood stringers) in railway trusses in the 1860s.

FIGURE 1.8 The Railway Suspension Bridge built in 1854 across the Niagara Gorge between New York, USA and Ontario, Canada. (From Niagara Falls Public Library.)

suited for use in iron bridges. Whipple, Warren, and post trusses were also used by US railroads in the 1870s. The Bollman truss bridge, patented in 1852 and used by the B&O and other railroads until 1873, was an example of the innovative* use of wrought iron in American railway bridge construction. Baltimore, Petit, or Pennsylvania truss spans were often used for longer wrought iron railway bridge spans (Figure 1.9).† The first use of a Baltimore truss (a Pratt truss with subdivided panels) was on the Pennsylvania Railroad in 1871.

Large railway viaduct bridges were also constructed using wrought iron. The 66 m (216 ft) high and 396 m (1300 ft) long Viaduc de la Bouble was built in France in 1871. In 1875, the Erie Railroad completed construction of a 249 m (818 ft) long wrought iron viaduct at Portage, New York‡ (Figure 1.10). This was followed in 1882 by the 92 m (300 ft) high and over 600 m (2000 ft) long wrought iron Kinzua Viaduct, PA (Figure 1.11).§ Also in France, Gustave Eiffel designed the wrought iron Garabit Viaduct, which opened to railroad traffic in 1884 (Figure 1.12).

A large number of iron railway bridges built after 1840 in the United States and the UK failed under train loads. It was estimated that about one-fourth of the railway bridges in the American railroad infrastructure were failing annually between 1875 and 1888. Most of these failures were related to fatigue and fracture, and the buckling instability of compression members (notably top chords of trusses). Although most of the failures were occurring in cast iron truss members and

* Bollman trusses used wrought iron tension members and cast iron compression members. The redundant nature of the truss form reduced the possibility of catastrophic failure.

† The Petit truss was used extensively by American railroad companies.

‡ The 1875 viaduct was designed by G. S. Morison and O. Chanute. It was extensively strengthened using steel in 1903 and is currently planned for replacement commencing in 2016 with a 147 m (483 ft) two-hinged spandrel braced steel arch (Irwin et al., 2013).

§ Both the Portageville Viaduct and the Kinzua Viaduct were designed by Morison and Chanute for the Erie Railroad.

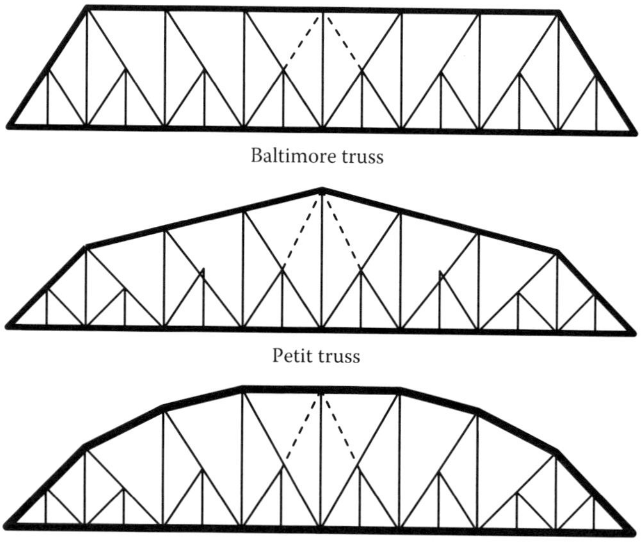

FIGURE 1.9 Baltimore trusses (inclined chord truss is also called Petit truss).

FIGURE 1.10 Portageville Viaduct 1875 (strengthened 1903) over Genesee River Gorge, New York. (From James N. Carter, Norfolk Southern Corp., Atlanta, GA. With permission.)

girders, by 1850 many American engineers had lost confidence in even wrought iron girder, truss, and suspension railway bridge construction.*

At this time, railway construction was not well advanced in Germany, and these failures interested Karl Culmann during construction of some major bridges for the Royal Bavarian Railroad. He proposed that American engineers use lower allowable stresses to reduce fatigue failures of iron truss railway bridges, and he recognized the issue of top chord compressive instability. Culmann also proposed the use of stiffening trusses for railroad suspension bridges after learning of the distress expressed by American bridge engineers concerning their flexibility under moving live loads.

* For example, following the collapse of an iron bridge in 1850, all metal bridges on the Boston and Albany Railroad were replaced with timber bridges.

FIGURE 1.11 Kinzua Viaduct 1882, Pennsylvania. (Courtesy of Historic American Engineering Record.)

FIGURE 1.12 The Garabit Viaduct built in 1884 over the Tuyere River, France. (From GFDL J. Thurion, http://fr.wikipedia, July 2005.)

A railroad Howe truss collapsed under a train at Tariffville, CT, in 1867 and a similar event occurred a decade later at Chattsworth, IL. However, the most significant railway bridge failure, due to the considerable loss of life associated with the incident, was the collapse of the cast iron Howe deck truss span on the Lake Shore & Michigan Southern Railroad at Ashtabula, OH, in 1876 (Figure 1.13a and b). The Ashtabula bridge failure provided further evidence that cast iron was not appropriate for heavy railway loading conditions and caused American railroad companies to abandon the use of cast iron elements for bridges.* This was, apparently, a wise decision as modern forensic analysis indicates the likely cause of the Ashtabula failure was a combination of fatigue and brittle fracture initiated at a cast iron flaw.

* With exception of cast iron bearing blocks at ends of truss compression members.

(a)

(b)

FIGURE 1.13 (a) The Ashtabula Bridge, OH, before 1876 collapse. (b) The Ashtabula Bridge, OH, after 1876 collapse. (Courtesy of Ashtabula Railway Historical Foundation.)

In addition, the collapse of the Tay Railway Bridge in 1879, only 18 months after completion, promoted a renewed interest in wind loads applied to bridges (Figure 1.14a and b). The Tay bridge collapse also reinforced the belief, held by many engineers, that light and relatively flexible structures are not appropriate for railway bridges.

These bridge failures shook the foundations of bridge engineering practice and created an impetus for research into new methods (for design and construction) and materials to ensure the safety and reliability of railway bridges. The investigation and specification of wind loads for bridges also emerged from research conducted following these railway bridge collapses. Furthermore, in both Europe and the United States, a new emphasis on truss analysis and elastic stability was developing in response to railway bridge failures.

A revitalized interest in the cantilever construction method occurred, particularly in connection with the erection of arch bridges. Early investigations by Stephenson, Brunel, and Eads had illustrated that the erection of long arch spans using the cantilever method* was feasible and precluded the requirement for falsework as temporary support for the arch. The cantilevered arms were joined to provide fixed or two-hinged arch action† or connected, allowing translation of members

* Often using guyed towers and cable stays as erection proceeds.

† Depending on whether fixed or pinned arch support conditions were used.

(a)

(b)

FIGURE 1.14 (a) The Tay River Bridge, Scotland, before 1879 collapse. (From http://en.wikipedia, January 2007). (b) The Tay River Bridge, Scotland, after 1879 collapse. (From http://en.wikipedia, January 2007.)

to provide a statically determinate structure. The cantilever construction method was also proposed for long-span truss erection, where the structure is made statically determinate after erection by retrofitting to allow appropriate members to translate. This creates a span suspended between two adjacent cantilever arms that are anchored by spans adjacent to the support pier, providing a statically determinate structure.* Alternatively, the cantilever arms may progress only partially across the main span and be joined by a suspended span erected between the arms.† Other benefits of cantilever construction are smaller piers (due to a single line of support bearings) and an economy of material for properly proportioned cantilever arms, anchor spans, and suspended spans.

Iron trusses continued to be built in conjunction with the rapid railroad expansion of the 1860s. However, in the second half of the 19th century, steel started to replace iron in the construction of railway bridges.‡ For example, the iron Kinzua Viaduct of 1882 was replaced with a similar structure of steel only 18 years after construction due to concerns about the strength of wrought iron bridges under increasing railroad loads.

---

* Statically indeterminate structures are susceptible to stresses caused by thermal changes and support settlements. Therefore, statically indeterminate cantilever bridges must incorporate expansion devices and be founded on unyielding foundations to ensure safe and reliable behavior.

† This was the method used in the 1917 reconstruction of the Quebec bridge.

‡ In 1895, steel completely replaced wrought iron for the production of manufactured structural shapes.

1.3 STEEL RAILWAY BRIDGES

Steel is stronger and creates lighter structures than wrought iron, but it was expensive to produce in the early 19th century. Bessemer developed the steelmaking process in 1856, and Siemens further advanced the steel industry with open-hearth steelmaking in 1867. These advances enabled the economical production of steel. These steelmaking developments, in conjunction with the demand for railway bridges following the American Civil War, provided remarkable stimulus to the extensive use of steel in the construction of railway bridges in the United States. In the latter part of the 19th century, North American and European engineers favored steel arches and cantilever trusses for long-span railway bridges which, due to their rigidity, were considered to better resist the effects of dynamic impact, vibration, and concentrated moving railway loads.

The first use of steel in a railway bridge* was during the 1869–1874 construction of two 152 m (500 ft) flanking spans and 158.5 m (520 ft) central span of the St. Louis Bridge (now named the Eads Bridge after its builder, James Eads†) across the Mississippi River at St. Louis, MO. Eads did not favor the use of a suspension bridge for railway loads‡ and proposed a cast steel arch bridge. Eads' concern for stiffness for railway loads is illustrated by the trusses built between the railway deck and main steel arches of the St. Louis Bridge (Figure 1.15). The Eads Bridge features not only the earliest use of steel but also other innovations in American railway bridge design and construction. The construction incorporated the initial use of the pneumatic caisson method§ and the first use

FIGURE 1.15 The St. Louis (Eads) Bridge built across the Mississippi River in 1874 at St. Louis, MO. (Courtesy of Historic American Engineering Record.)

---

* The first use of steel in any bridge was in the 1828 construction of a suspension bridge in Vienna, Austria, where open-hearth steel suspension chains were incorporated into the bridge.

† Eads was not an academically trained engineer and was assisted with design by Charles Pfeiffer and construction by Theodore Cooper.

‡ A suspension bridge was proposed for this site by John Roebling in 1864.

§ This method of pier construction was also used by Brunel in the construction of the Royal Albert Bridge at Saltash, UK, in 1859.

of the cantilever method of bridge construction in the United States.* It was also the first arch span over 150 m (500 ft) and incorporated the earliest use of hollow tubular chord members.† The plethora of innovations associated with this bridge caused considerable skepticism among the public and press. In response, before it was opened, Eads tested the bridge using 14 of the heaviest locomotives available. The construction of the Eads Bridge almost depleted the resources of the newly developed American steelmaking industry.

The initial growth of the American steel industry was closely related to the need for steel railway bridges, particularly those with long spans. The American railroads' demand for longer spans, and their use of increasingly heavier locomotives and freight cars encouraged Andrew Carnegie‡ and others to invest considerable resources toward the development of improved steels of higher strength and ductility. These improved steels elicited the construction of the first all-steel railway bridge (comprising Whipple trusses) by the Chicago and Alton Railway in 1879 at Glasgow, MO.

Despite concerns with suspension bridge flexibility under train and wind loads, some American bridge engineers continued to design and construct steel suspension railway bridges. The famous Brooklyn Bridge, when completed in 1883, carried two railway lines. However, lingering concerns with suspension bridge performance and increasing locomotive weights hastened the general demise of this relatively flexible type of railway bridge construction.

The structural and construction efficacy of cantilever-type bridges for carrying heavy train loads led to the erection of many long-span steel railway bridges of trussed cantilever design after 1876. The Cincinnati Southern Railway constructed the first cantilever, or Gerber§ type, steel truss railway bridge in the United States over the Kentucky River in 1877.¶ In 1883, the Michigan Central and Canada South Railway completed the construction of a counterbalanced cantilever deck truss bridge** across the Niagara Gorge parallel to Roebling's 1854 railway suspension bridge. Shortly afterward, in 1884, the Canadian Pacific Railway crossed the Fraser River in British Columbia with the first balanced cantilever steel deck truss (Figure 1.16). Cantilever bridges became customary for long-span railway bridge construction as they provided the rigidity required to resist dynamic train loads, may be made statically determinate, and require no main span (comprising cantilever arms and suspended span) falsework to erect. Table 1.5 summarizes some notable cantilever railway bridges constructed after 1876.

Theodore Cooper promoted the exclusive use of steel for railway bridge design and construction in his 1880 paper to the American Society of Civil Engineers (ASCE) titled "The Use of Steel for Railway Bridges." Following this, almost all railway bridges, and by 1895 many other bridges, in the United States were constructed of steel. Structural steel shape production was well developed for the bridge construction market by 1890.††

The British government lifted its ban on the use of steel in railway bridge construction in 1877. More than a decade later, Benjamin Baker reviewed precedent cantilever bridges constructed in North America (in particular, those on the Canadian Pacific Railway) and proposed a steel cantilever truss for the Firth of Forth Railway Bridge crossing in Scotland.‡‡ It was a monumental

* The cantilever method had been proposed in 1800 by Thomas Telford for a cast iron bridge crossing the Thames River at London and in 1846 by Robert Stephenson for construction of an iron arch railway bridge to avoid falsework in the busy channel of the Menai Straits. Eads used principles developed in the 17th century by Galileo to describe the principles of cantilever construction of arches to skeptics of the method.

† The tubular arch chords used steel with 1.5%–2% chromium content providing for a relatively high ultimate stress of about 100 ksi.

‡ Andrew Carnegie worked for the Pennsylvania Railroad prior to starting the Keystone Bridge Company (with J. H. Linville) and eventually going into the steelmaking business.

§ This type of bridge design and construction is attributed to the German engineer Heinrich Gerber who patented and constructed the first cantilever type bridge in 1867.

¶ At the location of an uncompleted suspension bridge by John Roebling.

** This was the first use of cantilever construction using a suspended span.

†† By 1895, structural shapes were no longer made with iron, and steel was used exclusively.

‡‡ Before this, Baker may not have known of the work of engineers C. Shaler Smith or C. C. Schneider who had already designed and constructed cantilever railway bridges in the United States.

FIGURE 1.16 Fraser River Bridge built in 1884, British Columbia, Canada. (Photo by J. A. Brock, Canadian Pacific Archives NS.11416.)

FIGURE 1.17 The Forth Rail Bridge built over the Firth of Forth in 1890, Scotland. (From GFDL Andrew Bell, http://en.wikipedia, January 2005.)

undertaking completed in 1890 (Figure 1.17). It is an example of steel truss cantilever-type railway bridge construction on a grand scale with cantilever arms of 207 m (680 ft) supporting a 107 m (350 ft) suspended span. Baker used the relatively new Bessemer steel in the bridge even though it was an untested material for such large structures, and some engineers thought it too susceptible to cracking. The bridge is very stiff and the 90 mm (3½ in.) deflection, measured by designer Baker

TABLE 1.5
Notable Steel Cantilever Railway Bridges Constructed during 1876–1917

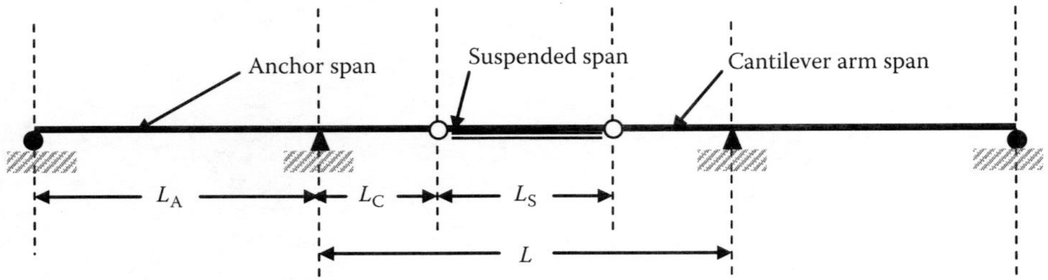

Location	Railroad	Engineer	Year	L_A m (ft)	L_C m (ft)	L_S m (ft)	L m (ft)
Posen, Poland	—	—	1876	23 (74)	—	—	45 (148)
Dixville, KY	Cincinnati Southern	C. Shaler Smith & G. Bouscaren	1877	~73(~240)	49.5 (162.5)	0	99 (325)
St. Paul, MN	Chicago, Milwaukee & St Paul	C. Shaler Smith	1880	74 (243)	49.5 (162)	0	99 (324)
Niagara Gorge, NY	Michigan Central	C. C. Schneider	1883	63 (207.5)	57 (187.5)	36.5 (120)	151 (495)
Fraser River, BC (Figure 1.16)	Canadian Pacific	C. C. Schneider	1884	32 (105)	32 (105)	32 (105)	96 (315)
St. John, NB	Canadian Pacific	G. H. Duggan	1885	43.5 (143), 58 (190)	43.5 (143), 58 (190)	43.5 (143)	145 (476)
Louisville, KY	—	—	1886	55 (180)	49 (160)	49 (160)	146 (480)
Point Pleasant, WV	—	—	1888	73 (240)	43.5 (142.5)	61 (200)	148 (485)
Tyrone, KY	Louisville and Southern	J. W. MacLeod	1889	~68.5 (~225)	—	—	551
Poughkeepsie, NY	Central New England	—	1889	80 (262.5)	~49 (~160)	~69.5 (~228)	167 (548)
Hoogly, India	East India	Sir B. Leslie	1890	—	—	—	—
Firth of Forth, Scotland (Figure 1.17)	North British	Sir B. Baker & Sir J. Fowler	1890	207 (680)	207 (680)	107 (350)	521 (1710)
Pecos River	Southern Pacific	A. Bonzano	1891	—	—	—	~61 (~200)
Red Rock, CO	—	J. A. L. Waddell	1892	50 (165)	50 (165)	100 (330)	200 (660)
Callao, Peru	Lima and Oroya	L. L. Buck	~1892	—	—	—	265
Cernavoda, Romania	—	—	~1892	71 (233.5)	50 (164)	90 (295)	190 (623)

(Continued)

TABLE 1.5 (*Continued*)
Notable Steel Cantilever Railway Bridges Constructed during 1876–1917

Location	Railroad	Engineer	Year	L_A m (ft)	L_C m (ft)	L_S m (ft)	L m (ft)
Memphis, TN	—	G. S. Morison	1892	69 (226) and 94.5 (310)	52 (170)	137 (450)	241 (790.5)
Ottawa, ON	Canadian Pacific	G. H. Duggan	1900	75 (247)	37.5 (123.5)	94 (308)	169 (555)
Loch Etive, Scotland	—	Sir J. W. Barry	1903	42.5 (139.5)	44.5 (146)	71 (232)	160 (524)
Pittsburgh, PA	Wabash	—	1904	106 (346)	69 (226)	110 (360)	248 (812)
Mingo Junction, OH	Wabash	—	1904	91 (298)	—	—	213 (700)
Thebes, IL	—	A. Noble & R. Modjeski	1905	79.5 (260.5)(½ of span)	46.5 (152.5)	111.5 (366)	204.5 (671)
Blackwell's Island (Queensboro), NY	City of New York (light rail)	G. Lindenthal	1907	143 (469.5) & 192 (630)	180 (591)	0	360 (1182)
Khushalgarth, India	—	Rendel & Robertson	1908	—	—	—	—
Westerburg, Prussia	Prussian State	—	1908	—	—	33.5 (110)	—
Daumer Bridge, China	Yunnan	—	1909	37.5 (123)	27.5 (90)	51 (168)	106 (348)
Beaver, PA	Pittsburgh & Lake Erie	—	1910	97.5 (320)	74 (242)	87 (285)	234.5 (769)
Quebec, QC (Figure 1.18)	Canadian Government	Duggan, Vautelet, Monsarrat, Modjeski, Schneider	1917	157 (515)	177 (580)	195 (640)	549 (1800)

under the heaviest locomotives available in the North British Railway, compared well with his estimate of 100 mm (4 in.). The bridge was further tested under extreme wind conditions with two long heavy coal trains, and the cantilever tip deflection was <180 mm (7 in.).

However, the Forth Railway Bridge used a large quantity of steel and was costly. This prompted engineers such as Theodore Cooper (who had worked with Eads on the St. Louis Bridge) to consider cantilever construction with different span types using relatively smaller members. Two such statically determinate railway bridges constructed were the 205 m (671 ft) main span bridge crossing the Mississippi River at Thebes, IL, and the 549 m (1800 ft) main span Quebec Bridge. The Thebes bridge, constructed in 1905, consists of five pin-connected through-truss spans, of which two spans are 159 m (521 ft) fixed double anchor spans [anchoring four 46.5 m (152.5 ft) cantilever arms] and three contain 111.5 m (366 ft) suspended spans. The Quebec Bridge, an example of economical long-span steel cantilevered truss construction for railroad loads, was completed in 1917 after two construction failures (Figure 1.18a and b). The initial 1907 failure was likely due to the calculation error in determining dead load compressive stresses in the bottom chord members during construction as the cantilever arms were increased in length. The bridge was redesigned,[*] and a new material, nickel steel,[†] was used in the reconstruction. In 1916, the suspended span truss fell while being hoisted into place. It was quickly rebuilt and the Quebec Bridge was opened to railway traffic in 1917 (Figure 1.18c). Another major cantilever-type bridge was not to be constructed until after 1930. The Quebec Bridge remains as the longest span cantilever bridge in the world.

Continuous spans were often used for long-span steel railway bridge construction in Europe but seldom in North America due to the practice of avoiding statically indeterminate structures. The first long-span continuous steel truss railway bridge was built by the Canadian Pacific Railway over the St. Lawrence River at Montreal in 1886 (Figure 1.19).[‡] The 124.5 m (408 ft) main spans were erected by the cantilever method with careful consideration of the deflections and stresses in the bottom chords of the truss. These were controlled during the cantilever erection procedure with cables and adjustment screws attached to the partially completed truss supported on the center pier of the continuous span. The Viaur Viaduct, built in 1898, was the first major steel railway bridge in France.[§]

Many iron and steel railway bridges were replaced in the first decades of the 20th century due to the development of substantially more powerful and heavier locomotives.[¶] Riveting was used extensively in Europe but only became a standard of American long-span steel railway bridge fabrication after about 1915[**] with construction of the Hell Gate and Sciotoville bridges. Hell Gate is a 298 m (978 ft) two-hinged steel trussed arch bridge in New York. It was built to carry four heavily loaded railroad tracks of the New England Connecting Railroad and Pennsylvania Railroad when it was completed in 1916 (Figure 1.20). It is the largest arch bridge in the world, and it was erected without the use of falsework. It was also the first major bridge to use high carbon steel members in its construction.[††] The Chesapeake & Ohio Railroad completed construction of two 236.5 m (775 ft)

---

[*] The original designers were Theodore Cooper and Peter Szlapka (Phoenixville Bridge Company). Following the collapse, a design was submitted by H. E. Vautelet; but the redesign of the bridge was tendered to various bridge companies and carried out by G. H. Duggan (St. Lawrence Bridge Company) under the direction of C. C. Schneider, R. Modjeski, and C. N. Monsarrat.

[†] Alloy nickel steel was first used in 1909 on the Blackwell's Island (now Queensboro) Bridge in New York. Nickel steel was also used extensively by J. A. L. Waddell for long-span railway bridge designs. A. N. Talbot conducted tests of nickel steel connections for the Quebec bridge reconstruction.

[‡] These spans were replaced in 1912 due to concern over performance under heavier train loads. The lead end of the simple span replacement trusses was supported by falsework on a movable barge during installation on an adjacent alignment.

[§] This cantilever truss arch bridge is unusual, in that it incorporates no suspended span, thereby rendering the structure statically indeterminate. Many engineers believe that the design was inappropriate for railroad loading.

[¶] Locomotive weights were typically about 40 tons in 1860, 70 tons in 1880, 100 tons in 1890, 125 tons in 1900, and 150 tons in 1910.

[**] Riveting was used on smaller spans earlier in the 20th century.

[††] Primarily, because of the high cost of alloy steel.

(a)

(b)

(c)

FIGURE 1.18 (a) The 1907 Quebec Bridge collapse, Canada. (b) The 1916 Quebec Bridge collapse, Canada. (c) The Quebec Bridge completed in 1917 across the St. Lawrence River at Quebec City, Canada. [(a) and (c) Courtesy of Carleton University Civil Engineering Exhibits; (b) Courtesy of A. A. Chesterfield, Library & Archives Canada.]

FIGURE 1.19 The St. Lawrence Bridge built in 1886 at Montreal, Canada. (Photo by J. W. Heckman Canadian Pacific Archives NS.1151.)

FIGURE 1.20 The Hell Gate Bridge built across the East River in 1916, New York. (Courtesy of Library of Congress from Detroit Publishing Co.)

span continuous steel trusses across the Ohio River at Sciotoville, OH, in 1917. This bridge remains the longest continuous span bridge in the world.

It has been estimated that in 1910 there were 80,000 iron and steel bridges* with a cumulative length of 2250 km (1400 miles) on about 300,000 km (190,000 miles) of track. Railroads were the

* The majority of the bridges were of steel construction by the beginning of the 20th century.

principal incentive for material and construction technology innovation in the latter half of the 19th century as the transition from wood and masonry to iron and steel bridges progressed in conjunction with the construction methods that minimized interference with rail and other traffic.* The art and science of bridge engineering was emerging from theoretical and experimental mechanical investigations prompted, to a great extent, by the need for rational and scientific bridge design to support a rapidly developing and expanding railroad infrastructure.

1.4 THE DEVELOPMENT OF RAILWAY BRIDGE ENGINEERING

1.4.1 Strength of Materials and Structural Mechanics

The early work of Robert Hooke (1678) concerning the elastic force and deformation relation, Jacob Bernoulli (1705) regarding the shape of deflection curves, Leonard Euler (1759) and C. A. Coulomb (1773) about elastic stability of compression members,† and Louis M. H. Navier (1826) on the subject of the theory of elasticity laid the foundation for the rational analysis of structures. France led the world in the development of elasticity theory and mechanics of materials in the 18th century, and produced well-educated engineers who, in many cases, became the leaders of American railway bridge engineering practice.‡ Railroad expansion continued at a considerable pace for another 80 years following inception in the 1820s. During that period, due to frequently increasing locomotive loads, it was not uncommon for railway bridges to be replaced at 10- to 15-year intervals. The associated demand for stronger, longer, and more reliable steel bridges, coupled with the in-service failures that were occurring, compelled engineers in the middle of the 19th century to engage in the development of a scientific approach to the design of iron and steel railway bridges.

American railway bridge engineering practice was principally experiential and based on the use of proven truss forms with improved tensile member materials. Many early Town, Long, Howe, and Pratt railway trusses were constructed without the benefit of a thorough and rational understanding of forces in the members. This lack of scientific approach was revealed by the many failures of railway bridge trusses between 1850 and 1870. This empirical practice had served the emerging railroad industry until heavier loads and longer span bridges, in conjunction with an increased focus on public safety,§ made a rational and scientific approach to the design of railway bridges imperative. In particular, American engineers developed a great interest in truss analysis because of the extensive use of iron trusses on US railroads. In response, Squire Whipple published the first rational treatment of statically determinate truss analysis (the method of joints) in 1847.

The rapid advancement of elasticity theory and engineering mechanics in Europe in the mid-19th century also encouraged French and German engineers to design iron and steel railway bridges using scientific methods. At this juncture, European engineers were also interested in the problems of truss analysis and elastic stability. B. P. E. Clapyron developed the three-moment equation in 1849 and used it in an 1857 postanalysis of the Britannia Bridge.¶ Concurrently, British railway bridge engineers were engaged in metals and bridge model testing for strength and stability. Following Whipple, two European railway bridge engineers, D. J. Jourawski** and Karl Culmann,

* For example, to not to interfere with railway traffic, the tubular spans of the Victoria Bridge at Montreal were replaced by extension of substructures and erecting steel trusses around the exterior of the tubular girders.

† Between 1885 and 1889, F. Engesser, a German railway bridge engineer, further developed a compression member stability analysis for general use by engineers.

‡ Charles Ellet (1830), Ralph Modjeski (1855), L. F. G. Bouscaren, Chief Engineer of the Cincinnati Southern Railroad (1873), and H. E. Vautelet, Bridge Engineer of the Canadian Pacific Railway (c. 1876) were graduates of early French engineering schools.

§ As a result of the considerable number of train accidents attributed to bridge structural failures.

¶ The design of the Britannia bridge was based on a simple span analysis even though Fairbairn and Stephenson had a good understanding of continuity effects on bending. The spans were erected simply supported, then sequentially jacked up at the appropriate piers and connected with riveted plates to attain span continuity.

** Jourawski was critical of Stephenson's use of vertical plate stiffeners in the Britannia Bridge.

provided significant contributions to the theory of truss analysis for iron and steel railway bridges. Karl Culmann, an engineer of the Royal Bavarian Railway, was a strong and early proponent of the mathematical analysis of trusses. He presented, in 1851, an analysis of the Howe and other proprietary trusses* commonly used in the United States. The Warren truss was developed in 1846,† and by 1850 W. B. Blood had developed a method of analysis for triangular trusses. Investigations, conducted primarily in the UK in the 1850s, into the effects of moving loads and speed were also being initiated. These investigations were preceded by theoretical work on the strength and vibration of railway bridges by Stokes and Willis in the late 1840s. Fairbairn also considered the effects of moving loads on determinate trusses as early as 1857.

J. W. Schwedler, a German engineer, presented the fundamental theory of bending moments and shear forces in beams and girders in 1862. Previously, he had also made a substantial contribution to truss analysis by introducing the method of sections. Also in 1862, A. Ritter improved truss analysis by simplifying the method of sections through development of the equilibrium equation at the intersection of two truss members. James Clerk Maxwell‡ and Culmann§ both published graphical methods for truss analysis. Culmann also developed an analysis for the continuous beams and girders that were often used in the 1850s by railroads. Later, in 1866, he published a general description of the cantilever bridge design method.¶ In subsequent years, Culmann also developed moving load analysis and beam flexure theories that were almost universally adopted by railroad companies in the Unites States and Europe. Developed by E. Winkler in 1867, bridge engineers were also given the powerful tool of influence lines for moving load analysis.

The effects of moving loads, impact (from track irregularities and locomotive hammer blow), pitching, nosing, and rocking of locomotives continued to be of interest to railway bridge engineers and encouraged considerable testing and theoretical investigation. Heavier and more frequent railway loadings were also creating an awareness of, and initiating research into, fatigue (notably by A. Wohler for the German railways).

North American engineers had recognized the need for rational and scientific bridge design, and, in response, J. A. L. Waddell published comprehensive books on steel railway bridge design in 1898 and 1916. Furthermore, Waddell and other engineers promoted independent bridge design in lieu of the usual proprietary bridge design and procurement practice of the American railroad companies. The Erie Railroad was the first to establish this practice and only purchased fabricated bridges from their own scientifically based designs. This soon became the usual practice of all American railroads.

1.4.2 Railway Bridge Design Specifications

Almost 40 bridges (about 50% of them iron) were collapsing annually in the United States during the 1870s. This was alarming as the failing bridges comprised about 25% of the entire American railway bridge inventory at that time. In particular, between 1876 and 1886, almost 200 bridge spans collapsed in the United States.

Proprietary railway bridges were failing, many due to a lack of rigidity and lateral stability. Most of these spans were built by bridge companies without the benefit of independent engineering design, and while some bridge companies had good specifications for design and construction, others did not. Therefore, without independent engineering design, railroad company officials required a good knowledge of bridge engineering to ensure public safety. This was not always the case, as demonstrated by the Ashtabula bridge collapse, where it was learned in the subsequent inquiry that the proprietary

* Culmann also analyzed Long, Town and Burr trusses using approximate methods for these statically indeterminate forms.

† The Warren truss was first used in a railway bridge in 1853 on the Great Northern Railway in the UK.

‡ Truss graphical analysis methods were developed and improved by J. C. Maxwell and O. Mohr between 1864 and 1874. Maxwell and W. J. M. Rankine were also among the first to develop theories for steel suspension bridge cables, lattice girders, bending force, shear force, deflection, and compression member stability.

§ Culmann published an extensive description of graphical truss analysis in 1866.

¶ Sir Benjamin Baker also outlined the principles of cantilever bridge design in 1867.

bridge design had been approved by a railroad company executive without bridge design experience.* To preclude further failures, American engineers were proposing the development and implementation of railroad company specifications that all bridge fabricators would use. Developments in the fields of materials and structural mechanics had supplied the tools for rational and scientific bridge design and provided the information required to establish specifications for iron and steel railway bridges.

The first specification for iron railway bridges was made by the Clarke, Reeves and Company (later the Phoenix Bridge Co.) in 1871. This was followed in 1873 by G. S. Morison's† "Specifications for Iron Bridges" for the Erie Railroad (formerly the New York, Lake Erie & Western Railroad). L. F. G. Bouscaren of the Cincinnati Southern Railroad published the first specifications with concentrated wheel loads in 1875.‡ Following this, in 1878, the Erie Railroad produced a specification (at least partially written by Theodore Cooper) with concentrated wheel loads that specifically referenced steam locomotive forces.

By 1876, the practice of bridge design by consulting engineers working on behalf of the railroad companies became more prevalent in conjunction with the expanding railroad business. In particular, Cooper's publications concerning railway loads, design specifications, and construction were significant contributions to the development of a rational basis for the design of steel railway bridges. Cooper produced specifications, intended for use by all railroad companies, for iron and steel railway bridges as early as 1884. These were updated until 1888, and Cooper delivered his first specification for steel railway bridges in 1890. Cooper's specifications for steel railway bridges were updated until 1905. Nevertheless, many railroad companies continued to use their own specifications.§ The multitude and variety of steel railway bridge specifications prepared by railroad companies, consulting engineers and fabricators, heralded the development of general specifications for steel railway bridges by the American Railway Engineering & Maintenance-of-way Association (AREMA) in 1906. This latter specification has been continuously updated and is the current recommended practice on which most North American railroad company design requirements are based. Other significant milestones in the development of general specifications for iron and steel railway bridges were as follows:

- 1867 St. Louis Bridge Co. specifications for Eads' steel arch¶
- 1873 Erie Railway Co. (G. S. Morison)
- 1877 Chicago, Milwaukee & St. Paul Railway Co. (C. Shaler Smith)
- 1877 Lake Shore & Michigan Southern Railway (C. Hilton)
- 1878 New York, Lake Erie, and Western Railroad (T. Cooper).
- 1880 Quebec Government Railways
- 1880 New York, Pennsylvania & Ohio Railroad
- 1886 Philadelphia and Reading Railroad
- 1890 Illinois Central Railroad
- 1894 Baltimore & Ohio Railroad (J. E. Greiner)

The large magnitude dynamic loads imposed on bridges by railroad traffic created a need for scientific design in order to ensure safe, reliable, and economical** construction. Railroad and consulting

* There were also material quality issues with the cast iron compression blocks, which were not discovered as the testing arranged by the Lake Shore and Michigan Southern Railroad company was inadequate.

† Morison and O. Chanute (an engineer educated in France) designed the 249-m (818 ft) long wrought iron viaduct for the Erie Railroad at Portage, New York, in 1875.

‡ However, it appears that the first use of concentrated wheel loads for bridge design was by the New York Central Railroad in 1862.

§ Cooper recommended a design live load of E40, but many railroads used their own specifications, which often specified design live loads equivalent to about Cooper's E50 and E60.

¶ This was not a general specification but was the first use of specification documents in the design and construction of railway bridges in the United States. The specification also included the first requirements for the inspection of material.

** This can be a critical consideration as most railway bridge construction projects are privately funded by railroad companies.

engineers engaged in iron and steel railway bridge design were the leaders in the development of structural engineering practice.* Evidence of this governance was the publication by AREMA, in 1906, of the first general structural design specification for steel bridges in the United States, where design loads and material stresses were specified.

Allowable stresses for materials were provided based on generally conservative estimates of the tensile and compressive strength of steel.† Allowable stresses for steel and fasteners have been continuously modified in subsequent editions of the Manual for Railway Engineering (MRE) reflecting the latest materials research and engineering.

In 1906, the design live load was specified as Cooper's E40. The design live loads were increased in 1920, 1935, and 1968 to Cooper's E60, E72, and E80, respectively. The 2015 AREMA Manual of Recommended Practice (MRE) indicates a minimum Cooper's E80 (SI equivalent is Cooper's EM360) live load with an alternate load consisting of four 100 kip (SI equivalent is 445 kN) axles.

The MRE has also specified various formulas for calculating steam and diesel locomotive impact forces (dynamic increment) in various editions of the AREMA specifications and recommended practices. Most North American railroads discontinued steam locomotive use in the early 1960s. The diesel and diesel-electric locomotives that followed, in conjunction with improved track design, construction, and maintenance practices, have allowed bridge designers to use smaller impact loads for design.‡ Figure 1.21 outlines the recommended dynamic increment (impact) in the AREMA and American Railway Engineering Association (AREA) specifications and recommended practices of 1906, 1920, 1935, and 1968.§

Well-maintained steel railway bridges designed prior to the 1960s with relatively conservative allowable stresses¶ for heavy steam locomotives** with large dynamic increment (steam locomotive impact) continue to safely and reliably carry modern railway traffic.†† Many of these bridges are over 100 years old, providing evidence of the exceptional design, fabrication, and erection skills of early steel railway bridge engineers, and the scientific methods and specifications that guided their work. Modern steel railway bridge design practice is able to continue this record of safety and reliability using cost-effective materials, analysis and design methods based on updated design specifications, guidelines, codes, and recommendations such as AREMA MRE Chapter 15—Steel Structures.

1.4.3 Modern Steel Railway Bridge Design

The basic forms of ordinary steel railway superstructures have not changed substantially since the turn of the 20th century. Steel arch, girder, and truss forms are still routinely designed. However, considerable improvements in materials, structural analysis and design, and fabrication and erection technology occurred during the 20th century.

---

* The advanced state of steel design and construction knowledge possessed by railway bridge engineers made them a greatly sought after resource by architects from about 1880 to 1900 during the rebuilding of Chicago after the Great Fire.

† The allowable tensile stress for steel was typically specified to be about 110 MPa (16,000 psi) in the AREMA specifications of the first quarter of the 20th century.

‡ Steam locomotive impacts were very large due to eccentric reciprocating wheel motion or "hammer blow."

§ Figure 1.21 is shown in US Customary or Imperial units only as the impact formulae of these older specifications and recommended practices were provided in only US Customary or Imperial units. The AREMA (2015) recommendations for impact loads in Chapter 15 are the same as the 1968 recommendations (see Chapter 4).

¶ Particularly for bridges designed in the early part of the 20th century.

** Steam locomotives used in the early part of the 20th century weighed about the same as modern diesel locomotives.

†† Modern rail car axle loads are typically not greater than modern diesel locomotive or older steam locomotive combined static and dynamic loads. However, older bridge design specifications did not consider fatigue as a design limit state (and did not need to because of the light rail cars pulled by few heavy locomotives). Nevertheless, older railway bridges generally perform well in the modern cyclical railway live load environment due to low allowable design stresses, internal redundancy of riveted connections, and the use of modern methods of fatigue life evaluation. Modern steel bridges must be designed considering fatigue due to the large number of high-magnitude tensile stress ranges experienced by some steel superstructure members (typically short members) and details.

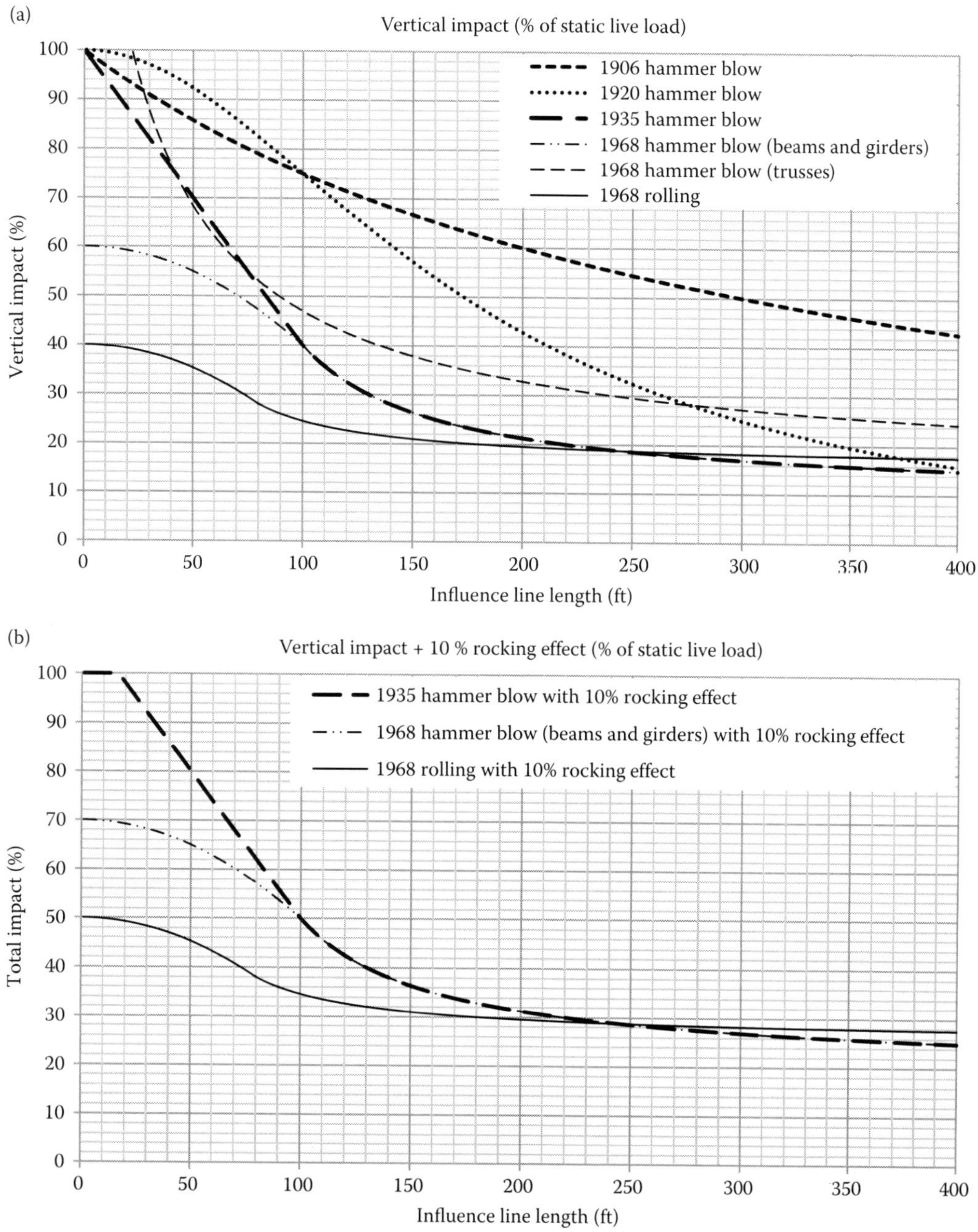

FIGURE 1.21 AREMA and AREA impact loads since 1905. (a) Vertical impact. (b) Total impact (with 10% rocking effect).

The strength, ductility, toughness, corrosion resistance, and weldability properties of structural steel have improved significantly since the middle of the 20th century. These material enhancements, combined with a greater understanding of planning considerations associated with modern bridge design and construction, have enabled the design of economical, reliable, and safe railway superstructures.

Modern structural analysis has also allowed considerable progress regarding the safety and economics of modern railway superstructures. Vast advancements in the theory of elasticity and structural mechanics were made in the 19th century as a result of railroad expansion. Today, the steel railway bridge engineer can take advantage of modern numerical methods, such as the matrix displacement (or stiffness) method, to solve difficult and complex structures. These methods of modern structural analysis have further evolved into multipurpose and specialized finite-element programs capable of linear elastic, nonlinear, static, dynamic (including seismic), stability, fracture, and other analyses using even small digital computers. In addition, modern methods of structural design, such as probabilistic (reliability) methods, that continue to enable the efficient and safe design of modern structures have ensued from recent research and practice.

Advances in manufacturing and fabrication technologies have permitted plates, sections, and members of large and complex dimensions to be fabricated and erected using superior fastening techniques such as welding and high-strength bolting. Modern fabrication with computer-controlled machines performing shop operations such as cutting, punching, drilling, bending, and welding has produced economical, expedient, and reliable steel railway superstructures. Advanced technologies such as radiographic and ultrasonic testing have enhanced modern fabrication quality control and quality assurance execution. Modern steel superstructure erection methods and procedures have also benefitted from technological advances in erection equipment such as large cranes, launching machinery, and transporter units.

BIBLIOGRAPHY

Akesson, B., 2008, *Understanding Bridge Collapses*, Taylor & Francis, London.

Baker, B., 1862, *Long-Span Railway Bridges*, Reprint from Original, BiblioBazaar, Charleston, SC.

Bennett, R. and Skinner, T., 1996, *Bridge Failures, Recent and Past Lessons for the Future*, American Railway Bridge and Building Association, Homewood, IL.

Billington, D.P., 1985, *The Tower and the Bridge: The New Art of Structural Engineering*, Princeton University Press, Princeton, NJ.

Buck, L.L., 1898, The Niagara railway arch, *Transactions of the American Society of Civil Engineers*, Vol. 40, No. 2, 125–150.

Chatterjee, S., 1991, *The Design of Modern Steel Bridges*, BSP Professional Books, Oxford, UK.

Clark, J.G., 1939, *Specifications for Iron and Steel Railroad Bridges prior to 1905*, Author, Urbana, IL.

Clary, J.N., N.D., *History of Early Bridge Specifications*, Virginia Department of Transportation, Structure and Bridge Division, Charlottesville, VA.

Cooper, T., 1889, *American Railroad Bridges*, Engineering News, New York.

Gasparini, D.A. and Fields, M., 1993, Collapse of Ashtabula bridge on December 29, 1876, *Journal of Performance of Constructed Facilities*, Vol. 7, No. 2, 109–125.

Ghosh, U.K., 2006, *Design and Construction of Steel Bridges*, Taylor & Francis, London.

Griggs, F.E., 2002, Kentucky high river bridge, *Journal of Bridge Engineering*, Vol. 7, No.2, 73–84.

Griggs, F.E., 2006, Evolution of the continuous truss bridge, *Journal of Bridge Engineering*, Vol. 12, No.1, 105–119.

Griggs, F., 2015, Historic structures: The Quebec bridge, *Structure Magazine*, December 2015.

Griggs, F., 2016, Historic structures: Roebling's Niagara river railroad suspension bridge—1885, *Structure Magazine*, June 2016.

Irwin, D.B., Johns, K.W., and Hauschildt, K.G., 2013, The design of the new Portageville bridge, *Proceedings of the American Railway Engineering and Maintenance-of-Way Association*, Indianapolis, IN.

Johnson, A., 2008, CPR high level bridge at Lethbridge, Occasional Paper No. 46, Lethbridge Historical Society, Lethbridge, AB.

Kurrer, K.-E., 2008, *The History of the Theory of Structures*, Ernst & Sohn Verlag, Berlin.

Kuzmanovic, B.O., 1977, History of the theory of bridge structures, *Journal of the Structural Division*, Vol. 103, No. 5, 1095–1111.

Marianos, W.N., 2008, George Shattuck Morison and the development of bridge engineering, *Journal of Bridge Engineering*, Vol. 13, No.3, 291–298.

Middleton, W.D., 2001, *The Bridge at Quebec*, Indiana University Press, Bloomington, IN.

Motley, P.B., 1914, *Double Tracking of the CPR's St. Lawrence River Bridge*, Canadian Railway and Marine World, Montreal, QC.

Petroski, H., 1996, *Engineers of Dreams*, Random House, New York.

Plowden, D., 2002, *Bridges: The Spans of North America*, W.W. Norton & Co., New York.
Ryall, M.J., Parke, G.A.R., and Harding, J.E., 2000, *Manual of Bridge Engineering*, Thomas Telford, London.
Stokes, G.G., 1849, Discussion of a differential equation relating to the breaking of railway bridges, *Transactions of the Cambridge Philosophical Society*, Vol. 8, No.5, 707–35.
Timoshenko, S.P., 1983, *History of Strength of Materials,* Dover Publications, New York.
Todhunter, I., 1885, *A History of the Theory of Elasticity and of the Strength of Materials*, Volume I, Dover Publications Reprint 1960, New York.
Todhunter, I. and Pearson, K., 1893, *A History of the Theory of Elasticity and of the Strength of Materials*, Volume II, Dover Publications Reprint 1960, New York.
Troitsky, M.S., 1994, *Planning and Design of Bridges,* John Wiley & Sons, New York.
Tyrrell, H.G., 1911, *History of Bridge Engineering*, H.G. Tyrell, Chicago, IL.
Unsworth, J.F., 2001, Evaluation of the load capacity of a rehabilitated steel arch railway bridge, *Proceedings of 3rd International Arch Bridges Conference*, Presses de L'ecole Nationale des Ponts et Chaussees, Paris.
Waddell, J.A.L., 1898, *De Pontibus*, John Wiley & Sons, New York.
Waddell, J.A.L., 1916a, *Bridge Engineering—Volume 1*, John Wiley & Sons, New York.
Waddell, J.A.L., 1916b, *Bridge Engineering—Volume 2*, John Wiley & Sons, New York.
Werry, S.D., 1997, Rails across the river: The story of the St. Lawrence Bridge (1881–1915), *Canadian Journal of Civil Engineering*, Vol.24, No. 3, 480–488.
Whipple, S., 1873, *Treatise on Bridge Building*, Reprint from original 2nd ed., University of Michigan, Ann Arbor, MI.
Willis, R., 1849, Application of iron to railway structures, Commissioner's report, William Clowes, London.

2 Steel for Modern Railway Bridges

2.1 INTRODUCTION

Steel development in the latter part of the 20th century has been remarkable. Modern steel is made of iron with small amounts of carbon, manganese, and traces of other alloy elements added to enhance physical properties. Chemical and physical metallurgical treatment has enabled improvements to many steel properties.

Mild carbon and high-strength low-alloy (HSLA) steels have been used for many years in railway bridge design and fabrication. Recent research and development related to high-performance steel (HPS) metallurgy has provided modern structural steels with even further enhancements to physical properties.

The important physical properties of modern structural bridge steels are:

- Strength
- Ductility
- Fracture resistance or toughness
- Weldability
- Corrosion resistance

These physical properties and general steel quality are controlled in the manufacturing process for structural steel shapes and plates used for superstructure fabrication.

2.2 MANUFACTURE OF STRUCTURAL STEEL

Significant advances in the art and science of steelmaking have occurred since the early part of the 20th century. Many of these advances have been related to the need for steels of increasingly higher strength with improved ductility, fracture toughness, corrosion resistance, and weldability properties. Modern high-strength structural steel shapes and plates are manufactured using chemistry* and process† to control these important physical and mechanical properties. Steel chemistry has the greatest influence on strength, ductility, fracture toughness, corrosion resistance, and weldability. Carbon and HSLA steels attain their mechanical properties through chemistry. However, increasing the strength of HSLA steel and HPS also requires supplemental heat treatment processes. HPS attains its mechanical properties through supplemental heat treatment in conjunction with chemistry manipulation.

Carbon and manganese are hardening and strengthening alloys. Carbon is the principal element controlling the mechanical properties of steel. The strength of steel may be increased by increasing the carbon content, but at the expense of ductility and weldability. Steel also contains deleterious elements, such as sulfur and phosphorous, that are present in the iron ore used to manufacture steel. Manganese also combines with sulfur to preclude the detrimental effects associated with the presence of elemental sulfur. Aluminum and silicon are alloyed to promote deoxidization and improve general steel quality. Chromium and copper are alloyed to increase atmospheric corrosion

* Chemical composition ranges for elements in various grades of structural steel are specified in ASTM and other applicable steel material specifications.

† Casting, rolling, and heat treatment operations.

resistance. Table 2.1 indicates the effects of various alloying elements on the physical and mechanical properties of steel.

The modern steelmaking process involves continuous casting of the molten steel (iron carbon, manganese, and other alloy elements) into slabs or blooms with relatively high cooling rates to discourage segregation of the elements.* Continuous casting provides plates with uniform physical properties at low production cost. Nevertheless, structural steel for railway superstructure fabrication requires steel mill process quality control to ensure that properties are appropriate in regard to fatigue and fracture performance. Specifically, measures are necessary to ensure that microscopic crack-like defects† do not occur due to trapped gasses and to minimize alloy element segregation

TABLE 2.1
Effects of Alloying Elements on Physical and Mechanical Properties of Steel

	Effect on Mechanical and Physical Properties	
Element	**Increase or Improve**	**Decrease or Reduce**
Aluminum (Al)	Toughness (with Si-killed steel)	Surface quality, hardness (aging)
Boron (B)	Hardenability (Q&T steels), strength (low-C Mo steels)	
Carbon (C)[a]	Strength, hardenability	Ductility, toughness, weldability
Chromium (Cr)[a]	Strength (high temperature), hardenability, toughness, abrasion resistance, corrosion resistance	Weldability
Columbium (Co)	Strength	Toughness
Copper (Cu)[a]	Corrosion resistance, strength, hardenability	Ductility, surface quality
Hydrogen (H)		Ductility (embrittlement)
Manganese (Mn)[a]	Strength, hardenability, sulfur control, toughness, corrosion resistance, ductility	Weldability
Molybdenum (Mo)[a]	Strength (high temperature), hardenability, abrasion resistance, corrosion resistance, weldability	Toughness, ductility
Nickel (Ni)[a]	Strength, toughness, hardenability, corrosion resistance, ductility	Weldability
Nitrogen (N)	Strength	Ductility
Oxygen (O)		Ductility, toughness
Phosphorus (P)[a]	Strength, hardenability, corrosion resistance	Ductility, weldability
Silicon (Si)[a] with other alloys	Strength, toughness, hardenability, ductility, deoxidation	Weldability, surface quality
Sulfur		Inclusions, weld porosity, and cracking
Titanium (Ti)	Strength, abrasion resistance, deoxidation, grain refinement	
Tungsten (W)	Strength (high temperature), hardenability, toughness, abrasion resistance	
Vanadium (V)[a]	Strength (high temperature, hardenability, abrasion resistance, deoxidation, grain refinement	

[a] Indicates the most common steel alloy elements.

* In particular, carbon segregation during casting may degrade steel uniformity, ductility, fracture toughness, and weldability. New HPSs with lower carbon content preclude excessive carbon segregation during casting.

† These defects occur at grain boundaries that are opened as trapped gasses escape.

during slab solidification and subsequent hot rolling operations. Atmospheric corrosion-resistant (weathering) steel chemistry also requires that production processes yield fine grain-size steel.

Degassing or "killing" steel involves alloying aluminum and/or silicone to reduce the oxygen available for the production of carbon dioxide. Aluminum alloying also promotes fine grain size. Low hydrogen processes such as vacuum degassing* can also be used to further protect against small crack-like defects caused by escaping hydrogen gases. Structural steel for railway superstructures must be killed or semikilled to reduce the creation of gases that affect fatigue strength and fracture resistance.

The cooled cast slabs are reheated and passed back and forth through a succession of rollers to create plates and shapes. Heat and roller pressure plastically deform the plate or shape to final dimensions for fabrication, but segregated alloy elements will tend to form planar inclusions.† Element segregation control is necessary to avoid the possibility of subsequent lamellar tearing.‡ Controlled cooling during the steel hot rolling process is often required to control element segregation, particularly for thicker plates such as those typically used for the flange plates of modern welded girders.

Nevertheless, hot-rolled structural shapes and plates may require postrolling heat treatment to improve physical and mechanical properties. Heat treatments such as normalizing, quenching and tempering (Q&T), and stress relieving may be used to enhance strength, ductility, and/or fracture toughness.

The quenching process following hot rolling increases strength, but at the expense of ductility and toughness.§ Normalizing¶ refines grain size and improves microstructure uniformity, providing increased ductility and fracture resistance.** Normalizing involves reheating the shape or plate between 900°C and 925°C (1650°F and 1700°F) and allowing the steel to cool slowly in air. However, because this postmanufacture heat treatment requires a furnace, shape and plate lengths for normalizing are often practically limited to about 15 m (50 ft).

Higher strength steel plates may be attained through the heat treatment of HSLA steel plates.†† These heat-treated low-alloy steel plates (Q&T steels) are not typically used for steel railway superstructure fabrication due to concerns with weldability.‡‡ Heat-affected zone (HAZ) strength may be detrimentally affected by welding, and welding consumables with equivalent yield and ultimate strength to that of the heat-treated low-alloy steel base metal are difficult to obtain. Thick plates and higher strength Q&T steels may also increase the propensity of the steel to hydrogen crack during welded fabrication.§§ Heat-treated low-alloy steel plates are produced by a Q&A process by reheating the plates to 900°C (1650°F) until an austenitic¶¶ microstructure is achieved. Subsequent rapid cooling provides increased hardness and strength, but at the expense of ductility and fracture toughness. Ductility and toughness may be improved through tempering by reheating between 425°C and 675°C (800°F and 1250°F) and slow cooling. Tempering results in a slight reduction in strength, but with greater ductility and fracture toughness. However, since a furnace is required, the production of heat-treated low-alloy steel (Q&T steels) may also be limited to lengths of about 15 m (50 ft).

Stress relieving is not typically required following the rolling process,*** but if necessary a specified heat can be applied followed by very slow cooling to relax internal stresses.

---

* Vacuum degassing is used for the production of modern HPS to further control fatigue strength and fracture resistance.

† Typically at mid-thickness of thicker plates due to lower cooling rate. Element segregation is potentially greater in copper-alloyed atmospheric resistant steels.

‡ Generally occurs due to loading and/or welding operations.

§ In particular, for thick plates.

¶ Normalizing is typically specified by bridge owners for plates thicker than about 38 mm (1–1/2 in.) or 50 mm (2 in.).

** Ductility and toughness are improved by tempering with only a small effect on strength.

†† Many modern 485 MPa (70 ksi) and 690 MPa (100 ksi) yield strength steels attain their increased strength through heat treatment of 345 MPa (50 ksi) steel chemistry.

‡‡ However, some Q&T steels have been developed with low carbon content and good weldability.

§§ Fabrication-induced hydrogen cracking may be precluded by using an under-matching strength filler metal, increasing the preheat or welding heat input (see Chapter 10).

¶¶ The crystal structure of the steel transforms from ferrite to austenite when heated above 900°C (1650°F).

*** Typically required following some welding, cold bending, cutting, or machining operations to relieve residual stresses (see Chapter 10).

Postrolling heat treatments, such as normalizing, may be precluded by controlled hot rolling. Controlled hot rolling involves regulating heating rates, cooling rates, and holding times during the rolling process. Modern controlled hot rolling of plates may be performed precisely using the thermomechanically controlled process (TMCP).* TMCP equalizes plate temperature by localized heating and variable cooling rate sprays. TMCP produces plates with a fine and uniform microstructure.†

Controlled rolling heat treatment is not limited by plate length, but by plate thickness. Plate thicknesses greater than 50 mm (2 in.) are precluded by the roll pressures required for thicker plates at the lower rolling temperatures used in portions of the controlled hot rolling process. However, in many cases,‡ controlled hot rolling may preclude the need to normalize and avoid limitations on plate length.§

In some cases, fabricators may need to understand the tempering temperatures used in production heat treatments to ensure that mechanical properties are not altered by shop heating above the tempering temperatures.

2.3 ENGINEERING PROPERTIES OF STEEL

2.3.1 Strength

2.3.1.1 Elastic Yield Strength of Steel

Strength may be defined in terms of tensile yield stress, F_y, which is the point where plastic behavior commences at almost constant stress (unrestricted plastic flow). Strength or resistance may also be characterized in terms of the ultimate tensile stress, F_U, which is attained after yielding and significant plastic behavior. An increase in strength is associated with plastic behavior (due to strain hardening) until the ultimate tensile stress is attained (Figure 2.1). The most significant properties of steel that are exhibited by stress–strain curves are the elastic modulus (linear slope of the initial portion of the curve up to the proportional limit), the existence of yielding, and plastic behavior, with some unrestricted flow and strain hardening, until the ultimate stress is attained.

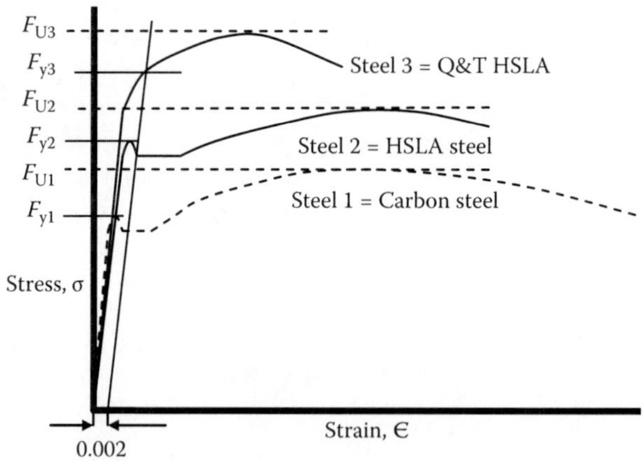

FIGURE 2.1 Engineering tensile stress–strain behavior of typical bridge structural steels.

* Not all steel mills have this technology.

† Grain size reduction and uniformity increase strength, ductility, and toughness.

‡ For economic and technical reasons (see Chapters 7 and 10), girder flange plate thickness is typically limited by bridge owners to less than about 65 mm (2–1/2 in.) or 75 mm (3 in.).

§ Limited plate lengths may require that flange plates of girders be spliced with shop butt welds. Such butt welds, particularly in tensile regions, must be carefully inspected (see Chapter 10).

Yield stress in tension can be measured by simple tensile tests (ASTM, 2015). Yield stress in compression is generally assumed to be equal to that in tension.* Yield stress in shear may be established from theoretical considerations of the yield criteria. Various yield criteria have been proposed, but most are in conflict with experimental evidence that yield stress is not influenced by hydrostatic (or octahedral normal) stress. However, two theories, the Tresca and von Mises yield criteria, meet the necessary requirement of being pressure independent. The von Mises criterion is most suitable for ductile materials with similar compression and tensile strength, and it also accounts for the influence of intermediate principal stress (Chen and Han, 1988; Chatterjee, 1991). It has also been shown by experiment that the von Mises criterion best represents the yield behavior of most metals (Chakrabarty, 2006).

The von Mises yield criterion is based on the octahedral shear stress, τ_h, attaining a critical value, τ_{hY}, at yielding. The octahedral shear stress, τ_h, in terms of principal stresses, $\sigma_1, \sigma_2, \sigma_3$, is

$$\tau_h = \frac{1}{3}\sqrt{(\sigma_1-\sigma_2)^2+(\sigma_1-\sigma_3)^2+(\sigma_2-\sigma_3)^2}. \quad (2.1)$$

Yielding in uniaxial tension will occur when $\sigma_1 = \sigma_Y$ and $\sigma_2 = \sigma_3 = 0$. Substitution of these values into Equation (2.1) provides

$$\tau_{hY} = \frac{\sqrt{2}}{3}\sigma_Y \quad (2.2)$$

or the criterion that, at yielding,

$$\sigma_Y = \frac{1}{\sqrt{2}}\sqrt{(\sigma_1-\sigma_2)^2+(\sigma_1-\sigma_3)^2+(\sigma_2-\sigma_3)^2}, \quad (2.3)$$

where σ_Y is the yield stress from the uniaxial tensile test.

It can also be shown that the octahedral shear stress at yield is (Hill, 1989)

$$\tau_{hY} = \sqrt{\frac{2}{3}}\tau_Y, \quad (2.4)$$

which when substituted into Equation (2.2) provides

$$\tau_Y = \frac{\sigma_Y}{\sqrt{3}}, \quad (2.5)$$

where τ_Y is the yield stress in pure shear. Therefore, a theoretical relationship is established between yield stress in shear and tension.

Example 2.1

Determine the allowable shear stress for use in design, f_v, if the allowable axial tensile stress, f_t, is specified as $0.55F_y$ and $0.60F_y$ (F_y is the axial tensile yield stress).

For $f_t = 0.55F_y$, f_v = allowable shear stress $= \dfrac{0.55F_y}{\sqrt{3}} = 0.32F_y$.

For $f_t = 0.60F_y$, f_v = allowable shear stress $= \dfrac{0.60F_y}{\sqrt{3}} = 0.35F_y$.

AREMA (2015) recommends an allowable shear stress for structural steel of $0.35F_y$.

* It is typically around 5% higher than the tensile yield stress.

2.3.1.2 Fatigue Strength of Steel

Localized material failure can occur when applied cyclical stresses* are greater than a threshold tensile stress range, but below the elastic yield stress. On a microscopic level, cyclical stresses may precipitate the movement of atomic dislocations creating slip bands and surface discontinuities,† particularly at grain boundaries.‡ Progressive microscopic material failure may involve a relatively long time to initiate a macroscopic crack, but some superstructure design details§ and fabrication imperfections¶ may cause more rapid fatigue crack initiation and propagation that could lead to failure.** The fatigue behavior of macroscopic detail stress raisers concerns the bridge design engineer. The macroscopic fatigue strength of steel railway superstructures is related to:

- Cyclical stress state (magnitude and number of cycles)
- Manufacturing residual stresses (casting and rolling)
- Design geometric details (welded attachments and stress concentrations)
- Fabrication quality and process residual stresses [rolling, cutting, welding (see Chapter 10)]
- In-service temperatures and atmospheric environment

A stress-life approach for the fatigue strength design of railway superstructures is appropriate for high-cycle fatigue at stress levels below the yield strength. The macroscopic fatigue behavior of common design details has been investigated by testing at nominal stress ranges that incorporate the stress concentration affects of the design detail. Analysis of the test data reveals a linear logarithmic relationship, with slope $-m$, between the number of cycles to failure and the constant amplitude stress range above a threshold or constant amplitude fatigue limit stress range as shown in Figure 2.2 (see also Chapter 5).

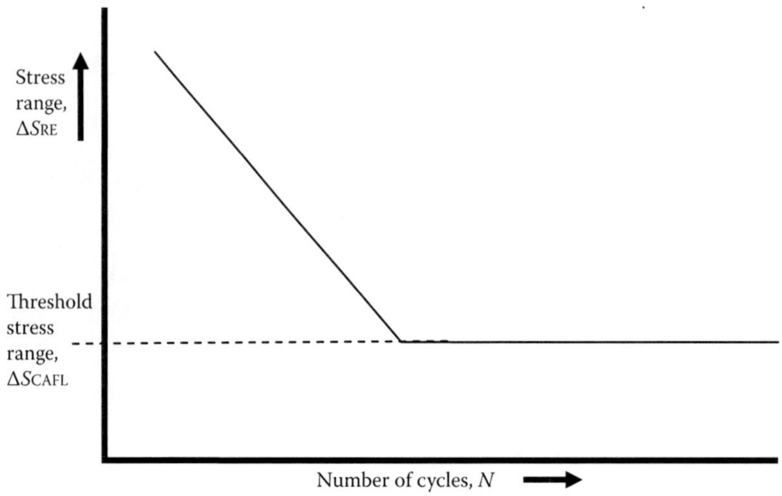

FIGURE 2.2 Fatigue strength behavior of typical bridge structural steels.

* Railway train loads are highly cyclical in nature creating a high-cycle fatigue regime, particularly on members with short influence lines (see Chapter 5).

† Essentially microscopic cracks.

‡ A principal reason for reduced grain size practice in steelmaking.

§ Design details such as welded attachments, intersecting welds, and copes have lower fatigue strength. The lower fatigue strength of design details is reflected by a lower value of the constant in the linear logarithmic relationship between stress range and number of cycles to failure (see Chapter 5).

¶ Many imperfections are avoided or mitigated during design, fabrication, and quality control/quality assurance (QC/QA) testing (see Chapter 10), but some may be unavoidable or undetected.

** Fatigue analysis is probabilistic and, therefore, fatigue "failure" is defined based on statistical criteria.

2.3.2 Ductility

Ductility is the ability of steel to withstand large strains after yielding and prior to fracture. Ductility is necessary in railway bridges and many civil engineering structures to provide advance warning of overstress conditions and potential failure. Ductility also enables the redistribution of stresses when a member yields in redundant systems, in continuous members, and at locations of stress concentrations (i.e., holes and discontinuities). Adequate ductility also assists in the prevention of lamellar tearing in thick elements.* Ductility is measured by simple tensile tests and specified as a minimum percentage elongation over a given gage length [usually 200 mm (8 in.)]. Only ductile steels are used in modern railway bridge fabrication.

2.3.3 Fracture Resistance

Brittle fracture occurs as cleavage failure with little associated plastic deformation. Once initiated, brittle fracture cracks can propagate at very fast rates as elastic strain energy is released (Fisher, 1984; Barsom and Rolfe, 1987). In steel railway bridges, this fracture can be initiated below the yield stress.

Fabrication-induced cracks, notches, discontinuities, or defects can create stress concentrations that may initiate brittle fracture in components in tension. Welding can also create hardened HAZ, hydrogen-induced embrittlement, and high residual tensile stresses near welds. All of these may be of concern with respect to brittle fracture. Rolled sections might contain rolling inclusions and defects that may also initiate brittle fracture. Thick plates are more susceptible to brittle fracture than thinner plates. Other factors that affect brittle fracture resistance are galvanizing (hot-dip), poor heat treatments, and the presence of nonmetallic alloy elements. Brittle fracture most often occurs from material effects in cold service temperatures, high load rates, and/or triaxial stress states (Figure 2.3).

Normal railway bridge strain rate application is relatively slow (in comparison to, e.g., machinery components or testing machines). Brittle fracture can, however, be caused by high strain rates associated with large impact forces from live loads.† Triaxial stress distributions and high stress

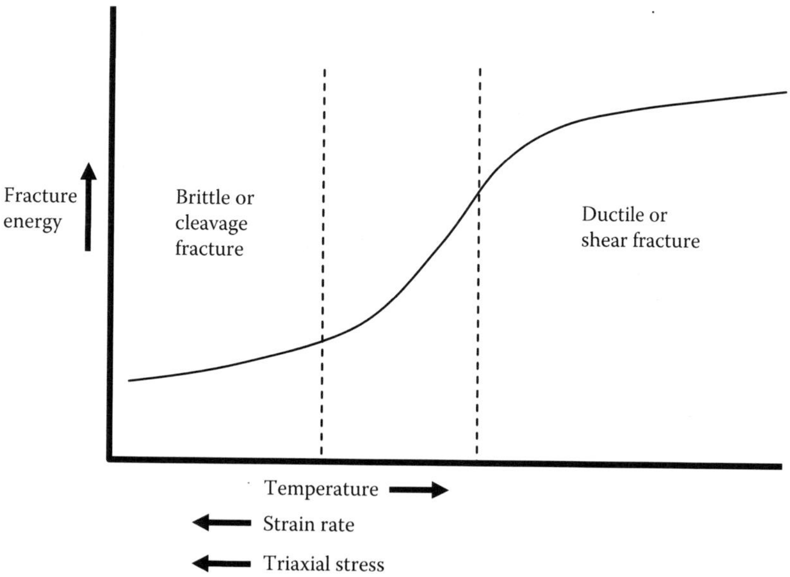

FIGURE 2.3 Fracture toughness behavior of typical bridge structural steels.

* Such as the relatively thick flange plates typically required for railway loads on long-span girders.

† Caused by poor wheel and/or rail conditions, derailment, or other vehicular collision.

concentrations can be avoided by good detailing and welding practice. Thick elements are often more susceptible to brittle fracture due to the triaxial stress state. Normalizing, a supplemental heat treatment, can be beneficial in improving material toughness through grain size reduction in thick elements (Brockenbrough, 2011). Adequate material toughness for the coldest service temperature likely to be experienced by the bridge (generally a few degrees cooler than the coldest ambient temperature) is critically important.

Temperature changes the ductile to brittle behavior of steel. A notch ductility measure, the Charpy V-notch (CVN) test, is used to ensure adequate material toughness against brittle fracture at intended service temperatures. A fracture control plan should ensure that weld metals have at least the same notch ductility as the specified base metal, and some specifications indicate even greater notch toughness requirements for welds in fracture critical members (FCMs). CVN testing is performed to establish notch ductility or material toughness based on energy absorbed at different test temperatures. CVN testing is done at a rapid load rate, so adjustments are made to the specified test temperature to account for the greater ductility associated with the slower strain rate application of railway traffic. For design purposes, temperature service zones are established with a specified minimum energy absorption at a specified test temperature for various steel types and grades. CVN requirements are often specified separately for FCM and non-FCM. Tables 2.2 and 2.3 show the specified CVN test requirements for non-FCM and FCM, respectively, for steel railway bridges recommended by AREMA (2015).

2.3.4 Weldability

If the carbon content of steel is less than 0.30%, it is generally weldable. Higher strength steels, where increased strength is attained through increased carbon and manganese content, will become hard and difficult to weld. The addition of other alloy elements to increase strength (Cr, Mo, and V) and corrosion resistance (Ni and Cu) will also reduce the weldability of steel.

The weldability of steel is estimated from an empirical carbon equivalency equation,[*] given as

$$CE = C + \frac{Mn + Si}{6} + \frac{Ni + Cu}{15} + \frac{Cr + Mo + V}{5}, \tag{2.6}$$

silicon, nickel, copper, chromium, molybdenum, and vanadium in the steel, respectively. Carbon equivalence, CE, of about 0.5% or greater indicates that special weld treatments may be required.

Weld cracking may result from resistance to weld shrinkage upon cooling. Thicker elements are more difficult to weld. Preheat and interpass temperature control, in conjunction with the use of low hydrogen electrodes, will prevent welding-induced hardening and cracking (see Chapter 10).

Modern high-strength structural steels have been developed with excellent weldability.[†] The increase in weldability enables limited preheat requirements and postweld treatments (translating into fabrication savings), and may eliminate hydrogen-induced weld cracking.

2.3.5 Corrosion Resistance

Atmospheric corrosion-resistant (weathering) steel chemistry (using chromium, copper, nickel, and molybdenum alloys) is such that a thin iron oxide film forms upon initial wetting cycles and prevents the further ingress of moisture. This type of corrosion protection works well where there are alternate wetting and drying cycles. It may not be appropriate in locations where deicing chemicals and salts are prevalent, in marine environments, or where there is a high level of sulfur content in the atmosphere.

---

[*] Other similar formulas also exist such as the Deardon and O'Neill equation and others formulated in Japan.

[†] For example, HPSs for bridges such as ASTM A709M (A709) HPS 345W (50W), 485W (70W), and 690W (100W).

Weldability is slightly compromised because carbon equivalence, CE (Equation 2.6), is raised through the addition of alloy elements for corrosion resistance. However, these steels have about four times the resistance to atmospheric corrosion as carbon steels (Kulak and Grondin, 2002), which makes their use in bridges economical from a life cycle perspective. Corrosion resistance can be estimated by a Corrosion index (CI), based on an empirical alloy content equation,*

$$CI = 26.01(Cu) + 3.88(Ni) + 1.20(Cr) + 1.49(Si) + 17.28(P) - 7.29(Cu)(Ni)$$
$$- 9.10(P)(Ni) - 33.39(Cu)^2, \quad (2.7)$$

where Cu, Ni, Cr, Si, and P are the percentage of elemental copper, nickel, chromium, silicon, and phosphorus in the steel, respectively. A CI of 6.0 or higher† is typically required for bridge weathering steels.

Nonweathering steels can be protected with paint or sacrificial coatings (hot-dip or spray-applied zinc or aluminum). Shop applied three-coat paint systems are commonly used by many North American railroads. Two, and even single, coat painting systems are being assessed by the steel coatings industry and bridge owners. An effective modern three-coat paint system consists of a zinc-rich primer, epoxy intermediate coat, and polyurethane top coat. For aesthetic purposes, steel with zinc or aluminum sacrificial coatings can be top-coated with epoxy or acrylic paints.

2.4 TYPES OF STRUCTURAL STEEL

2.4.1 Carbon Steels

Modern carbon steel contains only manganese, copper, and silicon alloys. Mild carbon steel has a carbon content of 0.15%–0.29%, and a maximum of 1.65% manganese (Mn), 0.60% copper (Cu), and 0.60% silicon (Si). Mild carbon steel is not of high strength, but it is very weldable and exhibits a well-defined upper and lower yield stress (Steel 1 in Figure 2.1). Shapes and plates of ASTM A36M (A36) and A709M (A709) Grade 250 (36) are mild carbon steels used in railway bridge fabrication.

2.4.2 High-Strength Low-Alloy Steels

Carbon content must be limited to preclude negative effects on ductility, toughness, and weldability. Therefore, it is not desirable to increase strength by increasing carbon content, and manipulation of the steel chemistry needs to be considered. HSLA steels have increased strength attained through the addition of many alloys.

Alloy elements can significantly change steel phase transformations and properties (Jastrewski, 1977). The addition of small amounts of chromium, columbium, copper, manganese, molybdenum, nickel, silicon, phosphorous, and vanadium in specified quantities results in improved mechanical properties. The total amount of these alloys is less than 5% in HSLA steels. These steels typically have a well-defined yield stress in the 300–415 MPa (44–60 ksi) range (Steel 2 in Figure 2.1). Shapes and plates of ASTM A572M (A572), A588M (A588), and A992M (A992) (rolled shapes only) and A709M (A709) Grade 345 (50), 345S (50S), and 345W (50W) are HSLA steels used in railway bridges.

A572M (A572) Grade 290 (42), 345 (50), and 380 (55) steels are used for bolted or welded construction. Higher strength A572M (A572) steel [Grades 415 (60) and 450 (65)] is used for bolted construction only, due to reduced weldability. A572M (A572), A588M (A588), and A992M (A992) steels are not

* This equation is given in ASTM G101. Other equations, such as the Townsend equation, have also been proposed and may be of greater accuracy.

† ASTM A588M (A588) steel has a CI of about 5.8 (Swanson, 2014), but it is considered acceptable as an atmospheric corrosion resistant steel for railway superstructures (Table 2.5).

material toughness graded at the mills and often require supplemental CVN testing to ensure adequate toughness, particularly for service in cold climates. A588M (A588) and A709M (A709) Grade 345W (50W) steels are atmospheric corrosion-resistant (weathering) steels. ASTM A709M (A709) Grade 345 (50), 345S (50S), and 345W (50W) steel is mill certified with a specific toughness in terms of the minimum CVN impact energy absorbed at a given test temperature (e.g., designations 345T2 (50T2) indicating non-FCM Zone 2 and 345WF3 (50WF3) indicating FCM Zone 3 toughness criteria).

Further increases in strength, ductility, toughness, and corrosion resistance through steel chemistry alteration have been made in recent years. HSLA steels with 485 MPa (70 ksi) yield stress have been manufactured with niobium, vanadium, nickel, copper, and molybdenum alloy elements. These alloys stabilize either austenite or ferrite so that martensite formation and hardening does not occur, as it may for higher strength steel attained by heat treatment. A concise description of the effects of various alloy and deleterious elements on steel properties is given in Brockenbrough (2011).

2.4.3 Heat-Treated Low-Alloy Steels

Higher strength steel plate [with yield stress in excess of 485 MPa (70 ksi)] is produced by heat treating HSLA steels. A disadvantage of higher strength steels is a decrease in ductility. Heat treatment restores loss of ductility through Q&A processes. The quenching of steel increases strength and hardness with the formation of martensite. Tempering improves ductility and toughness through temperature relief of the high internal stresses caused by martensite formation. However, after quenching, tempering, and controlled cooling, these steels will not exhibit a well-defined yield stress (Steel 3 in Figure 2.1). In such cases, the yield stress is determined at the 0.2% offset from the elastic stress–strain relation (Figure 2.4).

Use of these steels may result in considerable weight reductions and precipitate fabrication, shipping, handling, and erection cost savings. High-strength steel can also allow for design of shallower superstructures. ASTM A514M (A514), A852M (A852), and A709M (A709) Grade 485W (70W) and 690W (100W) are quenched and tempered low-alloy steel plates. However, none of these steels are typically used in ordinary railway superstructures due to weldability concerns.

2.4.4 High-Performance Steels

HPS plates have been developed in response to the need for enhanced toughness, weldability, and corrosion resistance of high-strength steels. HPS 485W (70W) and 690W (100W) steels are produced by a

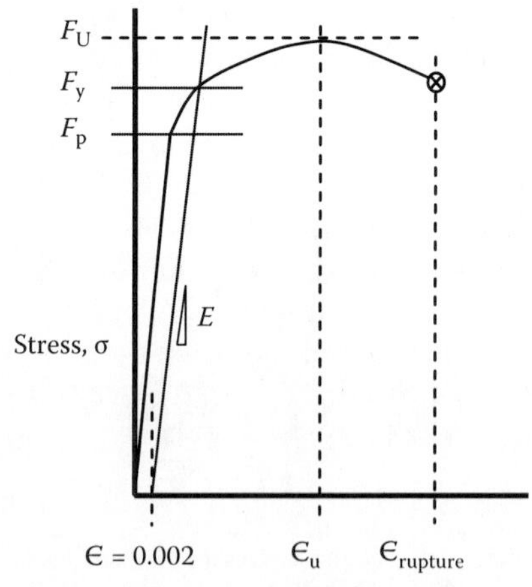

FIGURE 2.4 Engineering tensile stress–strain behavior of typical high strength bridge structural steel.

combination of chemistry manipulation and quench and temper operations or, for longer plates, TMCP. The first HPSs were produced with a yield stress of 485 MPa (70 ksi). However, HPS with 345 MPa (50 ksi) yield stress soon followed due to the weldability, toughness, and atmospheric corrosion resistance property improvements of HPS. HPS 345W (50W) is produced with the same chemistry as HPS 485W (70W), using conventional hot or controlled rolling techniques. HPS plates with 690 MPa (100 ksi) yield stress are also available. HPS 690W (100W) is considered an improvement to A514M (A514) steel plates (Lwin et al., 2005). HPS with 690 MPa (100 ksi) yield stress has been quench and temper heat treated to provide good ductility, weldability, and CVN toughness (Chatterjee, 1991).

Weldability is increased by lowering the carbon content [e.g., below 0.11% for HPS 485W (70W)], therefore, benefiting the carbon equivalence (Equation 2.6). This weldability increase results in the elimination of preheat requirements for thin members and limited preheat requirements for thicker members. Also, postweld treatments are reduced and hydrogen-induced cracking at welds eliminated (provided correct measures are taken to eliminate hydrogen from moisture, contaminants, and electrodes). Welding of HPSs using low hydrogen electrodes is done by submerged arc welding or shielded metal arc welding processes (see Chapters 9 and 10).

Toughness is significantly increased through reductions in sulfur content (0.006% max) and control of inclusions (by calcium treatment of steel). The fracture toughness of HPS is, therefore, much improved with the ductile to brittle transition occurring at lower temperatures (the curve shifts to the left in Figure 2.2). Higher toughness also translates into greater crack tolerance for fatigue crack detection and repair procedure development. HPSs meet or exceed the CVN toughness requirements specified for the coldest climates (Zone 3 in AREMA, 2015).

The corrosion-resistant properties of HPS are based on quenched and tempered ASTM A709M (A709) Grade 485W (70W) and 690W (100W) steels. Chromium, copper, nickel, and molybdenum are alloyed for improved weathering resistance. Improved weathering resistant steels are under development that might provide good service in even moderate chloride environments.

Hybrid* applications of HPSs with HSLA steels have proven technically and economically successful on a number of highway bridges (Lwin, 2002) and may be appropriate for some railway bridge projects.

2.5 STRUCTURAL STEEL FOR RAILWAY SUPERSTRUCTURES

There is no increase in stiffness associated with higher strength steels (deflections, vibrations, and elastic stability are proportional to the modulus of elasticity and not strength). Also, because fatigue strength depends primarily on applied stress range and detail (see Chapter 5), there is no appreciable increase in fatigue resistance for higher strength steels.† Therefore, the material savings associated with the use of higher strength steels [with greater than 345 MPa (50 ksi) yield stress] may not be available because deflection criteria and fatigue often govern critical aspects of ordinary steel railway superstructure design. The steel bridge designer must carefully consider all design limit states (strength, serviceability, fatigue, and fracture), procurement (availability and cost), and fabrication issues when selecting the materials for railway bridge projects.

2.5.1 Material Properties

The following material properties may be used for steel railway bridge design and construction:

- Density, $\gamma = 7850\,\text{kg/m}^3$ ($490\,\text{lb/ft}^3$)
- Modulus of elasticity (Young's modulus), $E = 200{,}000\,\text{MPa}$ ($29 \times 10^6\,\text{psi} = 29{,}000\,\text{ksi}$)
- Coefficient of thermal expansion, $\alpha = 12 \times 10^{-6}/°\text{C}$ ($6.5 \times 10^{-6}/°\text{F}$)

* An example is the use of HPSs for tension flanges in simple and continuous girders.

† Recent testing indicated that CVN requirements for HPS grades were only marginally better than current AREMA and AASHTO Zone 2 and Zone 3 specifications (Alstadt et al., 2014).

- Poisson's ratio, $\upsilon = 0.3$ (lateral to longitudinal strain ratio under load)
- In accordance with the theory of elasticity, shear modulus, $G = \dfrac{E}{2(1+\upsilon)} \sim 77{,}000$ MPa ($\sim 11.2 \times 10^6$ psi)

2.5.2 Structural Steel for Modern North American Railway Superstructures

Structural bridge steels have increased in strength and quality over the past century. Table 2.2 indicates the strength of some of the structural steels used in the past century in the United States and Canada (Canadian Institute of Steel Construction, 2004).

Modern structural bridge steels provide good ductility, weldability, and corrosion resistance. Structural steel for use in modern railway superstructures in North America is typically specified as ASTM A36M (A36), A572M (A572), A588M (A588), A709M (A709), and/or A992M (A992), depending on strength, ductility, welding, and corrosion-resistant requirements. Tables 2.3 and 2.4 indicate the toughness requirements for these steels for non-FCM and FCM applications, respectively. Table 2.5 outlines the strength of these steels for use in railway superstructures.

The AREMA (2015) recommendations do not include heat-treated low-alloy steels. The only steel with a yield stress greater than 345 MPa (50 ksi) currently recommended is A709M (A709)

TABLE 2.2
Structural Steel Used in North America Since 1900

			F_y		F_u	
Steel Designation	**Country**	**Date**	**(MPa)**	**(ksi)**	**(MPa)**	**(ksi)**
ASTM A7	USA	1900–1909	$0.5F_u$	$0.5F_u$	410–490	60–70
		1914	$0.5F_u$	$0.5F_u$	380–450	55–65
CSA A16	Canada	1924	$0.5F_u$	$0.5F_u$	380–450	55–65
ASTM A7	USA	1924	$0.5F_u \geq 210$	$0.5F_u \geq 30$	380–450	55–65
		1934	$0.5F_u \geq 230$	$0.5F_u \geq 33$	410–500	60–72
CSA S39	Canada	1935	210	30	380–450	55–65
CSA S40	Canada	1935	230	33	410–500	60–72
CSA G40.4 and G40.5	Canada	1950	230	33	410–500	60–72
CSA G40.6	Canada	1950	310	45	550–650	80–95
ASTM A242	USA	1955	350	50	480	70
ASTM A36	USA	1960	250	36	410–550	60–80
ASTM A440 and A441		1959 and 1960	350	50	480	70
CSA G40.8	Canada	1960	280[a]	40[a]	450–590	65–85
CSA G40.12	Canada	1964	300[b]	44[b]	450	65
ASTM A572 (Grade 50)	USA	1966	345	50	450	65
ASTM A588	USA	1968	345[c]	50[c]	485[c]	70[c]
CSA G40.21	Canada	1973	Incorporated all previous CSA G40 standards			
ATM A992	USA	1998	345–450	50–65	450	65

[a] Less for material thicker than 16 mm (5/8″)
[b] Less for material thicker than 40 mm (1–1/2″)
[c] Less for thicker material

TABLE 2.3
Fracture Toughness Requirements for Non-Fracture Critical Members (FCM)[e,f]

ASTM Designation	Thickness Inches (mm)	Minimum Average Energy, ft-lb(J), and Test Temperatures: Zone 1	Zone 2	Zone 3
A36/A36M	To 6(150) incl.	15(20)@70°F(21°C)	15(20)@40°F(4°C)	15(20)@10°F(−12°C)
A709/A709M, Grade 36T(250T)[a]	To 4(100) incl.	15(20)@70°F(21°C)	15(20)@40°F(4°C)	15(20)@10°F(−12°C)
A992/A992M[b]	To 2(50) incl.	15(20)@70°F(21°C)	15(20)@40°F(4°C)	15(20)@10°F(−12°C)
A709/A709M, Grade 50ST (Grade 345ST)[a,b]	Over 2(50)–4(100) incl.	20(27)@70°F(21°C)	20(27)@40°F(4°C)	20(27)@10°F(−12°C)
A588/A588M[b]				
A572/A572M, Grade 42 (Grade 290)[b]				
A572/A572M, Grade 50 (Grade 345)[b]				
A709/A709M, Grade 50T (Grade 345T)[a,b]				
A709/A709M, Grade 50WT (Grade 345WT)[a,b]				
A572/A572M, Grade 42 (Grade 290)[b]	Over 4(100)–6(150) incl.	20(27)@70°F(21°C)	20(27)@40°F(4°C)	20(27)@10°F(−12°C)
A588/A588M[b]	Over 4(100)–5(125) incl.	20(27)@70°F(21°C)	20(27)@40°F(4°C)	20(27)@10°F(−12°C)
A709/A709M, Grade HPS 50WT (Grade HPS 345WT)[a,b]	To 4(100) incl.	20(27)@10°F(−12°C)	20(27)@10°F(−12°C)	20(27)@10°F(−12°C)
A709/A709M, Grade HPS 70WT (Grade HPS 485WT)[a,c]	To 4(100) incl.	25(34)@−10°F(−23°C)	25(34)@−10°F(−23°C)	25(34)@−10°F(−23°C)
Minimum Service Temperature[d]		[d]0°F(−18°C)	−30°F(−34°C)	−60°F(−51°C)

Source: American Railway Engineering and Maintenance-of-Way Association (AREMA), Chapter 15—Steel Structures, *Manual for Railway Engineering*, Lanham, MD.

[a] The suffix T is an ASTM A709/A709M designation for non-fracture critical material requiring impact testing. A numeral 1, 2 or 3 should be added to the T marking to indicate the applicable service temperature zone.

[b] If the yield point of the material exceeds 65,000 psi (450 MPa) the test temperature for the minimum average energy required shall be reduced by 15 F (8°C) for each increment or fraction of 10,000 psi (70 MPa) above 65,000 psi (450 MPa).

[c] If the yield strength of the material exceeds 85,000 psi (585 MPa) the test temperature for the minimum average energy required shall be reduced by 15 F (8°C) for each increment or fraction of 10,000 psi (70 MPa) above 85,000 psi (585 MPa).

[d] Minimum service temperature of 0 F (−18°C) corresponds to Zone 1, −30 F (−34°C) to Zone 2, and −60 F (−51°C) to Zone 3.

[e] Impact tests shall be in accordance with the Charpy V-Notch (CVN) tests as governed by ASTM Specification A673/A673M with frequency of testing H for all grades except for A709/A709M, Grade HPS 70WT (Grade HPS 485WT), which shall be frequency of testing P.

[f] Impact test requirements for structural steel of Fracture Critical Members are specified in Table 2.4.

TABLE 2.4
Fracture Toughness Requirements for Fracture Critical Members (FCM)[f]

ASTM Designation	Thickness in. (mm)	Minimum Test Value Energy ft-lb(J)	Minimum Average Energy, ft-lb(J), and Test Temperatures		
			Zone 1	Zone 2	Zone 3
A36/A36M[a]	To 4(100) incl.	20(27)	25(34)@70°F(21°C)	25(34)@40°F(4°C)	25(34)@10°F(–12°C)
A709/A709M, Grade 36F (Grade 250F)[a,b]					
A992/A992M[c]	To 2(50) incl.	20(27)	25(34)@70°F(21°C)	25(34)@40°F(4°C)	25(34)@10°F(–12°C)
A709/A709M, Grade (Grade 345SF)[a,b,c]	Over 2(50)–4(100) incl.	24(33)	30(41)@ 70°F(21°C)	30(41)@40°F(4°C)	30(41)@10°F(–12°C)
A572/A572M, Grade 50 (Grade 345)[a,c]					
A709/A709M, Grade 50F (Grade 345F)[a,b,c]					
A588/A558[a,c]					
A709/A709M, Grade 50WF (Grade 345WF)[a,b,c]					
A709/A709M, Grade HPS 50WF (Grade HPS 345WF)[b,c]	To 4(100) incl.	24(33)	30(41)@10°F(–12°C)	30(41)@10°F(–12°C)	30(41)@10°F(–12°C)
A709/A709M, Grade HPS 70WF (Grade HPS 485WF)[b,d]	To 4(100) incl.	28(38)	35(48)@–10°F(–23°C)	35(48)@–10°F(–23°C)	35(48)@–10°F(–23°C)
Minimum Service Temperature[e]			0°F(–18°C)	–30°F(–34°C)	–60°F(–51°C)

Source: American Railway Engineering and Maintenance-of-Way Association (AREMA), 2015, Chapter 15—Steel Structures, *Manual for Railway Engineering*, Lanham, MD.

[a] Steel backing for groove welds joining steels with a minimum specified yield strength of 50,000 psi (345 MPa) or less may be base metal conforming to ASTM A36/A36M, A709/A709M, A588/A588M or A572/A572M, at the Contractor's option, provided the backing material is furnished as bar stock rolled to a size not exceeding 3/8 in. (10 mm) by 1–1/4 in. (32 mm). The bar stock so furnished need not conform to the Charpy V-Notch impact test requirements of this table.

[b] The suffix "F" is an ASTM A709/A709M designation for fracture critical material requiring impact testing. A numeral 1, 2, or 3 shall be added to the F marking to indicate the applicable service temperature zone.

[c] If the yield point of the material exceeds 65,000 psi (450 MPa), the test temperature for the minimum average energy and minimum test value energy required shall be reduced by 15 F(8°C) for each increment or fraction of 10,000 psi (70 MPa) above 65,000 psi (450 MPa). The yield point is the value given on the certified "Mill Test Report".

[d] If the yield strength of the material exceeds 85,000 psi (585 MPa), the test temperature for the minimum average energy and minimum test value energy required shall be reduced by 15 F (8°C) for each increment of 10,000 psi (70 MPa) above 85,000 psi (585 MPa). The yield strength is the value given on the certified "Mill Test Report".

[e] Minimum service temperature of 0 F (–18°C) corresponds to Zone 1, –30 F (–34°C) to Zone 2, –60 F (–51°C) to Zone 3.

[f] Impact tests shall be Charpy V-notch (CVN) impact testing, "P" plate frequency, in accordance with ASTM Designation A673/A673M except for plates of A709/A709M Grades 36F(250F), 50F(345F), 50WF(345WF), HPS 50WF (HPS 345WF) and HPS 70WF (HPS 485WF) and their equivalents in which case specimens shall be selected as follows: (1) as-rolled plates shall be sampled at each end of each plate-as-rolled; (2) normalized plates shall be sampled at one end of each plate-as-heat treated; and (3) quenched and tempered plates shall be sampled at each end of each plate-as-heat-treated.

TABLE 2.5
Structural Steel for Modern Railway Bridges[c]

ASTM Designation	F_y-min Yield Point or Yield Strength psi	F_u Ultimate Tensile Strength or Tensile Strength psi	Thickness Limitation: For Plates and Bars, in.	Thickness Limitation: Applicable to Shapes
A36	36,000 min	58,000 min 80,000 max	To 8 incl.	All[d]
A709, Grade 36	36,000 min	58,000 min 80,000 max	To 4 incl.	All[d]
A588[a] A709, Grade 50W[a] A709, Grade HPS 50W[a]	50,000 min	70,000 min	To 4 incl.	All
A588[a]	46,000 min	67,000 min	Over 4–5 incl.	None
A588[a]	42,000 min	63,000 min	Over 5–8 incl.	None
A992[b] A709, Grade 50S[b]	50,000 min 65,000 max[e]	65,000 min Yield to tensile ratio, 0.85 max	None	All
A572, Grade 50 A709, Grade 50	50,000 min	65,000 min	To 4 incl.	All
A572, Grade 42	42,000 min	60,000 min	To 6 incl.	All
A709, Grade HPS 70W[a]	70,000 min	85,000 min 110,000 max	To 4 incl.	None

Source: American Railway Engineering and Maintenance-of-Way Association (AREMA), 2015, Chapter 15—Steel Structures, *Manual for Railway Engineering*, Lanham, MD.

[a] ASTM A588 and A709, Grade 50W, Grade HPS 50W, and Grade HPS 70W have atmospheric corrosion resistance in most environments substantially better than that of carbon steels with or without copper addition. In many applications these steels can be used unpainted.

[b] The yield to tensile ratio shall be 0.87 or less for shapes that are tested from the web location; for all other shapes, the requirement is 0.85 maximum.

[c] These requirements are current as of May 2009. Refer to ASTM specifications for additional requirements.

[d] For wide flange shapes with flange thickness over 3 in., the 80.000 psi maximum tensile strength limit does not apply.

[e] A maximum yield strength of 70,000 psi (490 MPa) is permitted for structural shapes that are required to be tested from the web location.

HPS 485W (70W). Also, as seen in Table 2.5, AREMA (2015) recommends the use of weathering steels such as ASTM A588M (A588) and A709M (A709). Nonweathering steels such ASTM A36M (A36) and A572M (A572) are also indicated for use. Since A572M (A572) Grades 290 (42) and 345 (50) are recommended for welded and bolted construction, with higher grades used for bolted construction only, the AREMA (2015) recommendations for structural steel do not include A572M (A572) grades higher than Grade 345 (50).

Structural steel for use in modern Canadian steel railway superstructures may also be specified in accordance with CSA G40.21 Grades 260 (36), 300 (44), and 350 (50). Grade 350 is produced in five categories* of toughness for non-FCM and FCM applications. Typically, railway superstructure FCM use Grade 350WT or 350AT Category 5 steel with the toughness requirements† specified to be in accordance with Table 2.3 for ASTM A709M steel. Non-FCM applications may typically

* Category 5 is as specified by the designer or user. It may be used to specify CSA G40.21 Grade 350 steel with alternate toughness requirements.

† In terms of CVN impact testing.

specify Grade 350WT or 350AT Category 2, or Grade 350WT or 350AT Category 5 steel with the toughness requirements specified to be in accordance with Table 2.2 for ASTM A709M steel.

REFERENCES

Alstadt, S., Wright, W., and Connor, R., 2014, Proposed revisions to the current Charpy V-notch requirements for structural steel used in bridges, *Journal of Bridge Engineering*, Vol. 19, No. 1, 131–140.

American Railway Engineering and Maintenance-of-Way Association (AREMA), 2015, Chapter 15—Steel Structures, *Manual for Railway Engineering*, Lanham, MD.

American Society of Testing and Materials (ASTM), 2015, Standards Vol. 01.04, A36M (A36), A572M (A572), A588M (A588), A673M (A673), A709M (A709), A992M (A992); Vol. 03.01, E8; and Vol. 03.02, G101, *2015 Annual Book of ASTM Standards*, West Conshohocken, PA.

Barsom, J.M. and Rolfe, S.T., 1987, *Fracture and Fatigue Control in Structures*, 2nd ed., Prentice-Hall, Englewood Cliffs, NJ.

Brockenbrough, R.L., 2011, Properties of structural steels and effects of steelmaking and fabrication, in *Structural Steel Designer's Handbook*, 5th ed., Brockenbrough, R.L. and Merritt, F.S., Ed., McGraw Hill, New York.

Canadian Institute of Steel Construction (CISC), 2004, *Handbook of Steel Construction*, 8th ed., CISC, Rexdale, Markham, ON.

Chakrabarty, J., 2006, *Theory of Plasticity*, 3rd ed., Elsevier, Oxford, UK.

Chatterjee, S., 1991, *The Design of Modern Steel Bridges*, BSP Professional Books, Oxford, UK.

Chen, W.F. and Han, D.J., 1988, *Plasticity for Structural Engineers*, Springer-Verlag, New York.

Fisher, J.W., 1984, *Fatigue and Fracture in Steel Bridges*, John Wiley & Sons, New York.

Hill, R., 1989, *The Mathematical Theory of Plasticity*, Oxford University Press, Oxford, UK.

Jastrewski, Z.D., 1977, *The Nature and Properties of Engineering Materials*, 2nd ed., John Wiley & Sons, New York.

Kulak, G.L. and Grondin, G.Y., 2002, *Limit States Design in Structural Steel*, Canadian Institute of Steel Construction, Toronto, ON.

Lwin, M.M., 2002, *High Performance Steel Designer's Guide*, FHWA Western Resource Center, US Department of Transportation, San Francisco, CA.

Lwin, M.M, Wilson, A.D., and Mistry, V.C., 2005, *Use and Application of High-Performance Steels for Steel Structures—High-Performance Steels in the United States*, International Association for Bridge and Structural Engineering, Zurich, Switzerland.

Swanson, J.A., 2014, Steel design, in *Bridge Engineering Handbook, Fundamentals*, 2nd ed., Chen, W.-F. and Duan, L., Ed., CRC Press, Taylor & Francis Group, Boca Raton, FL.

3 Planning and Preliminary Design of Modern Steel Railway Bridges

3.1 INTRODUCTION

The primary purpose of railway bridges is to safely and reliably carry freight and passenger train traffic within the railroad operating environment. The majority of the railway bridges in the North American railway infrastructure have steel superstructures.*

It is estimated that, in terms of length, about 53% of the approximately 80,000 bridges [with an estimated total length of almost 2900 km (1800 miles)] in the US railroad bridge inventory are steel spans (Unsworth, 2003; Federal Railroad Administration (FRA), 2008).† Over 60,000 of these bridges are in the inventories of the four largest US railroads and, in terms of length, about 55% are steel. There are over 65,000 bridges in the inventories of the six largest North American freight railroads with an estimated 57% of their length constructed with steel.

Structural and/or functional obsolescence precipitates the regular rehabilitation and/or replacement of many of these steel railway superstructures. In addition, many of the steel bridges in the North American freight railroad bridge inventory are over 80 years old, and they may require replacement due to the effects of age, increases in freight equipment weight,‡ and the amplified frequency of the application of train loads.§ Bridge replacement requires careful planning with consideration of site conditions and transportation requirements in the modern freight railroad operating environment.

Site conditions relating to hydraulic or roadway clearances, as well as the geotechnical and physical environment (during and after construction), are important concerns during planning and preliminary bridge design. Railroad and other transportation entity operating practices also need careful deliberation. Interruption to traffic flow in rail, highway, or marine transportation corridors and safety (construction and public) are also of paramount concern. The planning phase should yield information concerning optimum bridge crossing geometry, layout, and anticipated construction methodologies (see Chapter 11). This information is required for the selection of span lengths, types, and materials for preliminary superstructure design. Preliminary bridge design concepts are often the basis of regulatory reviews, permit applications, and budget cost estimates. Therefore, planning and preliminary bridge design can be critical to successful project implementation and, particularly,

---

* By length, 17% of bridge superstructures are wood and 28% are concrete in the inventories of the four largest US railroads. When the inventories of short-line and regional railroads are considered, wood and concrete superstructures each comprises about 23% of US railway bridges in terms of length.

† In 1910, it was estimated that there were about 80,000 metal railway bridges with a cumulative length of about 2250 km (1400 miles) (see Chapter 1). In 2008, the cumulative length of US steel railway bridges was estimated as 1500 km (935 miles), of which about 1300 km (800 miles) were in the inventories of the four largest US railroads. In 2013, the cumulative length of steel bridges in the inventories of the six largest North American railroads was estimated to be 1450 km (900 miles).

‡ In 1910, locomotives typically weighed about 136,000 kg (300,000 lbs) (see Chapter 1). Over the next few decades the weight of some heavy service locomotives increased by over 50%. The weight of typical locomotives currently used on North American railroads approaches 200,000 kg (450,000 lbs).

§ Trains with loads causing many cycles of stress ranges that might accumulate significant fatigue damage did not occur until the latter half of the 20th century when typical train car weights increased from 80,000 kg (177,000 lbs) to over 120,000 kg (263,000 lbs) on a regular basis.

for large or complex bridges, warrants due deliberation. Detailed design of the superstructure for fabrication and erection can proceed following preliminary bridge and superstructure design.

3.2 PLANNING OF RAILWAY BRIDGES

Planning of railway bridges involves the careful consideration and balancing of multifaceted, and often competing, construction economics, business, public, and technical requirements.

3.2.1 Bridge Crossing Economics

In general, other issues not superseding, the bridge crossing should be close to perpendicular to the narrowest point of the river or flood plain. The economics of a bridge crossing depends on the relative costs of foundations, substructures, and superstructures.* Estimates related to the cost of foundation and substructure construction, and superstructure erection are often less reliable than those for superstructure fabrication. Superstructure fabrication cost estimates are often more dependable than erection estimates due to the inherently greater uncertainty and risk associated with field construction. Excluding, in particular, public and technical (hydraulic and geotechnical) constraints from the cost-estimating procedure enables the economical span length, l, to be estimated based on simple principles. Considering a fairly uniform bridge with similar foundations, substructures, and multiple equal length spans, the total estimated cost, CB, of a bridge crossing may be expressed as

$$\mathrm{CB} = n_s C_{\mathrm{sup}} w_s l + (n_s - 1) C_{\mathrm{pier}} + 2C_{\mathrm{abt}}, \tag{3.1}$$

where

n_s = number of spans = L/l

L = length of the bridge

C_{sup} = estimated cost of steel per unit weight (purchase, fabricate, and erect)

w_s = weight per unit length of the span elements that depend on span length, l, (e.g., girders and trusses). The weights of the span elements that are independent of span length, l (e.g., the floor system of a through span, which is dependent on panel length), are excluded from w_s.

C_{pier} = estimated average cost of one pier (materials, foundation, and construction)

C_{abt} = estimated average cost of one abutment (materials, foundation, and construction)

If $w_s = \alpha l + \beta$, where α and β are constants independent of span length and dependent only on span type and design live load, Equation 3.1 may be expressed as

$$\mathrm{CB} = C_{\mathrm{sup}} L(\alpha l + \beta) + C_{\mathrm{pier}} L\left(\frac{L - l}{lL}\right) + 2C_{\mathrm{abt}}, \tag{3.2}$$

which may be differentiated in terms of span length, l, to determine an expression for the minimum total estimated cost, CB, as

$$\frac{\mathrm{dCB}}{\mathrm{d}l} = C_{\mathrm{sup}} \alpha L - \frac{C_{\mathrm{pier}} L}{l^2} = 0. \tag{3.3}$$

Rearrangement of Equation 3.3 provides the economical span length, l, as

$$l = \sqrt{\frac{C_{\mathrm{pier}}}{C_{\mathrm{sup}} \alpha}}. \tag{3.4}$$

* A rule of thumb for economical, relatively uniform multispan bridges is that the cost of superstructure (fabrication and erection) equals the cost of foundation and substructure construction (Byers, 2009).

Again, considering a fairly uniform bridge with similar foundations, substructures, and multiple equal length spans, the total estimated cost per unit length, CBL, of a bridge crossing may be expressed as

$$\text{CBL} = \text{CL}_{\text{sup}} + \text{CL}_{\text{flr}} + \text{CL}_{\text{sub}}, \tag{3.5}$$

where

CL_{sup} = estimated cost per unit length of superstructures (e.g., girders and trusses) = $C_{\text{sup}} w_s = C_{\text{sup}} \alpha l$

CL_{flr} = estimated cost per unit length of the floor system, which is independent of span length, l

CL_{sub} = estimated cost of substructures per unit length of span (two piers or one pier and one abutment), which is approximately C_{pier}/l

and Equation 3.5 can be expressed as

$$\text{CBL} = C_{\text{sup}} \alpha l + \text{CL}_{\text{flr}} + \frac{C_{\text{pier}}}{l}, \tag{3.6}$$

which may be differentiated in terms of span length, l, to determine an expression for the minimum total estimated cost per unit length, CBL, as

$$\frac{d\text{CBL}}{dl} = C_{\text{sup}} \alpha - \frac{C_{\text{pier}}}{l^2} = \frac{\text{CL}_{\text{sup}}}{l} - \frac{\text{CL}_{\text{sub}}}{l} = 0, \tag{3.7}$$

which shows that the minimum total estimated cost per unit length, CBL, occurs when $\text{CL}_{\text{sup}} = \text{CL}_{\text{sub}}$ or when the estimated cost per unit length of superstructure fabrication and erection equals the estimated cost of foundation and substructure construction per unit length of span.

The cost of foundations, substructures, superstructures, floor systems (e.g., through plate girder, through truss, and some deck truss spans), and decks must be evaluated to estimate the economical cost of bridge construction. Budgetary costs can be obtained from previous similar projects, with appropriate adjustments for location, time, and other factors. Many bridge designers and owners have developed and updated the "average" unit costs for various bridge and/or span types based on historical data. These may also be used for budgetary construction cost estimates with appropriate adjustments for location and other factors.

Preliminary construction cost estimates should be based on quantity* take-offs from preliminary bridge design drawings. These quantities can be used with local unit costs and rates for materials, labor, and equipment to develop estimates for bridge construction (fabrication and erection).

Nevertheless, while Equation 3.4 provides a simple estimate of economical span length, and Equation 3.7 outlines a simplistic minimum cost criterion, the final general arrangement in terms of span lengths, l, may depend on other business, public, and technical requirements.

3.2.2 Railroad Operating Requirements

Most new freight railway bridges are constructed on existing routes on the same alignment. Construction methods that minimize the interference to normal rail, road, and marine traffic enable simple erection and are cost-effective, which must be carefully considered during the planning process. Often, to minimize interruption to railroad traffic, techniques such as laterally sliding spans into position from falsework, launching spans longitudinally on gantries, floated erection of spans from river barges (Unsworth and Brown, 2006), span installation with movable derricks

* Typically foundation, substructure, bearing, superstructure, floor system, deck, and walkway quantities.

and gantries, construction on adjacent alignment,* and the use of large cranes must be developed (see Chapter 11). These methodologies may add cost to the reconstruction project that are acceptable in lieu of the costs associated with extended interruption to railway† or marine traffic.

New rail lines are generally constructed in accordance with the requirements established by public agencies‡ and railroad business access. It is not often that bridge crossings are selected solely on the basis of localized bridge economics planning principles.§ Therefore, site reconnaissance (surveying and mapping) and route selection are typically performed on the basis of business, technical, and public considerations.

The railroad operating environment presents specific challenges for bridge design and construction. The design of steel railway bridges involves the following issues related to railroad operations:

- The magnitude, frequency, and dynamics of railroad live loads
- Other loads particular to railroad operations
- The location of the bridge (in relation to preliminary design of bridge type, constructability, and maintainability)
- Analysis and design criteria particular to railway bridges

3.2.3 Site Conditions (Public and Technical Requirements of Bridge Crossings)

Site conditions are of critical importance in the determination of location, form, type, length, height, and estimated cost of railway bridges. Existing records and drawings of previous construction are of considerable value during planning of railway bridges being reconstructed on the same, or nearby, alignment. In terms of the cost and constructability of bridges being built on a new alignment, the preferred bridge crossing is generally the shortest or shallowest crossing. However, regulatory, clearance (hydraulic, highway, railway, and marine), and foundation conditions may dictate crossing location, which will affect the form, type, length, height, and estimated cost of the bridge.

3.2.3.1 Regulatory Requirements

Bridge location, length, height, and, consequently, form are often governed by existing route location, pre-established design route locations, and/or regulation. Regulatory requirements relating to bridge crossing location and environmental protection may affect preliminary design and construction methods (see Chapter 11). Depending on location, environmental protection (vegetation, fish, and wildlife) and cultural considerations may be critical components of the bridge planning phase. Land ownership and use regulations also warrant careful review for potential new bridge crossing locations. Regulatory requirements vary by geographic location and jurisdiction. Railway bridge construction project managers and engineers must be well versed in the jurisdictional permitting requirements for bridge crossings. Regulatory concerns potentially affecting bridge location, length, height, and the construction method that may affect bridge form must be communicated to the railway bridge designer during the planning phase.

3.2.3.2 Hydrology and Hydraulics of the Bridge Crossing

Hydrological and hydraulic assessments are vital to establishing the required bridge opening at river crossings, and the ensuing form, type, length, height, and estimated cost. The bridge opening must

* Either the new bridge is constructed on an adjacent alignment or a temporary bridge is built on an adjacent alignment (shoo-fly) to not interrupt the flow of rail traffic. This may not always be feasible due to cost, site conditions, and/or route alignment constraints.

† North American Class 1 railroads often limit regular windows for construction to less than 6 h on mainline tracks. Longer windows are typically available only through advanced planning with the railroad's operating department.

‡ Generally, the requirements relate to environmental, fish and wildlife, land ownership and cultural considerations, and/or regulations.

§ The exception might be very long bridges.

safely pass the appropriate return frequency (probability of occurrence) water discharge,* ice and debris† past constrictions, and obstructions created by the bridge crossing substructures.

Typical flood discharge requirements for railway bridges are a 1:100 return frequency for mainlines and, in some cases, lesser frequencies‡ for other lines. However, recent flood damage occurrences and costs may precipitate the use of a single design discharge flood frequency for all river crossings (Byers, 2009).

At some crossing locations, the ability to pass drifting materials (debris and/or ice) will dictate bridge length and form. The length and form of the bridge may also depend on whether the channel or flood plain is stable. Stable channels and flood plains may be spanned with shorter spans unless shifting channel locations require the use of longer spans. Hydraulic studies must also consider the potential for scour at substructures, which may affect foundation location and design.

3.2.3.2.1 Bridge Crossing Hydraulics

A study of area hydrology and river hydraulics will provide information concerning the existing average channel velocity, V_u, at the required return frequency discharge, Q. If there are no piers to obstruct the river crossing and no constriction of the channel (Figure 3.1), the required area of the crossing is simply established as

$$A = \frac{Q}{V_u}. \tag{3.8}$$

3.2.3.2.1.1 Constricted Discharge Hydraulics Where abutments constrict the channel (Figure 3.2), the flow may become rapidly varied and exhibit a drop in water surface elevation (hydraulic jump) as a result of the increased velocity. Four types of constriction openings have been defined (Hamill, 1999) as follows:

- Type 1: Vertical abutments with and without wings walls with vertical embankments
- Type 2: Vertical abutments with sloped embankments
- Type 3: Sloped abutments with sloped embankments
- Type 4: Vertical abutments with wings walls and sloped embankments (typical of many railroad embankments at bridge crossings)

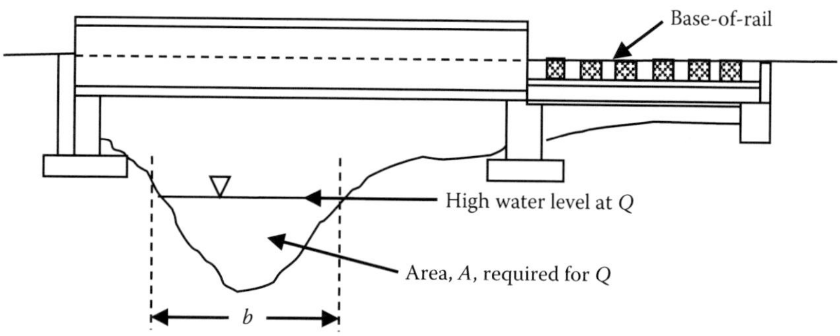

FIGURE 3.1 River crossing profile without constriction or obstruction.

* Developed from hydrological evaluations for the river crossing.

† In general, piers should not be skewed to river flow to avoid impact from ice, debris, or vessels. A "freeboard" is often required above the discharge frequency water elevation to assist with debris and ice flow through bridge constrictions and past obstructions. Freeboard requirements vary depending on location; however, typical freeboard requirements are between 300 (1 ft) and 600 mm (2 ft) for discharge frequencies of 1:50 or 1:100.

‡ Typically 1:50 for secondary mainlines and 1:50 or 1:25 for branchlines.

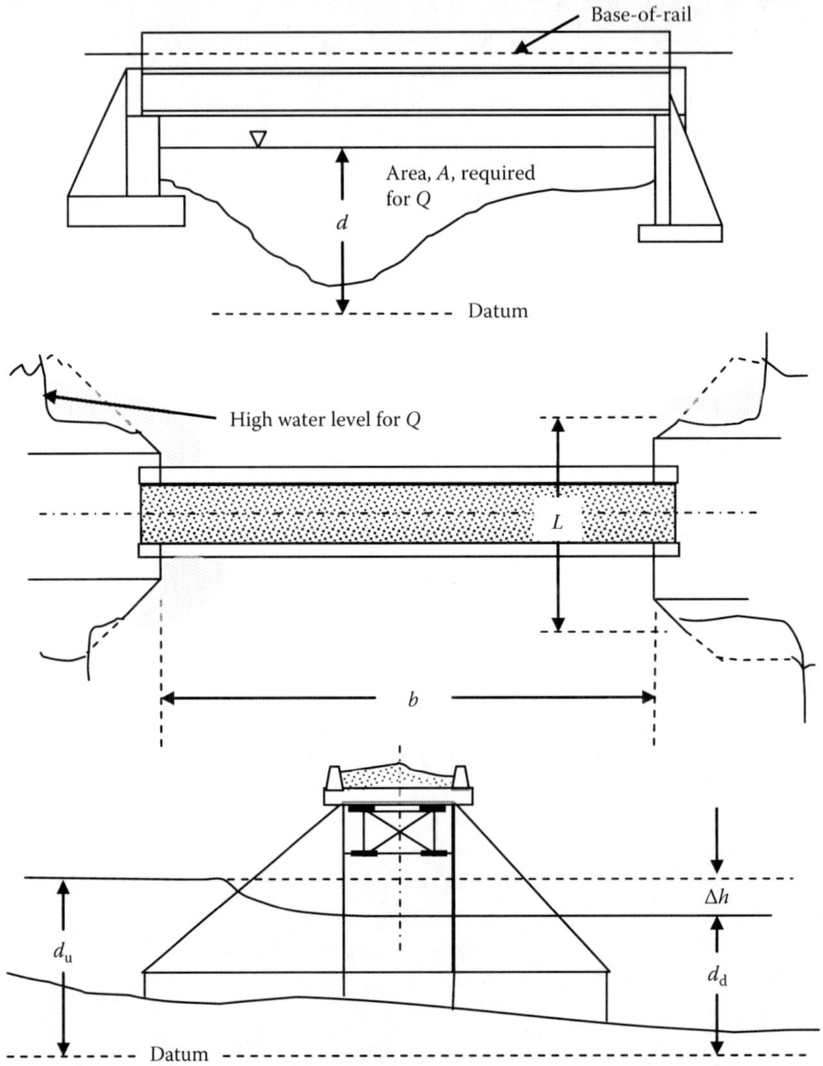

FIGURE 3.2 River crossing profile with constriction at Q.

The hydraulic design should strive for subcritical flow ($F < 1.0$ with a stable water surface profile). Discharge flows that exceed subcritical at, or even immediately downstream of, the bridge may also be acceptable with adequate abutment and channel scour protection. Supercritical flow ($F > 1.0$) is undesirable and may create an increase in water surface elevation at the bridge crossing. However, supercritical flow may be unavoidable for river crossings with steep slopes [generally greater than 0.5%–1% (Transportation Association of Canada (TAC), 2004)], and expert hydraulic design may be required. In the case of constrictions, the required area of the crossing is established as

$$A = \frac{Q}{C_c V_u}, \tag{3.9}$$

where

A = minimum channel area required under the bridge

Q = required or design return frequency (ex. 1:100) discharge

V_u = average velocity of existing (upstream) channel for discharge, Q
C_c = coefficient of contraction of the new channel cross section at the constriction

It can be shown that the coefficient of contraction, C_c, depends on the following:

- Contraction ratio
- Constriction edge geometry and angularity
- Submerged depth of abutment
- Slope of abutment face
- Eccentricity of the constriction relative to normal stream flow
- Froude number, which is

$$F = \frac{Q}{A\sqrt{gd}}, \tag{3.10}$$

where g is the acceleration due to gravity and d is the effective depth of the channel under the bridge, expressed as A/b, where b is the net or effective width of the bridge opening.

The theoretical determination of C_c is difficult, and numerical values are established experimentally. Published values of C_c for various constriction geometries are available in the literature of open-channel hydraulics (Chow, 1959).

One commonly used hydraulic analysis* is the US Geological Survey (USGS) method, which is based on extensive research. It determines a base coefficient of discharge, C', for four opening types in terms of the opening ratio, N, and constriction length ratio, L/b. The coefficient of discharge, C', is further modified by adjustment factors based on the opening ratio, Froude number, F, and detail abutment geometry to obtain the discharge coefficient, C. In terms of the USGS method, the coefficient of contraction is

$$C_c = \frac{C}{V_u}\sqrt{\left(2g\left(\Delta h + \frac{\alpha_u V_u^2}{2g} - h_f\right)\right)}, \tag{3.11}$$

where

C = USGS discharge coefficient, which depends on N, L, b, F, and other empirical adjustment factors based on the skew angle of the crossing. It also depends on conveyance, K, detail geometry and flow depth at the constriction

N = bridge opening ratio = Q_c/Q
Q_c = undisturbed flow that can pass the bridge constriction
Q = flow in the not constricted channel
L = length of the channel at the constricted bridge crossing

$$K = \frac{AR^{2/3}}{n}$$

$\Delta h = d_u - d_d$
d_u = depth of the channel upstream of the bridge
d_d = depth of the channel downstream of the bridge
h_f = friction loss upstream and through constricted opening, which is

* Other methods such as the US Bureau of Public Roads, Biery and Delleur, and UK Hydraulic Research methods are also used.

TABLE 3.1
Manning's Roughness Coefficients for Channels

Channel	Manning's Normal Roughness Coefficient, *n*
Natural streams or river channels	0.030–0.100
Natural flood plains	0.030–0.150
Lined streams or river channels	0.012–0.033
Excavated or dredged streams or river channels	0.018–0.100

$$h_f = L_u\left(\frac{Q^2}{K_u K_d}\right) + L\left(\frac{Q}{K_d}\right)^2, \tag{3.12}$$

where

L_u = length of upstream reach (from uniform flow to beginning of constriction)

K_u = upstream conveyance $= \dfrac{A_u R_u^{2/3}}{n_u}$

K_d = downstream conveyance $= \dfrac{A_d R_d^{2/3}}{n_d}$

A_u, A_d = upstream and downstream channel cross-sectional areas, respectively

R_u, R_d = upstream and downstream hydraulic radius, respectively = Area, A_u or A_d, divided by the channel wetted perimeter

n_u, n_d = upstream and downstream Manning's roughness coefficient, respectively. Typical normall* values of Manning's roughness coefficient, *n*, for common channel crossings are indicated in Table 3.1.†

3.2.3.2.1.2 Obstructed Discharge Hydraulics Due to the large live loads, long railway bridges often consist of many short- and medium-length spans, where the topography allows such construction. In these cases, many piers are required that may create an obstruction to the flow, and consideration of the contraction effects due to obstruction is also necessary (Equation 3.9 with C_c equals the coefficient of contraction of the new channel cross section at the obstruction). The flow past an obstruction is similar to the flow past a constriction but with more openings (Figure 3.3). The degree of contraction is usually less for obstructions than constrictions. Published values of C_c for various obstruction geometries are available in the literature of open-channel hydraulics (Yarnell, 1934; Chow, 1959).

The flow about an obstruction consisting of bridge piers was investigated extensively (Nagler, 1918) and Equation 3.13, which is similar in form to Equation (3.9), was derived from the results:

$$A = by = \frac{Q}{K_N\sqrt{\left(2gH_1 + \beta V_u^2\right)}}, \tag{3.13}$$

* Minimum and maximum values of Manning's roughness coefficient, *n*, are also given in AREMA Chapter 1 and in the literature of open channel hydraulics.

† Specific values of Manning's roughness coefficient, *n*, for various channel types and conditions are given in AREMA Chapter 1 and in the literature of open channel hydraulics.

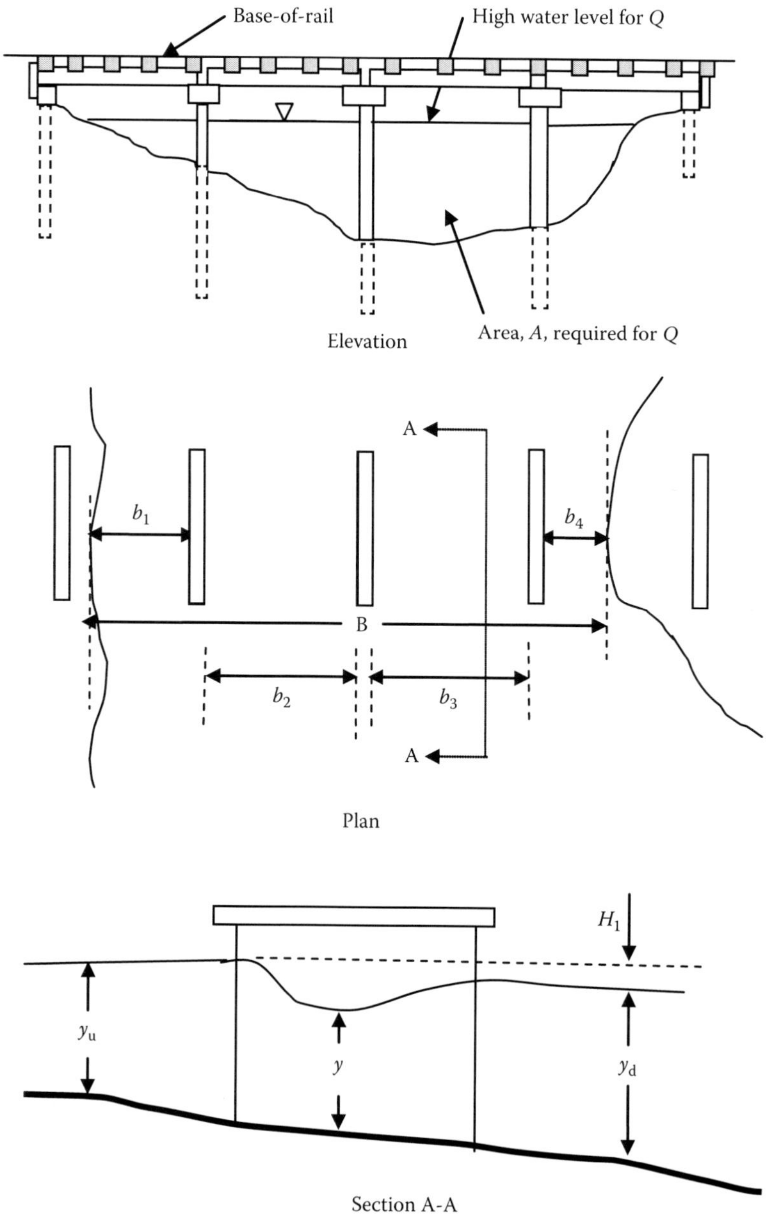

FIGURE 3.3 River crossing profile with obstructions at Q.

where

b = the effective width at the obstruction = $b_1 + b_2 + b_3 + b_4$ (Figure 3.3)

y = the depth at the pier or obstruction = $y_d - \varphi V_d^2/2g$

y_d = the depth downstream of the pier or obstruction

V_d = the average velocity of the downstream channel for discharge, Q

Φ = an adjustment factor (it has been evaluated from experiments that Φ is generally about 0.3)

K_N = the coefficient of discharge, which depends on the geometry of the pier or obstruction and the bridge opening ratio, $N = b/B$

H_l = the downstream afflux

β = a correction for upstream velocity, V_u, head where $N<0.6$ and $\beta \sim 2$

However, for subcritical flow the Federal Highway Administration (FHWA), 1990 recommends the use of the Energy equation (Schneider et al., 1977) or momentum balance methods (TAC, 2004) when pier drag is a relatively small proportion of the friction loss. When pier drag forces constitute the predominate friction loss through the contraction, the momentum balance or Yarnell equation methods are applicable. The momentum balance method yields more accurate results when pier drag becomes more significant.

The Yarnell equation is based on further experiments (summarized by Yarnell, 1934) with relatively large piers (typical of railway bridge substructures) that were performed to develop equations for the afflux for use with Equation 3.14 (the d'Aubuisson equation, which is applicable to subcritical flow conditions only).

$$A_d = by_d = \frac{Q}{K_A\sqrt{\left(2gH_l + V_u^2\right)}}, \tag{3.14}$$

where K_A is a coefficient from Yarnell's experiments, which depends on the geometry of the pier or obstruction and the bridge opening ratio, $N = b/B$, where B is the width of the channel without obstruction.

The afflux depends on whether the flow is subcritical or supercritical (Hamill, 1999). For subcritical flow conditions,

$$H_l = Ky_dF_d^2\left(K + 5F_d^2 - 0.6\right)\left((1-N) + 15(1-N)^4\right), \tag{3.15}$$

where K is Yarnell's pier shape coefficient (between 0.90 and 1.25 depending on pier geometry) and

$$F_d = \frac{Q}{A\sqrt{gy_d}} \le 1.0 = \text{the normal depth Froude number.}$$

The normal depth, y_d, is readily calculated from the usual open-channel hydraulics methods.

For supercritical flow conditions (which will cause downstream hydraulic jump), the analysis is more complex and design charts have been made to assist in establishing the discharge past obstructions (Yarnell, 1934).

3.2.3.2.1.3 Contraction at Constrictions and Obstructions Chapter 1 of American Railway Engineering and Maintenance-of-way Association (AREMA) (2015)* provides guidance on contraction coefficients, C_c, for typical railway bridge hydraulic analyses. Typical bridge constrictions and/or obstructions subjected to subcritical flow will have a contraction coefficient, $C_c \simeq 0.3$. C_c is typically between about 0.05 and 0.1 for supercritical flow† regimes.

* Recommended practices for the hydraulic design of railway bridges are developed and maintained by the American Railway Engineering and Maintenance-of-Way Association (AREMA). Recommended practice for the hydraulic design of railway bridges is outlined in Chapter 1. Many railroad companies establish railway bridge hydraulic design criteria based on, and incorporating portions of the AREMA, recommended practices.

† Supercritical flow velocity head is greater than that of subcritical flow resulting in a lower coefficient of contraction.

3.2.3.2.2 *Scour at Bridge Crossings*

It is not a primary concern to the superstructure designer, but once the required bridge opening and general geometry of the crossing is established, scour conditions at constrictions* and obstructions† must be investigated in order to ensure and sustain overall stability of the bridge foundations and substructure (Figure 3.4). Scour can occur when the streambed is composed of cohesive or cohesionless materials. However, scour generally occurs at a much faster rate for cohesionless materials, which is of crucial significance to hydraulic and bridge design engineers.

General scour occurs due to streambed degradation at the contraction (due to opening constriction and/or obstruction) of the waterway opening caused by the bridge substructures (abutments and piers). Contraction scour may occur under both live-bed‡ and clear-water§ conditions. General scour may also occur due to degradation, or adjustment of the river bed elevation, due to overall hydraulic changes not specifically related to the bridge crossing, such as lateral migration

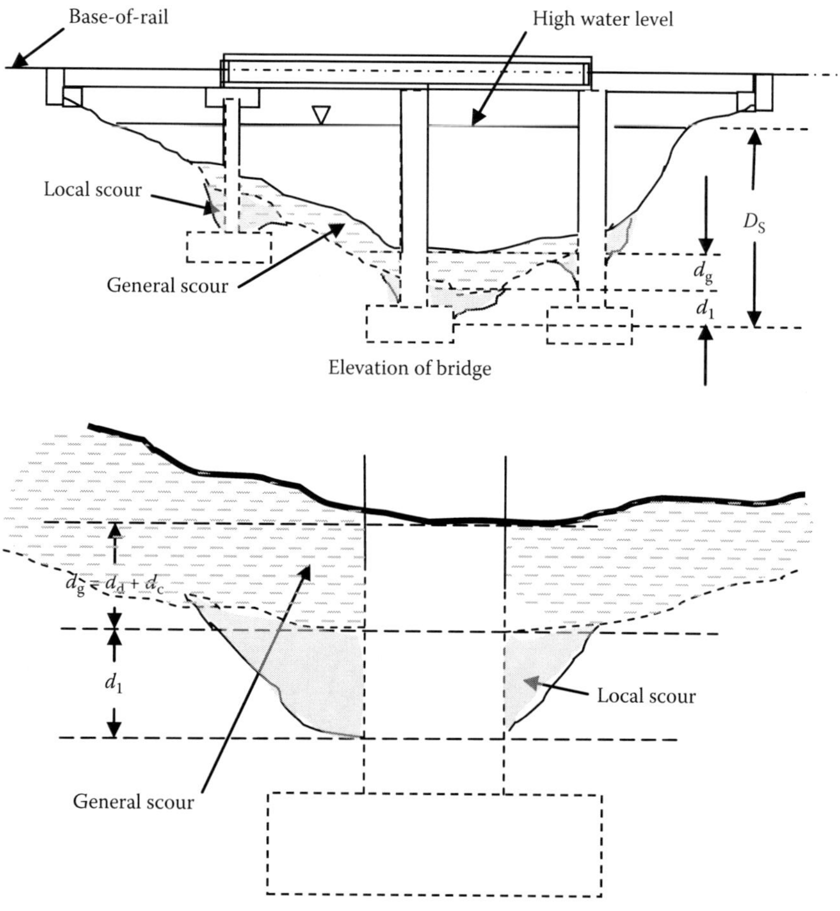

FIGURE 3.4 Scour at bridge crossings due to constrictions and obstructions.

---

* Typically abutments.

† Typically piers.

‡ The streambed material upstream of the bridge is moving.

§ The upstream streambed is at rest and there is no sediment in the water.

of the waterway. Degradation, general scour due to overall hydraulic changes may occur under live-bed scour conditions. For preliminary design,* and where overall hydraulic degradation is not a consideration, general scour can be estimated by assuming that the cross-sectional area of the scour is equal to the loss of cross-sectional discharge area due to the submerged substructures. The depth of general scour, d_g, can then be estimated based on the width of the channel bed. Local scour (due to obstruction) may also occur under both live-bed and clear-water conditions.

3.2.3.2.2.1 Contraction Scour During live-bed scour the contraction scour depths are affected by deposits of sediment from upstream. Scour will cease when the rate of sediment deposit equals the rate of loss by contraction scour. Under clear-water conditions, sediment is not transported into the contraction scour depth increase (creating channel bed depressions or holes). Scour will equilibrate and cease when the velocity reduction caused by the increased area becomes less than that required for continued contraction scour. Equation 3.16 provides an estimate of the approach streambed velocity at which live-bed scour will initiate, V_s, as follows (Laursen, 1963):

$$\text{In SI units: } V_s = 6.2 y_u^{1/6} D_{50}^{1/3}, \quad (3.16a)$$

$$\text{In US Customary or Imperial Units: } V_s = 11.2 y_u^{1/6} D_{50}^{1/3}. \quad (3.16b)$$

Here, y_u is the depth of channel upstream of the bridge crossing and D_{50} is the streambed material median diameter at which 50% by weight are smaller than that specified (the size which governs beginning of erosion in well-graded materials).

Contraction scour depth, d_c, for cohesionless materials under live-bed conditions can be estimated as (Laursen, 1962)

$$d_c = y_u \left(\left(\frac{Q_c}{Q} \right)^{6/7} \left(\frac{B}{b} \right)^{k_1} \left(\frac{n_c}{n_u} \right)^{k_2} - 1 \right), \quad (3.17)$$

where

Q_c = the discharge at the contracted channel (at bridge crossing)
b = the net or effective width of the bridge opening
B = the width of the channel without obstruction or constriction
n_c = Manning's surface roughness coefficient at the contracted channel
n_u = Manning's surface roughness coefficient at the upstream channel
k_1, k_2 are exponents that depend on $\sqrt{\frac{g y_u S_u}{V_{D50}}}$
S_u = upstream energy slope (often taken as streambed slope)
V_{D50} = median fall velocity of flow based on D_{50} median particle size

Contraction scour depth, d_c, for cohesionless materials under clear-water conditions can be estimated as (Laursen, 1962; Richardson and Davis, 2001)

* Final scour design should be based on the more accurate assessment methods for general scour available (TAC, 2004) (Richardson and Davis, 2001).

$$d_c = y_u\left(\left(\frac{B}{b}\right)^{6/7}\left(\frac{V_u^2}{42(y_u)^{1/3}(D_{50})^{2/3}}\right)^{3/7} - 1\right). \tag{3.18}$$

3.2.3.2.2.2 Local Scour Local scour occurs at substructures as a result of vortex flows induced by the localized disturbance to flow caused by the obstruction. The determination of local scour depths is complex, but there are published values relating local and general scour depths that are useful for preliminary scour evaluations. Procedures for establishing local scour relationships for abutments and piers are available (TAC, 2004; Richardson and Davis, 2001). For most modern bridges, local scour at abutments and piers can generally be precluded by the use of properly designed revetments and scour protection. Local scour depth for cohesionless materials at piers with width, b_{pier}, can be estimated as

$$d_l = 2yK_{1P}K_{2P}K_{3P}\left(\frac{b_{pier}}{y}\right)^{0.65}F^{0.43}. \tag{3.19}$$

Local scour depth for cohesionless materials at abutments can be estimated as (Richardson and Davis, 2001)

$$d_l = 4y\left(K_{1A}/0.55\right)K_{2A}F^{0.33}, \tag{3.20}$$

where

K_{1P} = pier nose geometry adjustment factor (from experimental values)
K_{2P} = angle of flow adjustment factor (from experimental values)
K_{3P} = bed configuration (dune presence) adjustment factor (from experimental values)
K_{1A} = abutment-type (vertical, sloped, wingwalls) adjustment factor (from experimental values)
K_{2A} = abutment skew adjustment factor (from experimental values)
y = depth at pier or abutment
F = Froude number calculated at pier or abutment

3.2.3.2.2.3 Total Scour The total scour depth, d_t, is then estimated as

$$d_t = d_g + d_l = d_d + d_c + d_l, \tag{3.21}$$

where

d_g = general scour
d_d = degradation depth related to global longer term channel bed adjustments
d_c = contraction scour (under live-bed or clear-water conditions)
d_l = local scour at substructure (abutment or pier)

In some cases, substructure depth must also be designed anticipating extreme natural scour and channel degradation events. Therefore, spread footings or the base of pile caps, for major bridges are often located such that the underside of the footing or cap is 1.5 m (5 ft) to 1.8 m (6 ft) below the estimated total scour depth, d_t. Also, it is often beneficial to consider the use of fewer long piles than a greater number of short piles when the risk of foundation scour is relatively great.

3.2.3.2.3 AREMA Scour Design Recommendations

Degradation depth is difficult to assess as it is generally a long-term phenomenon related to site geology and geomorphology considering channel, obstruction, and constriction geometries. Nevertheless, considering sediment transport as the principal cause of degradation, estimates can be made using computer models using long-term hydrographic flow information. AREMA (2015) provides recommendations for computer modeling of degradation depth related to global longer term channel bed adjustments.

AREMA (2015) also provides recommendations for the estimation of contraction scour depth, d_c, for live-bed conditions based on Equation 3.17 as

$$d_c = y_u\left(\frac{Q_c}{Q}\right)^{6/7}\left(\frac{B}{b}\right)^{k1}, \tag{3.22}$$

where k_1 varies from 0.59 to 0.69 depending on how much suspended bed material is in the flow.

AREMA (2015) also provides recommendations for the estimation of contraction scour depth, d_c, for clear-water conditions based on Equation 3.18 as

$$d_c = \left(\frac{K_u Q_c^2}{b^2 (D_{50})^{2/3}}\right)^{3/7}, \tag{3.23}$$

where $K_u = 0.025$ using SI units of meters and $K_u = 0.0077$ using US Customary or Imperial Units of feet.

The AREMA (2015) recommendations for estimating pier local scour depth for live-bed and clear-water conditions are based on Equation 3.19 as

$$d_l = 2b_{pier}K_{1P}K_{2P}K_{3P}\left(\frac{y_u}{b_{pier}}\right)^{0.35} F^{0.43}, \tag{3.24}$$

where

K_{1P} = pier nose geometry adjustment factor, which varies from 0.9 to 1.1 depending on the pier nose shape, K_{2P} = angle of flow adjustment factor $= \left(\cos Ø + \frac{L_{pier}}{b_{pier}}\sin Ø\right)^{0.65}$

Φ = angle of flow attack

L_{pier} = length of the pier

$L_{pier}/b_{pier} \leq 12$

K_{3P} = bed configuration adjustment factor = 1.1, except where large dunes are present.

The AREMA (2015) recommendation for estimating abutment local scour depth for live-bed and clear-water conditions, based on US Army Corps of Engineers (USACE) field and laboratory data, is

$$d_l = 7.3yK_{1A}K_{2A}F^{0.33}, \tag{3.25}$$

where K_{1A} is the abutment shape adjustment factor, which varies from 0.82 to 1.00 for vertical wall abutments and is 0.55 for spill through-type abutments, K_{2A} = angle (in degrees) of flow adjustment factor $= \left(\frac{\theta}{90}\right)^{0.13}$

θ = angle between centerline of bridge and waterway flow direction

The total scour depth, d_t, is then estimated as

$$d_t = d_d + d_c + d_l, \quad (3.26)$$

where

d_d = degradation depth

d_c = contraction scour depth

d_l = local scour depth at substructure

3.2.3.3 Highway, Railway, and Marine Clearances

Railway bridges crossing over transportation corridors must provide adequate horizontal and vertical clearance to ensure the safe passage of traffic under the bridge. Railway, highway, and navigable waterway minimum clearance requirements are prescribed by government agencies.* Provision for changes in elevation of the undercrossing (i.e., a highway or track raise) and widening should be considered during planning and preliminary design.

The minimum railway bridge clearance envelope recommended by AREMA (2015)† is generally 7.0 m (23 ft) from top of the rail and 2.75 m (9 ft) on each side of the track centerline. Chapters 15 and 28 of AREMA (2015) outline more detailed clearance requirements for railway bridges in various North American jurisdictions.‡ These dimensions must be revised to properly accommodate track curvature. Railroad companies may have additional clearance requirements relating to the safety of railroad operations.

3.2.3.4 Geotechnical Conditions

Geotechnical site conditions are often critical with respect to the location, foundation type, constructability, and cost of railway bridges. Soil borings should generally be taken at or near each proposed substructure location. Soil samples are submitted for laboratory testing and/or tested *in situ* to determine soil properties required for foundation design such as permeability, compressibility, and shear strength.

For the purposes of railway bridge design, the subsurface investigation should yield a report making specific foundation design recommendations. The geotechnical investigation should encompass:

- Foundation type and depth (spread footings, driven piles, drilled shafts, etc.)
- Construction effects on adjacent structures (pile driving, jetting, and drilling)
- Foundation settlement§
- Foundation scour analyses and protection design
- Foundation cost

---

* For example, Transport Canada and US State minimum legal clearances (see AREMA Chapter 28) in North America.

† Recommended practices for the design of railway bridges are developed and maintained by the American Railway Engineering and Maintenance-of-Way Association (AREMA). Recommended practice for the design of fixed railway bridges is outlined in Part 1—Design and for the design of movable railway bridges is outlined in Part 6—Movable Bridges, in Chapter 15—Steel Structures, of the AREMA MRE. Chapter 15—Steel Structures, provides detailed recommendations for the design of steel railway bridges for spans up to 120 m (400 ft) in length, standard gage track of 1435 mm (56.5″), and North American freight and passenger equipment at speeds up to 127 km/h (79 mph) and 145 km/h (90 mph), respectively. The recommendations may be used for longer span bridges with supplemental requirements. Clearance requirements are outlined in Chapter 28. Many railroad companies establish steel railway bridge design criteria based on, and incorporating portions of the AREMA, recommended practices.

‡ Clearances for highway and railway traffic under railway bridges and for railway traffic through railway bridges are outlined. In Canada, Transport Canada has established regulations concerning the clearances for structures and through railway bridges.

§ Generally simply supported spans and ballasted deck bridges tolerate greater settlements. However, tolerable settlements may depend on longitudinal and lateral track geometry (e.g., permissible variations in rail profile or cross-level) tolerance requirements.

Driven steel pipe, steel HP sections, and precast concrete piles are often cost-effective bridge foundations. Although typically more costly than driven piles, concrete piles may also be installed by boring when required by the site conditions. Geotechnical investigations for driven pile foundations should include the following:

- Recommended pile types based on design and installation criteria
- Pile capacities related to soil friction and/or end bearing
- Pile tip elevation estimation
- Allowable pile loads and factor of safety (FS) used
- Recommended test pile requirements and methods

Investigations for spread footing foundations should include the following:

- Footing elevation (scour and/or frost protection)
- Allowable soil bearing pressure and FS used
- Groundwater elevation
- Stability (overturning, sliding)
- Bedding materials and compaction

The design and construction of drilled shaft foundations should be based on geotechnical investigations and recommendations relating to the following:

- Friction and end bearing conditions (straight shaft, belled base)
- Construction requirements (support of hole)
- Allowable side shear and base bearing stresses and FS used for each
- Downdrag and uplift conditions

Also, specific dynamic soil investigations may be required in areas of high seismic activity to determine soil strength, foundation settlement, and stability under earthquake motions.

In some cases, it is possible that, due to geotechnical conditions, foundations are recommended to be relocated. This will result in significant changes in the proposed bridge arrangement and should be carefully and comparatively cost estimated. A geotechnical engineer experienced in shallow and deep bridge foundation design and construction should be engaged to manage geotechnical site investigations and provide recommendations for foundation design to the bridge designer.*

3.2.4 Geometry of the Track and Bridge

Railway horizontal alignments consist of simple curves (Figure 3.5) and tangent track connected by transition or spiral curves. Track profile, or vertical alignment, comprises constant grades connected by parabolic curves. Many high-density rail lines have grades of less than 1% and restrict curvature to safely operate at higher train speeds.

* The bridge designer should, in many cases, provide the geotechnical engineer with preliminary foundation design loads to assist with the foundation investigation and design recommendations. The bridge design engineer should inform the geotechnical engineer of final foundation design loads that differ substantially from the preliminary design loads.

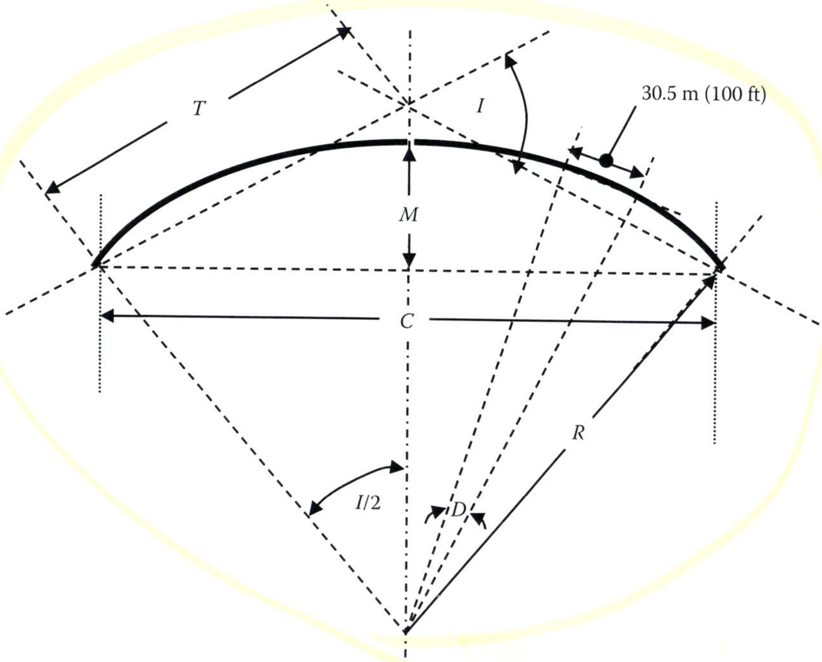

FIGURE 3.5 Simple curve geometry.

3.2.4.1 Horizontal Geometry of the Bridge

If curved tracks traverse a bridge, the consequences for steel superstructure design are effects due to the following:

- Centrifugal force created as the train traverses the bridge at speed, V (force effect)
- Offset or eccentricity of the track alignment with respect to the centerline of the span or centerline of supporting members (geometric effect)
- Offset or eccentricity of the center of gravity of the live load as it traverses the superelevated curved track (geometric effect)

The centrifugal force is horizontal and transferred to supporting members as a force couple. The magnitude of the force depends on the track curvature and live load speed, and it is applied at the center of gravity of the live load (see Chapter 4). Track alignment and superelevation also affect vertical live load forces (including impact) in supporting members based on geometrical eccentricity of the live load.

3.2.4.1.1 Route (Track) Geometrics

Train speed is governed by the relationship between curvature and superelevation. Railway bridge designers must have an accurate understanding of route geometrics to develop the horizontal geometry of the bridge, determine centrifugal forces, and ensure adequate horizontal and vertical clearances in through superstructures. The central angle subtended by a 30.5 m (100 ft) chord in a simple curve, or the degree of curvature, D, is used to describe the curvature of North American railroad track. Then the radius, R, and other simple curve data are as follows:

$$R = \frac{360(30.5)}{2\pi D} = \frac{1747.5}{D}\text{ m} = \frac{5729.6}{D}\text{ ft}, \tag{3.27}$$

$$I \approx \frac{L_c D}{30.5}(L_c = \text{m}) \approx \frac{L_c D}{100}(L_c = \text{ft}), \tag{3.28}$$

$$T = R \tan\frac{I}{2}, \tag{3.29}$$

$$C = 2R \sin\frac{I}{2}, \tag{3.30}$$

$$M = R\left(1-\cos\frac{I}{2}\right) = R - \sqrt{\left(R+\frac{C}{2}\right)\left(R-\frac{C}{2}\right)} = R - \sqrt{R^2 - \frac{C^2}{4}}. \tag{3.31}$$

Rearranging Equation 3.31 yields

$$R = \frac{C^2}{8M} + \frac{M}{2} \approx \frac{C^2}{8M}, \tag{3.32}$$

where R is greater than about 175 m (575 ft), which is typical of railway track curvature.

Rearranging Equation 3.32 yields

$$M = \frac{C^2}{8R} \approx \frac{L_c^2}{8R}, \quad \text{where } R \text{ is large}, \tag{3.33}$$

where

L_c = length of the curve [L_c >> 30.5 m (100 ft) for typical of railway track]

I = intersection angle

T = tangent distance

C = chord length

M = mid-ordinate of curve

The track is superelevated to accommodate the centrifugal forces that occur as the train traverses through a curved track (Figure 3.6).

For equilibrium, with weight equally distributed to both wheels, the superelevation, e, is (Hay, 1977)

$$e = \frac{\text{CF}(d)}{W}. \tag{3.34}$$

Also, since

$$\text{CF} = \frac{mV^2}{R}, \tag{3.35}$$

the superelevation may be expressed as

$$e = \frac{dV^2}{gR}, \tag{3.36}$$

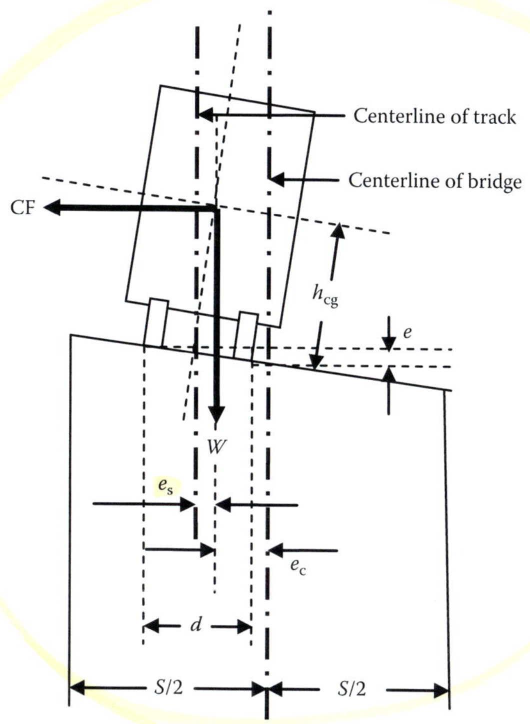

FIGURE 3.6 Railway track superelevation.

where

d = horizontal projection of track contact point distance = 1.5 m (4.9 ft) for North American standard gage track

CF = centrifugal force

W = weight of train

$m = W/g$ = mass of train (g = acceleration due to gravity)

V = speed of train

For SI system units of measurement, substitution of $R = 1750/D$, $d = 1.5$ m, and $g = 9.98$ m/s^2 into Equation 3.36 yields

$$e \approx 0.0066DV^2 \text{ (SI units)}, \tag{3.37a}$$

where

e = equilibrium superelevation, mm

V = speed of train, km/h

D = degree of curve (the central angle subtended by a 30.5 m chord in a simple curve)

For US Customary or Imperial units of measurement, substitution of $R = 5730/D$, $d = 4.9$ ft, and $g = 32.17$ ft/s^2 into Equation 3.36 yields

$$e \approx 0.0007DV^2 \text{ (US Customary or Imperial units)}, \tag{3.37b}$$

where

e = equilibrium superelevation, in.

V = speed of the train, miles/h

D = degree of the curve (the central angle subtended by a 100 ft chord in a simple curve)

Transition curves are required between tangent and curved track to gradually vary the change in lateral train direction. The cubic parabola is used by many freight railroads as a transition from tangent track to an offset simple curve. The length of the transition curve is based on the rate of change of superelevation. Safe rates of superelevation "run-in" are prescribed by regulatory authorities and railroad companies. For example, using the SI system units of measurement, with a rate of change of superelevation = 32 mm/s, the length of the transition curve, L_s, in m, is

$$L_s = 0.0089\text{eV}\left(\text{SI units}\right), \tag{3.38a}$$

where e is the equilibrium superelevation in mm and V is the speed of the train in km/h.

For US Customary or Imperial units of measurement and a rate of change of superelevation = 1.25 in./s, the length of transition curve, L_s, ft, is

$$L_s = 1.17\text{eV}\left(\text{US Customary or Imperial Units}\right), \tag{3.38b}$$

where e is the equilibrium superelevation in in., and V is the speed of the train in mph.

3.2.4.1.2 Bridge Geometrics

Track curvature can be accommodated by laying-out bridges using straight or curved spans. Straight spans must be laid-out on a chord to reflect the curved track alignment. The individual straight spans must be designed for the resulting eccentricities* as the curved track traverses the span (Figure 3.7). The

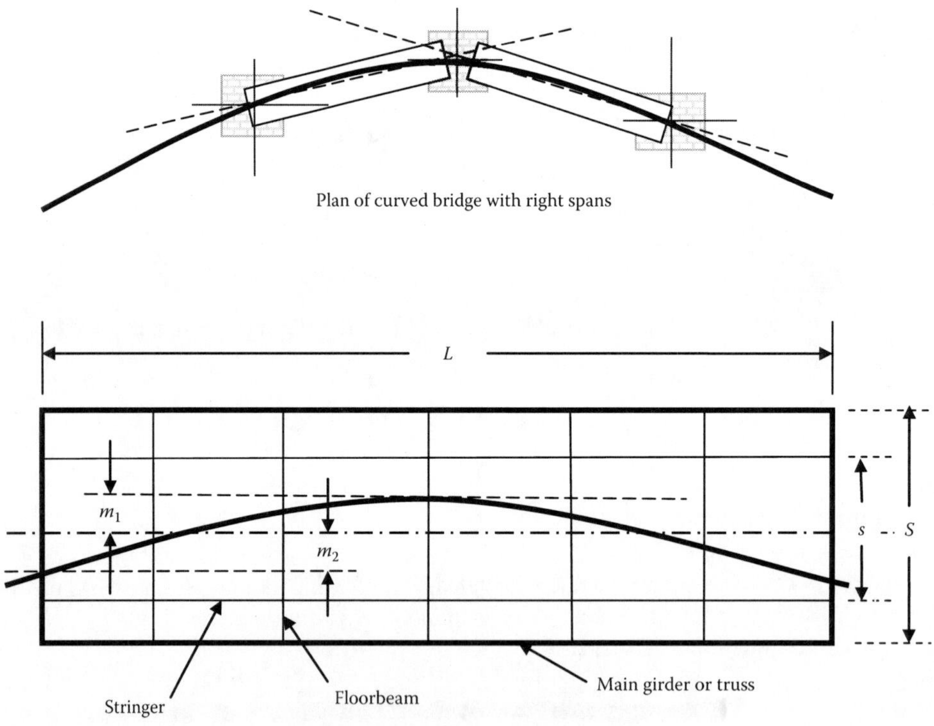

FIGURE 3.7 Horizontally curved bridges using right spans.

* For example, through girder or truss spacing and design forces are increased due to eccentricity of track (curvature effect) and superelevation (shift effect).

stringer spacing, s, may be adjusted to equalize eccentricities in the center and end panels ($m_1 = m_2$); and in some cases, such as sharp curves, it may be economical to offset the stringers equally in each panel. However, fabrication and erection effort and costs must be carefully considered prior to designing offset floor systems. It is common practice on freight railroads to use m_1 between $M/6$ and $M/2$,* where $M = m_1 + m_2$ is given by Equation 3.31.

The superelevated and curved track creates horizontal eccentricities based on the horizontal curve geometry (track curvature effect), e_c, and vertical superelevation (track shift effect), e_s. These eccentricities must be considered when determining the lateral distribution of live load forces (including dynamic effects) to members (stringers, floorbeams, and main girders or trusses). The shift effect eccentricity, e_s, is (Figure 3.6)

$$e_s = \frac{h_{cg} e}{d}. \tag{3.39}$$

Expressed as a percentage, this will affect the magnitude of forces to supporting members on each side of the track in the following proportion:

$$\frac{2e_s}{s}(100), \tag{3.40}$$

where

h_{cg} = the distance from the center of gravity of the rail car to the base of the track [AREMA (2015) recommends 2.45 m (8 ft) for the distance from the center of gravity of the car to the top of the rail]

e = the superelevation of the track

d = the horizontal projection of the track gage distance [generally taken as 1.5 m (4.9 ft)]

s = the distance between the center of gravity of the longitudinal members supporting the curved track

The eccentricity due to track curvature, e_c, is estimated by considering an equivalent uniform live load, W_{LL}, along the curved track across the square span length, L. The curvature effects on shear force and bending moment depend on the lateral shift of the curved track with respect to the centerline of the span and the degree of curvature. These effects are often negligible for short spans or shallow curvature (Waddell, 1916). However, if necessary, they can be determined in terms of the main member shear force and bending moment for tangent track across the span as (Figure 3.7)

$$V_0 = \frac{W_{LL} L}{4}\left(1 + \frac{2m_1}{S} - \frac{L^2}{12RS}\right), \tag{3.41}$$

$$V_i = \frac{W_{LL} L}{4}\left(1 - \frac{2m_1}{S} + \frac{L^2}{12RS}\right), \tag{3.42}$$

$$M_0 = \frac{W_{LL} L^2}{16}\left(1 + \frac{2m_1}{S} - \frac{L^2}{24RS}\right), \tag{3.43}$$

$$M_i = \frac{W_{LL} L^2}{16}\left(1 - \frac{2m_1}{S} + \frac{L^2}{24RS}\right), \tag{3.44}$$

* An eccentricity, $m_1 = M/3$, is often used, which provides for equal shear at the ends of the longitudinal members.

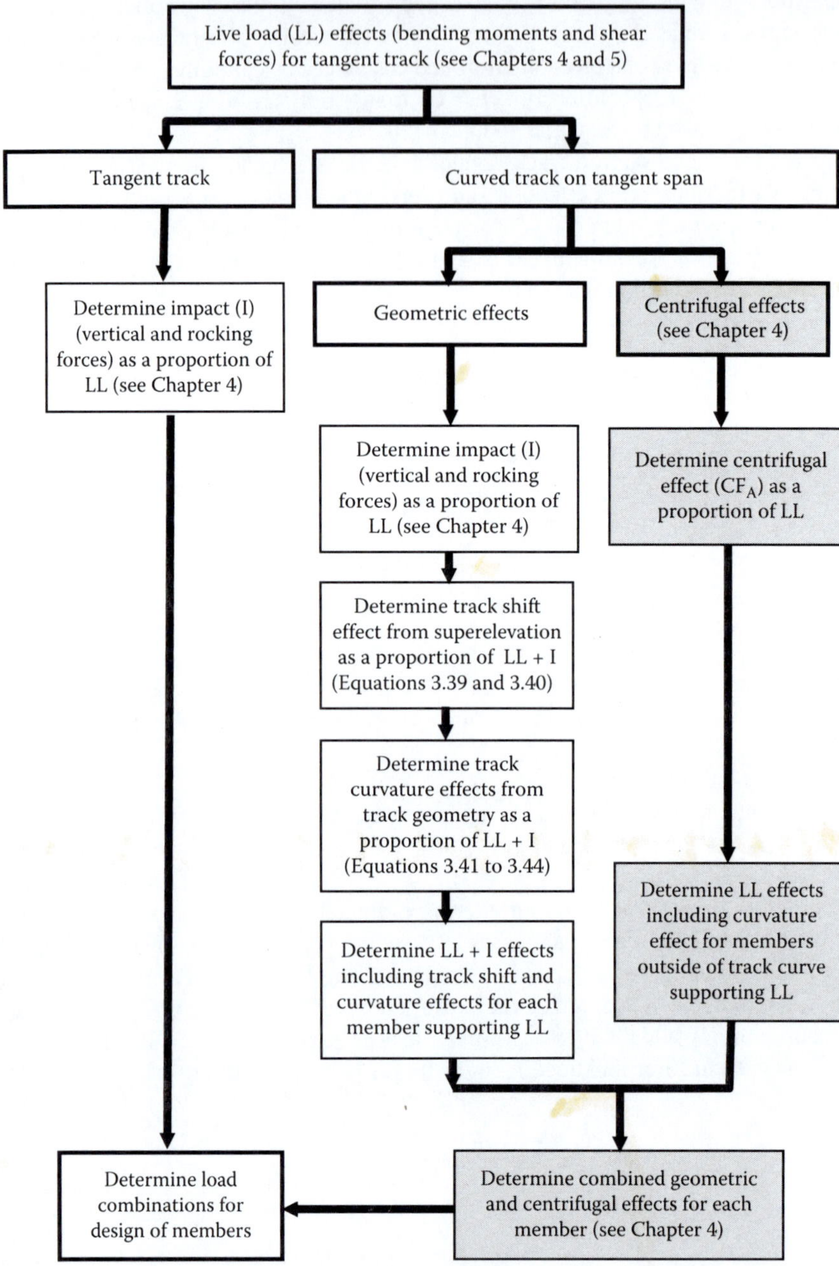

FIGURE 3.8 Flowchart for determination of geometric and centrifugal effects.

where

V_o = the shear force on the outer girder
V_i = the shear force on the inner girder
M_o = the bending moment on the outer girder
M_i = the bending moment on inner girder
W_{LL} = the equivalent uniform live load per track (see Chapter 5)
L = the length of span
S = the distance between the center of gravity of supporting longitudinal members

m_1 = the offset of the track centerline to the bridge centerline at the center of the span
R = the radius of the track curve

For the condition of equal shear (from Equations 3.31, 3.41, and 3.42) at the ends of the girders on each side of the track

$$m_1 = \frac{L^2}{24R} = \frac{M}{3}. \tag{3.45}$$

For the condition of equal moment (from Equations 3.31, 3.43, and 3.44) at the centers of the girders on each side of the track

$$m_1 = \frac{L^2}{48R} = \frac{M}{6}. \tag{3.46}$$

Figure 3.8 outlines the actions required for determination of the geometrical effects and centrifugal effects (shaded boxes, see Chapter 4) of the curved track on vertical live load forces. Example 3.1 outlines the calculation of the geometrical (shift and curvature) effects on vertical live load forces on straight superstructures.

Example 3.1a (SI Units)

A ballasted steel through plate girder railway bridge is to be designed with a 6° of curvature track across its 20 m span. The railroad has specified a 125 mm superelevation based on operating speeds and conditions. The track tie depth and rail height are each 175 mm, and the girders are spaced at 5 m. Determine the geometrical effects of the curvature on the design live load shear and bending moment for each girder.

The effect of the offset of the live load center of gravity (Equation 3.39) is

$$e_s = \frac{(2450+350)(125)}{1500} = 233 \text{ mm (from centerline of track)}.$$

The outside girder forces will be reduced by, and the inside girder forces increased by,

$$\frac{2(233)}{5000}(100) = 9.33\% = 0.093 \text{ (Equation 3.40)}.$$

The curve mid-ordinate over a 20 m span (Equation 3.33) is

$$M = \frac{(20)^2 6}{8(1747.5)} = 0.17 \text{ m} = 172 \text{ mm}.$$

Design for equal shear at girder ends and use a track offset at centerline of

$$m_1 = \frac{172}{3} = 57 \text{ mm}, \quad \text{use 60 mm}.$$

The effect of the curvature alignment (Equations 3.41 and 3.42) on girder shear forces is

$$\left(1 \pm \frac{2m_1}{S} \mp \frac{L^2}{12RS}\right) = \left(1 \pm \frac{2m_1}{S} \mp \frac{L^2 D}{12(1747.5)S}\right)$$

$$= \left(1 \pm \frac{2(60)}{5000} \mp \frac{(20)^2 6}{12(1747.5)5}\right) = (1 \pm 0.024 \mp 0.023) = 1.00.$$

The effect of the curvature alignment (Equations 3.43 and 3.44) on girder flexural forces is

$$\left(1 \pm \frac{2m_1}{S} \mp \frac{L^2}{24RS}\right) = \left(1 \pm \frac{2m_1}{S} \mp \frac{L^2 D}{24(1747.5)S}\right)$$

$$= \left(1 \pm \frac{2(60)}{5000} \mp \frac{(20)^2 6}{24(1747.5)5}\right) = (1 \pm 0.024 \mp 0.011).$$

The outside girder and inside girder forces are multiplied by $(1+0.024-0.011)=1.013$ and $(1-0.024+0.011)=0.987$, respectively, to account for the offset of curved track alignment.

Therefore, the following are determined for the shear, V_{LL+I}, and bending moment, M_{LL+I}, live load forces:

$$V_{out} = V_{LL+I}\,(1-0.093) = 0.907V_{LL+I}$$
$$V_{in} = V_{LL+I}\,(1+0.093) = 1.093V_{LL+I}$$
$$M_{out} = M_{LL+I}\,(1.013-0.093) = 0.920M_{LL+I}$$
$$M_{in} = M_{LL+I}\,(0.987+0.093) = 1.080M_{LL+I}$$

It should be noted that these shear and bending moment forces do not include the effects of the centrifugal force. The calculation of centrifugal forces is outlined in Chapter 4.

Example 3.1b (US Customary or Imperial Units)

A ballasted steel through plate girder railway bridge is to be designed with a 6° of curvature track across its 70 ft span. The railroad has specified a 5 in. superelevation based on operating speeds and conditions. The track tie depth and rail height are taken as 7 in. each and the girders are spaced at 16 ft. Determine the geometrical effects of the curvature on the design live load shear and bending moment for each girder.

The effect of the offset of the live load center of gravity (Equation 3.39) is

$$e_s = \frac{(8+(14/12))(5)}{(4.9)} = 9.35 \text{ in (from centerline of track)}.$$

The outside girder forces will be reduced by, and the inside girder forces increased by,

$$\frac{2(9.35)}{(16)(12)}(100) = 9.74\% = 0.097 \quad \text{(Equation 3.40)}.$$

The curve mid-ordinate over a 70 ft span (Equation 3.33) is

$$M = \frac{(70)^2 6}{8(5730)} = 0.64 \text{ ft} = 7.7 \text{ in.}$$

Design for equal shear at girder ends and use a track offset at centerline of

$$m_1 = \frac{7.7}{3} = 2.56, \text{ use } 2.5 \text{ in.}$$

The effect of the curvature alignment (Equations 3.41 and 3.42) on girder shear forces is

$$\left(1 \pm \frac{2m_1}{S} \mp \frac{L^2}{12RS}\right) = \left(1 \pm \frac{2m_1}{S} \mp \frac{L^2 D}{12(5730)S}\right)$$

$$= \left(1 \pm \frac{2(2.5)}{(16)(12)} \mp \frac{(70)^2 6}{12(5730)16}\right) = (1 \pm 0.026 \mp 0.027) = 1.00.$$

The effect of the curvature alignment (Equations 3.43 and 3.44) on girder flexural forces is

$$\left(1\pm\frac{2m_1}{S}\mp\frac{L^2}{24RS}\right)=\left(1\pm\frac{2m_1}{S}\mp\frac{L^2D}{24(5730)S}\right)$$

$$=\left(1\pm\frac{2(2.5)}{(16)(12)}\mp\frac{(70)^2 6}{24(5730)16}\right)=\left(1\pm 0.026\mp 0.013\right).$$

The outside girder and inside girder forces are multiplied by $(1+0.026-0.013)=1.013$ and $(1-0.026+0.013)=0.987$, respectively, to account for the offset of curved track alignment.

Therefore, the following are determined for the shear, V_{LL+I}, and bending moment, M_{LL+I}, live load forces:

$$V_{out} = V_{LL+I}\,(1-0.097) = 0.903V_{LL+I}$$
$$V_{in} = V_{LL+I}\,(1+0.097) = 1.097V_{LL+I}$$
$$M_{out} = M_{LL+I}\,(1.013-0.097) = 0.916M_{LL+I},$$
$$M_{in} = M_{LL+I}\,(0.987+0.097) = 1.084M_{LL+I}.$$

It should be noted that these shear and bending moment forces do not include the effects of the centrifugal force. The calculation of centrifugal forces is outlined in Chapter 4.

The bridge deck must be superelevated to accommodate the track curvature. The required superelevation is readily provided in ballasted deck bridges (Figure 3.9) but may also be developed in open-deck bridges by tie dapping, shimming, or varying the elevation of supporting superstructure or bearings (Figure 3.10). Varying the elevation of the members supporting the deck may be problematical from structural behavior, fabrication, and maintenance perspectives, and generally not recommended.

Curved spans must be designed for flexural and torsional effects. Dynamic behavior under moving loads is particularly complex for curved girders as flexural and torsional vibrations may be coupled. Even the static design of curved girder railway bridges requires careful consideration of torsional and distortional* warping stresses and shear lag considerations. These analyses are complex and often carried out using finite element analysis (FEA) software.† Curved girders are best suited to continuous span construction and, therefore, not often used for freight railroad bridges. Continuous construction is relatively rare for ordinary steel freight railway bridges due to remote location erection requirements (field splicing, falsework, and large cranes); and to preclude uplift that may occur due to the large railway live load to superstructure dead load ratio. Nevertheless, curved girders are often effectively utilized for light transit applications.

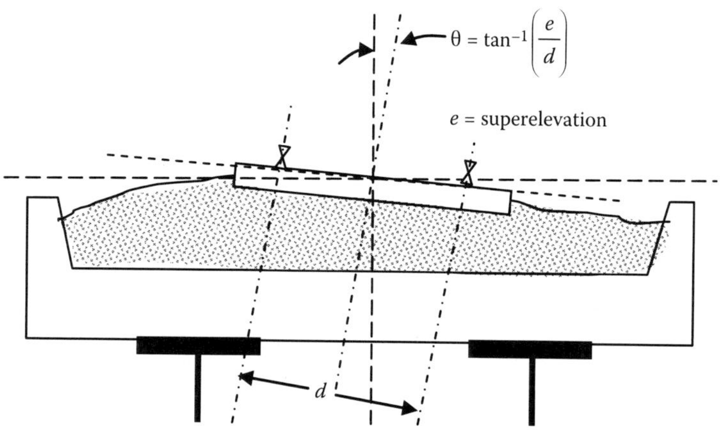

FIGURE 3.9 Track superelevation on ballasted deck bridges.

* In the case of box girders.

† The Federal Highway Administration (FHWA) has also performed extensive research on steel curved girders at the Turner-Fairbanks lab. A synthesis of this research is available from the FHWA.

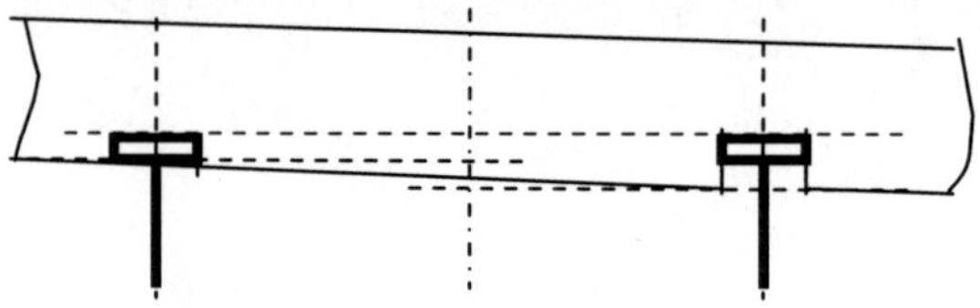

(a). Differential tie dapping (notched shear strength may govern depth of dap)

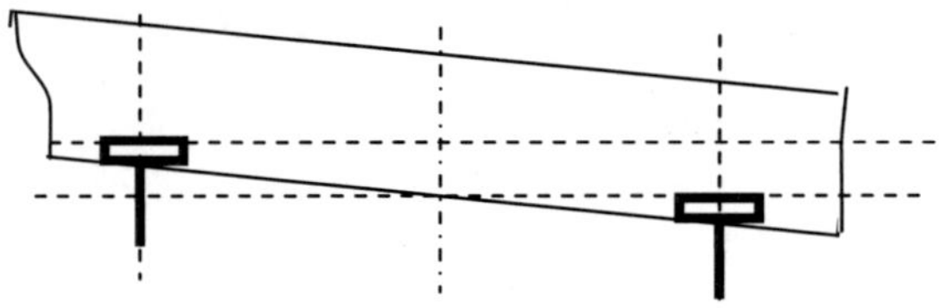

(b). Variable deck support elevation (dapped)

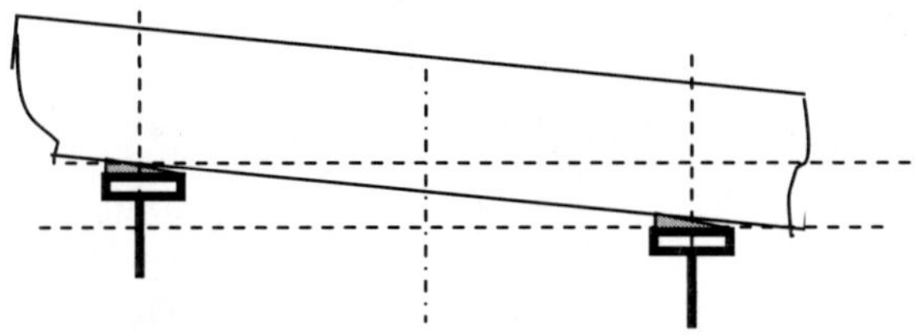

(c). Variable deck support elevation (shimmed, not generally recommended)

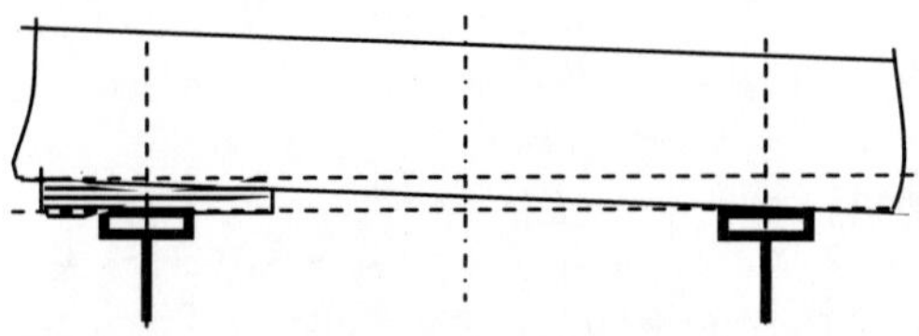

(d). Tie shimming (not generally recommended)

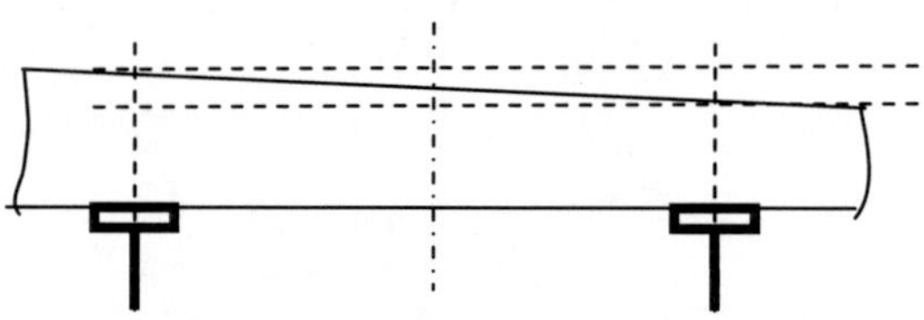

(e). Tie tapering

FIGURE 3.10 Track superelevation on open-deck bridges.

3.2.4.1.3 Skewed Bridges

Skewed bridges have been considered as a necessary inconvenience to an abomination (Waddell, 1916) by bridge designers. There are many salient design and construction reasons for avoiding skewed bridge construction. Torsional moments and unequal distribution of live load occur with larger skew angles and compromise performance. Also, skewed spans generally require more material than square spans and include details that increase fabrication cost. However, on occasion, and particularly in congested urban areas or where large skew crossings exist, skewed construction may be unavoidable.

Many railroads have specific design requirements regarding skew angle and type of construction for skewed railway bridges. Skewed connections and bent plates may be prohibited requiring that the track support at the ends of skewed spans be perpendicular to the track. This can be accommodated by many ways depending on bearing and span types. Figure 3.11 shows some variations for accommodating skew over a pier for ballasted- (closed-) and open-deck spans.

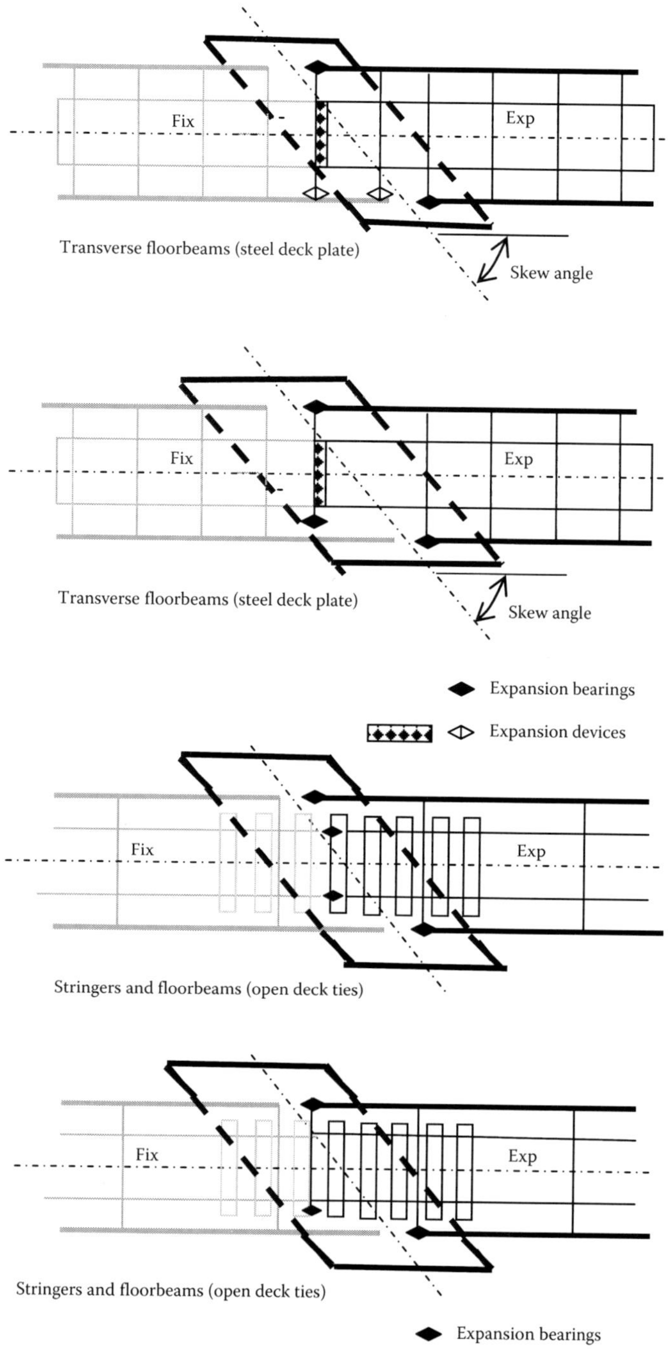

FIGURE 3.11 Square track support at skewed ends of steel railway spans over piers.

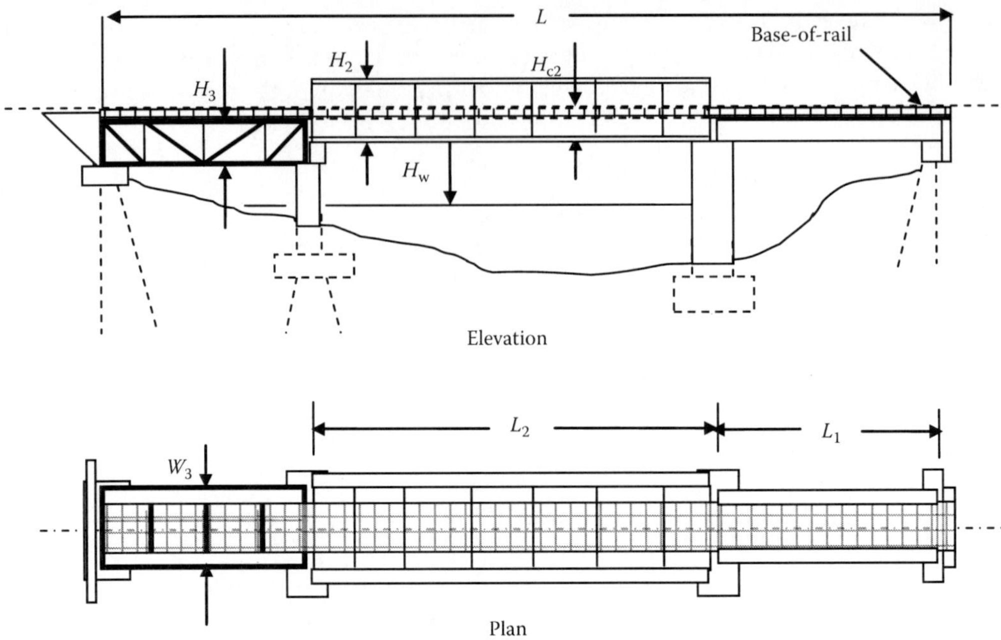

FIGURE 3.12 Basic dimensions of a railway bridge crossing.

3.2.4.2 Vertical Geometry of the Bridge

In addition to a ground profile survey at the crossing, the dimensions generally required to develop a preliminary general arrangement of a railway bridge crossing are shown in Figure 3.12.

These basic dimensions provide the information for preliminary design of the superstructure elements, where L = length of the bridge $= \sum_{i=1}^{ns} L_i$,

where

i = span number
ns = number of spans
L_i = length of span i
W_i = width of span i
H_i = height (depth) of span i
H_{ci} = construction depth of span $i = H_i$ for deck type spans
H_w = distance from base-of-rail to water level
g = grade of bridge

The length of spans is generally governed by site conditions, such as hydraulic or geotechnical considerations, or transportation corridor clearances (railroad, highway, or marine). Width is controlled by the number of tracks and the applicable railway company and regulatory clearances.

3.3 PRELIMINARY DESIGN OF STEEL RAILWAY BRIDGES

3.3.1 Bridge A Esthetics

Due to location, a esthetics is generally not of concern for most railway bridges. However, bridge a esthetics may be of importance in some urban or accessible natural environments. Bridge a esthetics may be considered in terms of the structure itself and/or its integration into the environment. Perception of beauty varies extensively among persons. However, there are some basic tenets of a esthetic bridge design that appear to be universal in nature (Leonhardt, 1984; Billington, 1985; Taly, 1998; Bernard-Gely and Calgaro, 1994).

Harmony is often of primary importance to the public who generally desire bridges to integrate and be compatible with their environment.* Therefore, the bridge should be of materials and form to achieve any necessary environmental and, in some cases, cultural congruence.

The bridge should also be expressive of function† and materials. In this manner the bridge will be a visual expression of the structural engineering effort involved in achieving safety, efficiency, and economy. Ornamentation that conceals function should generally be avoided. However, the economical proportioning of bridges does not necessarily produce a esthetic structures and other issues, in addition to harmony and expression of function, also warrant careful deliberation.

Proportion and scale are critically important. The dimensional relationships and relative size of components, elements, and/or parts of steel railway structures may affect public perception and support for the project. Slenderness, simplicity, and open space‡ often assist with public acceptance of railway structures constructed in urban environments.

The arrangement, rhythm, repetition, and order of members and/or parts of the structure are also essential considerations for a esthetic bridge design. Light, shade, color, and surface treatments are further means of a esthetic improvement in structures within urban or accessible rural environments.

3.3.2 Steel Railway Bridge Superstructures

Railway bridges transmit loads to substructures through decks, superstructures, and bearings. The superstructure carries loads and forces with members that resist axial, shear, and/or flexural forces.

The steel superstructure forms typically used in freight railway bridge construction are beams, girders, trusses, and arches. These superstructures have the strength and rigidity required to safely and reliably carry modern heavy railroad live loads, and the lightness required for transportation to, and erection at, remote locations. Beam, girder, truss, and arch bridges can be constructed as deck or through structures depending on the geometry of the crossing and clearance requirements (Figure 3.13). Steel frame and suspended structures (i.e., suspension and cable-stayed bridges) are less common but sometimes used in lighter passenger rail bridge applications. Simple span construction is prevalent on North American freight railroads for performance,§ rapid erection,¶ and maintenance considerations. AREMA (2015) recommends simple span types, based on length, for typical modern steel railway bridges as follows:

- Rolled or welded beams for spans up to about 15 m (50 ft) in length (cover plates may increase strength) (often used in floor systems of through plate girder and truss spans)
- Bolted or welded plate girders for spans between 15 m (50 ft) and 45 m (150 ft)
- Bolted or welded trusses for spans between 45 m (150 ft) and 120 m (400 ft).**

Steel freight railway bridge girder spans can be economically designed with a minimum depth to span ratio of about 1/15. Typically, depth to span ratios in the range of 1/10–1/12 are appropriate for modern short- and medium-span steel girder freight railway superstructures. Trusses may be

* The requirements related to environmental compatibility will vary depending upon whether the bridge is to be constructed in an urban, commercial, industrial, or rural environment.

† Sullivan's famous "form follows function" statement on architecture applies well to bridges, which are often most aesthetically pleasing when designed primarily for economy and strength.

‡ Structures that look enclosed or cluttered are often unacceptable from an aesthetic perspective.

§ The high railway live load to steel superstructure dead load ratio often precludes the use of continuous spans due to uplift considerations. Also, continuous spans may be susceptible to detrimental stresses from support settlements.

¶ Simple span construction is generally preferred by railroads due to relative ease of erection in comparison to continuous spans or spans requiring field splicing.

** Based on the upper limit of span length that the AREMA (2015, Chapter 15) recommendations consider.

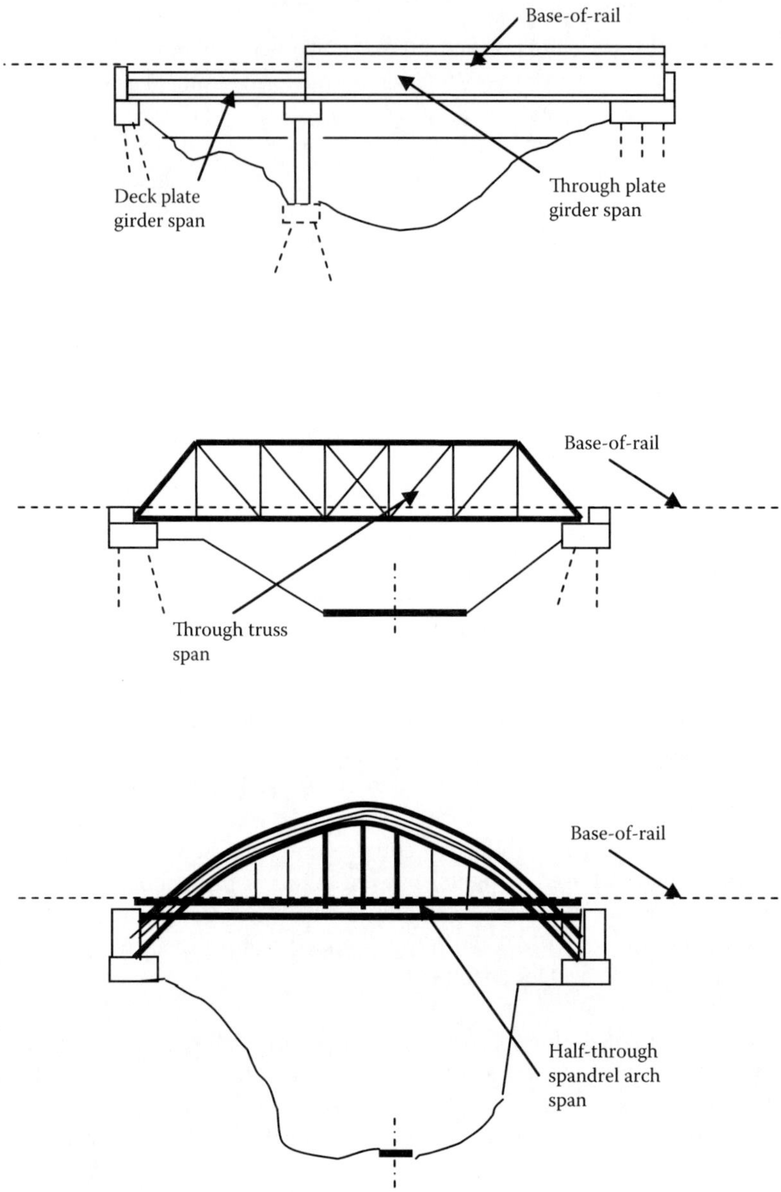

FIGURE 3.13 Basic forms of steel railway bridges—beams, trusses and arches.

economically designed with depth to span ratios in the range of 1/5–1/7. Beam, girder, truss, and arch railway bridges can be constructed with open or closed (i.e., ballasted) decks.

3.3.2.1 Bridge Decks for Steel Railway Bridges

3.3.2.1.1 Railway Track on Bridge Decks

Rails with elastic fasteners seated on steel plates fastened to wooden ties (sleepers) or embedded in prestressed concrete ties are typical of modern North American railroad freight track. On ballasted deck bridges the wood, steel, or concrete ties are bedded in compacted granular rock ballast for drainage and track stability. Steel ties have been used and preclude the need for steel tie plates.[*] Concrete ties, sometimes used on ballasted decks, may require damping, which may be achieved with rubber pads applied to the bottom of the ties with an adhesive (Akhtar et. al., 2006). Alternatively, well-designed rubber mats applied to the top of the deck have been successfully used with concrete ties to damp dynamic effects. The installation of concrete ties on open-deck superstructures is not typically permitted by railroads. Wooden ties are used in both open- and closed- or ballasted-deck construction. In recent years composite material ties[†] have been tested.[‡]

3.3.2.1.2 Open-Bridge Decks

Open decks using wooden ties[§] are still used in many instances on modern railway superstructures. Open-deck bridges are often used in situations when new superstructures are being erected on existing substructures where it is necessary to reduce dead weight to preclude substructure overloading and foundation creep. On open bridge decks the ties are directly supported on steel structural elements (i.e., stringers, beams, girders) (Figure 3.14a and b). Dead load is relatively small, but dynamic amplification may be increased because the track modulus is discontinuous.[¶] Bridge tie sizes can be large for supporting elements spaced far apart and careful consideration needs to be given to the deck fastening systems. Most North American railroads have open bridge deck standards based on the design criteria recommended by AREMA Chapter 7—Timber Structures.

Open bridge decks are often the least costly deck system and are free draining.[**] However, they generally require more maintenance during the deck service life. Continuous welded rail on long-span bridges can create differential movements causing damage and skewing in open bridge decks (see Chapter 4).

3.3.2.1.3 Ballasted Bridge Decks

Closed or ballasted steel plate and concrete slab deck bridges are common in new railway bridge construction. On ballasted or closed bridge decks, track ties (sleepers) are laid in stone ballast that is supported by steel or concrete decks (Figure 3.15a and b). The deck design may be composite or noncomposite with the superstructure. Dead load can be considerable, but dynamic effects are reduced and train ride quality is improved due to the relatively constant track stiffness in open track and across the bridge.

Steel plate decks are usually of isotropic design as orthotropic decks are often not economical for ordinary superstructures due to fabrication, welding, and fatigue design requirements. However, steel orthotropic plate modular deck construction is an effective means of rapid reconstruction of decks on existing steel railway bridges. Cast-in-place reinforced concrete and precast reinforced or prestressed concrete construction can also be used for deck slabs. Composite steel and concrete construction is structurally efficient (see Chapter 7 and Appendix B), but may not be feasible due to site constraints (i.e., need for falsework and site concrete supply). Noncomposite precast concrete deck systems[††] may be considered when site and installation time constraints exist in the particular railroad operating environment. However, precast concrete decks made composite with steel superstructures by casting recesses for shear connection devices and grouting after installation will provide superior performance. Concrete decks are not often used in BTPG superstructures due to deck

[*] Due to their shape, steel ties may also allow for a reduction in ballast depth on bridges.

[†] AREMA (2015, Chapter 30—Ties) has recommendations for the manufacturing of composite material ties for use on railway bridges.

[‡] The testing has been done at the Transportation Technology Center (TTC) of the Association of American Railroads (AAR) Facility for Accelerated Service Testing (FAST) test track and on a North American Class 1 railroad.

[§] Steel, concrete and composite ties may also be used but, due to their relatively large stiffness, may require a detailed analysis of structural behavior. Wooden ties may be used in accordance with the recommendations outlined in AREMA (2015, Chapter 7—Timber Structures and Chapter 15—Steel Structures).

[¶] Generally the open deck superstructure stiffness is considerably greater than the approach track stiffness.

[**] Free draining decks may not be appropriate over some roadways or environmentally sensitive waterways.

[††] Often precast elements are bedded on grout or suitable elastomer, keyed, bonding grouted, and nominally longitudinally post-tensioned. Some specifications may require match-casting to ensure rapid erection.

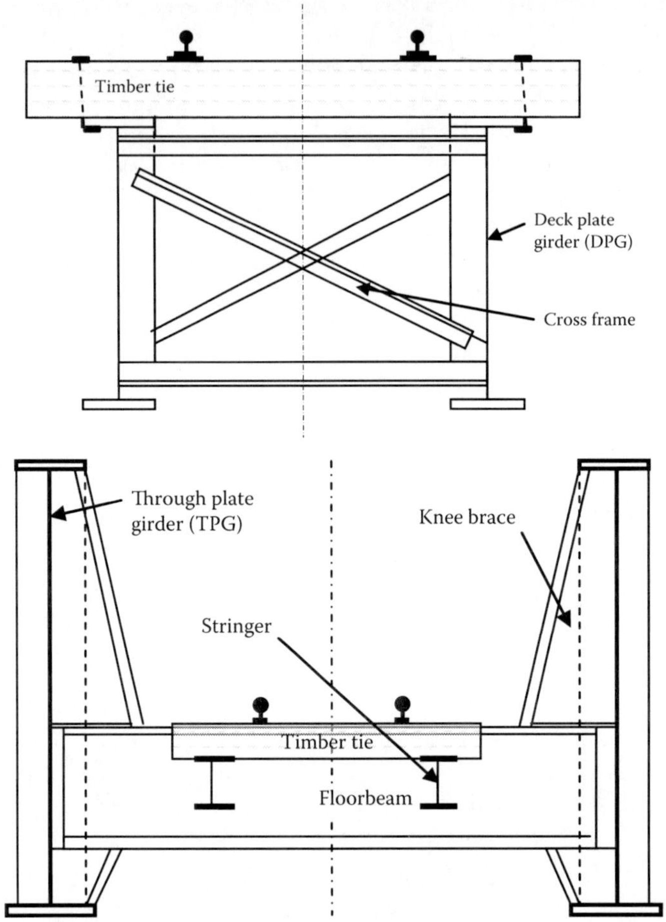

FIGURE 3.14 (a) Open-deck plate girder (DPG) span and (b) Open through plate girder (TPG) span.

cracking issues related to unintended partial composite behavior (Unsworth et al., 2005).* AREMA (2015) recommends a minimum deck thickness of 13 mm (1/2 in.) and 150 mm (6 in.) for steel plate and concrete slab decks, respectively.

Ballasted decks generally require less maintenance and are often used due to curved track geometry or when the bridge crosses over a roadway or environmentally sensitive waterway. Ballasted deck superstructures allow for easier track elevation changes, but drainage must be carefully considered. Drainage of the deck is often accomplished by sloping the deck surface to scuppers or through drains. In some cases, the through drains are connected to conduits to carry water to the ends of spans. In particular, deck drainage at the ends of spans using expansion plates under the ballast between decks must be carefully considered. Most railroads have standards for minimum ballast depth and waterproofing requirements. AREMA Chapter 8—Concrete Structures and Foundations—contains information on recommended deck waterproofing and protection systems.

3.3.2.1.4 Direct Fixation Decks

Rails may be fastened directly to the deck or superstructure where live loads are light and dynamic forces effectively damped. Direct fixation decks are most often used in passenger rail service with

* Concrete decks may be used if noncomposite behavior is ensured through design (Small and Ketler, 2006).

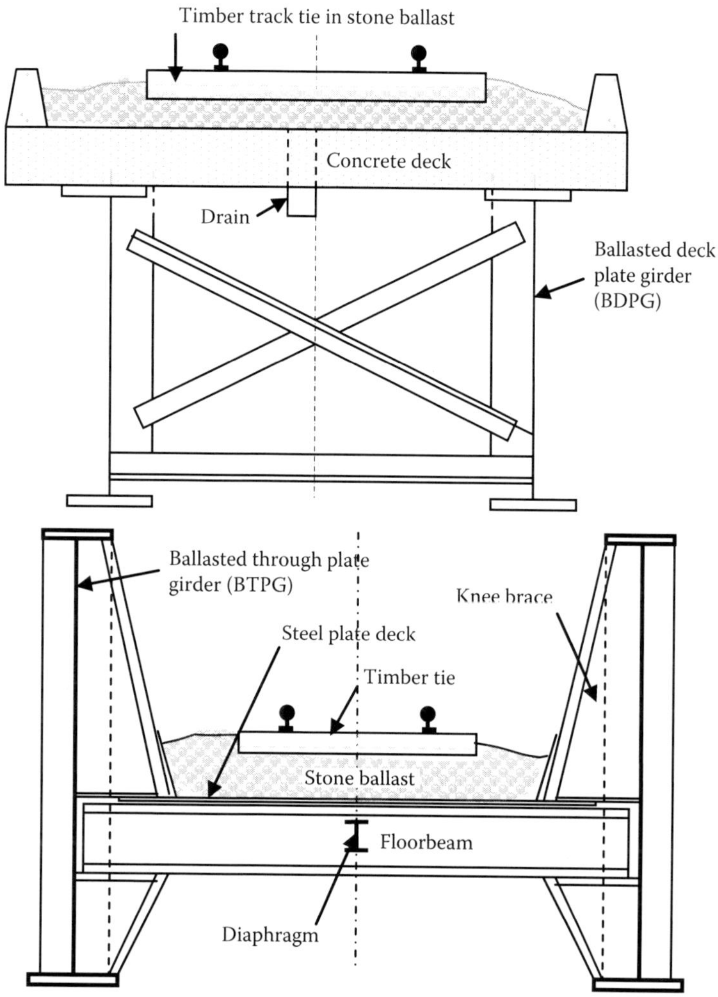

FIGURE 3.15 (a) Ballasted deck plate girder (BDPG) span. (b) Ballasted through plate girder (BTPG) span.

rails securely fastened to the steel or concrete decks. Dead load and structure depth are reduced, but dynamic forces can be large. Direct fixation decks are generally not recommended for modern freight railway superstructures and require careful design and detailing (Sorgenfrei et al., 2014).

3.3.2.2 Bridge Framing Details

Superstructures are framed to resist vertical (dead and live) and lateral (live,* wind, rail expansion and contraction, and seismic) loads.

The main vertical load carrying systems in typical steel railway superstructures are beams, girders, trusses, and arches. Open deck through plate girder, through truss, and some deck truss spans have floor systems consisting of longitudinal stringers and transverse floorbeams that transfer loads and forces to the main vertical load carrying system. Ballasted deck through plate girder spans generally have steel plate decks† supported on closely spaced transverse floorbeams framing into

* Such as braking, traction, centrifugal, and truck (bogie) hunting forces (nosing) from live load.

† BTPG spans with concrete decks have been used with more widely spaced floorbeams (with and without longitudinal stringers). However, detrimental composite behavior has, in some cases, caused premature deck cracking (Unsworth et al., 2005).

the main girder or truss (Figure 3.15b and Appendix A). In some cases, stringers with less closely spaced transverse floorbeams are used.*

Stringers should be placed parallel to the longitudinal axis of the bridge and transverse floorbeams should be perpendicular to main girders or trusses. Stringers are usually framed into the floorbeams and have intermediate cross frames or diaphragms. Floorbeams should frame into the main girders or trusses such that lateral bracing may be connected to both the floorbeam and main member. The end connections of stringers and floorbeams should generally be made with two angles† designed to ensure flexibility of the connection in accordance with the structural analysis used. Due to cyclical live load stresses, welded end connections should not be used on the flexing leg of connections (see Chapter 9). Freight railway bridge spans should have end floorbeams, or other members, designed to permit lifting and jacking of the superstructure without producing stresses in excess of 50% of the basic allowable stresses (see Chapter 4). Multiple beams, girders, and stringers should be arranged to equally distribute live load to all the members.

Redundancy, particularly for Fracture Critical Members (FCMs), is an important consideration in modern steel railway superstructure design. Although more costly from a fabrication perspective, internal redundancy can be achieved by the use of bolted built-up members. Structural (load path) redundancy can be achieved through establishing alternate load paths with additional members. For example, an open-deck steel girder or truss designer may elect to use two rolled beam stringers per rail instead of one stringer per rail. The nonredundant system of a single stringer per rail will require higher material toughness, and more stringent welding procedures and inspection if a built-up member is required.

The lateral load carrying systems in typical steel railway superstructures are horizontal truss bracing and vertical bracing. Horizontal truss bracing consists of lateral bracing members framed in the planes of the top‡ and bottom flanges of beams and girders, the top and bottom chords of trusses, and arch ribs. Vertical bracing consists of end and intermediate cross frames, diaphragms, and knee braces.

3.3.2.3 Bridge Bearings

Freight railway spans 15 m (50 ft) or greater in length should have fixed and expansion bearings that accommodate rotation due to live load and other span deflections.§ All spans should also have provision for expansion to accommodate horizontal movements due to temperature or other longitudinal effects.¶ In addition to these translations and rotations, the bearings must also transmit vertical, lateral horizontal, and, in the case of fixed bearings, longitudinal horizontal forces. Vertical forces are transferred through bearing plates directly to the substructure. Uplift forces may exist that require anchor bolts and many designers consider a nominal uplift force for the design of bearings, in any case. Horizontal forces are usually resisted by guide or key arrangements in bearing elements that transmit the horizontal forces to the substructure through anchor bolts.

Unconstrained elastomeric bearings may be used at the ends of short spans of usual form** to accommodate expansion and rotation. However, for longer spans, end rotation is permitted using spherical discs, curved bearing plates or hinges and expansion is enabled by low-friction sliding

* For example, when ballasted decks are used on through truss spans.

† Where floorbeams frame into girder webs at transverse web stiffener locations, it is often acceptable to use an angle connection on one side of the floorbeam and on the other side directly connect the floorbeam web to the outstanding leg or plate of the girder web stiffener. This requires careful coping (or blocking) of the top and bottom flanges of the floorbeam.

‡ Through girder and pony truss (through truss without top chord lateral bracing), spans use a vertical bracing system (knee braces) to resist lateral loads where top flange or chord horizontal bracing is not possible.

§ For example rotations due to bridge skew, curvature, camber, construction misalignments and loads, support settlements, and thermal effects.

¶ For example, translations due to braking and traction forces, construction misalignments and loads, support settlements, and thermal effects (particularly concerning Continuous Welded Rail (CWR) as outlined in Chapter 4).

** For example, elastomeric bearings might not be appropriate for spans greater than about 15 m (50 ft) or for heavily skewed bridges.

plates, rockers, or roller devices. Multirotational bearings may be required for long, skewed, curved, complex framed, and/or multiple track bridges; or in bridges where substructure settlement may occur.* Constrained elastomeric (pot) bearings have been used with success in highway bridge applications. However, they are not recommended for steel railway superstructure support due to experience with pot bearing component damage from the high-magnitude cyclical railway live loads.

Typical fixed bearing components used on North American freight railroad steel bridges that transmit vertical and horizontal forces while allowing for rotation between superstructure and substructure are as follows:

- Flat steel plates—this type of bearing component has limited application due to inability for rotation and should not be used in spans greater than 15 m (50 ft) and in any span without careful deliberation concerning long-term performance.
- Disc bearings—this spherical segment bearing component allows rotation in any direction (e.g., longitudinal rotations combined with horizontal rotations due to skew and/or radial rotations due to curvature).
- Fixed hinged bearings—this type of hinged bearing uses a pin and pedestal arrangement to resist vertical and horizontal forces and enable rotation at the pin.
- Elastomeric bearings—these plain or steel reinforced rubber, neoprene, or polyurethane bearing pads allow rotation through elastic compression of the elastomer. The design of elastomeric bearing pads is a balance between the required stiffness of the pad to carry vertical loads and that needed to allow rotation by elastic compression.

Typical expansion bearing components used on North American freight railroad steel bridges that transmit vertical forces while allowing for rotation and translation between superstructure and substructure are as follows:

- Flat steel plates—this type of bearing component has limited application due to a deficient ability for translation unless maintained with lubrication.
- Bronze, copper-alloy, or polytetrafluoroethylene (PTFE) flat, cylindrical, and spherical sliding plates—these bearing components enable translation on low friction surfaces. Bronze and copper-alloy sliding elements can be made self-lubricating by providing graphite or other solid lubricants in multiple closely spaced trepanned recesses. PTFE sliding plates should mate with stainless steel or other corrosion resistant surfaces and contain self-lubricating dimples containing a silicone grease lubricant (Figure 3.16).
- Roller bearings—these bearing elements allow translation through rotation of single or multiple cylindrical rollers.
- Linked bearings—this type of bearing uses a double pin and link arrangement between pedestals to allow for horizontal translation.
- Expansion hinged bearings—this type of hinged bearing uses a pin and rocker (segmental roller) arrangement with the pin allowing rotation and the rocker permitting translation.
- Elastomeric bearings—these plain or steel reinforced rubber, neoprene, or polyurethane bearing pads allow translation through shear deformation of the elastomer.

In addition to steel span expansion bearings, the bearings at the bases of columns in steel bents, and viaduct towers should be designed to allow for expansion and contraction of the tower or bent bracing system.

There are proprietary types of fixed and expansion bearings available to the steel railway bridge engineer. Most are similar, or combinations of, the basic elements described above (Stanton et al.,

* Multirotational bearings also accommodate construction tolerances for substructure construction and bearings installation.

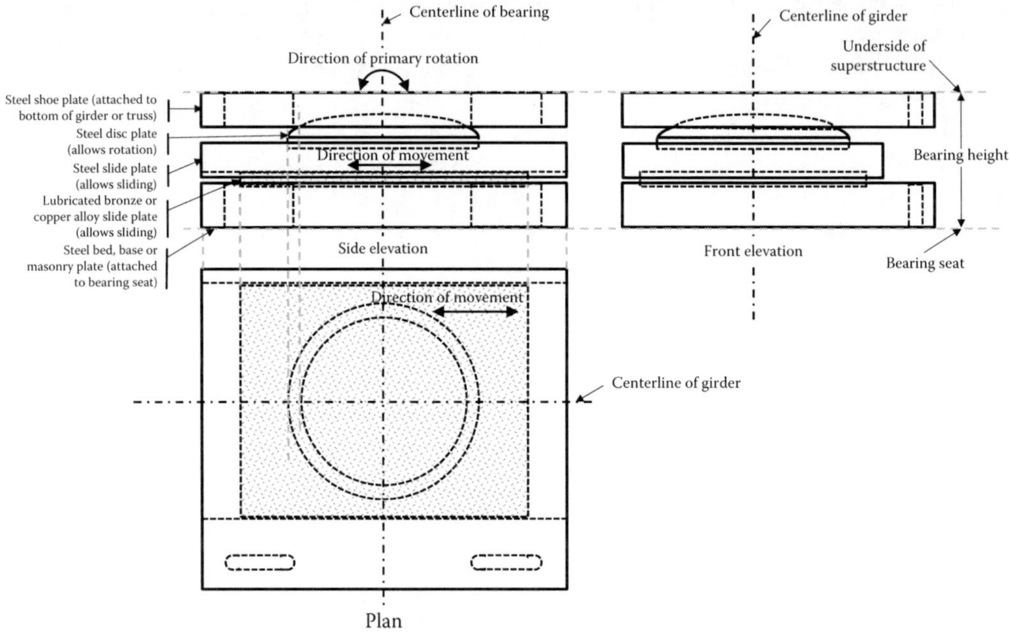

FIGURE 3.16 Multirotational steel expansion bearing.

1999; Ramberger, 2002). Bearings of mixed element types are not recommended (e.g., elastomeric fixed bearings with PTFE sliding bearings). Due to the large vertical cyclical loads and exposed environment of most railway bridges, bearing designs should generally produce simple, robust, and functional bearings that are readily maintained and replaced by jacking of the superstructure. Detailed recommendations on types, design, and fabrication of fixed and expansion bearings for steel freight railway bridge spans are found in AREMA (2015, Chapter 15).

3.3.3 Bridge Stability

Girders and trusses should be spaced to prevent overturning instability created by wind and equipment based lateral loads (centrifugal, wheel/rail interface, and train rocking). AREMA (2015) recommends that the spacing should be greater than 1/20 of the span length for through spans and greater than 1/15 of the span length for deck spans of freight railway bridges. The spacing between the center of pairs of beams, stringers, or girders should not be less than 2.0 m (6.5 ft).

3.3.4 Pedestrian Walkways

Pedestrian walkways on railroad bridges are typically for the use of only railroad employees and not the general public. The walkway must be outside regulatory or other prescribed railway clearances.* Most railroad companies have policies, generally based on government regulations, regarding walkway and guardrail† requirements for bridges. Width of walkways is often prescribed by the railroad company,‡ but should not be less than 600 mm (2 ft). Walkway surfaces should be a nonslip galvanized steel grating (checker plate) no less than 6 mm (1/4 in.) thick. High toe boards of dimension

* US State clearances are outlined in AREMA (2015) Chapters 15 and 28, and Canadian requirements are indicated in Transport Canada regulations, TC E-05, Standard Respecting Railway Clearances.

† Typically, consisting of posts and handrails of appropriate strength and spacing.

‡ Some companies require a minimum walkway width of 900 mm (3 ft).

100 mm (4 in.) to 150 mm (6 in.) are often installed on walkway surfaces, particularly on ballasted deck bridges over roadways.* Structural members (e.g., knees braces at floorbeams of through plate girder spans) should not be considered as obstructions on walkways designed for the use of railroad employees. Guardrail height is generally prescribed as a minimum of 1067 mm (3.5 ft), with clear distances between railings not exceeding 530 mm (21 in.), by North American regulatory authorities. Nevertheless, depending on the bridge location, greater guardrail heights might be required at some crossings. Posts are usually spaced considering the superstructure framing system, and are typically spaced at intervals of between about 2.5 m (8 ft) and 4 m (13 ft). Handrails and posts consisting of galvanized tubular, pipe, or angle sections not less than 6 mm (1/4 in.) thick are often used for railway bridge guardrails, where adequate strength and safety, without the need to consider a esthetics, are required. AREMA (2015) provides recommendations for the design of pedestrian walkways and guardrails (see Chapter 4). The designer should also consult with the railroad company and applicable regulations concerning specific safety appliances that may be required.

3.3.5 General Design Criteria

Engineers use the tools of analysis and design to ensure structural behavior that does not compromise established safety and performance limit states. Safety limits states (or ultimate limit states) to consider are strength, stability, fatigue, and fracture. Performance limit states are serviceability related. In some cases, extreme events limit states also require attention.

The strength limit state is capacity related and based on probabilities† of not exceeding the yield, plastic, or fracture strength of the member. Fracture is avoided through design by ensuring that member stresses do not exceed the ultimate tensile strength, F_u. The fracture limit state is also controlled by material toughness requirements and fabrication control plans (see Chapter 10). The stability limit state involves the avoidance of buckling by limiting member stresses to the critical buckling stress. The fatigue limit state defines the permissible live load stress ranges under normal service loads. Serviceability limit states define acceptable limits for deflections, vibrations, and durability‡ under normal service loads. Extreme event limit states require superstructure survivability during earthquake, flood, ice, or other severe environmental conditions.

3.3.5.1 Structural Analysis for Modern Steel Superstructure Design

Structural analysis is essentially scientific verification of suitable behavior regarding safety and performance limits states. It is based on the theories and applications of mechanics or strength of materials, linear and nonlinear elasticity and plasticity, elastic and inelastic stability, and structural dynamics. Structural analysis delivers the responses (in terms of member internal stresses and deformations) of the superstructure to actions (typically, loads and forces). For steel railway superstructures, the most significant actions include moving live loads, masses, or sprung masses (see Chapter 4). Many railway superstructures are statically determinate, which simplifies even moving load analysis (see Chapter 5). However, for complex, and most statically indeterminate superstructures, FEA and other specialized software are effective modern tools for moving load, mass, or sprung mass structural analysis.

Design method is closely coupled to the structural analysis (e.g., elastic, inelastic, or plastic). Allowable stress design (ASD) is based on elastic analysis to ensure that superimposed stresses calculated from various load combinations do not exceed allowable stresses calculated as a proportion of the elastic limit. Limit state design (LSD) or load and resistance factor design (LRFD) is based on superstructure member resistance (involving inelastic strength behavior) to load combinations.

---

* Installed to prevent ballast from damaging traffic crossing under the railway bridge. It is usually considered good practice to install toe boards on new superstructures.

† The probabilities may be expressed as safety factors or reliability indices directly related to probabilities of failure.

‡ Durability in steel structures is primarily related to environmental corrosion.

3.3.5.2 Structural Design for Modern Steel Superstructure Fabrication

The principal concerns of design are safety, performance, and constructability. Design essentially consists of conceptual, preliminary, and detailed design. Conceptual design is based on planning considerations such as economics, transportation operating requirements, site conditions, hydraulics, regulatory clearances, geotechnical conditions, and, in some cases, a esthetics. Preliminary design may be based on similar bridge superstructures, standard designs, experience, and/or initial calculations. Preliminary design plans are typically for construction budget cost estimating and planning purposes (see Chapters 10 and 11). Detailed design for fabrication must provide the size or dimension of members, attachments,[*] and connections required to resist stresses and deformations to within acceptable safety and performance limits established and recommended by codes, guidelines, and practices. Traditional ASD and modern design methods based on probabilistic analyses[†] consider safety and performance limit states and are used to design steel railway superstructures worldwide.

3.3.5.2.1 Allowable Stress Design

ASD is used for the design of steel railway superstructures that are analyzed on the premise of linear elastic behavior. The limiting stress is the yield stress,[‡] and allowable stresses are the limiting stress divided by a safety factor, SF. ASD is essentially the limit state for allowable stresses and deformations under regular service loads and can be expressed as

$$\frac{R}{\text{SF}} \geq \sum_{i} Q_i, \tag{3.47}$$

where R = the member or connection resistance

Q_i = service loads on the member or connection

ASD applies the SF, to the resistance, R, only. The use of a single SF[§] is unrealistic, particularly when load uncertainty is variable.[¶] A single SF implies the same average variability (or uncertainty) for resistance, R, and loads, Q_i, and does not explicitly recognize that strength may be governed by tension, compression, and flexural limit states.[**]

ASD does not identify the reserve strength in some members and cannot determine behavior at limit states related to stress redistribution[††] at localized yielding. ASD requires serviceability design for deflections and vibrations based on service loads, Q_i.

Nevertheless, elastic analysis and ASD allow for the simple superposition of stresses from various loads as no permanent deformations result by keeping allowable stresses well below the elastic limit.[‡‡] However, structural reliability methods, based on probabilistic or statistical analyses, are the basis of many modern steel bridge design codes and guidelines.

3.3.5.2.2 Probabilistic Structural Design

The consistent reliabilities and probabilities of failure determined from consideration of the various strength (resistance) and load uncertainties is the acumen of the probabilistic design approach.

---

[*] For example, gusset plates, cover plates, and stiffeners.

[†] The structural design method based on probabilistic analyses is known as Limit States Design (LSD) in many parts of the world and Load and Resistance Factor Design (LRFD) in the United States.

[‡] Steel has a well-defined yield stress.

[§] The SF in ASD has no mathematical basis, but is predicated on experience with successful structures designed based on this approach.

[¶] Different uncertainties are associated with various loads such as dead load (DL), live load (LL), impact load (I), and wind (W).

[**] These limit states reflect the various loads and associated load uncertainties.

[††] Available to ductile steel superstructures.

[‡‡] Appropriate for structural steel with a well-defined yield stress.

Probabilistic-based design considers the strength, fatigue, and serviceability limit states by factoring both loads and member strengths based on uncertainty models (Equation 3.48).

$$ØR \geq \sum_i \gamma_i Q_i. \tag{3.48}$$

The resistance factors, ϕ, account for uncertainties in analysis, material strength, size of mill produced sections, test result statistics, and consequence of failure. The load factors, γ_i, relate to maximum load magnitude uncertainties.

If R is the material resistance probability distribution and Q is the load distribution, acceptable safety and performance is defined by $Q \leq R$. However, due to the stochastic nature of loads and material strengths, $Q \leq R$ is not definitively predictable (Figure 3.17). A probabilistic approach to design, based on consistent safety margins (reliability indices) that account for the variable load and resistance uncertainties, and consequence of member failure,* is required. If the probability distributions for loads, Q, and resistances, R, are determined, failure can be defined as when some of the random variables for the resistance probability distribution are less than some of the random variables for the load probability distribution (Figure 3.17). The statistics of loads, Q, and resistances, R, are typically given by normal and/or lognormal probability distributions (Collins and Nowak, 2013).

3.3.5.2.2.1 Normal Probability Density The normal probability density is (Miller and Freund, 1972)

$$f(x) = \frac{1}{\sigma\sqrt{2\pi}} \exp\left[-\frac{1}{2}\left(\frac{x-\mu}{\sigma}\right)^2\right], \tag{3.49}$$

where

$f(x)$ = the probability of the occurrence of the normal random variable, x
μ = the mean of the data of the normal probability distribution
σ^2 = the variance of the data from the mean of the normal probability distribution
σ = the standard deviation (dispersion) of the data of the normal probability distribution
σ/μ = the coefficient of variation and is an indication of data scatter.

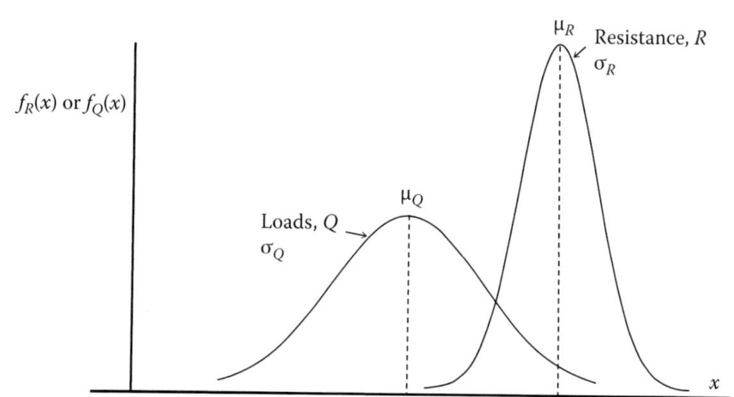

FIGURE 3.17 Load, Q, and resistance, R, probability distributions.

* Generally considered high for railway superstructures.

For normal probability distributions, the mean, $\mu_{(R-Q)}$, variance, $\sigma^2_{(R-Q)}$, and standard deviation, $\sigma_{(R-Q)}$, of the limit state function $(R–Q)$ are*

$$\mu_{(R-Q)} = \mu_R - \mu_Q, \tag{3.50}$$

$$\sigma^2_{(R-Q)} = \sigma^2_R + \sigma^2_Q, \tag{3.51}$$

$$\sigma_{(R-Q)} = \sqrt{\sigma^2_R + \sigma^2_Q}. \tag{3.52}$$

For normal probability distributions, the failure limit state function is $(R–Q)$ (Figure 3.18). To preclude failure, a safety margin or reliability index, β, related to the probability of failure, P_f, is required for the limit state function $(R–Q)$. A safety margin of β standard deviations ($\beta\sigma_{(R-Q)}$) from the mean to failure is considered for design as indicated in Figure 3.18.

$$\beta\sigma_{(R-Q)} = \mu_{(R-Q)}, \tag{3.53}$$

$$\beta = \frac{\mu_{(R-Q)}}{\sigma_{(R-Q)}} = \frac{\mu_R - \mu_Q}{\sqrt{\sigma^2_R + \sigma^2_Q}}. \tag{3.54}$$

The probability of failure for normal probability distributions, $P_f = P[(R-Q)<0]$, can be obtained from the limit state function, $(R–Q)$, as shown in Figure 3.18. The probability of failure, P_f, for normal probability distributions is

$$P_f = P\left(-\infty < (R-Q) < \beta\right) = 1 - \frac{1}{\sqrt{2\pi}} \int_{-\infty}^{\beta} \exp\left[-\frac{1}{2}(\beta)^2\right] = 1 - F_z(\beta), \tag{3.55}$$

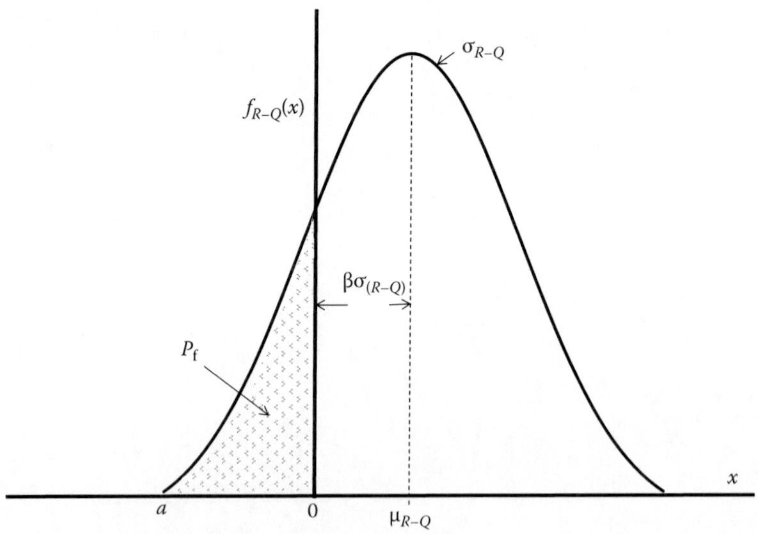

FIGURE 3.18 $(R–Q)$ probability distribution.

* In accordance with the central limit theorem for normal probability distributions.

TABLE 3.2
Reliability Index and Probability of Failure

β	P_f	P_f	β
1.0	1.59×10^{-1}	1.0×10^{-2}	2.32
2.0	0.23×10^{-1}	1.0×10^{-3}	3.09
2.5	0.62×10^{-2}	1.0×10^{-4}	3.72
3.0	1.35×10^{-3}	1.0×10^{-5}	4.27
3.5	2.33×10^{-4}	1.0×10^{-6}	4.75
4.0	3.17×10^{-5}	1.0×10^{-7}	5.20
4.5	3.40×10^{-6}	1.0×10^{-8}	5.62
5.0	2.87×10^{-7}	1.0×10^{-9}	6.00

where $F_z(z)$ is tabulated as the standard normal distribution.* The closed-form solution of the integral in Equation 3.55 is avoided by use of the standard normal distribution tables.† However, the random variables must be nondimensionalized to use the standard normal distribution tables. The nondimensional variable is $Z_{R-Q} = \dfrac{(R-Q)-\mu_{(R-Q)}}{\sigma_{(R-Q)}}$. Values of P_f for various β and the inverse‡ for normal distributions are shown in Table 3.2.

3.3.5.2.2.2 Lognormal Probability Density The lognormal frequency distribution is (Miller and Freund, 1972)

$$f(x) = \frac{1}{x\beta\sqrt{2\pi}}\exp\left[-\frac{(\ln x-\alpha)^2}{2\beta^2}\right] \quad \text{for } x>0,\ \beta>0, \tag{3.56}$$

$$= 0 \text{ for all other } x \text{ and } \beta,$$

where $f(x)$ is the probability of the occurrence of lognormal random variable, x.

The statistical properties of the lognormal distribution may be determined from the statistical properties of the normal distribution (Barker and Puckett, 1997) as

$$\mu_{\ln} = \exp\left(\alpha+\frac{\beta^2}{2}\right) = \ln\left[\frac{\mu}{\sqrt{1+\left(\frac{\sigma}{\mu}\right)^2}}\right] \tag{3.57}$$

= the mean of the data of the lognormal normal probability distribution,

* Tabulated standard normal distribution tables are available in many books concerning statistical analysis and reliability of structures.

† Normal distribution with $\mu = 0$ and $\sigma = 1$.

‡ The reliability index is directly and inversely related to the probability of failure

$$\sigma_{\ln}^2 = \exp\left(2\alpha+\beta^2\right)\left(\exp\left(\beta^2\right)-1\right) = \ln\left(1+\left(\frac{\sigma}{\mu}\right)^2\right) \tag{3.58}$$

= the variance of the data from the mean of the lognormal probability distribution.

For lognormal probability distributions, the failure limit state function is ($\ln R - \ln Q$) = $\ln(R/Q)$. To preclude failure, a safety margin or reliability index, β, related to the probability of failure, P_f, is required for the limit state function $\ln(R/Q)$. For lognormal probability distributions, the estimated-*safety margin of β standard deviations ($\beta\sigma_{(\ln(R/Q))}$) from the mean to failure for design is

$$\beta = \frac{\mu_{\left(\ln\left(\frac{R}{Q}\right)\right)}}{\sigma_{\left(\ln\left(\frac{R}{Q}\right)\right)}} = \frac{\mu_{\ln R} - \mu_{\ln Q}}{\sqrt{\left(\frac{\sigma_R}{\mu_R}\right)^2 + \left(\frac{\sigma_Q}{\mu_Q}\right)^2}} = \frac{\ln(\mu_R/\mu_Q)}{\sqrt{\left(\frac{\sigma_R}{\mu_R}\right)^2 + \left(\frac{\sigma_Q}{\mu_Q}\right)^2}}. \tag{3.59}$$

The probability of failure for lognormal probability distributions is more difficult to calculate but has been estimated for use in practice (Rosenblueth and Esteva, 1972).

3.3.5.2.2.3 LSD (LRFD) Calibration to ASD The single SF used in ASD implies the same average variability (or uncertainty) for resistance, R, and loads, Q_i. Uniform safety margins or reliability indices are unable to be realized using ASD. Therefore, if LSD (LRFD) methods are directly calibrated against existing superstructure designs based on ASD, consistent reliability indices are not possible. Nevertheless, a direct calibration of LSD (LRFD) to the ASD methodology can be made with Equations 3.47 and 3.48 as*†

$$\varnothing = \frac{\sum_i \gamma_i Q_i}{FS \sum_i Q_i}. \tag{3.60}$$

For $\sum_i Q_i = \gamma_{DL} Q_{DL} + \gamma_{LL} Q_{LL}$, Equation 3.60 is

$$\varnothing = \frac{\gamma_{DL} Q_{DL} + \gamma_{LL} Q_{LL}}{FS\left(Q_{DL} + Q_{LL}\right)} = \frac{\gamma_{DL}\left(\frac{Q_{DL}}{Q_{LL}}\right) + \gamma_{LL}}{FS\left(\frac{Q_{DL}}{Q_{LL}} + 1\right)}, \tag{3.61}$$

where γ_{DL} is the dead load factor applied to dead loads, Q_{DL}, and γ_{LL} is the live load factor applied to live loads, Q_{LL}.

Q_{DL}/Q_{LL} ratios for modern steel railway ballasted deck girder spans, based on bending moments, typically range from about 0.25 for short BTPG spans to 0.80 for long BTPG spans. For BDPG spans carrying railway traffic, typical Q_{DL}/Q_{LL} ratios are also less than about 0.80.

Equation 3.61 is plotted in Figure 3.19a for values of γ_{DL} = 1.25 and γ_{LL} = 1.75 (values typically used in highway bridge design codes) for various SF (typically between 1.67 and 1.82 for ASD design codes and recommendations). The plot outlines that reasonable resistance factors, ϕ, between

* The estimate is reasonable for small coefficients of variation, σ/μ, of less than about 20%.

† This calibration does not consider that resistance, R, For ASD is based on yield stress and resistance, R, For LSD (LRFD) is based on ultimate stress.

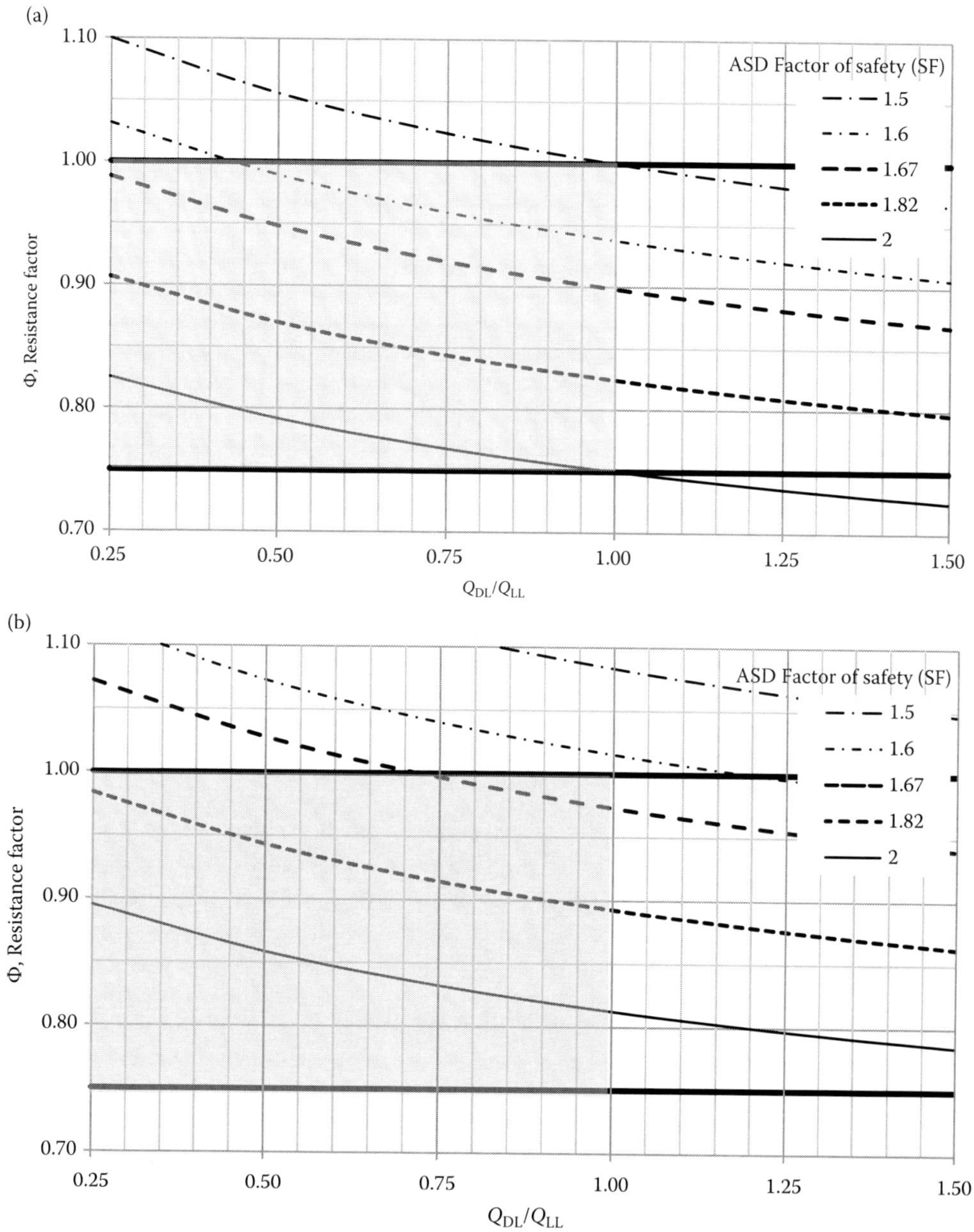

FIGURE 3.19 (a) LRFD resistance factor and ASD safety factor calibrations for $\gamma_{DL} = 1.25$ and $\gamma_{LL} = 1.75$. (b) LRFD resistance factor and ASD safety factor calibrations for $\gamma_{DL} = 1.35$ and $\gamma_{LL} = 1.90$.

0.75 and 1.00, can be estimated when calibrated against ASD using a SF of greater than about 1.65. Equation 3.61 is also plotted in Figure 3.19b for values of $\gamma_{DL} = 1.35$ and $\gamma_{LL} = 1.90$ for various SF. The plot outlines that reasonable resistance factors, φ, between about 0.75 and 1.00, can be estimated when calibrated against ASD using a SF of greater than about 1.80.

Nevertheless, direct calibration to existing ASD codes, practices, and recommendations does not provide the consistent reliabilities and probabilities of failure determined from consideration of the resistance and load uncertainties. Therefore, calibration of the LSD (LRFD) methodology is based on reliability methods and the judgment of experienced bridge designers engaged in the preparation of codes, practices, and recommendations for LSD (LRFD) development.

3.3.5.2.2.4 LSD (LRFD) Calibration Using Reliability Methods The load and resistance distributions are often modeled as normal and lognormal distributions, respectively (Nowak, 1999). Load and resistance factors used for design are derived from the statistical parameters of the load and resistance probability distributions at strength, fatigue, serviceability, and extreme event limit states.

A significant testing program was conducted on six steel superstructures* traversed by over 35,000 locomotives and rail cars† with a total gross load of just over 3.7×10^6 tonnes (4.1×10^6 tons) (Tobias et al., 1996). Measured live load spectra for locomotives and their axles best fit gamma and lognormal distributions with coefficients of variation, $C_v = \sigma/\mu$, of between 5% and 10%‡. Measured live load spectra for hopper cars and their axles typically best fit normal and lognormal distributions with coefficients of variation, $C_v = \sigma/\mu$, of between 2% and 10%. Measured live load spectra for intermodal cars and their axles best fit gamma, beta, and normal distributions with coefficients of variation, $C_v = \sigma/\mu$, of up to 42%. The maximum stress responses to these live load spectra fit a normal distribution (Rakoczy and Nowak, 2013).

Fatigue loading has been modeled using a normal probability distribution (Tobias and Foutch, 1997) and reliability based analysis has demonstrated that accumulated fatigue damage follows a normal probability distribution (Rakoczy et al., 2016).

The axial, shear, and bending strength resistances of steel follow a lognormal probability distribution with relatively small variations.§ Fatigue resistance has been modeled using lognormal (Tobias and Foutch, 1997) and normal (Rakoczy et al., 2016) probability distributions. A reliability analysis for the serviceability limit state of deflection response is a normal distribution (Ghasemi and Nowak, 2016).

These load and resistance probability distributions yield the statistical parameters required to determine load and resistance factors for design codes, practices, and recommendations. Load and resistance factors may be determined based on the selected target reliability index, β_T, and Equation 3.54¶ as

$$\mu_R-\mu_Q=\beta\sqrt{\sigma_R^2+\sigma_Q^2}\simeq 0.75\beta(\sigma_R+\sigma_Q), \tag{3.62}$$

which can be presented as

$$\begin{aligned}\mu_R-0.75\beta\sigma_R=\mu_Q+0.75\beta\sigma_Q &= R\left(\frac{\mu_R}{R}\right)\left(1-0.75\beta_T\left(\frac{\sigma_R}{\mu_R}\right)\right),\\ &= Q\left(\frac{\mu_Q}{Q}\right)\left(1+0.75\beta_T\left(\frac{\sigma_Q}{\mu_Q}\right)\right),\\ &= \varphi R+\gamma Q,\end{aligned} \tag{3.63}$$

* Consisting of open deck plate girder (DPG), ballasted deck plate girder (BDPG), through plate girder (TPG), and truss superstructures.

† Just over 4% of the equipment traversing the superstructures were locomotives.

‡ With the larger values of C_v attributed to axle statistics.

§ Which provides a steeper probability distribution curve enabling strength to be evaluated with relatively small quantities of test data.

¶ This approximation results in a maximum error of less than 6% (Barker and Puckett, 1997).

where

$$\varphi = \left(\frac{\mu_R}{R}\right)\left(1 - 0.75\beta_T\left(\frac{\sigma_R}{\mu_R}\right)\right) = \text{the resistance factor,} \tag{3.64}$$

$$\gamma = \left(\frac{\mu_Q}{Q}\right)\left(1 + 0.75\beta_T\left(\frac{\sigma_Q}{\mu_Q}\right)\right) = \text{the load factor.} \tag{3.65}$$

Therefore, load factors, γ, and resistance factors, φ, may be calculated from the target reliability index and load and resistance distribution statistics. The values of φ and β calculated by reliability theory are adjusted through experience and judgment* for inclusion in codes, practices, and recommendations for steel superstructure design.

3.3.5.2.3 AREMA (2015) Recommendations Relating to the Analysis and Design of Steel Superstructures

The superposition of stresses from various load combinations is applicable for linear elastic behavior as no permanent deformations occur with design stresses well below the elastic limit or yield stress.† Elastic analysis also simplifies the strength and serviceability investigation of other load effects such as creep in concrete, thermal, and/or due to substructure displacements. Also, the iterative updating of member sizes is uncomplicated in an elastic analysis, even for statically indeterminate superstructures. Inelastic analysis is not considered for steel railway superstructures due to uncertainties in models (e.g., boundary conditions) and analytical complexity (e.g., stiffness, strength, strain hardening behavior, and other properties are required).‡ Therefore, strength, serviceability, and fatigue limit states are based on elastic analyses with unfactored service loads, which is the ASD methodology.§

Serviceability and fatigue limit states are evaluated within the elastic range of behavior. Deflections are calculated with unfactored loads and the fatigue limit state is evaluated based on allowable fatigue stresses. Elastic behavior for the strength limit state is also appropriate for railway structures, where failure consequences are high (e.g., safety, economic, and environmental considerations), and even localized yielding is not permitted.

LSD (LRFD) based on reliability theory requires load and resistance statistics. Dead load, environmental load, and steel member resistance statistics are well established. However, for steel railway superstructures, live load and impact (dynamic effect) statistics developed by testing and simulation are required for further development of steel railway superstructure reliability methods. Dynamic locomotive, car, truck (bogie), and axle load statistics have been developed based on measurements at six steel spans traversed by a total gross load of about 3.7×10^6 tonnes (4.1×10^6 tons) (Tobias et al., 1996). These statistics are an important contribution to the development of railway traffic loading statistics required to develop suitable target reliability indices and live load factors.

Dynamic loads (impact) affect the strength, serviceability, and fatigue limit states. The statistical properties of impact measurements (Ruble, 1955) were shown to follow a normal probability distribution (Byers, 1970) but with relatively large variability. Low values of impact¶ with large

* For example, the target reliability index must be chosen to reflect the consequence of failure.
† Ensured by using a safety factor (SF).
‡ In contrast, only stiffness properties are required for the elastic analysis of even statically indeterminate models.
§ In ASD, the safety limit state is validated by ensuring that stresses are below an allowable stress.
¶ As low as 10%.

coefficients of variation,* C_V, have been used in some reliability studies (Tobias and Foutch, 1997; Rakoczy and Nowak, 2013). However, these low impact values are based on only a limited number of observations and may not be appropriate for current design practice.

It is clear that further research and development is required regarding the modern live load and impact spectra, and their statistical parameters, to develop reliability design methods for steel railway bridges. Relevant target reliability indices and load factors could then be developed based on a comprehensive reliability analysis. The target reliability indices must also be calibrated against existing successful ASD designs and simulations (typically by the Monte Carlo method). However, because many recent steel railway superstructure designs are controlled by the serviceability or fatigue limit states, calibration for strength design may be onerous.

When serviceability or fatigue limit states govern superstructure design, and because LSD (LRFD) target reliability indices and load factors are calibrated to the statistical parameters of existing ASD designs, it is not expected that many LSD (LRFD) designs would be substantially different than ASD designs. Therefore, since there does not appear to be a clear and pressing need for steel railway superstructure design based on probabilistic design methods in North America, elastic structural analysis continues to be used for freight railway steel superstructure design based on the ASD methods of AREMA (2015, Chapter 15—Steel Structures).

AREMA (2015) outlines recommended practices relating to materials, type of construction, loads, strength, serviceability, and fatigue design of steel railway bridges. Strength criteria for tensile yielding, tensile fracture, and compressive instability are provided in AREMA (2015). Serviceability (vibrations and deflections), fatigue, and fracture requirements are important aspects of steel railway superstructure design and also comprehensively considered. Dynamic amplification of live load (commonly referred to as impact) may be very large in freight railway superstructures (see Chapter 4) and vibration control is primarily exercised through span live load deflection limitations. Some railroads and owners may impose more stringent live load deflection limits to control vibrations and concrete deck cracking based on operating conditions. Railway equipment, such as long unit trains (some with over 150 cars), can create a significant number of stress cycles on high traffic density rail lines, particularly on bridge members with relatively small influence lines (see Chapter 5).

AREMA (2015) recommends a performance-based approach to seismic design. A performance-based approach is valid considering that steel railway bridges have suffered little damage or displacement during many recent large magnitude earthquake events (Byers, 2006). Steel freight railway bridges have performed well in seismic events due to the type of construction usually employed. Steel freight railway bridges have relatively light superstructures, stiff substructures, large bridge seat dimensions, and substantial bracing and anchor bolts (used to resist the considerable longitudinal and lateral forces associated with train operations).

3.3.6 Fabrication Considerations

The steel superstructure fabrication process commences with shop (or detail) drawings produced by the fabricator from engineering design drawings and specifications. The approved shop drawings are then used for material procurement, cutting, drilling, punching, bolting, bending, welding, surface finishing, coatings application, shop assembly, and quality control (QC). Tolerances from dimensions on engineering drawings concerning straightness, length, cross section, connection geometry, clearances, and surface contact must be respected during fabrication.† Design and shop drawings should indicate all FCMs since these members require specific material (see Chapter 2) and fabrication (see Chapter 10) requirements.

* C_v of up to 50% have been used in some load statistical analyses.

† These tolerances are outlined in AREMA 2015, Chapter 15, Part 3.

Fabricators must make joints and connections with high-strength steel bolts in accordance with ASTM F3125 (includes specifications for A325M, A325, A490M,* A490, and "twist-off" type structural bolts F1852 and F2280). Steel freight railway bridges are designed with slip critical connections and pretensioning is required for bolt installation (see Chapter 9). Bolts should be installed with a minimum tension† by the turn-of-nut, tension-control, direct-tension-indicator, or calibrated wrench methods.

Welding process procedures, preparation, workmanship, qualification, and inspection requirements for steel railway bridges (dynamically loaded structures) should conform to the requirements of the American Welding Society AWS D1.5—Bridge Welding Code or other applicable welding code.‡ In particular, for FCMs, additional provisions concerning welding processes, procedures and inspection merit careful attention during fabrication.§ Welding processes typically used for steel railway bridges are shielded metal arc welding, submerged arc welding, and flux cored arc welding (see Chapters 9 and 10). Railroad companies may prescribe supplementary limitations concerning acceptable welding processes for superstructure fabrication.

Steel railway bridge fabrication, particularly for FCMs, should be accompanied by testing of materials, fasteners, and welding. Material mill certifications should be reviewed to confirm material properties such as ductility, strength, fracture toughness, corrosion resistance, and weldability. Bolted joints and connections should be inspected by turn, tension, and torque tests to substantiate adequate joint-strength. QC¶ and Quality assurance (QA)** inspection of welding procedures, equipment, welder qualification, and nondestructive testing (NDT) are also required to validate the fabrication. NDT of welds is typically performed by magnetic particle, ultrasonic, and/or radiographic testing by qualified personnel. Railroad companies often have specific QA requirements for the testing of fillet, complete joint penetration, or partial joint penetration welds.

Steel bridges fabricated with modern atmospheric corrosion resistant (weathering) steels are often not coated, with exception of specific areas that may be galvanized, metalized, and/or painted for localized corrosion protection.†† Nevertheless, modern multiple-coat painting systems are typically used for steel railway bridge protection. Many modern steel railway bridges are protected with a three-coat system consisting of a zinc rich primer, epoxy intermediate coat, and polyurethane topcoat. However, railroads often develop their own cleaning and painting guidelines or specifications for shop and field coating of steel bridges.

Recommendations related to the fabrication of steel freight railway bridges are included in AREMA (2015).‡‡ Engineers should consult with experienced fabricators early in the design process concerning shape (rolled section) and plate size and grade availability from steel mills or supply companies (steel service centers). After the completion of fabrication, marking, loading, and shipping are arranged. Steel railway superstructures are often most economically shipped from fabrication shop§§ to erection site by rail. Large steel railway spans may also require that experienced engineers provide loading and shipping arrangements to ensure stability of the fabricated structure and the safety of railroad operations.

---

* ASTM A490 bolts are often discouraged or prohibited by bridge owners due to brittleness concerns.

† For example, AREMA (2015) recommends a minimum tension force of 175 kN (39,000 lb) for 22 mm (7/8 in.) diameter ASTM F3125 A325 bolts.

‡ For example, Canadian Standards Association (CSA) W59-13—Welded steel construction (metal arc welding), British Standards BS 5400 Part 3—Code of Practice for the Design of Steel Bridges or Eurocode 3—Design of steel structures.

§ Fracture Control Plans are usually specified to ensure FCM fabrication is performed in accordance with the additional requirements indicated by AREMA (2015) and AWS D1.5 (2015) or other applicable welding code.

¶ QC inspection is carried out by the fabricator in accordance with an approved Fracture Control Plan (FCP).

** QA inspection is carried out by the owner (typically, railroad company) or representative (typically, the design engineer) in accordance with the owner's requirements.

†† For example, at bearing areas or top flange surfaces supporting ties (sleepers) in open deck spans.

‡‡ Recommended practice for fabrication of steel railway bridges is outlined in AREMA 2015, Chapter 15, Part 3.

§§ Fabrication shops or facilities with freight rail access are quite common in North America.

3.3.7 Erection Considerations

Erection complexity, location, cost, schedule, equipment requirements, and erection experience will typically dictate whether the erection of steel railway bridges is performed by steel fabricator, general contractor, specialty erection contractor or railroad construction forces. Erection procedures typically depend primarily on site conditions, contractor experience, and equipment availability. Nevertheless, designer engineers should consider typical erection methods for steel railway bridge superstructure erection (see Chapter 11) during conceptual, preliminary, and detailed superstructure design. Steel span erection procedures and drawings should be in conformance with the engineering design drawings, specifications, special provisions, shop drawings, camber diagrams, match marking diagrams, fastener bills of material, and all other information related to erection planning requirements. Erection procedures should always be made with due consideration of safety* and transportation corridor interruption.† Recommendations related to the erection of steel freight railway bridges are included in AREMA (2015).‡

3.3.8 Detailed Design of the Superstructure

Detailed design of the superstructure may proceed following all deliberations related to the railroad operating environment, site conditions, geometrics, a esthetics, superstructure type, deck type, preliminary design of framing systems, fabrication, and erection are completed to an appropriate level.§ Detailed design will proceed from load development through structural analysis¶ and design of members and connections to prepare structural steel design drawings and specifications for fabrication and erection of the superstructure (see Chapters 4 through 11, and Appendices A and B).

REFERENCES

Akhtar, M.N., Otter, D., and Doe, B., 2006, *Stress-State Reduction in Concrete Bridges Using Under-Tie Rubber Pads and Wood Ties*, Transportation Technology Center, Inc. (TTCI), Association of American Railroads (AAR), Pueblo, CO.

American Railway Engineering and Maintenance-of-way Association (AREMA), 2015, *Manual for Railway Engineering*, Lanham, MD.

American Society of Testing and Materials (ASTM), 2015, *Standard F3125—Standard Specification for High Strength Structural Bolts, Steel and Alloy Steel, Heat Treated, 120 ksi (830 MPa) and 150 ksi (1040 MPa) Minimum Tensile Strength, Inch and Metric Dimensions*, *Annual Book of ASTM Standards*, West Conshohocken, PA.

American Welding Society (AWS), 2015, *Bridge Welding Code*, AWS D1.5, Miami, FL.

Barker, R.M. and Puckett, J.A., 1997, *Design of Highway Bridges*, John Wiley & Sons, New York.

Bernard-Gély, A. and Calgaro, J.-A., 1994, *Conception des Ponts*, Presses de l'Ecole Nationale des Ponts et Chaussées, Paris, France.

Billington, D.P., 1985, *The Tower and the Bridge, The New Art of Structural Engineering*, Princeton University Press, Princeton, NJ.

Byers, W.G., 1970, Impact from railway loading on steel girder spans, *Journal of the Structural Division, ASCE*, Vol. 96, No. 6, 1093–1103.

---

* The safety of the public, railroad operations, and railroad and contractor personnel must be carefully considered in erection procedures.

† Typically, railroad, highway, and/or marine traffic.

‡ Recommended practice for the erection of steel railway bridges is outlined in AREMA 2015, Chapter 15, Part 4.

§ The level of planning and preliminary design effort is related to the scope, magnitude, and complexity of the proposed bridge.

¶ Structural analysis may range from simple manual calculations based on strength of materials theory to complex dynamic and/or nonlinear finite element methods using digital computers, depending on the complexity of loading and/or geometry of the structure (Fu and Wang (2015)).

Byers, W.G., 2006, Railway bridge performance in earthquakes that damaged railroads, *Proceedings of the 7th International Conference on Short and Medium Span Bridges*, Canadian Society for Civil Engineering (CSCE), Montreal, QC.

Byers, W.G., 2009, Overview of bridges and structures for heavy haul operations, in *International Heavy Haul Association (IHHA) Best Practices*, Chapter 7, R. & F. Scott, North Richland Hills, TX.

CSA W59-13, *Welded Steel Construction (Metal Arc Welding)*, Canadian Standards Association, Toronto, Canada.

Chow, V.T., 1959, *Open Channel Hydraulics*, McGraw-Hill, New York.

Nowak, A.S. and Collins, K.R., 2013, *Reliability of Structures*, 2nd ed., CRC Press, Boca Raton, FL.

Federal Highway Administration (FHWA), 1990, *User Manual for WSPRO*, Publication IP-89-027, Washington, DC.

Federal Railroad Administration (FRA), 2008, *Railroad Bridge Working Group Report to the Railroad Safety Advisory Committee*, Washington, DC.

Fu, C.C. and Wang, S., 2015, *Computational Analysis and Design of Bridge Structures*, CRC Press, Boca Raton, FL.

Ghasemi, H.S. and Nowak, A.S., 2016, Reliability analysis for serviceability limit state of bridges concerning deflection criteria, *Structural Engineering International*, Vol. 26, No. 2, 168–175.

Hamill, L., 1999, *Bridge Hydraulics*, E & FN Spon, London, UK.

Hay, W.W., 1977, *Introduction to Transportation Engineering*, John Wiley & Sons, New York.

Laursen, E.M., 1962, Scour at bridge crossings, *Transactions of the ASCE*, Vol. 127, No. 1, 166–179.

Laursen, E.M., 1963, Analysis of relief bridge scour, *Journal of Hydraulics Division*, ASCE, Vol. 89, No. HY3, 93–118.

Leonhardt, F., 1984, *Bridges*, MIT Press, Cambridge, MA.

Miller, I., and Freund, J.E., 1972, *Probability and Statistics for Engineers*, 2nd ed., Prentice-Hall Inc., Englewood Cliffs, NJ.

Nagler, F.A., 1918, Obstruction of bridge piers to the flow of water, *Transactions of American Society of Civil Engineers (ASCE)*, Vol. 82, No. 1, 334–363.

Nowak, A.S., 1999, Calibration of LRFD bridge design code, NCHRP Report 368, Transportation Research Board, National Research Council, Washington, DC.

Rakoczy, A.M. and Nowak, A.S., 2013, Reliability-based strength limit state for steel railway bridges, *Structure and Infrastructure Engineering*, Vol. 58, No. 4, 81–92.

Rakoczy, A.M., Nowak, A.S., and Dick, S.M., 2016, Fatigue reliability model for steel railway bridges, *Structure and Infrastructure Engineering*, Vol. 12, No. 12, 1602–1613.

Ruble, E.J., 1955, Impact in railroad bridges, *Proceedings of the American Society of Civil Engineers*, Vol. 81, No. 7, 1–36.

Ramberger, G., 2002, *Structural Bearings and Expansion Joints for Bridges*, Structural Engineering Document 6, International Association for Bridge and Structural Engineering (IABSE), Zurich, Switzerland.

Richardson, E.V. and Davis, S.R., 2001, *Evaluating Scour at Bridges*, Hydraulic Engineering Circular 18, FHWA, McLean, VA.

Rosenblueth, E. and Esteva, L., 1972, *Reliability Basis for Some Mexican Codes*, ACI Publication SP-31, American Concrete Institute, Detroit, MI.

Schneider, V.R., Board , J.W., Colson, B.E., Lee, F.N., and Druffel, L., 1977, Computation of back-water and discharge at width constrictions of heavily vegetated floodplains, Water Resources Investigations 76-129, US Geological Survey.

Small, G. and Ketler, L., 2006, Recycling of an open deck half through plate girder span and conversion to a concrete ballast deck through plate girder span, *Proceedings of the AREMA 2006 Annual Conference*, AREMA, Lanham, MD.

Sorgenfrei, D.F., Marianos, W.N., and Sweeney, R.A.P., 2014, Railroad bridge design specifications, in *Bridge Engineering Handbook*, Chen, W.-F. and Duan, L., CRC Press, Boca Raton, FL.

Stanton, J.F., Roeder, C.W., and Campbell, T.I., 1999, *High-Load Multi-Rotational Bridge Bearings*, NCHRP Report 432, National Academy Press, Washington, DC.

Taly, N., 1998, *Design of Modern Highway Bridges*, McGraw-Hill, New York.

Tobias, D.H., Foutch, D.A., and Choros, J., 1996, Loading spectra for railway bridges under current operating conditions, *Journal of Bridge Engineering*, Vol. 1, No. 4, 127–134.

Tobias, D.H. and Foutch, D.A., 1997, Reliability-based method for fatigue evaluation of railway bridges, *Journal of Bridge Engineering*, Vol. 2, No. 2, 53–60.

Transportation Association of Canada (TAC), 2004, *Guide to Bridge Hydraulics*, Thomas Telford, London, UK.

Unsworth, J.F., 2003, *Heavy Axle Load Effects on Fatigue Life of Steel Bridges*, TRR 1825, Transportation Research Board, Washington, DC.

Unsworth, J, Small, G., and Afhami, S., 2005, Service load investigation of the composite behavior of a ballasted through plate girder span, *Proceedings of the 2005 AREMA Annual Conference, Chicago, IL*, AREMA, Lanham, MD.

Unsworth, J.F. and Brown, C.H., 2006, *Rapid Replacement of a Movable Steel Railway Bridge*, TRR 1976, Transportation Research Board, Washington, DC.

Waddell, J.A.L., 1916, *Bridge Engineering, Volume 1*, John Wiley & Sons, New York.

Yarnell, D.L., 1934, *Bridge Piers as Channel Obstructions*, Technical Bulletin 442, US Department of Agriculture, Washington, DC.

4 Loads and Forces on Steel Railway Bridges

4.1 INTRODUCTION

The loads and forces on steel railway bridge superstructures are vertical (gravity), longitudinal, or lateral in nature.

Gravity loads comprising dead, live, and impact loads are the principal loads to be considered for steel railway bridge design. Live load impact (dynamic effect) is included due to the relatively rapid application of railway live loads. However, longitudinal forces (due to live load, wind forces, and/or thermal forces) and lateral forces (due to live load, wind forces, thermal forces,* and/or seismic activity) also warrant careful consideration in steel railway bridge design. A resource for the review of load effects on structures, in general, is American Society of Civil Engineers (ASCE)/Structural Engineering Institute (SEI).

Railway bridges are subjected to unique forces related to railroad moving loads. These are live load impact from vertical and rocking effects, longitudinal forces from the acceleration or deceleration of railroad equipment, lateral forces caused by irregularities at the wheel-to-rail interface (commonly referred to as "truck hunting" or "nosing"), and centrifugal forces due to track curvature.

4.2 DEAD LOADS

Superstructure dead load consists of the weight of the superstructure itself, track, deck (open or ballasted), utilities (conduits, pipes, and cables), walkways (some engineers also include walkway live load as a component of superstructure dead load), permanent formwork, snow, ice, and anticipated future dead loads (e.g., larger deck ties, increases in ballast depth, additional utilities). However, snow and ice loads are generally excluded from consideration due to their relatively low magnitude. For ordinary steel railway bridges, dead load is often a comparatively small proportion of the total superstructure load (steel railway bridges typically have a relatively high live load to dead load ratio).

Curbs, parapets, and sidewalks may be poured after the deck slab in reinforced concrete construction. This superimposed dead load may be distributed according to superstructure geometry (e.g., by tributary widths). However, it is common practice to equally distribute superimposed dead loads to all members supporting the hardened deck slab. This is appropriate for most superstructure geometries, but may require refinement in multibeam spans where exterior beams may be subjected to a greater proportion of any superimposed dead load.

At the commencement of design, dead load must be estimated from experience or review of similar superstructure designs. This estimated design load must be reviewed against the actual dead load calculated after the final design of the superstructure (see Appendices A and B). Small differences between the estimated and actual dead load are not important, provided the dead load is a reasonably small constituent of the total design load. Dead loads typically used for ordinary steel railway bridge design are shown in Table 4.1. Steel railway bridge engineers will often include an allowance of 10%–15% of estimated steel superstructure weight to account for bolts, gusset plates, stiffeners, and other appurtenant steel components. Temporary construction dead loads and the transfer of dead load during shored or unshored construction of steel and concrete composite deck spans should also be considered during design (see Chapter 7 and Appendix B).

* On curved track.

TABLE 4.1
Dead Loads on Steel Railway Bridges

Item	Dead Load	
Track (rails and fastenings)	200 lb/ft	300 kg/m
Steel	490 lb/ft^3	7850 kg/m^3
Reinforced and prestressed concrete	150 lb/ft^3	2400 kg/m^3
Plain (unreinforced) concrete	145 lb/ft^3	2320 kg/m^3
Timber	35–60 lb/ft^3	560–960 kg/m^3
Sand and gravel, compacted (railroad ballast)	120 lb/ft^3	1920 kg/m^3
Sand and gravel, loose	100 lb/ft^3	1600 kg/m^2
Permanent formwork (incl. concrete in valleys)	15 lb/ft^2	75 kg/m^2
Waterproofing on decks	10 lb/ft^2	50 kg/m^2

4.3 RAILWAY LIVE LOADS

Railroad locomotives and equipment (box and flat cars, commodity gondolas, hopper, and tank cars) vary greatly with respect to weight, number of axles, and axle spacing.

Modern freight locomotives have two three-axle sets (trucks or bolsters) with a spacing between axles of between 1.96 m (6.42 ft) and 2.08 m (6.83 ft), and a spacing between axle sets of between 13.91 m (45.62 ft) and 16.66 m (54.63 ft). These modern generation locomotives weight up to 200,000 kg (435,000 lb). There are, however, many four- and six-axle locomotives of weight between about 115,000 kg (250,000 lb) and 180,000 kg (400,000 lb), and with lengths between about 15 m (50 ft) and 25 m (80 ft) operating on the railroad infrastructure.

Axle spacing is typically between 1.52 m (5 ft) and 1.78 m (5.83 ft) for North American four-axle freight car equipment. Truck spacing may vary from about 5 m (17 ft) to 20 m (66 ft) (Dick, 2002). Gross car weights of up to 130,000 kg (286,000 lb)* are common on North American railroads and some railroad lines carry 145,000 kg (315,000 lb) cars.

This variability in railroad equipment weight and geometry requires a representative live load model for design that provides a safe and reliable estimate of railroad operating equipment characteristics within the design life of the bridge.

4.3.1 Static Freight Train Live Load

The railway bridge design live load currently recommended in AREMA (2015) is Cooper's EM360† (E80) load. This design load is based on two consolidation-type steam locomotives with trailing cars represented by a uniformly distributed load (Cooper, 1894). The maximum locomotive axle load is 360 kN (80 kips‡), and freight equipment is represented by a uniform load of 120 kN/m of the track (8 kips per ft of track). An alternate live load, consisting of four 445 kN (100 kips) axles, is also recommended in AREMA (2015, Chapter 15) to represent the stress range effects of adjacent heavy rail cars on short spans. These design live loads are shown in Figure 4.1a and b.

* Cars with weight not exceeding 130,000 kg (286,000 lb) and lengths over car couplers greater than 12.78 m (41′–11″) are considered to be in free interchange between most lines of North American Class 1 railroads.

† Cooper's EM360 is the recommended live load in units of the International System (Système International or SI), the modern metric system. AREMA (2015, Chapter 15—Steel Structures) does not explicitly indicate Cooper's EM360; however, AREMA (2015, Chapter 8—Concrete Structures and Foundations) does outline the Cooper's EM360 live load as shown in Figure 4.1a.

‡ 1 kip = 1000 lb_f (see also Appendix C).

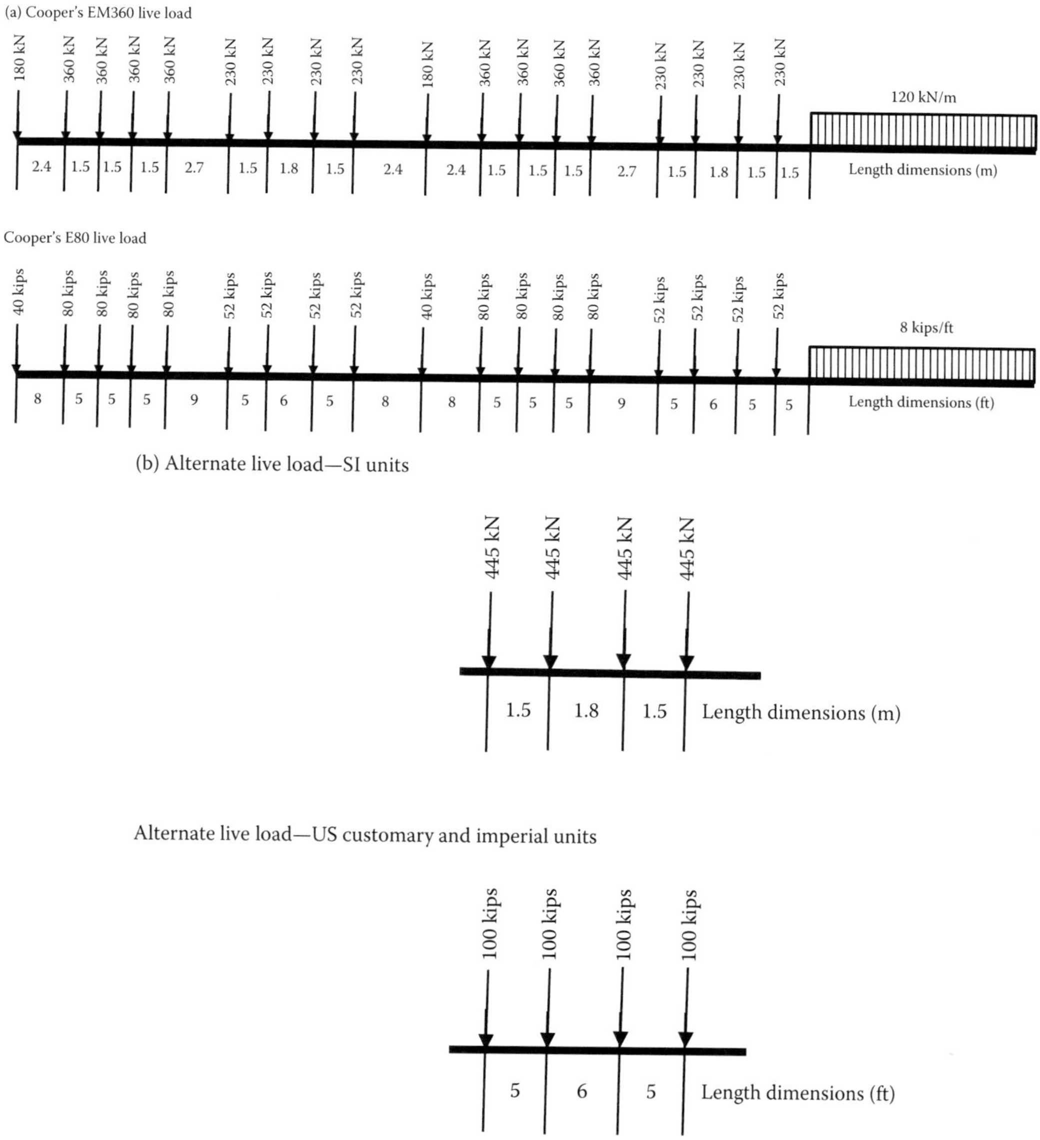

FIGURE 4.1 (a) Cooper's Axle Loads and (b) Alternate live load—axle loads. (After AREMA, *Manual for Railway Engineering*, Lanham, MD, 2015.)

This design load appears antiquated, particularly with respect to the use of steam locomotive geometry.* However, it is a reasonably good representation of the maximum load effects of modern freight traffic as illustrated in Figure 4.2. The figure is plotted from a moving load analysis (see Chapter 5) of end shear and midspan flexure on simple spans for continuous and uniform strings of various heavy freight equipment vehicles. For a design live load of Cooper's EM360 (E80) with the alternate live load of four 445 kN (100 kips) axles, Figure 4.2 shows that the alternate live load governs for spans less than about 16 m (53 ft) in bending and less than about 14 m (46 ft) in shear. For a design live load of Cooper's EM400 (E90) with the alternate load of 445 kN (100 kips) axles,

* Alternate live loads have been proposed, such as the M live load system proposed by D.B. Steinman (1922), but Cooper's E live load has been maintained as the railway bridge design load by most railroads.

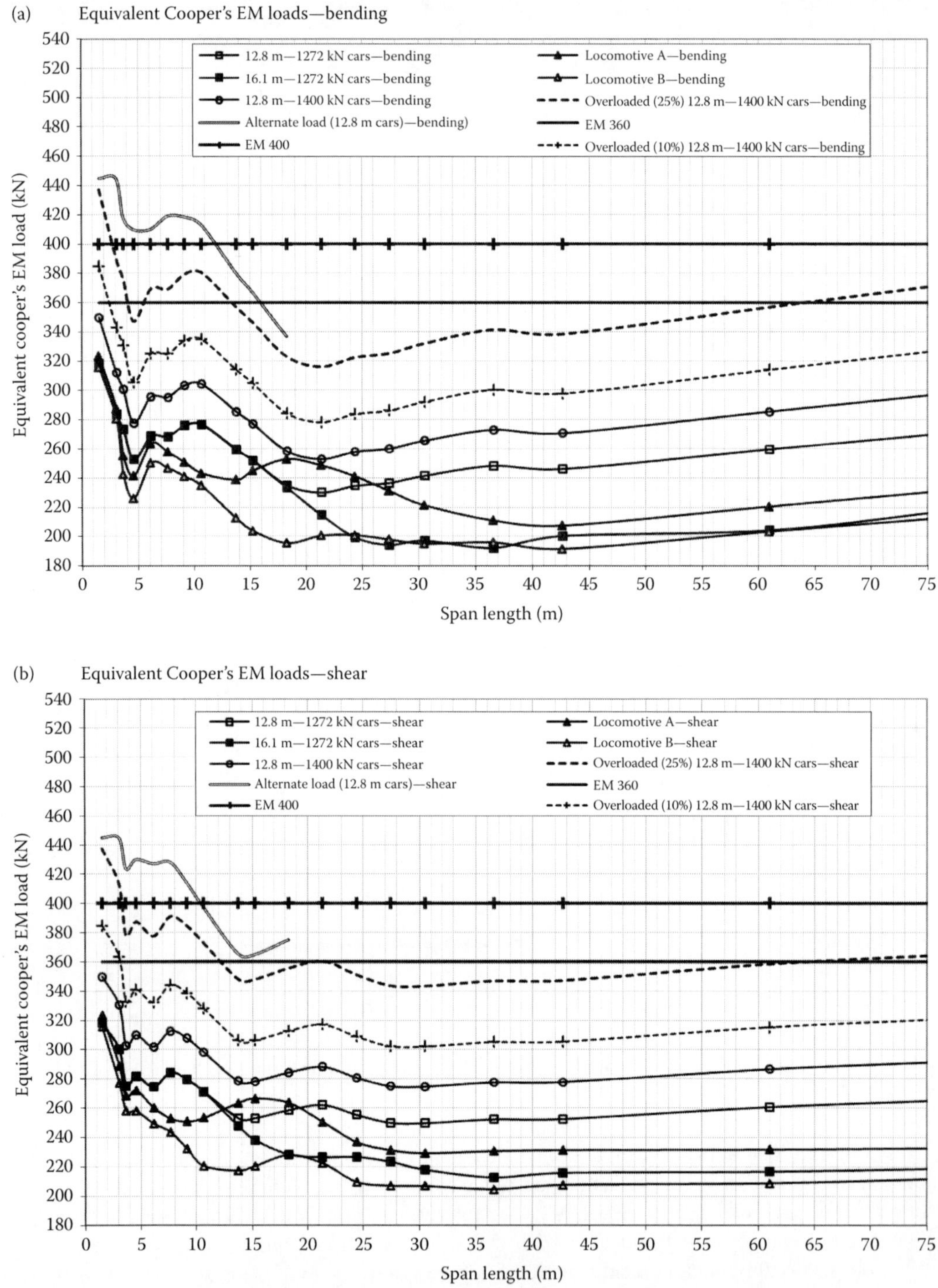

FIGURE 4.2 (a) Equivalent Cooper's E loads in bending for some modern railway freight locomotives and equipment on simply supported spans up to 75 m long. (b) Equivalent Cooper's E loads in shear for some modern railway freight locomotives and equipment on simply supported spans up to 75 m long.

(Continued)

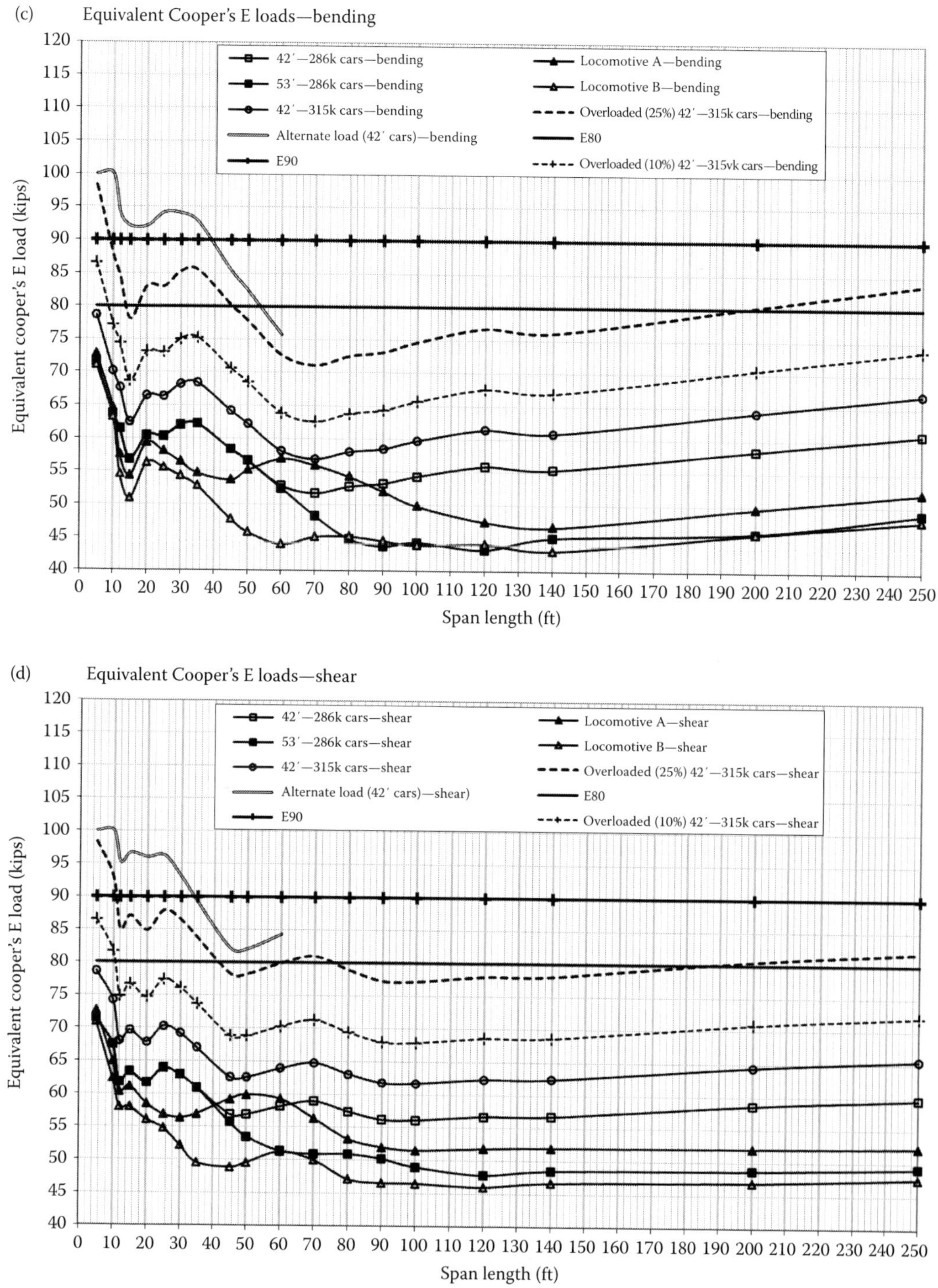

FIGURE 4.2 (CONTINUED) (c) Equivalent Cooper's E loads in bending for some modern railway freight locomotives and equipment on simply supported spans up to 250 ft long. (d) Equivalent Cooper's E loads in shear for some modern railway freight locomotives and equipment on simply supported spans up to 250 ft long.

Figure 4.2 shows that the alternate load governs for spans less than about 12 m (40 ft) in bending and less than about 10 m (33 ft) in shear. Figure 4.2 also indicates that the alternate load is a good representation of the static bending and shear forces imposed on short spans by overloaded cars (with overload weights of 10% and 25% of the total railcar load), which may occur during the life of the superstructure.*

For the car weight and configurations investigated in Figure 4.2, the overloaded cars† exceed the Cooper's EM360 (E80) design load for bending for spans less than about 14 m (45 ft) in length and longer than 56 m (185 ft) when overloaded 25%. The 10% overloaded cars exceed the Cooper's EM360 (E80) design load for bending for spans less than about 2.5 m (8 ft) in length. The 25% overloaded cars also exceed the Cooper's EM400 (E90) design load for bending for spans less than about 2.5 m (8 ft) in length. The Cooper's EM360 (E80) design live load appears to be generally adequate for current equipment weights and load control practices. However, in addition to reflecting current equipment weight and overloads, the design live load must also consider anticipated future freight equipment weight.‡ Some railroad companies may vary from the Cooper's EM360 (E80) bridge design load based on their current and/or projected operating practice.§ It is usual that the magnitude of the design axle loads is changed [e.g., 320 kN (72 kips) or 400 kN (90 kips)], but the axle spacing is unaltered. Therefore, different bridge designs can be readily compared.

Increasing the AREMA design live load was also recommended by Tobias et al. (1996), following a statistical study of the load spectra from over 500 North American freight trains. Average four-axle car loads up to 1253 kN (282 kips) and average six-axle locomotive loads up to 1838 kN (413 kips) were measured. Also measured were maximum four-axle car loads of 1508 kN (339 kips) and maximum six-axle locomotive loads of 2157 kN (485 kips). For the same equipment, average car axle loads of up to 313 kN (79 kips) and average locomotive axle loads of up to 306 kN (69 kips) were measured. In addition, maximum car axle loads of 471 kN (106 kips) and maximum locomotive axle loads of 427 kN (96 kips) were measured. In 1996, much of the North American freight railroad industry was transitioning from 1170 kN (263 kips) cars to 1272 kN (286 kips) cars. With average and maximum measured car loads of 1253 kN (282 kips) and 1508 kN (339 kips), respectively, the guidance to increase the recommended design live load appears warranted. Cooper's EM400 (E90) design live load provides a load margin of 26% for 1272 kN (286 kips) cars and 14% for 1400 kN (315 kips) cars. If maximum car loads for 1272 kN (286 kips) cars are estimated as 1530 kN (344 kips),¶ the design live load margin is only 5%. Cooper's EM400 (E90) may be appropriate for current bridge design projects to anticipate occasionally overloaded cars, high dynamic (impact) forces,** and projected rail car weights.

Nevertheless, the flexural cyclical stress ranges created by Cooper's design load do not necessarily accurately reflect the cyclical stress ranges created by modern railway freight equipment. Figure 4.3 shows the variation in midspan bending moment as 7.6 m (25 ft), and 18.3 m (60 ft) simply supported spans, respectively, are traversed by various train configurations. The Cooper's EM360 (E80) design live load appears to conservatively represent the design stress range magnitude for both the 7.6 m (25 ft) and 18.3 m (60 ft) spans.†† However, the alternate live load better represents

* Depending on the rail line, scale availability, and freight being shipped, overloaded cars may occur with some frequency. Overloaded rail cars may occur where high-density freight (e.g., metals) is being transported in cars with volumes intended for lighter weight materials or products.

† These cars are very short and heavy, and not typical of those used routinely on North American railroads. They are, however, representative of equipment currently used on some specific routes and, in terms of weight, the potential direction for future freight equipment.

‡ In particular, four-axle rail car weights, which have been increasing at a greater rate than locomotive weights.

§ Some North American railroads currently design new bridges for Cooper's EM400 (E90) live load.

¶ A preliminary estimate based on the ratio of the measured maximum car load [1508 kN (339 kips)] to average car load [1253 kN (282 kips)].

** Typically from flat rail wheels and/or, if present, joints in the rail traversing the bridge.

†† This is accounted for with adjustments to the number of equivalent constant-amplitude cycles based on the ratio of typical train loads to the Cooper's EM360 (E80) design load.

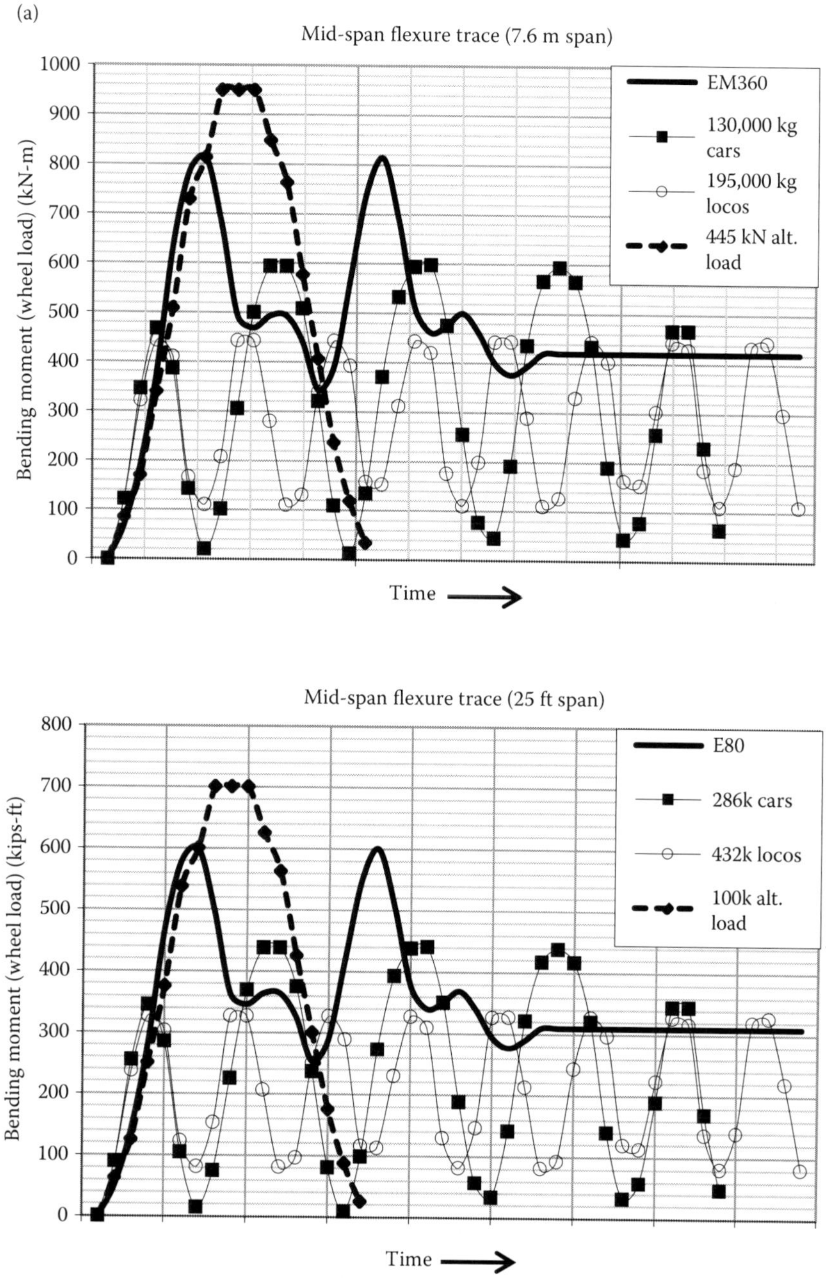

FIGURE 4.3 (a) Midspan bending moments for Cooper's EM 360 on a 7.6 m simply supported span and Cooper's E80 load on a 25 ft simply supported span with midspan bending moments for typical modern heavy freight equipment.

(Continued)

(b)

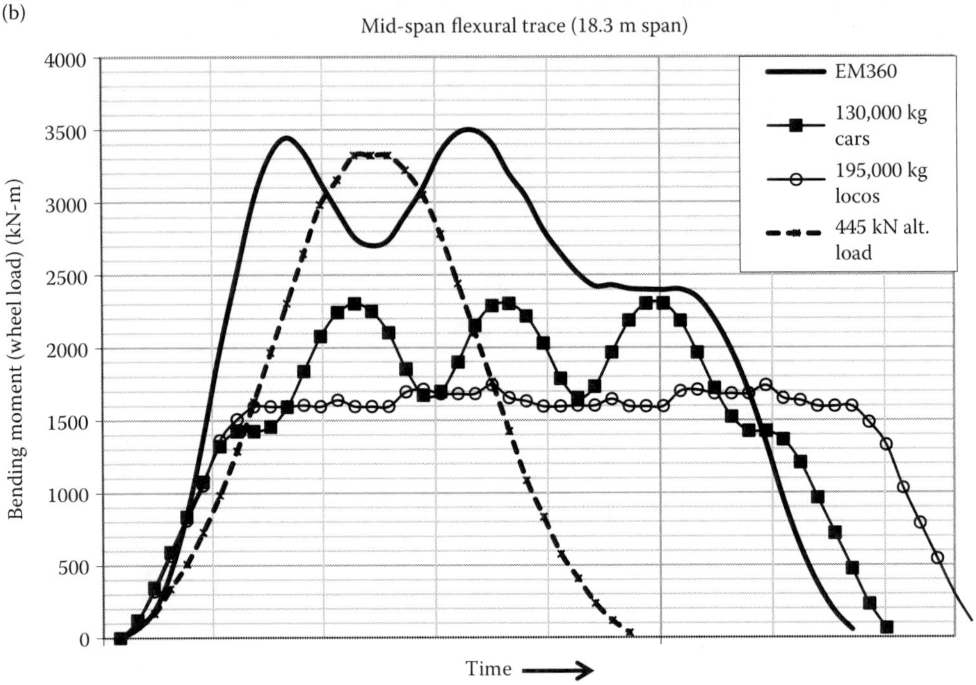

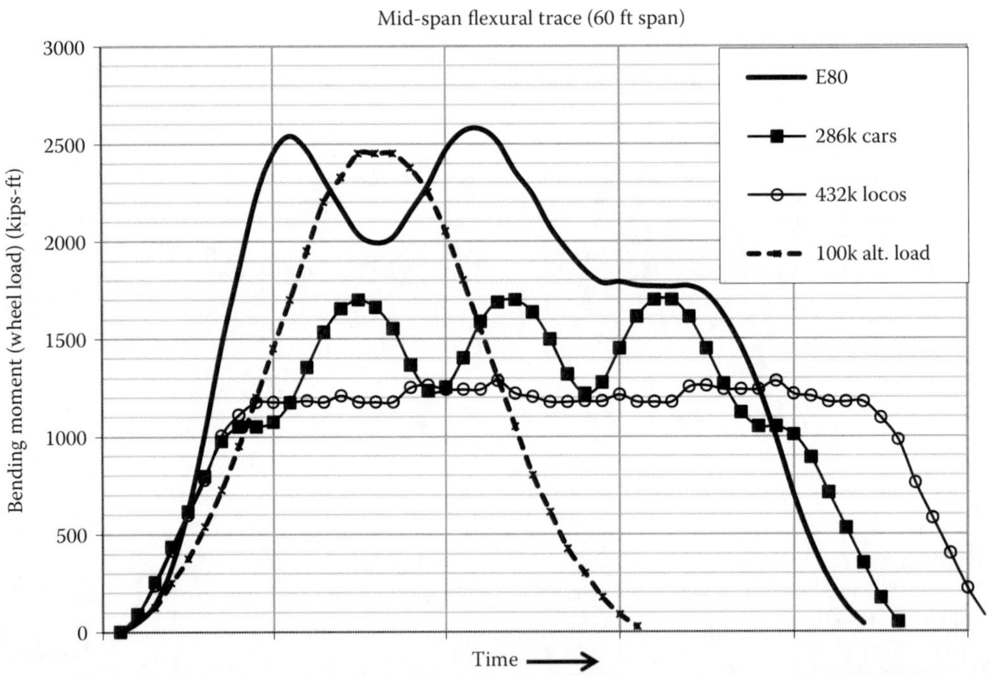

FIGURE 4.3 (CONTINUED) (b) Midspan bending moments for Cooper's EM 360 on an 18.3 m simply supported span and Cooper's E80 load on a 60 ft simply supported span with midspan bending moments for typical modern heavy freight equipment.

(*Continued*)

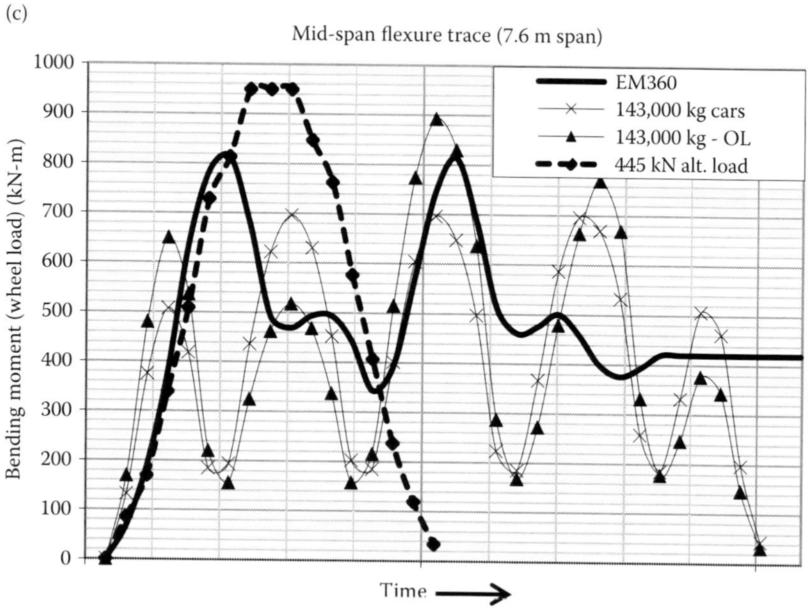

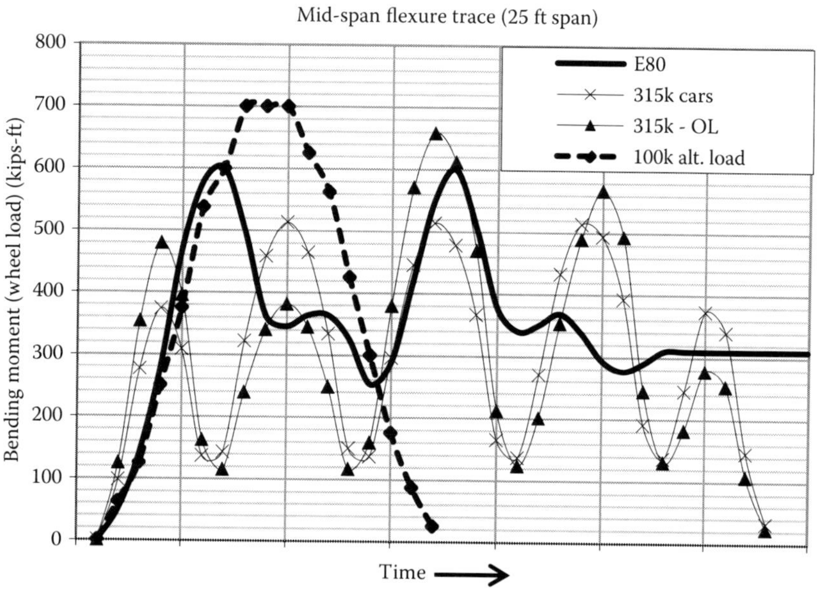

FIGURE 4.3 (CONTINUED) (c) Midspan bending moments for Cooper's EM 360 on a 7.6 m simply supported span and Cooper's E80 load on a 25 ft simply supported span with midspan bending moments for modern very heavy freight equipment.

(*Continued*)

(d)

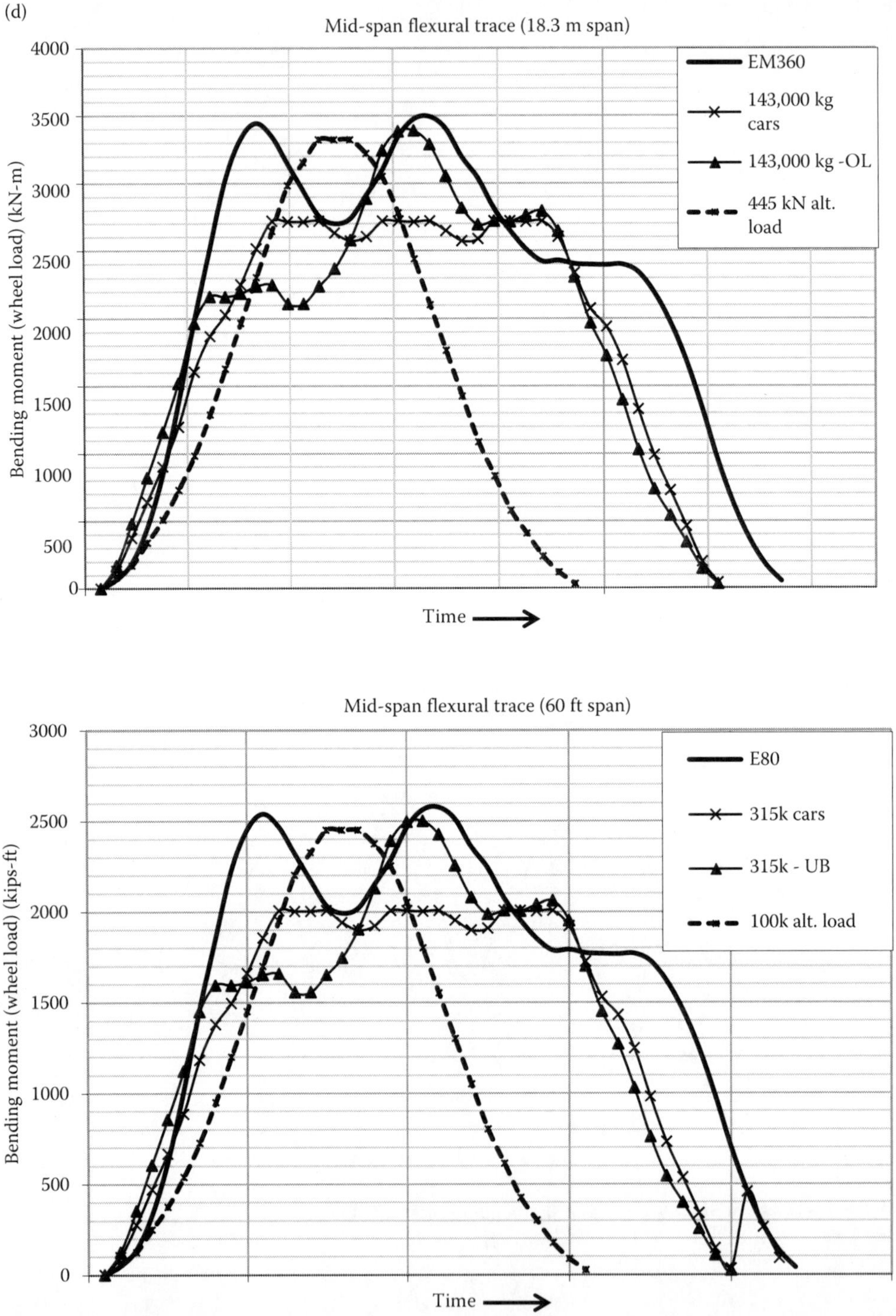

FIGURE 4.3 (CONTINUED) (d) Midspan bending moments for Cooper's EM 360 on an 18.3 m simply supported span and Cooper's E80 load on a 60 ft simply supported span with midspan bending moments for modern very heavy freight equipment.

the cyclical behavior of the various train configurations, particularly on short spans. Therefore, the allowable fatigue stress ranges, S_{Rfat}, recommended for design in AREMA (2015) are based on equivalent constant-amplitude stress cycles from projected variable-amplitude stress cycles due to typical railroad traffic (see Chapter 5).

4.3.1.1 Cooper's Design Live Load for Projected Railway Equipment

A Cooper's design live load appropriate for projected railroad operating practice may be considered, based on the development of design live loads in conjunction with railroad operating practice of the past.

In the latter part of the 19th century and the early part of the 20th century, North American steel railway bridges were designed for the heaviest locomotives that would be used on the specific railroad that owned the bridge. However, as railroads merged and expanded, the rapid increase in locomotive weights and growth of interchange traffic created the need for a specified design live load that would represent operating railroad locomotives and equipment. This was achieved in 1906 with the publication of the first steel railway bridge design specifications by AREMA[*] (see Chapter 1). Locomotive weights and representative design loads continued to increase, and many steel railway bridges were replaced in the early 20th century to accommodate the increasing weights of steam locomotives.[†]

Increases in steam locomotive power were accompanied by considerable increases in locomotive weight. The increased hauling power also enabled rapid increases in rail car weights from about 75,000 kg (170,000 lb) in 1910 to 95,000 kg (210,000 lb) just a decade later. High-power modern diesel locomotives have been developed since the middle of the 20th century without substantial increase in weight. Modern six-axle locomotives with weights of less than 200,000 kg (435,000 lb) can attain a tractive effort of up to 35%. Modern rail cars have increased in weight from about 100,000 kg (220,000 lb) to 130,000 kg (286,000 lb) in the past 40–50 years, and some rail lines currently carry 145,000 kg (315,000 lb) rail cars. As shown in Figure 4.2, the four-axle 130,000 kg (286 kips) cars create greater equivalent Cooper's EM (E) loads than the six-axle diesel locomotives, except for influence line lengths between about 12 m (40 ft) and 26 m (85 ft). Figure 4.2 also indicates that the four-axle 145,000 kg (315 kips) cars create greater equivalent Cooper's EM (E) loads than the six-axle diesel locomotives. Many steel railway spans are designed in the 12 m (40 ft) and 26 m (85 ft) length range and warrant attention to locomotive effects, which may be in the order of 10%–12% greater than four-axle 130,000 kg (286 kips) car effects. Nevertheless, because the American Railway Engineering Association (AREA) and AREMA Cooper's design live load trailing car weights are proportional to the locomotive axle loads, a review of Cooper's design live load in terms of rail car weight will provide useful insight into the design live load margins used since the early part of the 20th century.

From the beginning of the 20th century, recommended design live loads have increased in conjunction with locomotive and car weight increases and projections for future traffic. Figure 4.4 shows the increases in weight for rail cars commonly used in North American railroad interchange from 1920 to 1995 when the current 130,000 kg (286,000 lb) rail cars were typically introduced. These 130,000 kg (286,000 lb) rail cars have 320 kN (71,500 lb) axle loads, which is also typical of modern locomotive axle loads. Figure 4.4 also shows a linear regression of rail car weight data and the Cooper's design live loads specified by AREA and AREMA since early in the 20th century.[‡] A maximum projected rail car weight of 160,000 kg (360,000 lb) is assumed, which might be considered a reasonable future upper limit based on other technical and operational criteria. Future design live loads may be investigated by projecting currently operating 130,000 kg (286,000 lb) and 145,000 kg (315,000 lb)[§] rail cars with load margins similar to those successfully used since the early part of the 20th century.

---

[*] The live load specified in the 1906 AREMA specifications was Cooper's E40 (AREMA, 1906).

[†] The live load specified in the 1920 AREA specifications was Cooper's E60 (AREA, 1920).

[‡] Cooper's design live loads representing rail cars trailing locomotives are uniformly distributed.

[§] 145,000 kg (315,000 lb) four-axle rail cars are currently used on some heavy haul railroads.

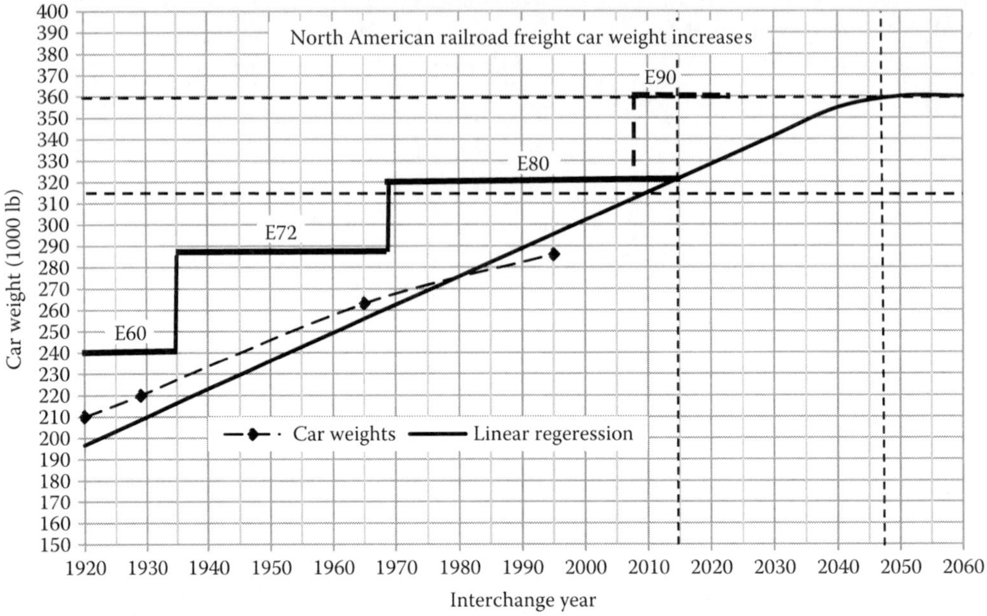

FIGURE 4.4 Car weight increases and AREMA Cooper's design live loads.

Figure 4.4 shows that, since the early part of the 20th century, the Cooper's design live load has exceeded actual rail car weights by a margin intended to anticipate occasional overloaded cars, and, most essentially, future heavier axle loads.

Actual loads may exceed design loads during the life of a bridge. Actual rail cars with 25% increase in individual axle loads are not uncommon (Tobias et al., 1996) due to flat wheels, rail and track conditions, overloading, and/or unbalanced loading. In particular, these large axle loads can influence the maximum effects on members with short influence lines (e.g., stringers in through span floor systems). However, these large axle loads may not be detrimental to performance, provided they are infrequent (AREMA, 2015) with an adequate minimum live load margin.

The service life of steel railway superstructures designed for the appropriate loads and correctly fabricated with appropriate materials is generally considered to be about 80 years (AREMA, 2015). However, steel superstructures that are well maintained after erection can safely and reliably carry live loads for considerably longer service lives.* The design live load must anticipate the maximum likely effects created by the actual traffic over a significant proportion of the expected design life of the superstructure. It is often considered that if large increases in equipment weights occur,† smaller superstructures may require replacement and longer superstructures rehabilitation (strengthening) at about 50% of their service lives. In that case, a design live load appropriate for about 40–60 years may be considered.

Historical live load margins from Figure 4.4, at the beginning and end of each period in which the Cooper's design live load was in effect, assuming a linear increase in actual car weight are shown in Table 4.2. This table indicates that, from 1920 to 2015, a minimum live load margin of about 1.10 has been maintained. The lowest live load margins occur at members with short influence lines, where only a few axles affect bending and shear forces. This is typical behavior for spans less than about 15 m (50 ft), where axle loads affect bending and shear forces. Car load effects typically dominate shear and bending in spans greater than about 27 m (90 ft).

* Many steel railway bridges are over 80 years old and some are over 100 years old.

† For example, the Commentary to Part 1 of AREMA (Chapter 15—Steel Structures) indicates that design loads must anticipate trends toward heavier locomotives and some heavy rail cars that currently produce loads equivalent to Cooper's EM360 (E80) or greater.

The live load margin for Cooper's E80 for various rail car weights and member influence line lengths is shown in Figure 4.5. This figure shows that, for spans less than about 12 m (40 ft) in length, a minimum Cooper's E80 live load margin of about 1.10 is compromised for 145,000 kg (315,00 lb) rail car traffic. Table 4.3 indicates the live load margins for Cooper's E80 design live load if used with actual rail cars of 130,000 kg (286,000 lb) and 145,500 kg (315,000 lb).

TABLE 4.2
Live Load Margins for Historical and Future Cooper's Design Live Loads

Cooper's Design Live Load	Period	Years	Live Load Margin: Initial	Live Load Margin: Minimum
E60	1920–1935	15	1.20	1.09 (245 kN (55k) axle, linear regression)
E72	1935–1968	33	1.20	1.11 (290 kN (65k) axle, linear regression)
E80	1968–2015	47	1.11	1.12 (1275 kN, 286k traffic) 1.02 (1400 kN, 315k traffic)
E90	2015–2045	30	1.13	1.26 (1275 kN, 286k traffic) 1.00 (1400 kN, 315k traffic)

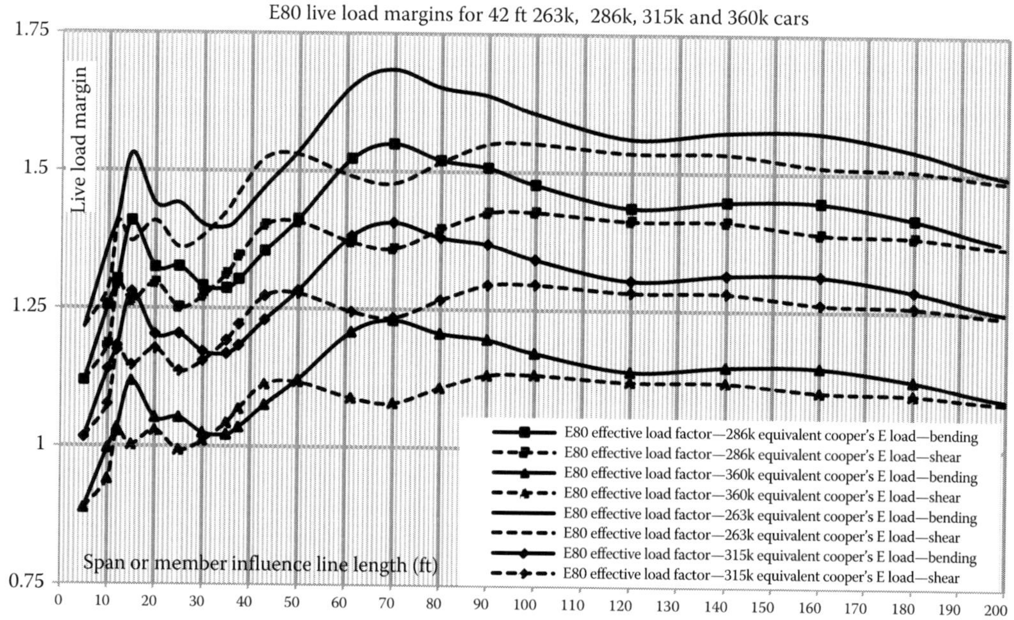

FIGURE 4.5 Cooper's E80 design live load margins.

TABLE 4.3
Live Load Margins for Cooper's E80 Design Live Load

Cooper's Design Live Load	Initial Live Load Margin (130,000 kg (286 kips) cars)	Minimum Live Load Margin (145,000 kg (315 kips) Cars)
E80	1.12	1.02

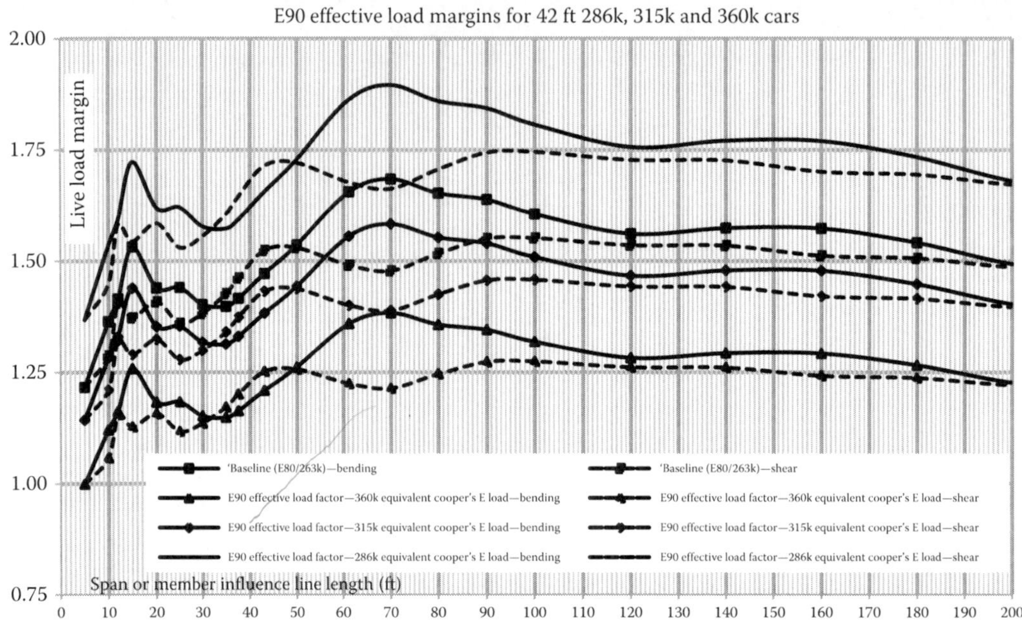

FIGURE 4.6 Cooper's E90 design live load margins.

TABLE 4.4
Live Load Margins for Cooper's E90 Design Live Load

Cooper's Design Live Load	Initial Live Load Margin (130,000 kg (286 kips) Cars)	Minimum Live Load Margin (145,000 kg (315 kips) Cars)
E90	1.26	1.14

The live load margin for Cooper's E90 for various rail car weights and member influence line lengths is shown in Figure 4.6. This figure shows that, only for spans less than about 3 m (10 ft) in length,* a minimum Cooper's E90 live load margin of about 1.10 is compromised for 145,000 kg (315,00 lb) rail car traffic. Table 4.4 indicates the live load margins for Cooper's E90 design live load if used with actual rail cars of 130,000 kg (286,000 lb) and 145,500 kg (315,000 lb). With Cooper's E90 design live load, a minimum live load margin of about 1.10 is maintained for rail car weights of up to about 147,500 kg (325,000 lb). On rail lines where rail car weights exceeding 130,000 kg (286,000 lb) are operating or planned, bridge owners may wish to consider adopting the Cooper's EM400 (E90) design live load for new bridge design projects to anticipate occasional overloaded cars, high impacts, and to consider future rail car weights.

4.3.1.2 Fatigue Design Live Load for Railway Equipment

As equipment (locomotives and rail cars) traverse the length of an influence line or span, cyclical variable-amplitude stress ranges are induced in the superstructure members carrying the live load. The magnitude of the stress ranges may be less than that of the maximum stresses created by the static train load located at the position along the length of the influence line or span that creates the

* Spans less than 3 m (10 ft) are not considered bridges in accordance with US Federal Railroad Authority (FRA) Bridge Safety Regulations and Transport Canada (TC) Bridge Safety Guidelines.

greatest stresses in the member under consideration. However, if large enough, the cyclical* nature of the stress ranges may induce progressive fatigue damage that may result in cracking, particularly in areas of the superstructure subject to stress concentrations.†

The AREMA (2015) fatigue design methodology uses the Cooper's EM360 (E80) live load as the base for a fatigue design load. The stress-life approach, recommended for the design of steel superstructures by AREMA (2015), requires that the variable-amplitude cyclical railway live load be developed as an effective, or equivalent, constant-amplitude cyclical design load. This is necessary because fatigue strength (see Chapter 5) is established by the constant-amplitude cyclical stress testing of typical steel superstructure member and detail specimens. This equivalent constant-amplitude cyclical fatigue design load must accumulate the same damage as the variable-amplitude cyclical load over the total number of stress range cycles to failure. The stress ratio, $R = S_{remin}/S_{remax}$, and the stress range, $\Delta S_{re} = S_{remax} - S_{remin}$, may be used to describe constant-amplitude loading (Figure 4.7). The constant-amplitude loads used for fatigue testing are often performed with $R = 0$ (cyclical tension with $S_{min} = 0$) or $R = -1$ (fully reversed with $S_{max} = -S_{min}$). The mean stress‡ is $S_{remean} = (S_{remax} + S_{remin})/2$.

The variable-amplitude cyclical load history for the AREMA (2015) design load midspan bending moment on a 7.6 m (25 ft) and 18.3 m (60 ft) span is shown in Figure 4.8a and b, respectively. The uniformly distributed load (120 kN/m for Cooper's EM360 and 8000 lb/ft for Cooper's E80 design live load) creates no change in stress and is shown truncated in Figure 4.8a and b. The number of stress range cycles and their magnitudes can be determined directly from relatively simple load traces in elastic structures. However, actual freight rail traffic stress ranges measured on in-service bridges are typically more complex. Areas near the ¼ span length and locations of change in section may also be important for the determination of the maximum number of stress cycles and their magnitude.§

The variable-amplitude load and stress cycles from actual freight trains are typically irregular and require a method of reducing the complex spectra to simple cyclic loads and stresses. There are many techniques for counting the significant cycles of variable-amplitude load or stress range spectra. For many structures, the rainflow cycle counting method provides very good results (Dowling, 1999). The rainflow method counts the number of full cycles and their load or stress range magnitudes from half cycles. A frequency distribution histogram for the numbers of stress range cycles,

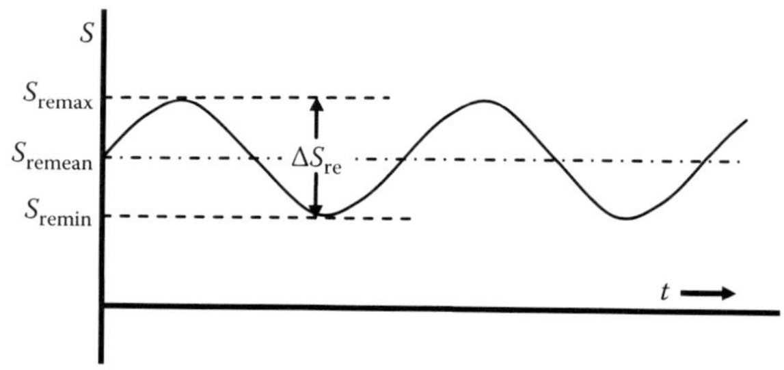

FIGURE 4.7 Constant-amplitude cyclical loading.

* Due to wheel configuration, typical railway traffic creates a substantially greater number of cycles on shorter spans than on longer spans.

† At a microscopic level, cyclical stresses may precipitate movement of atomic dislocations and create slip lines resulting in surface stress raisers and crack initiation. However, in typical steel superstructures, stress concentrations due to poor design details and/or fabrication practices are more prevalent.

‡ Since constant-amplitude cyclical fatigue testing of members and details includes the effects of stress concentrations and residual stresses (present from rolling, forming, fabricating, and welding operations) (see Chapter 10), the fatigue life is not influenced by mean stress effects and only the range of stress is significant for fatigue.

§ Because of the relationship between span length and car length during the cycling of spans, locations around the ¼ point may govern the maximum number of stress range cycles and magnitude (Dick, 2002).

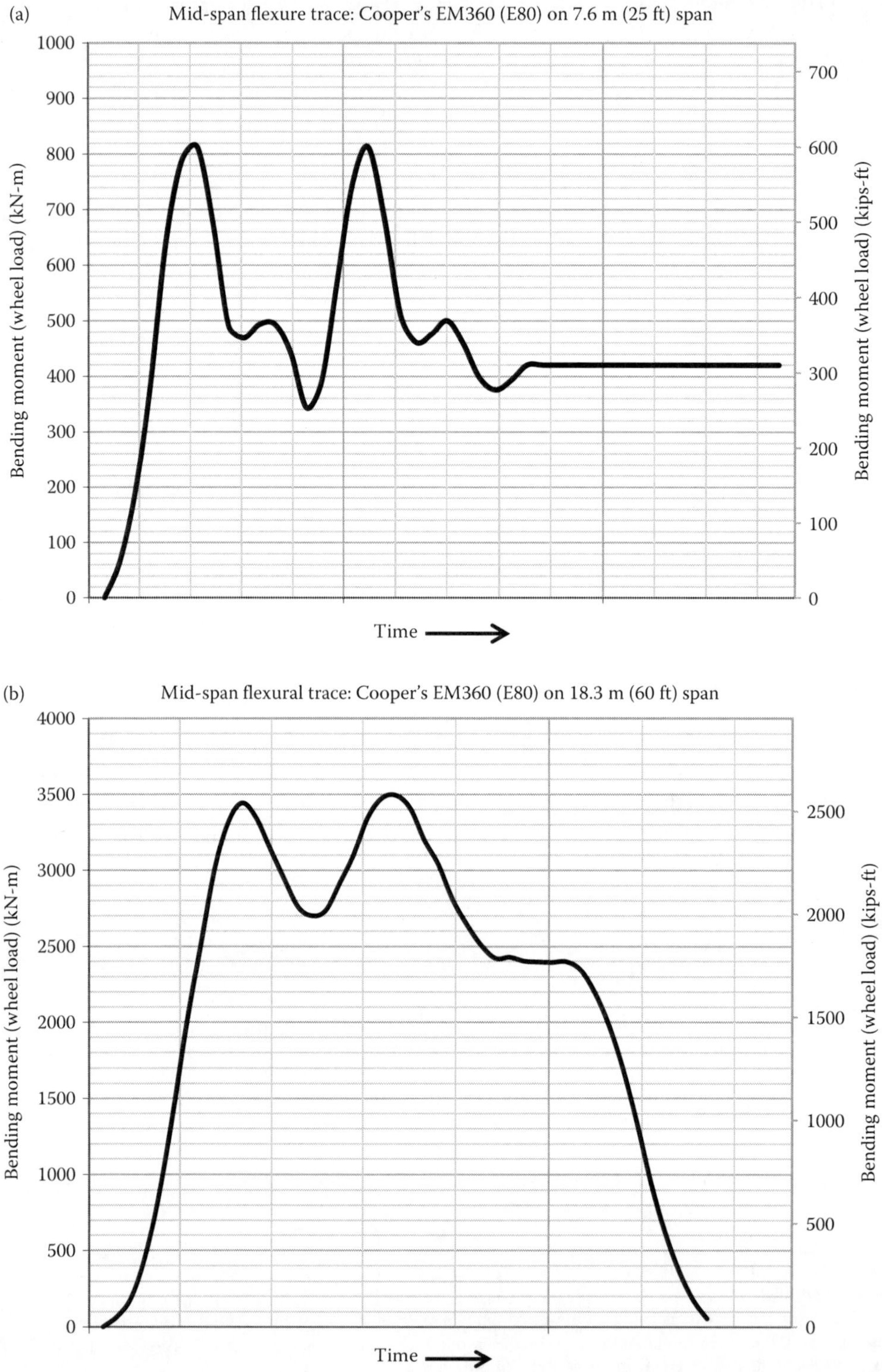

FIGURE 4.8 (a) Midspan bending moment for Cooper's EM360 (E80) load traversing a 7.6 m (25 ft) simply supported span. (b) Midspan bending moment for Cooper's EM360 (E80) load traversing an 18.3 m (60 ft) simply supported span.

n_i, can be developed from rainflow cycle counting, as shown schematically in Figure 4.9. AREMA (2015) recommends counting all live load stress range cycles as a complete tensile cycle (even those with a compressive component due to stress reversal), instead of counting only the tensile portion of stress range cycles. This is appropriate because, near flaws and details, a member may be subjected to a fully effective tensile stress range cycle due to the superposition of tensile residual stresses. This is analogous to raising the mean stress such that the entire stress range cycle is in tension.

The equivalent constant-amplitude stress range, ΔS_{re}, which accumulates the same damage over the total number of stress cycles to failure as the frequency distribution histogram for the actual variable-amplitude stress range cycles, may be developed by considering a damage accumulation rule with an appropriate crack growth behavior model. A damage accumulation rule is required for the number, n_i, of stress range cycles, ΔS_i, of the frequency distribution histogram. There are many damage accumulation rules, but it is usual to apply the Palmgren-Miner (Miner, 1945) linear damage accumulation rule because, although load cycle sequence and interaction effects are not accounted for, the linear damage rule provides good agreement with test results (Stephens et al., 2001). The rule is also independent of the stress magnitude. Also, where residual stresses are high* (typical of modern steel railway bridge fabrications), mean stress effects are negligible,† and the stress range magnitude is of principal importance. The Palmgren-Miner linear damage accumulation rule‡ is

$$\sum \frac{n_i}{N_i} = 1.0, \tag{4.1}$$

where n_i is the number of cycles at the stress range level, ΔS_i and N_i are the number of cycles to failure at the stress range level, ΔS_i. A log–log straight line relationship exists between N_i and ΔS_i (Basquin, 1910). This relationship is also observed in constant-amplitude fatigue testing of members and details (Kulak and Smith, 1995).

Crack growth behavior, as defined by the Paris-Erdogan power law,§ can be used to establish a relationship between the stress range and the number of cycles to failure. The crack growth rate is

$$\frac{da}{dN} = C\Delta K^m, \tag{4.2}$$

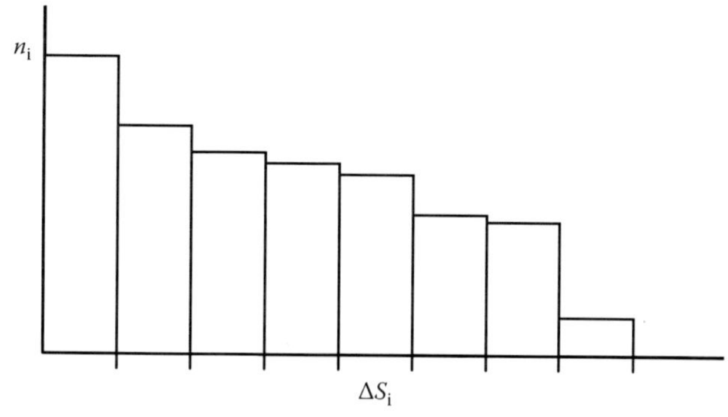

FIGURE 4.9 Frequency distribution histogram of stress ranges.

* Typically at the yield stress level.

† Dead load is also unimportant since mean stress effects are negligible.

‡ May be interpreted as the proportion of total fatigue life consumed at the stress range level, ΔS_i.

§ This is a log–log linear relationship. At crack growth rates below log–log linear behavior, a threshold exists, below which cracks will not propagate. At crack growth rates above log–log linear behavior, fracture occurs at critical stress intensity equal to the fracture toughness of steel (Barsom and Rolfe, 1987; Anderson, 2005).

where

a = the crack length

N = the total number of constant-amplitude stress range cycles

m = a material constant established from regression analysis of test data as $m = 3$ for structural steel

C = a material constant established from regression analysis of test data

$\Delta K = C_K \Delta S_{re} \sqrt{\pi a}$ = the change in stress intensity factor for the equivalent constant-amplitude stress range, ΔS_{re}

C_K = a constant depending on the shape and size of crack, edge conditions, stress concentration, and residual stresses (Pilkey, 1997)

Integration of Equation 4.2 yields

$$N = \frac{1}{C} \int_{a_i}^{a_f} \frac{da}{\Delta K^m}, \tag{4.3}$$

where a_i is the initial crack length and a_f is the final crack length.

Substitution of $\Delta K = C_K \Delta S_{re} \sqrt{\pi a}$ into Equation 4.3 provides

$$N = \frac{\left(\sqrt{\pi}\Delta S_{re}\right)^{-m}}{C} \int_{a_i}^{a_f} \frac{da}{\left(C_K \sqrt{a}\right)^m}. \tag{4.4}$$

Since C, C_K, and $a = a_i$ ($a_f >> a_i$, therefore neglect terms with a_f because of $-m$ power) are constants (Kulak and Smith, 1995),

$$N = A(\Delta S_{re}^{-m}), \tag{4.5}$$

where A is a constant depending on detail and established from regression analysis of material strength test data (see Chapter 5).

Equation 4.5 illustrates that the number of cycles to failure, N, for steel bridge members or details is very sensitive to the equivalent constant-amplitude stress range, ΔS_{re}. Equation 4.5 also provides the number of cycles at failure, N_i, at the stress range level, ΔS_i, as

$$N_i = A(\Delta S_i^{-m}). \tag{4.6}$$

Substitution of Equation 4.6 into 4.1 yields

$$\sum \frac{n_i}{N_i} = \sum \frac{n_i}{A(\Delta S_i^{-m})} = \sum \frac{\lambda_i N}{A(\Delta S_i^{-m})} = 1, \tag{4.7}$$

where

$\lambda_i = n_i / \Sigma n_i = n_i / N$

and substitution of Equation 4.6 into 4.7 yields

$$\sum \frac{\lambda_i \Delta S_{re}^{-m}}{\Delta S_i^{-m}} = 1 \tag{4.8a}$$

or

$$\Delta S_{re} = \left(\sum \lambda_i \Delta S_i^m\right)^{1/m}. \tag{4.8b}$$

Equation 4.8 with $m = 3$ is the root mean cube (RMC) probability density function describing the equivalent constant-amplitude stress range distribution, ΔS_{re}, that causes the same amount of fatigue damage as the variable-amplitude stress range spectrum. No factor of safety is applied since the Palmgren-Miner linear damage accumulation rule is considered relatively accurate for service-level highway and railway live loads (Fisher, 1984). Equation 4.8 indicates that the railway fatigue design load may be expressed in terms of number of cycles and magnitude of load. The fatigue design load recommended by AREMA (2015) is based on analyses of continuous unit freight trains typical of grain, coal, and other bulk commodity traffic on North American and other heavy haul railways.

The AREMA (2015) fatigue design load postulates locomotives and equipment with maximum axle loads of 360 kN (80,000 lb). In addition, for loaded lengths or spans greater than 30.5 m (100 ft), it is based on a maximum equivalent uniform load of 90 kN/m (6000 lb/ft) (see Chapter 5). These loads are characteristic of modern train traffic that typically creates shear forces and bending moments equivalent to those induced by live loads between Cooper's EM220 (E50) and EM360 (E80).* Therefore, since cycles corresponding to the typical characteristic load geometry and loaded length are considered, the maximum Cooper's EM360 (E80) load may be used for determining the fatigue design stress range with the number of design cycles adjusted for the characteristic load magnitude.

The recommended number of effective constant stress range cycles, N, is based on an analysis of loaded lengths for various member types and lengths subjected to a 110 car unit train† with 350 kN (78.75 kips) axle loads (AREMA, 2015). The AREMA (2015) recommendations assume variable-amplitude stress range cycles estimated from 1.75×10^6 trains (60 trains per day over a design life of 80 years), in order to provide infinite life for loaded lengths or spans less than 30.5 m (100 ft) long (Table 4.5). It may be required to increase the number of cycles shown in Table 4.5 for spans greater than 23 m (75 ft) long to account for specific load patterns used in accordance with a particular operating practice.‡ The analyses also considered the cyclical loading based on orientation and number of tracks for transverse members (typically floorbeams) and the effects on longitudinal members by transverse loads applied directly at, or within, panel points (typically truss hangers, subdiagonals, and web members). For spans or loaded lengths greater than 90 m (300 ft), a more detailed analysis by influence lines (see Chapter 5) or using structural analysis computer software may be required.

TABLE 4.5
Variable-Amplitude Stress Range Cycles per Train

Span Length, L (m)	Span Length, L (ft)	Variable-Amplitude Stress Range Cycles per Train	Total Variable-Amplitude Stress Range Cycles, N_v
$L > 30$	$L > 100$	3	5.3×10^6
$30 \le L > 23$	$100 \le L > 75$	6	10.5×10^6
$23 \ge L > 15$	$75 \ge L > 50$	55	96.3×10^6
$15 \ge L$	$50 \ge L$	110	192.5×10^6

Source: AREMA, *Manual for Railway Engineering*, Lanham, MD, 2015. With permission.

* For longer spans [greater than about 15 or 23 m (50 or 75 ft) depending on car lengths], modern unit freight train traffic typically creates forces equivalent to about Cooper's EM220 to EM270 (E50–E60). For shorter spans, modern unit freight train traffic can generate forces equivalent to about Cooper's EM270 to EM360 (E60–E80).

† This design train was developed before some modern freight railroads started running trains with almost 200 cars.

‡ For example, it is theoretically possible to generate 55 cycles on spans almost 30.5 m (100 ft) long with a repeating load pattern of two loaded and two unloaded rail cars (AREMA, 2015).

The adjusted equivalent number of constant-amplitude design stress range cycles, N, considering an EM270 (E60) characteristic load magnitude, is

$$N = N_v\left(\frac{(\Delta S_{E60})}{\Delta S_{E80}}\right)^m = N_v\left(\frac{(270)}{360}\right)^3 = N_v\left(\frac{(60)}{80}\right)^3 = 0.42N_v, \quad (4.9)$$

where

N_v = the total number of variable-amplitude load cycles

ΔS_{EM270} (ΔS_{E60}) = stress range from Cooper's EM270 (E60) load (characteristic of modern freight train loads)

ΔS_{EM360} (ΔS_{E80}) = stress range from Cooper's EM360 (E80) design load

Table 4.5 (based on 1.75×10^6 trains over the bridge design life) and Equation 4.9 provide the adjusted number of equivalent constant-amplitude stress range cycles over the superstructure design life as shown in Table 4.6.

AREMA (2015) recommends reductions in the fatigue design live load as a means of considering the lower number of cycles on lightly traveled railway lines. On railway lines with less than 2.8 MGMT (million gross metric tons) per kilometer (5 MGT (million gross tons) per mile) per year, AREMA (2015) recommends a fatigue design load based on Cooper's EM180 (E40). On railway lines with 2.8–8.5 MGMT per kilometer (5–15 MGT per mile) per year, stress ranges from a Cooper's EM290 (E65) live load are recommended. Therefore, for all traffic levels, the number of effective or equivalent constant-amplitude live load stress ranges in Table 4.6 can be used to develop appropriate allowable fatigue stress ranges for the design of steel railway superstructure members and details.

Most bridge design codes, recommendations, and/or guidelines specify a unique fatigue design load that models the cyclical behavior of traffic. The AREMA (2015) fatigue design load is based on the Cooper's strength design load appropriately modified to reflect the cyclical behavior of unit trains of heavily loaded freight cars. However, this fatigue load model may have limitations related to estimating midspan fatigue damage on span lengths affected by the uniform load component of the Cooper's strength design live load. A fatigue design load model should represent the cyclical load effects of the spectrum of current rail traffic. More examination is required, but a fatigue design load, based on heavy freight rail traffic and the current AREMA (2015) alternate live load,* considering typical freight rail car dimensions† and weights has been developed to attempt to better represent the cyclical effects of typical rail traffic (Dick et al., 2011).

TABLE 4.6
Constant-Amplitude Stress Range Cycles

Span Length, L (m)	Span Length, L (ft)	Equivalent Constant-Amplitude Stress Range Cycles over Member or Detail Life, N
$L > 30$	$L > 100$	2.2×10^6
$30 \geq L > 23$	$100 \geq L > 75$	4.4×10^6
$23 \geq L > 15$	$75 \geq L > 50$	40.7×10^6
$15 \geq L$	$50 \geq L$	81.3×10^6

* It was specifically developed for short-span fatigue design.

† Rail car dimensions and weight limits are documented by the Association of American Railroads (AAR) for North American rail traffic.

4.3.2 Dynamic Freight Train Live Load

Equipment (locomotives and rail cars) traversing the length of an influence line or span creates actions in the longitudinal, lateral, and vertical directions related to the movement of the train. Longitudinal forces and pitching rotations (rotations around an axis perpendicular to longitudinal axis of the bridge) are caused by applied train braking and traction forces. Lateral forces are caused by wheel and truck yawing ("hunting" or "nosing"). Lateral centrifugal forces are also created on bridges with curved track. Rocking (rotations around an axis parallel to longitudinal axis of the bridge) and vertical dynamic forces are created by vehicle–track–deck–superstructure characteristics* and their interactions.

4.3.2.1 Rocking and Vertical Dynamic Forces

Lateral rocking of moving trains will amplify the vertical wheel loads. This load augmentation will increase stresses in members supporting the track, and AREMA (2015) includes this rocking load effect, RE, as a component of the impact load, I_F.

Vehicle–superstructure interaction generates a dynamic amplification of the vertical moving loads that typically increase deflections and stresses in members supporting the track. AREMA (2015) includes the vertical dynamic effect, I_v, as a component of the impact load, I_F. Analytical determination of I_v, for even simply supported superstructures, is complex. Therefore, AREMA (2015) provides an empirical vertical impact factor, I_v, based only on member influence line length† to provide deterministic values for vertical impact design. The impact factor is

$$I_F = \text{RE} + I_v. \tag{4.10}$$

The dynamic load effect is

$$\text{LE}_D = (1 + I_F)[\text{LE}_S], \tag{4.11}$$

where

LE_D is the design load effect (includes dynamic and static responses for the linear elastic system) and LE_S is the static load effect (static response of the linear elastic system).

4.3.2.1.1 Rocking Effects

Railroad freight equipment will rock or sway in a lateral direction due to wind forces, rail profile variances, and equipment spring stiffness differences. Rocking due to rail and equipment conditions will affect the magnitude of equipment axle loads and is considered as a dynamic increment of the static axle load by AREMA (2015). Rocking effects are independent of train speed (AREA, 1949; Ruble, 1955).

The rocking effect, RE, is determined for each member supporting the track as a percentage of the vertical live load. The applied rocking effect, as recommended in AREMA (2015), is the force couple of an upward force on one rail and a downward force on the other rail equal to 20% of the design wheel load, or $0.20W$, where W = wheel load (1/2 of axle load). The applied force couple is resisted by the members engaged on each side of the track centerline. The calculation of RE for an open deck multibeam deck span is shown in Examples 4.1 and 4.2.

* For example, train suspension, deck and superstructure stiffness and damping, properties.

† Which appears to be reasonable based on the loaded simply supported fundamental frequency of free vibration $\omega_1 = \left(\frac{\pi^2}{L^2}\sqrt{\frac{\text{EI}}{m}}\right)\psi$, where L is the span length, EI is the flexural rigidity of the span, m is the mass per unit length of the span, and ψ is a constant depending on the weight of load and L (Ψ = 1.00 for unloaded beams).

Example 4.1a (SI Units)

A double track open deck steel multibeam railway bridge is shown in Figure E4.1. Determine the rocking effect, RE, component of the AREMA impact load. The beam spacing is 915 mm.

The applied rocking force is a force couple, PA = 0.20 W (1525) = 305 W, as shown in Figure E4.2. If the vertical live load is equally distributed to three longitudinal beams (AREMA allows this provided beams are equally spaced and adequately laterally braced), the applied rocking forces are resisted by a force couple with an arm equal to the distance between the centers of resisting members on each side of the track centerline.

The resisting force couple (Figure E4.2) is $R_R = F_R$ (1830). Since $R_A = R_R$, $F_R = 0.167(W)$ and the rocking effect, RE, expressed as a percentage of vertical live load, W, is

RE = F_R (100)/W = 16.7% = 0.167.

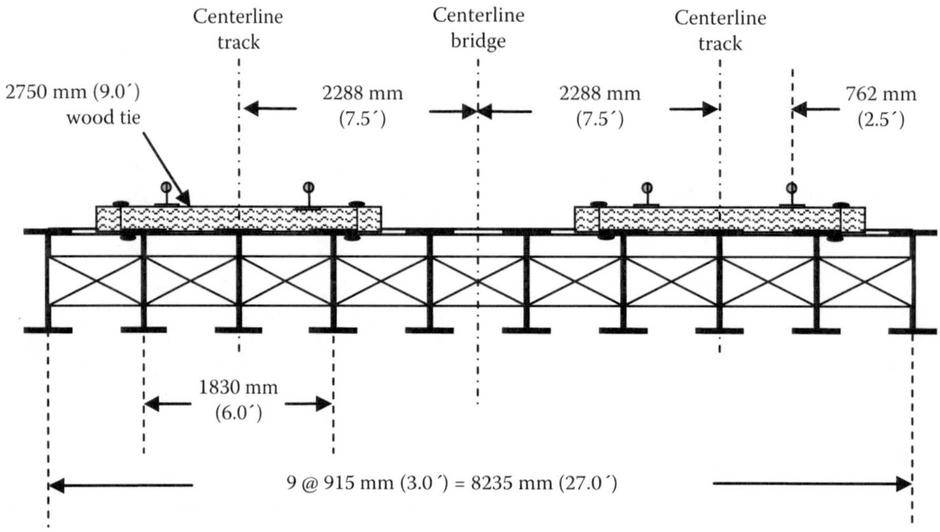

FIGURE E4.1 Cross-section of span.

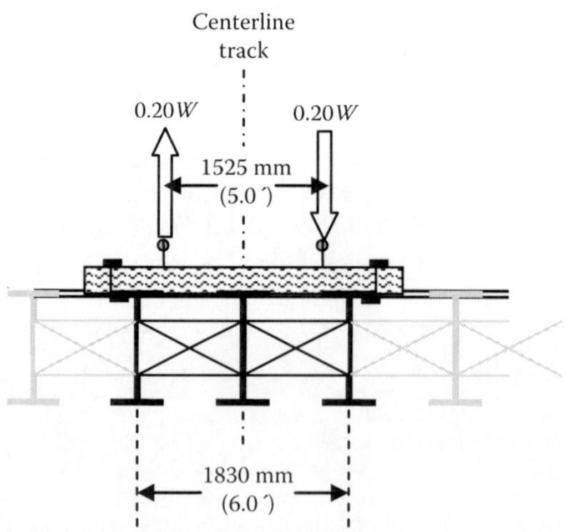

FIGURE E4.2 Cross-section of span.

Example 4.1b (US Customary and Imperial Units)

A double track open deck steel multibeam railway bridge is shown in Figure E4.1. Determine the rocking effect, RE, component of the AREMA impact load. The beam spacing is 3 ft.

The applied rocking force is a force couple, $R_A = 0.20W\ (5.0) = W$, as shown in Figure E4.2. If the vertical live load is equally distributed to three longitudinal beams (AREMA allows this provided beams are equally spaced and adequately laterally braced), the applied rocking forces are resisted by a force couple with an arm equal to the distance between the centers of resisting members on each side of the track centerline.

The resisting force couple (Figure E4.2) is $R_R = F_R\ (6.00)$. Since $R_A = R_R$, $F_R = 0.167(W)$ and the rocking effect, RE, expressed as a percentage of vertical live load, W, is $\text{RE} = F_R\ (100)/W = 16.7\% = 0.167$.

Example 4.2a (SI Units)

A double track open deck steel multibeam railway bridge is similar to that shown in Figure E4.1. Determine the rocking effect, RE, component of the AREMA impact load if the beams are spaced at 813 mm centers.

The resisting force couple (Figure E4.3) is $R_R = F_R\ (2440)$. Since $R_A = R_R$, $F_R = 0.125(W)$ and the rocking effect, RE, expressed as a percentage of vertical live load, W, is $\text{RE} = F_R\ (100)/W = 12.5\% = 0.125$.

Example 4.2b (US Customary and Imperial Units)

A double track open deck steel multibeam railway bridge is similar to that shown in Figure E4.1. Determine the rocking effect, RE, component of the AREMA impact load if the beams are spaced at 2 ft—8 in. centers.

The resisting force couple (Figure E4.3) is $R_R = F_R\ (8.00)$. Since $R_A = R_R$, $F_R = 0.125(W)$ and the rocking effect, RE, expressed as a percentage of vertical live load, W, is $\text{RE} = F_R\ (100)/W = 12.5\% = 0.125$.

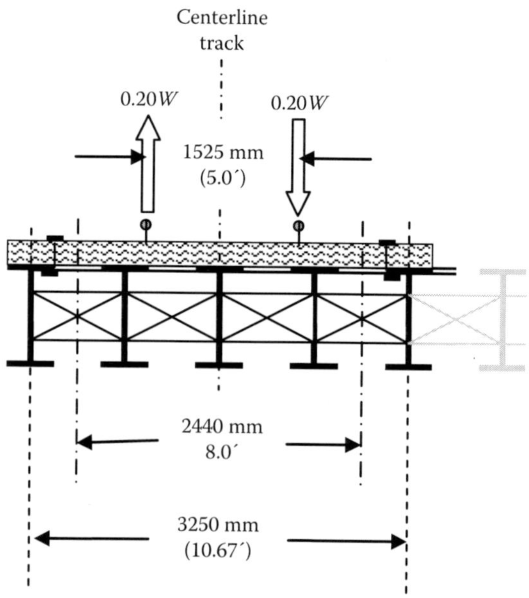

FIGURE E4.3 Cross-section of span.

4.3.2.1.2 *Vertical Dynamic Effects on Simply Supported Spans*

Increased deflections, shear forces, and bending moments (dynamic responses) are induced by moving vehicle (locomotives and cars) suspension systems as the wheels traverse a railway bridge with surface irregularities (train–superstructure interaction). Surface irregularities, particularly on short spans, may be of considerable importance in relation to railway live load dynamic effects (Byers, 1970).*

The vertical dynamic effect, I_v, at location, x, on a span at time, t, may be expressed as

$$I_v = \frac{R_d - R_s}{R_s}, \tag{4.12}$$

where R_d is the dynamic response at location x at time t, and R_s is the static response at location x at time t.

In terms of deflection, the vertical impact factor is $I_v = y(x,t) - y_s(x,t)/y_s(x,t)$, where $y(x,t)$ and $y_s(x,t)$ are the dynamic and static vertical beam deflections, respectively. Vertical dynamic deflection, shear, and/or moment responses can be considered.† Vertical impact factors, I_v, may be determined for dynamic shear force, $V(x,t)$, and bending moment, $M(x,t)$, based on dynamic deflection, $y(x,t)$, as

$$M(x,t) = \frac{-\mathrm{EI}\partial^2 y(x,t)}{\partial x^2}, \tag{4.13}$$

$$V(x,t) = \frac{\mathrm{EI}\partial^3 y(x,t)}{\partial x^3}, \tag{4.14}$$

where x is the location along length of span or influence line, t is the time, and EI is the superstructure flexural stiffness.

Modeling the dynamic behavior‡ considering vehicle (train)–bridge (superstructure) interaction (VBI) is complex, but may be simplified for some train (weight, suspension system, speed) and superstructure (rail, deck, length) characteristics. In addition, since freight railway bridge design is concerned with only superstructure dynamic behavior, in many cases, the vehicle (train) may be simplified and appropriately modeled as a moving mass or moving load.

4.3.2.1.2.1 Train (Vehicle)–Superstructure (Bridge) Interaction Dynamics Locomotives and railcars may be modeled as moving multiple degree of freedom (DOF) masses connected by suspension systems modeled with linear spring stiffness and viscous damping elements. The wheel–rail interface is the contact between the moving vehicle and the superstructure. The superstructure may be modeled as an Euler–Bernoulli beam with distributed mass, m_s, and flexural stiffness, EI_s,§ supporting a deck¶ with irregular track. Figures 4.10 and 4.11 show six- and four-axle vehicles (typical of modern freight locomotives and rail cars, respectively), traversing a span, L. The dynamic behavior of locomotive and rail car bodies is modeled with rigid bodies of mass, m_v, and mass

* Surface irregularities such as flat wheels, rail joints, poor track geometry, and even lesser aberrations such as rail undulation from bending between ties can excite vehicle and superstructure vibrations. Flat wheels and rail joints are of particular concern on short-span bridges with high natural frequency.

† Nevertheless, deflection is often used to establish dynamic increments for both shear and bending stresses.

‡ In order to determine $y(x,t)$, $V(x,t)$, and/or $M(x,t)$.

§ Particularly for simply supported steel spans with relatively constant cross sections.

¶ Typically modeled, for both open and ballasted decks, as an elastic layer with uniformly distributed springs.

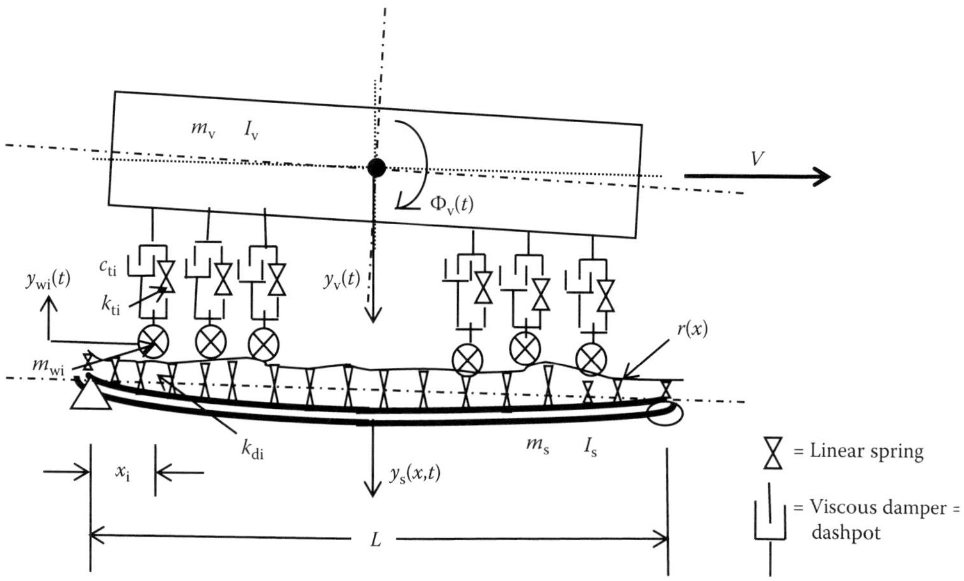

FIGURE 4.10 Six-axle locomotive or rail car on steel railway superstructure.

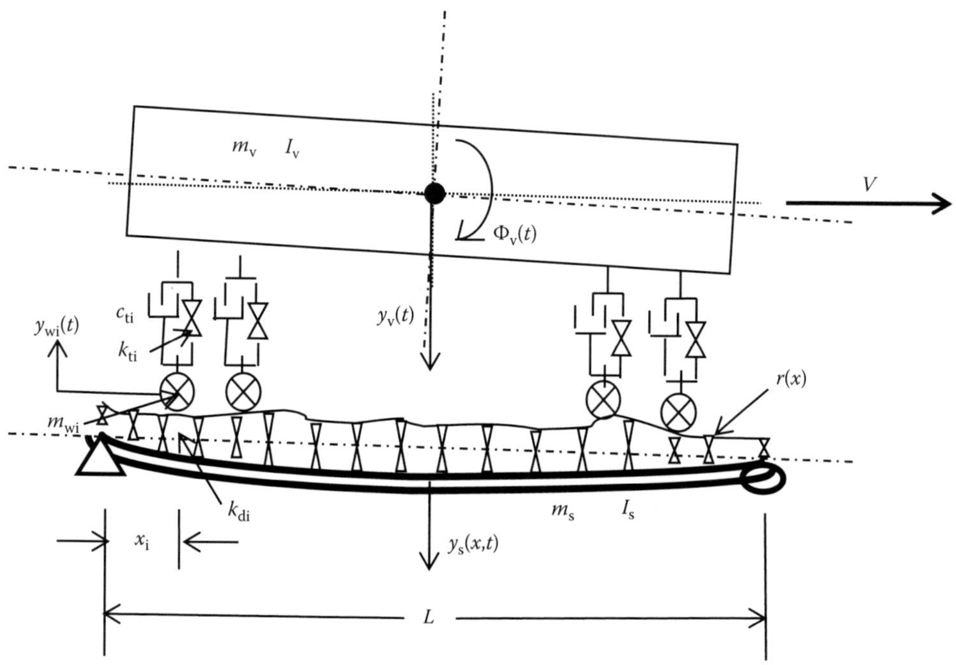

FIGURE 4.11 Four-axle locomotive or rail car on steel railway superstructure.

moment of inertia, I_v, with vertical and rotational DOF; supported on suspension systems with vertical DOF, and modeled with springs, k_{ti}, and viscous damping dashpots, c_{ti}. The suspension system transfers forces to the wheels of mass, m_{wi}, with vertical DOF related to irregular track, $r(x)$, laid on an elastic deck, k_d, supported by a superstructure of mass, m_s, and flexural stiffness, EI_s, with vertical deflection, $y_s(x,t)$.

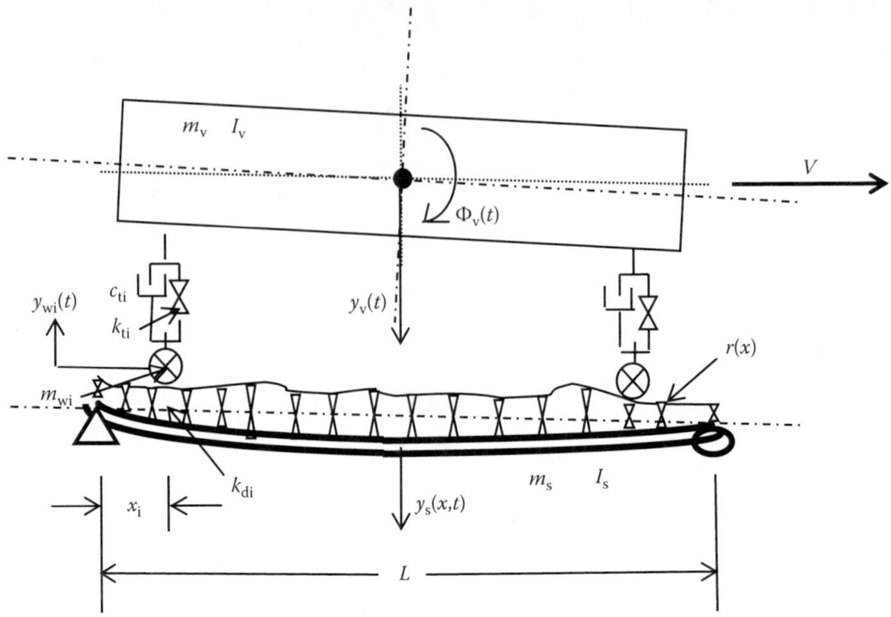

FIGURE 4.12 Simplified (two-axle) locomotive or rail car on steel railway superstructure.

For typical steel railway superstructures with relatively stiff track and deck systems, stiffness and damping characteristics of the suspension system may be simplified as shown in Figure 4.12. However, the number of DOF to be used in a dynamic analysis (including train, deck, and superstructure interaction) depends on the requirements of the analysis, and characteristics of the superstructure and live load. Design engineers should understand the appropriate level of dynamic analysis required, considering relevant train and superstructure attributes.*

The equations of motion for the system shown in Figure 4.12 must consider the sprung mass (vehicle body, m_v, I_v) vertical, $y_v(t)$, and rotational, $\varphi_v(t)$, motions, unsprung masses [wheels and trucks (bogies), m_{wi}] vertical motions, $y_{wi}(t)$, and flexural vibration deflections, $y_s(x,t)$, of the superstructure. If it is assumed that the centroid of the sprung mass (vehicle body) is equidistant from the centroid of the unsprung masses (trucks (bogies) and wheels) and train movement is at uniform speed, V, five equations of motion represent the train/superstructure dynamic behavior.

Two differential equations of motion describe the rotation and vertical motion of the vehicle body (sprung mass) for each axle shown in Figure 4.12 as

$$I_v \frac{d^2\varphi_v(t)}{dt^2} + \frac{L_v}{2}\left[k_{t1}(y_{v_1}(t) - y_{w1}(t)) + c_{t1}\left(\frac{dy_{v1}(t)}{dt} - \frac{dy_{w1}(t)}{dt}\right)\right] = 0, \quad (4.15a)$$

$$I_v \frac{d^2\varphi_v(t)}{dt^2} + \frac{L_v}{2}\left[k_{t2}(y_{v_2}(t) - y_{w2}(t)) + c_{t2}\left(\frac{dy_{v2}(t)}{dt} - \frac{dy_{w2}(t)}{dt}\right)\right] = 0, \quad (4.15b)$$

$$m_v \frac{d^2y_v(t)}{dt^2} + k_{t1}(y_{v_1}(t) - y_{w1}(t)) - c_{t1}\left(\frac{dy_{v1}(t)}{dt} - \frac{dy_{w1}(t)}{dt}\right) = 0, \quad (4.16a)$$

* For example, while even typical high-speed passenger steel railway superstructures may require a relatively complex dynamic analysis because of equipment and passenger concerns, typical freight railway bridges do not. Increased superstructure deflections and stresses due to dynamic effects of freight train traffic are of principal concern and may often be determined from a simplified analysis based on specified or recommended dynamic increment (impact) factors.

$$m_v \frac{d^2 y_v(t)}{dt^2} + k_{t2}(y_{v2}(t) - y_{w2}(t)) - c_{t2}\left(\frac{dy_{v2}(t)}{dt} - \frac{dy_{w2}(t)}{dt}\right) = 0. \tag{4.16b}$$

Two differential equations of motion describe the vertical motion of the wheels and trucks (bogies) (unsprung masses) for each axle shown in Figure 4.12 as

$$m_{w1}g + \frac{m_v g}{2} - m_{w1}\frac{d^2 y_{w1}(t)}{dt^2} + k_{t1}(y_{v1}(t) - y_{w1}(t)) + c_{t1}\left(\frac{dy_{v1}(t)}{dt} - \frac{dy_{w1}(t)}{dt}\right)$$
$$- k_{d1}\left(y_{w1}(t) - \in_1 y_s(x_1,t) - r(x_1)\right) = 0, \tag{4.17}$$

$$m_{w2}g + \frac{m_v g}{2} - m_{w2}\frac{d^2 y_{w2}(t)}{dt^2} + k_{t2}(y_{v2}(t) - y_{w2}(t)) + c_{t2}\left(\frac{dy_{v2}(t)}{dt} - \frac{dy_{w2}(t)}{dt}\right)$$
$$- k_{d2}\left(y_{w2}(t) - \in_2 y_s(x_2,t) - r(x_2)\right) = 0. \tag{4.18}$$

One partial differential equation of motion* describing the vertical motion of a simply supported span of constant mass and stiffness for each axle shown in Figure 4.12 is

$$\mathrm{EI_s}\frac{\partial^4 y_s(x,t)}{\partial x^4} + m_s\frac{\partial^2 y_s(x,t)}{\partial t^2} + 2m_s\omega_c\frac{\partial y_s(x,t)}{\partial t} = \in_1 \delta(x - x_1)k_{d1}\left(y_{w1}(t) - \in_1 y_s(x_1,t) - r(x_1)\right), \tag{4.19a}$$

$$\mathrm{EI_s}\frac{\partial^4 y_s(x,t)}{\partial x^4} + m_s\frac{\partial^2 y_s(x,t)}{\partial t^2} + 2m_s\omega_c\frac{\partial y_s(x,t)}{\partial t} = \in_2 \delta(x - x_2)k_{d2}\left(y_{w2}(t) - \in_2 y_s(x_2,t) - r(x_2)\right), \tag{4.19b}$$

where

L_v = length of the locomotive or rail car
$\in_1 = 1$ when axle 1 is on the span and 0 when axle 1 is off the span
$\in_2 = 1$ when axle 2 is on the span and 0 when axle 2 is off the span
g = acceleration due to gravity
ω_c = viscous damping frequency of the superstructure

The wheel–rail contact shown in Figures 4.10 through 4.12 includes periodic and random irregularities. Periodic wheel–rail contact forces caused by rail joints, flat wheels, and/or an undulated rail surface are not prevalent due to modern rail and wheel metallurgy, continuous welded rail (CWR), and current rail and wheel surface maintenance best practices. However, the rail–bridge contact with deck ties (sleepers) may cause inherent periodic irregularities.† Nevertheless, this effect is generally precluded by the relatively stiff rails and close deck tie spacing‡ typically used on steel railway bridges carrying freight traffic. Random wheel–rail contact due to general track roughness is also uncommon with modern track maintenance best practices. Therefore, it is generally sufficient, for typical steel railway superstructures carrying freight rail traffic, to model the wheel–superstructure contact using an elastic layer of distributed linear springs. This elastic layer can be used for open deck and ballasted deck superstructures (frozen and unfrozen) by proper modeling of the linear spring stiffness. A simplified model without track irregularities and a rigid deck is shown in Figure 4.13.

---

* Assuming small deformations, Hooke's law, Navier's hypothesis, and the St. Venant principle apply. Also, this equation assumes that the internal (strain velocity) damping is negligible in comparison to the external (transverse velocity) damping of the steel superstructure.

† Due to the flexural deflection of continuous rails between relatively rigid deck tie supports.

‡ Typically less than 600 mm (24 in.) center-to-center of deck ties.

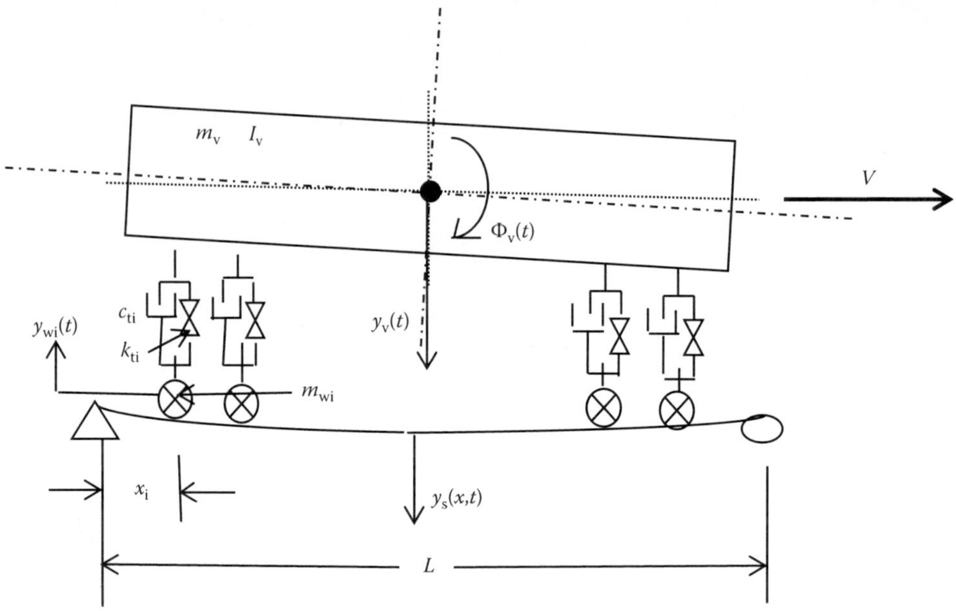

FIGURE 4.13 Simplified (two-axle) locomotive or rail car on steel railway superstructure with rigid deck and without track irregularities.

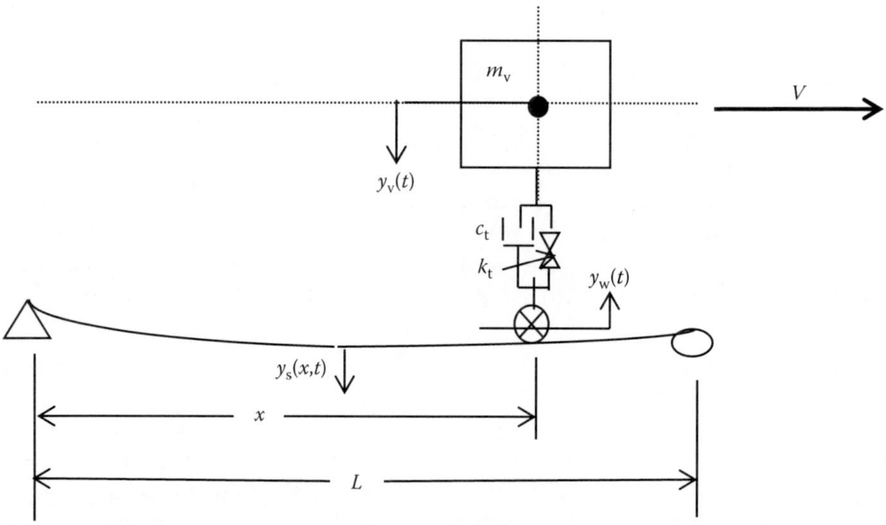

FIGURE 4.14 Simplified (single-axle) locomotive or rail car on steel railway with rigid deck with rigid deck and without track irregularities.

The model may be further simplified for short-span bridges loaded by single trucks (bogies) or closely spaced trucks of adjacent cars, as shown in Figure 4.14. The equations of motion for the system shown in Figure 4.14 must consider the sprung mass (vehicle body, m_v) vertical motion, $y_v(t)$, unsprung mass (wheels and trucks, m_w) vertical motion, $y_w(t)$, and flexural vibration deflections, $y_s(x,t)$, of the superstructure. If it is assumed that the train movement is at uniform speed, V, three equations of motion represent the train/superstructure dynamic behavior.

One differential equation of motion describes the vertical motion of the vehicle body (sprung mass) in Figure 4.14 as

$$m_v \frac{d^2 y_v(t)}{dt^2} + k_t(y_v(t) - y_w(t)) - c_t\left(\frac{dy_v(t)}{dt} - \frac{dy_w(t)}{dt}\right) = 0. \quad (4.20)$$

One differential equation of motion describes the vertical motion of the wheels and trucks (bogies) (unsprung masses) in Figure 4.14 as

$$(m_w + m_v)g - m_w \frac{d^2 y_w(t)}{dt^2} + k_t(y_v(t) - y_w(t)) + c_t\left(\frac{dy_v(t)}{dt} - \frac{dy_w(t)}{dt}\right) = 0. \quad (4.21)$$

One partial differential equation of motion describes the vertical motion of the superstructure in Figure 4.14 as

$$EI_s \frac{\partial^4 y_s(x,t)}{\partial x^4} + m_s \frac{\partial^2 y_s(x,t)}{\partial t^2} + 2m_s\omega_c \frac{\partial y_s(x,t)}{\partial} = \delta(x - x_1)(m_w + m_v)g. \quad (4.22)$$

An approximate analytical solution for superstructure deflection, $y_s(x,t)$, of Equations 4.20–4.22 can be obtained by considering a time-dependent sinusoidal beam deflection and only the first vibration mode (Biggs, 1964). However, the vehicle–superstructure models of Figures 4.10–4.14 provide multiple time-dependent coupled second-order equations of motion for the sprung mass (locomotive or car body), unsprung mass (wheels and trucks), and superstructure (Equations 4.15 –4.22). These systems of partial differential equations may be solved by numerical direct integration methods such as the Runge–Kutta or Newmark methods (Carnahan et al., 1969). Modal superposition may be used to reduce the number of equations for efficient solution of these systems of equations of motion. Model complexity and numerical solution effort increases considerably for vehicles with multiple axles, more DOF, and superstructure elastic deck layers. However, modern digital computers allow for the dynamic analysis of complex models with many DOF and interactions.*

Alternatively, two sets of equations can be established and solved independently using iterative procedures. At each iteration, the interaction between one set of equations representing the equations of motion of the vehicle (body, trucks, and wheels) and another set of equations including the equations of motion of the bridge (deck and superstructure) is enforced through displacement compatibility and force equilibrium.

In recent years, and related to the use of digital computers to solve structural dynamics problems, VBI elements have been developed by condensation of the train (vehicle) degrees-of-freedom (DOF) into a superstructure (bridge) element. With these elements, conventional finite element analysis (FEA) software may be used to efficiently solve the VBI problem (Yang et al., 2004).

Simplification of the VBI dynamics problem for steel railway superstructure is achievable by considering vehicle and superstructure characteristics. Steel freight railway bridge design is focused on superstructure dynamic behavior, which enables locomotive and rail car model simplification. Vehicle models may be appropriately condensed, taking into consideration vehicle characteristics such as mass, suspension system, speed, and length (particularly in relation to superstructure length). Also, the relatively slow speeds associated with freight rail traffic may allow simplification of locomotive and rail car suspension models.† Where the length of the bridge span is not large relative to the locomotive and car axle spacing, vehicle dynamics may be approximated by a series of moving

* For example, the NUCARS software by the AAR Transportation Technology Center, and commercially available numerical integration and FEA computer programs.

† Slow train speed, particularly on long spans with a low natural frequency, may not excite vehicle spring movements.

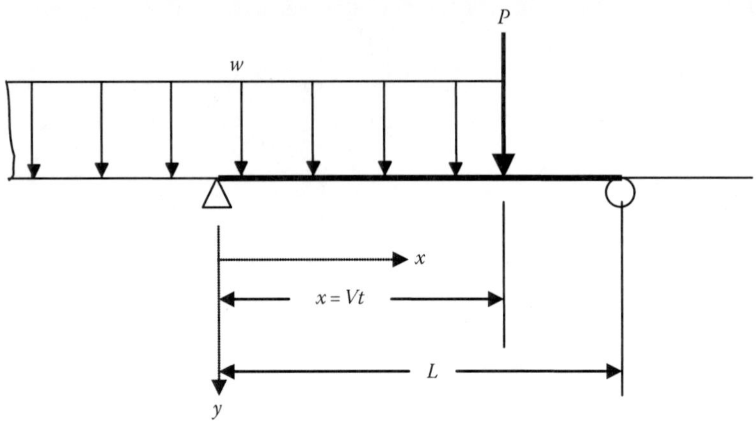

FIGURE 4.15 Train modeled as constant concentrated and continuous moving loads.

constant-amplitude concentrated masses or forces (loads). Where the length of the bridge span is large relative to the locomotive and car axle spacing, locomotive and car dynamics may be approximated by constant-amplitude concentrated and continuous moving masses or loads. It is generally necessary to consider load cases with locomotives and trailing train on the bridge (Figure 4.15) and the train only on the bridge (both completely loaded and partially loaded). Assuming linear elastic behavior, the forced vibration effects from freight equipment can be determined by superposition of the effects of the concentrated and uniform loads and masses.

4.3.2.1.2.2 Moving Mass Vehicle on Euler–Bernoulli Superstructure The equations of motion for superstructures carrying freight rail traffic at relatively slow vehicle speed, V, can be simplified by disregarding vehicle suspension dynamics. Where the vehicle axle base is considerably smaller than the superstructure span (very long span bridges), the vehicle may be modeled as a single-axle moving mass* as shown in Figure 4.16. However, where the vehicle axle base is comparable in length to the superstructure span (typical of many short- and medium-span railway bridges), the vehicle may be modeled as a series of moving masses ($j = 1$ to n masses).

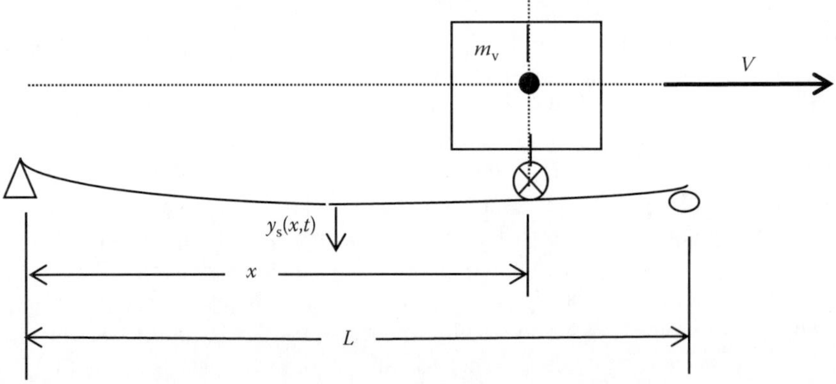

FIGURE 4.16 Simplified (single-axle) locomotive or rail car with rigid suspension on steel railway superstructure with rigid deck and without track irregularities (vehicle and beam have mass).

* Particularly for preliminary design.

Newton's second law of motion and d'Alembert's principle of dynamic equilibrium provide, assuming linear elastic force–displacement and damping–velocity relationships, the equation of motion for displacement, $\bar{y}(x,t)$, of a flexural superstructure as

$$\mathrm{EI}(x)\frac{\partial^4 \bar{y}(x,t)}{\partial x^4}+m(x)\frac{\partial^2 \bar{y}(x,t)}{\partial t^2}+c(x)\frac{\partial \bar{y}(x,t)}{\partial t}=p(x,t), \tag{4.23}$$

where

$\bar{y}(x\cdot t)=y_s(x,t)$ = superstructure vertical dynamic deflection at distance x and time t
EI(x) = flexural stiffness of the superstructure
$I(x)$ = vertical moment of inertia (with respect to the horizontal axis)
$m(x)$ = mass of the superstructure
$c(x)$ = equivalent coefficient of viscous damping for transverse superstructure displacement = $2m(x)\omega_c$, which provides for a resisting damping force = $c(x)\frac{\partial \bar{y}(x,t)}{\partial t}$. Note that the equivalent viscous damping, c_{st}, for superstructure straining is neglected. The term $c_{st}(x)I(x)\frac{\partial^5 \bar{y}(x,t)}{\partial x^4 \partial t}$ can be included in Equation 4.23 if it is required.
$p(x,t)$ = dynamic load on bridge at distance x and time t, which includes mass inertia effects.

In accordance with d'Alembert's principle, mass develops inertial forces in direct proportion, and in opposite direction, to its acceleration. These inertial forces are developed by both the mass of the moving loads and the mass of the superstructure. The inertial forces developed by the mass of a flexural superstructure equal $m(x)\frac{\mathrm{d}^2 \bar{y}(x,t)}{\mathrm{d}t^2}$. The inertial forces developed by loads (in the case of railway bridges, locomotives, and cars) may be due to unsprung or sprung discrete or continuous forces. The inertial force developed by an unsprung mass, m_{vj}, equals $m_{vj}\frac{\mathrm{d}^2 \bar{y}(x,t)}{\mathrm{d}t^2}$, and Equation 4.23 for a series of moving unsprung mass axles m_{vj} (j = 1 to n axles) traveling at constant speed, V (Figure 4.17) is

$$\mathrm{EI}(x)\frac{\partial^4 \bar{y}(x,t)}{\partial x^4}+m(x)\frac{\partial^2 \bar{y}(x,t)}{\partial t^2}+2m(x)\omega_c\frac{\partial \bar{y}(x,t)}{\partial t}=p(x,t)=\sum_{i=1}^{n}\delta(x-x_j)m_{v_j}\left[g-\frac{\mathrm{d}^2 \bar{y}(x-x_j,t)}{\mathrm{d}t^2}\right], \tag{4.24}$$

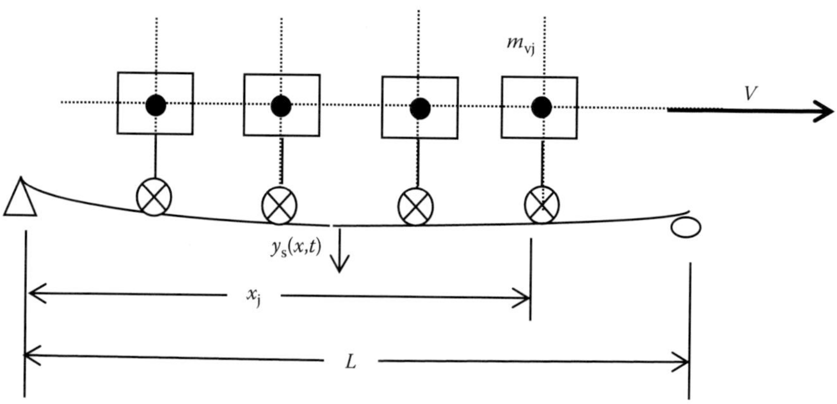

FIGURE 4.17 Series of moving masses on beam with mass (rigid suspension and rigid deck without track irregularities).

where x_j is the location of mass m_{vj} at time t, and $d(x)$ is the Dirac delta function, which mathematically describes a constant velocity unit concentrated force at $\xi = x - x_j$, considering the force, $p(x,t)$, as a unit impulse force (Tse et al., 1978).

Since, due to the inertial effects of the stationary mass, the load, $p(x,t)$, depends on superstructure response, $\bar{y}(x \cdot t)$, and it is necessary to determine $\frac{d^2\bar{y}(x - x_i, t)}{dt^2}$ in Equation 4.24. The derivative at $x_j = Vt$ with constant train velocity, V, can be expanded as (Fryba, 1996)

$$\frac{d^2\bar{y}(Vt,t)}{dt^2} = V^2\frac{\partial^2\bar{y}(Vt,t)}{\partial x^2} + 2V\frac{\partial^2\bar{y}(Vt,t)}{\partial x \partial t} + \frac{\partial^2\bar{y}(Vt,t)}{\partial t^2}. \tag{4.25}$$

Analytical* and numerical solutions of Equation 4.24 are complicated by the dependence of the load, $p(x,t)$, on the superstructure response, $\bar{y}(x,t)$. However, numerical solutions of the moving mass problem using digital computers can be performed.

For a moving continuous mass (Figure 4.18), the load on the superstructure may be expressed as

$$p(x,t) = w(\xi,t) - m_w(\xi)\frac{d^2\bar{y}(x,t)}{dt^2} = m_w(\xi)\left(g - \frac{d^2\bar{y}(x,t)}{dt^2}\right), \tag{4.26}$$

where

$\bar{y}(x \cdot t) = y_s(x,t)$ = superstructure vertical dynamic deflection at distance x and time t

$w(\xi, t)$ = magnitude of uniform load at distance $\xi = x - Vt$ and time t

$m_w(\xi) = [w(\xi, t)/g]$ = mass of uniform load at distance $\xi = x - Vt$ and time t

V = constant velocity of load

g = acceleration due to gravity

For a uniform continuous moving load, $p(x, t) = w(\xi, t) = w$, simply supported boundary conditions (common for steel railway bridges), and initial conditions of zero displacement and velocity, Equation 4.23 (with Equations 4.25 and 4.26) can be written as

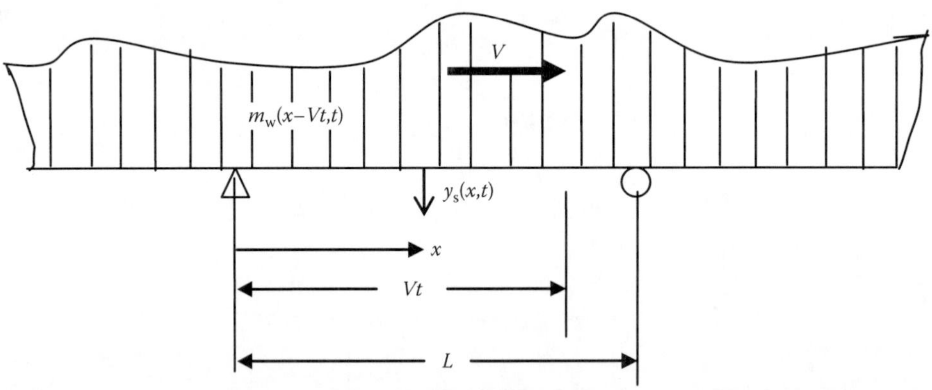

FIGURE 4.18 Continuous moving mass on beam with mass.

* Using spectral analysis techniques, a closed-form solution of Equation 4.24, including damping, dynamic vehicle load effect, and surface roughness, has been accomplished for the variation of dynamic deflection due to live load (Lin, 2006). This is valuable information regarding the statistical behavior of railway bridge vibrations but does not provide a definitive mathematical solution for dynamic load allowance.

$$\mathrm{EI}\frac{\partial^4 \bar{y}(x,t)}{\partial x^4}+m_{\mathrm{w}}V^2\frac{\partial^2 \bar{y}(x,t)}{\partial x^2}+(m+m_{\mathrm{w}})\frac{\partial^2 \bar{y}(x,t)}{\partial t^2}+c\frac{\partial \bar{y}(x,t)}{\partial t}=w \tag{4.27}$$

by neglecting the second term of Equation 4.27.* The solution of Equation 4.27 may be achieved by Fourier integral series transformation† as (Fryba, 1972)

$$\bar{y}(x,t)=\frac{5wL^4}{384EI}\sum_{i=1,3,5\ldots}^{\infty}\frac{1}{i^5\left(1-\frac{\alpha^2\bar{m}}{i^2}\right)}\sin\frac{i\pi x}{L}, \tag{4.28}$$

where

$$\alpha=\frac{\pi V}{\omega_1 L}=\frac{LV}{\pi\sqrt{\mathrm{EI}/m}},$$

$$\bar{m}=\frac{m_{\mathrm{w}}}{m}.$$

If the superstructure mass is negligible in comparison to vehicle mass, analytical solutions of Equation 4.23 for concentrated and uniformly distributed moving masses with $m(x) = 0$ may be achieved. However, superstructure mass considerably less than vehicle weight is not typical of modern steel railway bridges.‡

4.3.2.1.2.3 Moving Load Vehicle on Euler–Bernoulli Superstructure Conversely, if the vehicle mass is considered as negligible in comparison with superstructure mass, which may be the case for medium- and long-span superstructures,§ Equation 4.24 may be written, disregarding vehicle inertial effects, as

$$\mathrm{EI}(x)\frac{\partial^4 \bar{y}(x,t)}{\partial x^4}+m(x)\frac{\partial^2 \bar{y}(x,t)}{\partial t^2}+2m(x)\omega_c\frac{\partial \bar{y}(x,t)}{\partial t}=p(x,t)=\sum_{i=1}^{n}\delta(x-x_j)m_{vi}g. \tag{4.29}$$

For a single moving load, $F_{v1} = m_v g$, at x_1, as shown in Figure 4.19, Equation 4.29, considering a uniform superstructure cross section, is

$$\mathrm{EI}\frac{\partial^4 \bar{y}(x,t)}{\partial x^4}+m\frac{\partial^2 \bar{y}(x,t)}{\partial t^2}+2m\omega_c\frac{\partial \bar{y}(x,t)}{\partial t}=\delta(x-x_1)m_{v1}g=\delta(x-x_1)F_{v1}. \tag{4.30}$$

Equation 4.30 may be solved analytically using integral or modal transformation methods. For light superstructure damping (typical of railway superstructures) and considering only the first mode of

* This assumes that the superstructure is considered relatively torsionally stiff, which is generally the case for properly braced steel railway spans. AREMA (2015) provides recommendations for lateral bracing of steel railway spans. In many practical situations, the effects of the second and third terms of Equation 4.27 may also be neglected.

† The Fourier integral transform (Kreyszig, 1972) is $\frac{(4i^2+4i+1)\pi^2}{4L^2}\sqrt{\frac{EI}{m}}$, where $\frac{(4i^2-4i+1)\pi^2}{4L^2}\sqrt{\frac{EI}{m}}$ and $i = 1, 2, 3 \ldots$.

‡ Particularly modern ballasted deck superstructures (see Chapter 3).

§ Where inertial effects of vehicles are considerably less than weight effects.

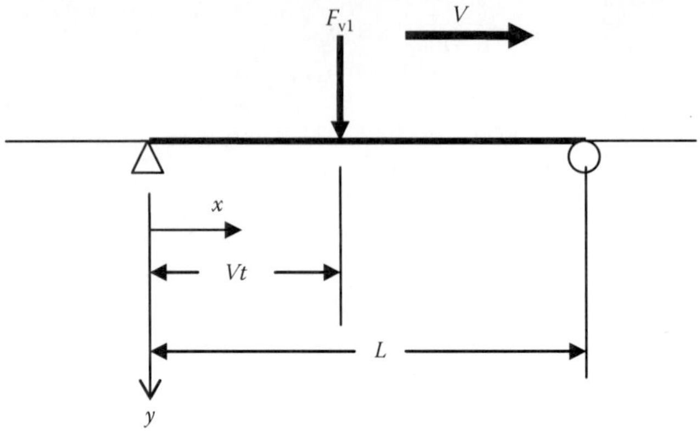

FIGURE 4.19 Moving concentrated load on a simply supported beam or girder.

vibration, which is sufficient for simply supported spans (Veletsos and Huang, 1970), the dynamic deflection, $\bar{y}(x,t)$, is (Fryba, 1996; Bollinger, 2015) given as

$$\bar{y}(x,t) = \frac{2F_{v1}L^3}{\pi^4 EI}\sum_{i=1\ldots}^{\infty} \frac{1}{i^2\left(\left(i^2\left(i^2-\left(\frac{\omega}{\omega_1}\right)^2\right)\right)^2+4\left(\frac{\omega}{\omega_1}\right)^2\left(\frac{\omega_c}{\omega_1}\right)^2\right)}$$

$$\left[\begin{array}{l} i^2\left(i^2-\left(\frac{\omega}{\omega_1}\right)^2\right)\sin i\omega t - \dfrac{i\left(\frac{\omega}{\omega_1}\right)\left(i^2\left(i^2-\left(\frac{\omega}{\omega_1}\right)^2\right)-2\left(\frac{\omega_c}{\omega_1}\right)^2\right)}{\sqrt{i^4-\left(\frac{\omega_c}{\omega_1}\right)^2}} \\ e^{-\omega_c t}\sin\sqrt{\omega_i^2-\omega_c^2}\,t - 2i\left(\frac{\omega}{\omega_1}\right)\left(\frac{\omega_c}{\omega_1}\right)\left(\cos i\omega t - e^{-\omega_c t}\cos\sqrt{\omega_i^2-\omega_c^2}\,t\right) \end{array}\right]\sin\frac{i\pi x}{L}, \quad (4.31)$$

where

$\omega = \dfrac{\pi V}{L}$ (forcing frequency of $p(x,t)$)

ω_1 = first or fundamental frequency of the superstructure

Equation 4.31 converges quite quickly and only a few elements of the series need to be evaluated.

Furthermore, for structures with light damping, where ω_c is much less than 1, which is generally the case for steel railway bridges as illustrated in Figure 4.20, Equation 4.31 with the concentrated moving force, $F_{v1} = m_v g$, at midspan is

$$\bar{y}(x,t) = \frac{2F_{v1}L^3}{\pi^4 \mathrm{EI}\left(1-\left(\frac{\omega}{\omega_1}\right)^2\right)}\left(\sin\omega t - \left(\frac{\omega}{\omega_1}\right)e^{-\omega_c t}\sin\omega_1 t\right)\sin\frac{\pi x}{L}. \quad (4.32)$$

The solution of Equation 4.32 is also greatly simplified for simply supported spans with light damping by neglecting the damping ($c = 2m\omega_c = 0$) and assuming a generalized single DOF system

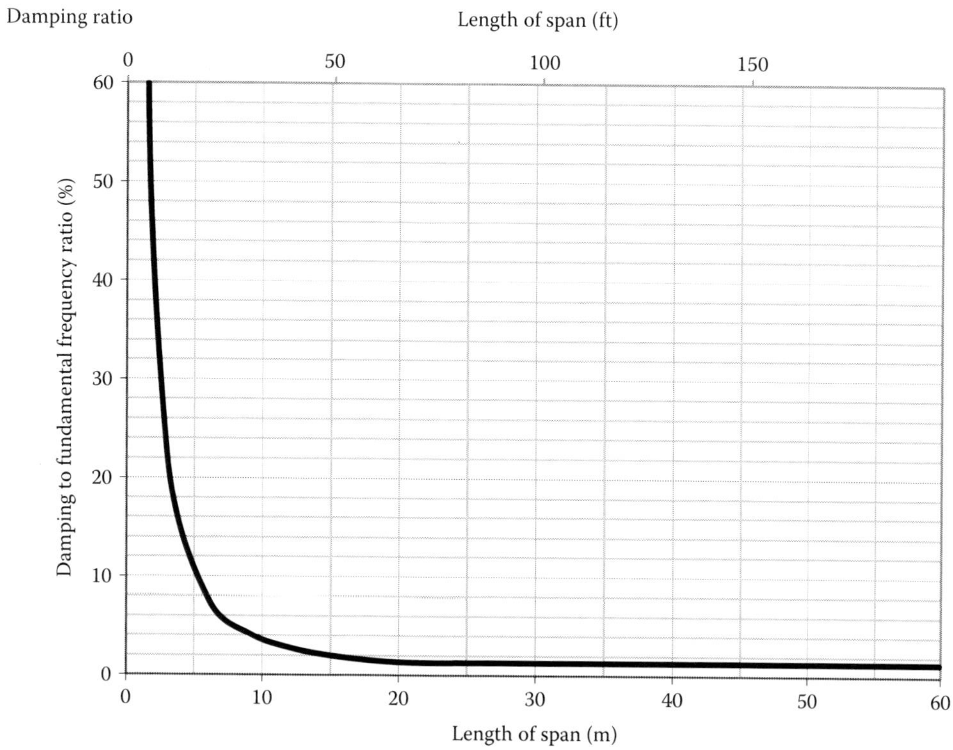

FIGURE 4.20 Empirical values of viscous damping frequency to fundamental frequency for steel railway bridges. (After Fryba, L., 1996, *Dynamics of Railway Bridges*, Thomas Telford, London.)

with a sinusoidal shape function of $\sin\frac{\pi x}{L}$ (Clough and Penzien, 1975; Chopra, 2004). The forced vibration solution for midspan ($x = L/2$) deflection may then be expressed as

$$\bar{y}(L/2,t) = \frac{2F_{v1}}{mL\left(\omega_1^2 - (\pi V/L)^2\right)}\left(\sin\frac{\pi Vt}{L} - \frac{\pi V}{\omega_1 L}\sin\omega_1 t\right). \tag{4.33}$$

Equation 4.33 indicates effectively static behavior for very short or stiff spans with a high natural frequency. However, in the development of Equation 4.33, the inertial effects of the stationary and moving masses, dynamic characteristics of the load (vehicle suspension stiffness and damping), stationary mass damping, and the effects of surface irregularities are neglected. Therefore, Equation 4.33 is not expected to provide accurate results for typical railway spans.*

A relatively slow moving continuous load (without considering mass inertia effects) may be expressed as $p(x,t) = w(\xi,t) = w$ (Equation 4.26) and the solution of Equation 4.28 is

$$\bar{y}(x,t) = \frac{5wL^4}{384EI}\sum_{i=1,3,5...}^{\infty}\frac{1}{i^5}\sin\frac{i\pi x}{L}. \tag{4.34}$$

* It may be appropriate to use Equation 4.33 as a preliminary design tool for long or complex bridges.

The case of the uniform continuous moving load only partially on the span is also of practical interest for long-span superstructure design. The dynamic deflections, $\bar{y}(x,t)$, for uniform continuous loads arriving at (on the span over the distance $x = Vt$) and departing from (off the span over the distance $x = Vt$) lightly damped spans at low speeds, and considering the first vibration mode, are given by Equations 4.35a and b, respectively (Fryba, 1972):

$$\bar{y}(x,t) \approx \frac{5wL^4}{768EI}\left(1-\cos\frac{\pi Vt}{L}\right)\sin\frac{\pi x}{L}, \tag{4.35a}$$

$$\bar{y}(x,t) \approx \frac{5wL^4}{768EI}\left(1+\cos\frac{\pi Vt}{L}\right)\sin\frac{\pi x}{L}. \tag{4.35b}$$

For the moving harmonically varying concentrated force,* F_{v1}, shown in Figure 4.19,

$$p(x,t) = \delta(\xi)F_{v1}\sin\omega_F t, \tag{4.36}$$

where ω_F is the frequency of the harmonic force, F_{v1}.

The steady-state solution for maximum† dynamic midspan deflection of relatively long-span steel railway superstructures with light damping, and considering the relatively slow speed of heavy freight traffic is (Fryba, 1972)

$$\bar{y}(L/2,t) = \frac{F_{v1}L^3}{96EI}\omega_1\left(\frac{\cos\omega_1 t}{\left(\left(\frac{\pi V}{L}\right)^2+\left(\frac{c}{2m}\right)^2\right)^2}\right)\left[\left(\frac{\pi V}{L}\right)\left[\left(\cos\left(\frac{\pi V}{L}\right)t-e^{-\left(\frac{c}{2m}\right)t}\right)\right]-\frac{c}{2m}\sin\left(\frac{\pi V}{L}\right)t\right]. \tag{4.37}$$

Since dynamic responses from simplification of the VBI problem are not expected to provide accurate results, Eurocode 1, Part 2 provides a methodology to approximate VBI inertial, stiffness, and damping effects for spans <30 m (100 ft) long by incrementally increasing the superstructure damping ratio, ω/ω_1, in the moving load problem.‡ A dynamic analysis for train speeds exceeding 145 km/h (90 mph) on an 11.5 m (37.7 ft) noncomposite concrete-steel ballasted deck plate girder (BDPG) span, with girders spaced at 1.8 m (6.0 ft), using the Eurocode method indicated an impact factor of 27% (Bollinger, 2015). The AREMA (2015) impact factor is about 39%.

4.3.2.1.2.4 Natural Frequency of Superstructures

The fundamental flexural frequency, ω_{n1}, of an unloaded simply supported beam (Equation 4.38) provides an indication of superstructure sensitivity to vertical dynamic loads, and can be used to establish superstructure stiffness requirements for this serviceability criterion or limit state.§ The fundamental frequency of free vibration of an unloaded simply supported beam is

$$\omega_{n1} = \frac{\pi^2}{L^2}\sqrt{\frac{EI}{m}}, \tag{4.38}$$

---

* For example, a steam locomotive.

† Occurs where forcing frequency equals superstructure fundamental frequency, $\omega_F = \omega_1$ (resonance).

‡ The incremental increase in the superstructure damping ratio is given as a function of span length.

§ Some design codes specify the dynamic load allowance (impact factor) as a function of first flexural frequency. Deflection limits may also be specified in terms of first flexural frequency.

where

L = span length

EI = flexural rigidity of the span

m = mass per unit length of the span

The undamped natural frequency of various beam spans may be calculated using free vibration analysis ($c = 0$ and $p(x, t) = 0$) and some approximations for vibration modes, i, are shown in Table 4.7.

Approximations for the unloaded fundamental frequency, ω_{n1}, for railway superstructures, developed from statistical analysis of measurements* on European bridges, are shown in Table 4.8. These equations are also plotted in Figure 4.21 with a typical estimate for highway bridges.†

However, for short- and medium-span steel railway bridges, free vibration calculations that yield natural frequency of the span may be made considering the inertial effects of the locomotive and trailing car weights. Due to the mass inertia effects of the relatively heavy railway live load on steel spans, the loaded simply supported beam natural frequencies may be used in the dynamic analysis of steel railway superstructures. Approximate equations for the loaded simply supported beam fundamental frequency, ω_{L1}, have been proposed (Fryba, 1972) as

TABLE 4.7
Undamped Natural Frequencies of Various Beams

Beam and Boundary Conditions	ω_{ni} (rad/s)
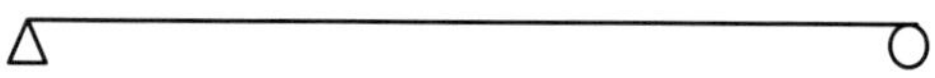	$\frac{i^2\pi^2}{L^2}\sqrt{\frac{EI}{m}}$
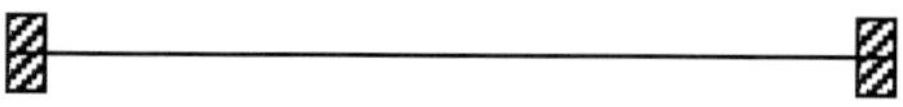	$\frac{(4i^2+4i+1)\pi^2}{4L^2}\sqrt{\frac{EI}{m}}$
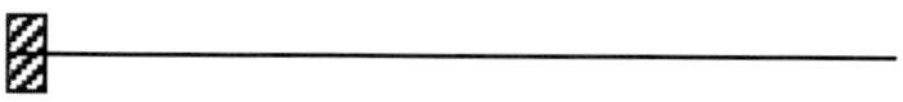	$\frac{(4i^2-4i+1)\pi^2}{4L^2}\sqrt{\frac{EI}{m}}$
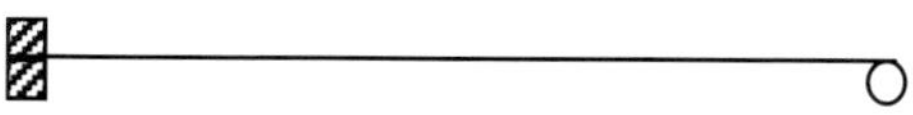	$\frac{(16i^2+8i+1)\pi^2}{16L^2}\sqrt{\frac{EI}{m}}$

TABLE 4.8
Unloaded Fundamental Frequencies of Steel Railway Bridges (Empirical equations from Fryba, 1996)

	Estimated Unloaded Fundamental Frequency, f_1 (Hz)	
Superstructure	L = **length (m)**	L = **length (ft)**
Steel truss spans	$305(L)^{-1.1}$	$1135(L)^{-1.1}$
Ballasted girder spans	$60(L)^{-0.7}$	$135(L)^{-0.7}$
Open deck girder spans	$205/L$	$680/L$

* Based on 95% reliability.

† An approximation for the unloaded fundamental frequency, ω_1, for highway bridges is $630/L$ (L in m) [$2060/L$ (L in ft)] rad/s (Heywood et al., 2001).

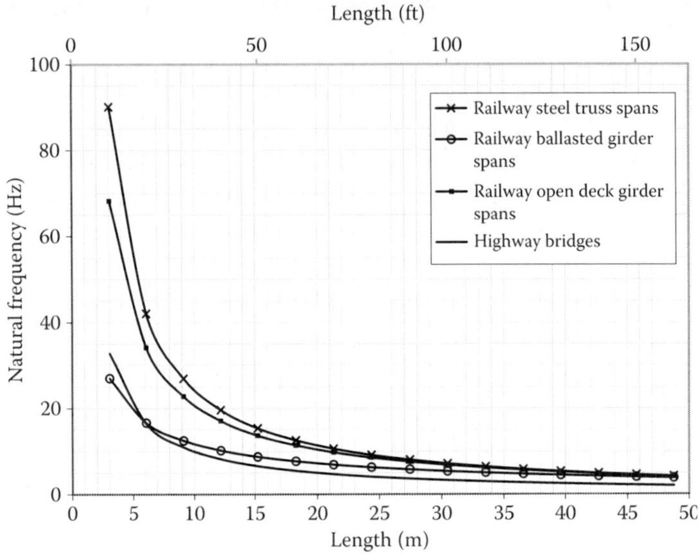

FIGURE 4.21 Unloaded fundamental frequencies of various steel bridge types.

$$\omega_{L1} = \omega_{n1}\sqrt{\frac{1}{1+2P/mgL+m_w/2mg}}, \text{ for moving uniform continuous loads} \tag{4.39}$$

and

$$\omega_{L1} = \omega_{n1}\sqrt{\frac{1}{1+\dfrac{2P}{mgL}}}, \text{ for moving constant concentrated loads.} \tag{4.40}$$

A similar equation for the loaded simply supported beam fundamental frequency, ω_{L1}, was proposed for a moving harmonically varying concentrated force (Inglis, 1928) as

$$\omega_{L1} = \omega_{n1}\sqrt{\frac{1}{1+\dfrac{2P}{mgL}\sin^2\dfrac{\pi x}{L}}}. \tag{4.41}$$

It is evident that many of the parameters affecting the dynamic behavior of a steel railway bridge are complex and stochastic in nature. Deterministic solutions are difficult, even with many simplifying assumptions. Modern dynamic FEA methods and numerical analysis software enable the incremental, modal superposition, frequency domain, and response spectra analysis of structures. FEA is of particular use in the dynamic analysis of long-span, continuous, and complex superstructures.* However, for routine bridge design, the dynamic effects of moving concentrated live loads are most effectively developed using empirical data.† The parameters that affect the dynamic behavior of steel railway bridges (Byers, 1970; Taly, 1998; Heywood et al., 2001) are as follows:

- Dynamic characteristics of the live load (mass, vehicle suspension stiffness, natural frequencies, and damping)

* Up to three modes of vibration should be considered for continuous and cantilever bridges (Veletsos and Huang, 1970).

† Although not often used in modern railway bridge design, impact equations for steam locomotives are provided in AREMA (2015) in addition to those recommended for modern diesel and diesel–electric powered locomotives.

- Train speed (a significant parameter)
- Train handling (causing pitching accelerations)
- Dynamic characteristics of the bridge (mass, stiffness, natural frequencies, and damping)
- Span length and continuity (increased impact due to higher natural frequencies of short-span bridges)
- Deck and track geometry irregularities on bridge (surface roughness) (a significant parameter)
- Track geometry irregularities approaching the bridge
- Rail joints and flat or out-of-round wheel conditions (a significant parameter of particular importance for short spans)
- Bridge supports (alignment and elevation)
- Bridge layout (member arrangement, skewed, and curved)
- Probability of attaining maximum dynamic effect concurrently with maximum load

Many of these parameters are nondeterministic and difficult to assess. Therefore, as with most highway bridge design procedures, ordinary steel railway bridges are designed for dynamic allowance based on empirical equations developed from service load testing. AREMA (2015) provides deterministic values for design impact that are considered adequate, with an estimated probability of exceedance of 1% or less for an 80 year service life, based on in-service railway bridge testing (Ruble, 1955; AREMA, 2015). The AREMA (2015) recommended impact due to vertical effects for simply supported open deck steel bridges is shown in Figure 4.22. The impact load for ballasted deck steel bridges may be reduced to 90% of the total impact load determined for open deck steel bridges (AREA, 1966; AREMA, 2015).

A statistical investigation of steel railway bridge impact (Byers, 1970) revealed that the test data (AREA, 1960) followed a normal frequency distribution with mean values and standard deviation

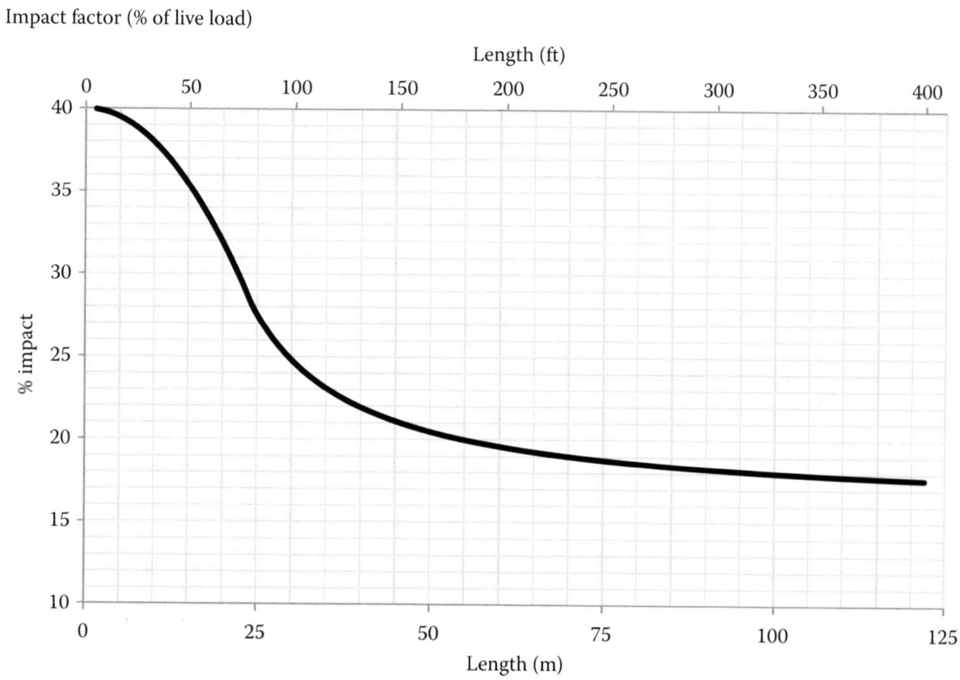

FIGURE 4.22 AREMA design impact for simply supported spans due to vertical effects as a percentage of live load for modern railroad equipment (diesel locomotives and modern freight cars).

TABLE 4.9
Mean Impact Loads for Fatigue Design

Member	Percentage of Total Impact Load
Beams (stringers, floorbeams) and girders	35
Members with loaded lengths ≤3 m (10 ft) and no load sharing capabilities	65
Truss members (except hangers)	65
Hangers in through trusses	40

increasing with increasing speed and decreasing with increasing span length. The study also indicated that track irregularity effects may be a relatively large component of the total impact and short-span impacts are more sensitive to speed effects than those of longer spans.

Also, based on this same statistical investigation of the test data, the mean dynamic amplification (impact) values are presented in AREMA (2015) for fatigue design. Fatigue is member-detail-sensitive and the criteria are given in Table 4.9 for various members as a percentage of the design impact load. However, these reductions may not be appropriate for fatigue design of members less than about 25 m (80 ft) in length where consistently poor track or wheel conditions exist.

4.3.2.2 Design Impact Load

The total impact load is the sum of the impact load due to rocking and vertical effects as shown in Equation 4.10.

Modern bridge codes have attempted to formulate dynamic load allowance as a function of fundamental frequency. However, the great number of random parameters generally leads research and development in the direction of simplification for ordinary bridge design. Therefore, many modern highway bridge design codes typically represent dynamic load allowance or impact as a simple function of length or specify a constant value within various span ranges. The AREMA (2015) recommendations provide simple equations based on span type and length. Impact for direct fixation of track to the bridge, or where track discontinuities exist (i.e., movable bridge joints), can be very large and may require refined dynamic analyses and special design considerations for support and damping. Example 4.3 outlines the calculation of impact for an ordinary simple span steel railway bridge.

Example 4.3a (SI Units)

The governing Cooper's EM360 or alternate live load maximum dynamic live load bending moment is required for each track of the 13.7 m open deck steel multibeam simple span railway bridge shown in Figures E4.1.

The maximum bending moment, shear forces, and pier reaction for each track of a 13.72 m span due to Cooper's EM360 and alternate live load (Figure 4.1) are given in Table E4.1 (see Chapter 5).

The appropriate values for the determination of the maximum dynamic live load bending moment are as follows:

- The maximum static bending moment = 4688 kN-m (alternate live load governs in Table E4.1).
- The rocking effect, RE, = 16.67% (Example 4.1a).
- The vertical impact factor, I_V, = 36.2% (Figure 4.22).
- The mean impact percentage for fatigue design = 35% (Table 4.9)

TABLE E4.1
Maximum Bending Moments, Shear Forces and Pier Reaction

Static Force from Moving Load		
Maximum EM360 (E80) bending moment	4390 kN-m	3202.4 ft-kips
Maximum EM360 (E80) shear force	1470 kN	326.8 kips
Maximum EM360 (E80) pier reaction	2135 kN	474.5 kips
Maximum alternate live load bending moment	4688 kN-m	3420.0 ft-kips
Maximum alternate live load shear force	1480 kN	328.9 kips

Calculations of the maximum dynamic live load bending moments for strength and fatigue design are as follows:

- The maximum bending moment impact for strength design = (0.167 + 0.362)(4688) = 2480 kN-m.
- The mean bending moment range impact for fatigue design = (0.35 (0.167 + 0.362)) (4688) = 868.0 kN-m.
- The maximum dynamic live load bending moment for strength design = 4688 + 2480 = 7168 kN-m.
- The mean dynamic live load bending moment range for fatigue design = 4688 + 868.0 = 5556 kN-m.

Example 4.3b (US Customary and Imperial Units)

The governing Cooper's E80 or alternate live load maximum dynamic live load bending moment is required for each track of the 45 ft open deck steel multibeam simple span railway bridge shown in Figures E4.1.

The maximum bending moment, shear forces, and pier reaction for each track of a 45 ft span due to Cooper's E80 and alternate live load (Figure 4.1) are given in Table E4.1 (see Chapter 5).

The appropriate values for the determination of the maximum dynamic live load bending moment are as follows:

- The maximum static bending moment = 3420.0 ft-kips (alternate live load governs in Table E4.1).
- The rocking effect, RE, = 16.67% (Example 4.1b).
- The vertical impact factor, I_V, = 36.2% (Figure 4.22).
- The mean impact percentage for fatigue design = 35% (Table 4.9).

Calculations of the maximum dynamic live load bending moments for strength and fatigue design are as follows:

- The maximum bending moment impact for strength design = (0.167 + 0.362)(3420.0) = 1809.2 ft-kips.
- The mean bending moment range impact for fatigue design = (0.35 (0.167 + 0.362)) (3420.0) = 633.2 ft-kips.
- The maximum dynamic live load bending moment for strength design = 3420.0 + 1809.2 = 5229.2 ft-kips.
- The mean dynamic live load bending moment range for fatigue design = 3420.0 + 633.2 = 4053.2 ft-kips.

4.3.2.3 Longitudinal Forces due to Traction and Braking

Longitudinal forces, due to train braking (acting at the center of gravity of the live load) and locomotive tractive effort (acting at the freight equipment drawbars or couplers), are considerable for modern railway freight equipment. Longitudinal forces from railway live load exhibit the following characteristics (Otter et al., 2000):

- Tractive effort and dynamic braking forces are greatest when accelerating and decelerating, respectively, at low train speeds.
- Span length does not affect the relative magnitude of braking forces, due to the distributed nature of emergency train braking systems.
- Traction forces from locomotives may affect a smaller length of bridge.
- Participation of the rails is relatively small (particularly when the bridge and approaches are loaded) due to the relatively stiff elastic rail fastenings used in modern bridge deck construction.
- The ability of the approach embankments to resist longitudinal forces at the bearings is reduced when the bridge and approaches are loaded.
- Grade-related traction is relatively insignificant for modern high adhesion locomotives.

The locomotive and car wheels may be modeled as accelerating or decelerating rolling* masses that do not slide (complete adhesion†) as they traverse the bridge superstructure. The forces created by the vertical, horizontal, and rotational translation of the rolling mass are shown in Figure 4.23a–c.

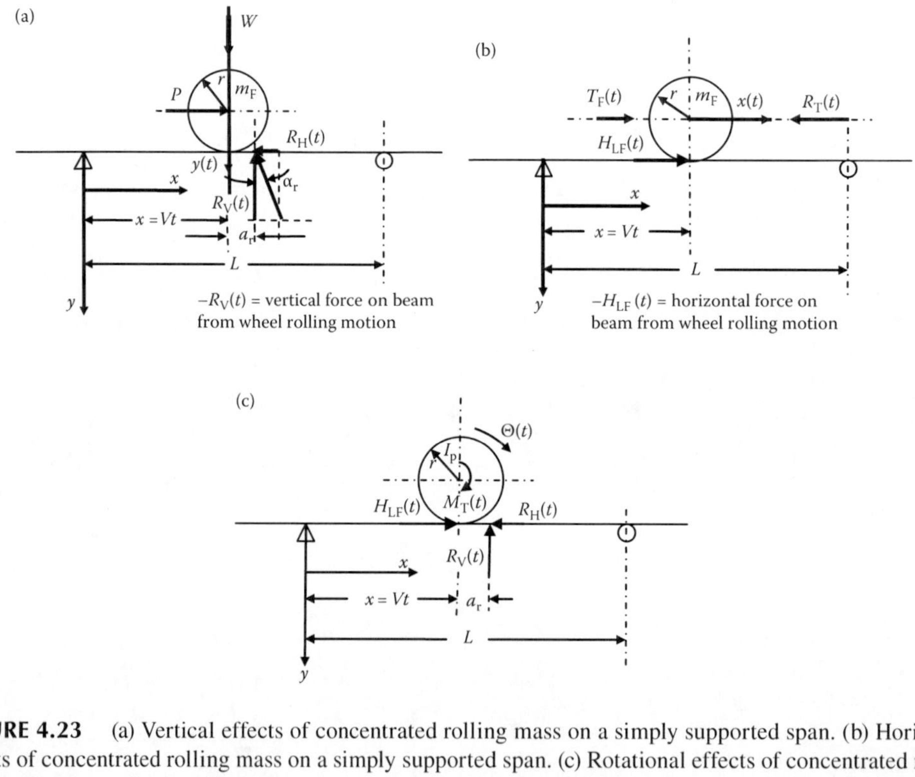

FIGURE 4.23 (a) Vertical effects of concentrated rolling mass on a simply supported span. (b) Horizontal effects of concentrated rolling mass on a simply supported span. (c) Rotational effects of concentrated rolling mass on a simply supported span.

* Rolling is the superposition of translation and rotation (Beer and Johnston, 1976).

† Nonuniform speed (acceleration for starting and deceleration for braking) and adhesion must exist between the wheel and the rail interface to start and stop trains.

Neglecting axle bearing and wheel rim friction,* force equilibrium for the vertical effects of rolling motion, considering complete adhesion (no sliding), provides (Figure 4.23a)

$$W - m_F \frac{d^2 y(t)}{dt^2} - R_V(t) = 0. \tag{4.42}$$

The horizontal reaction at the wheel axle, P, is

$$P = R_H(t) = \frac{W a_r}{r}, \tag{4.43}$$

where

$W = m_F g$ = weight of the concentrated force

m_F = mass of the concentrated force

$R_V(t)$, $R_H(t)$ = vertical and horizontal components, respectively, of reaction force due to rolling friction. The resultant reaction force, $R(t)$, is located at a horizontal distance, a_r, from the wheel centroid as a result of rolling friction (McLean and Nelson, 1962). Accordingly, the distance a_r is often referred to as the coefficient of rolling resistance. Rolling friction is small at constant train speed and greater at nonuniform train speeds. The horizontal component of the reaction, $R_H(t)$, is generally small because the applied vertical forces significantly exceed applied horizontal forces.

r = wheel radius, typically of 420 mm (16.5 in.) or 460 mm (18 in.), on North American locomotives and heavy freight cars.

Also, neglecting axle bearing and wheel rim friction, force equilibrium for the horizontal effects of rolling motion, considering complete adhesion, yields (Figure 4.23b)

$$H_{LF}(t) - R_T(t) + T_F(t) - m_F \frac{d^2 x(t)}{dt^2} + P = 0, \tag{4.44}$$

where

$H_{LF}(t)$ = longitudinal force transferred to rails- deck-superstructure

$R_T(t)$ = resistance to horizontal movement (primarily air resistance or vehicle drag forces since axle bearing and wheel flange friction is considered negligible). $R_T(t)$ is generally relativity small in comparison to other horizontal forces and it is not too conservative to neglect this force.

$T_F(t)$ = locomotive traction force = $\frac{M_T(t)c'}{r}$

$M_T(t)$ = driving torque applied to wheel

c' = constant depending on locomotive engine characteristics and gear ratio

Accordingly, Equation 4.44 may be simplified to

$$H_{LF}(t) + T_F(t) - m_F \frac{d^2 x(t)}{dt^2} = 0. \tag{4.45}$$

Again neglecting axle bearing and wheel rim friction, force equilibrium relating to the rotational effects of rolling motion, considering complete adhesion, is (Figure 4.23c)

$$-rH_{LF}(t) + M_T(t) - a_r R_V(t) + rR_H(t) - I_p \frac{d^2\theta(t)}{dt^2} = 0, \tag{4.46}$$

where I_p is the rotational moment of inertia of mass, m_F.

Since the distance, a_r, is small, the moment from rolling friction, $a_r R_V(t)$, may be neglected. In addition, because $R_H(t)$ is relatively small, Equation 4.46 may be simplified to

* Axle bearing and wheel rim friction are very small in comparison to rolling friction.

$$-rH_{\mathrm{LF}}(t)+M_{\mathrm{T}}(t)-I_{\mathrm{p}}\frac{\mathrm{d}^2\theta(t)}{\mathrm{d}t^2}=0. \tag{4.47}$$

For the condition of no slippage (complete adhesion), $\theta(t)=\frac{x(t)}{r}$. Substitution of $\frac{\mathrm{d}^2\theta(t)}{\mathrm{d}t^2}=\frac{\mathrm{d}^2x(t)}{r\mathrm{d}t^2}$ into Equation 4.47 yields

$$\frac{\mathrm{d}^2x(t)}{\mathrm{d}t^2}=\frac{I_{\mathrm{p}}}{r}\left(M_{\mathrm{T}}(t)-rH_{\mathrm{LF}}(t)\right). \tag{4.48}$$

Substitution of Equation 4.48 into Equation 4.44 provides

$$H_{\mathrm{LF}}(t)=\frac{1}{1-m_{\mathrm{F}}I_{\mathrm{p}}}\left(\frac{m_{\mathrm{F}}I_{\mathrm{p}}}{r}\left(M_{\mathrm{T}}(t)+T_{\mathrm{F}}(t)\right)\right)\leq\mu R_{\mathrm{V}}(t), \tag{4.49}$$

where μ is the coefficient of adhesion between locomotive wheels and rail without slippage (can be as high as 0.35 for modern locomotives with software-controlled wheel slip).

Equation 4.49 allows for the numerical solution of longitudinal force, $H_{\mathrm{LF}}(t)$. Numerical solutions are, however, too arduous and costly for ordinary or routine superstructure design. The longitudinal forces described by Equation 4.49 (including the effects of axle bearing, wheel rim friction, air resistance, rolling friction, and other effects) have been observed and recorded by field testing in both Europe and the United States. The longitudinal forces exhibit almost static behavior since maximum traction and braking forces occur at low speeds when starting and at the cessation of braking, respectively (Figure 4.24). Therefore, a static analysis can be performed with $H_{\mathrm{LF}}=\mu R_{\mathrm{V}}=\mathrm{LF}=\mu W$.

For a static longitudinal force analysis, the rails may be modeled as a series of longitudinal elastic bars (with independent longitudinal and flexural deformations) on horizontal elastic foundations represented* by equivalent horizontal springs with stiffness, k_i. The superstructures may also be modeled as elastic bars supported on expansion bearings at one end and fixed at the other end. The static longitudinal equilibrium equations for a system of bars, $i = 1, 2, \ldots, N_{\mathrm{B}}$, resisting N_i longitudinal forces, LF_{in}, in each of the rail bars can be written for the approach rails, rails on the superstructure, and the superstructure independently. For the longitudinal forces in the rails on the approaches to bridges [$i = 1$ and $(N_{\mathrm{S}}+2)$]:

$$-E_iA_i\frac{\mathrm{d}^2u_i(x)}{\mathrm{d}x^2}+k_iu_i(x)=\sum_{n=1}^{N_i}\delta(x-S_{in})\mathrm{LF}_{in}. \tag{4.50}$$

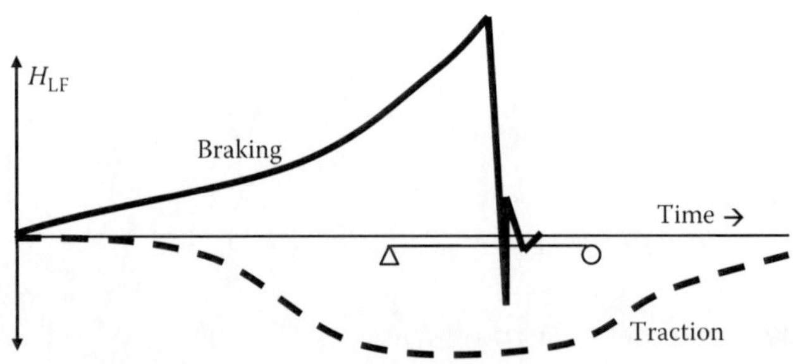

FIGURE 4.24 Time history of braking and traction forces (at a fixed bearing) from railroad equipment.

* Other bridge deck and approach track models that incorporate unique longitudinal restraint models for the rail-to-deck and deck-to-superstructure interfaces may be used to provide greater accuracy.

For the longitudinal forces in the rails on simply supported superstructures [$i = 2$ to $(N_B/2)$]:

$$-E_iA_i\frac{d^2u_i(x)}{dx^2}+k_i\left(u_i(x)-u_{\left(i+\left(\frac{N_B}{2}\right)\right)}(x)\right)=\sum_{n=1}^{N_i}\delta(x-S_{in})LF_{in} \tag{4.51}$$

and for the longitudinal forces in the superstructures [$i = (N_B/2 + 2)$ to N_B]:

$$-E_iA_i\frac{d^2u_i(x)}{dx^2}+k_i\left(u_i(x)-u_{\left(i-\left(\frac{N_B}{2}\right)\right)}(x)\right)=0. \tag{4.52}$$

For cases with uniform deck types,* $k_s = k_2 = k_3 = k_4 = k_5$ and Equation 4.52 is

$$-E_iA_i\frac{d^2u_i(x)}{dx^2}+k_s\left(u_i(x)-u_{\left(i-\left(\frac{N_B}{2}\right)\right)}(x)\right)=0, \tag{4.53}$$

where

E_iA_i = axial stiffness of the member (rail or span)
N_S = number of spans
N_B = number of bars in the model
N_i = number of longitudinal forces LF_{in} at location S_{in} in bar i

This system of equations may be solved for the longitudinal displacements, $u_i(x)$, and forces, $N_i(x) = E_iA_i\left(du_i(x)/dx\right)$, in the bars using transformation techniques (Fryba, 1996) with the appropriate boundary conditions. The longitudinal traction and braking forces transferred to the bearings may be determined from equilibrium following computation of the rail and span axial forces, $N_i(x)$. However, as seen in Example 4.4, even relatively simple bridge models with four simply supported spans will involve considerable calculation to determine the longitudinal displacements and forces in the superstructure.

Example 4.4

Develop the equations for longitudinal displacements and the boundary conditions for the open deck steel railway bridge shown in Figure E4.4.

The appropriate boundary conditions are shown in Table E4.2.

The equations for longitudinal displacements are

$$-E_iA_i\frac{d^2u_i(x)}{dx^2}+k_iu_i(x)=q_i(x),\quad \text{for } i=1 \text{ to } 6,$$

$$-E_iA_i\frac{d^2u_i(x)}{dx^2}+k_i\left[u_i(x)-u_{i+5}(x)\right]=q_i(x),\quad \text{for } i=2 \text{ to } 5,$$

$$-E_iA_i\frac{d^2u_i(x)}{dx^2}+k_{i-5}\left[u_i(x)-u_{i-5}(x)\right]=0,\quad \text{for } i=7 \text{ to } 10,$$

$$q_i(x)=\sum_{n=1}^{N_i}\delta(x-S_{in})LF_{in},\quad \text{for } i=1 \text{ to } 6.$$

* Typical of many steel railway superstructures with standardized deck types [rail to tie (sleeper) fasteners, tie sizes and spacing, and tie to deck fasteners].

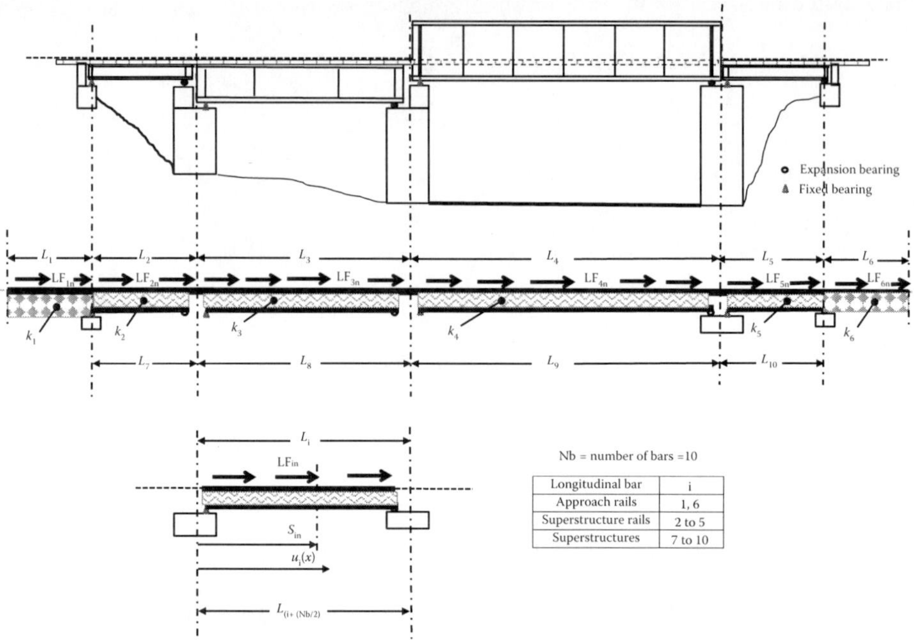

FIGURE E4.4 Elevation of bridge.

TABLE E4.2
Bar Element Boundary Conditions

Bridge Model Element (Rails, Decks, Superstructures)	**Boundary Conditions** **Force and Displacement Conditions**
Rails (i = 1–6)	
Forces at ends of rails	$N_1(0) = N_6(L_6) = 0$
Force equilibrium	$N_i(L_i) - \mathrm{LF}_{\mathrm{L}i} = N_{i+1}(0)$
Displacement compatibility	$u_i(L_i) = u_{i+1}(0)$
Expansion joints at ends of bridge	$L_1 = L_6 = 0$
Rail continuity (CWR across bridge)	$L_1 = L_6$
Decks (i = 2–5)	
No longitudinal rail restraint (free rails)	$k_i = 0$
Rails fixed (direct fixation)	$k_1 = \infty$
Superstructures (i = 7–10)	
Expansion bearings	$N_i(L_i) = 0$
Fixed bearings	$u_i(0) = 0$

Extensive testing and analytical work has been performed (Foutch et al., 1996, 1997; Otter et al., 1996, 1997, 1999, 2000; LoPresti and Otter, 1998; LoPresti et al., 1998; Tobias et al., 1998, 1999; Uppal et. al., 2001) to overcome the theoretical model complexities and numerical modeling efforts. This work has established relationships for braking and traction dependent on the length of the portion of the bridge under consideration. Testing in the United States has provided longitudinal forces for Cooper's EM360 (E80) design live load that are shown in Figure 4.25 and Equations 4.54

and 4.55. For loaded lengths less than about 110 m (350 ft), longitudinal force due to traction governs. However, locomotive tractive effort occurs over a relatively small length, and braking forces in each span from a loaded length consisting of the entire bridge may exceed the tractive effort (see Examples 4.5 and 4.6).* The force due to traction governs for short- and medium-length bridges. The longitudinal braking, LF_B, and traction, LF_T, forces for Cooper's EM360 (E80) design live load are

$$LF_B = 200 + 17.5L \text{ (SI units)}, \tag{4.54a}$$

$$LF_T = 200\sqrt{L} \text{ (SI units)}, \tag{4.55a}$$

where LF_B is the longitudinal force due to train braking (kN), LF_T is the longitudinal force due to locomotive traction (kN), and L is the length of the portion of bridge under consideration (m);

$$LF_B = 45 + 1.2L \text{ (US customary and imperial units)}, \tag{4.54b}$$

$$LF_T = 25\sqrt{L} \text{ (US customary and imperial units)}, \tag{4.55b}$$

where LF_B is the longitudinal force due to train braking (kips), LF_T is the longitudinal force due to locomotive traction (kips), and L is the length of the portion of bridge under consideration (ft).

However, while an estimate of the magnitude of the applied longitudinal traction and braking forces appropriate for design is readily available, the distribution of longitudinal forces for the design of span bracing, bearings, substructures, and foundations needs careful consideration. The distribution and path of longitudinal forces between their point of application and the superstructure supports depends on the arrangement, orientation, and relative stiffness of

- Members in the load path
- Bearing type (fixed or expansion)
- Substructure characteristics.

For very long or complex steel bridges, a numerical analysis using FEA software that can model longitudinal loads due to locomotive traction and train braking should be used. This software can also generally accommodate the analysis of thermal loads.

Example 4.5a (SI Units)

The longitudinal design force for Cooper's EM360 loading is required for each track of the two equal span 27.4 m long open deck steel multibeam railway bridge shown in Figure E4.1. From Figure 4.25, it is determined that

- The longitudinal force due to train braking is $LF_B = 680$ kN per track on the entire bridge, which because of relative span lengths and bearing arrangement may be equally distributed to each span as 340 kN.
- The longitudinal force due to train braking is $LF_B = 440$ kN per track on one span. However, this is an unlikely scenario considering the bridge length, train length, and distributed nature of train brake application.
- The longitudinal force due to locomotive traction is $LF_T = 1047$ kN per track on the entire bridge, which because of relative span lengths and bearing arrangement may be equally distributed to each span as 523 kN.
- The longitudinal force due to locomotive traction is $LF_T = 740$ kN per track on one span.

The longitudinal force due to locomotive traction, $LF_T = 740$ kN per track, may be used for superstructure design. The longitudinal forces are distributed through the superstructure to the bearings and substructures. Bearing component and substructure design will require consideration of these longitudinal forces. However, in this multibeam span, longitudinal forces of this magnitude will result in only small axial stresses in the longitudinal beams or girders, which may be disregarded in the design.

* As illustrated by Figure 4.26 showing the ratio of the longitudinal force transmitted to the bearings, H_B, to the applied longitudinal force for bridges with continuous welded rail and steel bearings (based on European tests reported by Fryba, 1996).

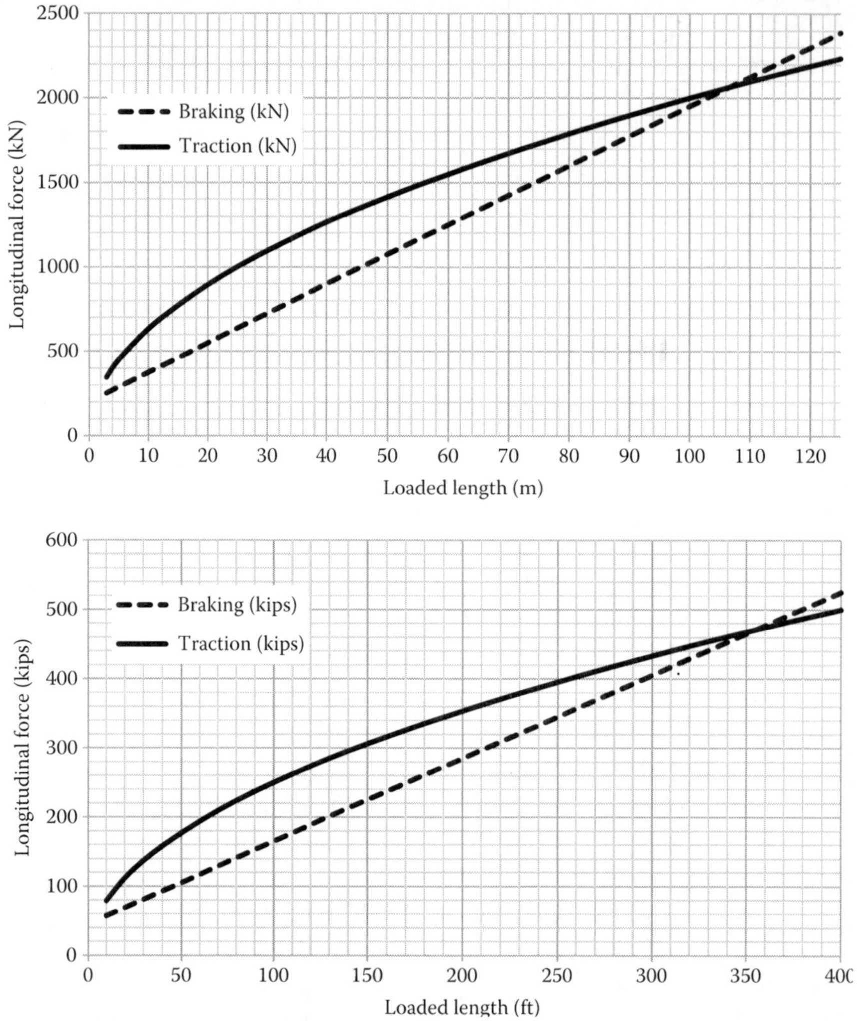

FIGURE 4.25 AREMA design longitudinal forces.

Example 4.5b (US Customary and Imperial Units)

The longitudinal design force for Cooper's E80 loading is required for each track of the two equal span 90 ft long open deck steel multibeam railway bridge shown in Figure E4.1. From Figure 4.25, it is determined that

- The longitudinal force due to train braking is LF_B = 153.0 kips per track on the entire bridge, which because of relative span lengths and bearing arrangement may be equally distributed to each span as 76.5 kips.
- The longitudinal force due to train braking is LF_B = 99.0 kips per track on one span. However, this is an unlikely scenario considering the bridge length, train length, and distributed nature of train brake application.
- The longitudinal force due to locomotive traction is LF_T = 237.2 kips per track on the entire bridge, which because of relative span lengths and bearing arrangement may be equally distributed to each span as 118.6 kips.
- The longitudinal force due to locomotive traction is LF_T = 167.7 kips per track on one span.

TABLE E4.3
Length of Spans

Span	Type	Length (m)	Length (ft)
1	Through plate girder	30	100
2–6	Deck plate girder	25 each	80 each
7	Deck truss	125	400
8–10	Deck plate girder	30 each	100 each
Total		370	1200

The longitudinal force due to locomotive traction, $LF_T = 167.7$ kips per track, may be used for superstructure design. The longitudinal forces are distributed through the superstructure to the bearings and substructures. Bearing component and substructure design will require consideration of these longitudinal forces. However, in this multibeam span, longitudinal forces of this magnitude will result in only small axial stresses in the longitudinal beams or girders, which may be disregarded in the design.

Example 4.6a (SI Units)

The longitudinal design force for Cooper's EM400 loading is required for the deck truss of the 370 m long single track ten span steel bridge outlined in the data in Table E4.3. Each span has fixed and expansion bearings. All substructures have spans with adjacent fixed and expansion bearings.

- The longitudinal force due to train braking is $LF_B = (400/360)6675 = 7417$ kN on the entire 370 m long bridge and is distributed to the deck truss span as (125/370) (7417) = 2506 kN.
- The longitudinal force due to train braking is $LF_B = (400/360)2388 = 2653$ kN on the 125 m long deck truss span. However, this is an unlikely scenario considering the bridge length, train length, and distributed nature of train brake application. Therefore, other portions of the bridge should be investigated for train braking. For example, the longitudinal force due to train braking is $LF_B = (400/360)(125/215)(3963) = 2560$ kN on the deck truss span when the train is on spans 7–10 (215 m long) only and (400/360)(125/280)(5100) = 2530 kN when the train is on spans 1–7 (280 m long) only.
- The longitudinal force due to locomotive traction is $LF_T = (400/360)3847 = 4275$ kN on the entire bridge. However, this is not likely (unless a string of powered accelerating/decelerating locomotives traverses the bridge) and other portions of the bridge should be investigated. For example, the longitudinal force due to locomotive traction is $LF_T = (400/360)(125/215)(2933) = 1894$ kN on the deck truss span when the train is on spans 7–10 only and (400/360)(125/280)(3347) = 1660 kN when the train is on spans 1–7 only.
- The longitudinal force due to locomotive traction is $LF_T = (400/360)2236 = 2485$ kN on the 125 m deck truss span.

The longitudinal force due to train braking, $LF_B = 2560$ kN, is likely appropriate for the design of the deck truss span.

Example 4.6b (US Customary and Imperial Units)

The longitudinal design force for Cooper's E90 loading is required for the deck truss of the 1200 ft long single track ten span steel bridge outlined in the data of Table E4.3. Each span has fixed and expansion bearings. All substructures have spans with adjacent fixed and expansion bearings.

- The longitudinal force due to train braking is $LF_B = (9/8)1485 = 1671$ kips on the entire 1200 ft long bridge and is distributed to the deck truss span as (400/1200) (1671) = 557 kips.
- The longitudinal force due to train braking is $LF_B = (9/8)525 = 591$ kips on the 400 ft deck truss span. However, this is an unlikely scenario considering the bridge length, train

length, and distributed nature of train brake application. Therefore, other portions of the bridge should be investigated for train braking. For example, the longitudinal force due to train braking is $LF_B = (9/8)(400/700)(885) = 569$ kips on the deck truss span when the train is on spans 7–10 (700 ft long) only and (9/8)(400/900)(1125) = 563 kips when the train is on spans 1–7 (900 ft long) only.

- The longitudinal force due to locomotive traction is $LF_T = (9/8)886 = 974$ kips on the entire bridge. However, this is not likely (unless a string of powered accelerating/decelerating locomotives traverses the bridge) and other portions of the bridge should be investigated. For example, the longitudinal force due to locomotive traction is $LF_T = (9/8)(400/700)(661) = 425$ kips on the deck truss span when the train is on spans 7–10 only and (9/8)(400/900)(750) = 375 kips when the train is on spans 1–7 only.
- The longitudinal force due to locomotive traction is $LF_T = (9/8)500 = 563$ kips on the 400 ft deck truss span.

The longitudinal force due to train braking, $LF_B = 569$ kips, is likely to be used for the design of the deck truss span.

As noted in Example 4.5, the distribution of longitudinal forces in the superstructure may be of little concern for some span types (e.g., multiple longitudinal beam and deck plate girder spans). However, for other types of superstructures, the longitudinal force path from rails to bearings is of considerable importance (e.g., floorbeams with direct fixation of track and some through span floor systems). The resistance of steel deck plates, due to diaphragm behavior, to horizontal axial forces may preclude the need for bracing elements to carry longitudinal forces to the main girders or trusses. Nevertheless, in some open deck spans, specific consideration of the lateral bracing (traction bracing) requirements is necessary to adequately transfer longitudinal forces to the main girders or trusses for transmission to the substructures through the bearings. (Figure 4.26 illustrates the bearing force relationship to LF_T and LF_B from European testing.) A typical instance where traction bracing may be required is within the panel adjacent to the fixed bearings in an open deck span with a stringer and floorbeam system supported each side of the track by long-span main girders or trusses. In order to preclude the torsional and/or lateral bending of floorbeams that might

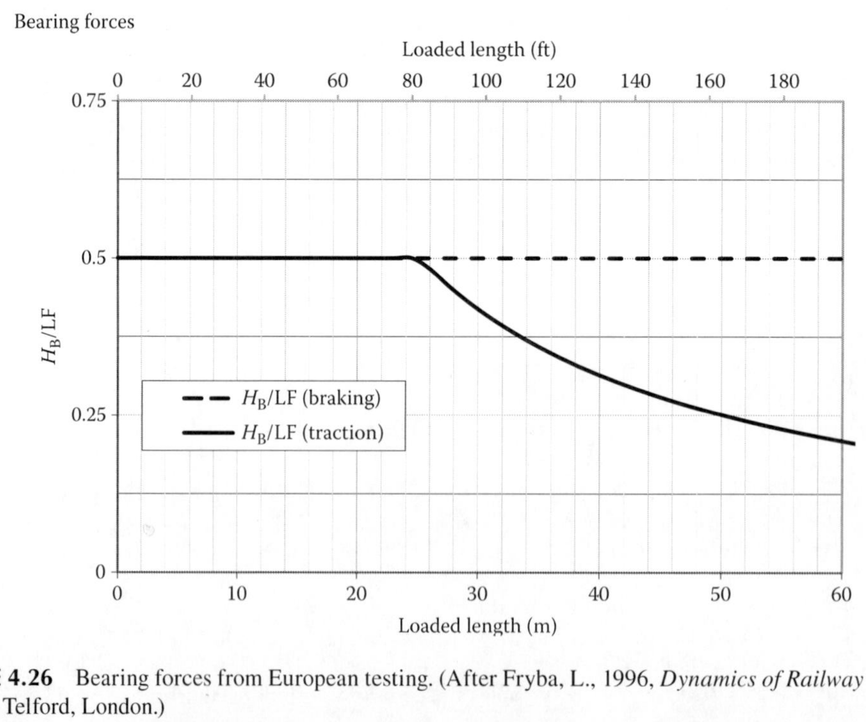

FIGURE 4.26 Bearing forces from European testing. (After Fryba, L., 1996, *Dynamics of Railway Bridges*, Thomas Telford, London.)

result from longitudinal forces transmitted by floor systems without connection to the lateral bracing (Figure 4.27a), traction bracing is used (Figure 4.27b). Traction bracing is provided through connection of the stringers to the lateral bracing and addition of a new transverse member (shown dashed in Figure 4.27b) between the stringers at the bracing connections. Provided the main girder or truss fixed bearings are adequate to transfer the longitudinal forces to the substructure, the traction bracing truss (Figures 4.27b and 4.28) will avoid lateral loading of floor beams (member 1–1 in Figure 4.28) since the stringers (members 2–3 in Figure 4.28) can carry no longitudinal force. Other traction bracing arrangements may be used in a similar manner at the fixed end of long single and multiple track spans to properly transmit longitudinal traction and braking forces to the bearings.

4.3.2.4 Centrifugal Forces

Centrifugal forces acting horizontally at the vehicle center of gravity (recommended as 2.4 m (8 ft) above the top of the rails in AREMA, 2015) act on the moving live load as it traverses the curved track on a bridge, as shown in Figure 4.29. The centrifugal force corresponding to each axle load is

$$CF_A = \frac{m_A V^2}{R}, \tag{4.56}$$

where

$m_A = A/g$
A = axle load
g = acceleration due to gravity
V = speed of train
R = radius of curve
which, in SI units, may be expressed as

$$CF_A = \left(4.42 \times 10^{-6}\right) AV^2 D \text{ (SI units)}, \tag{4.57a}$$

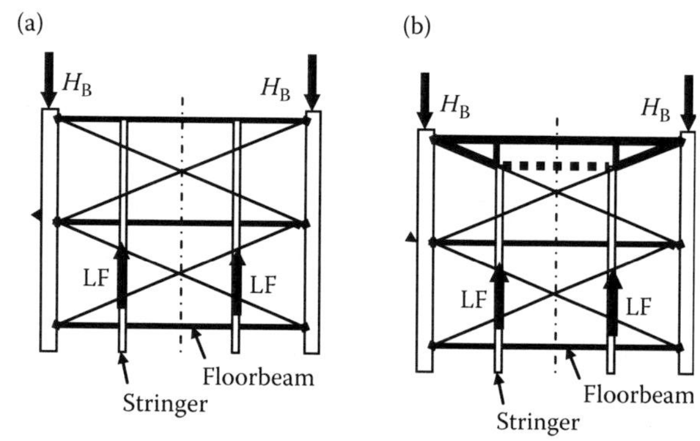

FIGURE 4.27 Plan of floor system: (a) without traction bracing and (b) with additional member (dashed line) to create traction bracing.

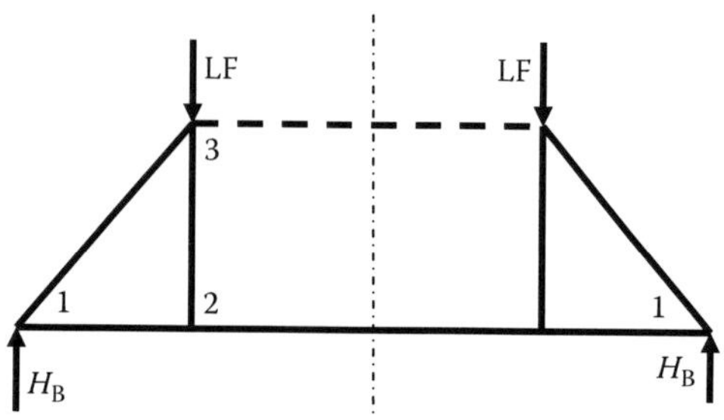

FIGURE 4.28 Traction frame truss for a single track span with two stringers per track.

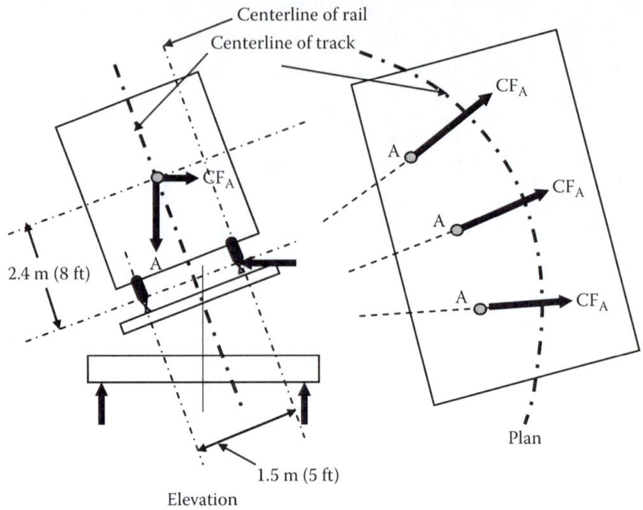

FIGURE 4.29 Centrifugal forces from a curved track.

where

CF_A = centrifugal force at each axle, kN
A = axle load, kN
V = speed of train, km/h
R = 1747.5/D, m
D = degree of curve (the central angle subtended by a 30.5 m chord—see Chapter 3)
or, in US Customary and imperial units, expressed as

$$CF_A = (11.7\times10^{-6})AV^2D = 0.0000117AV^2D \text{ (US customary or imperial units),} \quad (4.57b)$$

CF_A = centrifugal force at each axle, kips
A = axle load, kips
V = speed of train, mph
R = 5730/D, ft
D = degree of curve (the central angle subtended by a 100 ft chord—see Chapter 3)

Due to the rail–wheel interface contact as trains traverse through curved track, the entire centrifugal force will be transmitted at the outer or high rail and, therefore, centrifugal effects are not considered for longitudinal superstructure members inside the curved rack. Figure 4.30 outlines the actions required for the determination of the geometrical (shaded boxes, see Chapter 3) and centrifugal effects on vertical live load forces.

It should be noted that the maximum centrifugal effect can be developed from consideration of practical values of maximum superelevation of the track. For example, it is typical that North American railways specify a maximum superelevation of 150 mm (6 in) with a maximum underbalance of 75 mm (3 in.). This equates to an equilibrium superelevation of 225 mm (9 in.). Therefore, Equation 3.34 (See Chapter 3) with the horizontal distance between rails of 1.5 m (4.91 ft) provides a centrifugal force to train weight ratio of 225/1500 = 0.15 (9/58.8 = 0.15). The maximum practical centrifugal force is 15% of the train weight.

Example 4.7 outlines the calculation of the centrifugal effect and its combination with geometrical effects (see Chapter 3) to determine the total effect of track curvature on vertical live load forces.

Example 4.7a (SI Units)

A ballasted steel through plate girder railway bridge is to be designed with a 6 degree of curvature track across its 20 m span. The railroad has specified a 125 mm superelevation based on a

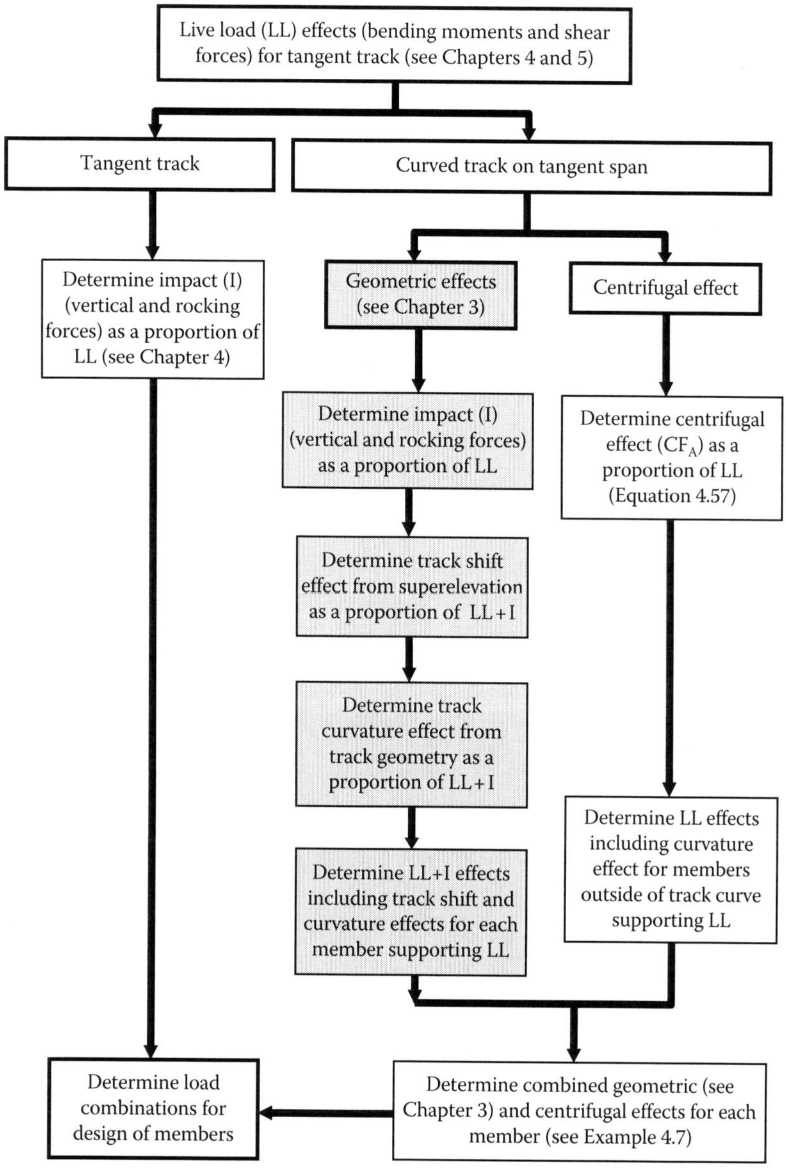

FIGURE 4.30 Determination of geometrical and centrifugal effects on vertical live load forces.

65 km/h operating speed. The track tie depth and rail height are 175 mm each. The girders are spaced at 5 m and the plane at the top of the rail elevations is 800 mm above the deck surface at track centerline. Determine the effects of the curvature on the design live load shear and bending moment for each girder.

Geometrical effects (see Example 3.1a in Chapter 3):

$V_{out} = 0.907V_{LL+I}$

$V_{in} = 1.093V_{LL+I}$

$M_{out} = 0.920M_{LL+I}$

$M_{in} = 1.080M_{LL+I}$.

Centrifugal effects:

$$CF_A/A = (4.42\times10^{-6})V^2D = (4.42\times10^{-6})(65)^2(6) = 0.112$$

$\frac{M_{CF}}{M_{LL}} = \frac{V_{CF}}{V_{LL}} = \pm 0.112\left(\frac{(2.4+0.80)}{(5/2)}\right) = \pm 0.143$ (see Figure 4.29). (Note that girder design forces due to centrifugal effects are independent of impact.)

Combined geometrical and centrifugal effects:

Since centrifugal forces are not affected by impact (rocking and dynamic vertical effects), the centrifugal effect on the girder design forces, which include impacts, must be reduced as follows:

$I = 30.5/5 + 32 = 38\%$

$$\frac{M_{CF(out)}}{M_{LL+I}} = \frac{V_{CF(out)}}{V_{LL+I}} = \pm\frac{0.112}{(1+I)}\left(\frac{(2.4+0.80)}{(5/2)}\right) = \pm\frac{0.143}{(1+0.38)} = \pm 0.104.$$

Therefore, the following is determined for the shear, V_{LL+I}, and bending moment, M_{LL+I}, live load forces (combining impact, geometric, and centrifugal force effects):

$V_{out} = V_{LL+I}\,(0.907+0.104) = 1.011V_{LL+I}$

$V_{in} = 1.093V_{LL+I}$

$M_{out} = M_{LL+I}\,(0.920+0.104) = 1.024M_{LL+I}$

$M_{in} = 1.080M_{LL+I}$.

Example 4.7b (US Customary and Imperial Units)

A ballasted steel through plate girder railway bridge is to be designed with a 6 degree of curvature track across its 70 ft span. The railroad has specified a 5 in. superelevation based on a 40 mph operating speed. The track tie depth and rail height are 7 in. each. The girders are spaced at 16 ft and the plane at the top of the rail elevations is 2′-8″ above the deck surface at track centerline. Determine the effects of the curvature on the design live load shear and bending moment for each girder.

Geometrical effects (see Example 3.1b in Chapter 3):

$V_{out} = 0.903V_{LL+I}$

$V_{in} = 1.097V_{LL+I}$

$M_{out} = 0.916M_{LL+I}$

$M_{in} = 1.084M_{LL+I}$.

Centrifugal effects:

$$CF_A/A = 0.0000117V^2D = 0.0000117(40)^2(6) = 0.112$$

$\frac{M_{CF}}{M_{LL}} = \frac{V_{CF}}{V_{LL}} = \pm 0.112\left(\frac{(8+2.67)}{(16/2)}\right) = \pm 0.150$ (see Figure 4.29). (Note that girder design forces due to centrifugal effects are independent of impact.)

Combined geometrical and centrifugal effects:

Since centrifugal forces are not affected by impact (rocking and dynamic vertical effects), the centrifugal effect on the girder design forces, which include impacts, must be reduced as follows:

$I = 100/16 + 31 = 37\%$

$$\frac{M_{CF(out)}}{M_{LL+I}} = \frac{V_{CF(out)}}{V_{LL+I}} = \pm\frac{0.112}{(1+I)}\left(\frac{(8+2.67)}{(16/2)}\right) = \pm\frac{0.150}{(1+0.37)} = \pm 0.110.$$

Therefore, the following is determined for the shear, V_{LL+I}, and bending moment, M_{LL+I}, live load forces (combining impact, geometric, and centrifugal force effects):

$V_{out} = V_{LL+I}\,(0.903+0.110) = 1.013V_{LL+I}$

$V_{in} = 1.097V_{LL+I}$

$M_{out} = M_{LL+I}\,(0.916+0.110) = 1.026M_{LL+I}$

$M_{in} = 1.084M_{LL+I}$.

4.3.2.5 Lateral Forces from Moving Freight Equipment

In addition to the centrifugal lateral forces due to track curvature, lateral forces caused by track irregularities at the wheel–rail interface must be considered in the design of steel railway bridges. The differential equation of motion for lateral deflection, $z(x,t)$, of a simply supported beam with i = 1 to N moving lateral forces, $H_i(t)$, is

$$\mathrm{EI_y}\frac{\partial^4 z(x,t)}{\partial x^4}+m\frac{\partial^2 z(x,t)}{\partial t^2}+2m\omega_c\frac{\partial z(x,t)}{\partial t}=\sum_{i=1}^{N}\delta(x+s_i-Vt)H_i(t), \tag{4.58}$$

where I_y is the lateral moment of inertia of superstructure (with respect to vertical axis) and s_i is the distance from load $H_i(t)$ to the first load $H_1(t)$.

Neglecting viscous damping ($\omega_c = 0$), and assuming lateral, vertical, and torsional vibrations are uncoupled, the solution of Equation 4.58 at $z(L/2,t)$ for lateral loads of equal magnitude, $H(t)$, is (see Equation 4.33) given as

$$z(L/2,t)=\frac{2H(t)}{mL\left(\omega_{y1}^2-\left(\frac{\pi V}{L}\right)^2\right)}\left(\sin\frac{\pi Vt}{L}-\frac{\pi V}{\omega_{y1}L}\sin\omega_{y1}t\right), \tag{4.59}$$

where

$$\omega_{y1}=\frac{\pi^2}{L^2}\sqrt{\frac{\mathrm{EI_y}}{m}}.$$

Clearly, even relatively simple analytical solutions are often unsuitable for routine or ordinary design, and it is most suitable to determine design lateral forces from tests conducted on in-service bridges. Tests concerning the dynamic lateral forces on in-service bridges (Otter et al., 2005) have confirmed that the AREMA (2015) design recommendation of a single moving lateral force of 25% of the heaviest axle of Cooper's EM360 (E80) load is a suitable representation of these effects.

The magnitude of lateral forces is of particular importance regarding the design of span lateral and cross bracing.* Therefore, in addition to the recommendation of a single moving lateral force of 25% of the heaviest axle of the Cooper's EM360 (E80) live load, a notional vibration load of 2.9 kN/m (200 lb/ft) applied to the loaded chord or flange (e.g., the top flange of a deck plate girder span) and 2.2 kN/m (150 lb/ft) on the unloaded chord or flange (e.g., the top chord of a through truss span) is recommended. This notional vibration load is recommended by AREMA (2015) to ensure a minimum lateral bracing stiffness to resist vibration from live load. It is not to be combined with other loads and forces (Waddell, 1916) and, therefore, it is to be applied to the lateral bracing as an alternative to the wind load (see Section 4.4.1) on a loaded railway bridge.

4.3.3 Distribution of Live Load

Unlike highway live loads that may move laterally across the bridge deck, railway live loads are generally fixed in lateral position. However, as a longitudinal series of large magnitude concentrated wheel loads, their longitudinal and lateral distribution to the deck and supporting members must be considered.

4.3.3.1 Distribution of Live Load for Open Deck Steel Bridges

For open deck bridges, no longitudinal distribution is made and lateral distribution to supporting members is based on span cross-sectional geometry and type of lateral bracing system. Lateral

* Lateral loads from freight rail equipment are considered to be applied directly to bracing members (see Chapter 5) without producing lateral bending of supporting member flanges or chords.

bracing between longitudinal beams should be made with cross frames, or for spans with shallow beams, rolled beams, and/or close beam spacing, solid diaphragms.* The cross frames and diaphragms should not have a spacing exceeding 5.5 m (18 ft). In some cases, AREMA (2015) recommends that diaphragms and cross bracing be fastened to the beam or girder flanges. When the lateral bracing system meets these criteria and is properly designed for the lateral forces (see Chapters 5 through 7), all beams or girders supporting the track are considered as equally loaded.

4.3.3.2 Distribution of Live Load for Ballasted Deck Steel Bridges

For ballasted deck bridges, the longitudinal and lateral distribution of design live load to the deck, and the longitudinal and transverse members supporting the deck, is based on tests performed by the Association of American Railroads (AAR) (Sanders and Munse, 1969).

For deck design, axle loads are distributed as shown in Figure 4.31. The longitudinal deck distribution width, 915 mm + $d_b + h$ (3 ft + $d_b + h$), should not exceed 1525 mm (5 ft) nor the minimum axle spacing of the design live load. The lateral deck distribution width (length of tie + $d_b + h$) should not exceed 4.25 m (14 ft), or the distance between adjacent track centerlines or width of the deck. Timber, concrete, and steel decks of thickness, h, supporting rock ballast with depth, d_b, should be designed in accordance with the recommended practices of AREMA (2015) Chapters 7, 8, and 15,

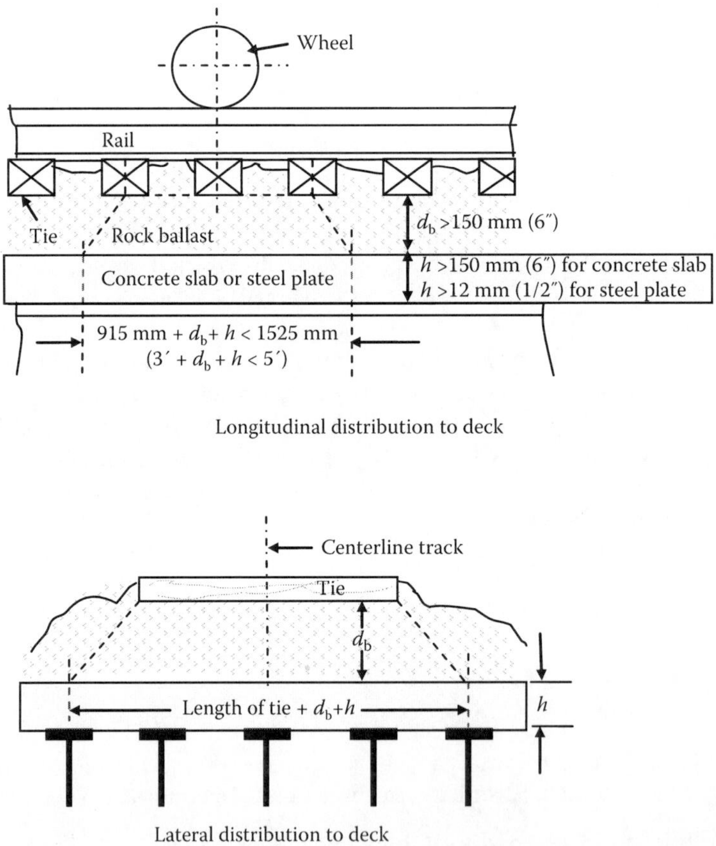

FIGURE 4.31 Longitudinal and lateral distribution of live load to deck on ballasted deck bridges.

* Channels, coped wide flange shapes, and plate/angle fabrications are often used for diaphragms between longitudinal beams. Plates alone are generally not used due to the absence of flanges and low bending strength.

respectively (see Appendices A and B for examples of steel plate and concrete slab deck design, respectively) .

For transverse beam design, the longitudinal distribution of live load is given in terms of an effective beam spacing, which is dependent on deck material, beam span, beam spacing, and for concrete decks, stiffness of beams, deck rigidity, and width of deck (Figure 4.32). The portion of the axle load, P, on a transverse beam is

$$P = \frac{1.15AD}{S}, \tag{4.60}$$

A = axle load

S = axle spacing, ft

D = effective beam spacing, ft, which can be calculated as

- For bending moment calculations with a concrete deck*

$$D = d\left[\frac{1}{1+\frac{d}{aH}}\right]\left(0.4+\frac{1}{d}+\frac{\sqrt{H}}{12}\right), \text{ but } D \le d \text{ or } S, \tag{4.61a}$$

where the first term of the equation reflects the effect of the transverse steel beams and the second part indicates the relative effect of a concrete slab in the longitudinal distribution of axle loads to the transverse beams. Equation 4.61a is valid provided that the concrete slab extends over at least 75% of the transverse beam length. If the slab width is less than 75% of the transverse beam length, then $D = d$.

- For bending moment calculations with a steel plate deck, or for end shear with both concrete and steel decks

$$D = d \tag{4.61b}$$

where d is the transverse beam spacing, ft, when $d \le S$ (if $d > S$, then assume the deck as simply supported between transverse beams for the longitudinal distribution of axle loads to supporting transverse beams).

In Equation 4.61a, a is the transverse beam span, ft, and H is given by

$$H = \frac{nI_b}{ah^3}, \text{ in./ft}, \tag{4.62}$$

where n is the steel to concrete modular ratio, I_b is the transverse beam moment of inertia, in.^{4}, and h is the concrete slab thickness, in. Equations 4.60–4.62 were empirically developed based on US Customary and imperial units. No lateral distribution of live load is made for transverse beams supporting ballasted decks (Figure 4.32).

For longitudinal beam and girder design, axle loads are distributed as shown in Figure 4.33. Axle loads are distributed equally to all beams or girders within the lateral distribution width, length of tie + $2(d_b + h)$. No longitudinal distribution of live load is made for longitudinal members spaced equally about the centerline of the track.

* This equation was empirically developed based on testing, and US Customary or Imperial units are inherent within its development.

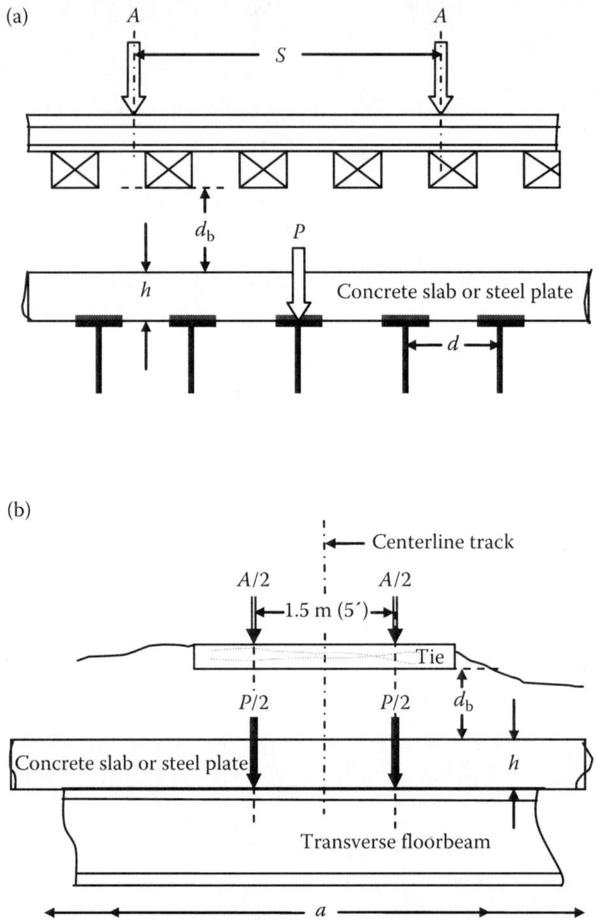

FIGURE 4.32 (a) Longitudinal and (b) lateral distribution of live load on ballasted deck bridges with transverse floorbeams.

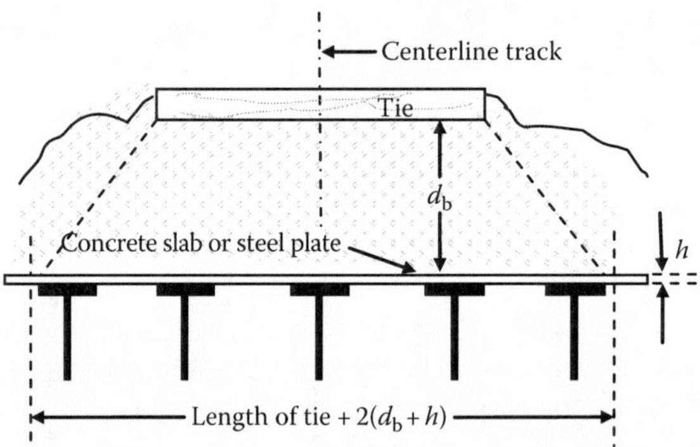

FIGURE 4.33 Lateral distribution of live load to longitudinal members on ballasted deck bridges.

4.3.3.3 Distribution of Live Load for Direct Fixation Deck Steel Bridges

For longitudinal members supporting the deck, axle loads are distributed laterally based on a structural analysis of the deck and supporting beam system considering the location of wheel loads, the location and spacing of supporting members, and deck and longitudinal member materials and properties. No longitudinal distribution of live load is made for longitudinal beams supporting direct fixation decks.

AREMA (2015) indicates that, for bridges with rails directly fixed to concrete or steel decks, the longitudinal and lateral distribution of design live load to the transverse members supporting the deck may also be based on the tests of ballasted deck bridges conducted by the AAR. For transverse members supporting the deck, the longitudinal distribution of live load is given in terms of the effective beam spacing of Equation 4.61a or 4.61b, and the portion of the axle load, *P*, on a transverse beam due to longitudinal distribution through the deck is given by Equation 4.60. No lateral distribution of live load is made for transverse beams supporting direct fixation decks.

Example 4.8a (SI Units)

The longitudinal distribution of Cooper's EM360 axle loads to 5 m long transverse *W* 920 × 223 floorbeams (Figure E4.5) spaced at 750 mm supporting a 175 mm thick reinforced concrete deck slab is required.

Some SI units require conversion in order to use Equation 4.61a:

$I_b = 3770 \times 10^6$ mm^4 = 9057 in.^{4},
$n = 8$,
$a = 5$ m = 16.40 ft,
$h = 175$ mm = 6.9 in.,
$H = (8)(9057)/(16.40(6.9^3)) = 13.45$ in./ft,
$d = 750$ mm − 2.46 ft,

$$D_M = \frac{2.46}{1+\left(\dfrac{2.46}{16.4(13.45)}\right)}\left(0.4+\frac{1}{2.46}+\frac{\sqrt{13.45}}{12}\right) = 2.71 \text{ ft, use } 2.46 \text{ ft,}$$

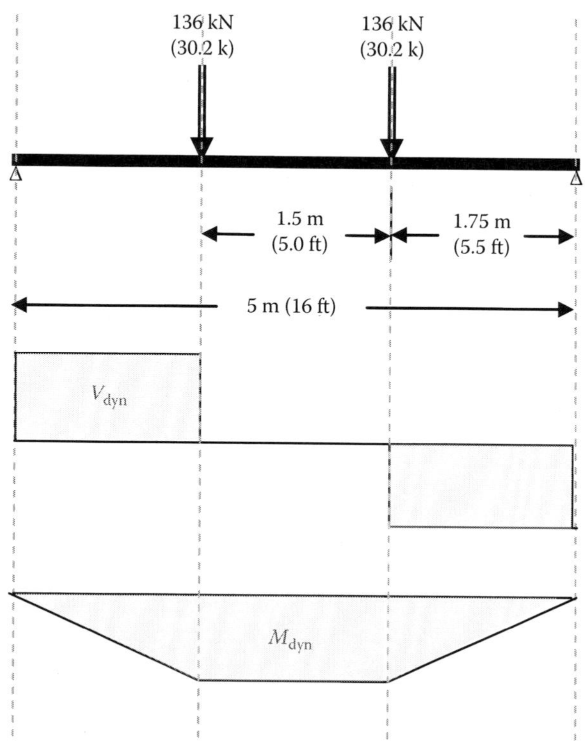

FIGURE E4.5 Longitudinal distribution to floorbeams.

$D_V = d = 2.46$ ft,
S = design load axle spacing = 1.5 m = 4.92 ft, and
$P = 1.15(360)(2.46)/4.92 = 207$ kN (for both shear and moment calculations) (58% of axle load).
Considering dynamic effects:
RE = $0.20W(1.5)(100)/5W = 6.00\%$,
$I_V = 39.5\%$ (Figure 4.22),
$P_{LL+I} = 0.90(1 + 0.060 + 0.395)207 = 271$ kN (136 kN each rail),
$V_{LL+I} = 136$ kN, and
$M_{LL+I} = (1.75)\ 136 = 237$ kN-m.

Example 4.8b (US Customary and Imperial Units)

The longitudinal distribution of Cooper's E80 axle loads to 16 ft long transverse W 36 × 150 floorbeams (Figure E4.5) spaced at 2.5 ft supporting a 7 in. thick reinforced concrete deck slab is required.
$I_b = 9040$ in.^{4},
$n = 8$,
$H = (8)(9040)/(16(7^3)) = 13.18$ in./ft,

$$D_M = \frac{2.50}{1+\left(\dfrac{2.50}{16(13.18)}\right)}\left(0.4+\frac{1}{2.50}+\frac{\sqrt{13.18}}{12}\right) = 2.72 \text{ ft, use } 2.50 \text{ ft.,}$$

$D_V = d = 2.50$ ft, and
$P = 1.15(80)(2.5)/5 = 46.0$ kips (for shear and moment calculations)(58% of axle load).
Considering dynamic effects:
RE = $0.20W(5)(100)/16W = 6.25\%$,
$I_V = 39.5\%$ (Figure 4.22),
$P_{LL+I} = 0.90(1 + 0.063 + 0.395)46.0 = 60.3$ kips (30.2 kips each rail),
$V_{LL+I} = 30.2$ kips, and
$M_{LL+I} = (5.5)\ 30.2 = 166.1$ ft-kips.

4.4 ENVIRONMENTAL AND OTHER STEEL RAILWAY BRIDGE DESIGN FORCES

In addition to dead and live load effects, environmental forces (wind, thermal, and seismic events) and other forces related to serviceability and overall stability criteria must be considered in the design of steel railway bridges.

4.4.1 Wind Forces on Steel Railway Bridges

In contrast to long-span or flexible bridges (such as suspension or cable-stayed bridges), typical steel railway bridges (such as those comprising of beam, girder, truss, and arch spans) need not consider aerodynamic effects* of the wind in design.† However, the aerostatic effects of the wind on the superstructure and moving train must be considered, particularly in regard to lateral bracing design.

A steady wind with uniform upstream velocity, V_u, flowing past a bluff body (such as the bridge cross section of Figure 4.34a) will create a maximum steady-state local or dynamic pressure, p_m, in accordance with Bernoulli's fluid mechanics equation as

* The effects from dynamic behavior and buffeting.

† An equivalent static wind pressure is appropriate since the natural or fundamental frequency of the superstructure is substantially greater than the frequency of localized gust effects.

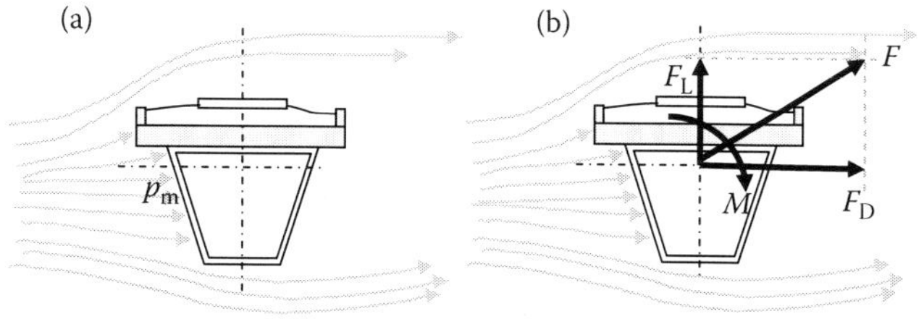

FIGURE 4.34 (a) Wind flow past a bluff body and (b) wind forces on a bluff body.

$$p_m = p_{amb} + \frac{1}{2}\rho V_u^2, \tag{4.63}$$

where p_{amb} is the ambient air pressure which is equal to 0 at atmospheric pressure, ρ is the air density, and V_u is the upstream air speed.

However, the average dynamic pressure on the bridge span will be less than the maximum dynamic pressure given by Equation 4.63. Therefore, the dynamic pressure, p, at any point on the bluff body can be expressed as

$$p = C_p p_m = C_p\left(\frac{1}{2}\rho V_u^2\right), \tag{4.64}$$

where C_p is a dimensionless mean pressure coefficient that depends on the shape of the obstruction.

For example, if we assume a 160 km/h (100 mph) wind speed (which may occur during gale and hurricane events), Equation 4.63 yields a maximum dynamic pressure of 1.13 kPa (23.7 psf).

Design wind forces must be based on average dynamic wind pressures (i.e., reduced by use of an appropriate pressure coefficient) calculated over an appropriate cross-sectional area. The design wind forces must also consider the effects of wind gusts.* It is beneficial, from a design perspective, to calculate design wind forces based on the maximum dynamic pressure, a characteristic area, and a dimensionless coefficient that includes the effects of bridge cross-sectional shape as well as the wind flow characteristics.† These coefficients are determined from tests and applied to the design process. If the dynamic pressure, p, is integrated over the surface of the bluff body, it will create a force, F, and moment, M, as shown in Figure 4.34b. The force is resolved into horizontal (drag), F_D, and vertical (lift), F_L, forces. The equations for the forces and moment can then be expressed in a form similar to Equation 4.64 as

$$F_D = C_D\left(\frac{1}{2}\rho V_u^2\right)A_{RD}, \tag{4.65}$$

$$F_L = C_L\left(\frac{1}{2}\rho V_u^2\right)A_{RL}, \tag{4.66}$$

$$M = C_M\left(\frac{1}{2}\rho V_u^2\right)A_{RM}^2, \tag{4.67}$$

* Gust factors are generally between 2 and 3 for tall structures (Liu, 1991).

† Wind flow characteristics are described by the Reynolds number on a characteristic geometry, which is dependent on wind velocity and viscosity.

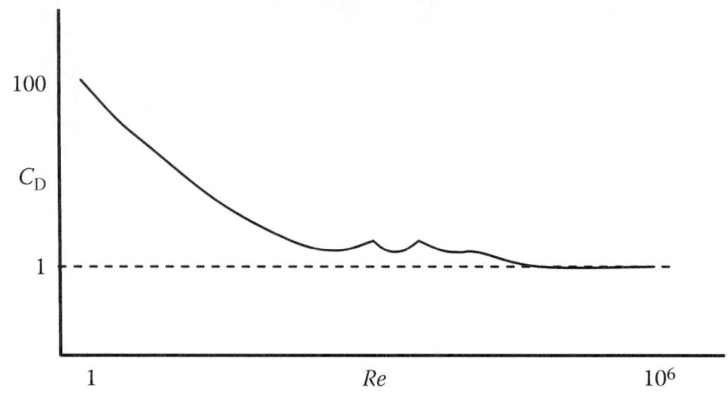

FIGURE 4.35 Typical relationship between drag coefficient, C_D, and Reynolds number, *Re*.

where C_D is a dimensionless drag coefficient that depends on span geometry and Reynolds number, *Re*. The Reynolds number is indicative of wind flow patterns related to inertial effects (*Re* large and C_D small) and viscous effects (*Re* small and C_D large). Figure 4.35 illustrates the typical relationship between C_D and *Re*. The Reynolds number, *Re*, is

$$Re = \frac{\rho V_u L_D}{\mu}, \tag{4.68}$$

L_D = a characteristic length of the bridge or object for drag
C_L = dimensionless lift coefficient
C_M = dimensionless moment coefficient
A_{RD} = a characteristic area of the bridge or object for drag
A_{RL} = a characteristic area of the bridge or object for lift
A_{RM} = a characteristic area of the bridge or object for moment
μ = dynamic wind viscosity

The drag force or total wind thrust on a bluff body, such as a bridge cross section, created by wind flow is of primary interest for the design of bridges. Therefore, drag coefficients are established by wind tunnel tests, which incorporate the effects of geometry and flow characteristics (as described by the Reynolds number), and which may be used for design purposes. Drag coefficients for a solid element, such as a plate girder, are generally no greater than about 2.0 and drag coefficients for a truss are typically about 1.70 (Simiu and Scanlon, 1986). The difference is related primarily to the characteristic dimension of effective area, generally taken as the area projected onto a plane normal to the wind flow. The solidity ratio, φ, is defined as the ratio of the effective area to the gross area.

The effects of the usual pairing of girders, trusses, and arches in steel railway bridges must also be considered. Figure 4.36 illustrates the typical relationship between the drag coefficient relating to the total wind force on two girders or trusses, C_{DT}, and the drag coefficient for a single girder or truss, C_D, in terms of the solidity ratio, φ, for spans with girder or truss spacing, *s*, no greater than the girder or truss height, *h*. Examples 4.9 and 4.10 outline the use of these drag coefficients.

Example 4.9a (SI Units)

A 40 m long ballasted deck steel deck plate girder span is shown in Figure E4.6a. Determine the design wind force for a wind speed of 120 km/h.

ρ =1.134 kg/m3
The solidity ratio φ =1.00 (plate girder)
$s/h = 0.67$
$C_{DT} = 1.15\ (C_D) = 1.15\ (2.0) = 2.3$ (Figure 4.36)

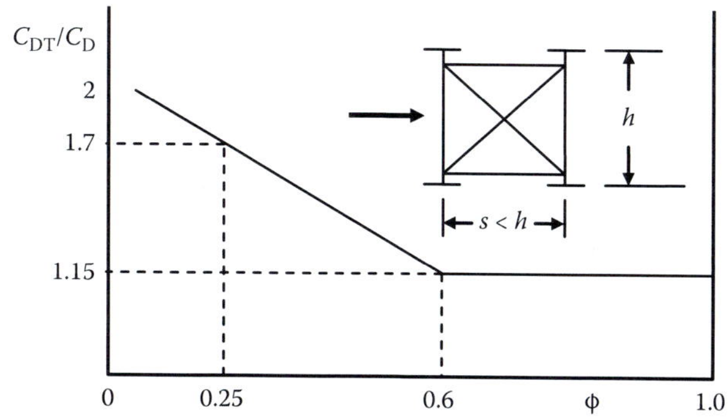

FIGURE 4.36 Typical relationship between the total drag coefficient on two girders or trusses, C_{DT}, and the drag coefficient for a single girder or truss, C_D.

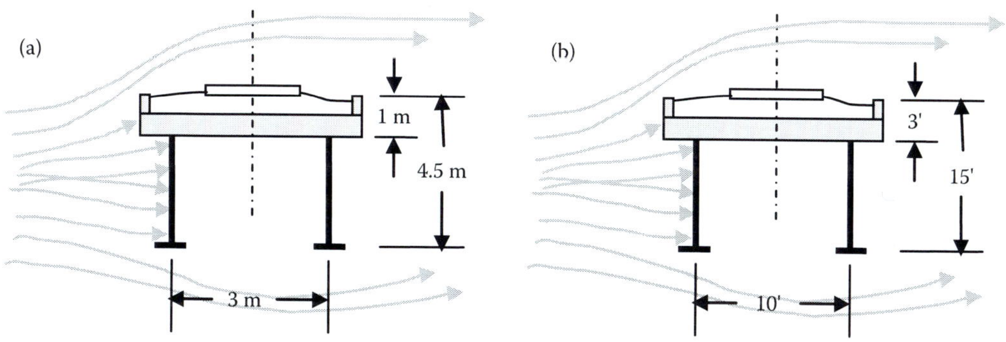

FIGURE E4.6 Cross-section of span.

$F_D = 2.3\left(\frac{1}{2}(1.134)(33.3)^2\right)A_{RD} = \frac{1449(40)(4.5)}{1000} = 261$ kN, not including gust factor. If we assume a typical gust factor of 2.0, the design wind force = 2.0(261) = 522 kN.

For a wind speed of 160 km/h,

$F_D = 2.3\left(\frac{1}{2}(1.134)(44.4)^2\right)A_{RD} = \frac{2576(40)(4.5)}{1000} = 464$ kN. With a gust factor of 2, $F_D = 927$ kN.

Example 4.9b (US Customary and Imperial Units)

A 125 ft long ballasted deck steel deck plate girder span is shown in Figure E4.6b. Determine the design wind force for a wind speed of 75 mph.

$\rho = 0.0022$ slug/ft^3

The solidity ratio $\varphi = 1.00$ (plate girder)

$s/h = 0.67$

$C_{DT} = 1.15\ (C_D) = 1.15\ (2.0) = 2.3$ (Figure 4.36)

$F_D = 2.3\left(\frac{1}{2}(0.0022)(110)^2\right)A_{RD} = \frac{30.6(125)(15)}{1000} = 57.4$ kips, not including gust factor. If we assume a typical gust factor of 2.0, the design wind force = 2.0(57.4) = 114.8 kips.

For a wind speed of 100 mph,

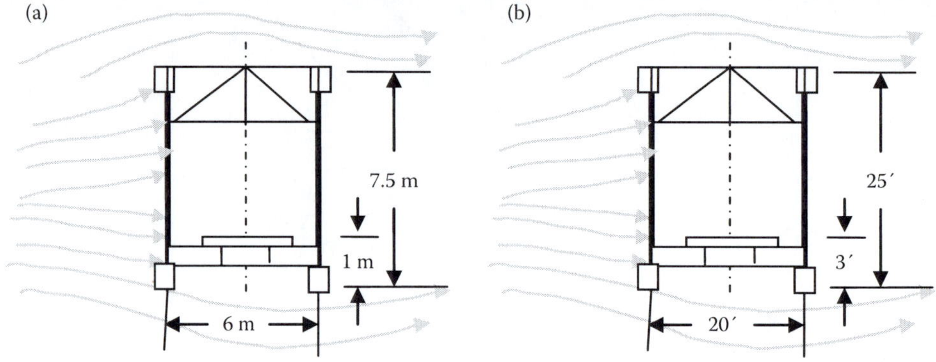

FIGURE E4.7 Cross-section of span.

$$F_D = 2.3\left(\frac{1}{2}(0.0022)(147)^2\right)A_{RD} = \frac{54.4(125)(15)}{1000} = 102 \text{ kips. With a gust factor of 2, } F_D = 204 \text{ kips.}$$

Example 4.10a (SI Units)

A 60 m steel through truss railway span is shown in Figure E4.7a. The solidity ratio, φ, for this truss is 0.25. Determine the design wind force for a wind speed of 120 km/h.

$s/h = 0.80$

$C_{DT} = 1.70\ (C_D) = 1.70\ (1.7) = 2.9$ (from Figure 4.36)

$F_D = 2.9\left(\frac{1}{2}(1.134)(33.3)^2\right)A_{RD} = \frac{1823(60)(7.5)(0.25)}{1000} = 205$ kN, not including gust factor. If we assume a typical gust factor of 2.0, the design wind force = 2.0(205) = 410 kN.

For a wind speed of 160 km/h,

$F_D = 2.3\left(\frac{1}{2}(1.134)(44.4)^2\right)A_{RD} = \frac{2576(60)(7.5)(0.25)}{1000} = 290$ kN. With a gust factor of 2, $F_D = 580$ kN.

Example 4.10b (US Customary and Imperial Units)

A 200 ft steel through truss railway span is shown in Figure E4.7b. The solidity ratio, φ, for this truss is 0.25. Determine the design wind force for a wind speed of 75 mph.

$s/h = 0.80$

$C_{DT} = 1.70\ (C_D) = 1.70\ (1.7) = 2.9$ (from Figure 4.36)

$F_D = 2.9\left(\frac{1}{2}(0.0022)(110)^2\right)A_{RD} = \frac{38.6(200)(25)(0.25)}{1000} = 48.3$ kips, not including gust factor. If we assume a typical gust factor of 2.0, the design wind force = 2.0(48.3) = 96.5 kips.

For a wind speed of 100 mph,

$F_D = 2.3\left(\frac{1}{2}(0.0022)(147)^2\right)A_{RD} = \frac{54.4(200)(25)(0.25)}{1000} = 68$ kips. With a gust factor of 2, $F_D = 136$ kips.

The AREMA (2015) design recommendations for wind force on a loaded steel railway bridge superstructure assume that the maximum wind velocity at which trains can safely operate* will produce a wind pressure of 1.44 kPa (30 psf). The AREMA (2015) design recommendations for wind load on an unloaded steel railway bridge superstructure assume a maximum wind velocity corresponding

* To avoid the overturning of empty cars.

to a typical hurricane event with a wind pressure of 2.39 kPa (50 psf) (see Examples 4.11 and 4.12). In order to account for the effects of paired or multiple girders, these wind pressures are to be applied to a surface area 50% greater than the projected surface area of a girder span. For truss spans the area is taken as the projected surface area of the windward truss plus the projected surface area of the leeward truss not shielded by the floor system.

The AREMA (2015) design recommendations also indicate that the load on the moving train is to be taken as 4.38 kN/m (300 lb/ft) at a distance 2.4 m (8 ft) above the top of the rails. Designers of railway bridges that carry high loads (e.g., double-stack rail cars) should review this recommendation.

Example 4.11a (SI Units)

Determine the AREMA (2015) recommended design wind force for the unloaded girder span of Figure E4.6a.

$$F_D = 2.39(1.5)(4.5)(40) = 645 \text{ kN}.$$

Example 4.11b (US Customary and Imperial Units)

Determine the AREMA (2015) recommended design wind force for the unloaded girder span of Figure E4.6b.

$$F_D = \frac{50(1.5)(15)(125)}{1000} = 140.6 \text{ kips}.$$

Example 4.12a (SI Units)

Determine the AREMA-recommended design wind force for the unloaded truss span of Figure E4.7a.

$$F_D = 2.39(0.25)((7.5 - 1.0) + 7.5)(60) = 502 \text{ kN}.$$

Example 4.12b (US Customary and Imperial Units)

Determine the AREMA-recommended design wind force for the unloaded truss span of Figure E4.7b.

$$F_D = \frac{50(0.25)((25 - 3) + 25)(200)}{1000} = 117.5 \text{ kips}.$$

A notional vibration load of 2.9 kN/m (200 lb/ft) applied to the loaded chord or flange and 2.2 kN/m (150 lb/ft) applied to the unloaded chord or flange is recommended in AREMA (2015) as an alternative to the wind pressure of 1.44 kPa (30 psf) on a loaded railway superstructure.* Example 4.13 outlines the calculation of wind forces on a ballasted steel deck plate girder span.

Example 4.13a (SI Units)

Determine the design wind force (including the notional vibration load) for the top and bottom lateral bracing of the 40 m long ballasted steel deck plate girder railway span shown in Figure E4.6a.

Unloaded span:

$W_T = (1+1.75)(2.39)(1.5) = 9.86$ kN/m wind at the top lateral bracing

$W_B = (1.75)(2.39)(1.5) = 6.27$ kN/m wind at the bottom lateral bracing

* The notional vibration load can occur only when the bridge is loaded by live load.

Loaded span:

$W_T = (1+1.75)(1.44)(1.5) + 4.38 = 10.32$ kN/m wind at the top lateral bracing

$W_B = (1.75)(1.44)(1.5) = 3.78$ kN/m wind at the bottom lateral bracing

$V_T = 2.90$ kN/m notional vibration load at the top lateral bracing

$V_B = 2.20$ kN/m notional vibration load at the bottom lateral bracing

Top lateral bracing design is based on $W_T = 10.32$ kN/m in addition to other lateral loads such as those due to live load. Bottom lateral bracing design is based on $W_B = 6.27$ kN/m.

Example 4.13b (US Customary and Imperial Units)

Determine the design wind force (including the notional vibration load) for the top and bottom lateral bracing of the 125 ft long ballasted steel deck plate girder railway span shown in Figure E4.6b.

Unloaded span:

$W_T = (3+6)(50)(1.5) = 675$ lb/ft wind at the top lateral bracing

$W_B = (6)(50)(1.5) = 450$ lb/ft wind at the bottom lateral bracing

Loaded span:

$W_T = (3+6)(30)(1.5) + 300 = 705$ lb/ft wind at the top lateral bracing

$W_B = (6)(30)(1.5) = 270$ lb/ft wind at the bottom lateral bracing

$V_T = 200$ lb/ft notional vibration load at the top lateral bracing

$V_B = 150$ lb/ft notional vibration load at the bottom lateral bracing

Top lateral bracing design is based on $W_T = 705$ lb/ft in addition to other lateral loads such as those due to live load. Bottom lateral bracing design is based on $W_B = 450$ lb/ft.

4.4.2 Thermal Forces from Continuous Welded Rail on Steel Railway Bridges

Continuously welded rail is used in modern track construction because it diminishes dynamic effects (no impact forces due to joints in the rail), provides a smoother ride, and results in reduced rail maintenance and increased tie life. The rail may be fastened to the deck to provide lateral and longitudinal restraint.* The deck is also fastened to the superstructure to provide lateral and longitudinal restraint.†

The longitudinal forces‡ generated due to restraint of thermal expansion and contraction of the rail and superstructure may need to be considered in the design of some steel railway superstructures. Longitudinal forces in the rail are due to thermal expansion and contraction restrained by rail-to-deck interface mechanical fasteners and the stone ballast. However, on open deck bridges, there are additional longitudinal forces created by thermal expansion and contraction of the superstructure with their magnitude dependent on the stiffness of rail-to-deck and deck-to-superstructure longitudinal restraint. The distribution of longitudinal forces through the superstructure may be of little concern for some superstructure types (e.g., multiple longitudinal beam spans and deck plate girder spans). However, for other types of superstructures and long spans, the longitudinal force path from rails to bearings may be of importance (e.g., floorbeams with direct fixation of track and some span floor systems). In addition to thermal actions, the CWR may also experience internal stresses due to bridge span movements from live load bending. The magnitude of the CWR–superstructure thermal interaction is governed by the following conditions:

* Longitudinal restraint by elastic hold-down fasteners, friction, and/or rail anchors applied at the base-of-rail against the ties. Lateral restraint by fasteners and tie plate seats.

† Ballasted decks are generally rigidly connected to the superstructure. However, open deck spans may have various degrees of longitudinal restraint depending on deck-to-superstructure connection (see Chapter 3) and the deck support surface. Open decks are often fastened to the superstructure with bolts or "hook bolts" installed at regular intervals (e.g., every third tie). Lateral deck restraint is typically controlled by fasteners and daps in ties.

‡ These longitudinal forces caused by the thermal expansion and contraction of restrained rails and superstructure are in addition to the longitudinal forces in the rails and superstructures due to traction and braking.

- Movement of the bridge spans, in particular the maximum span length, which may freely expand in the bridge
- The rail laying temperature (neutral temperature) and ambient temperature extremes at the bridge (the temperature ranges experienced by the rail and superstructure depend on neutral temperature, and maximum and minimum ambient temperatures)
- The type of bridge (open deck, ballasted*) and bridge materials
- The connection at rail-to-deck and deck-to-superstructure interfaces
- The cross-sectional area of the rail and coefficient of thermal expansion of the bridge
- The location of fixed and expansion bearings (spans with adjacent expansion bearings on the same pier create a long expansion length and generally provide the governing condition for design)

The partial differential equation of horizontal motion from force equilibrium on a simply supported span bridge of constant mass and stiffness is

$$-\mathrm{EA}\frac{\partial^2 \bar{x}(x,t)}{\partial x^2}+m\frac{\partial^2 \bar{x}(x,t)}{\partial t^2}+c_x\frac{\partial \bar{x}(x,t)}{\partial t}=h(x,t), \tag{4.69}$$

where

$\bar{x}(x,t)$ = superstructure horizontal deflection at distance x and time t
EA = axial stiffness of the span
c_x = superstructure equivalent longitudinal viscous damping coefficient
$h(x,t) = -k_d\bar{x}(x,t)$ is the distributed longitudinal force due to thermal movements transferred through an elastic rail-to-deck-to-superstructure system represented by an equivalent horizontal spring stiffness, k_d.† Longitudinal movement will typically occur primarily at the rail-to-deck or deck-to-superstructure interface depending on their respective degrees of longitudinal restraint.‡ Longitudinal restraint at the rail-to-deck interface is generally much greater than that at the deck-to-superstructure interface.§ Tests have indicated normal-strain-rate longitudinal restraint for smooth top deck supports of 0.70 N/mm (4 lb/in.) and 8.6 N/mm (49 lb/in.) for unanchored and fully anchored track, respectively.¶ Intermediate values for longitudinal restraint were observed for partially anchored track. High-strain-rate load application indicated longitudinal restraint about 50% lower than that resulting from normal-strain-rate load application. Longitudinal restraint at the smooth top deck support interfaces was about 25% of that at the rail-to-deck interface. The tests also showed that about 80% of the longitudinal displacement occurred at the smooth top deck support interface with the remainder at the rail-to-deck interface (Joy et al., 2007, 2009). The longitudinal restraint at the rail-to-deck and deck-to-superstructure interfaces affects the stress in the CWR, relative displacements between rail-to-deck and deck-to-superstructure at the interfaces, and the extent of rail travel upon rail fracture at cold temperatures (resulting in a gap in the rail).

---

* For ordinary ballasted deck bridges, the differential thermal movements are generally accommodated by the relatively flexible ballast section between the rail and superstructure.

† A linear elastic spring is assumed for all levels of displacement in this model. Rail-to-deck and deck-to-superstructure interfaces may be more accurately modeled using bilinear springs which, following initial elastic behavior, act perfectly plastically during steady-state sliding friction displacement.

‡ For example, longitudinal movement may occur at the deck-to-superstructure interface for open deck beams and girders with smooth tops, and at the rail-to-deck interface for girders with substantial longitudinal resistance at the deck-to-superstructure interface (e.g., by restraint angles and bars,) and positive deck connection. Some modern elastic rail fasteners allow for longitudinal movement (no "hold down" forces) at the rail-to-deck interface.

§ This is the case for modern superstructures designed with smooth deck support surfaces, even with tightened deck-to-superstructure connections. Tests of the deck-to-superstructure interface with riveted girder flanges and tightened deck-to-superstructure connections indicated less longitudinal restraint than that of the rail-to-deck interface (Joy et al., 2009). Conventional rail anchors and elastic fasteners provide considerable longitudinal restraint at the rail-to-deck interface.

¶ CWR specifications in China typically use 7.0 N/mm (40 lb/in.) for normal strain rate longitudinal restraint.

This model oversimplifies the rail-to-deck and deck-to-superstructure interaction with a single elastic horizontal stiffness, which, nevertheless, is appropriate for usual bridge design purposes. For more complex superstructures, sophisticated models may be developed* that use different elastic horizontal stiffness at the rail-to-deck and deck-to-superstructure interfaces.

Assuming negligible longitudinal viscous damping and neglecting superstructure longitudinal inertial effects (acceptable for ordinary steel railway superstructures), Equation 4.69 may be expressed as

$$-\mathrm{EA}\frac{\mathrm{d}^2\bar{x}(x)}{\mathrm{d}x^2}+k_\mathrm{d}\bar{x}(x)=0, \tag{4.70}$$

which may be solved considering various failure criteria, such as:

- Safe rail gap (separation) on a bridge after fracture of CWR. Rail fracture,† and the sudden release of tensile stresses in the rail, is a high-strain-rate event that may occur due to cold weather contraction, particularly at weld flaws and metallurgical discontinuities in the rail. The impulsive release of tensile rail force will result in a rail gap and transfer of forces to the deck and superstructure with magnitudes that depend on the stiffness of the longitudinal restraint at the rail-to-deck interface. The safe rail gap depends on individual railroad temporary operating practices, but it is generally considered to be between about 50 mm (2 in.) and 100 mm (4 in.) for relatively slow speed freight trains. Rigid longitudinal restraint may increase the risk of cold weather rail fracture (due to high rail stresses). The risk and consequences of rail fracture must be carefully assessed in conjunction with the safety concerns associated with CWR buckling and deck and fastener damage.‡
- Safe stress in the CWR to preclude buckling.§ Rail buckling, particularly at the typically weaker¶ bridge approach track, is a normal-strain-rate event that may occur when rails on the bridge are highly longitudinally restrained such that large rail forces are created during hot weather rail and superstructure expansion.
- Acceptable relative displacement between rail-to-deck and deck-to-superstructure to preclude damage to the deck and/or fasteners. Damage may occur due to excessive longitudinal movements at either rail-to-deck or deck-to-superstructure interfaces.
- Avoidance of bearing component damage.

4.4.2.1 Safe Rail Separation Criteria

If a steel bridge is modeled as a series of spans with a distributed longitudinal force, due to thermal expansion of rail, transferred through an elastic deck system, the magnitude of the axial force in the CWR, $N(x)$, is

$$N(x)=\mathrm{EA}_\mathrm{r}\left(\frac{\mathrm{d}\bar{x}(x)}{\mathrm{d}x}-\alpha\Delta t_\mathrm{c}\right), \tag{4.71}$$

* Usually used in conjunction with computer-based FEA.

† Modern North American heavy freight railroad CWR is considered to have a fracture strength of about 1300 kN (300 kips).

‡ Deck and fastener damage is typically not a safety concern for bridge decks inspected, maintained, and retrofitted in accordance with the appropriate regulatory requirements and/or guidelines (e.g., US FRA and TC Bridge Safety Management programs). Buckling of the rail is sudden and a critical safety condition may cause train derailments.

§ Modern North American heavy freight railroad CWR is considered to have a safe buckling strength of about 650 kN (150 kips).

¶ Weaker lateral restraint behind abutment backwalls and approach track.

where

EA_r = axial stiffness of CWR

A_r = cross-sectional area of CWR (about 8500 mm² (13 in.²) on typical North American heavy freight railroads)

α = coefficient of thermal expansion of CWR ~ 12×10^{-6}/°C (6.5×10^{-6}/F)

Δt_c = cold weather rail temperature change (with respect to neutral temperature)

Assuming zero displacement far from the rail break, $\bar{x}(\infty) = 0$, and zero force at the rail break, $N(0) = 0$, Equation 4.71 yields

$$\bar{x}(x) = \frac{-\alpha \Delta t_c}{\lambda} e^{-\lambda x}, \tag{4.72}$$

$$N(x) = -EA_r \alpha \Delta t_c (1 - e^{-\lambda x}), \tag{4.73}$$

where

$$\lambda = \sqrt{\frac{k_l}{EA_r}}$$

k_l = longitudinal stiffness associated with a high-strain-rate event such as a rail breaking. It is generally about ½ of the normal-strain-rate event (such as rail thermal expansion and contraction) stiffness.

The separation of the CWR at fracture (likely to occur over the expansion bearings) is

$$\Delta \bar{x}_S = -\alpha \Delta t_c \left(\frac{1}{\lambda_d} + \frac{1}{\lambda_t} \right), \tag{4.74}$$

where

$$\lambda_d = \sqrt{\frac{k_d}{EA_r}}$$

$$\lambda_t = \sqrt{\frac{k_t}{EA_r}}$$

k_d = the equivalent high-strain-rate event horizontal spring constant for bridge deck

k_t = the equivalent high-strain-rate event horizontal spring constant for track approach

Figure 4.37 outlines the relationship of Equation 4.74, where $Fk = 1 + \sqrt{k_d / k_t}$.

4.4.2.2 Safe Stress in the CWR to Preclude Buckling

Assuming a multiple span bridge with n spans of equal length, L, with alternating fixed and expansion bearings on substructures, and the boundary conditions of (Figure 4.38):

- Zero displacement of the CWR away from the bridge (e.g., $\bar{x}_5(\infty) = 0$)
- Compatibility of displacements in the rail over bearings (e.g., $\bar{x}_2(L_2) = \bar{x}_3(0)$)
- Zero displacement at fixed bearings (e.g., $\bar{x}_7(0) = 0$)
- Rail force compatibility over bearings (e.g., $N_2(L_2) = N_3(0)$).
- Zero forces at expansion bearings (e.g., $N_6(L_6) = 0$).

Rail separation criteria

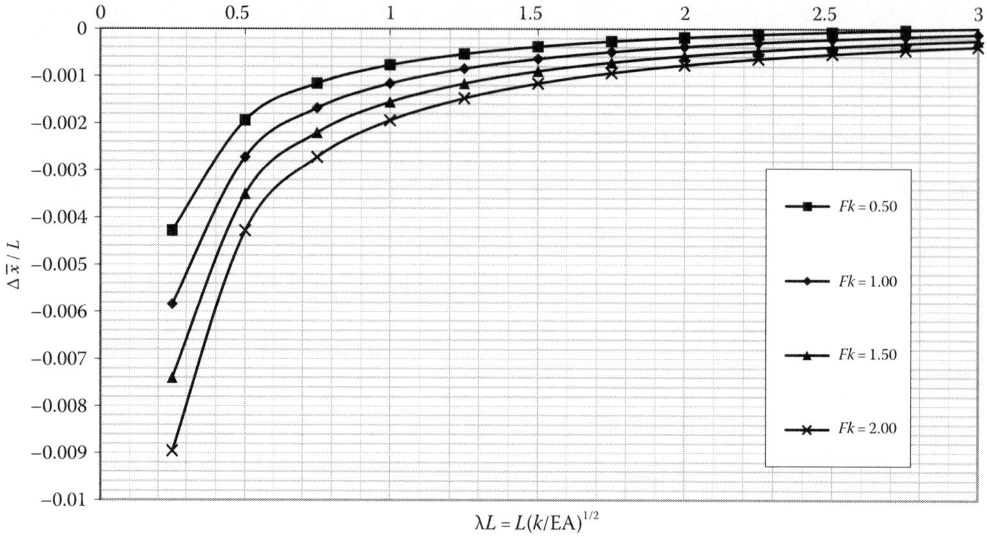

FIGURE 4.37 Typical relationships between rail separation, length of span, deck and track stiffness, and rail size for four stiffness ratios.

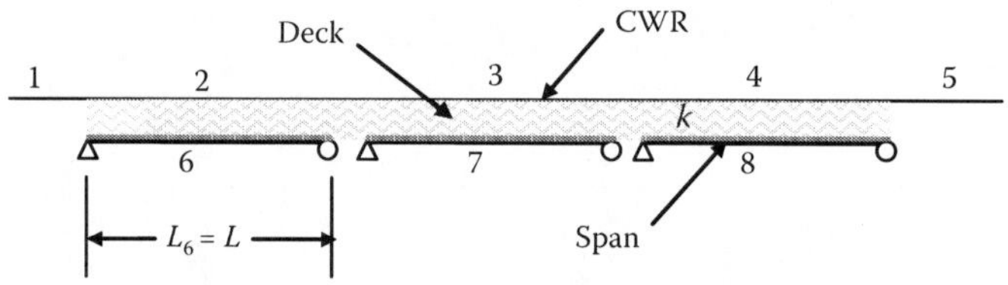

FIGURE 4.38 Three-span bridge model.

Equation 4.70 may be solved to yield

$$\sigma_{\mathrm{cwr}} = \frac{N_{n+1}(L)}{A_{\mathrm{r}}} = -E\alpha\Delta t_{\mathrm{h}}\left(1 + \frac{\alpha_0 \Delta T_{\mathrm{h}}}{2\alpha\Delta t_{\mathrm{h}}}\left(\lambda L - 1 + C_n\right)\right), \tag{4.75}$$

where

$$C_1 = e^{-\lambda L}$$

$$C_n = \left(\lambda L + C_{n-1}\right)e^{-\lambda L} \quad \text{for } n \geq 2$$

α_0 = coefficient of thermal expansion of the bridge

Δt_{h} = the hot weather rail temperature change with respect to neutral temperature

ΔT_{h} = the hot weather bridge temperature change with respect to construction temperature

$$\lambda = \sqrt{\frac{k}{\mathrm{EA_r}}}$$

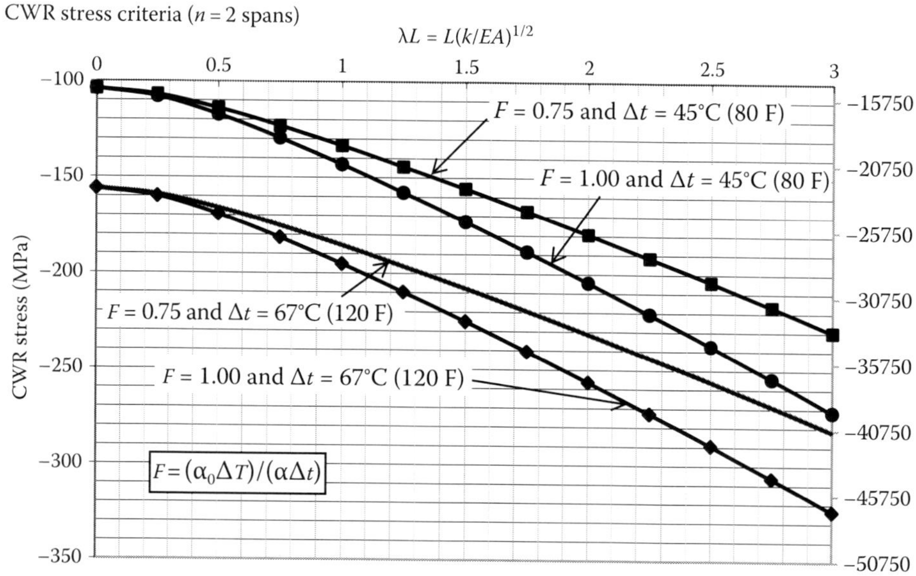

FIGURE 4.39 Typical relationships between stress in CWR, length of span, deck and track stiffness, and rail size for two expansion/contraction ratios.

k = the equivalent normal-strain-rate event horizontal spring constant modeling rail-to-deck-to-superstructure interaction

Figure 4.39 shows the relationships of Equation 4.75 for a two-span bridge with two expansion ratios, $F = \dfrac{\alpha_0 \Delta T_h}{\alpha \Delta t_{h,}}$, at $\Delta t_{h} = 67°C$ (120°F) and $\Delta t_h = 45°C$ (80°F).

4.4.2.3 Acceptable Relative Displacement between Rail-to-Deck and Deck-to-Span

Assuming a multiple span bridge with n spans of equal length, L, and alternating fixed and expansion bearings on substructures, Equation 4.70 with the boundary conditions outlined in Section 4.4.2.2 may be solved to yield

$$\Delta\bar{x} = \bar{x}_{2(n+1)}(L) - \bar{x}_{n+1}(L) = \frac{\alpha_0 \Delta T}{2\lambda}\left(1 + \lambda L - C_n\right), \tag{4.76}$$

where ΔT is the change in bridge temperature with respect to construction temperature.

Examples 4.14 and 4.15 illustrate the calculation of rail gap on the superstructure following cold weather fracture, the axial stress in the CWR due to thermal expansion and contraction of the rails and superstructure, the relative displacements between rail-to-deck and deck-to-superstructure,* and the magnitude of bearing forces for relatively short and long spans, respectively.

Example 4.14a (SI Units)

The double track open deck steel multibeam railway bridge shown in Figure E4.8 consists of two 13.7 m simple spans. CWR with elastic rail fastenings is used on the friction-bolt fastened†

* Combined into a rail-to-superstructure relative displacement in Examples 4.14 through 4.18. This is appropriate for the routine analysis of typical open deck steel railway bridges where the rail-to-deck interface is considerably longitudinally stiffer than the deck-to-superstructure interface.

† Often referred to as hook bolts in North America.

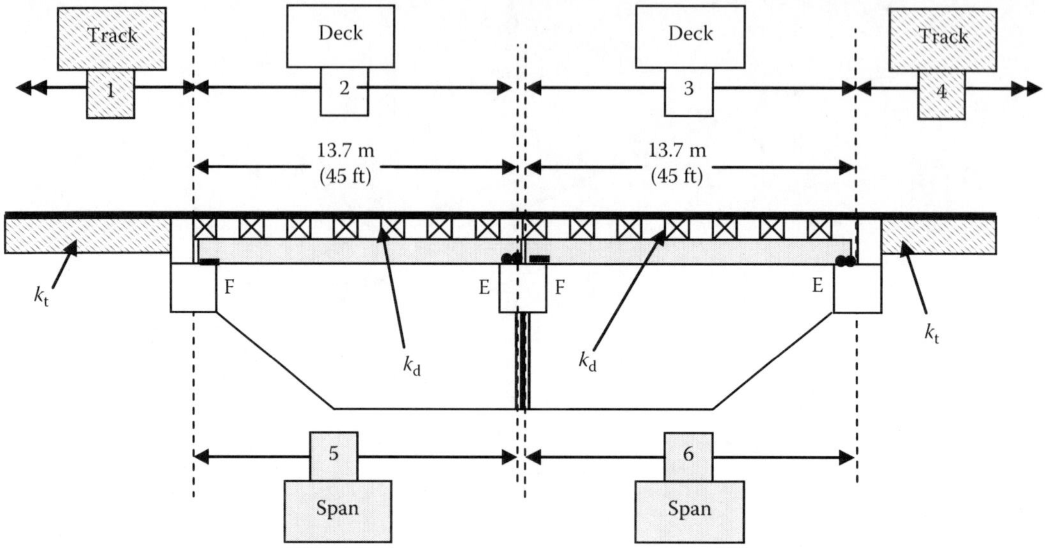

FIGURE E4.8 Elevation of bridge.

timber deck supported on smooth top flange surfaces. Determine the maximum stress in the CWR, relative displacement between the rail and superstructure, rail separation, and longitudinal bearing force at the pier. The following are the characteristics of the bridge:

$\Delta T_c = \Delta t_c = 55°C$,
$\Delta t_h = 25°C$,
$\Delta T_h = 20°C$,
$\alpha_0 \Delta T_c = -4.95 \times 10^{-4}$ (bridge),

$\alpha \Delta t_c = -6.60 \times 10^{-4}$(CWR),

$\alpha_0 \Delta T_h = 1.80 \times 10^{-4}$(bridge),

$\alpha \Delta t_h = 3.00 \times 10^{-4}$ (CWR),

$EA_r = 200{,}000(16{,}800) = 3.36 \times 10^9$ N per unit length of track = 3.36×10^6 kN per unit length of track (two typical CWRs),

$k_d = 3$ N/mm per unit length of track (normal strain rate),
$k_t = 5$ N/mm per unit length of track (normal strain rate).

Maximum stress in the CWR:

$$\lambda_d = \sqrt{\frac{k_d}{EA_r}} = 2.97 \times 10^{-5} \text{mm}^{-1},$$

$$\lambda_t = \sqrt{\frac{k_t}{EA_r}} = 3.83 \times 10^{-5} \text{mm}^{-1},$$

$$\lambda_d L = 0.41.$$

Substitution into Equation 4.75 with $n = 2$ yields

$$\sigma_{cwr} = -60.0\left(1 + 0.30\left(\lambda_d L - 1 + \left(\lambda_d L + e^{-\lambda_d L}\right)e^{-\lambda_d L}\right)\right),$$

$$\sigma_{cwr} = -60.0\left(1+0.30\left(0.41-1+\left(0.41+0.67\right)0.67\right)\right) = -62.4 \text{ MPa} \quad \text{for two rails.}$$

Force in each rail = 62.4 (8400)/1000 = 524 kN compression, OK.

Rail separation:

k_d = 1.5 N/mm per unit length of track (rapid strain rate)

k_t = 2 N/mm per unit length of track (rapid strain rate)

$$\lambda_d = \sqrt{\frac{k_d}{EA_r}} = 2.10\times10^{-5}\text{mm}^{-1},$$

$$\lambda_t = \sqrt{\frac{k_t}{EA_r}} = 2.44\times10^{-5}\text{mm}^{-1}.$$

Substitution into Equation 4.74 yields

$\overline{\Delta x_s} = -6.60\times10^{-4}\left(\dfrac{1}{2.10\times10^{-5}}+\dfrac{1}{2.44\times10^{-5}}\right) = -58$ mm, which may be marginally acceptable.

Relative displacement:

Substitution into Equation 4.76 with n = 2 yields

$$\overline{\Delta x} = 8.3\left(1+\lambda L-\left(\lambda L+e^{-\lambda L}\right)e^{-\lambda L}\right),$$

$\overline{\Delta x} = 8.3\left(1+0.41-\left(0.41+0.67\right)0.67\right) = 5.7$ mm, which is likely OK and will not cause deck fastener damage.

The fixed bearing force at the pier is

$$X_F = N_4(0)-N_3(0) = -EA_r\alpha\Delta t\left(\frac{\alpha_0\Delta T}{2\alpha\Delta t}(C_3-C_2)\right) = -2244(0.375)(C_3-C_2),$$

$\lambda_d L = 0.41,$

$C_1 = e^{-\lambda L} = 0.67,$

$C_2 = (\lambda_d L + C_1)e^{-\lambda L} = 0.72,$

$C_3 = (\lambda_d L + C_2)e^{-\lambda L} = 0.75,$

$X_F - 842(0.75-0.72) = 25.3$ kN for both bearings.

Example 4.14b (US Customary and Imperial Units)

The double track open deck steel multibeam railway bridge shown in Figure E4.8 consists of two 45 ft simple spans. CWR with elastic rail fastenings is used on the friction-bolt fastened timber deck supported on smooth top flange surfaces. Determine the maximum stress in the CWR, relative displacement between the rail and superstructure, rail separation, and longitudinal bearing force at the pier. The following are the characteristics of the bridge:

$\Delta T_c = \Delta t_c = 100°F,$

$\Delta t_h = 50°F,$

$\Delta T_h = 40°F,$

$\alpha_0\Delta T_c = -5.00\times10^{-4}$ (bridge),

$\alpha\Delta t_c = -6.50\times10^{-4}$ (CWR),

$\alpha_0 \Delta T_h = 2.00 \times 10^{-4}$ (bridge),
$\alpha \Delta t_h = 3.25 \times 10^{-4}$ (CWR),
$EA_r = 29 \times 10^6(26) = 7.5 \times 10^8$ lb per unit length of track (two typical CWRs),
$k_d = 400$ lb/in. per unit length of track (normal strain rate),
$k_t = 550$ lb/in. per unit length of track (normal strain rate).
Maximum stress in the CWR:

$$\lambda_d = \sqrt{\frac{k_d}{EA_r}} = \sqrt{\frac{400}{(7.5 \times 10^8)}} = 7.30 \times 10^{-4} \text{ in.}^{-1},$$

$$\lambda_t = \sqrt{\frac{k_t}{EA_r}} = \sqrt{\frac{550}{(7.5 \times 10^8)}} = 8.56 \times 10^{-4} \text{ in.}^{-1},$$

$$\lambda_d L = 0.39.$$

Substitution into Equation 4.75 with $n = 2$ yields

$$\sigma_{cwr} = -9425\left(1 + 0.308\left(\lambda_d L - 1 + \left(\lambda_d L + e^{-\lambda_d L}\right) e^{-\lambda_d L}\right)\right),$$

$$\sigma_{cwr} = -9425\left(1 + 0.308\left(0.39 - 1 + (0.39 + 0.68)0.68\right)\right) = -9766 \text{ psi for two rails.}$$

Force in each rail = 9766 (13)/1000 = 127 kips compression, OK.

Rail separation:

$k_d = 200$ lb/in. per unit length of track (rapid strain rate),
$k_t = 300$ lb/in. per unit length of track (rapid strain rate),

$$\lambda_d = \sqrt{\frac{k_d}{EA_r}} = 5.16 \times 10^{-4} \text{ in.}^{-1},$$

$$\lambda_t = \sqrt{\frac{k_t}{EA_r}} = 6.32 \times 10^{-4} \text{ in.}^{-1}.$$

Substitution into Equation 4.74 yields

$$\Delta \bar{x}_s = -6.50 \times 10^{-4}\left(\frac{1}{5.16 \times 10^{-4}} + \frac{1}{6.32 \times 10^{-4}}\right) = -2.3 \text{ in.,}$$ which may be marginally acceptable.

Relative displacement:

Substitution into Equation 4.76 with $n = 2$ yields

$$\Delta \bar{x} = 0.34\left(1 + \lambda L - \left(\lambda L + e^{-\lambda L}\right) e^{-\lambda L}\right),$$

$\Delta \bar{x} = 0.34\left(1 + 0.39 - (0.39 + 0.68)0.68\right) = 0.23$ in., which is likely OK and will not cause deck fastener damage.

The fixed bearing force at the pier is

$$X_F = N_4(0) - N_3(0) = -EA_r \alpha \Delta t\left(\frac{\alpha_0 \Delta T}{2\alpha \Delta t}(C_3 - C_2)\right) = -487{,}500(0.385)(C_3 - C_2),$$

$$\lambda_d L = 0.39,$$

$$C_1 = e^{-\lambda L} = 0.68,$$

$$C_2 = (\lambda_d L + C_1)e^{-\lambda L} = 0.73,$$

$$C_3 = (\lambda_d L + C_2)e^{-\lambda L} = 0.76,$$

$$X_F - 187{,}688(0.76 - 0.73) = 6081 \text{ lb for both bearings.}$$

Example 4.15a (SI Units)

The double track open deck steel multibeam railway bridge of Example 4.14a is rehabilitated with stiffer rail/tie and tie/superstructure interface stiffness. Determine the maximum stress in the CWR, relative displacement between the rail and superstructure and rail separation.

k_d = 14 N/mm per unit length of track (normal strain rate),

k_t = 5 N/mm per unit length of track (normal strain rate).

Maximum stress in the CWR:

$$\lambda_d = \sqrt{\frac{k_d}{EA_r}} = 6.45 \times 10^{-5} \text{mm}^{-1},$$

$$\lambda_t = \sqrt{\frac{k_t}{EA_r}} = 3.83 \times 10^{-5} \text{mm}^{-1},$$

$$\lambda_d L = 0.88.$$

Substitution into Equation 4.75 with $n = 2$ yields

$$\sigma_{cwr} = -60.0\left(1 + 0.30\left(0.88 - 1 + (0.88 + 0.41)0.41\right)\right) = -67.4 \text{ MPa} \quad \text{for two rails.}$$

Force in each rail = 67.4 (8400)/1000 = 566 kN compression, OK.

Rail separation:

k_d = 8 N/mm per unit length of track (rapid strain rate),

k_t = 2 N/mm per unit length of track (rapid strain rate),

$$\lambda_d = \sqrt{\frac{k_d}{EA_r}} = 4.88 \times 10^{-5} \text{ mm}^{-1},$$

$$\lambda_t = \sqrt{\frac{k_t}{EA_r}} = 2.44 \times 10^{-5} \text{ mm}^{-1}.$$

Substitution into Equation 4.74 yields

$\overline{\Delta x}_s = -6.60 \times 10^{-4}\left(\frac{1}{4.88 \times 10^{-5}} + \frac{1}{2.44 \times 10^{-5}}\right) = -41 \text{ mm}$, which is generally considered as acceptable.

Relative displacement:

Substitution into Equation 4.76 with $n = 2$ yields

$\overline{\Delta x} = 8.3\left(1 + 0.88 - (0.88 + 0.41)0.41\right) = 11.2$ mm, which may cause some deck fastener damage.

Example 4.15b (US Customary and Imperial Units)

The double track open deck steel multibeam railway bridge in Example 4.14b is rehabilitated with stiffer rail/tie and tie/superstructure interface stiffness. Determine the maximum stress in the CWR, relative displacement between the rail and superstructure and rail separation.

k_d = 1800 lb/in. per unit length of track (normal strain rate),

k_t = 550 lb/in. per unit length of track (normal strain rate).

Maximum stress in the CWR:

$$\lambda_d = \sqrt{\frac{k_d}{EA_r}} = \sqrt{\frac{1800}{(7.5\times10^8)}} = 1.55\times10^{-3}\ \text{in.}^{-1},$$

$$\lambda_t = \sqrt{\frac{k_t}{EA_r}} = \sqrt{\frac{550}{(7.5\times10^8)}} = 8.56\times10^{-4}\ \text{in.}^{-1},$$

$$\lambda_d L = 0.84.$$

Substitution into Equation 4.75 with $n = 2$ yields

$\sigma_{cwr} = -9425\big(1+0.308\big(0.84-1+\big(0.84+0.43\big)0.43\big)\big) = -10,520$ psi for two rails.

Force in each rail = 10520 (13)/1000 = 137 kips compression, OK.

Rail separation:

k_d = 1000 lb/in. per unit length of track (rapid strain rate),

k_t = 300 lb/in. per unit length of track (rapid strain rate),

$$\lambda_d = \sqrt{\frac{k_d}{EA_r}} = 1.15\times10^{-3}\ \text{in.}^{-1},$$

$$\lambda_t = \sqrt{\frac{k_t}{EA_r}} = 6.32\times10^{-4}\ \text{in.}^{-1}.$$

Substitution into Equation 4.74 yields

$\Delta\bar{x}_s = -6.50\times10^{-4}\left(\frac{1}{1.15\times10^{-3}}+\frac{1}{6.63\times10^{-4}}\right) = -1.5$ in., which is generally considered acceptable.

Relative displacement:

Substitution into Equation 4.76 with $n = 2$ yields

$\Delta\bar{x} = 0.34\big(1+0.84-\big(0.84+0.43\big)0.43\big) = 0.44$ in., which may cause some deck fastener damage.

For bridges with short spans, the amount of thermal movement per span is small and generally easily accommodated by the normal tolerances of railroad track on bridges. It should be noted that the longitudinal stiffness values assumed for the rail-to-deck and deck-to-superstructure interfaces may not reflect actual values at a particular bridge. Appropriate values for particular applications may be developed from research documents, obtained from codes and standards, and/or by relatively simple testing of suitable existing bridge decks.*

Example 4.16a (SI Units)

An open deck steel deck truss bridge consists of a single 70 m span. CWR with elastic rail fastenings is used on the friction-bolt fastened timber deck supported on smooth top flange surfaces. Determine the maximum stress in the CWR, relative displacement between the rail and

* Testing may be most appropriate for long span and/or complex superstructures.

superstructure, rail separation, and longitudinal bearing force at the abutment. The following are the characteristics of the bridge:

$\Delta T_c = \Delta t_c = 40°C$,
$\Delta t_h = 25°C$,
$\Delta T_h = 35°C$,
$\alpha_0 \Delta T_c = -3.60 \times 10^{-4}$ (bridge),
$\alpha \Delta t_c = -4.80 \times 10^{-4}$ (CWR),
$\alpha_0 \Delta T_h = 3.15 \times 10^{-4}$ (bridge),
$\alpha \Delta t_h = 3.00 \times 10^{-4}$ (CWR),
$EA_r = 200{,}000(16800) = 3.36 \times 10^{9\text{N}}$ per unit length of track (two typical CWRs),
$k_d = 5$ N/mm per unit length of track (normal strain rate),
$k_t = 10$ N/mm per unit length of track (normal strain rate).
Maximum stress in the CWR:

$$\lambda_d = \sqrt{\frac{k_d}{EA_r}} = \sqrt{\frac{5}{(3.36 \times 10^9)}} = 3.86 \times 10^{-5} \text{ mm}^{-1},$$

$$\lambda_t = \sqrt{\frac{k_t}{EA_r}} = \sqrt{\frac{10}{(3.36 \times 10^9)}} = 5.46 \times 10^{-5} \text{ mm}^{-1},$$

$$\lambda_d L = 2.70.$$

Substitution into Equation 4.75 with $n = 1$ yields

$$\sigma_{cwr} = -60.0\left(1 + 0.375\left(\lambda_d L - 1 + e^{-\lambda_d L}\right)\right),$$

$\sigma_{cwr} = -60.0\left(1 + 0.375(2.70 - 1 + 0.067)\right) = -96.7$ MPa for both rails.

Force in each rail = 96.7(8400)/1000 = 813 kN compression; the rail may buckle at weak locations such as bridge approaches.

Rail separation:
$k_d = 2.5$ N/mm per unit length of track (rapid strain rate),
$k_t = 5$ N/mm per unit length of track (rapid strain rate),

$$\lambda_d = \sqrt{\frac{k_d}{EA_r}} = 2.73 \times 10^{-5} \text{mm}^{-1},$$

$$\lambda_t = \sqrt{\frac{k_t}{EA_r}} = 3.86 \times 10^{-5} \text{mm}^{-1}.$$

Substitution into Equation 4.74 yields

$\overline{\Delta x_s} = -4.80 \times 10^{-4}\left(\frac{1}{2.73 \times 10^{-5}} + \frac{1}{3.86 \times 10^{-5}}\right) = 30.0$ mm, which is acceptable.

Relative displacement:
Substitution into Equation 4.76 with $n = 1$ yields

$$\overline{\Delta x} = 4.66\left(1 + \lambda L - e^{-\lambda L}\right),$$

$\Delta\bar{x} = 4.66(1+2.70-0.067) = 17$ mm, which may be excessive.

The fixed bearing force at the abutment is

$$X_F = N_3(0) - N_2(0) = -EA_r\alpha\Delta t\left(\frac{\alpha_0\Delta T}{2\alpha\Delta t}(C_2 - C_1)\right) = -529.2(C_2 - C_1),$$

$$C_1 = e^{-\lambda L} = 0.15,$$

$$C_2 = (\lambda_d L + C_1)e^{-\lambda L} = 0.31,$$

$$X_F - 529.2(0.31 - 0.15) = 84.7 \text{ kN for both bearings.}$$

Example 4.16b (US Customary and Imperial Units)

An open deck steel deck truss bridge consists of a single 225 ft span. CWR with elastic rail fastenings is used on the friction-bolt fastened timber deck supported on smooth top flange surfaces. Determine the maximum stress in the CWR, relative displacement between the rail and superstructure, rail separation, and longitudinal bearing force at the abutment. The following are the characteristics of the bridge:

$\Delta T_c = \Delta t_c = 70$ °F,

$\Delta t_h = 50$ °F,

$\Delta T_h = 40$ °F,

$\alpha_0\Delta T_c = -3.50\times10^{-4}$ (bridge),

$\alpha\Delta t_c = -4.55\times10^{-4}$ (CWR),

$\alpha_0\Delta T_h = 2.00\times10^{-4}$ (bridge),

$\alpha\Delta t_h = 3.25\times10^{-4}$ (CWR),

$EA_r = 29 \times 10^6(26) = 7.5 \times 10^8$ lb per unit length of track (two typical CWRs),

$k_d = 650$ lb/in. per unit length of track (normal strain rate),

$k_t = 1300$ lb/in. per unit length of track (normal strain rate).

Maximum stress in the CWR:

$$\lambda_d = \sqrt{\frac{k_d}{EA_r}} = \sqrt{\frac{650}{(7.5\times10^8)}} = 9.31\times10^{-4} \text{ in.}^{-1},$$

$$\lambda_t = \sqrt{\frac{k_t}{EA_r}} = \sqrt{\frac{1300}{(7.5\times10^8)}} = 13.17\times10^{-4} \text{ in.}^{-1},$$

$$\lambda_d L = 2.51.$$

Substitution into Equation 4.75 with $n = 1$ yields

$$\sigma_{cwr} = -9425\left(1+0.38\left(\lambda_d L - 1 + e^{-\lambda_d L}\right)\right),$$

$$\sigma_{cwr} = -9425(1+0.38(2.51-1+0.081)) = -14,995 \text{ psi} \quad \text{for both rails.}$$

Force in each rail = 14,995(13)/1000 = 195 kips compression; the rail may buckle at weak locations such as bridge approaches.

Rail separation:
$k_d = 300$ lb/in. per unit length of track (rapid strain rate),
$k_{t\,=}$ 650 lb/in. per unit length of track (rapid strain rate),

$$\lambda_d = \sqrt{\frac{k_d}{EA_r}} = 6.32 \times 10^{-4} \text{ in.}^{-1},$$

$$\lambda_t = \sqrt{\frac{k_t}{EA_r}} = 9.31 \times 10^{-4} \text{ in.}^{-1}.$$

Substitution into Equation 4.74 yields

$\overline{\Delta x_s} = -4.55 \times 10^{-4}\left(\dfrac{1}{6.32 \times 10^{-4}} + \dfrac{1}{9.31 \times 10^{-4}}\right) = 1.21$ in., which is acceptable.

Relative displacement:
Substitution into Equation 4.76 with $n = 1$ yields

$$\overline{\Delta x} = 0.19\left(1 + \lambda L - e^{-\lambda L}\right),$$

$\overline{\Delta x} = 0.19\left(1 + 2.51 - 0.081\right) = 0.65$ in., which may be excessive.

The fixed bearing force at the abutment is

$$X_F = N_3(0) - N_2(0) = -EA_r\alpha\Delta t\left(\frac{\alpha_0\Delta T}{2\alpha\Delta t}(C_2 - C_1)\right) = -131,250\left(C_2 - C_1\right),$$

$$C_1 = e^{-\lambda L} = 0.15,$$

$$C_2 = \left(\lambda_d L + C_1\right)e^{-\lambda L} = 0.31,$$

$$X_F - 131,250\left(0.31 - 0.15\right) = 20,556 \text{ lb for both bearings.}$$

For bridges with long spans, the amount of thermal movement per span is larger. CWR stresses that exceed buckling strength may occur due to high longitudinal restraint at the rail-to-deck interface. However, excessive relative displacements between the rail and superstructure that can cause deck and fastener damage may occur due to inadequate longitudinal restraint at either the rail-to-deck or deck-to-superstructure interface. Examples 4.17 and 4.18 illustrate the calculation of rail gap on the superstructure following cold weather fracture, the axial stress in the CWR due to thermal expansion and contraction of the rails and superstructure, and the relative displacements between rail-to-deck and deck-to-superstructure for relatively stiff and flexible longitudinal restraint, respectively.

Example 4.17a (SI Units)

In order to reduce the relative displacements at the rail-to-deck-to-superstructure system in Example 4.16a, a fastening system on the bridge with greater horizontal elastic spring stiffness is proposed. Determine the maximum stress in the CWR, relative displacement between the rail and superstructure and rail separation.

$k_d = 25$ N/mm per unit length of track (normal strain rate),
$k_{t\,=}$ 10 N/mm per unit length of track (normal strain rate).
Relative displacement:

$$\lambda_d = \sqrt{\frac{k_d}{EA}} = 8.63 \times 10^{-5} \text{ mm}^{-1},$$

$$\lambda_d L = 6.04.$$

Substitution into Equation 4.76 with $n = 1$ yields

$\Delta\bar{x} = 2.1(1+6.04-0.0036) = 14.7$ mm; the relative displacement is reduced by about 15% but remains fairly large.

Maximum stress in the CWR:

Substitution into Equation 4.75 with $n = 1$ yields

$\sigma_{cwr} = -60.0\left(1+0.375(6.04-1+0.003)\right) = -218$ MPa for both rails.

Force in each rail = 218 (8400)/1000 = 1835 kN compression; the rail may buckle.

Rail separation:

$k_d = 15$ N/mm per unit length of track (rapid strain rate),

$k_t = 5$ N/mm per unit length of track (rapid strain rate),

$$\lambda_d = \sqrt{\frac{k_d}{EA_r}} = 6.68\times10^{-5}\text{ in.}^{-1},$$

$$\lambda_t = \sqrt{\frac{k_t}{EA_r}} = 3.86\times10^{-5}\text{ in.}^{-1}.$$

Substitution into Equation 4.74 yields

$$\Delta\bar{x}_s = -4.80\times10^{-4}\left(\frac{1}{6.68\times10^{-5}}+\frac{1}{3.86\times10^{-5}}\right) = 19.6\text{ mm, OK.}$$

The longitudinally stiffer deck results in a reduction in relative displacement and rail separation upon rail fracture, but will create forces in the rails that exceed buckling strength.

Example 4.17b (US Customary and Imperial Units)

In order to reduce the relative displacements at the rail-to-deck-to-superstructure system in Example 4.16b, a fastening system on the bridge with greater horizontal elastic spring stiffness is proposed. Determine the maximum stress in the CWR, relative displacement between the rail and superstructure and rail separation.

$k_d = 2600$ lb/in. per unit length of track (normal strain rate),

$k_t = 1300$ lb/in. per unit length of track (normal strain rate).

Relative displacement:

$$\lambda_d = \sqrt{\frac{k_d}{EA}} = 18.62\times10^{-4}\text{ in.}^{-1},$$

$$\lambda_d L = 5.03.$$

Substitution into Equation 4.76 with $n = 1$ yields

$\Delta\bar{x} = 0.09(1+5.03-0.007) = 0.57$ in.; the relative displacement is reduced by about 15% but remains fairly large.

Maximum stress in the CWR:

Substitution into Equation 4.75 with $n = 1$ yields

$$\sigma_{cwr} = -9425\left(1+0.38\left(\lambda_d L-1+e^{-\lambda_d L}\right)\right),$$

$$\sigma_{cwr} = -9425\left(1+0.38(5.03-1+0.007)\right) = -31{,}000\text{ psi}\quad\text{for both rails.}$$

Force in each rail = 31,000 (13)/1000 = 403 kips compression; the rail may buckle.
Rail separation:
k_d = 1500 lb/in. per unit length of track (rapid strain rate),
k_t = 650 lb/in. per unit length of track (rapid strain rate),

$$\lambda_d = \sqrt{\frac{k_d}{EA_r}} = 14.14 \times 10^{-4} \text{ in.}^{-1},$$

$$\lambda_t = \sqrt{\frac{k_t}{EA_r}} = 9.31 \times 10^{-4} \text{ in.}^{-1}.$$

Substitution into Equation 4.74 yields

$$\Delta\bar{x}_s = -4.55\times10^{-4}\left(\frac{1}{14.14\times10^{-4}} + \frac{1}{9.31\times10^{-4}}\right) = 0.81 \text{ in., OK.}$$

The longitudinally stiffer deck results in a reduction in relative displacement and rail separation upon rail fracture, but will create forces in the rails that exceed buckling strength.

Examples 4.16 and 4.17 illustrate that, for long open deck spans, there are conflicting design requirements that the rail-to-deck-to-superstructure connection be flexible enough to avoid excessive compressive stress in CWR (that could precipitate buckling), and rigid enough to reduce rail separation (and the associated high-strain-rate forces suddenly imparted to the superstructure) and relative displacements at rail-to-deck and deck-to-superstructure interfaces. These conflicting design requirements must be balanced in order to reduce both CWR stresses and relative displacements at the rail-to-deck and deck-to-superstructure interfaces.

Therefore, in order to allow for unrestrained movement between the rail and superstructure while providing sufficient longitudinal rail restraint (anchoring) precluding excessive relative displacements, CWR may be anchored to only a portion of the deck length. The portion of the deck length to which the CWR is anchored should be adjacent to the fixed bearings to allow the necessary movement between the rail and superstructure (anchored CWR over the expansion-bearing areas will resist the thermal movements of the span). The effect of this is illustrated in Example 4.18.

Example 4.18a (SI Units)

In order to preclude rail buckling and reduce the relative displacement between the rail and superstructure in Example 4.16a, anchoring the CWR to only a portion of the rail is proposed. If only one-third of the span length (from fixed bearings) has the CWR anchored to the deck, determine the maximum stress in the CWR and relative displacement between the rail and superstructure.

Maximum stress in the CWR:

$$\lambda_d = \sqrt{\frac{k_d}{EA_r}} = \sqrt{\frac{5}{(3.36\times10^9)}} = 3.86 \times 10^{-5} \text{ mm}^{-1},$$

$$\lambda_t = \sqrt{\frac{k_t}{EA_r}} = \sqrt{\frac{10}{(3.36\times10^9)}} = 5.46 \times 10^{-5} \text{ mm}^{-1},$$

$$\lambda_d L = (70(1000/3))(3.86 \times 10^{-5}) = 0.90.$$

Substitution into Equation 4.75 with $n = 1$ yields

$\sigma_{cwr} = -60.0\left(1+0.375(0.90-1+0.41)\right) = -67.0$ MPa for both rails.

Force in each rail = 67.0 (8400)/1000 = 563 kN compression, OK. The force is reduced about 30% from 813 kN.

Relative displacement:

Substitution into Equation 4.76 with $n = 1$ yields

$$\Delta\bar{x} = 4.66\left(1+\lambda L - e^{-\lambda L}\right),$$

$\Delta\bar{x} = 4.66\left(1+0.90-0.41\right) = 6.9$ mm, OK, reduced about 60% from 17 mm.

The stress in the CWR and the relative displacement between the rail and superstructure are substantially reduced through partial anchoring of the rails to deck.

Example 4.18b (US Customary and Imperial Units)

In order to preclude rail buckling and reduce the relative displacement between the rail and superstructure in Example 4.16b, anchoring the CWR to only a portion of the rail is proposed. If only one-third of the span length (from fixed bearings) has the CWR anchored to the deck, determine the maximum stress in the CWR and relative displacement between the rail and superstructure.

Maximum stress in the CWR:

$$\lambda_d = \sqrt{\frac{k_d}{EA_r}} = \sqrt{\frac{650}{(7.5\times10^8)}} = 9.31\times10^{-4}\ \text{in.}^{-1},$$

$$\lambda_t = \sqrt{\frac{k_t}{EA_r}} = \sqrt{\frac{1300}{(7.5\times10^8)}} = 13.17\times10^{-4}\ \text{in.}^{-1},$$

$$\lambda_d L = (225/3)(12)(9.31\times10^{-4}) = 0.84.$$

Substitution into Equation 4.75 with $n = 1$ yields

$$\sigma_{cwr} = -9425\left(1+0.38\left(0.84-1+0.43\right)\right) = -10{,}394 \text{ psi for both rails.}$$

Force in each rail = 10,394(13)/1000 = 135 kips compression, OK. The force is reduced about 30% from 195 kips.

Relative displacement:

Substitution into Equation 4.76 with $n = 1$ yields

$\Delta\bar{x} = 0.19\left(1+0.84-0.43\right) = 0.27$ in., OK, reduced about 60% from 0.69 in.

The stress in the CWR and the relative displacement between the rail and superstructure are substantially reduced through partial anchoring of the rails to deck.

4.4.2.4 Design for CWR on Steel Railway Bridges

Based on similar considerations, AREMA (2015) and many railway companies establish standard practices for longitudinal anchoring of CWR to long open deck steel spans. In general, the recommended practice is to use longitudinal rail anchors or elastic rail fasteners on bridge approaches and near the fixed ends of spans, while allowing some movement near expansion ends of spans.* For very long or complex steel bridges, a numerical analysis using FEA software that can model thermal loads in rails, deck, and superstructure should be used. This software can also generally accommodate the analysis of longitudinal loads due to locomotive traction and train braking.

* Movement at the expansion end of spans is accommodated by unanchored rail or zero-longitudinal-force elastic fasteners. Rail expansion joints may be required for very long or complex bridges, or bridges with unusual bearing configurations (i.e., adjacent expansion bearings on long spans).

4.4.3 Seismic Forces on Steel Railway Bridges

The level of seismic dynamic analysis required depends on the location and characteristics of the bridge.

An equivalent static analysis is often used for the analysis of ordinary steel railway bridges where the response to seismic forces is depicted primarily by the first or fundamental vibration mode. Steel railway bridges that may be analyzed by an equivalent static analysis are typically simply supported, not (or only slightly) skewed or curved, and have spans of almost equal length and supporting substructures of almost equal stiffness. Seismic forces in an equivalent static analysis are developed based on a period-dependent coefficient and the weight of the bridge. AREMA (2015) recommends the use of a seismic response coefficient and the uniform load method.*

The seismic forces on complex steel railway bridges are generally determined for use in a dynamic structural analysis.† These loads are typically represented by an elastic design seismic response spectrum. AREMA (2015) recommends the use of a normalized response spectrum based on the seismic response coefficient.

4.4.3.1 Equivalent Static Lateral Force

The equivalent static distributed lateral force, $p(x)$, applied to the steel superstructures of a railway bridge is

$$p(x)=C_n w(x), \tag{4.77}$$

where

$C_n=\dfrac{1.2ASD}{T_n^{2/3}}\le 2.5AD$ = seismic response coefficient for the nth mode of vibration and a 5% damping ratio

$w(x)$ = distributed weight of the superstructure

A = base acceleration ratio determined from appropriate geological sources‡ for the design return period§

S = site coefficient between 1.0 and 2.0 depending on foundation soil conditions¶

$D=\left(\dfrac{1.5}{0.4\xi+1}+0.5\right)$ = damping adjustment factor to account for the actual superstructure percentage of critical damping, ξ**

T_n = natural period of the nth mode of vibration = $\dfrac{2\pi}{\omega_n}$

ω_n = natural frequency of the nth mode of vibration (see Tables 4.7 and 4.8, and Figure 4.21)

The equivalent static lateral distributed force, $p(x)$, is calculated in two orthogonal directions (longitudinal and transverse for ordinary bridges). Following a linear elastic analysis†† in each direction, forces are distributed to superstructure members based on load path, support conditions, and stiffness. Since these member loads are orthogonal and uncorrelated, they must be combined‡‡ for

* For some bridges, it may be appropriate to consider the multimode dynamic analysis method.

† Chapter 9 of AREMA indicates that a modal analysis is appropriate for such railway bridges.

‡ For example, the US Department of the Interior Geological Survey Maps.

§ The design return period depends on the earthquake event frequency and the limit state under consideration (serviceability, ultimate, or survivability).

¶ Rock, soil type, stratigraphy, depth, soil stiffness, and shear wave velocity are considered in the site coefficient.

** Established from tests or other sources in the literature of structural dynamics. The percentage of critical damping for steel superstructures is often less than 5% and depends on materials, structural system/foundation, deck type, and whether the structural response is linear elastic or post yield.

†† Linear elastic analysis is used for the equivalent lateral force method at the serviceability limit state.

‡‡ These combined forces account for the directional uncertainty and simultaneous occurrence of the seismic design forces in members.

design purposes. AREMA (2015) recommends the method often referred to as the 100%–30% rule (Equations 4.78a and b) to combine the seismic loads for member design.

$$\mathrm{EQ} = 1.00F_\mathrm{T} + 0.30F_\mathrm{L}, \tag{4.78a}$$

$$\mathrm{EQ} = 0.30F_\mathrm{T} + 1.00F_\mathrm{L}, \tag{4.78b}$$

where

EQ = combined seismic design force

F_T = absolute value of the seismic force in the transverse direction

F_L = absolute value of the seismic force in the longitudinal direction

However, in some cases, the development of the equivalent static distributed lateral force based on the seismic response coefficient is inappropriate and consideration of loading based on site-specific information is required.*

4.4.3.2 Response Spectrum Analysis of Steel Railway Superstructures

The response spectrum used to represent the seismic loading of more complex steel superstructures is a plot of the peak value of the response as a function of the natural period of vibration of the superstructure. These are typically graphed for a particular damping ratio† and response (deformation, velocity, or acceleration). AREMA (2015) recommends the use of a normalized spectral response based on the seismic response coefficient. This is essentially a pseudo-acceleration‡ response spectrum normalized by the natural period of vibration, T_n. The actual pseudo-acceleration response spectrum for a given earthquake and the design pseudo-acceleration response spectrum will typically look like the charts of Figures 4.40 and 4.41, respectively.

AREMA (2015) recommends the following with respect to the normalized design response spectra: T_r is the maximum natural vibration period for essentially rigid response, $T_0 = 0.096\mathrm{S}$, and $T_\mathrm{s} = (0.48\mathrm{S})^{3/2}$.

However, dynamic analyses of railway bridges typically underestimate the actual natural vibration period and, therefore, the response of the bridge for low period of vibration structures. AREMA (2015) recommends a design response spectrum without reduced response (or C_n) below T_0 (Figure 4.42) unless the effects of foundation flexibility, foundation rotational movement, and lateral span flexibility were included in the dynamic analysis.

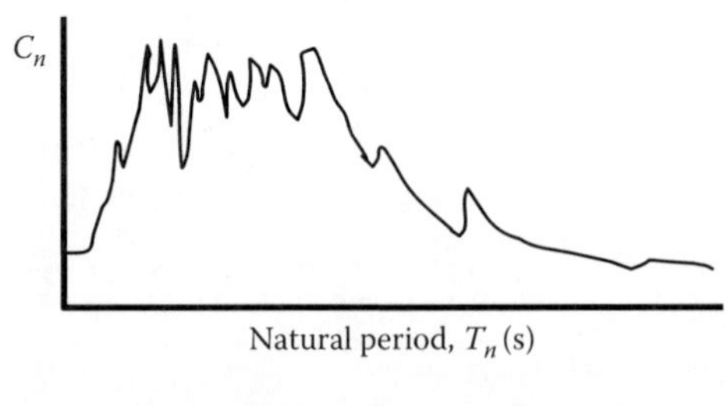

FIGURE 4.40 Typical actual response spectrum.

* For example, some bridges on soft clays and silts where vibration modes greater than the fundamental mode have periods of less than 0.3 s and bridges near faults or in areas of high seismicity. In these cases, alternative equations, available in seismic design standards and guidelines, for C_n may apply.

† Often established for a damping ratio (percentage of critical damping) of 5%.

‡ For steel bridge superstructures with low damping and short vibration periods, the pseudo-acceleration response is a close approximation to the actual acceleration response.

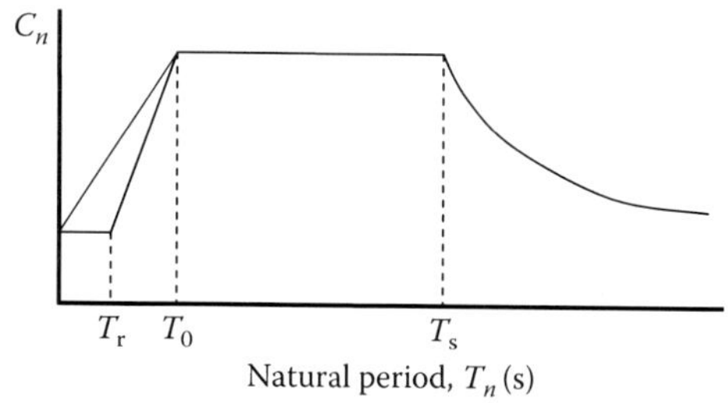

FIGURE 4.41 Typical design response spectrum.

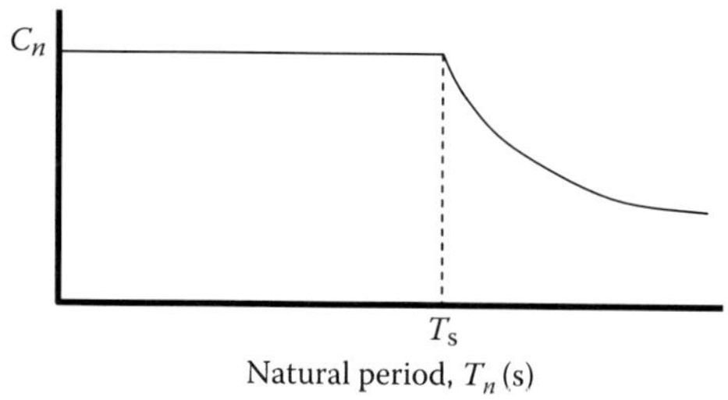

FIGURE 4.42 Typical AREMA design response spectrum used with simple dynamic analyses.

The response spectrum must be calculated in each orthogonal and uncorrelated direction (longitudinal and transverse) and, therefore, must be combined for design purposes. AREMA (2015) recommends either the square root sum of squares method (Equation 4.79) or the 100%–30% rule (Equations 4.78a and b) to combine the seismic loads:

$$F = \sqrt{F_T^2 + F_L^2}. \tag{4.79}$$

However, in some cases, the development of the response spectra from the seismic response coefficient is inappropriate and consideration of loading based on site specific response spectra is required.*

Example 4.19

The normalized response spectrum is required for a 30.5 m (100 ft) long steel girder span with the following properties:

Weight: 4465 kg/m (3000 lb/ft),
$I_x = 41620 \times 10^6$ mm^4 (100,000 in.^{4}),
$I_y = 2080 \times 10^6$ mm^4 (5000 in.^{4}),
$\in = 3\%$,

$$D = \left(\frac{1.5}{0.4(3)+1} + 0.5\right) = 1.18.$$

* For example, where $A \geq 0.2$ and $T_n \geq 0.7$ for bridges on very soft clays and silts, and for bridges on soft clays and silts where vibration modes greater than the fundamental mode have periods of less than 0.3 s and bridges near faults or in areas of high seismicity. In these cases, alternative equations, available in seismic design standards and guidelines, for C_n may apply.

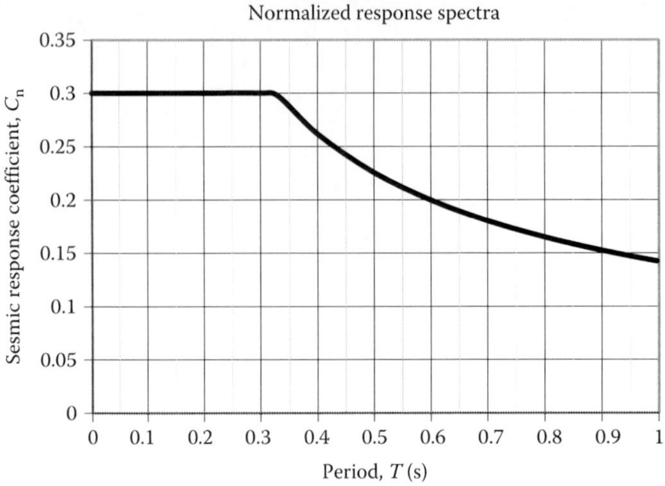

FIGURE E4.9 Normalized response spectrum.

The bridge is located where $A = 10\%$ and founded on material with $S = 1.0$ (rock), $T_s = (0.48(1.0))^{3/2} = 0.33$ s,

$$C_n = \frac{1.2(0.10)(1.0)1.18}{T_n^{2/3}} = \frac{0.142}{T_n^{2/3}} \leq 2.5(0.10)1.18 \leq 0.30.$$

The normalized response spectrum is shown in Figure E4.9.

4.4.4 Loads Relating to Overall Stability of the Superstructure

4.4.4.1 Derailment Load

Events such as train derailments on bridges are relatively infrequent. However, particularly on long bridges, train derailments can occur and create overall instability of individual spans. AREMA (2015) recommends an eccentric derailment load be used to ensure stability of spans. This derailment load, Q, is a single line of wheel loads, equal to the design live load including impact, at a 1.5 m (5 ft) eccentricity from the track centerline (Figure 4.43). It is used as a load case for the design of cross frames and diaphragms in beam and girder spans requiring lateral bracing.* AREMA (2015) recognizes that damage to some span elements may occur in these relatively extreme, but infrequent, events. Therefore, a 50% increase in allowable stress is permitted when determining stresses in cross frames, diaphragms, anchor rods, or other members resisting overall instability of the span.

Example 4.20a (SI Units)

Determine the Cooper's EM360 derailment load forces in the brace frame modeled in Figure E4.10a. The brace frames are spaced at 2.5 m intervals and the calculated impact factor for the span is 40%. The derailment force applied to the cross frame at (a) is assumed to be transferred to the opposite girder through the cross frame members ab and ac.
The derailment force, Q, at the brace frame is

* The tendency for the span to "roll over" is prevented by lateral bracing between beams or girders.

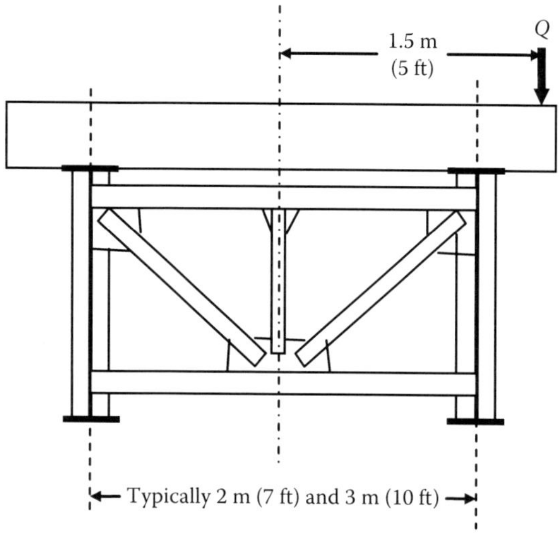

FIGURE 4.43 Derailment load.

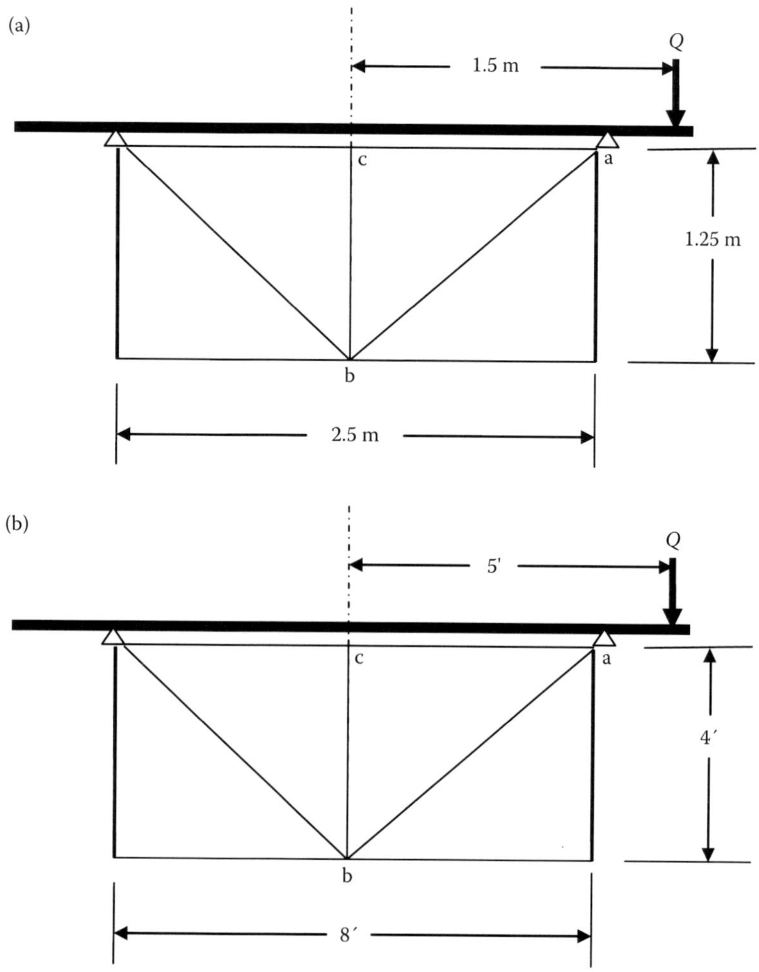

FIGURE E4.10 Cross-section of span.

$$Q = 1.40\left(\frac{360}{2(S_a)}\right)L_c = 1.40\left(\frac{360}{2(1.5)}\right)2.5 = 420 \text{ kN},$$

where

S_a = Live load axle spacing = 1.5 m
L_c = cross frame spacing = 2.5 m
R_a = Reaction at a from derailment load = −420((1.5–1.25)/2.5) = −42.0 kN
R_b = Reaction at b from derailment load = +420 + 42.0 = 462 kN
From free-body analysis:

$$P_{ab} = \frac{462}{\cos 45°} = 653 \text{ kN compression},$$

$$P_{ac} = 653 \sin 45° = 462 \text{ kN tension}.$$

These forces are checked against the usual allowable stresses increased by 50%.

Example 4.20b (US Customary and Imperial Units)

Determine the Cooper's E80 derailment load forces in the brace frame modeled in Figure E4.10b. The brace frames are spaced at 8 ft intervals and the calculated impact factor for the span is 40%.

The derailment force applied to the cross frame at (a) is assumed to be transferred to the opposite girder through the cross frame members ab and ac. The derailment force, Q, at the brace frame is

$$Q = 1.40\left(\frac{80}{2(S_a)}\right)L_c = 1.40\left(\frac{80}{2(5)}\right)8 = 89.6 \text{ kips},$$

where

S_a = live load axle spacing = 5 ft
L_c = cross frame spacing = 8′
R_a = Reaction at a from derailment load = −89.6((5–4)/8) = −11.2 kips
R_b = Reaction at b from derailment load = +89.6 + 11.2 = 100.8 kips
From free-body analysis:

$$P_{ab} = \frac{100.8}{\cos 45°} = 142.6 \text{ kips compression},$$

$$P_{ac} = 142.6 \sin 45° = 100.8 \text{ kips tension}.$$

These forces are checked against the usual allowable stresses increased by 50%.

4.4.4.2 Other Loads for Overall Lateral Stability

The overall stability of the superstructure against wind, nosing, and centrifugal forces must also be ensured. The stability of spans and towers should be calculated using a live load, without impact, of 18 kN/m (1200 lb per ft).* On multiple track bridges this live should be placed on the most leeward track on the bridge. A 50% increase in allowable stress is permissible when determining stresses in members resisting overall instability of the span (such as anchor rods at bearings).

4.4.5 Pedestrian Loads

Walkway surfaces† and supporting members must be designed for a uniformly distributed live load of 4.1 kPa (85 psf). Walkway components must not deflect more than 1/160 of the applicable span

* This represents a uniform load of empty rail cars.

† Typically galvanized nonslip steel grating for walkways used by railroad employees.

for a load of 1.1 kN (250 lb) applied at the location creating the greatest deflection. Railings and their connections should be designed for the greatest effects from a lateral or vertical force of 890 N (200 lb) applied at any location along the span. Posts must be designed to resist the greatest effects from a force of 890 N (200 lb) applied at the top railing connection.

The walkway dead and live loads may often be considered as a component of the girder or truss dead load in the applicable design load combinations.

4.5 LOAD AND FORCE COMBINATIONS FOR DESIGN OF STEEL RAILWAY SUPERSTRUCTURES

Table 4.10 outlines the load combinations that apply to steel superstructure design found in the various recommendations of AREMA (2015). Loads and forces related to LL (or LL_T) that affect the stresses in the superstructure members are *I*, CF, *N*, and LF. Other loads and forces related to LL are LV, BF, and DF. Loads and forces related to LL that affect the overall stability of the superstructure are SL, *N*, CF, and *Q*. Environmentally induced loads and forces such as W_L, W_{UL}, CWR, and OF must also be considered in load combinations. These loads and forces must be appropriately combined for design to reflect the probability of occurrence at superstructure members. To comprise these probabilities, AREMA (2015) applies factors to the allowable axial, shear, and bending stresses associated with various load combinations.

Load cases D1 and D2 are load combinations applying to all superstructure members loaded by live load. Load case D3 applies to truss web members loaded by live load. It is based on the AREMA (2015) recommendation that truss web members and their connections be designed for a live load, LL_T, that increases the total stress by 33% over the design stress in the most highly stressed chord of a truss. This live load ensures that web members attain their safe capacity at about the same increased live load as other truss members due to the observation that, in steel railway trusses, the web members typically reach capacity prior to other members in the truss (Hardesty, 1935).* Load cases D1, D2, and D3 presuppose that truss members may be designed as axial members (see Chapter 6). This is appropriate provided that secondary forces from truss distortion,† force eccentricity, end conditions (unsymmetrical connections), or other effects do not create excessive bending stresses in the members. AREMA (2015) recommends that truss members be designed as axial members where secondary forces do not create stresses in excess of 27.5 MPa (4000 psi) in tension members and 20.7 MPa (3000 psi) in compression members. For secondary stresses in excess of 27.5 MPa (4000 psi) or 20.7 MPa (3000 psi) for tension or compression members, respectively, the excess stress is superimposed on main primary stresses and the member designed as a combined axial and flexural member (see Chapter 8).

Load case D4 outlines the load combinations applicable to superstructure members loaded by wind only. Load case D5 provides load combinations for secondary members typically used in the bracing systems of members carrying LL. Load case E1 indicates the appropriate load combinations when seismic design of superstructures is necessary. Seismic design recommendations for steel railway superstructures are outlined in Chapter 9 of AREMA (2015). Load case S1 summarizes the load combinations to consider for investigations of the overall stability of the superstructure. Load cases C1, C2, and C3 apply to superstructure erection activities (see Chapter 11).

---

* However, in parametric studies of some recent railway truss designs, the effect was found not to govern design (Conway, 2003) and, when tensile stress ranges are present, fatigue criterion often governs truss web member design, in any case.

† Truss distortion effects on a member are generally negligible for relatively slender members where the width of the member parallel to the plane of distortion is less than 10% of the member length.

TABLE 4.10
Load Combinations for Steel Railway Superstructure Design

Load Case	Load Combination	Member	F_L
D1-A	DL + LL + I + CF	All members subjected to LL	1.00
D1-B	LL + I + CF	All members with tensile stress ranges from LL	S_{rfat}
D2-A	DL + LL + I + CF + W_L + LF + N + CWR + OF	All members subjected to LL, excluding floorbeam hangers	1.25
D2-B	DL + LL + I + CF + W + LF + N + CWR + OF	Floorbeam hangers	1.00
D3	DL+ LL_T+ I_T+ CF_T	Truss web members and connections	1.33
D4-A	W_L or LV	Members loaded by wind only	1.00
D4-B	W_{UL}	Members loaded by wind only	1.00
D5-A	CF + BF + N + W_L + CWR	Bracing between compression members	1.25
D5-B	DF	Cross frames, diaphragms, and anchor rods	1.50
E1-A	DL + EQ	All members in active seismic zones	1.50
E1-B	DL+ LL + I + CF + EQ	Members in active seismic zones in long bridges only	1.50
S1-A	SL + N + CF	Members resisting overall instability	1.50
S1-B	Q	Members in deck spans resisting overall instability	1.50
C1	DL	Members stressed during lifting or jacking	1.50
C2-A	DL	Members stressed during erection	1.25
C2-B	DL + W_{UL}	Members stressed during erection	1.33

F_L, allowable stress factor (multiplier for basic allowable stresses); DL, dead loads (self weight and superimposed dead loads) (see Section 4.2). For load case C2 includes erection loads such as temporary dead loads and/or temporary live loads (equipment); LL, live loads (see Section 4.3.1); I, impact load due to LL (due to dynamic amplification) (see Section 4.3.2.1); CF, centrifugal force due to LL (see Section 4.3.2.3); S_{rfat}, allowable stress based on member loaded length and fatigue detail category; W_L, wind forces on loaded superstructure (see Section 4.4.1); LF, longitudinal forces from equipment (braking and locomotive traction) (see Section 4.3.2.2); N, lateral forces from equipment (nosing) (see Section 4.3.2.4); CWR, thermal forces from CWR (lateral and/or longitudinal) (see Section 4.4.2); OF, other forces caused by thermal changes, support settlements and/or conditions particular to the superstructure; LL_T, live load that creates a total stress increase of 33% over the design stress (computed from load case D1-A) in the most highly stressed chord member of the truss (see Section 4.4.6); I_T, impact load due to $LL_T = (LL_T/LL)(I)$; CF_T, centrifugal force due to $LL_T = (LL_T/LL)(CF)$; LV, "notional" lateral vibration load (see Section 4.3.2.4); W_{UL}, wind forces on unloaded superstructure (see Section 4.4.1); BF, 2.5% of the total axial compressive force in adjacent panels of truss chords, girder flanges, and viaduct tower posts; DF, lateral forces from out-of-plane bending and from load distribution effects (see Section 4.4.4.1); EQ, forces from earthquake (combined transverse and longitudinal) (see Section 4.4.3); SL, live load on leeward track of 18 kN/m (1200 lb/ft) without impact, I (see Section 4.4.4.2); Q, derailment load (see Section 4.4.4.1).

REFERENCES

American Railway Engineering and Maintenance-of-Way Association (AREMA), 1906, General specifications for steel railroad bridges, *AREMA Proceedings*, Vol. 7, Chicago, IL.

American Railway Engineering and Maintenance-of-Way Association (AREMA), 2015, Chapter 15—Steel structures, *Manual for Railway Engineering*, Lanham, MD.

American Railway Engineering Association (AREA), 1920, General specifications for steel railroad bridges, *AREA Proceedings*, Vol. 21, Chicago, IL.

American Railway Engineering Association (AREA), 1949, Test results on relation of impact to speed, *AREA Proceedings*, Vol. 50, Chicago, IL.

American Railway Engineering Association (AREA), 1960, Summary of tests on steel girder spans, *AREA Proceedings*, Vol. 61, Chicago, IL.

American Railway Engineering Association (AREA), 1966, Reduction of impact forces on ballasted deck bridges, *AREA Proceedings*, Vol. 67, Chicago, IL.
American Society of Civil Engineers (ASCE)/Structural Engineering Institute (SEI), 2013, *Minimum Design Loads for Buildings and Other Structures*, ASCE/SEI 7-10, Reston, VA.
Anderson, T.L., 2005, *Fracture Mechanics*, 3rd ed., CRC Press, Boca Raton, FL.
Barsom, J.M. and Rolfe, S.T., 1987, *Fatigue and Fracture Control in Steel Structures*, 2nd ed., Prentice-Hall, Englewood Cliffs, NJ.
Basquin, O.H., 1910, The exponential law of endurance tests, *Proceedings of the American Society for Testing Materials*, Vol. 10, No. 2, 625–630.
Beer, F.P. and Johnston, E.R., 1976, *Statics and Dynamics*, 3rd ed., McGraw-Hill, New York.
Biggs, J.M., 1964, *Introduction to Structural Dynamics*, McGraw-Hill, New York.
Bollinger, K., 2015, Railroad bridge dynamics and rating, Transportation Research Board, Final Report for Rail Safety IDEA Project 24, Washington, DC.
Byers, W.G., 1970, Impact from railway loading on steel girder spans, *Journal of the Structural Division*, Vol. 96, No. 6, 1093–1103.
Carnahan, B., Luther, H.H., and Wilkes, J.O., 1969, *Applied Numerical Methods*, John Wiley & Sons, New York.
Chatterjee, P.K., Datta, T.K., and Surana, C.S., 1994, Vibration of continuous bridges under moving vehicles, *Journal of Sound and Vibration*, Vol. 169, No. 5, 619–632.
Chopra, A.K., 2004, *Dynamics of Structures*, 2nd ed., Prentice-Hall, Upper Saddle River, NJ.
Clough, R.W. and Penzien, J., 1975, *Dynamics of Structures*, McGraw-Hill, New York.
Conway, W.B., 2003, Article 1.3.16 of Chapter 15, Communication with AREMA Committee 15.
Cooper, T., 1894, Train loadings for railroad bridges, *Transactions of the American Society of Civil Engineers*, Vol. 31, 174–184.
Dick, S.M., 2002, Bending moment approximation analysis for use in fatigue life evaluation of steel railway girder bridges, PhD Thesis, University of Kansas.
Dick, S.M., Otter, D.E. and Connor, R.J., 2011, Comparison of railcar and bridge design loadings for development of a railroad bridge fatigue loading, *Proceedings of the 2011 AREMA Conference*, Lanham, MA.
Dowling, N.E., 1999, *Mechanical Behavior of Materials*, 2nd ed., Prentice-Hall, Upper Saddle River, NJ.
Fisher, J.W., 1984, *Fatigue and Fracture in Steel Bridges*, John Wiley & Sons, New York.
Foutch, D.A., Tobias, D., and Otter, D., 1996, Analytical investigation of the longitudinal loads in an open-deck through-plate-girder bridge, Report R-894, Association of American Railroads.
Foutch, D.A., Tobias, D.H., Otter, D.E., LoPresti, J.A., and Uppal, A.S., 1997, Experimental and analytical investigation of the longitudinal loads in an open-deck plate-girder railway bridge, Report R-905, Association of American Railroads.
Fryba, L., 1972, *Vibration of Solids and Structures under Moving Loads*, Noordoff International, Groningen, the Netherlands.
Fryba, L., 1996, *Dynamics of Railway Bridges*, Thomas Telford, London.
Hardesty, S., 1935, Live loads and unit stresses, *AREA Proceedings*, Vol. 36, 770–773.
Heywood, R., Roberts, W., and Boully, G., 2001, Dynamic loading of bridges, *Journal of the Transportation Research Board*, Vol. 1770, 58–66.
Inglis, C.E., 1928, *Impact in Railway-Bridges*, Minutes of Proceedings of the Institution of Civil Engineers, London.
Joy, R., Read, D., and Otter, L.D., 2007, Continuous welded rail restraint on an open-deck girder bridge, Technology Digest 07-026, Association of American Railroads, Pueblo, CO.
Joy, R., Read, D., and Otter, L.D., 2009, Thermal forces on open deck steel bridges, Research Report R-996, Association of American Railroads, Pueblo, CO.
Kreyszig, E., 1972, *Advanced Engineering Mathematics*, John Wiley & Sons, New York.
Kulak, G. L. and Smith, I.F.C., 1995, Analysis and design of fabricated steel structures for fatigue, University of Alberta Structural Engineering Report No. 190, Edmonton, AB.
Lin, J.H., 2006, Response of a bridge to a moving vehicle load, *Canadian Journal of Civil Engineering*, Vol. 33, No. 1, 49–57.
Liu, H., 1991, *Wind Engineering: A Handbook for Structural Engineers*, Prentice-Hall, Englewood Cliffs, NJ.
LoPresti, J.A., Otter, D.A., Tobias, D.H., and Foutch, D.A., 1998, Longitudinal forces in an open-deck steel bridge, Technology Digest 98-007, Association of American Railroads, Pueblo, CO.
LoPresti, J.A. and Otter, D.A., 1998, Longitudinal forces in a two-span open-deck steel bridge at FAST, Technology Digest 98-020, Association of American Railroads, Pueblo, CO.
McLean, W.G. and Nelson, E.W., 1962, *Engineering Mechanics*, McGraw-Hill, New York.
Miner, M.A., 1945, Cumulative damage in fatigue, *Journal of Applied Mechanics*, Vol. 67, A159–A164.

Otter, D.E., Doe, B., and Belport, S., 2005, Rail car lateral forces for bridge design and rating, Technology Digest 05-002, Association of American Railroads, Pueblo, CO.

Otter, D.E., Joy, R., and LoPresti, J.A., 1999, Longitudinal forces in a single-span, ballasted-deck, plate-girder bridge, Report R-935, Association of American Railroads, Pueblo, CO.

Otter, D.E., LoPresti, J., Foutch, D.A., and Tobias, D.H., 1996, Longitudinal forces in an open-deck steel plate-girder bridge, Technology Digest 96-024, Association of American Railroads, Pueblo, CO.

Otter, D. E., LoPresti, J., Foutch, D.A., and Tobias, D.H., 1997, Longitudinal forces in an open-deck steel plate-girder bridge, *AREA Proceedings*, Vol. 98, 101–105.

Otter, D.E., Sweeney, R.A.P., and Dick, S.M., 2000, Development of design guidelines for longitudinal forces in bridges, Technology Digest 00-018, Association of American Railroads, Pueblo, CO.

Pilkey, W.D., 1997, *Stress Concentration Factors*, 2nd ed., John Wiley & Sons, New York.

Ruble, E.J., 1955, Impact in railroad bridges, *Proceedings of the American Society of Civil Engineers*, Vol. 81, No. 736, 1–36.

Sanders, W.W. and Munse, W.H., 1969, Load distribution in steel railway bridges, *Journal of the Structural Division*, Vol. 95, No. ST12, 2763–2781.

Simiu, E. and Scanlon, R.H., 1986, *Wind Effects on Structures*, John Wiley & Sons, New York.

Stephens, R.I., Fatemi, A., Stephens, R.R., and Fuchs, H.O., 2001, *Metal Fatigue in Engineering*, 2nd ed., John Wiley & Sons, New York.

Steinman, D. B., 1922, Locomotive loadings for railway bridges, Transactions, Paper No. 1520, ASCE, New York.

Taly, N., 1998, *Design of Modern Highway Bridges*, McGraw-Hill, New York.

Tobias, D.H., Foutch, D.A., and Choros, J., 1996, Loading spectra for railway bridges under current operating conditions, *Journal of Bridge Engineering*, Vol. 1, No. 4, 127–134.

Tobias, D.H., Foutch, D.A., Lee, K., Otter, D.E., and LoPresti, J.A., 1999, Experimental and analytical investigation of the longitudinal loads in a multi-span railway bridge, Report R-927 Association of American Railroads, Pueblo, CO.

Tobias, D.H., Otter, D.E., and LoPresti, J.A., 1998, Longitudinal forces in three open-deck steel bridges, *Proceedings of the 1998 AREMA Technical Conference*.

Tse, F.S., Morse, I.E., and Hinkle, R.T., 1978, *Mechanical Vibrations: Theory and Applications*, Allyn and Bacon, Boston, MA.

Uppal, A.S., Otter, D.E., and Joy, R.B., 2001, Longitudinal forces in bridges due to revenue service, Report R-950, Association of American Railroads, Pueblo, CO.

Veletsos, A.S. and Huang, T, 1970, Analysis of dynamic response of highway bridges, *Journal of Engineering Mechanics*, 96, No. EM5, 593–620.

Waddell, J.A.L., 1916, *Bridge Engineering*, Vols. 1 and 2, John Wiley & Sons, New York.

Yang, Y.B., Yau, J.D., and, Wu, Y.S., 2004, *Vehicle-Bridge Interaction Dynamics*, World Scientific, Singapore.

5 Structural Analysis and Design of Steel Railway Bridges

5.1 INTRODUCTION

Elastic structural analysis procedures are used for steel railway bridge design based on the allowable stress design methods of the AREMA (2015) *Manual for Railway Engineering**. Strength (yield, ultimate, and stability), serviceability (deflection and vibration), fatigue, and extreme events criteria (or limit states) must be considered for safe and reliable steel railway bridge design.

The strength limit state concerns the axial, bending, shear, and torsional resistance of members to tensile yielding†, tensile fracture (ultimate), and/or compressive instability (buckling).

Fatigue, or the failure of steel at nominal cyclical stresses lower than yield stress, is a phenomenon that occurs due to the fluctuating nature of railway traffic and the presence of stress concentrations in the superstructure (see Chapter 4). The fatigue limit state governs the design of many typical railway superstructures. Ordinary steel railway superstructure design may also be controlled by deflection criteria, typically established to ensure adequate stiffness for vibration control. Since live load deflection and the fatigue strength of details are evaluated at service loads, allowable stress design is appropriate for typical steel railway superstructures. The design service life of railroad bridges is generally considered to be about 80 years.

5.2 STRUCTURAL ANALYSIS OF STEEL RAILWAY SUPERSTRUCTURES

Railway live loads are a longitudinal series of moving concentrated axle or wheel masses or loads that are fixed with respect to lateral position‡. The maximum elastic static normal stresses, shear stresses, and deformations in a steel superstructure member depend on the global longitudinal position of the railway live load. In addition, these maximum elastic static stresses are amplified due to dynamic effects as illustrated in Figure 5.1. The local longitudinal and lateral distribution of the moving loads to the deck and supporting members, as well as their dynamic effects are considered in Chapter 4.

5.2.1 Live Load Analysis of Steel Railway Superstructures

The static analysis of railway superstructures involves the determination of the deformations and stresses in members caused by the moving loads. These effects are influenced by the position of the moving load. Maximum effects are of primary interest, but the designer must also carefully consider effects of the moving load at other locations on the superstructure, where stress reversal, changes of cross section, splices, fatigue effects,§ and other considerations may require investigation.

* Recommended practices for the design of railroad bridges are developed and maintained by the American Railway Engineering and Maintenance-of-Way Association (AREMA, 2008). Chapter 15—Steel structures, provides detailed recommendations for the design of steel railway bridges for spans up to 120 m (400 ft) in length, standard gage track of 1435 mm (56.5″), and North American freight and passenger equipment at speeds up to 127 km/h (79 mph) and 145 km/h (90 mph), respectively. The recommendations may be used for longer span bridges with supplemental requirements. Many railroad companies establish steel railway bridge design criteria based on these recommended practices.

† Shear and compressive yielding is related to tensile yielding (see Chapter 2).

‡ By necessity, due to the steel wheel flange and rail head interface.

§ For example, for some span lengths traversed by railway cars, stress ranges are greatest near the ¼ point of the simple span length (Dick, 2002).

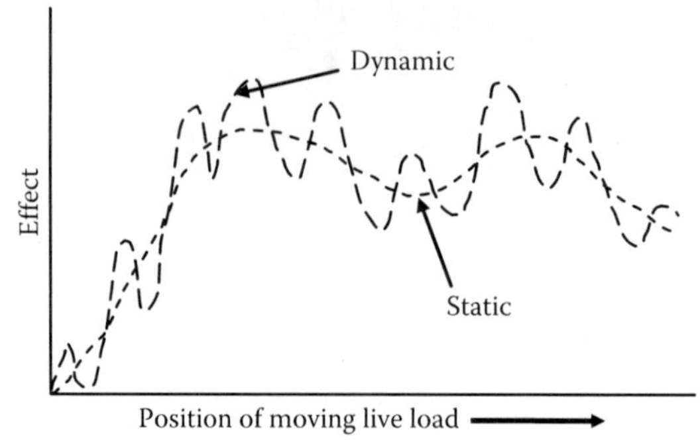

FIGURE 5.1 Static and dynamic effects on steel railway superstructures.

Structural analysis is required for multiple load positions to determine the maximum, or other significant, effects for design of members and connections. The necessary analytical effort may be reduced by careful consideration of the load configuration and the use of influence lines, computer software, and analytical experience. Furthermore, if the concentrated load configuration remains constant (typical of Cooper's E and other moving design loads), the analyses may be carried out and prepared in tables, in equations, and as equivalent uniform loads.

Some modern structural engineering software has the ability to perform dynamic interaction analyses to determine elastic deformations and forces in the members. Many steel railway bridge spans are simply supported* and, therefore, statically determinate†. This permits the use of relatively simple computer programs and spreadsheets for moving load analysis to determine the deformations and forces. For more complex superstructures (e.g., very long span or statically indeterminate superstructures‡), it may be necessary to utilize more sophisticated finite-element analysis (FEA) software that enables moving sprung mass, mass, or load analysis (see Chapter 4). In any case, for many superstructures, digital computing has made the analysis of the effects of moving loads a routine component of the superstructure design process. Nevertheless, it is often necessary that only individual members of a superstructure be investigated (e.g., during retrofit design or quality assurance design reviews), or that relatively simple superstructures be designed. In these cases, and in general, a basic understanding of classical moving load analysis is beneficial to the design engineer§.

Therefore, some principles of moving load analysis for shear force and bending moment are developed in this chapter. The analyses are performed for beam and girder spans with loads applied directly to the longitudinal members or at discrete locations via transverse members (typically floorbeams in through girder and truss spans). The maximum shear force and bending moment in railway truss¶ and arch** spans are also briefly outlined.

---

* Reasons for this are given in Chapter 3.

† The equations of static equilibrium suffice to determine forces in the structure.

‡ Typical of continuous and some movable bridge steel superstructures.

§ For example, influence line analysis is often the analytical foundation of algorithms used in moving load analysis software.

¶ The influence lines for simple span shear force and bending moment are useful for the construction of influence lines for axial force in truss web and chord members, respectively.

** Two-hinged arches (hinged at bases) are statically indeterminate and many steel railway arch superstructures are designed and constructed as three-hinged arches to create a statically determinate structure. For statically determinate arches, influence lines for axial forces in members may be constructed by superposition of horizontal and vertical effects. The influence lines for simple span bending moment are useful for the construction of the influence lines for the vertical components of axial force in arch members.

5.2.1.1 Maximum Shear Force and Bending Moment due to Moving Concentrated Loads on Simply Supported Spans

5.2.1.1.1 Criteria for Maximum Shear Force (with Loads Applied Directly to the Superstructure)

A series of concentrated loads applied directly to the steel beam or girder is typically assumed in the design of open deck beam and plate girder (DPG) spans, ballasted deck plate girder (BDPG) superstructures, and ballasted through plate girder (BTPG) spans with closely spaced transverse floorbeams.

The maximum shear force, V_C, at a location, C, in a simply supported span of length, L, traversed by a series of concentrated loads with resultant force at a distance, x_T, from one end of the span is (Figure 5.2)

$$V_C = P_T \frac{x_T}{L} - P_L, \tag{5.1}$$

where

P_T = total load on the span

P_L = load to left of location, C

Equation 5.1 indicates that V_C will be a maximum at a location where P_T (x_T/L) is a maximum and P_L a minimum. If $P_L = 0$, the absolute maximum shear in the span occurs at the end of the span and is

$$V_A = P_T \frac{x_T}{L}. \tag{5.2}$$

For any span length, L, the maximum end shear, V_A, will be largest when the product $P_T x_T$ is greatest. Therefore, for a series of concentrated loads (such as Cooper's E loading), the maximum end shear, V_A, must be determined with the heaviest loads included in P_T and these heavy loads should be close to the end of the beam (to maximize the distance, x_T).

This information assists in determination of the absolute maximum value of end shear force, which can be determined by a stepping the load configuration across the span (by each successive load spacing) until $P_T x_T$ causes a decrease in V_A. With the exception of end shear in spans between

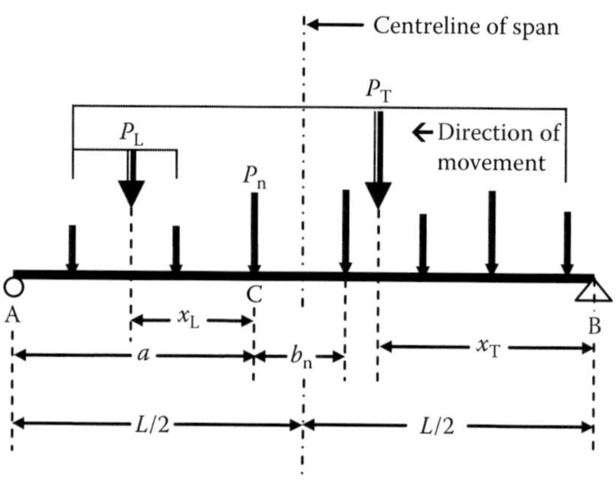

FIGURE 5.2 Concentrated moving loads applied directly to the superstructure.

$L = 7.0$ m (23 ft) and $L = 8.3$ m (27.3 ft), this occurs when the second axle* of Cooper's design load configuration is placed at the end of the beam (location A in Figure 5.2). For spans between $L = 7.0$ m (23 ft) and $L = 8.3$ m (27.3 ft), the maximum shear occurs with the fifth axle at the end of the span.

The maximum shear force at other locations, C, may be determined in a similar manner by considering a constant P_T moving from x_T to $x_T + b_n$ (where b_n = the successive load spacing). In that case, the change in shear force, ΔV_c, at location, C, is

$$\Delta V_C = P_T \frac{b_n}{L} - P_L. \tag{5.3}$$

The relative changes in shear given by Equation 5.3 can be examined to determine the location of the concentrated loads for maximum shear at any location, C, in the span.

5.2.1.1.2 Criteria for Maximum Shear Force (with Loads Applied to the Superstructure through Transverse Members)

A series of concentrated loads applied through longitudinal members (stringers) to transverse members (floorbeams) of the steel beam or girder span is typically assumed in the design of open deck through spans†.

The maximum shear force, V_{BC}, in panel BC in a simply supported span of length, L, traversed by a series of concentrated loads (with resultant force at a distance, x_T) is (Figure 5.3)

$$V_{BC} = P_T \frac{x_T}{L} - \left(P_L + P_{BC} \left(\frac{c}{s_p} \right) \right). \tag{5.4}$$

The maximum shear force in panel BC may be determined by considering a constant P_T moving from x_T to $x_T + \Delta x_T$, where Δx_T is a small increment of movement of load assuming no change in concentrated forces on the span or within panel BC. In that case, the change in shear force, ΔV_{BC}, in panel BC, is

$$\Delta V_{BC} = \left(\frac{P_T}{L} - \frac{P_{BC}}{s_p} \right) \Delta x_T. \tag{5.5}$$

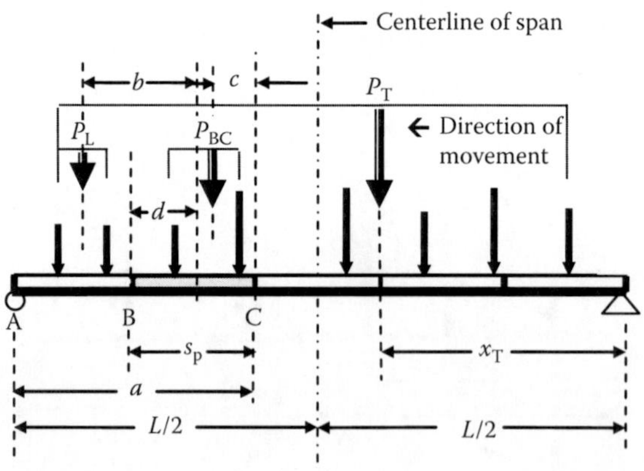

FIGURE 5.3 Concentrated moving loads applied to the superstructure at transverse members.

---

* The first driving wheel of the configuration.

† Such as open deck through plate girder (TPG) and through truss (TT) spans.

When $P_T/L = P_{BC}/s_p$ maximum shear in panel BC occurs as the change in shear changes sign (positive to negative). Therefore, the maximum shear occurs in panel BC when the average distributed load on the span, $P(T)/L$, equals the average distributed load in panel BC, P_{BC}/s_p.

When $s_p = L/n$ (n = number of equal length panels)

$$\Delta V_{BC} = \left(\frac{P_T}{n} - P_{BC}\right)\frac{\Delta x_T}{s_p} \tag{5.6}$$

and $\frac{P_T}{n} - P_{BC} = 0$ for maximum shear in the panel. Therefore, the maximum shear in spans with equal length panels occurs in panel BC when the average panel load on the span, P_T/n, equals the load in panel BC, P_{BC}.

The relative changes in shear given by Equation 5.6 can be examined to determine the location of the concentrated loads for maximum shear in any panel on the span.

5.2.1.1.3 *Criteria for Maximum Bending Moment (with Loads Applied Directly to the Superstructure)*

A series of concentrated loads applied directly to the steel beam or girder is typically assumed in the design of DPG spans, BDPG superstructures, and ballasted through plate girder (BTPG) spans with closely spaced transverse floorbeams.

The maximum bending moment, M_C, at a location, C, in a simply supported span, of length, L, traversed by a series of concentrated loads with resultant at a distance, x_T, from one end of the span is (Figure 5.2)

$$M_C = \left(P_T \frac{x_T}{L}\right)a - (P_L)x_L, \tag{5.7}$$

where

P_T = total load on the span

P_L = load to left of location, C

The change in bending moment, ΔM_C, at location C as constant force P_T moves from $x_T + \Delta x_T$ is

$$\Delta M_C = \left(P_T \frac{a}{L} - P_L\right)\Delta x_T, \tag{5.8}$$

where

Δx_T = a small increment of movement of load assuming no change in concentrated forces on the span.

When $\frac{P_T}{L} = \frac{P_L}{a}$, maximum bending moment at location C occurs as the change in bending moment changes sign (positive to negative). Therefore, the maximum bending moment occurs at location C when the average distributed load on the span, P_T/L, equals the average distributed load to the left of location C, P_L/a.

The relative changes in bending moment given by Equation 5.8 can be examined to determine the location of the concentrated loads for maximum bending moment at any location, C, in the span.

5.2.1.1.4 *Criteria for Maximum Bending Moment (with Loads Applied to the Superstructure through Transverse Members)*

A series of concentrated loads applied through longitudinal members (stringers) to transverse members (floorbeams) to the steel beam or girder is typically assumed in the design of open deck through spans.

The maximum bending moment, M_{BC}, in panel BC in a simply supported span of length, L, traversed by a series of concentrated loads (with resultant force at x_T) transferred to the span by stringers and transverse floorbeams is (Figure 5.3)

$$M_{BC} = \left(P_T \frac{x_T}{L}\right)a - (P_L)b - (P_{BC})\left(\frac{c}{s_p}\right)d. \tag{5.9}$$

The change in bending moment, ΔM_{BC}, in panel BC is

$$\Delta M_{BC} = \left(P_T \frac{a}{L} - \left(P_L + P_{BC}\frac{d}{s_p}\right)\right)\Delta x_T. \tag{5.10}$$

When $P_T \frac{a}{L} - \left(P_L + P_{BC}\frac{d}{s_p}\right) = 0$, the maximum bending moment occurs in panel BC.

The relative changes in bending moment given by Equation 5.10 can be examined to determine the location of the concentrated loads for maximum bending moment at any panel in the span. The maximum bending moment will occur at the panel point nearest the center of the span.

5.2.1.1.5 Maximum Bending Moment with Cooper's EM360 (E80) Load

The criteria for maximum shear force and bending moment in a simply supported span illustrate that loads may be stepped across the span and their effects investigated at the location of interest. In particular, the load position for maximum bending moment is of interest to superstructure designers.

For live load configurations, such as Cooper's EM360 (E80) that are expressed as a series of concentrated loads (with or without uniform load segments), the wheel load under which the maximum bending moment occurs may not be readily known by inspection, particularly on longer spans. In such cases, the development of a moment table or chart is beneficial for determining the maximum bending moments at any location along the span. For example, to determine the maximum moment at location C in Figure 5.4, the load configuration would have to be moved in many successive increments across the span. The construction of a moment table, for the particular live load configuration, makes such an iterative analysis unnecessary.

The bending moment, M_C, at any location, C, due to moving concentrated and uniform loads as shown in Figure 5.4 is

$$M_C = R_A\left(\frac{L}{2} - x\right) - \sum P_i x_i, \tag{5.11}$$

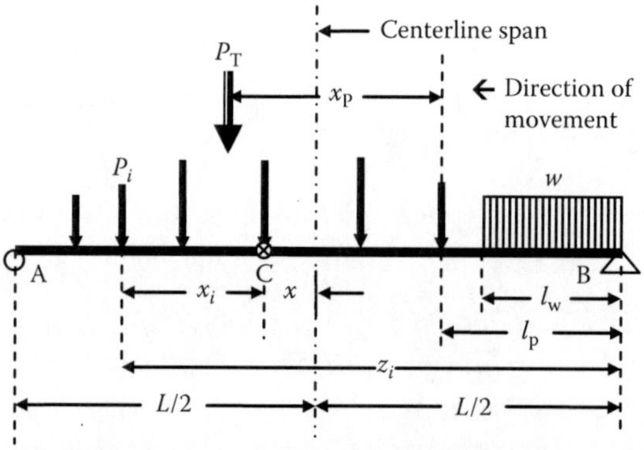

FIGURE 5.4 Concentrated and uniform moving loads applied directly to the superstructure.

where

$\sum P_i x_i$ is the sum of moments due to loads to left of C.

The left reaction, R_A, is

$$R_A = \frac{\sum P_i z_i + \frac{w l_w^2}{2}}{L}. \tag{5.12}$$

Substitution of Equation 5.12 into Equation 5.11 yields

$$M_C = \left(\frac{\sum P_i z_i + \frac{w l_w^2}{2}}{L} \right) \left(\frac{L}{2} - x \right) - \sum P_i x_i. \tag{5.13}$$

From Figure 5.4, the sum of the moments of concentrated loads about B is

$$\sum P_i z_i = P_T(x_P) + \left(\sum P_i \right) l_P. \tag{5.14}$$

Substitution of Equation 5.14 into Equations 5.12 and 5.13 yields

$$R_A = \frac{P_T(x_P) + \left(\sum P_i \right) l_P + \frac{w l_w^2}{2}}{L} \tag{5.15}$$

and

$$M_C = \left(\frac{P_T(x_P) + \left(\sum P_i \right) l_P + \frac{w l_w^2}{2}}{L} \right) \left(\frac{L}{2} - x \right) - \sum P_i x_i. \tag{5.16}$$

Equations 5.15 and 5.16 illustrate that to determine the end shear force and bending moment at any location in the simple span due to moving concentrated and uniform loads (such as Cooper's EM360 (E80) load) the following is required:

- The sum of the bending moments of all concentrated loads in front of, and about, the last concentrated load (at l_p from B in Figure 5.4) on the span, $P_T(x_P)$
- The sum of all concentrated loads on the span, $\sum P_i$
- The negative bending moment or the sum of the moments about C of all concentrated loads in front of C, $\sum P_i x_i$

Since Cooper's load pattern is constant, it is possible to develop charts and tables to readily determine the bending moment for various simple span lengths using Equation 5.16. Tables 5.1a*

* It should be noted that the bending moments in Table 5.1a are calculated based on the values in Table 5.1b but only small errors (typically less than 2%) will result from using Table 5.1a with the distance (S) and load (P) values shown.

TABLE 5.1a
Moment Table for Cooper's EM360 Wheel Load

N	P	S_1	S_W	SP_1	SP_{18}	SM	SM_{18}	SM_{17}	SM_{16}	SM_{15}	SM_{14}	SM_{13}	SM_{12}	SM_{11}	SM_{10}	SM_9	SM_8	SM_7	SM_6	SM_5	SM_4	SM_3	SM_2	SM_1
18	115	31.2	1.5	2541	115	40040	178																	
17	116	29.7	3.0	2426	231	36293	535	178																
16	115	27.9	4.8	2310	346	32782	1105	570	214															
15	115	26.4	6.3	2195	461	29436	1854	1141	606	178														
14	180	23.7	9.0	2080	641	26328	3499	2512	1703	946	494													
13	180	22.2	10.5	1900	821	21995	5419	4158	3074	1988	1262	274												
12	180	20.7	12.0	1720	1001	17936	7613	6077	4720	3305	2304	823	274											
11	180	19.2	13.5	1540	1181	14151	10081	8271	6640	4895	3620	1645	823	274										
10	90	16.8	15.9	1360	1271	10641	11535	9588	7819	5910	4498	2276	1316	631	219									
9	115	14.4	18.3	1270	1386	9105	13710	11584	9637	7514	5924	3381	2243	1379	790	285								
8	115	12.9	19.8	1155	1501	7387	16063	13759	11634	9297	7528	4665	3349	2306	1539	749	178							
7	115	11.1	21.6	1040	1616	5859	18630	16148	13844	11293	9346	6162	4668	3447	2501	1426	570	214						
6	115	9.6	23.1	925	1731	4543	21375	18715	16233	13468	11343	7838	6165	4766	3642	2282	1141	606	178					
5	180	6.9	25.8	810	1911	3401	26092	23157	20401	17308	14908	10910	8962	7289	5891	4092	2512	1703	946	494				
4	180	5.4	27.3	630	2091	2139	31083	27874	24844	21421	18748	14255	12034	10087	8414	6176	4158	3074	1988	1262	274			
3	180	3.9	28.8	450	2271	1152	36349	32866	29561	25809	22861	17875	15380	13158	11211	8535	6077	4720	3305	2304	823	274		
2	180	2.4	30.3	270	2451	439	41889	38131	34552	30472	27249	21770	19000	16504	14283	11167	8271	6640	4895	3620	1645	823	274	
1	90	0.0	32.7	90	2541	0	44878	40984	37267	33022	29663	23936	21029	18397	16038	12703	9588	7819	5910	4498	2276	1316	631	219

N, wheel number;
P, Cooper's EM360 wheel load, kN, (1/2 axle load);
S_1, distance, m, from wheel *N* to wheel 1;
S_W, distance, m, from wheel *N* to uniform train load of 60 kN/m;
$S_1 + S_w = 32.7$ m;
SP_1, sum of wheel loads, kN, between and including wheel 1 to wheel *N*;
SP_{18}, sum of wheel loads, kN, between and including wheel *N* to wheel 18;
SM, sum of the moments, kNm, about wheel 1 of wheel loads between and including wheel 2 to wheel *N*;
SM_{18}, sum of the moments, kNm, about the beginning of the uniform load of wheel loads between and including wheel *N* to wheel 18;
SM_k, sum of the moments, kNm, about wheel load $(k + 1)$ of wheel loads between and including wheel *N* to wheel *k*.

TABLE 5.1b

Moment Table for Cooper's E80 Wheel Load

N	P	S_1	S_W	SP_1	SP_{18}	SM	SM_{18}	SM_{17}	SM_{16}	SM_{15}	SM_{14}	SM_{13}	SM_{12}	SM_{11}	SM_{10}	SM_9	SM_8	SM_7	SM_6	SM_5	SM_4	SM_3	SM_2	SM_1
18	26	104	5	568	26	29200	130																	
17	26	99	10	542	52	26467	390	130																
16	26	93	16	516	78	23907	806	416	156															
15	26	88	21	490	104	21467	1352	832	442	130														
14	40	79	30	464	144	19200	2552	1832	1242	690	360													
13	40	74	35	424	184	16040	3952	3032	2242	1450	920	200												
12	40	69	40	384	224	13080	5552	4432	3442	2410	1680	600	200											
11	40	64	45	344	264	10320	7352	6032	4842	3570	2640	1200	600	200										
10	20	56	53	304	284	7760	8412	6992	5702	4310	3280	1660	960	460	160									
9	26	48	61	284	310	6640	9998	8448	7028	5480	4320	2466	1636	1006	576	208								
8	26	43	66	258	336	5387	11714	10034	8484	6780	5490	3402	2442	1682	1122	546	130							
7	26	37	72	232	362	4273	13586	11776	10096	8236	6816	4494	3404	2514	1824	1040	416	156						
6	26	32	77	206	388	3313	15588	13648	11838	9822	8272	5716	4496	3476	2656	1664	832	442	130					
5	40	23	86	180	428	2480	19028	16888	14878	12622	10872	7956	6536	5316	4296	2984	1832	1242	690	360				
4	40	18	91	140	468	1560	22668	20328	18118	15622	13672	10396	8776	7356	6136	4504	3032	2242	1450	920	200			
3	40	13	96	100	508	840	26508	23968	21558	18822	16672	13036	11216	9596	8176	6224	4432	3442	2410	1680	600	200		
2	40	8	101	60	548	320	30548	27808	25198	22222	19872	15876	13856	12036	10416	8144	6032	4842	3570	2640	1200	600	200	
1	20	0	109	20	568	0	32728	29888	27178	24082	21632	17456	15336	13416	11696	9264	6992	5702	4310	3280	1660	960	460	160

N, wheel number;

P, Cooper's E80 wheel load, kips, (1/2 axle load);

S_1, distance, ft, from wheel *N* to wheel 1;

S_W, distance, ft, from wheel *N* to uniform train load of 4000 lb/ft;

$S_1 + S_w = 109$ ft;

SP_1, sum of wheel loads, kips, between and including wheel 1to wheel *N*;

SP_{18}, sum of wheel loads, kips, between and including wheel *N* to wheel 18;

SM, sum of the moments, kips-ft, about wheel 1 of wheel loads between and including wheel 2 to wheel *N*;

SM_{18}, sum of the moments, kips-ft, about the beginning of the uniform load of wheel loads between and including wheel N to wheel 18;

SM_k, sum of the moments, kips-ft, about wheel load $(k + 1)$ of wheel loads between and including wheel N to wheel *k*.

and 5.1b are developed for the wheel load (1/2 of axle load) of Cooper's EM360 and E80 live load. The legend to Tables 5.1a and 5.1b outlines the methods used to determine the values shown. The use of Tables 5.1a and 5.1b for determining maximum bending moments due to Cooper's EM360 and E80 live load, respectively, is outlined in Examples 5.1 and 5.2.

Example 5.1a (SI Units)

Determine the maximum bending moment per rail for Cooper's EM360 live load on an 18 m long DPG span. The moment at the center, C, is assumed to be near to the maximum bending moment in both location and magnitude.

A review of Cooper's load configuration indicates that the maximum moment will likely occur under axles, NP = 3, 4, 5, 12, 13, or 14 (Figure E5.1).

With NP = 3 (Cooper's load configuration wheel number 3)
From Table 5.1a:
$x1 = (S_1)_3 = 3.9\,\text{m}$
Since $x1 \le L/2 \le 9\,\text{m}$; $N1 = 1$
Support B is $(L/2 + x1) = 9 + 3.9 = 12.9\,\text{m}$ from N1
Since $(S_1)_8 = 12.0\,\text{m}$, NL = 8 and is over support B
$xL = 12.9 - 12.9 = 0$
NE = last wheel on span = NL − 1 = 7 when NL is over support B

$$R_B = \frac{\Sigma M_B}{L} = \frac{\Sigma M_{(NL-1),N1} + \Sigma P_{N1,NE}(xL)}{18} = \frac{\Sigma M_{7,1} + \Sigma P_{1,7}(xL)}{18} = \frac{7819 + 1040(0)}{18} = 434.4\text{ kN},$$

$$M_C = R_B(L/2) - \sum M_{(NP-1),N1} = R_B(L/2) - \sum M_{2,1} = 434.4(9) - 631 = 3279\text{ kNm}.$$

With NP = 4 (Cooper's load configuration wheel number 4)
From Table 5.1a:
$x1 = (S_1)_4 = 5.4\,\text{m}$
Since $x1 \le L/2 \le 9\,\text{m}$, N1 = 1
Support B is 9 + 5.4 = 14.4 m from N1
Since $(S_1)_9 = 14.4\,\text{m}$, NL = 9 and is over support B
$xL = 0$
NE = last wheel on the span = NL − 1 = 8 when NL is over support B.

$$R_B = \frac{\Sigma M_B}{L} = \frac{\Sigma M_{8,1} + \Sigma P_{1,8}(xL)}{18} = \frac{9588 + 1155(0)}{18} = 532.7\text{ kN},$$

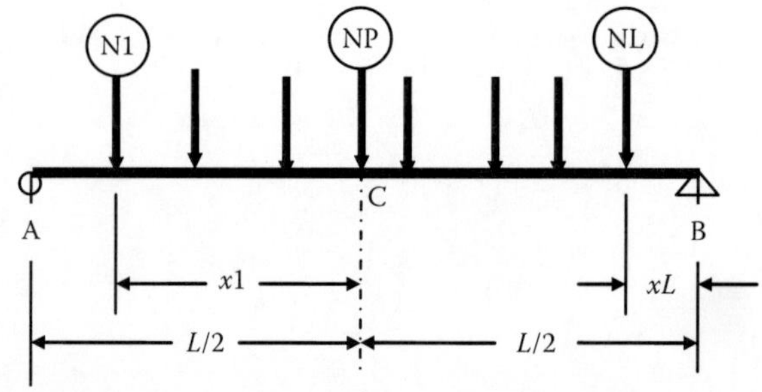

FIGURE E5.1 Position of load.

$$M_c = R_B(L/2) - \sum M_{3,1} = 532.7(9) - 1316 = 3478 \text{ kNm}.$$

With NP = 5 (Cooper's load configuration wheel number 5)
From Table 5.1a:
$x1 = (S_1)_5 = 6.9$ m
Since $x1 \leq L/2 \leq 9$ m; N1 = 1
Support B is 9 + 6.9 = 15.9 m from N1
Since $(S_1)_9 = 14.4$ m, NL = 9 and is $xL = 15.9 - 14.4 = 1.5$ m from support B
NE = last wheel on span = NL = 9 when NL is not over support B

$$R_B = \frac{\Sigma M_B}{L} = \frac{\Sigma M_{8,1} + \Sigma P_{1,9}(xL)}{18} = \frac{9588 + 1270(1.5)}{18} = 638.5 \text{ kN},$$

$$M_c = R_B(L/2) - \sum M_{4,1} = 638.5(9) - 2276 = 3471 \text{ kNm}.$$

With NP = 12 (Cooper's load configuration wheel number 12)
From Table 5.1a:
With NP = 12, the first wheel on the span = N1 − 8
$x1 = (S_1)_{12} - (S_1)_8 = 20.7 - 12.9 = 7.8$ m
Support B is 9 + 7.8 = 16.8 m from N1 = 8
With NP = 12, the last wheel on the span = NL = 17
$xL = (L/2) - [(S_1)_{17} - (S_1)_{12}] = 9 - (29.7 - 20.7) = 0$ m and N17 is over support B
NE = last wheel on span = NL − 1 = 16 when NL is over support B

$$R_B = \frac{\Sigma M_B}{L} = \frac{\Sigma M_{16,8} + \Sigma P_{8,16}(xL)}{18} = \frac{11{,}634 + (1501 - 346)(0)}{18} = 646.3 \text{ kN},$$

$$M_c = R_B(L/2) - \sum M_{11,8} = 646.3(9) - 2306 = 3511 \text{ kNm}$$

With NP = 13 (Cooper's load configuration wheel number 13)
From Table 5.1a:
With NP = 13, the first wheel on the span = N1 = 9
$x1 = (S_1)_{13} - (S_1)_9 = 22.2 - 14.4 = 7.8$ m
Support B is 9 + 7.8 = 16.8 m from N1 = 9
With NP = 13, the last wheel on the span = NL = 18
$xL = (L/2) - [(S_1)_{18} - (S_1)_{13}] = 9 - (31.2 - 22.2) = 0$ m and N18 is over support B
NE = last wheel on span = NL − 1 = 17 when NL is over support B

$$R_B = \frac{\Sigma M_B}{L} = \frac{\Sigma M_{17,9} + \Sigma P_{9,17}(xL)}{18} = \frac{11{,}584 + (1386 - 231)(0)}{18} = 643.6 \text{ kN},$$

$$M_c = R_B(L/2) - \sum M_{12,9} = 643.6(9) - 2243 = 3549 \text{ kNm}.$$

With NP = 14 (Cooper's load configuration wheel number 14)
From Table 5.1a:
With NP = 14, the first wheel on the span = N1 = 10
$x1 = (S_1)_{14} - (S_1)_{10} = 23.7 - 16.8 = 6.9$ m
Support B is 9 + 6.9 = 15.9 m from N1 = 10
With NP = 14, the last wheel on the span, NL, is the beginning of the uniform load, w
$xL = (L/2) - [(S_1)_w - (S_1)_{14}] = 9 - (31.2 + 1.5 - 23.7) = 0$ m from beginning of uniform load, w, to support B

NE = last wheel on span = beginning of uniform load, w

$$R_B = \frac{\Sigma M_B}{L} = \frac{\Sigma M_{18,10} + \Sigma P_{10,w}(xL)}{18} = \frac{11535+(1271)(0)}{18} = 640.8 \text{ kN},$$

$$M_c = R_B(L/2) - \sum M_{13,10} = 640.8(9) - 2276 = 3492 \text{ kNm}.$$

With NP = 15 (Cooper's load configuration wheel number 15)

From Table 5.1a:

With NP = 15, the first wheel on the span = N1 = 11

$x1 = (S_1)_{15} - (S_1)_{11} = 26.4 - 19.2 = 7.2$ m

Support B is 9 + 7.2 = 16.2 m from N1 = 11

With NP = 15, the last wheel on the span, NL, is the end of 2.7 m of the uniform load, w

$xL = (L/2) - [(S_1)_w - (S_1)_{15}] = 9 - (31.2 + 1.5 - 26.4) = 2.7$ m from beginning of uniform load, w, to support B

NE = last wheel on span = 2.7 m of uniform load, w

$$R_B = \frac{\Sigma M_B}{L} = \frac{\Sigma M_{18,11} + \Sigma P_{11,w}(xL)}{18} = \frac{10081+[(1181)(2.7)+60(2.7)(2.7/2)]}{18} = 749.4 \text{ kN},$$

$$M_c = R_B(L/2) - \sum M_{14,11} = 749.5(9) - 3620 = 3124 \text{ kNm}.$$

The maximum bending moment is 3549 kNm (NP = 13).

Example 5.1b (US Customary and Imperial Units)

Determine the maximum bending moment per rail for Cooper's E80 load on a 60 ft long DPG span. The moment at the center, C, is assumed to be near to the maximum bending moment in both location and magnitude.

A review of Cooper's load configuration indicates that the maximum moment will likely occur under axles, NP = 3, 4, 5, 12, 13, or 14 (Figure E5.1).

With NP = 3 (Cooper's load configuration wheel number 3)

From Table 5.1b:

$x1 = (S_1)_3 = 13$ ft

Since $x1 \le L/2 \le 30$ ft; N1 = 1

Support B is $(L/2 + x1) = 30 + 13 = 43$ ft from N1

Since $(S_1)_8 = 43$ ft, NL = 8 and is over support B

$xL = 43 - 43 = 0$

NE = last wheel on span = NL − 1 = 7 when NL is over support B

$$R_B = \frac{\Sigma M_B}{L} = \frac{\Sigma M_{(NL-1),N1} + \Sigma P_{N1,NE}(xL)}{60} = \frac{\Sigma M_{7,1} + \Sigma P_{1,7}(xL)}{60} = \frac{5702+232(0)}{60} = 95.03 \text{ kips},$$

$$M_c = R_B(L/2) - \sum M_{(NP-1),N1} = R_B(L/2) - \sum M_{2,1} = 95.03(30) - 460 = 2391 \text{ kips-ft}$$

With NP = 4 (Cooper's load configuration wheel number 4)

From Table 5.1b:

$x1 = (S_1)_4 = 18$ ft

Since $x1 \le L/2 \le 30$ ft; N1 = 1

Support B is 30 + 18 = 48 ft from N1
Since $(S_1)_9 = 48$ ft, NL = 9 and is over support B
$xL = 0$
NE = last wheel on span = NL − 1 = 8 when NL is over support B

$$R_B = \frac{\Sigma M_B}{L} = \frac{\Sigma M_{8,1} + \Sigma P_{1,8}(xL)}{60} = \frac{6992 + 258(0)}{60} = 116.5 \text{ kips},$$

$$M_c = R_B(L/2) - \sum M_{3,1} = 116.5(30) - 960 = 2536 \text{ kips-ft}.$$

With NP = 5 (Cooper's load configuration wheel number 5)
From Table 5.1b:
$x1 = (S_1)_5 = 23$ ft
Since $x1 \le L/2 \le 30$ ft; N1 = 1
Support B is 30 + 23 = 53 ft from N1
Since $(S_1)_9 = 48$ ft, NL = 9 and is $xL = 53 - 48 = 5$ ft from support B
NE = last wheel on span = NL = 9 when NL is not over support B

$$R_B = \frac{\Sigma M_B}{L} = \frac{\Sigma M_{8,1} + \Sigma P_{1,9}(xL)}{60} = \frac{6992 + 284(5)}{60} = 140.2 \text{ kips},$$

$$M_c = R_B(L/2) - \sum M_{4,1} = 140.2(30) - 1660 = 2546 \text{ kips-ft}.$$

With NP = 12 (Cooper's load configuration wheel number 12)
From Table 5.1b:
With NP = 12, the first wheel on the span = N1 = 8
$x1 = (S_1)_{12} - (S_1)_8 = 69 - 43 = 26$ ft
Support B is 30 + 26 = 56 ft from N1 = 8
With NP = 12, the last wheel on the span = NL = 17
$xL = (L/2) - [(S_1)_{17} - (S_1)_{12}] = 30 - (99 - 69) = 0$ ft and N17 is over support B
NE = last wheel on span = NL − 1 = 16 when NL is over support B

$$R_B = \frac{\Sigma M_B}{L} = \frac{\Sigma M_{16,8} + \Sigma P_{8,16}(xL)}{60} = \frac{8484 + (336 - 78)(0)}{60} = 141.4 \text{ kips},$$

$$M_c = R_B(L/2) - \sum M_{11,8} = 141.4(30) - 1682 = 2560 \text{ kips-ft}.$$

With NP = 13 (Cooper's load configuration wheel number 13)
From Table 5.1b:
With NP = 13, the first wheel on the span = N1 = 9
$x1 = (S_1)_{13} - (S_1)_9 = 74 - 48 = 26$ ft
Support B is 30 + 26 = 56 ft from N1 = 9
With NP = 13, the last wheel on the span = NL = 18
$xL = (L/2) - [(S_1)_{18} - (S_1)_{13}] = 30 - (104 - 74) = 0$ ft and N18 is over support B
NE = last wheel on span = NL − 1 = 17 when NL is over support B

$$R_B = \frac{\Sigma M_B}{L} = \frac{\Sigma M_{17,9} + \Sigma P_{9,17}(xL)}{60} = \frac{8448 + (310 - 52)(0)}{60} = 140.8 \text{ kips},$$

$$M_c = R_B(L/2) - \sum M_{12,9} = 140.8(30) - 1636 = 2588 \text{ kips-ft}.$$

With NP = 14 (Cooper's load configuration wheel number 14)

From Table 5.1b:

With NP = 14, the first wheel on the span = N1 = 10

$x1 = (S_1)_{14} - (S_1)_{10} = 79 - 56 = 23$ ft

Support B is 30 + 23 = 53 ft from N1 = 10

With NP = 14, the last wheel on the span, NL, is the beginning of the uniform load, *w*

$xL = (L/2) - [(S_1)_w - (S_1)_{14}] = 30 - (104 + 5 - 79) = 0$ ft from beginning of uniform load, *w*, to support B

NE = last wheel on span = beginning of uniform load, *w*

$$R_B = \frac{\Sigma M_B}{L} = \frac{\Sigma M_{18,10} + \Sigma P_{10,w}(xL)}{60} = \frac{8412 + (284)(0)}{60} = 140.2 \text{ kips},$$

$$M_c = R_B(L/2) - \sum M_{13,10} = 140.2(30) - 1660 = 2546 \text{ kips-ft}.$$

With NP = 15 (Cooper's load configuration wheel number 15)

From Table 5.1b:

With NP = 15, the first wheel on the span = N1 = 11

$x1 = (S_1)_{15} - (S_1)_{11} = 88 - 64 = 24$ ft

Support B is 30 + 24 = 54 ft from N1 = 11

With NP = 15, the last wheel on the span, NL, is the end of 9 ft of the uniform load, *w*

$xL = (L/2) - [(S_1)_w - (S_1)_{15}] = 30 - (104 + 5 - 88) = 9$ ft from beginning of uniform load, *w*, to support B

NE = last wheel on span = 9 ft of uniform load, *w*

$$R_B = \frac{\Sigma M_B}{L} = \frac{\Sigma M_{18,11} + \Sigma P_{11,w}(xL)}{60} = \frac{7352 + [(264)(9) + 4(9)(9/2)]}{60} = 164.8 \text{ kips},$$

$$M_c = R_B(L/2) - \sum M_{14,11} = 164.8(30) - 2640 = 2305 \text{ kips-ft}.$$

The maximum bending moment is 2588 kips-ft (NP = 13).

Example 5.2a (SI Units)

Determine the bending moment per rail at location C under axles NP = 3, 4, and 13 of Cooper's EM360 load on a 18 m long through plate girder span with a floor system comprising floorbeams and 6 m long stringers (Figure E5.2).

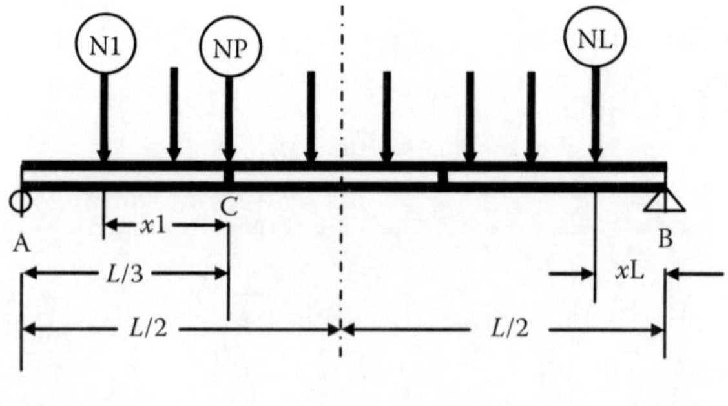

FIGURE E5.2 Position of load.

With NP = 3 (Cooper's load configuration wheel number 3)
From Table 5.1a:
$x1 = (S_1)_3 = 3.9\,\text{m}$
Since $x1 \leq L/2 \leq 9\,\text{m}$; N1 = 1
Support B is $(2L/3 + x1) = 12.0 + 3.9 = 15.9\,\text{m}$ from N1
Since $(S_1)_9 = 14.4\,\text{m}$, NL = 9 and $xL = 15.9 - 14.4 = 1.5\,\text{m}$ from NL = 9 to support B
NE = last wheel on span = NL = 9

$$R_B = \frac{\Sigma M_B}{L} = \frac{\Sigma M_{(NL-1),N1} + \Sigma P_{N1,NE}(xL)}{18} = \frac{\Sigma M_{8,1} + \Sigma P_{1,9}(xL)}{18} = \frac{9588 + 1270(1.5)}{18} = 638.5 \text{ kN},$$

$$M_c = R_B(L/3) - \sum M_{(NP-1),N1} = R_B(L/3) - \sum M_{2,1} = 638.5(6) - 631 = 3200 \text{ kNm}.$$

The superstructure with transverse floorbeams has a 100[1 − (3200/3279)] = 2.4% decrease in bending moment with NP = 3 at location of maximum moment.
With NP = 4 (Cooper's load configuration wheel number 4)
From Table 5.1a:
$x1 = (S_1)_4 = 5.4\,\text{m}$
Since $x1 \leq L/2 \leq 9\,\text{m}$; N1 = 1
Support B is $(2L/3 + x1) = 12.0 + 5.4 = 17.4\,\text{m}$ from N1
Since $(S_1)_{10} = 16.8\,\text{m}$, NL = 10 and $xL = 17.4 - 16.8 = 0.60\,\text{m}$ from NL = 10 to support B
NE = last wheel on span = NL = 10

$$R_B = \frac{\Sigma M_B}{L} = \frac{\Sigma M_{(NL-1),N1} + \Sigma P_{N1,NE}(xL)}{18} = \frac{\Sigma M_{9,1} + \Sigma P_{1,10}(xL)}{18} = \frac{12{,}703 + 1360(0.60)}{18} = 751.1 \text{ kN},$$

$$M_c = R_B(L/3) - \sum M_{(NP-1),N1} = R_B(L/3) - \sum M_{3,1} = 751.1(6) - 1316 = 3190 \text{ kNm}.$$

The superstructure loaded with transverse floorbeams has a 100[1 − (3190/3478)] = 8.3% decrease in bending moment with NP = 4 at location of maximum moment.
With NP = 13 (Cooper's load configuration wheel number 13)
From Table 5.1a:
With NP = 13, the first wheel on the span = N1 = 10
$x1 = (S_1)_{13} - (S_1)_{10} = 22.2 - 16.8 = 5.4\,\text{m}$
Support B is $12.0 + 5.4 = 17.4\,\text{m}$ from N1 = 10
With NP = 13, the last wheel on the span, NL, is the end of 1.5 m of the uniform load, w
$xL = (2L/3) - [(S_1)_w - (S_1)_{13}] = 12.0 - (31.2 + 1.5 - 22.2) = 1.5\,\text{m}$ from beginning of uniform load, w, to support B
NE = last wheel on span = 1.5 m of uniform load, w

$$R_B = \frac{\Sigma M_B}{L} = \frac{\Sigma M_{18,10} + \Sigma P_{10,w}(xL)}{18} = \frac{11{,}535 + [(1271)(1.5) + 60(1.5)(1.5/2)]}{18} = 750.5 \text{ kN},$$

$$M_c = R_B(L/3) - \sum M_{12,10} = 750.5(6) - 1316 = 3187 \text{ kNm}.$$

The superstructure with transverse floorbeams has a 100[1 − (3187/3549)] = 10.2% decrease in bending moment with NP = 13 at location of maximum moment.

Example 5.2b (US Customary and Imperial Units)

Determine the bending moment per rail at location C under axles NP = 3, 4, and 13 of Cooper's E80 load on a 60 ft long through plate girder span with a floor system comprising floorbeams and 20 ft long stringers (Figure E5.2).

With NP = 3 (Cooper's load configuration wheel number 3)
From Table 5.1b:
$x1 = (S_1)_3 = 13$ ft
Since $x1 \times L/2 \times 30$ ft; N1 = 1
Support B is $(2L/3 + x1) = 40 + 13 = 53$ ft from N1
Since $(S_1)_9 = 48$ ft, NL = 9 and $xL = 53 - 48 = 5$ ft from NL = 9 to support B
NE = last wheel on span = NL = 9

$$R_B = \frac{\Sigma M_B}{L} = \frac{\Sigma M_{(NL-1),N1} + \Sigma P_{N1,NE}(xL)}{60} = \frac{\Sigma M_{8,1} + \Sigma P_{1,9}(xL)}{60} = \frac{6992 + 284(5)}{60} = 140.20 \text{ kips},$$

$$M_c = R_B(L/3) - \sum M_{(NP-1),N1} = R_B(L/3) - \sum M_{2,1} = 140.2(20) - 460 = 2344 \text{ kips-ft}.$$

The superstructure with transverse floorbeams has a 100[1 – (2344/2391)] = 2.0% decrease in bending moment with NP = 3 at location of maximum moment.

With NP = 4 (Cooper's load configuration wheel number 4)
From Table 5.1b:
$x1 = (S_1)_4 = 18$ ft
Since $x1 \times L/2 \times 30$ ft; N1 = 1
Support B is $(2L/3 + x1) = 40 + 18 = 58$ ft from N1
Since $(S_1)_{10} = 56$ ft, NL = 10 and $xL = 58 - 56 = 2$ ft from NL = 10 to support B
NE = last wheel on span = NL = 10

$$R_B = \frac{\Sigma M_B}{L} = \frac{\Sigma M_{(NL-1),N1} + \Sigma P_{N1,NE}(xL)}{60} = \frac{\Sigma M_{9,1} + \Sigma P_{1,10}(xL)}{60} = \frac{9264 + 304(2)}{60} = 164.5 \text{ kips},$$

$$M_c = R_B(L/3) - \sum M_{(NP-1),N1} = R_B(L/3) - \sum M_{3,1} = 164.5(20) - 960 = 2331 \text{ kips-ft}.$$

The superstructure loaded with transverse floorbeams has a 100[1 – (2331/2536)] = 8.1% decrease in bending moment with NP = 4 at location of maximum moment.

With NP = 13 (Cooper's load configuration wheel number 13)
From Table 5.1b:
With NP = 13, the first wheel on the span = N1 = 10
$x1 = (S_1)_{13} - (S_1)_{10} = 74 - 56 = 18$ ft
Support B is 40 + 18 = 58 ft from N1 = 10
With NP = 13, the last wheel on the span, NL, is the end of 5 ft of the uniform load, *w*
$xL = (2L/3) - [(S_1)_w - (S_1)_{13}] = 40 - (104 + 5 - 74) = 5$ ft from beginning of uniform load, *w*, to support B
NE = last wheel on span = 5 ft of uniform load, *w*

$$R_B = \frac{\Sigma M_B}{L} = \frac{\Sigma M_{18,10} + \Sigma P_{10,w}(xL)}{60} = \frac{8412 + [(284)(5) + 4(5)(5/2)]}{60} = 164.7 \text{ kips},$$

$$M_c = R_B(L/3) - \sum M_{12,10} = 164.7(20) - 960 = 2334 \text{ kips-ft}.$$

The superstructure with transverse floorbeams has a 100[1 – (2334/2588)] = 9.8% decrease in bending moment with NP = 13 at location of maximum moment.

5.2.1.2 Influence Lines for Maximum Effects of Moving Loads on Superstructures

Influence lines for effects (shear force, bending moment, axial force, and deformations) at any location along the length of a superstructure element illustrate the variation in the effect at the location as a unit load traverses the superstructure. In this manner, influence lines facilitate both the appropriate placement of loads and determination of the maximum effects in steel beam and girder superstructures (shear forces and bending moments), trusses (axial forces), and arches (axial forces, shear forces, and bending moments).

Influence lines may be constructed for moving load analysis of statically determinate superstructures by moving a unit concentrated force across the superstructure and determining the value of the effect of the unit load at each location in the superstructure. The construction of influence lines may be simplified by determining the value of the effect at locations where changes in the influence line will occur (i.e., supports, panel points, hinges, etc.) and joining those locations with straight lines*.

Influence lines may also be constructed for moving load analysis of statically indeterminate superstructures by use of the Muller-Breslau principle. This principle enables the construction of influence line ordinates for effects as the ordinate of the deflection curve. The ordinate is determined by releasing the restraint compatible with the effect and introducing a corresponding unit displacement in the remaining length of the superstructure.

5.2.1.2.1 Influence Lines for Maximum Shear Force and Bending Moment in Simply Supported Beam and Girder Spans

5.2.1.2.1.1 Maximum Shear Force (with Loads Applied Directly to the Superstructure) The influence lines for shear force at location C and at the end (location A) of a simple span are shown in Figure 5.5. They are developed by determining the shear force at location, C, and reaction at the end (location A) of the simple span with a unit load placed at locations A, B, and C.

5.2.1.2.1.2 Maximum Shear Force (with Loads Applied to the Superstructure through Transverse Members) The influence line for shear in panel BC of a simply supported span is shown in Figure 5.6.

It is developed by determining the shear force at locations B and C with a unit load placed at locations A, B, C, and D:

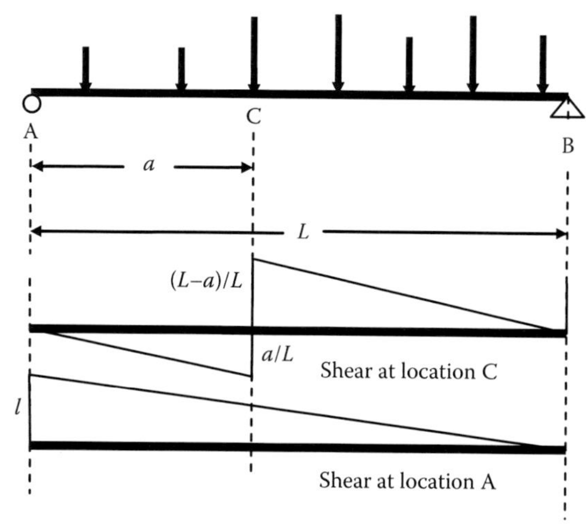

FIGURE 5.5 Influence lines for shear at locations C and A for concentrated moving loads applied directly to the superstructure.

* For axial force, shear force, and bending moment in statically determinate structures, influence lines are comprised of straight-line segments. However, for deflections this is not the case.

$$d_1 = \frac{\left(\frac{n_R}{n_L}\right)s_p}{\left(1+\left(\frac{n_R}{n_L}\right)\right)} = \frac{s_p n_R}{2\left(\frac{n_L}{n_R}+1\right)}, \quad (5.17)$$

where, in Figure 5.6, n is the number of panels, n_L is the number of panels left of panel BC, and n_R is the number of panels right of panel BC.

$$L = n(s_p) = (n_L + n_R + 1)\,(s_p).$$

5.2.1.2.1.3 Maximum Bending Moment (with Loads Applied Directly to the Superstructure) The influence lines for bending at location C and at the center of a simple span are shown in Figure 5.7. They are developed by determining the bending moment at location C and at center span with a unit load placed at locations A, B, and C.

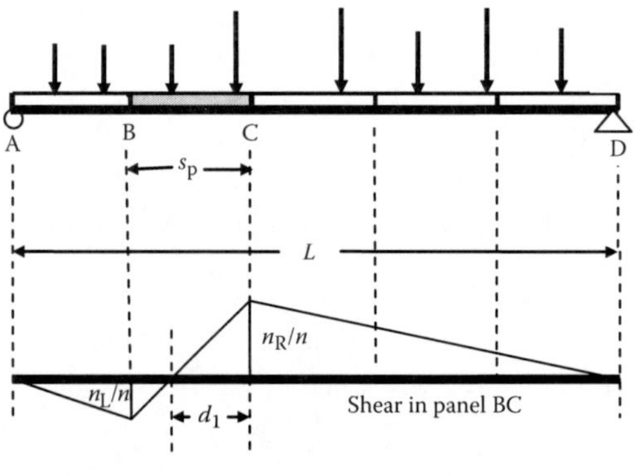

FIGURE 5.6 Influence line for shear in panel BC for concentrated moving loads applied to the superstructure at transverse members.

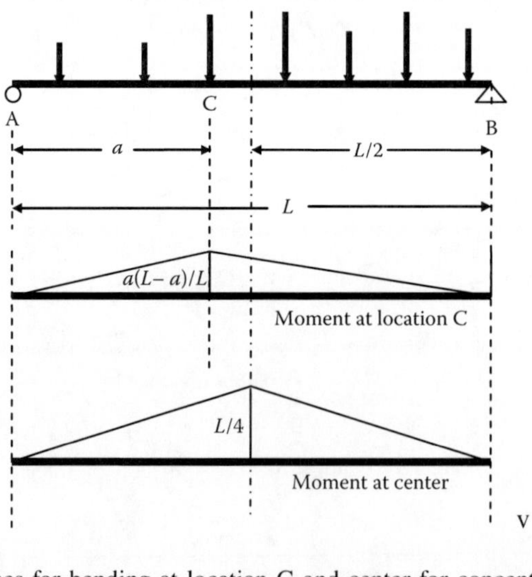

FIGURE 5.7 Influence lines for bending at location C and center for concentrated moving loads applied directly to the superstructure.

5.2.1.2.1.4 Maximum Bending Moment (with Loads Applied to the Superstructure by Transverse Members) The influence lines for moment in panel BC (at distance d_2 from B) and at location C of a simple span are shown in Figure 5.8. They are developed by determining the bending moments at locations B and C with a unit load placed at locations A, B, C, and D. As shown in Figure 5.8, a reduction in bending moment occurs for beams and girders loaded through transverse members.

5.2.1.2.1.5 Maximum Floorbeam Reactions for Loads on Simply Supported Stringers The influence line for floorbeam reaction at location C assuming simply supported stringer spans is shown in Figure 5.9. It is developed by determining the shear forces at locations B, C, and D with a unit load placed at locations B, C, and D. Since stringer spans are generally relatively short, the location of concentrated loads for maximum floorbeam reaction is usually quite obvious by inspection.

Influence lines may be readily constructed for moving load analysis of statically determinate superstructures.

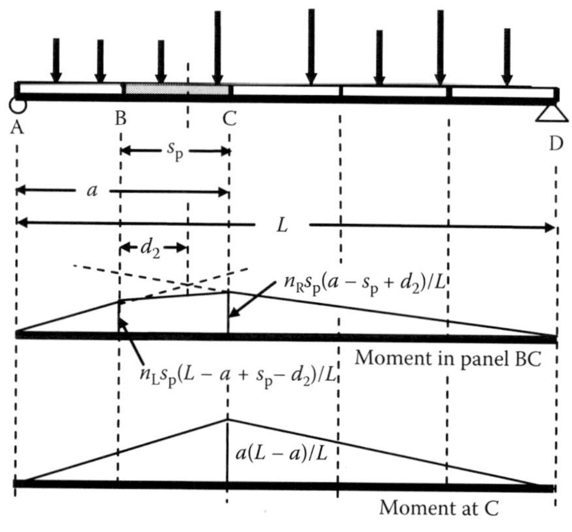

FIGURE 5.8 Influence lines for moment in panel BC and at location C for concentrated moving loads applied at transverse members.

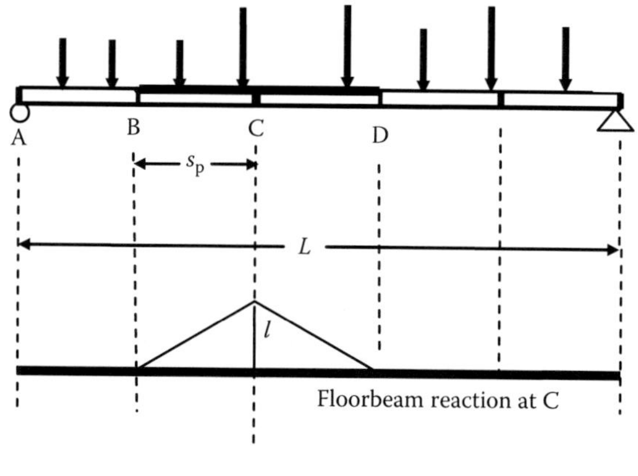

FIGURE 5.9 Influence line for floorbeam reaction at location C.

Example 5.3a (SI Units)

Determine the maximum bending moment per rail under axle NP = 5 of Cooper's EM360 load on an 18 m long DPG span. The moment at the center, *C*, is assumed to be near to the maximum bending moment in both location and magnitude.

The influence line for center span bending moments is shown in Figure E5.3a.

The ordinates of the influence lines are as follows:

$a = (2.1/9)4.50 = 1.05$
$b = (4.5/9)4.50 = 2.25$
$c = (6.0/9)4.50 = 3.00$
$d = (7.5/9)4.50 = 3.75$
$e = 18/4 = 4.50$
$f = (6.3/9)4.50 = 3.15$
$g = (4.8/9)4.50 = 2.40$
$h = (3.0/9)4.50 = 1.50$
$i = (1.5/9)4.50 = 0.75$
$M_C = 90(1.05) + 180(2.25 + 3.00 + 3.75 + 4.50) + 115(3.15 + 2.40 + 1.50 + 0.75) = 3422$ kNm.

Example 5.3b (US Customary and Imperial Units)

Determine the maximum bending moment per rail under axle NP = 5 of Cooper's E80 load on a 60 ft long DPG span. The moment at the center, *C*, is assumed to be near to the maximum bending moment in both location and magnitude.

The influence line for center span bending moments is shown in Figure E5.3b.

The ordinates of the influence lines are as follows:

$a = (7/30)15 = 3.50$
$b = (15/30)15 = 7.50$
$c = (20/30)15 = 10.00$
$d = (25/30)15 = 12.50$

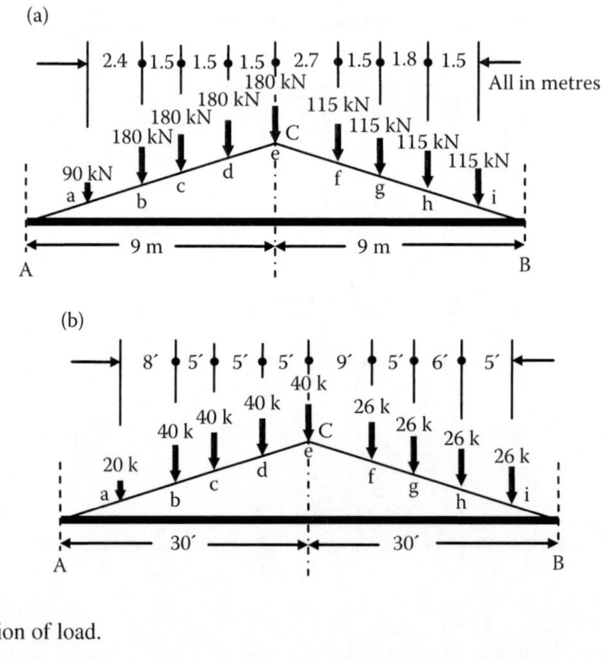

FIGURE E5.3 Position of load.

$e = 60/4 = 15.00$
$f = (21/30)15 = 10.50$
$g = (16/30)15 = 8.00$
$h = (10/30)15 = 5.00$
$i = (5/30)15 = 2.50$
$M_C = 20(3.50) + 40(7.50 + 10.00 + 12.50 + 15.00) + 26(10.50 + 8.00 + 5.00 + 2.50) = 2546$ kips-ft.

Example 5.4a (SI Units)

Determine the bending moment per rail at location C under axle NP = 3 of Cooper's EM360 load on a 18 m long through plate girder span with a floor system comprising floorbeams and 6 m long stringers.

The influence line for bending moments at location C is shown in Figure E5.4a.

The ordinates of the influence lines are as follows:

$a = (2.1/6.0)4.00 = 1.40$
$b = (4.5/6.0)4.00 = 3.00$
$c = 12.0(6.0)/18 = 4.00$
$d = (10.5/12.0)4.00 = 3.50$
$e = (9.0/12.0)4.00 = 3.00$
$f = (6.3/12.0)4.00 = 2.10$
$g = (4.8/12.0)4.00 = 1.60$
$h = (3.0/12.0)4.00 = 1.00$

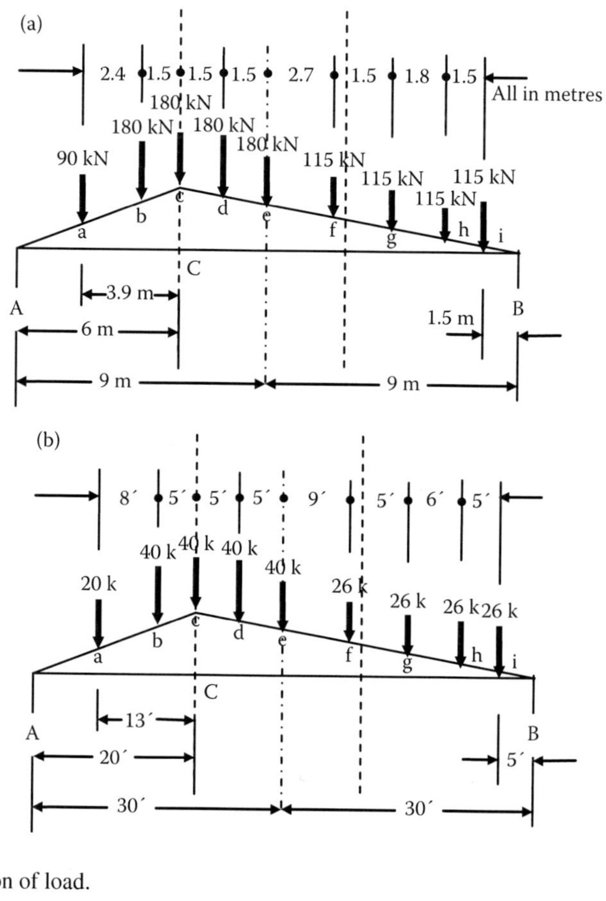

FIGURE E5.4 Position of load.

$i = (1.5/12)4.00 = 0.50$
$M_C = 90(1.40) + 180(3.00 + 4.00 + 3.50 + 3.00) + 115(2.10 + 1.60 + 1.00 + 0.50) = 3154$ kNm.

Example 5.4b (US Customary and Imperial Units)

Determine the bending moment per rail at location C under axle NP = 3 of Cooper's E80 load on a 60 ft long through plate girder span with a floor system comprising floorbeams and 20 ft long stringers.

The influence line for bending moments at location C is shown in Figure E5.4b.

The ordinates of the influence lines are as follows:

$a = (7/20)13.33 = 4.67$
$b = (15/20)13.33 = 10.00$
$c = 40(20)/60 = 13.33$
$d = (35/40)13.33 = 11.67$
$e = (30/40)13.33 = 10.00$
$f = (21/40)13.33 = 7.00$
$g = (16/40)13.33 = 5.33$
$h = (10/40)13.33 = 3.33$
$i = (5/40)13.33 = 1.67$
$M_C = 20(4.67) + 40(10.00 + 13.33 + 11.67 + 10.00)) + 26(7.00 + 5.33 + 3.33 + 1.67) = 2345$ kips-ft.

5.2.1.2.2 Influence Lines for Maximum Axial Forces in Statically Determinate Truss Spans

The influence lines developed for shear force and bending moment in simply supported spans are useful in the construction of axial force influence lines for truss web and chord members, respectively. In addition, consideration of the moving load effect at panel points simplifies the construction of axial force influence lines for statically determinate truss spans.

Influence lines for truss chord members may be constructed by considering free body diagrams and equilibrium of moments. Influence lines for truss web members may be constructed by considering free body diagrams and equilibrium of forces. The construction of influence lines for axial forces in the members of simply supported truss spans is illustrated by examples a Pratt truss and a Parker truss, respectively, in Examples 5.5 and 5.6*.

Example 5.5a (SI Units)

Construct influence lines for the 80 m long eight panel Pratt through truss in Figure E5.5aa. The influence lines are constructed by locating unit loads at appropriate locations and using the method of sections or method of joints.

Determine influence lines for the reactions and members U1–U2, U3–L3, L1–L2, L3–L4, U1–L1, and U1–L2.

Section 1–1 may be isolated to determine the forces in members U1–U2 (Figure E5.5aa), L1–L2 (Figure E5.5ab) and U1–L2 (Figure E5.5ac).

Member U1–U2:
In Figure E5.5aa: with unit load at L1 and taking moments about L2, the force in U1–U2 = [(1/8)(6)(10.0)]/12.0 = −0.63 (compression direction to balance reaction moment about L2).

In Figure E5.5aa: with unit load at L2 and taking moments about L2, the force in U1–U2 = [(2/8)(6)(10.0)]/12.0 = −1.25 (compression direction to balance reaction moment about L2).

Member L1–L2:
In Figure E5.5ab: with unit load at L1 and taking moments about U1, the force in L1–L2 = [1/8)(7)(10.0)]/12.0 = +0.73 (tension direction to balance reaction moment about U1).

* These truss forms are often used for medium span steel railway bridges.

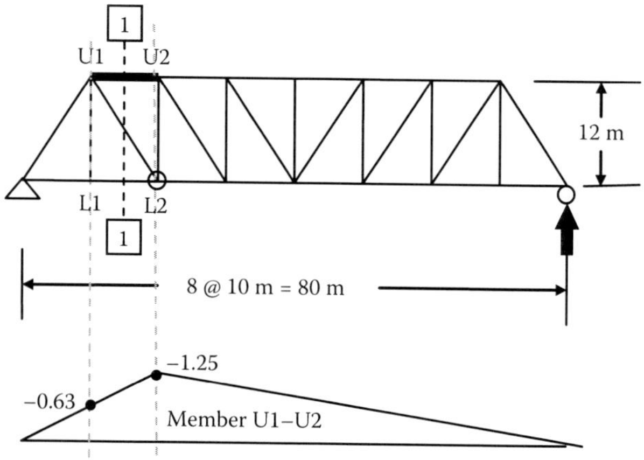

FIGURE E5.5aa Influence line for member U1–U2.

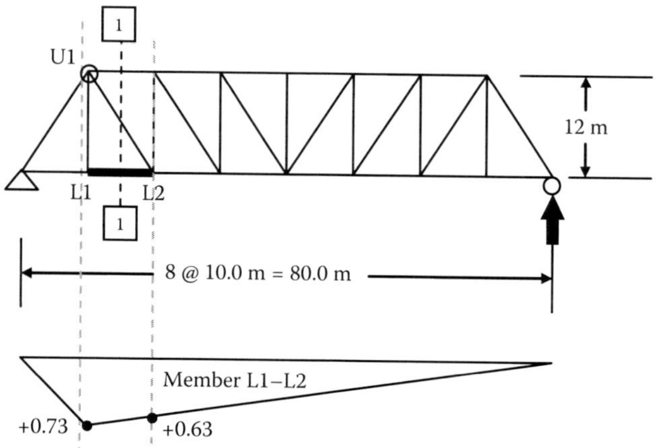

FIGURE E5.5ab Influence line for member L1–L2.

In Figure E5.5ab: with unit load at L2 and taking moments about U1, the force in L1–L2 = [(2/8)(7)(10.0) – (1)(10.0)]/12.0 = + 0.63 (tension direction to balance reaction moment about U1).

Member U1–L2:

In Figure E5.5ac: with unit load at L1 and summing horizontal forces in panels 1–2, the force in U1–L2 = –(–0.63 + 0.73)$(10.0^2 + 12.0^2)^{1/2}$/10.0 = – 0.16.

In Figure E5.5ac: with unit load at L2 and summing horizontal forces in panels 1–2, the force in U1–L2 = –(–1.25 + 0.63) $(10.0^2 + 12.0^2)^{1/2}$/10.0 = +0.97.

Member L3–L4:

Section 2–2 may be isolated to determine the forces in member L3–L4 (Figure E5.5ad).

In Figure E5.5ad: with unit load at L3 and taking moments about U3, the force in L3–L4 = [3/8)(5)(10.0)]/12.0 = +1.56 (tension direction to balance reaction moment about U1).

In Figure E5.5ad: with unit load at L4 and taking moments about U3, the force in L3–L4 = [(4/8)(5)(10.0) – (1)(10.0)]/12.0 = +1.25 (tension direction to balance reaction moment about U1).

Member U3–L3:

Section 3–3 may be isolated to determine the forces in member U3–L3 (Figure E5.5ae).

In Figure E5.5ae: with unit load at L3 and summing vertical forces in panels 3–4, the force in U3–L3 = +3/8 = +0.38.

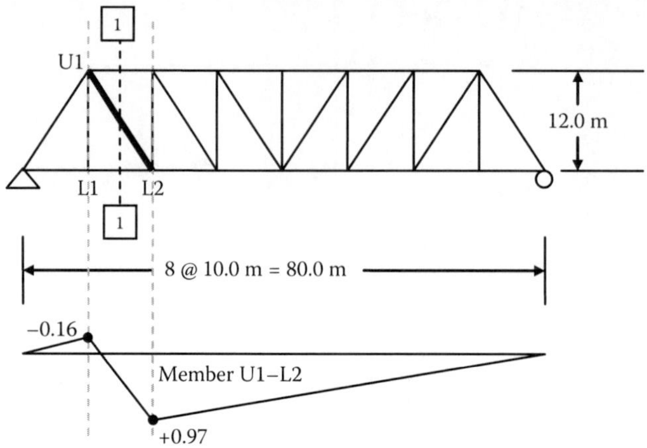

FIGURE E5.5ac Influence line for member U1–L2.

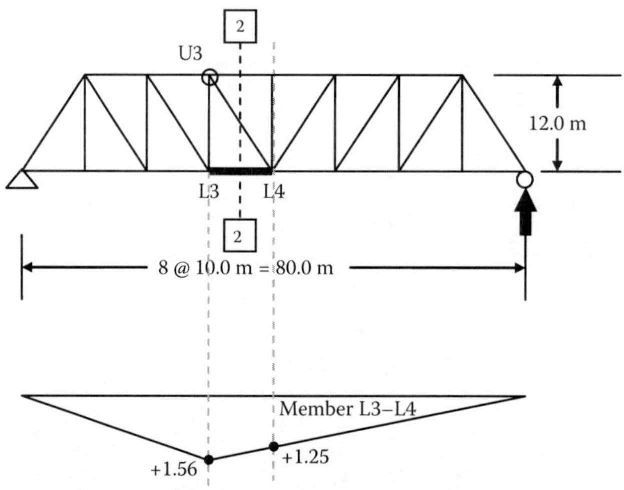

FIGURE E5.5ad Influence line for member L3–L4.

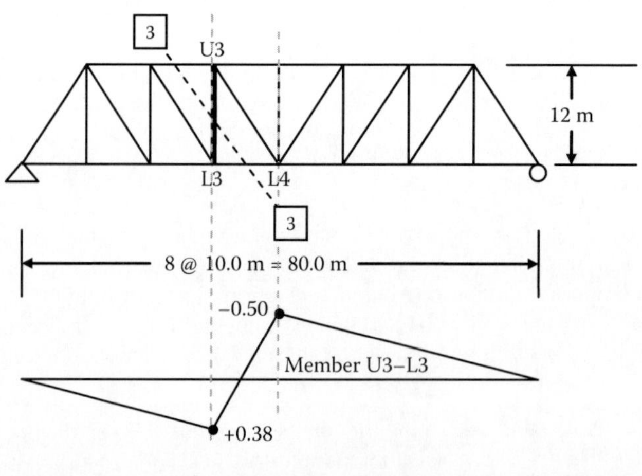

FIGURE E5.5ae Influence line for member U3–L3.

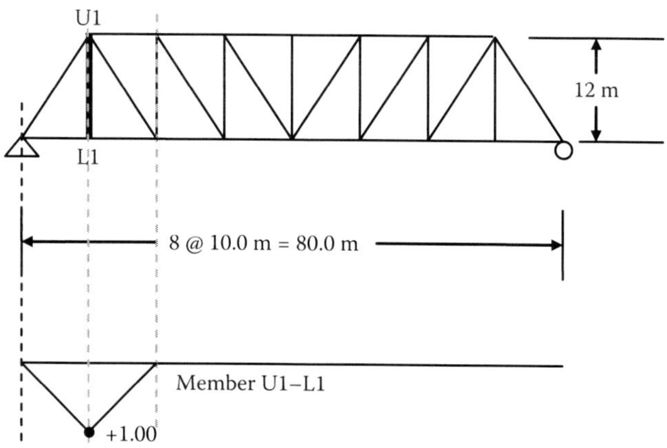

FIGURE E5.5af Influence line for member U1–L1.

In Figure E5.5ae: with unit load at L4 and summing vertical forces in panels 3–4, the force in U3–L3 = +1/2 − 1 = −0.50.

Member U1–L1:
The forces in member U1–L1 (Figure E5.5af) may be determined by the method of joints by locating unit loads at L0, L1, and L2.

The hanger U1–L1 is loaded only when moving loads are in adjacent panels of the hanger*.

Example 5.5b (US Customary and Imperial Units)

Construct influence lines for the 156.38 ft eight panel Pratt through truss in Figure E5.5ba. The influence lines are constructed by locating unit loads at appropriate locations and using the method of sections or method of joints.

Determine influence lines for the reactions and members U1–U2, U3–L3, L1–L2, L3–L4, U1–L1, and U1–L2.

Section 1–1 may be isolated to determine the forces in members U1–U2 (Figure E5.5ba), L1–L2 (Figure E5.5bb), and U1–L2 (Figure E5.5bc).

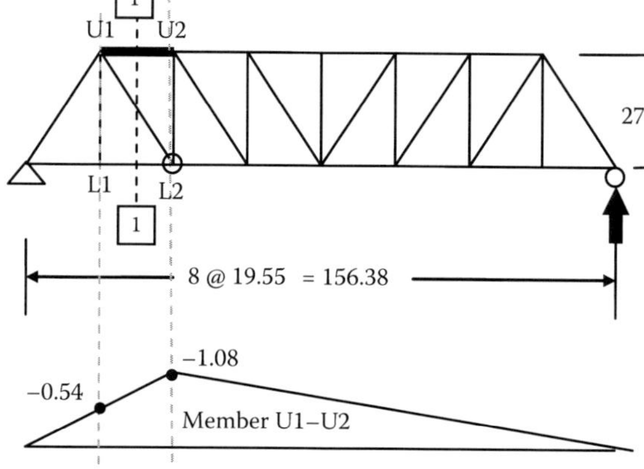

FIGURE E5.5ba Influence line for member U1–U2.

* There are also increased impact effects for through truss hangers due to the short live load influence line (see Chapter 4).

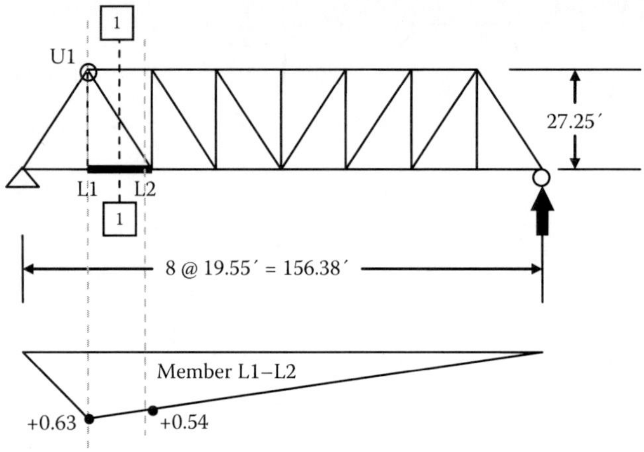

FIGURE E5.5bb Influence line for member L1–L2.

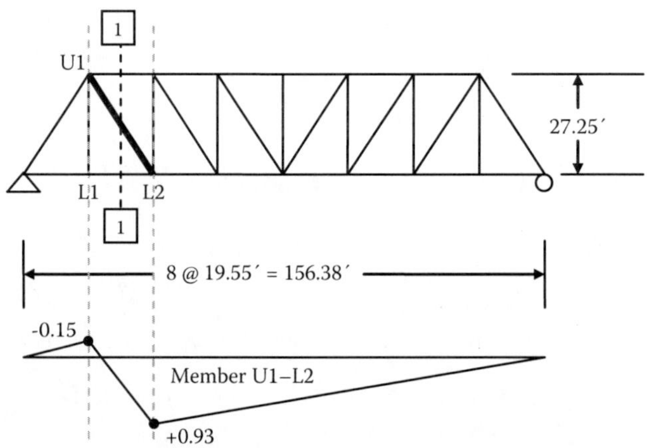

FIGURE E5.5bc Influence line for member U1–L2.

Member U1–U2:
In Figure E5.5ba: with unit load at L1 and taking moments about L2, the force in U1–U2 = [(1/8)(6)(19.55)]/27.25 = −0.54 (compression direction to balance reaction moment about L2).

In Figure E5.5ba: with unit load at L2 and taking moments about L2, the force in U1–U2 = [(2/8)(6)(19.55)]/27.25 = −1.08 (compression direction to balance reaction moment about L2).

Member L1–L2:
In Figure E5.5bb: with unit load at L1 and taking moments about U1, the force in L1–L2 = [1/8)(7)(19.55)]/27.25 = +0.63 (tension direction to balance reaction moment about U1).

In Figure E5.5bb: with unit load at L2 and taking moments about U1, the force in L1–L2 = [(2/8)(7)(19.55) − (1)(19.55)]/27.25 = + 0.54 (tension direction to balance reaction moment about U1).

Member U1–L2:
In Figure E5.5bc: with unit load at L1 and summing horizontal forces in panels 1–2, the force in U1–L2 = $-(-0.54 + 0.63)(19.55^2 + 27.25^2)^{1/2}/19.55 = -0.15$.

In Figure E5.5bc: with unit load at L2 and summing horizontal forces in panels 1–2, the force in U1–L2 = $-(-1.08 + 0.54)(19.55^2 + 27.25^2)^{1/2}/19.55 = +0.93$.

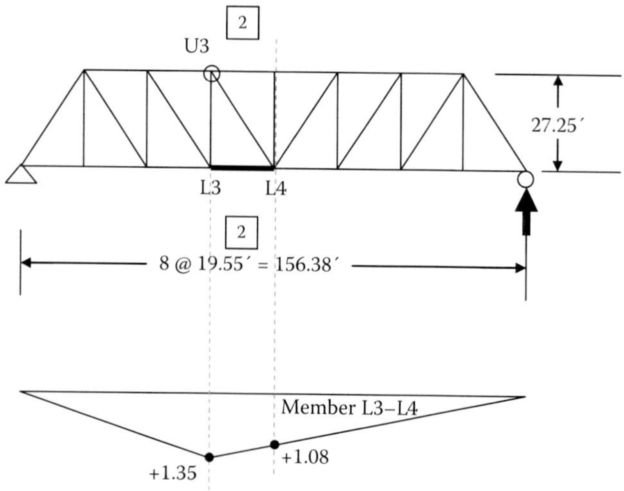

FIGURE E5.5bd Influence line for member L3–L4.

Member L3–L4:

Section 2–2 may be isolated to determine the forces in member L3–L4 (Figure E5.5bd).

In Figure E5.5bd: with unit load at L3 and taking moments about U3, the force in L3–L4 = [3/8)(5)(19.55)]/27.25 = +1.35 (tension direction to balance reaction moment about U1).

In Figure E5.5bd: with unit load at L4 and taking moments about U3, the force in L3–L4 = [(4/8)(5)(19.55) − (1)(19.55)]/27.25 = +1.08 (tension direction to balance reaction moment about U1).

Member U3–L3:

Section 3–3 may be isolated to determine the forces in member U3–L3 (Figure E5.5be).

In Figure E5.5be: with unit load at L3 and summing vertical forces in panels 3–4, the force in U3–L3 = +3/8 = +0.38.

In Figure E5.5be: with unit load at L4 and summing vertical forces in panels 3–4, the force in U3–L3 = +1/2 − 1 = −0.50.

Member U1–L1:

The forces in member U1–L1 (Figure E5.5bf) may be determined by the method of joints by locating unit loads at L0, L1, and L2.

The hanger U1–L1 is loaded only when moving loads are in adjacent panels of the hanger.

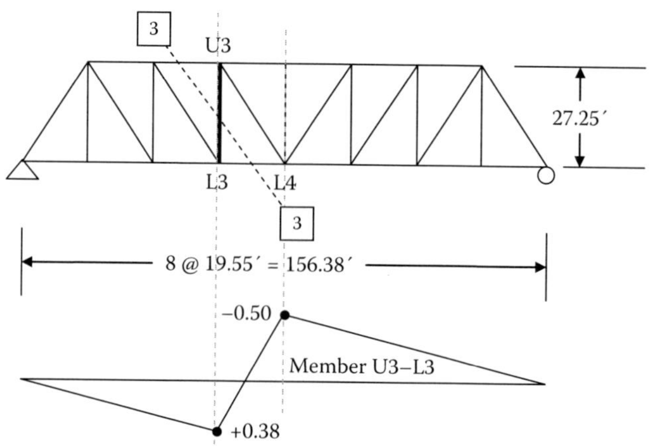

FIGURE E5.5be Influence line for member U3–L3.

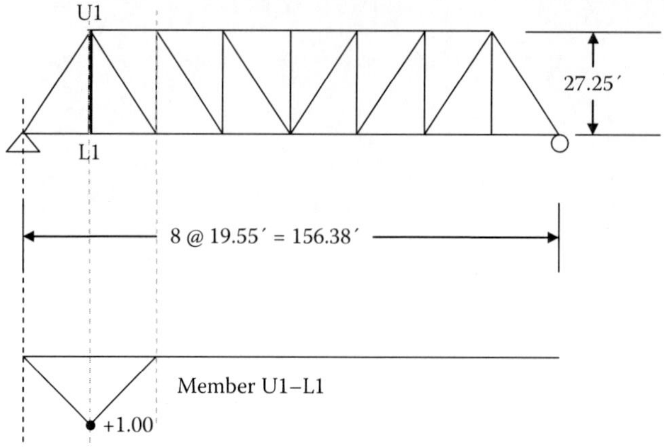

FIGURE E5.5BF Influence line for member U1–L1.

Example 5.6a (SI Units)

Construct influence lines for members U1–U2, U1–L2, and U2–L2 in the 90 m six panel curved-chord Parker through truss in Figure E5.6aa. The influence lines are constructed by using the method of sections and locating unit loads at appropriate locations:

$$a_p = \frac{15}{\left(\frac{12}{8} - 1\right)} - 15 = 15 \text{ m}$$

$$h_p = (a_p + 2(15))\left(\frac{8}{\sqrt{(a_p + 15)^2 + 8^2}}\right) = 22.7 \text{ m}$$

$$\beta_p = 45^\circ - \tan^{-1}\frac{8}{a_p + 15} = 14.9^\circ$$

$$L_p = \sqrt{(a_p + 15)^2 + 8^2}\cos(\beta_p) = 30.0 \text{ m}.$$

Considering section 1–1 in Figure E5.6aa: with unit load at L2 and taking moments about L2, the force in U1–U2 = [−(4/6)(2)(15)]/22.7 = −0.88 (compression direction to balance reaction moment about L2) (Figure E5.6ab).

Considering section 1–1 in Figure E5.6aa: with unit load at L1 and taking moments about P0, the force in U1–L2 = (−(1/6)(90 + 15))/30.0 = −0.58.

Considering section 1–1 in Figure E5.6aa: with unit load at L2 and taking moments about P0, the force in U1–L2 = (4/6)(15.0)/30.0 = 0.33 (Figure E5.6ab).

Considering section 2–2 in Figure E5.6aa: with unit load at L2 and taking moments about P0, the force in U2–L2 = 2/6(15 + 90)/30.0 = 1.17.

Considering section 2–2 in Figure E5.6aa: With unit load at L3 and taking moments about P0, the force in U2–L2 = −1/2(15.0)/30.0 = −0.25 (Figure E5.6ab).

The distance, h_p, in Figure E5.6aa illustrates the effect of the "modified panel shear" created by the sloped chord, which participates in resisting the panel shear force.

Influence lines for other chord and web members of the truss may be constructed in a similar manner by applying unit loads at panel points and determining axial forces in members by the method of sections or method of joints.

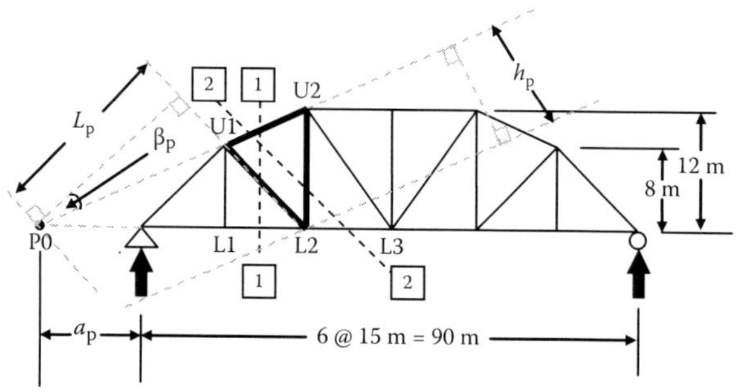

FIGURE E5.6AA Sections for influence lines for members U1–U2, U1–L2 and U2–L2.

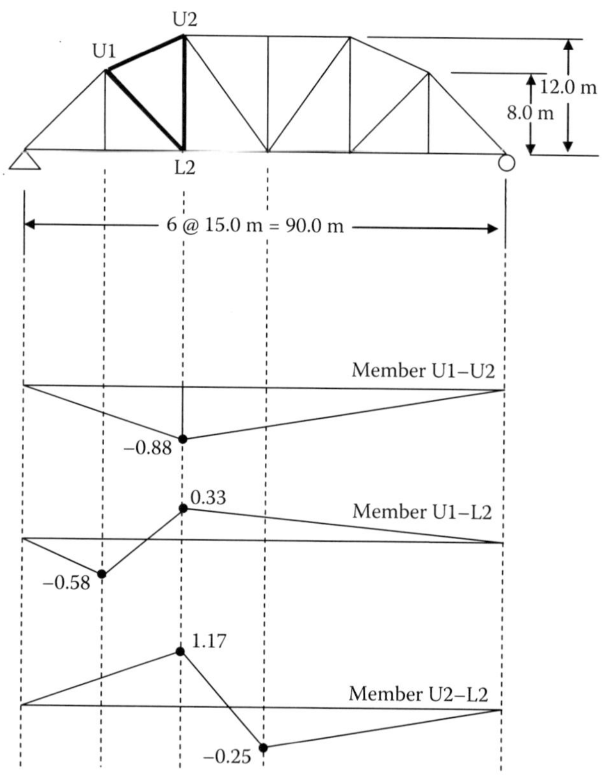

FIGURE E5.6AB Influence lines for members U1–U2, U1–L2 and U2–L2.

Example 5.6b (US Customary and Imperial Units)

Construct influence lines for members U1–U2, U1–L2, and U2–L2 in the 240 ft six panel curved-chord Parker through truss in Figure E5.6ba. The influence lines are constructed by using the method of sections and locating unit loads at appropriate locations:

$$a_p = \frac{40}{\left(\frac{36}{28}-1\right)} - 40 = 100\,\text{ft},$$

$$h_p = (a_p + 2(40))\left(\frac{28}{\sqrt{(a_p+40)^2 + 28^2}}\right) = 35.3\,\text{ft},$$

$$\beta_p = 45^\circ - \tan^{-1}\frac{28}{a_p + 40} = 33.7^\circ,$$

$$L_p = \sqrt{(a_p+40)^2 + 28^2}\,\cos(\beta_p) = 118.8\,\text{ft}.$$

Considering section 1–1 in Figure E5.6ba: with unit load at L2 and taking moments about L2, the force in U1–U2 = [−(4/6)(2)(40)]/35.3 = −1.51 (compression direction to balance reaction moment about L2) (Figure E5.6bb).

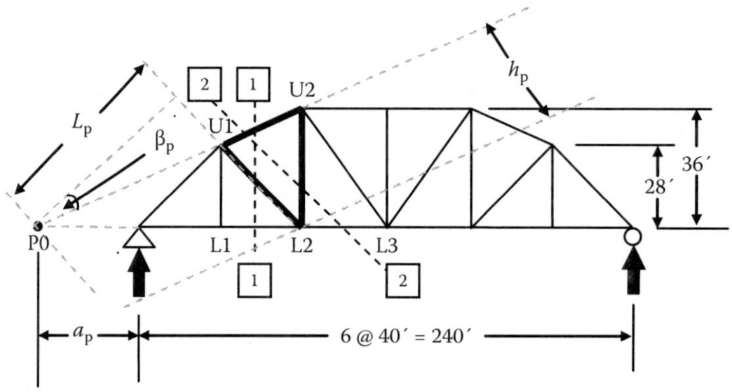

FIGURE E5.6BA Sections for influence lines for members U1–U2, U1–L2 and U2–L2.

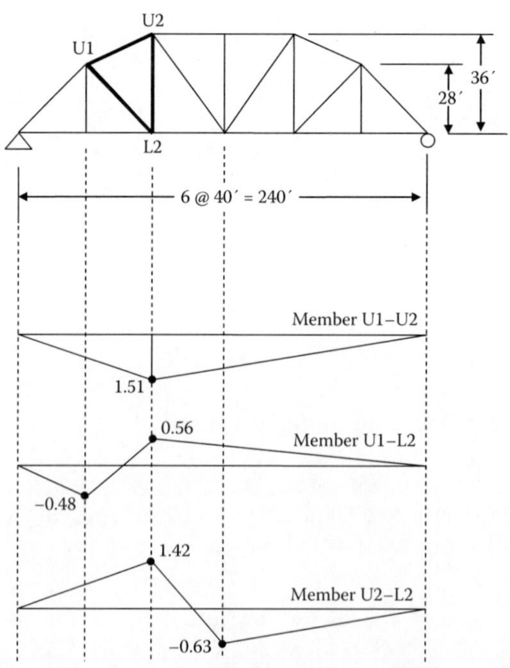

FIGURE E5.6BB Influence lines for members U1–U2, U1–L2 and U2–L2.

Considering section 1–1 in Figure E5.6ba: with unit load at L1 and taking moments about P0, the force in U1–L2 = (–(1/6)(240 + 100))/118.8 = –0.48.

Considering section 1–1 in Figure E5.6ba: with unit load at L2 and taking moments about P0, the force in U1–L2 = (4/6)(100)/118.8 = 0.56 (Figure E5.6bb).

Considering section 2–2 in Figure E5.6ba: with unit load at L2 and taking moments about P0, the force in U2–L2 = 2/6(100 + 240)/80 = 1.42.

Considering section 2–2 in Figure E5.6ba: With unit load at L3 and taking moments about P0, the force in U2–L2 = –1/2(100)/80 = –0.63 (Figure E5.6bb).

The distance, h_p, in Figure E5.6ba illustrates the effect of the "modified panel shear" created by the sloped chord, which participates in resisting the panel shear force.

Influence lines for other chord and web members of the truss may be constructed in a similar manner by applying unit loads at panel points and determining axial forces in members by the method of sections or method of joints.

5.2.1.2.3 *Influence Lines for Maximum Effects in Statically Determinate Arch Spans*

Many steel railway arches are designed as three-hinged to impose statically determinate conditions (Figure 5.10a). Statically determinate arches are typically simpler to fabricate and erect; and are not subjected to temperature or support displacement induced stresses. The construction of influence lines for statically determinate arches can be made efficient by understanding the relationships between arch reactions, internal forces (shear, bending, and axial), and the influence lines obtained in Section 5.2.1.2.1 for shear and bending in simply supported spans.

5.2.1.2.3.1 Maximum Bending Moment, Shear Force, and Axial Force (with Moving Loads Applied Directly to the Arch) For the moving concentrated load, $P = 1$, a distance x_p from support A in Figure 5.10a:

$$R_a = \frac{L - x_p}{L}, \tag{5.18a}$$

$$R_B = \frac{x_p}{L}. \tag{5.18b}$$

Therefore, the influence line for the vertical components of the arch reactions, R_A and R_B, will be the same as those for a simply supported beam of length, L, as shown in Figure 5.10a.

If moments are taken about the arch crown pin (point C)*,

$$H_A(h) = R_A(L/2). \tag{5.19}$$

Since $R_A(L/2)$ is the bending moment at point C in a simply supported span, the influence line for horizontal thrust reaction, H_A, is proportional (by the arch rise, h) to this simple span bending moment as shown in Figure 5.10a. Therefore, the criteria for the position of Cooper's load for maximum bending moment (see Section 5.2.1.1.3) can be used for the determination of maximum horizontal thrust in a statically determinate arch.

The arch reactions may be used to determine the internal shear force, bending moment, and axial force influence lines for the arch rib. From Figure 5.10b, the bending moment, M_D, at a location, D, is

* It is the inclusion of the crown pin that enables this equilibrium equation to be written; thereby illustrating the benefits of statically determinate design and construction.

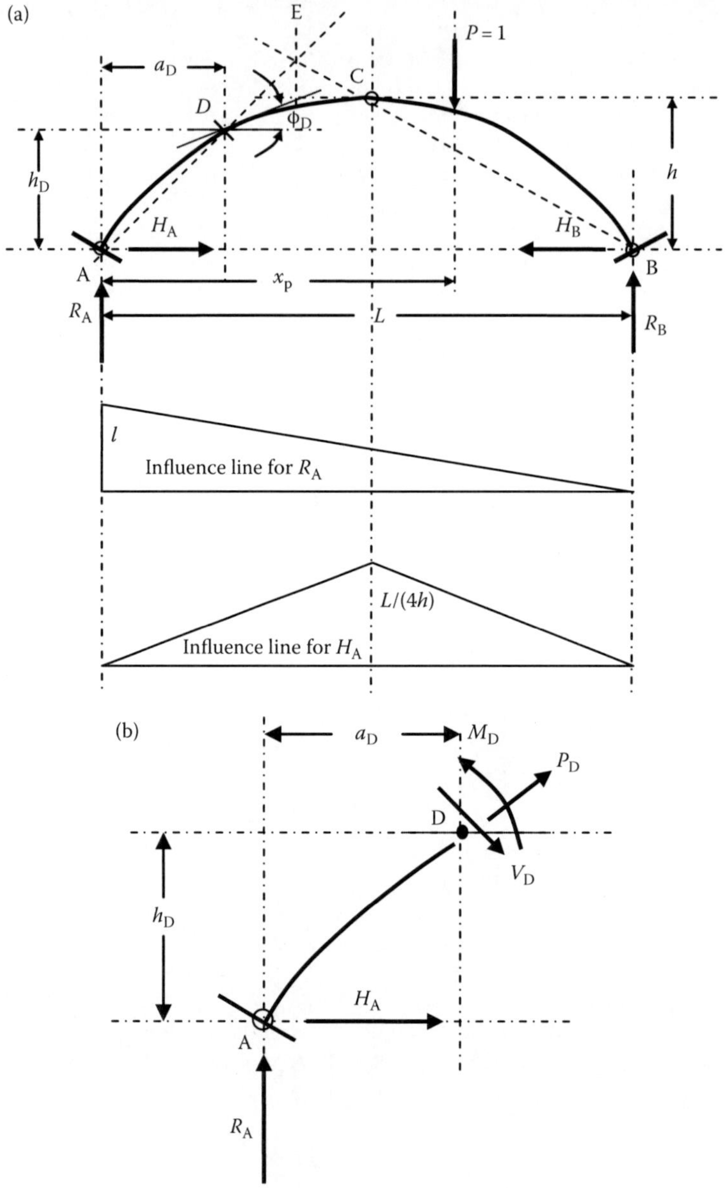

FIGURE 5.10 (a) Three-hinged arch rib with concentrated moving loads applied directly to the rib. (b) Free body diagram of arch rib from support A to point D.

$$M_D = R_A(a_D) - H_A(h_D). \tag{5.20}$$

Equation 5.20 indicates that the influence line for bending moment in the arch rib at location D can be obtained by subtracting the ordinates for the influence line for H_A (Figure 5.10a) multiplied by the distance h_D from the ordinates for simple beam bending at location, D, described by $R_A(a_D)$. The construction of this influence line is shown in Figure 5.11. The ordinates (shaded areas) may be plotted on a horizontal line for ease of use in design.

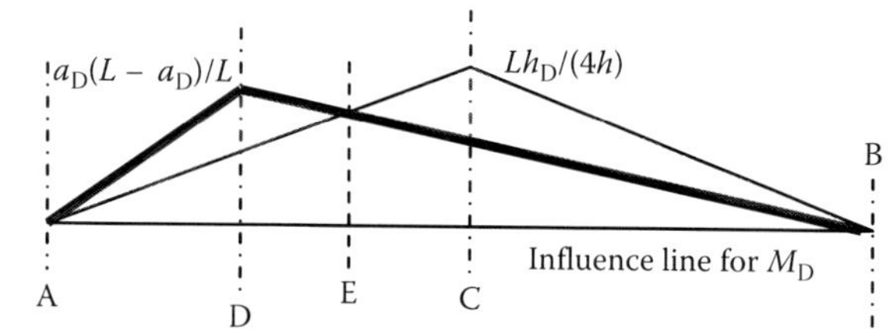

FIGURE 5.11 Influence line for bending moments at location D in three-hinged arch rib.

From Figures 5.10a and b, the shear force, V_D, at a location, D, is

$$V_D = R_A \cos \phi_D - H_A \sin \phi_D. \tag{5.21}$$

Equation 5.21 indicates that the influence line for shear force in the arch rib at location D can be obtained by subtracting the ordinates for the influence line for H_A multiplied by $\sin\phi_D$ from the ordinates for simple beam shear at D multiplied by $\cos\phi_D$. The construction of this influence line is shown in Figure 5.12. Again, the ordinates (shaded areas) may be plotted on a horizontal line for ease of use in design. Location E is the position of the moving load that creates no shear force or bending moment in the arch at location D (Figure 5.10a).

From Figures 5.10a and 5.10b, the axial force, F_D, at a location, D, is

$$F_D = -R_A \sin \phi_D - H_A \cos \phi_D. \tag{5.22}$$

Equation 5.22 indicates that the influence line for axial force at location D in the arch rib can be obtained by adding the ordinates for the influence line for H_A multiplied by $\cos\phi_D$ to the ordinates for simple beam shear at D multiplied by $\sin\phi_D$. The construction of this influence line is shown in Figure 5.13. Again, the ordinates (shaded areas) may be plotted on a horizontal line for ease of use in design.

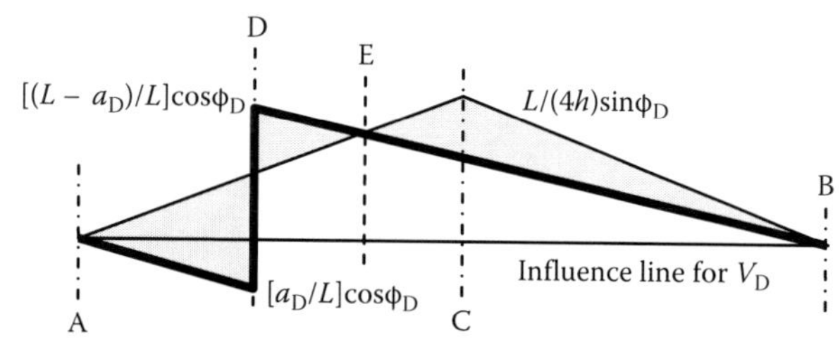

FIGURE 5.12 Influence line for shear forces at location D in three-hinged arch rib.

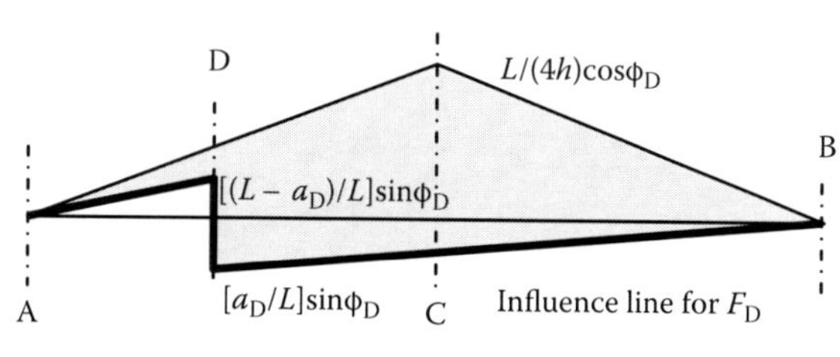

FIGURE 5.13 Influence line for axial force at location D in three-hinged arch rib.

5.2.1.2.3.2 Maximum Bending Moment, Shear Force, and Axial Force [with Loads Applied to the Arch by Transverse Members (Spandrel Columns or Walls)] Medium and long-span steel railway bridges can be economically constructed of three-hinged arches with the arch rib loaded by vertical spandrel members (Figure 5.14). Influence lines for arch spans with spandrel columns or vertical posts can be developed from influence lines for directly loaded arches in a manner analogous to simple spans with transverse members (floorbeams) (see Sections 5.2.1.2.1.2 and 5.2.1.2.1.4).

For example, with a pin at location C, the influence line for bending moment at D will be of the general form shown in Figure 5.15. The influence lines for other internal forces can be determined in a similar manner.

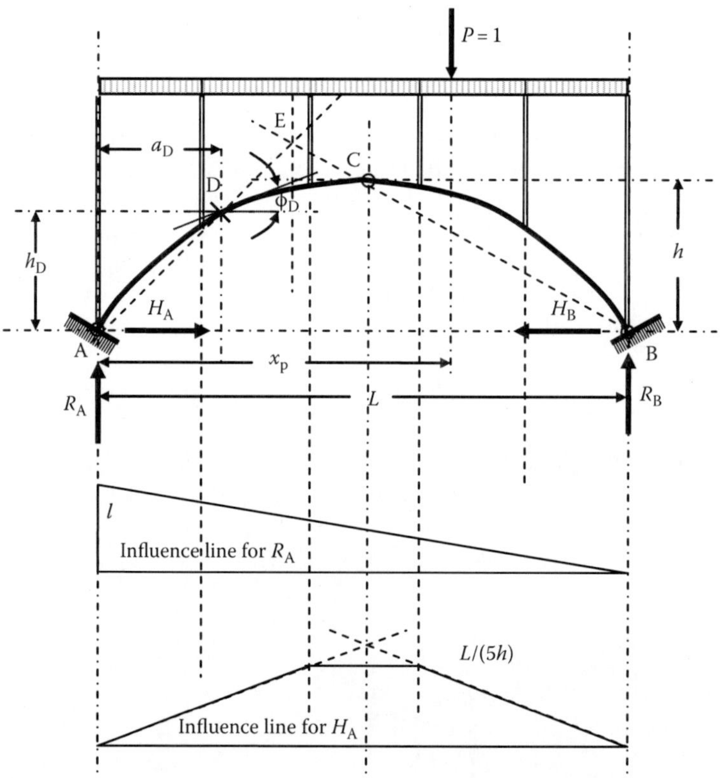

FIGURE 5.14 Three-hinged arch rib with concentrated moving loads applied to the rib at transverse members e.g., spandrel columns).

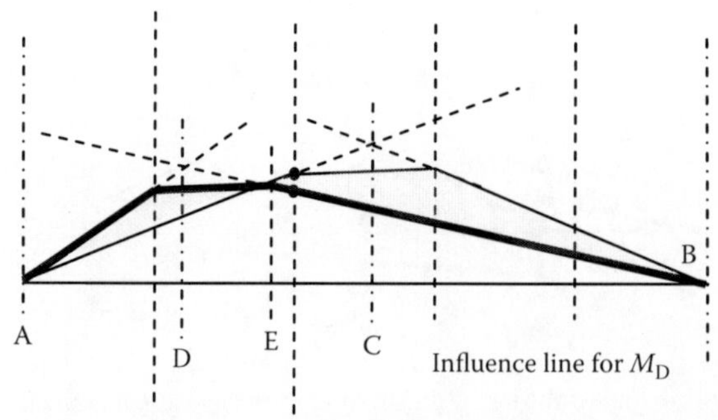

FIGURE 5.15 Influence line for bending moments at location D in three-hinged arch rib.

5.2.1.2.3.3 Maximum Axial Forces with Moving Loads on a Statically Determinate Trussed Arch Long-span steel railway bridges can be economically constructed of three-hinged arches with the arch rib replaced by a truss. The techniques used in Section 5.2.1.2.3.1 to determine maximum effects are useful for construction of influence lines for trussed arches. The crown hinge is designed to achieve static determinacy with a bottom chord pin and top chord sliding arrangement* as shown in Example 5.7 and Figures E5.7 and E5.8.

Example 5.7a (SI Units)

Determine the influence line for member U1–U2 in the 100 m long eight-panel deck trussed arch in Figure E5.7.

The force in the chord U1–U2 can be determined using Equation (5.20) by considering section 1–1 and taking moments about L2:

$$F_{U1-U2} = \frac{M_D}{y_D} = \frac{R_A(a_D) - H_A(h_D)}{y_D} = \frac{R_A(a_D) - \left(L/4h\right)(h_D)}{y_D}.$$

The ordinate of the influence line at L2 provides $R_A(a_D) = (6/8)(25) - 18.75$. This component of the influence line is related to vertical reaction, R_A.

The ordinate of the influence line at L4 provides $(L/4h)(h_D) = (100/(4(44))(15 + 10) = 14.20$. This component of the influence line is related to horizontal thrust reaction, H_A.

With $y_D = 50 - 10 - 15 = 25$ m, the influence line for axial force in chord U1–U2 (shown by the shaded area in Figure E5.7) can be determined by the superposition of the influence lines for R_A and H_A.

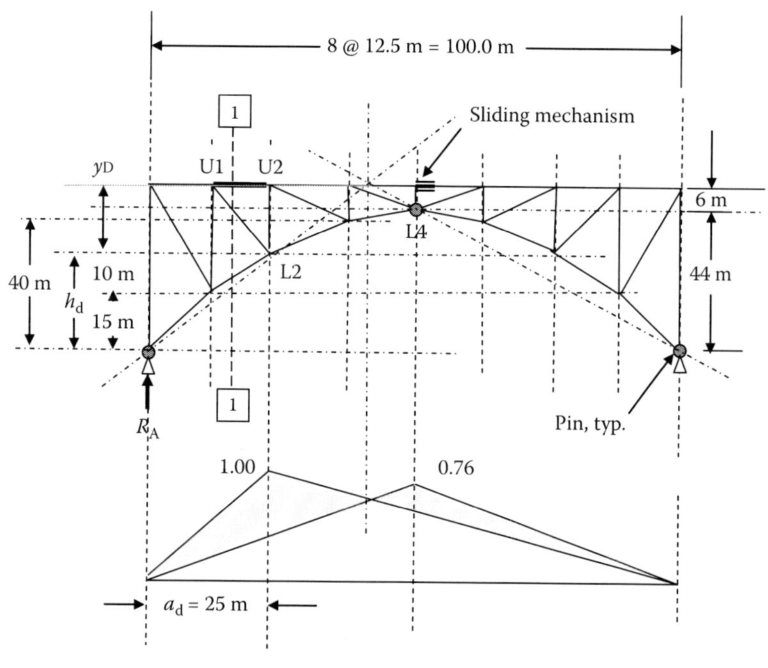

FIGURE E5.7 Influence line for member U1–U2.

* Thereby, rendering the force in one top chord member as zero.

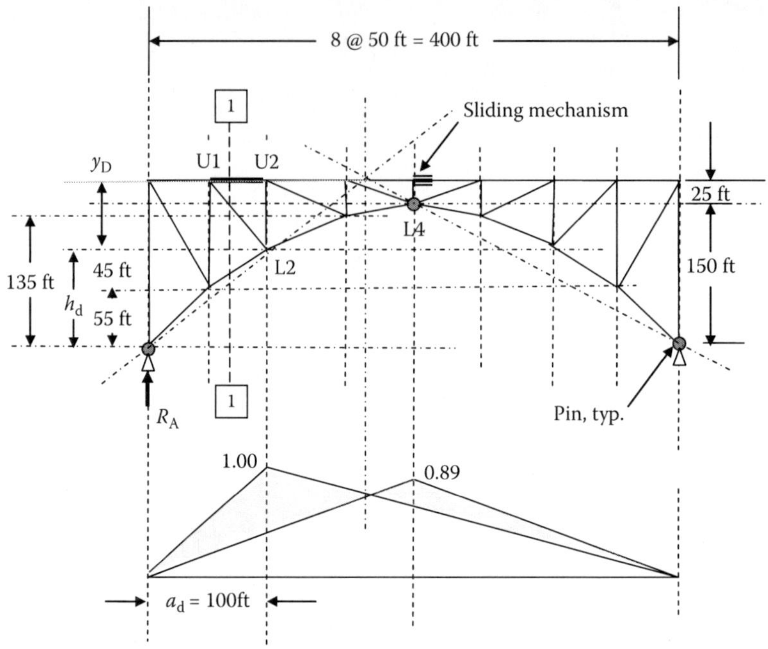

FIGURE E5.8 Influence line for member U1–U2.

Example 5.7b (US Customary and Imperial Units)

Determine the influence line for member U1–U2 in the 400 ft eight-panel deck trussed arch in Figure E5.8.

The force in the chord U1–U2 can be determined using Equation (5.20) by considering section 1–1 and taking moments about L2:

$$F_{U1\text{-}U2} = \frac{M_D}{y_D} = \frac{R_A(a_D) - H_A(h_D)}{y_D} = \frac{R_A(a_D) - \left(L/4h\right)(h_D)}{y_D}.$$

The ordinate of the influence line at L2 provides $R_A(a_D) = (6/8)(100) = 75$. This component of the influence line is related to vertical reaction, R_A.

The ordinate of the influence line at L4 provides $(L/4h)(h_D) = (400/(4(150))(45 + 55) = 66.67$. This component of the influence line is related to horizontal thrust reaction, H_A.

With $y_D = 175 - 45 - 55 = 75$ ft, the influence line for axial force in chord U1–U2 (shown by the shaded area in Figure E5.8) can be determined by the superposition of the influence lines for R_A and H_A.

5.2.1.2.4 Influence Lines for Maximum Effects in Statically Determinate Cantilever Bridge Spans

Long-span steel railway bridges may be economically constructed as cantilever bridges (see Chapters 1 and 11). The economical relative lengths of the cantilever arms, L_c, anchor spans, L_a, and suspended spans, L_s, will vary with live to dead load bending moment ratio. For the relatively high live to dead load bending moment ratios of steel railway superstructures, typical L_a/L_c values of 1–2 are used, depending on the suspended span length, L_s. In steel railway superstructures, L_c/L_s values typically range from 0.4 to 2. The relative lengths of the cantilever arm, anchor, and suspended spans may also be governed by site conditions that dictate the location of piers at a crossing (see Chapter 3). Influence lines for cantilever superstructures may also be constructed by consideration of unit loads traversing the bridge. The ordinates of the influence lines are readily determined by calculation of the reaction, bending moment and shear due to unit loads at locations where the influence lines change direction.

5.2.1.2.4.1 Cantilever Bridge Span Influence Lines (with Loads Applied Directly to the Superstructure) Influence lines for reactions at locations A and B, bending moment in the anchor span at location E and at location F in the cantilever span may be constructed by considering effects of unit loads placed at locations A, B, and C as shown qualitatively* in Figure 5.16.

5.2.1.2.4.2 Cantilever Bridge Span Influence Lines [with Loads Applied to the Superstructure by Transverse Members (Floorbeams)] Influence lines for shear force and bending moment in the anchor span panel points A1–A2 and in the cantilever span panel points C2–C3 can be constructed by considering effects of unit loads placed at locations A, A1, A2, B, C2, C3, and C as shown qualitatively in Figure 5.17.

For long-span railway superstructures, it is further efficient to utilize truss spans in cantilever bridges. The influence lines developed in Figure 5.17, in conjunction with those developed for

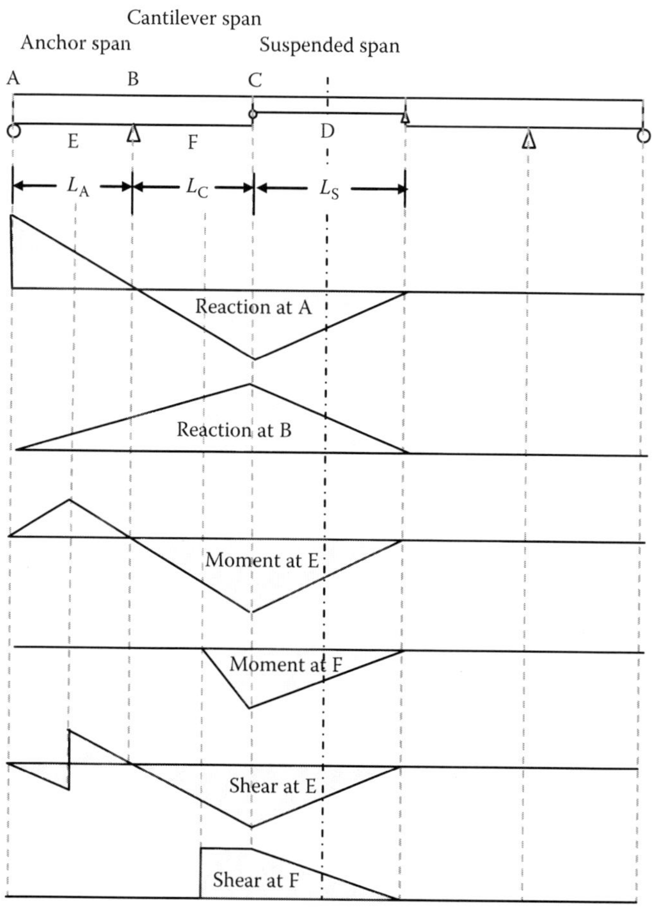

FIGURE 5.16 Influence lines for reactions, bending moments, and shear forces in anchor and cantilever spans with loads applied directly to superstructure.

* Qualitative influence lines are useful in both manual and electronic calculations of maximum or minimum effects to determine the approximate location of live load.

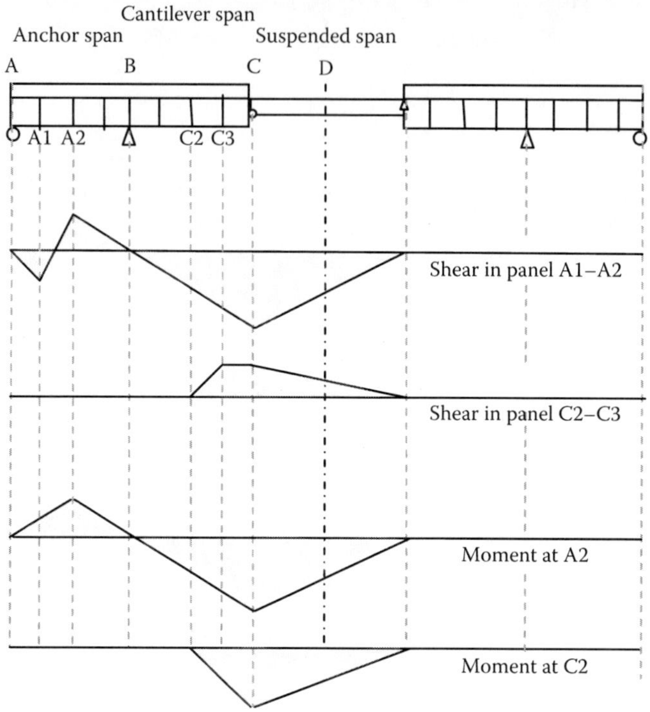

FIGURE 5.17 Influence lines for shear forces and bending moment in anchor and cantilever spans with panel points.

beam and girder (Section 5.2.1.2.1) and truss (Section 5.2.1.2.2) superstructures, are useful in the construction of axial force influence lines for cantilever bridge truss web and chord members. Also, as usual, the consideration of the moving load effect at panel points simplifies the construction of axial force influence lines. Influence lines constructed in this manner for axial forces in cantilever bridge truss members are shown in Example 5.8.

Example 5.8

Construct influence lines for members in panel points 2–3 in the anchor arm of the cantilever truss bridge in Figure E5.9.

By inspection and placement of unit loads at L0, L2, L3, and L6 and considering the hinge at the end of the cantilever and suspended spans, the influence lines for axial forces in L2–L3, U2–U3, and U2–L3 are shown in Figure E5.9. Influence lines for axial force in other members of the trusses may be constructed in a similar manner.

5.2.1.3 Equivalent Uniform Loads for Maximum Shear Force and Bending Moment in Simply Supported Spans

The methods outlined in Sections 5.2.1.1 and 5.2.1.2 require iteration which can be readily digitally programmed. However, for concentrated design loads used on many bridge spans (e.g., Cooper's configuration), it is often beneficial* to determine an equivalent uniform load, w_e, that represents the effects of the concentrated design loading.

* Equivalent uniform loads are particularly useful for preliminary design.

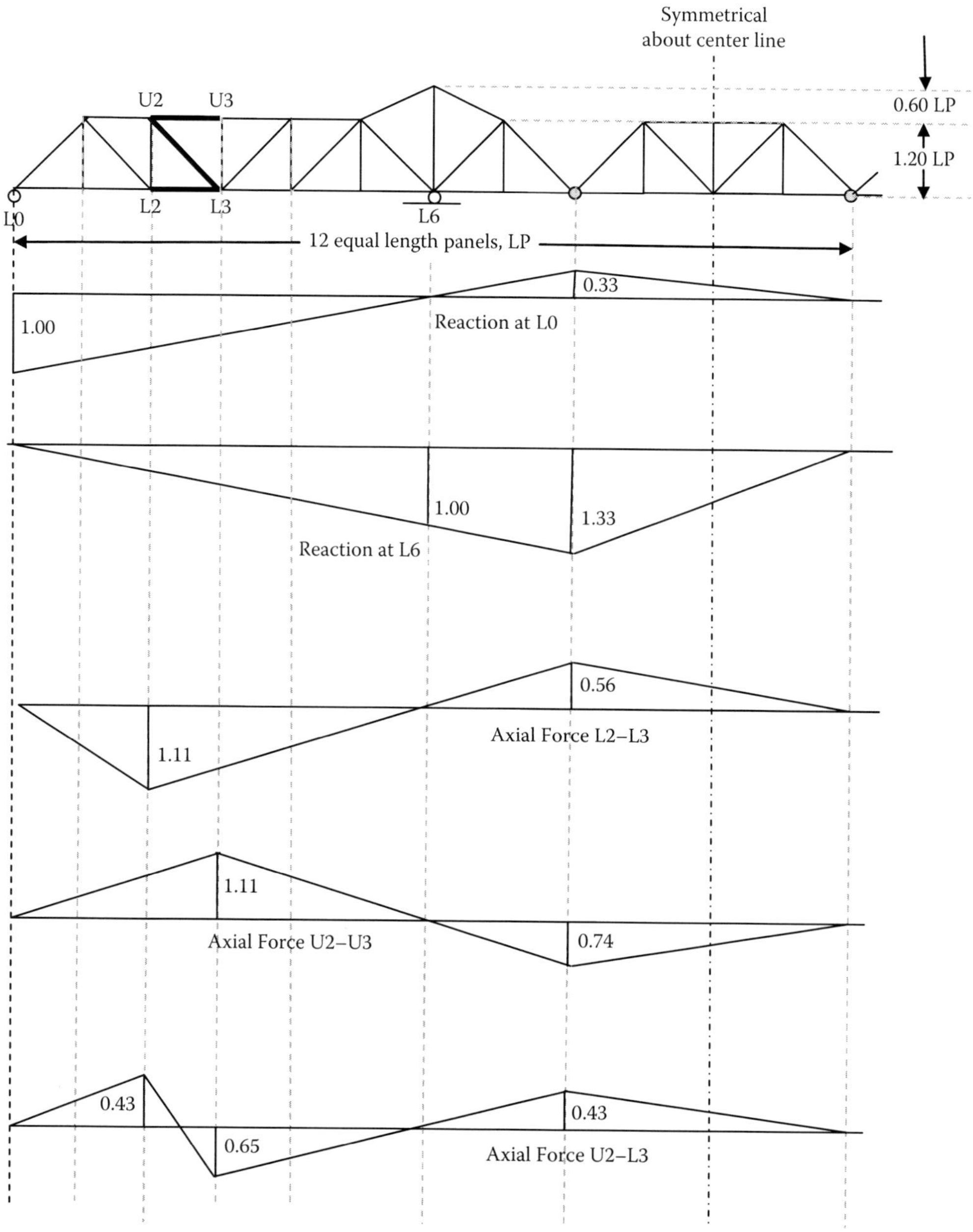

FIGURE E5.9 Influence lines for reactions and forces in members L2–L3, U2–U3 and U2–L3.

5.2.1.3.1 Maximum Shear Force in Simply Supported Spans [with Concentrated Moving Loads Applied Directly to the Superstructure (Figure 5.18)]

Equating maximum shear force, V_C, from Equation 5.1 with the shear force, V_{Ce}, at location C from an equivalent uniform load, w_{ev}, yields

$$w_{ev} = \left(P_T\left(\frac{x_T}{L}\right) - P_L\right)\left(\frac{2L}{(L-a)^2}\right). \tag{5.23}$$

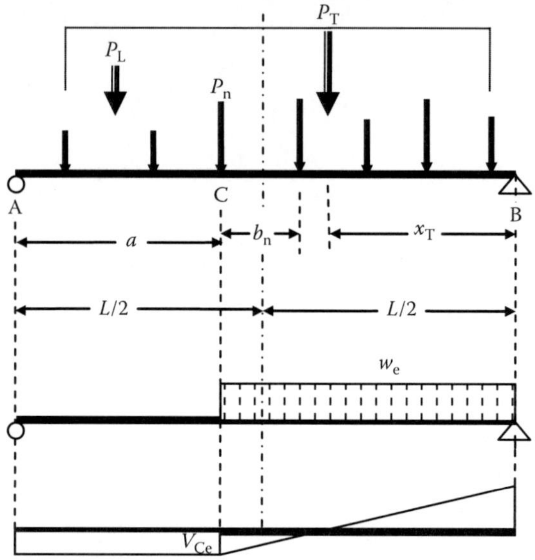

FIGURE 5.18 Equivalent uniform load for shear force for concentrated moving loads applied directly to the superstructure.

Equation 5.23 can be plotted for different P_T and P_L (which are dependent on load configuration and span length) at locations C on the span. Figure 5.19 shows the equivalent uniform load for shear force at the end, 1/4 point and center of span for a Cooper's EM360 (and E80) series of concentrated moving wheel loads applied directly to the superstructure.

5.2.1.3.2 Maximum Shear Force in Simply Supported Spans [with Concentrated Moving Loads Applied to the Superstructure by Transverse Members (Figure 5.20)]

The location in the panel BC where the shear due to an equivalent uniform load, $V_{BCe} = 0$, is

$$d = \frac{(L-a)s_p}{L-s_p} = \frac{(L-a)}{n_p - 1}, \tag{5.24}$$

where

$n_p = L/s_p$ = number of equal length panels.

Equating the maximum shear force, V_{BC}, from Equation 5.4 with the shear force, V_{BCe} in panel BC, from an equivalent uniform load, w_{ev}, yields

$$w_{ev} = \left(P_T \frac{x_T}{L} - \left(P_L + P_{BC}\left(\frac{c}{s_p}\right)\right)\right)\left(\frac{2L}{(L-a+d)^2}\right). \tag{5.25}$$

Equation 5.27 can be plotted for different P_T and P_L (which are dependent on load configuration and span length) and P_{BC} (which is dependent on load configuration and panel length) int different panels on the span (described by distances a and d). For a specific design load such as Cooper's configuration, the value of $P_T(x_T/L)-(P_n+P_{BC}(c/s_p))$ can be calculated for various values of x_T and the equivalent uniform load for shear can be determined in various panels along the span. The equivalent uniform load for maximum shear in the panels will have the general form shown in Figure 5.19.

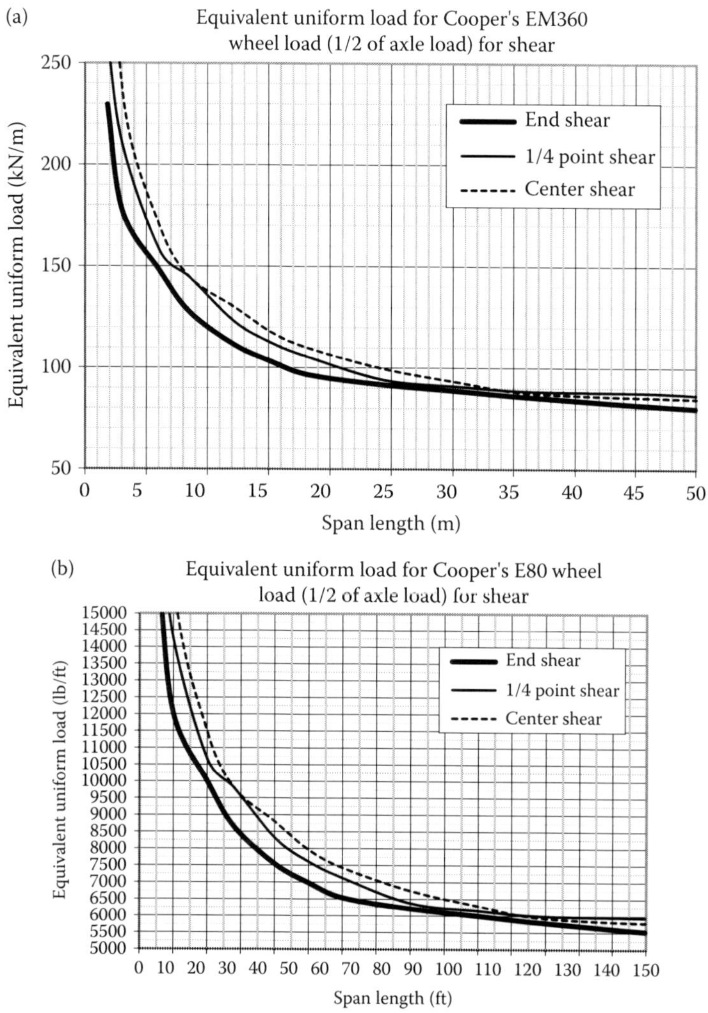

FIGURE 5.19 (a) Equivalent uniform load for shear force for a Cooper's EM360 series of concentrated moving wheel loads applied directly to the superstructure and (b) Equivalent uniform load for shear force for a Cooper's E80 series of concentrated moving wheel loads applied directly to the superstructure.

5.2.1.3.3 Maximum Bending Moment in Simply Supported Spans [with Concentrated Moving Loads Applied Directly to the Superstructure (Figure 5.21)]

Equating maximum bending moment, M_C, from Equation 5.7 with the bending moment, $M_{Ce} = w_e a(L-a)/2$, at location C from an equivalent uniform load, w_{em}, yields

$$w_{em} = \frac{2\left(P_T\left(\frac{x_T}{L}\right)a - P_L x_L\right)}{a(L-a)}. \tag{5.26}$$

Equation 5.28 can be plotted for different P_T and P_L at locations C on the span. Figure 5.22 shows the equivalent uniform load for bending moment at the 1/4 point and center of span for a Cooper's E80 series of concentrated moving wheel loads applied directly to the superstructure.

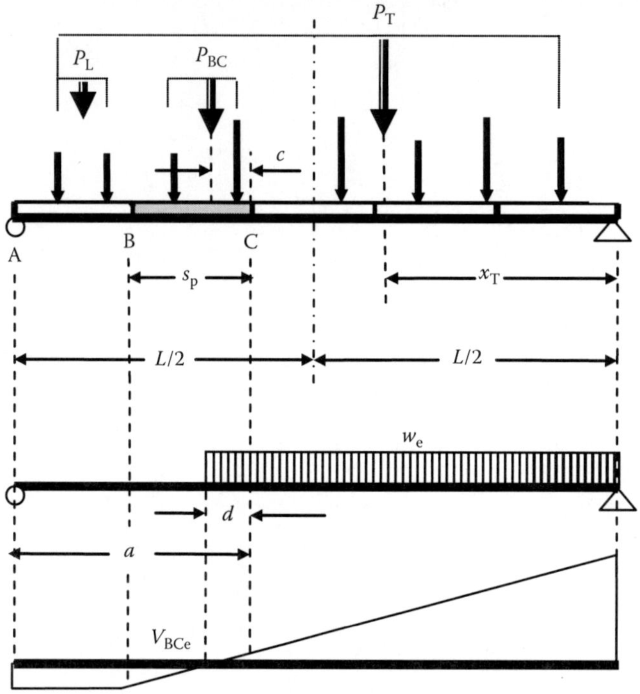

FIGURE 5.20 Equivalent uniform load for shear force for concentrated moving loads applied at transverse members to the superstructure.

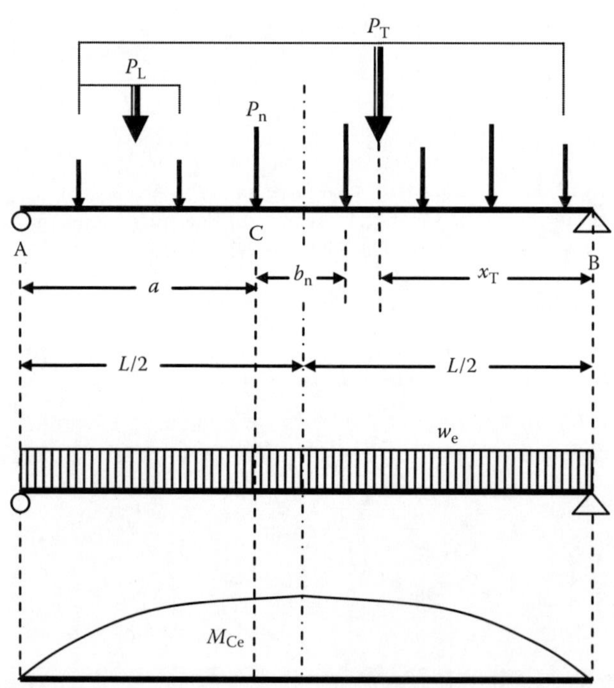

FIGURE 5.21 Equivalent uniform load for bending moment for concentrated moving loads applied directly to the superstructure.

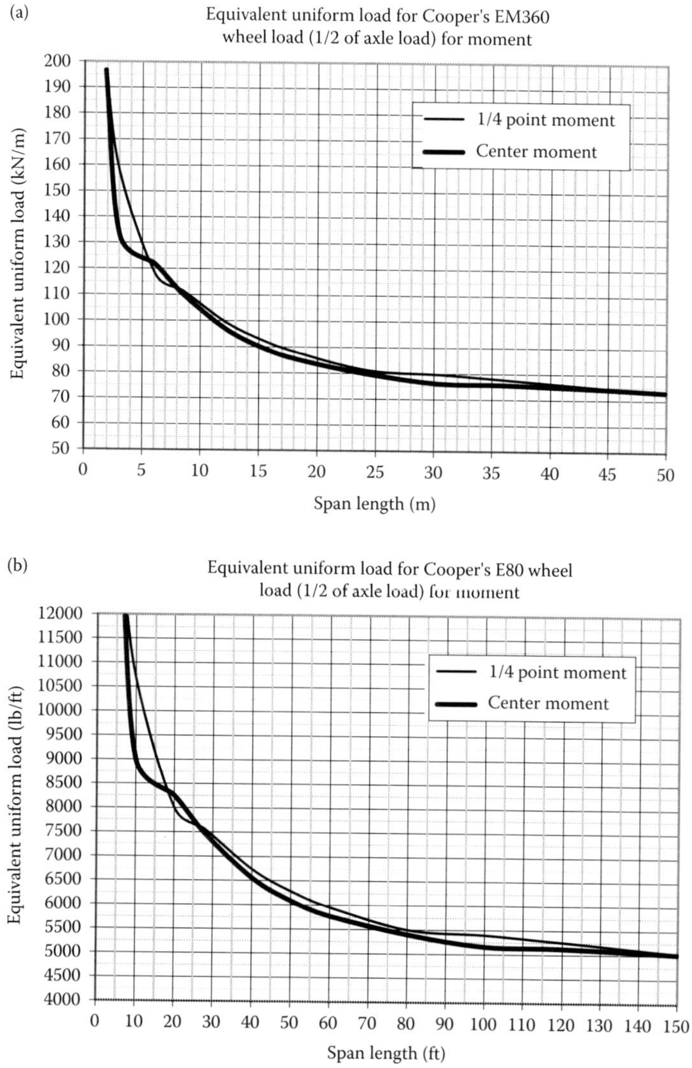

FIGURE 5.22 (a) Equivalent uniform load for bending moment for a Cooper's EM360 series of concentrated moving wheel loads applied directly to the superstructure and (b) Equivalent uniform load for bending moment for a Cooper's E80 series of concentrated moving wheel loads applied directly to the superstructure.

5.2.1.3.4 Maximum Bending Moment in Simply Supported Spans [with Concentrated Moving Loads Applied at Panel Points to the Superstructure (Figure 5.23)]

Equating the maximum bending moment, M_{BC}, from Equation 5.9 with the bending moment, $M_{BCe} = w_e s_p (s_p + a)$, in panel BC from an equivalent uniform load, w_{em}, yields

$$w_{em} = \frac{\left(\left(P_T \frac{x_T}{L}\right) a - (P_L) b - (P_{BC}) \left(\frac{c}{s_p}\right) d\right)}{s_p (s_p + a)}. \tag{5.27}$$

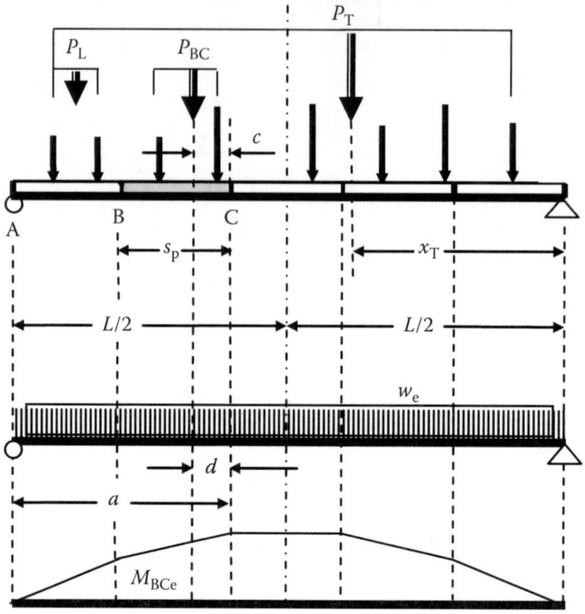

FIGURE 5.23 Equivalent uniform load for bending moment for concentrated moving loads applied at transverse members to the superstructure.

Equation 5.29 can be plotted for different P_T and P_L and P_{BC} for various panels on the span. For a specific design load such as Cooper's configuration the value of $((P_T(x_T/L))a-(P_L)b-(P_{BC})(c/s_p)d)$ can be calculated for various values of x_T and the equivalent uniform load for bending moment can be determined in various panels along the span. The equivalent uniform load for maximum bending moments in the panels will have the general form shown in Figure 5.22.

Example 5.9a (SI Units)

Determine the maximum shear forces and bending moment per rail for Cooper's EM360 live load on an 18 m long DPG span.

From Figure 5.19a, w_{ev} = 98 kN/m and V_{LL} = 98(18)/2 = 882 kN

From Figure 5.22a, w_{em} = 85 kN/m and M_{LL} = 86(18)²/8 = 3483 kNm.

This is within 2% of the M_{LL} calculated in Example 5.1a.

Example 5.9b (US Customary and Imperial Units)

Determine the maximum shear forces and bending moment per rail for Cooper's E80 live load on a 60 ft long DPG span.

From Figure 5.19b, w_{ev} = 6550 lb/ft and V_{LL} = 6.50(60)/2 = 195 kips.

From Figure 5.22b, w_{em} = 5760 lb/ft and M_{LL} = 5.76(60)²/8 = 2592 kips-ft.

This is well within 1% of the M_{LL} calculated in Example 5.1b.

Equivalent uniform loads provide the design engineer with an efficient technique for determining the effects of moving loads on simply supported superstructures. Equivalent uniform loads for Cooper's and other locomotive and train live loads were presented often in early railway bridge design literature (Waddell, 1916; Ketchum, 1924). However, their use for bridge design did not gain favor among North American railway bridge engineers who preferred to use concentrated loads that more closely reflected actual locomotive and train wheel loads.

5.2.1.3.5 Shear Force and Bending Moment at any Location in Simply Supported Spans [with Concentrated Moving Loads Applied Directly to the Superstructure (Figure 5.24)]

The use of uniform loads can be generalized for shear, bending, and floorbeam reaction at any location, C, on a simple span. The area under the shear influence line in Figure 5.24 is $b^2/2L$ and the area under the bending moment influence line in Figure 5.24 is $ab/2$. Therefore, the equivalent uniform load for Cooper's live load shear and bending moments, respectively, are

$$w_{ev} = V_{LL}\left(\frac{2L}{b^2}\right), \tag{5.28}$$

$$w_{em} = M_{LL}\left(\frac{2}{ab}\right). \tag{5.29}$$

The equivalent uniform load, w_{ev} or w_{em}, can be calculated for various span lengths, $L = a + b$, at location C (with $a < b$) and plotted to provide curves for use by design engineers. The curves will be of the general form shown in Figure 5.25. Curves such as these were prepared by the bridge engineer David. B. Steinman* in 1915. The curves (referred to as Steinman's charts) are available in many early bridge design handbooks, manuals, and texts (e.g., Grinter, 1942).

5.2.1.3.6 Shear Force and Bending Moment at any Location in Simply Supported Spans [with Concentrated Moving Loads Applied at Panel Points in the Superstructure (Figure 5.26)]

The area under the shear influence line in Figure 5.26 is $[(a + b)(n_R/n)/]2$. Therefore, the equivalent uniform load for Cooper's live load shear is

$$w_{ev} = V_{LL}\left(\frac{2n}{(a+b)\,n_R}\right). \tag{5.30}$$

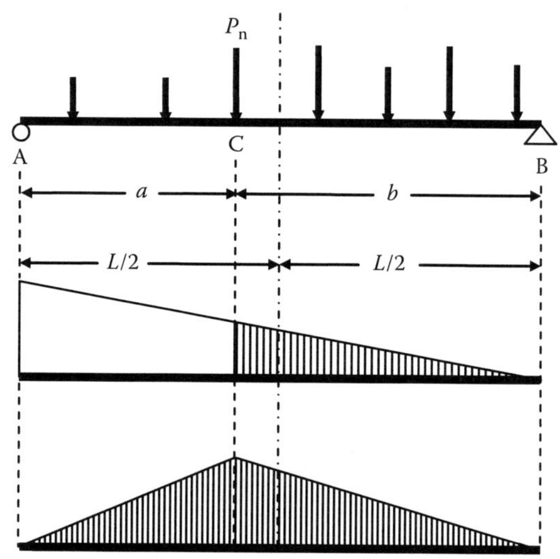

FIGURE 5.24 Determination of equivalent uniform loads for simple span shear and bending at location C.

* David B Steinman also designed long span suspension bridges and further developed J. Melan's "deflection theory" for suspension bridge design (Steinman, 1953; Petroski, 1995).

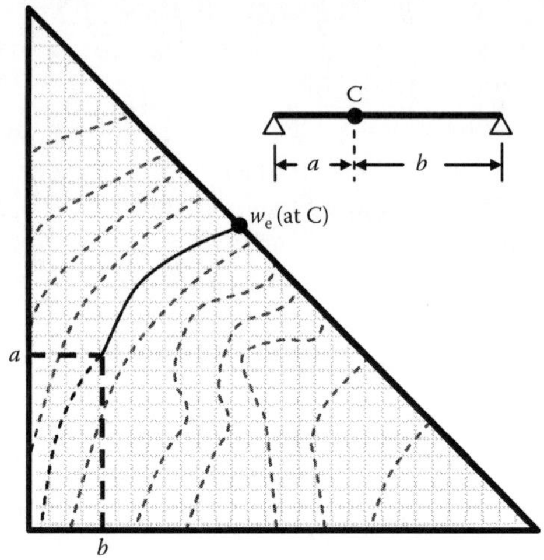

FIGURE 5.25 Schematic of generalized equivalent uniform loads for design live load shear, V_{LL}, bending moment, M_{LL}, and floorbeam reaction, R_{LL}.

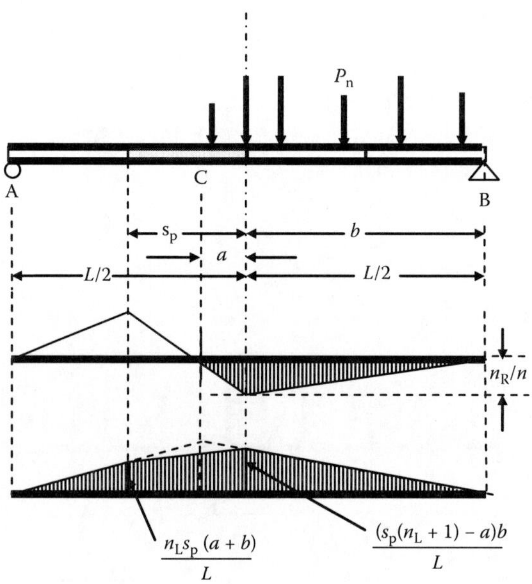

FIGURE 5.26 Determination of equivalent uniform loads for simple span shear and bending at location C.

From Equation (5.17),

$$a = \frac{s_p n_R}{2\left(\frac{n_L}{n_R}+1\right)}, \tag{5.31}$$

and from Figure 5.26,

$$b = n_R s_p, \tag{5.32}$$

where

n = number of panels ($n = 4$ in Figure 5.26)
n_L = number of panels left of panel under consideration ($n_L = 1$ in Figure 5.26)
n_R = number of panels right of panel under consideration ($n_R = 2$ in Figure 5.26)
s_p = uniform panel spacing ($s_p = L/n = L/4$ in Figure 5.26)

Substitution of Equations 5.31 and 5.32 into Equation 5.30 yields

$$w_{ev} = V_{LL}\left(\frac{2s_p}{ab}\right). \tag{5.33}$$

The area under the bending moment influence line in Figure 5.26 is $\dfrac{m_1 s_p(n_L+1)+m_2(s_p+b)}{2}$, where

$$m_1 = (L-b-s_p)\left(\frac{a+b}{L}\right), \tag{5.34a}$$

$$m_2 = (L-b-a)\left(\frac{b}{L}\right). \tag{5.34b}$$

Therefore, the equivalent uniform load for Cooper's live load moment is

$$w_{em} = \frac{2M_{LL}}{m_1 s_p(n_L+1)+m_2(s_p+b)}, \tag{5.35}$$

and substitution of Equations 5.34a and 5.34b in Equation 5.35 yields

$$w_{em} = \frac{2LM_{LL}}{(a+b)[(L-b-s_p)(L-b)]+(L-b-a)[b(s_p+b)]}. \tag{5.36}$$

The equivalent uniform load, w_{ev}, for shear (Equation 5.33)* and, w_{em}, for bending moment (Equation 5.36)† can be calculated within various panels at location C ($a + b$ from the right side in Figure 5.26) using the same charts plotted for simple spans shown in Figure 5.25.

5.2.1.3.7 *Floorbeam Reaction at any Location in Simply Supported Spans [with Concentrated Moving Loads Applied at Panel Points (at Transverse Floorbeams) in the Superstructure (Figure 5.27)]*

The area under the shear influence line in Figure 5.27 is $(a + b)/2$. Therefore, the equivalent uniform load for Cooper's live load reaction is

$$w_{eR} = V_{LL}\left(\frac{2}{(a+b)}\right). \tag{5.37}$$

---

* Equation 5.33 is similar in form to Equation 5.28.

† Equation 5.36 is similar in form to Equation 5.29 with L, b, and s_p being constant. If the constants are included together as K_1 and K_2, Equation 5.36 is $w_e = 2M_{LL}/(a+b)K_1+(L-b-a)K_2$ and the similarity with Equation 5.29 is clear.

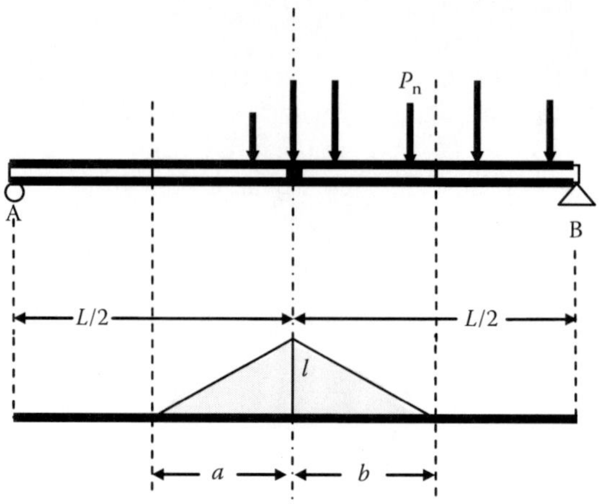

FIGURE 5.27 Determination of equivalent uniform loads for floorbeam reaction.

The equivalent uniform load, w_{eR}, can be calculated at various floorbeam locations with adjacent panel lengths, a and b, as shown in Figure 5.27, using the same charts plotted for simple spans shown in Figure 5.25.

The generalized equivalent uniform loads presented in Steinman's charts are useful in the design of usual steel girder and truss railway spans. However, despite the appeal of ease in design, the use of equivalent uniform live loads has never been prevalent in North America and most engineers develop shear forces and bending moments from an analysis of concentrated loads.

In order to encourage efficiency in the design process and avoid the use of charts and influence lines, digital computers, equations, and tables are useful. For usual bridge design projects (e.g., simply supported beam, girder and truss superstructures), equations and tables have been prepared for Cooper's load configuration for the determination of maximum shearing forces, axial forces, and bending moments.

5.2.1.4 Maximum Shear Force and Bending Moment in Simply Supported Spans from Equations and Tables

Tabulated values for shear and bending moment at the end, 1/4 point, and center of simple beam and girder spans from 1.5 m (5 ft) to 120 m (400 ft) are given in AREMA (2015). AREMA (2015) also provides equations for some span lengths for the shear and bending moment at the end, 1/4 point and center of simple beam and girder spans.

Shear in panels and moments at panel points for Pratt trusses with various panel lengths and number of panels have also been tabulated by railroad bridge engineers and are available in the railway bridge design literature (e.g., Ketchum, 1924). For typical railway truss span design, these tables can save considerable computational effort.

For the design of complex steel bridges, such as continuous and cantilever steel spans, the use of influence lines and/or modern FEA software may be required.

Example 5.10a (SI Units)

Determine the maximum shear forces and bending moment per rail for Cooper's EM360 live load on an 18 m long DPG span.

Using the tables and equations in AREMA Chapter 15,

$V_{LL} = (194)(4.448)(360/356) = 872$ kN

$M_{LL} = 349(L) - 2826 + 245.7/L = 349(18) - 2826 + 245.7/18 = 6282 - 2826 + 19.4 = 3475$ kNm.

These values are calculated from converted US Customary unit tables and equations in AREMA (2015) (see Example 5.10b).

These are within 2% of the V_{LL} and M_{LL} calculated in Examples 5.1a and 5.9a.

Example 5.10b (US Customary and Imperial Units)

Determine the maximum shear forces and bending moment per rail for Cooper's E80 live load on a 60 ft long DPG span.

Using the equations in AREMA Chapter 15,

$V_{LL} = 196$ kips

$M_{LL} = 77.5(L) - 2062 + 585.4/L = 77.5(60) - 2062 + 585.4/60 = 4650 - 2062 + 9.8 = 2598$ kips-ft.

These are within 1% of the V_{LL} and M_{LL} calculated in Examples 5.1b and 5.9b.

5.2.1.5 Modern Structural Analysis

The analysis of structures based on the classical theories and methods of applied elasticity and mechanics of materials are used for the determination of stresses and deformations in typical steel railway superstructures. However, for complex superstructures, and for efficiency in the design of even usual superstructure types, the computational capabilities of modern computers and availability of inexpensive computer software* based on these classical methods has been of great benefit to bridge engineers. When effectively utilized, such software enables the engineer to avoid many long and tedious calculations and attain greater speed and accuracy, and enhances the ability to perform multiple analyses for optimization purposes.

Usual steel railway superstructure types are beams, girders, and trusses. The analysis of typical simply supported deck beam and girder (DPG and BDPG) superstructures may be based on simplified methods using equations, charts and tables as shown in this chapter. In many cases, simplified methods are also appropriate for typical simply supported through beam and girder (TPG and BTPG) superstructures. A line-girder type of analysis using the load distribution criteria in AREMA (2015) with influence lines† is appropriate for most typical beam and girder superstructures. However, the structural analysis of more complex superstructures may require modeling the superstructure for a grillage analysis (Bakht and Jaeger, 1985; Jaeger and Bakht, 1989) or, more frequently, a 2D or 3D FEA. In some cases, typical simply supported steel railway truss superstructures may also be analyzed using simplified methods with charts and tables‡. However, an analysis using influence lines for the various truss members is usually necessary. Influence line generation for relatively simple truss span members is not onerous, but for more complex truss superstructures, influence line generation with software§ may be required. Influence line analysis is a component of most commercially available FEA software for bridge analysis, making railway truss analysis by 2D FEA or, if necessary, 3D FEA an effective structural engineering design tool. Modern complex, long span, and/or structures that require specialized analysis (e.g., dynamic structural analysis for wind or seismic effects) generally require the use of FEA software. FEA also enables more accurate modeling of structural behavior such as the participation of truss floor systems in chord resistance, effects of secondary bracing on behavior, end conditions of beams and truss members, and dynamic load-structure interaction.

* For example, commercially available spreadsheets are relatively easy to program and are used extensively for structural analyses (Christy, 2006). Also, many bridge design engineers and offices have internally developed software, often using spreadsheet tools.

† For complex superstructures, computer software generated influence lines based on the Muller-Breslau principle may be required.

‡ Some early books on steel railway bridge design contained tables of shear in panels and moment at joints of commonly used pin-connected statically determinate truss spans (typically for the commonly used Pratt truss).

§ Typically, such software utilizes the Muller-Breslau principle (Fu and Wang, 2015).

There are many proprietary specialized and general-purpose FEA programs available. For the analysis of steel railway superstructures, FEA software that incorporates moving loads* and dynamic analysis† is required. Efficient FEA applications, specifically for bridge analysis, that do not require the review and interpretation of a large quantity of data are particularly useful and available. Many commercially available FEA applications also routinely include linear and nonlinear elements. Engineers using FEA software should be familiar with the theory and approximations used in the software and how the software realizes the appropriate loading, material property, and boundary conditions. There are many standard textbooks that provide fundamental information regarding the theory and applications of FEA (e.g., Martin and Carey (1973), Zienkiewicz (1983), Cook (1981), Weaver and Johnston (1984), and Wilson (2004)).

Modern trends in structural engineering software are towards integrated structural analysis, design, drafting, and fabrication. Some proprietary systems successfully integrate many of these functions; and it is likely that, as such integrated systems become more "user-friendly" and reliable; they will be used even more frequently in structural engineering practice.

5.2.2 Lateral Load Analysis of Steel Railway Superstructures

The analysis of railway superstructures also involves the determination of the maximum deformations and stresses caused by lateral effects such as those due to moving loads (centrifugal and nosing), wind, and earthquakes‡.

For usual steel railway bridge superstructures, lateral load effects may be determined by simplified analyses. This enables the use of manual calculations, relatively simple computer programs and spreadsheets to determine the deformations and forces. For more complex superstructures, more sophisticated computerized frame analysis or FEA software may be employed.

5.2.2.1 Lateral Bracing Systems

Lateral forces on steel railway superstructures from wind, nosing,§ and centrifugal forces are generally transferred to the bearings and then substructures via bracing members in horizontal truss systems. Components of the horizontal bracing systems may also resist the buckling propensity of compression members, such as, girder top flanges or truss top chords in simply supported spans. Forces from horizontal truss systems that are not in the plane of the bearings are transferred to the substructures by end vertical (DPG spans and some deck truss (DT) spans) or portal (through truss (TT) and some DT spans) bracing systems. Knee braces are used to provide resistance to buckling of the compression flange and transfer wind forces from the top flanges to the bearings in through plate girder spans.

5.2.2.1.1 Horizontal Truss Bracing

Since, for usual steel railway bridges, the determination of lateral loads is approximate, it is reasonable to utilize simplifications regarding load distribution in horizontal bracing systems. It is generally adequate to use a horizontal Pratt or Warren truss and apply lateral forces at the windward side of the lateral truss panel points. For bracing systems (horizontal trusses) with two diagonals in each panel (Pratt type cross-bracing), the lateral shear can be assumed to be transferred equally between diagonals, and the members are designed for both maximum tension and compression forces. When the double bracing is not connected to the floor system or otherwise supported¶, the diagonals can be assumed to act as tension only members with the transverse members (struts) in compression. For bracing systems with only a single diagonal in each panel, the diagonals are also assumed to act as tension-only members.

The approximate determination of forces in lateral bracing systems is shown in Example 5.10.

---

* Moving load analysis is often achieved using influence lines or surfaces. Some FEA applications also include moving mass and moving sprung-mass vehicle loads, which may be required for the dynamic analysis of complex superstructures.

† Typically performed using modal superposition and solving for eigenvectors and eigenvalues.

‡ Wind and earthquake forces may also have longitudinal components.

§ Truck hunting or lateral movements due to variations at the wheel flange to rail head interface.

¶ Therefore, relatively long and slender with a low critical buckling load and compressive force capacity.

5.2.2.1.1.1 Members in Top Lateral Systems In addition to lateral forces from wind, the top lateral system in through spans* requires bracing members that resist a transverse shear force of 2.5% of the total compressive axial force in the chord or flange at that panel point. The top lateral system in deck spans requires bracing to resist a transverse shear force of 2.5% of total compressive axial force in the chord or flange at that panel point in addition to other lateral forces from wind, nosing, and centrifugal forces. Deck span top lateral systems are usually the most robust bracing system required in steel railway superstructures. Concrete slab and steel plate decks may be effective in behaving as a diaphragm for resisting the lateral forces in deck spans.

5.2.2.1.1.2 Members in Bottom Lateral Systems Lower lateral bracing is generally required when the span supports are located at the bottom chord of a truss or the bottom flange of a girder span. When the span supports are at the top chord of a truss† only struts at the bottom panel points are strictly required. However, a nominal lateral bracing system is often employed to ensure adequate overall lateral rigidity of the span.

The bottom lateral system in through spans may use the floorbeams as struts of the bracing system. The bracing is designed to resist lateral wind, nosing, and centrifugal forces. Depending on location relative to the bottom flange or chord, concrete slab,‡ and steel decks may act as full or partial bottom lateral bracing, through diaphragm behavior, for relatively short ballasted deck spans§.

The bottom lateral bracing system in deck spans is lightly loaded by wind and, for short spans in particular, may not be required. However, in order to ensure overall rigidity of longer spans, a light bracing system (based on the maximum slenderness ratio for compression members) is often used. At a minimum, struts should be installed at each panel point in the bottom chord or flange. AREMA (2015) recommends bottom lateral bracing for all deck spans greater than 15 m (50 ft) long.

Example 5.10a (SI Units)

The forces in the top and bottom lateral bracing system members of the TT span in Figure E5.10 are required.

The lateral wind and nosing forces, and compression forces required to be resisted for bracing of main compression members are as follows:

Wind load at top chord = 5.0 kN/m

Wind load at top lateral bracing panels = 6.0(5.0) = 30.0 kN per panel

Wind load at bottom chord = 3.0 kN/m

Wind load on train = 4.5 kN/m

Wind load at bottom lateral bracing panels = 6.0(7.5) = 45.0 kN per panel

Cooper's EM400 nosing load (lateral equipment load) at bottom lateral bracing panels = 400/4 = 100.0 kN at any panel

Bracing forces required to resist top chord buckling are shown in Table E5.1a.

Top lateral bracing:

Due to their slenderness, top lateral bracing compressive members are assumed inactive and tension members only resist the panel forces. The top lateral bracing member forces are shown in Table E5.2a.

---

* Through spans, such as plate girder and pony truss spans, without room for horizontal top lateral bracing generally utilize vertical knee brace frames to resist the transverse shear force of 2.5% of total compressive axial force in the chord or flange, and the lateral forces from wind. An analysis of knee-braced through span transverse frames is outlined in Section 5.2.2.1.4 (also see Appendix A).

† Such as in a "fish-bellied" deck truss span.

‡ Concrete slab decks acting compositely with steel through girder or truss spans are not recommended (see Chapter 4).

§ It is generally good practice to consider bottom lateral bracing in addition to any deck diaphragm behavior for spans longer than about 15 m (50 ft).

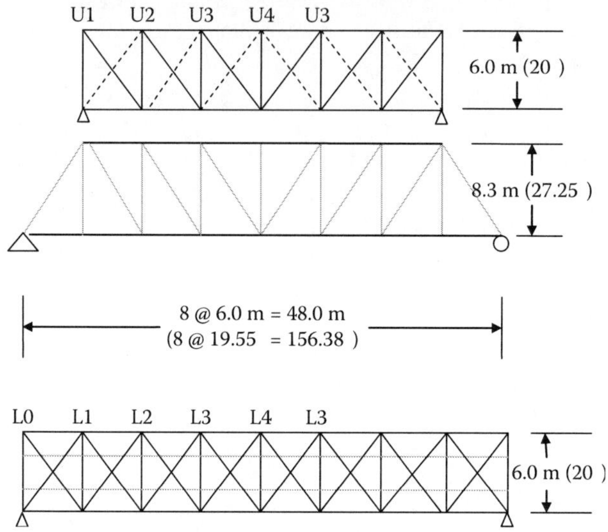

FIGURE E5.10 Top and bottom lateral bracing of through truss span.

TABLE E5.1A
Top Chord Bracing Force

Panel Point	Total Axial Compression in Top Chord (kN)	Bracing Force (kN) (2.5% of Main Member Compressive Force)
U1	1650	41.3
U2	2850	71.3
U3	3550	88.8
U4	3800	95.0

TABLE E5.2A
Force in Diagonal Bracing

Panel	Shear (Wind) (kN)	Shear (Top Chord Compression) (kN)	Total Panel Shear (kN)	Force in Each Diagonal (kN)
U1–U2	(30.0) (5 + 0.5 + 0.5)/2 = 90.0	41.3	131.3	+185.7
U2–U3	90.0 − 30.0 = 60.0	71.3	131.3	+185.6
U3–U4	60.0 − 30.0 = 30.0	88.8	118.8	+168.0
U4–U3′	0	95.0	95.0	+134.4

Bottom lateral bracing:
Since the bracing members are connected to the floor system, both are assumed to equally participate in resisting panel shear forces. Therefore, each member is required to resist 50% of the panel shear force in both tension and compression. The bottom lateral bracing member forces are shown in Table E5.3a.

Example 5.10b (US Customary and Imperial Units)

The forces in the top and bottom lateral bracing system members of the TT span in Figure E5.10 are required. The lateral wind and nosing forces; and compression forces required to be resisted for bracing of main compression members are as follows:

TABLE E5.3A
Force in Diagonal Bracing

Panel	Shear (Wind) (kN)	Shear (Nosing) (kN)	Total Panel Shear (kN)	Force in Each Diagonal (kN)
L0–L1	45.0 (7 + 0.5 + 0.5)/2 = 180.0	100.0	280.0	±198.0
L1–L2	180.0 – 45.0 = 135.0	100.0	235.0	±166.2
L2–L3	135.0 – 45.0 = 90.0	100.0	190.0	±134.4
L3–L4	90.0 – 45.0 = 45.0	100.0	145.0	±102.5
L4–L3′	0	100.0	100.0	±70.7

TABLE E5.1B
Top Chord Bracing Force

Panel Point	Total Axial Compression in Top Chord (kips)	Bracing Force (kips) (2.5% of Main Member Compressive Force)
U1	370	9.3
U2	640	16.0
U3	800	20.0
U4	850	21.3

TABLE E5.2B
Force in Diagonal Bracing

Panel	Shear (Wind) (kips)	Shear (Top Chord Compression) (kips)	Total Panel Shear (kips)	Force in Each Diagonal (kips)
U1–U2	(6.8) (5 + 0.5 + 0.5)/2 = 20.4	9.3	29.7	+41.6
U2–U3	20.4 – 6.8 = 13.6	16.0	29.6	+41.4
U3–U4	13.6 – 6.8 = 6.8	20.0	26.8	+37.5
U4–U3′	0	21.3	21.3	+29.8

Wind load at top chord = 350 lb/ft
Wind load at top lateral bracing panels = 19.55(0.35) = 6.8 kips per panel
Wind load at bottom chord = 200 lb/ft
Wind load on train = 300 lb/ft
Wind load at bottom lateral bracing panels = 19.55(0.5) = 9.8 kips per panel
Cooper's E90 nosing load (lateral equipment load) at bottom lateral bracing panels = 90/4 = 22.5 kips at any panel

Bracing forces required to resist top chord buckling are shown in Table E5.1b.

Top lateral bracing:
Due to their slenderness, top lateral bracing compressive members are assumed inactive and tension members only resist the panel forces. The top lateral bracing member forces are shown in Table E5.2b.

Bottom lateral bracing:
Since the bracing members are connected to the floor system, both are assumed to equally participate in resisting panel shear forces. Therefore, each member is required to resist 50% of the panel shear force in both tension and compression. The bottom lateral bracing member forces are shown in Table E5.3b.

TABLE E5.3B
Force in Diagonal Bracing

Panel	Shear (Wind) (kips)	Shear (Nosing) (kips)	Total Panel Shear (kips)	Force in Each Diagonal (kips)
L0–L1	9.8(8)/2 = 39.2	22.5	61.7	±43.1
L1–L2	39.2 − 9.8 = 29.4	22.5	51.9	±36.3
L2–L3	29.4 − 9.8 = 19.6	22.5	42.1	±29.4
L3–L4	19.6 − 9.8 = 9.8	22.5	32.3	±22.6
L4–L3′	0	22.5	22.5	±15.7

5.2.2.1.2 End Vertical and Portal Bracing

A supplemental structural system is required to transfer lateral forces from horizontal bracing members that are not located in the plane of the bearings to the horizontal lateral systems in the plane of the bearings. In through spans, the lateral forces may be transferred through a system of knee braces, or via sway and end portal frames. In deck spans a system of vertical cross frames or diaphragms may serve this purpose*. The vertical end cross frame bracing and portal frame bracing are required to carry the entire reaction from lateral loads to the substructures via the bearings.

TT end portal frames must be designed to transfer the total wind force reaction of the top lateral truss system, P_L, through flexure of end posts. The end posts are also required to resist additional axial forces from the portal frame action. In order to estimate the end portal frame effects on the end posts, it is assumed that horizontal reactions are equal at the bottom of the end posts and at an inflection point located midway between the bottom of the end post and the bottom of the portal bracing frame (often cross-braced or knee-braced in modern steel trusses) at a distance, $(h_e - h_p)/2$, as shown in Figure 5.28.

In this case, the vertical load, R_e, and horizontal shear, H_e, are

$$R_e = \frac{P_L((h_e + h_p)/2)}{b_t}, \tag{5.38}$$

$$H_e = \frac{P_L}{2}. \tag{5.39}$$

and the end post bending moment, M_e, due to the force, P_L, is estimated as

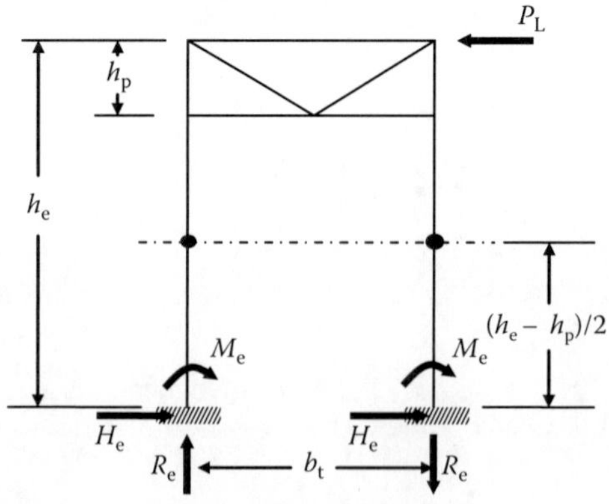

FIGURE 5.28 End post forces from through truss portal action.

* In deck truss spans supported at the top chord, end portal bracing is typically used in the plane of the end diagonal member.

$$M_e = \frac{P_L((h_e - h_p)/2)}{2}. \tag{5.40}$$

The end portal bracing member forces can then be determined from free body equilibrium equations for one leg of the portal. The end portal bracing member forces for various portal configurations are shown in Examples 5.11 through 5.14. Example 5.15 outlines the analysis of a typical railway TT span end portal.

Example 5.11

The axial forces in lattice portal frame members (Figure E5.11) are

$$P_A = \frac{\pm P_L\left(\frac{h_e + h_p}{2}\right)}{2b_t}\left(\frac{\sqrt{h_p^2 + b_p^2}}{h_p}\right),$$

$$P_B = \frac{-P_L}{8h_p}\left(3\left(\frac{h_e + h_p}{2}\right) + 4h_p\right),$$

$$P_C - \frac{+3P_L}{8h_p}\left(\frac{h_e + h_p}{2}\right).$$

Example 5.12

The axial forces in cross-braced portal frame members (Figure E5.12) are

$$P_A = \frac{-P_L\left(\frac{h_e + 3h_p}{2}\right)}{2h_p},$$

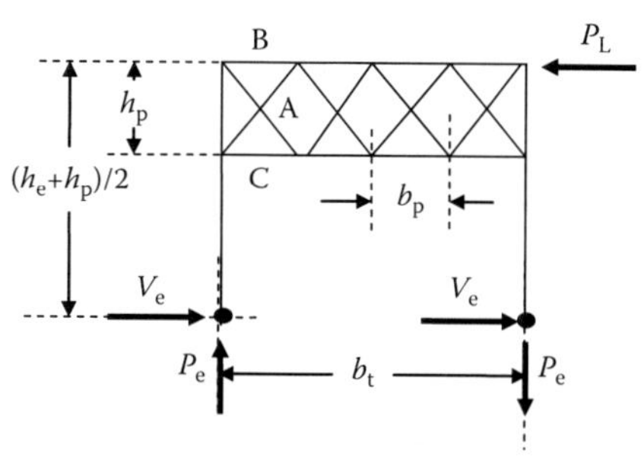

FIGURE E5.11 Lattice portal frame.

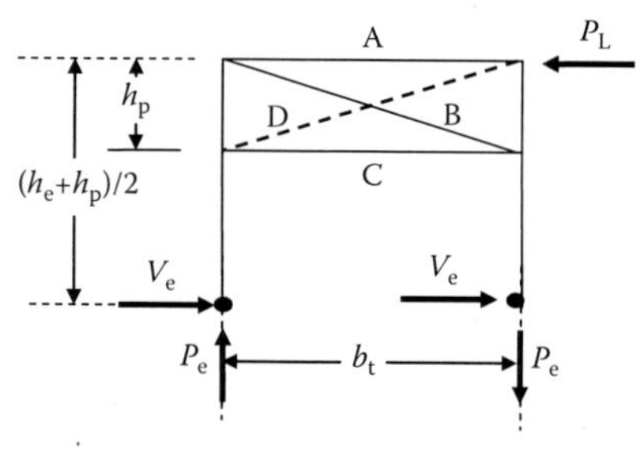

FIGURE E5.12 Cross braced portal frame.

$$P_B = \frac{+P_L}{b_t}\left(\frac{h_e + h_p}{2}\right)\left(\frac{\sqrt{h_p^2 + b_t^2}}{h_p}\right),$$

$$P_C = \frac{-P_L}{2h_p}\left(\frac{h_e + h_p}{2}\right).$$

Example 5.13

The axial, shear, and bending forces in knee-braced portal frame members (Figure E5.13) are

$$P_A = -P_B = \frac{+P_L\left(\dfrac{h_e + h_p}{2}\right)}{2h_p\left(\dfrac{b_t - d_t}{2}\right)}\sqrt{h_p^2 + \left(\frac{b_t - d_t}{2}\right)^2},$$

$$P_C = \frac{+P_L}{2}\left(\left(\frac{h_e + h_p}{2h_p}\right) - 1\right),$$

$$P_D = \frac{-P_L}{2},$$

$$P_E = \frac{-P_L}{2}\left(\left(\frac{h_e + h_p}{2h_p}\right) + 1\right),$$

$$V_C = V_E = \frac{-P_L\left(\dfrac{h_e + h_p}{2}\right)}{b_t}\left(\left(\frac{b_t}{b_t - d_t}\right) - 1\right),$$

$$V_D = \frac{+P_L\left(\dfrac{h_e + h_p}{2}\right)}{b_t},$$

$$M_E = \frac{P_L\left(\dfrac{h_e + h_p}{2}\right)}{2b_t}d_t.$$

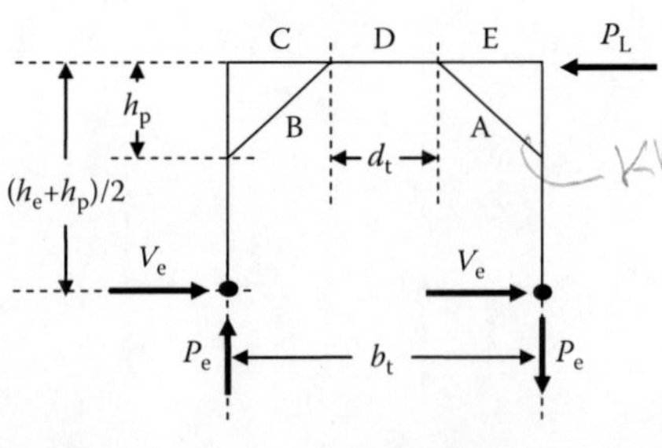

FIGURE E5.13 Knee braced portal frame.

Example 5.14

The axial forces in triangular portal frame members (Figure E5.14) are

$$P_A = -P_B = \frac{+P_L\left(\dfrac{h_e + h_p}{2}\right)}{b_t h_p}\sqrt{h_p^2 + \left(\dfrac{b_t}{2}\right)^2},$$

$$P_C = \frac{+P_L}{2}\left(\left(\frac{h_e + h_p}{2h_p}\right) - 1\right),$$

$$P_D = \frac{-P_L}{2}\left(\left(\frac{h_e + h_p}{2h_p}\right) + 1\right).$$

Example 5.15a (SI Units)

The forces in the end portal bracing system (Figure E5.15) of the TT span in Example 5.10a are required.

$h_e = 10.0$ m (in-plane height of portal)
$h_p = 2.5$ m (in-plane height of portal bracing system)

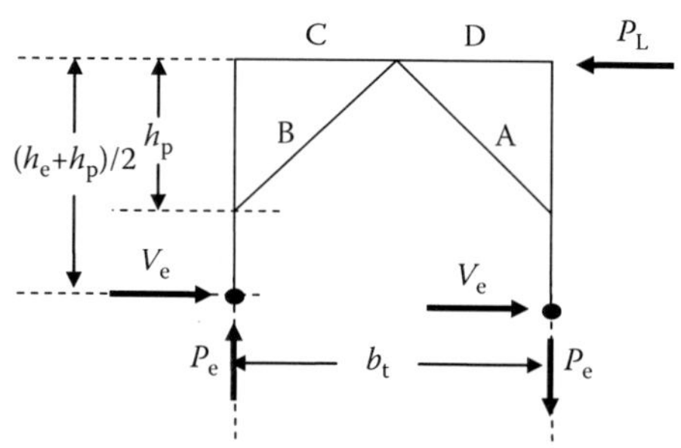

FIGURE E5.14 Triangular portal frame.

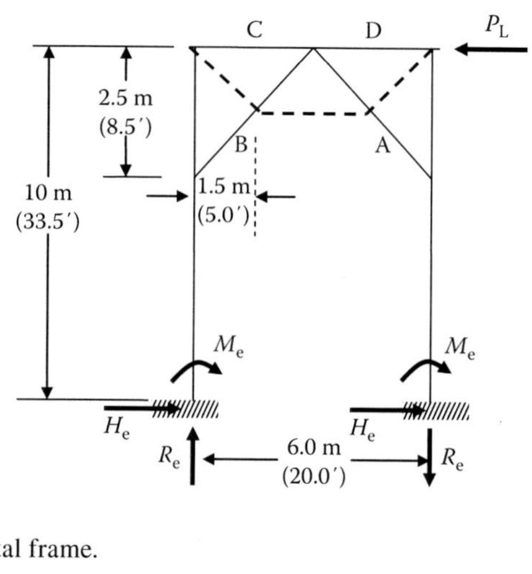

FIGURE E5.15 End portal frame.

$b_t = 6.0$ m (spacing of trusses)

P_L = (6)(6.0)(5.0)/2 = 90.0 kN (top lateral truss wind force reactions transferred to end portal frame)

$$R_e = \frac{90.0((10.0+2.5)/2)}{6.0} = 93.8 \text{ kN} \text{ additional compression in end post due to portal action}$$

$$H_e = \frac{90.0}{2} = 45.0 \text{ kN} \text{ (small shear force that is generally neglected)}$$

$$M_e = \frac{90.0((10.0-2.5))/2)}{2} = 168.8 \text{ kNm} \text{ bending moment at bottom of portal frame end post.}$$

The portal is of the triangular type (Example 5.14) and member forces are

$P_A = -P_B = 90.0(1.63) = 146.4$ kN

$P_C = 90.0(0.75) = 67.5$ kN

$P_D = -90.0(1.74) = -157.5$ kN.

The members shown as dotted lines may be designed for 2.5% of the compressive force in members A and B. However, as this force will be small (3.7 kN in this example), design based on compression member slenderness ratio criteria ($r_{min} \leq$ member length/120) will likely govern.

Example 5.15b (US Customary and Imperial Units)

The forces in the end portal bracing system (Figure E5.15) of the TT span in Example 5.10b are required.

$h_e = 33.5$ ft (in-plane height of portal)

$h_p = 8.5$ ft (in-plane height of portal bracing system)

$b_t = 20.0'$ (spacing of trusses)

P_L = (6)19.55(0.35)/2 = 20.4 kips (top lateral truss wind force reactions transferred to end portal frame)

$$R_e = \frac{20.4((42.0)/2)}{20} = 21.4 \text{ kips} \text{ additional compression in end post due to portal action}$$

$$H_e = \frac{20.4}{2} = 10.2 \text{ kips} \text{ (small shear force that is generally neglected)}$$

$$M_e = \frac{20.4((25.0)/2)}{2} = 127.5 \text{ kips-ft} \text{ bending moment at bottom of portal frame end post.}$$

The portal is of the triangular type (Example 5.14) and member forces are

$P_A = -P_B = 20.4(1.62) = 33.0$ kips

$P_C = 20.4(0.74) = 15.1$ kips

$P_D = -20.4(1.74) = -35.5$ kips.

The members shown as dotted lines may be designed for 2.5% of the compressive force in members A and B. However, as this force will be small (825 lb in this example), design based on compression member slenderness ratio criteria ($r_{min} \leq$ member length/120) will likely govern.

The member forces in other portal frame arrangements may be determined in a similar approximate manner or by a more rigorous frame analysis. AREMA (2015) indicates that TT spans should have portal bracing with knee braces (e.g., members A and B in Figure E5.15) as deep as clearances (see Chapter 3) will allow.

Cross-frame members at the end of deck spans must transfer the reaction of the top lateral truss to the bearings and substructure. AREMA (2015) indicates that diaphragms may be used in lieu of cross frames for closely spaced shallow girders. Example 5.16 outlines the analysis of a typical DPG vertical end brace frame.

Example 5.16a (SI Units)

Determine the forces in the members of the end brace frame shown in Figure E5.16.

P_L = 158 kN (top lateral truss wind force and nosing reactions transferred to end portal frame).

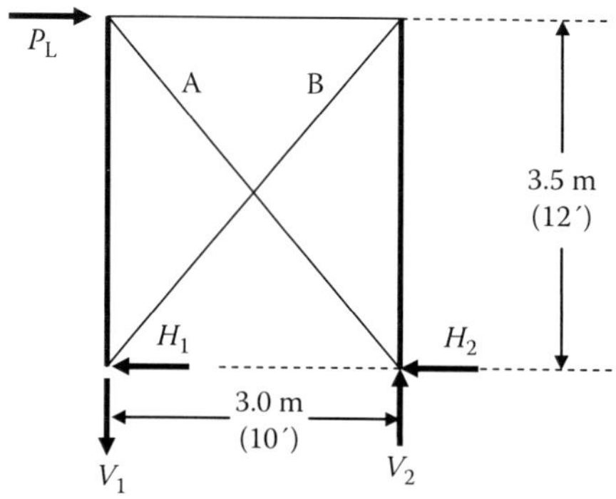

FIGURE E5.16 End brace frame.

Each brace is assumed to resist 1/2 of the horizontal shear. The force in each brace is estimated as

$$F_A = -F_B = -\frac{(158)}{2}\frac{\sqrt{3.5^2+3.0^2}}{3.0} = -121.4\text{ kN}.$$

Example 5.16b (US Customary and Imperial Units)

Determine the forces in the members of the end brace frame shown in Figure E5.16.

$P_L = 35.5$ kips (top lateral truss wind force and nosing reactions transferred to end portal frame).

Each brace is assumed to resist 1/2 of the horizontal shear. The force in each brace is estimated as

$$F_A = -F_B = -\frac{(35.5)}{2}\frac{\sqrt{12^2+10^2}}{10} = -27.7\text{ kips}.$$

5.2.2.1.3 *Intermediate Vertical and Sway Bracing*

Intermediate vertical cross-frame bracing in deck spans and sway bracing in TT spans are required to provide compression flange or chord stability, resist wind forces, and ensure adequate stiffness for load distribution and serviceability conditions.

In TT spans, intermediate vertical sway bracing carries only small forces because of the negligible difference in relative lateral deformation of top and bottom lateral systems in typical TT spans*. It is often estimated that 50% of panel load due to wind in addition to 2.5% of the total compressive axial force in the chord at the panel point is carried by the sway bracing. The analysis of forces may then proceed in a similar manner to that for the end portal frames of TT spans (see Examples 5.11 through 5.14). Intermediate sway bracing is often designed as the knee-braced frame type (see Figure E5.13). Where estimated lateral forces are small, it may be sufficient to proportion members based on maximum slenderness criteria for buckling. AREMA (2015) provides recommendations regarding the types and geometry of TT span sway bracing.

In deck spans, the intermediate cross frames or diaphragms provide for proper load distribution between main girders or trusses and therefore, in addition to the wind and stability-related forces, must be designed to resist the forces induced by the differential vertical deflections of trusses or girders†. AREMA (2015) indicates that, for deck spans, diaphragms may be used in lieu of cross frames for closely spaced shallow girders. AREMA (2015) also provides the guidelines shown in Table 5.2 for the recommended spacing of intermediate vertical brace frames.

---

* Provided that there are no substantial live load eccentricities. Track eccentricity can create additional forces in the bracing members that may be determined by the simple tension member only assumption or by a more rigorous analysis.

† These forces can be particularly large in skewed spans (see Chapter 3) or spans with a substantial track eccentricity.

TABLE 5.2
Maximum Spacing of Intermediate Vertical Brace Frames in Deck Spans

Type of Bridge Deck	Maximum Vertical Brace Frame Spacing (m)	Maximum Vertical Brace Frame Spacing (ft)
Open deck construction (see Chapter 3)	5.5	18
Noncomposite steel-concrete ballasted decks (precast concrete, steel plate, solid timber) with top lateral bracing	5.5	18
Noncomposite steel-concrete ballasted decks (precast concrete, steel, timber) without top lateral bracing	3.65	12
Cast-in-place composite concrete decks	7.30	24

5.2.2.1.4 Knee Bracing in Through Spans

The members (knee braces) which provide intermittent lateral bracing* to pony truss compression chords and through plate girder compression flanges must have adequate transverse elastic frame stiffness to ensure that the overall chord or flange has panel lengths with appropriate stiffness to attain the buckling load, P_c (Figure 5.29). Nodal points are created at each knee brace transverse frame (panel point) location if the transverse frame stiffness is very large. Conversely, if the transverse frame is too flexible, the entire compression chord or flange may buckle in a single half-wave. For typical through superstructures, the buckled shape of the compression chord or flange usually comprises many half-waves of length less than the distance between panel points (Bleich, 1952).

The lateral forces associated with resisting the compression chord or flange deformations can be estimated as the product of the buckling deformation and transverse frame elastic stiffness. The transverse frame elastic stiffness is expressed as an equivalent spring constant, C, developed by considering the stiffness contributions of the girders/knee brace, EI_c, and floorbeam/deck, EI_b as (Galambos, 1988)

$$C = \frac{E}{h^2\left(\left(h/3I_c\right)+\left(S/2I_b\right)\right)}. \tag{5.41a}$$

If the floorbeam is very stiff in comparison to the vertical members of the transverse frame† ($I_b >> I_c$),

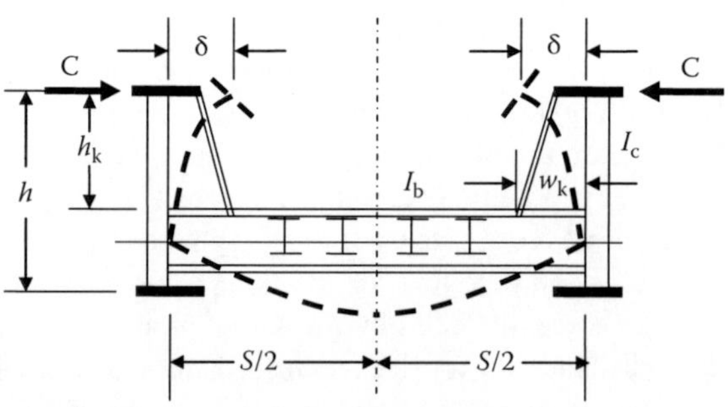

FIGURE 5.29 Through plate girder transverse frame behavior.

* Bracing of the compression chord or flange in the vertical direction is provided by truss and girder web members, respectively.

† Which might be the case for some pony truss spans and girders without substantial knee braces.

$$C = \frac{3EI_c}{h^3}. \tag{5.41b}$$

Assuming that the compression chord or flange is rigidly connected* at the ends and elastically supported at equally spaced transverse frames (girder/knee brace and floorbeam/deck)†, the force (reaction), R_F, at the transverse frames is

$$R_F = C\delta. \tag{5.42}$$

Furthermore, assuming that the span buckles in a half-wave with continuously distributed elastic intermediate supports between ends of the span (Figure 5.30a), Engesser provided the solution for the required spring constant, C_{req}, as (Bleich, 1952)

$$C_{req} = \frac{F_{cr}^2 l}{4EI} = \frac{\pi^2 F_{cr}}{4k^2 l}, \tag{5.43}$$

where

F_{cr} = compression chord or flange critical buckling force (for either the entire chord or flange supported by transverse elastic frames or the length between the transverse frames with elastic end supports)

kl = effective panel length.

However, because Equation 5.43 is only accurate when the half-length of the buckled chord or flange is greater than about $1.8l$, it is not applicable to short spans or spans with only a few panel points. A considerably larger spring constant, $\pi^2 F_{cr}/k^2 l$, is required if it is assumed that the ends of the girder or pony truss are laterally unsupported (Figure 5.30b) (Davison and Owens, 2003). This condition is unlikely and an analysis performed by Holt (1952, 1956) that provides for end supports modeled as cantilever springs is applicable to short spans (Figure 5.30c). The results of this analysis and an associated design procedure are given in (Galambos, 1988). In such analyses, the compression area for through plate girders is generally taken as the area of the top flange and 1/3 of the web compression area.

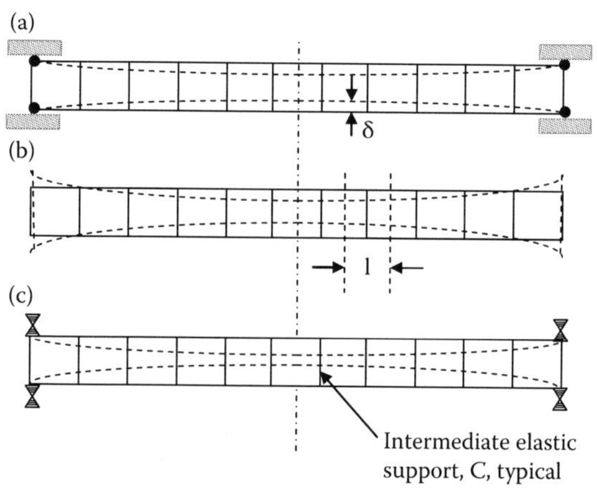

FIGURE 5.30 End restraint with intermediate transverse elastic frames: (a) pinned ends, (b) unrestrained ends, and (c) elastic ends.

* The assumption of pin connected span ends will result in a non-conservative analysis for short spans.

† Also assuming that the chord or flange has a constant cross-sectional area and moment of inertia.

AREMA (2015) recommends that the lateral bracing of compression chords and flanges be designed for a transverse shear force, R_F, equal to 2.5% of the total axial force in both members in the panel. This "notional" force is recommended to ensure that intermediate transverse frames with knee braces are designed with adequate stiffness to prevent buckling failure. The "notional" force can be used to determine the axial and bending-induced forces for the design of knee brace web and flange elements that provide adequate frame stiffness. However, these "notional" forces developed for knee brace design are not to be transferred to the end of the floorbeams supporting the knee braces. An analytical study by FEA and field testing (Paterson et al., 2016) illustrated that through spans behave as a system of connected frames with the girder top flanges relatively free to rotate. The investigation confirmed the following:

- Knee braces do not affect floorbeam design and the AREMA (2015) guidance that floorbeams should be designed as pinned supported members, even at knee braces, is appropriate.
- It is not relevant to consider floorbeam bending induced by the knee brace and bending from the directly applied live load (see Chapter 4) concurrently for the same load position.
- The "notional" transverse force does not apply bending moment to the floorbeam and is resolved through knee brace frame action.

The transverse shear force due to restraint of compression flange or chord buckling, R_F, can then be determined for through plate girder (Figure 5.29) or pony truss spans, as

$$R_F = 0.025 A_f f_c, \tag{5.44}$$

where

A_f = area of compression chord or flange

f_c = compressive stress in the chord or flange

Bracing members may also have to be designed considering the shear force in the panel from lateral wind loads.

Example 5.17a (SI Units)

Determine the AREMA recommended bracing design force for the knee braces (at a 3:1 slope) of the through plate girder span of Figure 5.29 with the following data:

$h = 2550$ mm

$h_k = 1900$ mm

$w_k = 885$ mm

$S = 6600$ mm

$I_c = 62.5 \times 10^6$ mm^4

$I_b = 4150 \times 10^6$ mm^4

$A_f = 30.0 \times 10^3$ mm^2

$A_w = 32.5 \times 10^3$ mm^2 (web plate area)

$f_c = 138$ MPa

$L = 30$ m

N_p = number of panels = 10

$l = 30/10 = 3$ m

$R_F = 0.025 A_f f_c = 0.025(30,000/1000)(138) = 103.5$ kN

Column force in knee brace = (103.5)(3) = 310.5 kN

Bending moment induced force in knee brace flange = 103.5(1900/885) = 222 kN.

Maximum bending moment in knee brace = 103.5(1.9) = 196.7 kNm

Knee brace design forces and moments from "notional" transverse shear forces are not transferred to the floorbeams.

Example 5.17b (US Customary and Imperial Units)

Determine the AREMA recommended bracing design force for the knee braces (at a 3:1 slope) of the through plate girder span of Figure 5.29 with the following data:

$h = 100$ in
$h_k = 75$ in
$w_k = 35''$
$S = 260$ in.
$I_c = 150$ in.^{4}
$I_b = 10{,}000$ in.^{4}
$A_f = 45$ in.^{2}
$A_w = 50$ in.^{2} (web plate area)
$f_c = 20$ ksi
$L = 100$ ft $= 1200$ in.
N_p = number of panels = 10
$l = 1200/10 = 120$ in.
$R_F = 0.025 A_f f_c = 0.025(45)(20) = 22.5$ kips
Column force in knee brace = (22.5)(3) = 67.5 kips
Bending moment induced force in knee brace flange = 22.5(75/35) = 48.2 kips
Maximum bending moment in knee brace = 22.5(75/12) = 140.6 kips-ft.

Knee brace design forces and moments from "notional" transverse shear forces are not transferred to the floorbeams.

Example 5.18a (SI Units)

Determine the transverse stiffness of the through plate girder span compression flange bracing of Example 5.17a using the Engesser approach and the lateral deflection associated with the AREMA recommended design force.

$$C = \frac{200{,}000(10^6)}{(2550)^2((2550/3(62.5))+(6600/2(4150)))} = 2137 \text{ N/mm}$$

$F_{cr} \sim 1.80(30.0 + (32.5/2/3))(138) \sim 8798$ kN (using safety factor of 1.80)

$$C_{req} = \frac{\pi^2(8798)(10^3)}{4(3000)} = 7236 \text{ N/mm.}$$

Since $C < C_{req}$, the transverse frames are not stiff enough to preclude excessive buckling deformations.

If, for example, the vertical member stiffness, I_c, was increased by addition of a substantial knee brace so that $I_c = 250 \times 10^6\,\text{mm}^4$,

$$C = \frac{200{,}000(10^6)}{(2550)^2((2550/3(250))+(6600/2(4150)))} = 7332 \text{ N/mm.}$$

Since $C > C_{req}$, the critical buckling load of the flange, assuming a factor of safety of 1.80, is $F_{cr} \sim 1.80(30.0)(138) \sim 7452$ kN.

Using the shear force, $R_F = 103.5$ kN from Example 5.17a (AREMA recommended force):

$$\delta = \frac{R_f}{C} = \frac{103.5(1000)}{7332} = 14 \text{ mm,}$$

which is (span/2143) and appears reasonable for a 30.0 m long-span lateral deflection and a 2.05 m (assuming 300 mm deep floorbeams) frame wall (girder web/knee brace) cantilever tip deflection of (height/146).

Example 5.18b (US Customary and Imperial Units)

Determine the transverse stiffness of the through plate girder span compression flange bracing of Example 5.17b using the Engesser approach and the lateral deflection associated with the AREMA recommended design force.

$$C = \frac{29,000}{100^2((100/3(150))+(260/2(10,000)))} = 12.3 \text{ kips/in.}$$

$F_{cr} \sim 1.80(45 + (50/2/3))(20) \sim 1920$ kips (using safety factor of 1.80)

$$C_{req} = \frac{\pi^2(1920)}{4(1200/10)} = 39.5 \text{ kips/in.}$$

Since $C < C_{req}$, the transverse frames are not stiff enough to preclude excessive buckling deformations.

If, for example, the vertical member stiffness, I_c, was increased by addition of a substantial knee brace so that $I_c = 600$ in.^{4},

$$C = \frac{29,000}{100^2((100/3(600))+(260/2(10,000)))} = 42.3 \text{ kips/in.}$$

Since $C > C_{req}$, the critical buckling load of the flange, assuming a factor of safety of 1.80, is $F_{cr} \sim 1.80(45)(20) \sim 1620$ kips.

Using the shear force, $R_F = 22.5$ kips from Example 5.17b (AREMA recommended force):

$$\delta = R_f/C = 22.5/42.3 = 0.53 \text{ in.}$$

which is (span/2256) and appears reasonable for a 100 ft long-span lateral deflection and a 6 ft – 9 (assuming 12 in. deep floorbeams) in frame wall (girder web/knee brace) cantilever tip deflection of (height/153).

5.3 STRUCTURAL DESIGN OF STEEL RAILWAY SUPERSTRUCTURES

The structural design of members and connections in the superstructure may proceed once the bridge design engineer has determined the loads on the superstructure (Chapter 4) and the internal member forces from structural analysis with the appropriate load combinations.

Preliminary structural analyses use superstructure models developed through the planning and preliminary design process that are refined, as necessary, through the structural analysis process. The structural analysis may range from the routine analysis of statically determinate superstructures (reactions and internal forces determined from equilibrium) to continuous and more complex statically indeterminate structures (additional equations required such as compatible displacement equations, which require section properties and dimensions). Structural design for strength, serviceability, and fatigue criteria (or limit states) involves material selection and determination of the dimensions or section properties of the members and connections in the superstructure. For statically indeterminate structures, iterative analyses and design are typically required. Strength, serviceability, and fatigue design of the superstructure require examination of material (yielding and fracture) and member (instability, cyclical stresses, deformation and vibration) behavior.

5.3.1 Failure Modes of Steel Railway Superstructures

Strength failure by yielding, instability, or fracture must be precluded. The von-Mises yield criterion (see Chapter 2) is appropriate for use in elastic strength design (Armenakas, 2006). Therefore, tension, compression, and shear yielding failure are based on the yield criterion of this failure theory. For allowable stress design (ASD), allowable tension, compression, and shear stresses are based on the tensile yield stress (the yield stress is divided by a safety factor to obtain the allowable stress). Compression members and elements (e.g., the top flange of a simply supported girder) may become unstable prior to yielding and this effect is incorporated into ASD elastic strength design procedures as an effective reduction in the allowable compression stress (usually expressed as parabolic transition equations) (see Chapters 6 and 7). In addition to yielding, tension members must also be designed considering the ultimate stress fracture condition. The strength design of axial members, flexural members, and connections is discussed further in Chapters 6 through 9.

Serviceability failures occur as excessive elastic deformations, vibrations, and/or cracking of concrete decks in steel-concrete composite construction (see Chapter 7). Allowable live load deflection criteria, based on length of span, are recommended by AREMA (2015), which will affect the stiffness design of the superstructure. The live load deflection criteria are intended to control dynamic behavior. Live load vibration effects on stresses are included in the empirically developed dynamic load or stress increment (see Chapter 4) and vibration from wind is generally not a concern for the usually relatively stiff steel railway superstructures*. The deflection design of steel railway superstructures is discussed in greater detail later in this chapter.

In addition to ultimate stress tensile fracture, failure by tensile fracture can, under certain conditions, be sudden or caused by accumulated fatigue damage. Sudden or brittle fracture is typically related to pre-existing flaws (e.g., cracks, notches, and weld discontinuities) and members with details where triaxial stresses are constrained creating stress concentrations with high mean normal tensile stresses, that can precipitate failure prior to yielding†. Therefore, the failure may be a sudden fracture without evidence of yielding. Fracture susceptibility is generally more severe with dynamic loads, thick plates,‡ and low service temperatures§.

Failure by fracture may also result from the accumulated damage from the cyclical application of tensile stresses from live loads over time (fatigue). Fracture through accumulated fatigue damage is of primary concern in the design of steel railway superstructure members and connections. The fatigue life, or number cycles to failure¶, depends on the frequency and number of load cycles, load magnitude (in particular, tensile stress range), member size, and member details. A fracture mechanics approach to fatigue design is not generally used for ordinary steel bridge design (Fisher, 1984; Kulak and Smith, 1995; Dexter, 2005) and the stress-life approach, recommended for the design of steel superstructures by AREMA (2015), is outlined further in this Chapter.

The extreme events limit state, such as, seismic activity or derailment, considers that some damage to the superstructure may occur due to the event, but it will survive for rehabilitation, if necessary, to the pre-event capacity.

Table 5.3 outlines the strength, serviceability, fatigue and extreme events limit state criteria for typical members and corresponding load combinations (see Table 4.10).

---

* Wind vibration is implicitly considered in the design of steel railway spans in accordance with AREMA (2015) by recommendation of a notional lateral load (see Chapter 4) that ensures sufficiently stiff lateral bracing systems.

† The von-Mises yield criterion is independent of mean normal (or hydrostatic) stresses.

‡ At a given temperature, thicker plates exhibit lower fracture toughness in elastic-plastic regions (crack tips) due to plane strain conditions.

§ Conditions in which steel railway bridges are often required to perform.

¶ Generally considered through-thickness fracture of a component.

TABLE 5.3
Limit State Criteria for Typical Members and Corresponding Load Combinations

Limit State	Failure Criteria	Internal Forces	Typical Members	Load Cases (See Table 4.10)
Strength	Tensile yielding	Axial	Truss members	D1-A, D2-A, D2-B, D3, S1-A, C1, C2-A, C2-B
			Bracing members	D4-A, D4-B, D5-A, D5-B
		Flexural and shear	Beams and girders	D1-A, D2-A, S1-A, C1, C2-A, C2-B
	Tensile fracture (ultimate)	Axial	Truss members	D1-A, D2-A, D2-B, S1-A, C1, C2-A, C2-B
			Bracing	D4-A, D4-B, D5-A, D5-B
	Compressive instability (buckling)	Axial	Truss members	D1-A, D2-A, D3, S1-A, C1, C2-A, C2-B
			Bracing members	D4-A, D4-B, D5-A, D5-B
		Flexural	Beams and girders	D1-A, D2-A, S1-A, C1, C2-A, C2-B
Serviceability	Vertical deflection		Trusses, beams, and girders	LL + I
	Vibration		Trusses, beams, and girders	LL + I
	Lateral deflection		Trusses, beams, and girders	CF + W_L + N + other lateral forces
	Concrete deck cracking (composite steel-concrete)	Flexural	Beams and girders	see AREMA Chapter 8
Fatigue	Tensile fracture (cyclical)	Axial	Truss members	D1-B
		Flexural	Beams and girders	D1-B
Extreme events	Earthquake	Axial, flexural, and shear	Truss members, beams, girders, and bracing members	E1-A, E1-B
	Derailment	Axial	Bracing members (diaphragms and cross frames)	S1-B

5.3.2 Steel Railway Superstructure Design

5.3.2.1 Strength Design

The strength design of members and connections, as recommended by AREMA (2015), is performed through elastic structural analyses and the allowable stress design (ASD) method. The allowable stress design methodology divides the ultimate and yield stress of the steel by a factor of safety (FS) to determine allowable stresses. Yield stress is associated with plastic deformation and ultimate stress with fracture. Internal stresses in members and connections must not be greater than the allowable yield or fracture criteria. As indicated in Chapter 2, the allowable stresses for tension, compression, and shear are all expressed in terms of the material tensile yield and ultimate stresses.

The allowable stress approach ensures all members behave elastically, which is appropriate for steel with its well defined elastic behavior and tensile yield stress. Also, since stresses from loads are maintained within the elastic region of behavior, design load combinations may be based on load and stress superposition. However, the use of a single safety factor against yielding for the many different loads within a load combination is a shortcoming of allowable stress design. In addition, ASD elastic design methods do not fully consider the localized yielding and load redistribution attributes of steel structures at failure.

AREMA (2015) recommends modification of the factor of safety (modification of allowable stresses) for design load combinations based on the probability of the loads being applied concurrently to the member*. However, design load combinations with ASD based FS do not consider the real uncertainties associated with different loads or combinations of loads. Therefore, methods based on a probabilistic approach to the estimation of loads and member strength have been adopted by many international building and bridge design guidelines, recommendations, codes, and specifications.

Nevertheless, the use of a single safety factor in ASD is not a significant shortcoming for ordinary steel railway superstructure design due to the relatively high live load to dead load ratio, and the importance of the serviceability (deflection) and fatigue limit states, which are both evaluated at service loads. In addition, while ASD does not consider the localized yielding and load redistribution of steel structures at failure, it is a valid design methodology in regards to the acceptability of failure (yielding, fracture, and stability) of superstructures in the railroad operating environment (see Chapter 3).

The FS for axial and flexural tensile stresses recommended by AREMA (2015) (9/5 = 1.80 ~ 1/0.55) is greater than the typical allowable tensile stress FS (5/3 = 1.67 = 1/0.60) used in building or highway bridge ASD because of the high-magnitude variable-amplitude cyclical dynamic live load spectra on steel railway superstructures (see Chapter 4). Further considerations relating to the use of a larger tensile factor of safety for steel railway superstructures are fracture (cold weather service), corrosion (industrial and wet environments), and damage susceptibility due to location (railway, highway, or marine vehicle contact with tension chords or flanges).

The factor of safety for ASD design of axial compression members is generally between 1.9 and 2.0 because of stability issues related to unintended load eccentricities and the initial curvature of compression members. However, for short axial compression members that will yield prior to buckling, the FS corresponding to compressive yielding (related to tensile yield stress, see Chapter 2) of 9/5 = 1.8 can be used. A cubic polynomial equation (representing a quarter sine wave) is an appropriate transition function for an axial compression member FS (Salmon & Johnson, 1980) and can be applied to the AREMA (2015) recommended FS for axial compression stresses as shown in Figure 5.31, where

K = effective length factor (depends on compression member end condition) (see Chapter 6)

L = the length of the member

r = the radius of gyration of the member

C_{cr} = the limiting or critical value of (KL/r) at the proportional limit $(0.50F_y)$ to preclude instability in the elastic range (Euler buckling) (see Chapter 6)

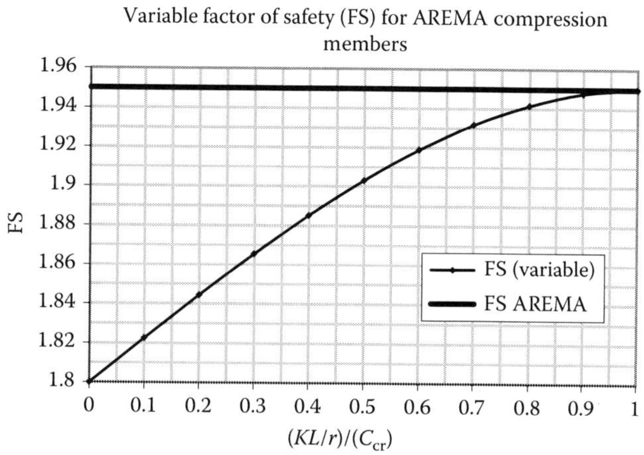

FIGURE 5.31 Factor of safety for compression members.

* Generally based on steel bridge performance and the engineering experience of AREMA Committee 15 members.

Nevertheless, AREMA (2015) recommends a FS of 1.95 for axial compression members of all slenderness ratios. This is appropriate unless the fabrication and erection of axial compression members can be carefully controlled to avoid eccentricities or other unintended secondary effects.

Beams, girders, trusses, arches, and frames are subjected to internal normal and shear stresses across cross sections caused by internal axial forces, shearing forces, torsional moments, and bending moments. Steel beam and girder design is based on internal elastic shearing and normal stresses caused by bending moments and shearing forces. Steel arch design is based on elastic shearing and normal stresses caused by bending moments, axial forces, and shearing forces. Steel truss design is concerned primarily with axial forces causing elastic normal stresses, although eccentricities and secondary effects (e.g., due to truss deflections) might create additional normal (due to bending moments) and shearing stresses. Members and connections with internal stresses not greater than the allowable tension, compression, or shear stresses recommended by AREMA (2015) are considered to be of safe and reliable design.

Design against brittle fracture is accomplished through the use of steel with adequate notch toughness* for the design service temperature (which depends on geographical location†) (see Chapter 2) and fabrication quality controls (see Chapter 10). AREMA (2015) recommends fracture toughness requirements‡ for steel members considered as primary and FCM. FCM are those members in tension whose failure would result in failure of the entire superstructure (e.g., nonredundant structural members such as welded girders and trusses of many typical steel railway superstructures). The fracture toughness requirements for ordinary steel superstructure design are based on relatively simple and standardized Charpy V-notch testing§ in lieu of the more complex methods available by fracture mechanics testing and analysis (Hertzberg, 1989; Anderson, 2005).

Table 5.3 outlines the strength limit state criteria for typical members and corresponding load combinations (see Table 4.10).

5.3.2.2 Serviceability Design

Serviceability criteria (or limit states) of deflection and vibration are important aspects of the structural design of steel railway superstructures. Deflection criteria may be established to ensure adequate flexural stiffness to preclude excessive deflections and vibrations.

5.3.2.2.1 Deflection Criteria

5.3.2.2.1.1 Vertical Flexural Deflections (Beams and Girders) Flexural deflections are calculated at the location of the maximum live load bending moment in a span. AREMA (2015) recommends that the maximum flexural deflection from live load including impact not exceed 1/640 of the span length¶. Railroad companies, authorities, and designers may further limit deflections based on span types (trusses, girders, and composite girder or beam spans**) and other operating practices.

The maximum flexural deflection in an ordinary simply supported span from live load including impact, Δ_{LL+I}, can be estimated considering an equivalent uniform load, $w_{e\Delta}$, as

$$M_{LL+I} = \frac{(w_{e\Delta})a(L-a)}{2} = \frac{w_{e\Delta}L^2}{8}, \text{ at } a = L/2. \tag{5.45}$$

---

* Toughness can be interpreted as the energy required to cause fracture at a given temperature.

† Indicated as Zones 1, 2 and 3 in AREMA (2015).

‡ Material with adequate toughness to initiate yielding prior to brittle fracture.

§ In North America the CVN tests are generally carried out in accordance with ASTM A673.

¶ This deflection limitation was derived (based on allowable unit stresses and design loads) from earlier AREA recommendations controlling the span to depth ratio to no greater than $L/12$.

** Often to limit cracking and improve behavior of concrete decks.

Therefore,

$$w_{e\Delta} = \frac{8M_{LL+I}}{L^2}, \tag{5.46}$$

where

a = distance to location of interest (see Figure 5.21)

M_{LL+I} = maximum bending moment due to live load and impact (at $a = L/2$)

L = length of simply supported span

Substitution of Equation (5.46) into the equation for the maximum deflection from a uniformly distributed load on a simple beam provides an estimate of the maximum flexural deflection due to live load including impact as

$$\Delta_{LL+I} = \frac{5w_{e\Delta}L^4}{384EI} = \frac{0.104M_{LL+I}L^2}{EI}, \tag{5.47}$$

where

E = modulus of elasticity

I = gross moment of inertia (used for flexural member deflection calculations)

Similar to other guidelines, codes, and specifications, AREMA (2015) recommends that the maximum flexural deflection from live load including impact not exceed $L/f\Delta$ of the span, where $f\Delta$ is an integer established based on structural behavior and experience. Therefore, the minimum gross moment of inertia, I, of a simple span required to meet the deflection criteria is

$$I \geq 520(M_{LL+I}Lf_{\Delta})\ \text{mm}^4, \tag{5.48a}$$

where

M_{LL+I} = live load including impact bending moment for span, L, kNm

L = length of span, m

or

$$I \geq \frac{M_{LL+I}Lf_{\Delta}}{1934}\ \text{in.}^4, \tag{5.48b}$$

where

M_{LL+I} = live load including impact bending moment for span, L, kips-ft

L = length of span, ft

If the AREMA (2015) recommended $f\Delta$ of 640 is used in Equations (5.48), the minimum gross moment of inertia is

$$I \geq 333\times10^3 M_{LL+I}L\ \text{mm}^4 \tag{5.49a}$$

or

$$I \geq 0.33M_{LL+I}L\ \text{in}^4 \tag{5.49b}$$

for a simply supported beam or girder span. The relationship shown in Equations 5.49a and 5.49b for f_{Δ} = 640, 800, and 1000 (corresponding to deflection criteria of $L/640$, $L/800$, and $L/1000$, respectively)*, are shown in Figure 5.32. It should be noted that the rolling impact used in Figure 5.32 varies from 5% for long spans to 12.5% for short spans and may require amendment for particular span designs. However, Figure 5.32 provides a preliminary estimate of minimum gross moment of inertia required to meet various deflection criteria for simply supported steel railway beam and girder spans.

5.3.2.2.1.2 Vertical Truss Deflections AREMA (2015) recommends that truss members may be designed as axial members provided that secondary forces do not create stresses in excess of 27.5 MPa (4000 psi) in tension members and 20.7 MPa (3000 psi) in compression members. Secondary stresses in excess of this are superimposed on the primary stresses and the member is designed as a combined axial and flexure member (see Chapter 8). In order to simplify the analysis of trusses, designers may maintain truss deformations that do not create secondary stresses in excess of 27.5 MPa (4000 psi) in tension members and 20.7 MPa (3000 psi) in compression members.

The maximum deflection of a simply supported truss from live load including impact, Δ_{LL+I}, can be determined by calculating truss joint horizontal and vertical translations by the method of virtual work or through graphical means (Utku, 1976; Armenakas, 1988). However, modern computer software based on matrix methods (stiffness or flexibility) and FEA enable the routine calculation of truss deflections and member forces. The analysis of trusses for joint translation should use the gross area of truss members not designed with perforated cover plates†. For truss members designed with perforated cover plates, the gross area should be reduced by the area determined by dividing the volume of a perforation by the spacing of perforations.

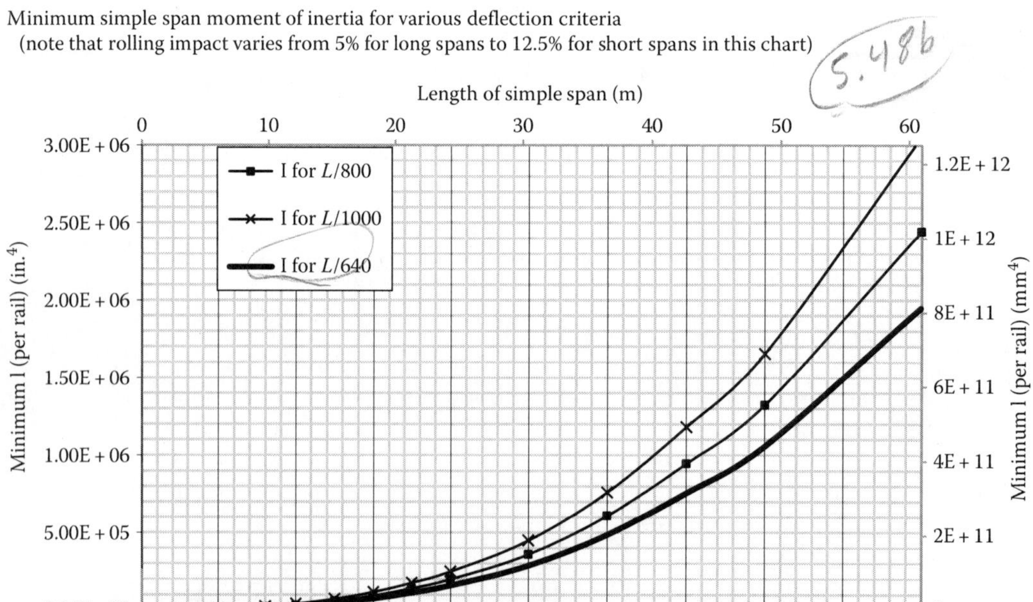

FIGURE 5.32 Stiffness design of simply supported spans for deflection criteria.

* Bridge design guidelines, specifications and codes typically specify deflection criteria of *L*/800 to control vibrations and *L*/1000 to control cracking of steel-concrete composite beam or girder decks. The AREMA (2015) criteria of *L*/640 will allow for greater superstructure vibration, which is typically of less concern for freight rail traffic.

† Cover plates on truss members are usually used in compression chords and end posts.

5.3.2.2.1.3 Lateral Deflections of Beams, Girders, and Trusses The lateral deflection of bridge superstructures must be limited to avoid misalignments of the track that could be detrimental to train operations.

AREMA (2015) recommends that lateral deflection of the superstructure (including lateral deflections imposed by the substructures) at track level be determined for wind (on the loaded superstructure), W_L, track curvature induced lateral forces (centrifugal, CF, and other lateral forces such as those due to lateral components of longitudinal forces, LF, and continuous welded rail CWR), lateral forces from equipment (nosing), *N*, and all other lateral forces excluding those due to earthquake, EQ. Depending on the type of structure and lateral deflection created by the applicable lateral forces, the maximum deflection at track level is referenced to an appropriate fixed or restrained location on the bridge. Flexible substructures may deflect and lateral deflection at track level is referenced to a vertical plane from the base of the substructures. Rigid substructures may deflect negligibly and lateral deflection at track level is referenced to a vertical plane from the superstructure bearings. Lateral forces creating differential lateral deflection throughout the height of a girder or truss are referenced to a vertical plane from the bottom flange of the girder or the bottom chord of the truss.

AREMA (2015) recommends lateral deflection limits that consider the usual construction tolerances for track alignment and deterioration before track maintenance is required (e.g., due to rail and/or fastener wear and movement). The guidance in AREMA (2015) Chapter 15 applies to freight trains with speeds not greater than 129 km/h (80 mph) and passenger trains with speeds not greater than 145 km/h (90 mph). These train speeds correspond to the United States Department of Transportation (USDOT) Federal Railroad Authority (FRA) Track Safety Standards (FRA, 2014) Class 5 track, which has allowable deviations from alignment for tangent track of 20 mm (3/4 in.) at the mid-ordinate of a 19 m (62 ft) chord, 12 mm (1/2 in.) at the mid-ordinate of a 9.5 m (31 ft) chord for curved track or 16 mm (5/8 in.) at the mid-ordinate of a 19 m (62 ft) chord for curved track. The AREMA (2015) recommendations consider track construction tolerances and deterioration by recommending lateral deflection limits that are 50% of the FRA track alignment deviation limits.

Therefore, the AREMA (2015) recommended lateral deflection limits are 10 mm (3/8 in.) at the mid-ordinate of a 19 m (62 ft) chord for tangent track, 6 mm (1/4 in.) at the mid-ordinate of a 9.5 m (31 ft) chord for curved track, or 8 mm (5/16 in.) at the mid-ordinate of an 19 m (62 ft) chord for curved track.

The allowable lateral deflection is presented in terms of the mid-ordinate of a 19 m (62 ft) chord because track curvature has been traditionally measured by this method. The use of a 19 m (62 ft) chord is convenient as track curvature is directly determined by the mid-ordinate. In Chapter 3 it was shown that

$$M = \frac{C^2}{8R}, \tag{5.50}$$

where

M = the mid-ordinate between chord, C, and the rail

R = the radius of the curve

If C = 19 m (62 ft) and R = 1747.5/D (5730/D), Equation 5.50 yields M = 25.4D (1.0D), which illustrates that the degree of curve may be readily and directly measured at the mid-ordinate of a 19 m (62 ft) chord.

Table 5.3 outlines the serviceability limit state criteria for typical members and corresponding load combinations (see Table 4.10).

5.3.2.2.2 *Fatigue Analysis and Design of Steel Railway Superstructures*

The fracture* of steel by fatigue may be caused by modern high-magnitude cyclical railway live loads. Fatigue cracks may initiate and then propagate at nominal tensile† cyclical stresses below the tensile yield stress at stress concentrations in the superstructure. The cyclical railway loading accumulates damage (which may be manifested as plastic deformation, crack initiation and crack extension) at stress concentrations, which may precipitate fracture, leading to unserviceable deformations or failure at a certain number of cycles, N_f.

The high cycle‡ fatigue life, $N \times N_f$, of a member or detail is determined by constant-amplitude cyclical stress testing of specimens typical of steel superstructure members and details§. The testing of representative specimens makes the determination of stress concentration factors and consideration of residual stresses unnecessary for ordinary steel bridge design¶. Therefore, fatigue analysis and design may be performed at nominal stresses.

5.3.2.2.2.1 Fatigue Loading of Steel Railway Superstructures The load and resulting stress cycles from moving freight trains are typically irregular and of variable amplitude. In Chapter 4, the variable-amplitude cyclical railway live load stress ranges were developed as an effective or equivalent constant-amplitude stress range, ΔS_{re}, which accumulates the same damage as the variable-amplitude cyclical stress ranges over the total number of stress range cycles to failure. The resulting expression for ΔS_{re}, the root mean cube (RMC) of the probability density function, is the foundation of the fatigue design load recommended by AREMA (2015). It was developed through stress-life analyses of continuous unit freight trains on heavy haul freight railways. The fatigue design load recommended by AREMA (2015), to determine the design stress range, is Cooper's EM360 (E80) live load with the number of design cycles adjusted for the characteristic load** magnitude.

The stress-life approach is also appropriate for determining the fatigue strength of members and details for high cycle stress range magnitudes that are generally low enough to preclude the need for considering yield effects (predominantly elastic strains with no, or small, plastic deformations).

5.3.2.2.2.2 Fatigue Strength of Steel Railway Superstructures Fatigue damage accumulation occurs at stress concentrations in tension zones†† making location and detail characteristics of prime importance regarding strength. These characteristics are compiled in AREMA (2015) within various Fatigue Detail Categories‡‡ based on the number of cycles to failure§§, N, from constant-amplitude stress range, ΔS_{re}, tests. Since railway live load is applied as a high cycle (long life) load, testing must also be conducted at high cycle constant-amplitude stress ranges. The allowable fatigue stress for the design of a particular detail is based on a probabilistic analysis (without a Factor of Safety) of high cycle test data, and, therefore, it is appropriate to perform fatigue design

---

* The fracture limit state can be defined in various ways, such as crack propagation to some critical length or number of cycles to appearance of a visible crack (generally considered to be in the order of 1 to 5 mm in the stress-life approach to fatigue). It is typically defined as when initiated fatigue cracks propagate through the thickness of the component, member or detail.

† At members and details with a net applied tensile stress, since there is no fatigue cracking in purely compression regions that never experience tensile stress.

‡ High cycle or long life fatigue analysis is appropriate for steel railway superstructure design.

§ These tests reveal considerable data scatter indicating that a probabilistic approach to fatigue strength may be appropriate.

¶ However, the use of nominal stresses without stress concentration factors should be carefully reviewed in areas of high stress gradients.

** Equivalent Cooper's loads typical for lines with various freight traffic densities (see Chapter 4).

†† Nevertheless, the presence of residual tensile stresses from rolling or welding processes may only be important in some cases (see Chapters 9 and 10).

‡‡ These are designated as A, B, B′, C, C′, D, E, E′, and F details according to the number of constant amplitude stress cycles to "failure" at a given stress range.

§§ "Failure" in terms of fatigue design does not mean failure as typically defined by the strength limit state. Fatigue "failure" is a criteria based on data at some standard deviation (generally, 2 or 2.5) from the mean of test data for the member or detail (AREMA (2015) uses a standard deviation of 2.5).

at service load levels. Also, since stress concentration effects are accounted for within the various Fatigue Detail Categories, a nominal applied stress approach for fatigue design is recommended in AREMA (2015).

The allowable fatigue stress range for design, ΔS_{rall}, depends on the number of equivalent constant-amplitude stress range cycles over the member or detail life, N, as*

$$\Delta S_{\text{rall}} = \left(\frac{A}{N}\right)^{1/m} \tag{5.51a}$$

or

$$\log(N) = \log(A) - m\log(\Delta S_{\text{rall}}), \tag{5.51b}$$

where

A = a constant depending on detail and established from regression analysis of material strength test data

Equation 5.51b is plotted in Figure 5.33 for $m = 3$ and various values of constant A (as shown in Table 5.4 for Fatigue Detail Categories A, B, B′, C, C′, D, E, E′, and F†). The constant A is established from regression analysis of test results such that Equation 5.51 describes S-N behavior for details with 95% confidence limits for 97.5% survival (2.5% probability of failure). Testing has also indicated that there is a constant-amplitude fatigue limit (CAFL) stress range, ΔS_{CAFL}, below which no fatigue damage accumulates‡. The CAFL is also shown in Table 5.4 and by the horizontal lines in Figure 5.33.

A total of 2.0×10^6 cycles is considered an infinite life condition in terms of fatigue testing (Taly, 1998). In Table 5.5, the number of applied equivalent constant-amplitude stress range cycles over the member or detail life, N, clearly exceeds 2.0×10^6 cycles for loaded lengths less than 30.5 m (100 ft). Therefore, the allowable fatigue stress range for loaded lengths or spans less than 30.5 m (100 ft,) is limited to the CAFL stress range (Table 5.4), which provides for infinite life.

A review of Tables 5.4 and 5.5 indicates that limiting the allowable fatigue stress range to the CAFL stress range for Category D, E and E′ details appears to be conservative for spans between 23.0 m (75 ft) and 30.5 m (100 ft) long (see Figure 5.34). However, because it is relatively easy to obtain many more than 6 stress range cycles for some load conditions§ on spans between 23.0 m (75 ft) and 30.5 m (100 ft) long, the apparent conservatism may not be fully realized and limiting the allowable stress range to the CAFL stress range is acceptable for routine bridge design.

Table 5.6 shows the maximum number of variable-amplitude stress cycles per train at the CAFL for various Fatigue Detail Categories. Table 5.6 indicates that limiting the allowable fatigue stress range to the CAFL is conservative for Fatigue Category Details B and B′ (at EM220 (E50) characteristic load), C (at EM245 (E55) and EM220 (E50) characteristic loads), and D, E, and E′ (at EM270 (E60), EM245 (E55) and EM220 (E50) characteristic loads) in spans between 23.0 m (75 ft) and 30.5 m (100 ft) long with an applied number of effective constant-amplitude cycles of 4.4×10^6 (based on six variable-amplitude stress range cycles per train). The analysis is performed for various typical characteristic loads. It is not unreasonable, for example, that trains with a characteristic load of about EM245 (E55) may cause 39 cycles of live load with some frequency on spans between 23.0 m (75 ft) and 30.5 m (100 ft) long. Therefore, the AREMA (2015) recommendation to limit the

---

* See Chapter 4.

† Category F is for the allowable shear stress range on the throat of a fillet weld.

‡ However, even a small number of cycles exceeding the CAFL may effectively render it as nonexistent. Therefore, fatigue design ensures all design live load stress ranges are below the CAFL.

§ Typically occurs for trains with adjacent empty and loaded cars.

(a)

S-N Curves for fatigue detail categories A, B, B′, C, C′, D, E, E′ and F

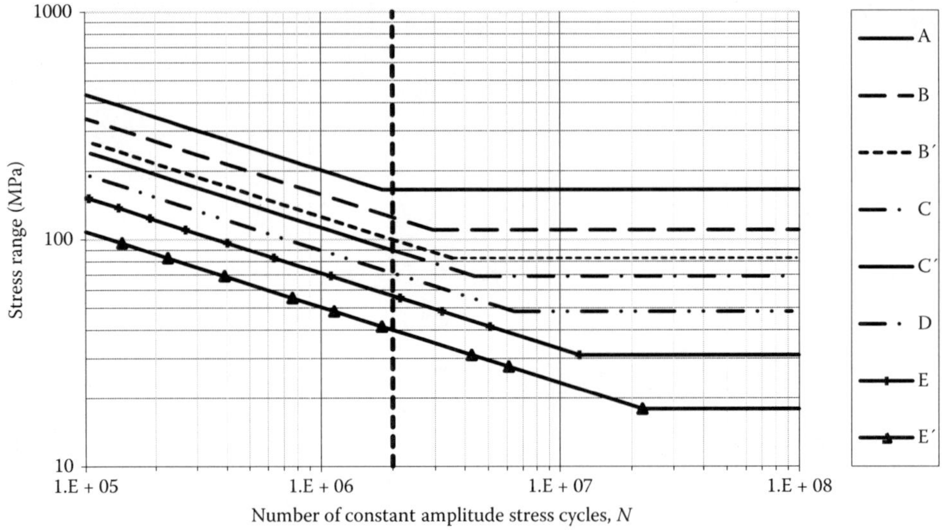

(b)

S-N Curves for fatigue detail categories A, B, B′, C, C′, D, E, E′ and F

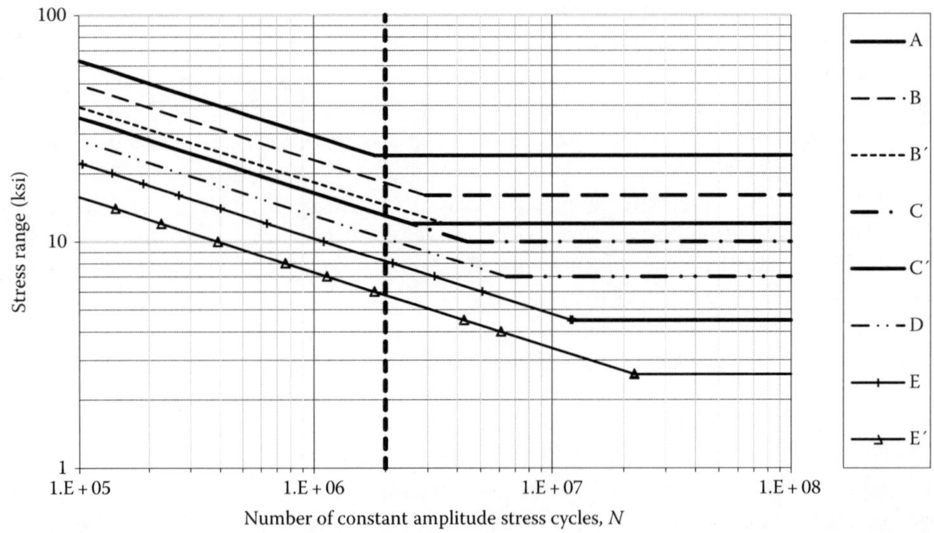

FIGURE 5.33 Constant amplitude S-N curves for fatigue detail categories.

allowable fatigue stress range to the appropriate Fatigue Category Detail CAFL is appropriate for loaded lengths or spans less than 30.5 m (100 ft) long.

Table 5.5 indicates that, for loaded lengths or spans greater than 30.5 m (100 ft), the allowable fatigue stress range may be based on the detail strength at 2.0×10^6 constant-amplitude stress range cycles as

TABLE 5.4
Number of Constant-Amplitude Stress Range Cycles at CAFL for Fatigue Detail Categories

Fatigue Detail Category	A (MPa³)	A(ksi³)	CAFL (MPa) (Allowable Fatigue Stress Range for $L \leq 30.5$ m)	CAFL (ksi) (Allowable Fatigue Stress Range for $L \leq 100'$)	N, Constant-Amplitude Cycles to Failure at CAFL
A	8.2×10^{12}	2.5×10^{10}	165	24	1.8×10^{6}
B	3.9×10^{12}	1.2×10^{10}	110	16	2.9×10^{6}
B′	2.0×10^{12}	6.1×10^{9}	83	12	3.5×10^{6}
C	1.4×10^{12}	4.4×10^{9}	69	10	4.4×10^{6}
C′	1.4×10^{12}	4.4×10^{9}	83	12	2.5×10^{6}
D	7.2×10^{11}	2.2×10^{9}	48	7	6.4×10^{6}
E	3.6×10^{11}	1.1×10^{9}	31	4.5	1.2×10^{7}
E′	1.3×10^{11}	3.9×10^{8}	18	2.6	2.2×10^{7}

TABLE 5.5
Constant-Amplitude Stress Range Cycles

Span Length, L (m)	Span Length, L (ft)	Equivalent Constant-Amplitude Stress Range Cycles over Member or Detail Life, N
$L > 30.5$	$L > 100$	2.2×10^{6}
$30.5 \geq L > 23.0$	$100 \geq L > 75$	4.4×10^{6}
$23.0 \geq L > 15.0$	$75 \geq L > 50$	40.7×10^{6}
$15.0 \geq L$	$50 \geq L$	81.3×10^{6}

$$\Delta S_{\text{rall}} = \left(\frac{A}{2\times10^{6}}\right)^{1/3}. \tag{5.52}$$

Using Table 5.4 and Equation 5.52, the allowable fatigue stress range, S_{rall}, for loaded lengths or spans greater than 30.5 m (100 ft) long is shown in Table 5.7.

AREMA (2015) also recommends Fatigue Detail Category F for shear stress range on the throat of fillet welds. The allowable fatigue stress range is 62 MPa (9 ksi) for loaded lengths or spans greater than 30.5 m (100 ft) and 55 MPa (8 ksi) for loaded lengths or spans less than 30.5 m (100 ft)*. It may be acceptable to consider this as Fatigue Detail Category E, where adequate weld throat is provided by recommended minimum sizes or strength requirements, because cracking will occur in the base metal at the weld toe (Dexter, 2005).

Mechanical fasteners designed in accordance with AREMA (2015) (see Chapter 9) will generally not experience shear stress range induced fatigue failure prior to the connection or member base metal. Therefore, AREMA (2015) contains no recommendations concerning allowable fatigue shear stress ranges for fasteners. Tensile stress ranges in mechanical fasteners from combined external loads (see Chapter 4) and prying forces† (see Chapter 9) must not exceed the allowable tensile stress ranges of 215 MPa (31 ksi) and 262 MPa (38 ksi) on the tensile stress area‡ for ASTM F3125 Grade A325 and Grade A490 bolts, respectively.

* These allowable fatigue shear stress ranges for fillet welds are determined from an S-N curve line with a slope, m, greater than 3.
† Prying forces must not exceed 20% of the combined external forces.
‡ Cross-sectional area at thread root.

(a)

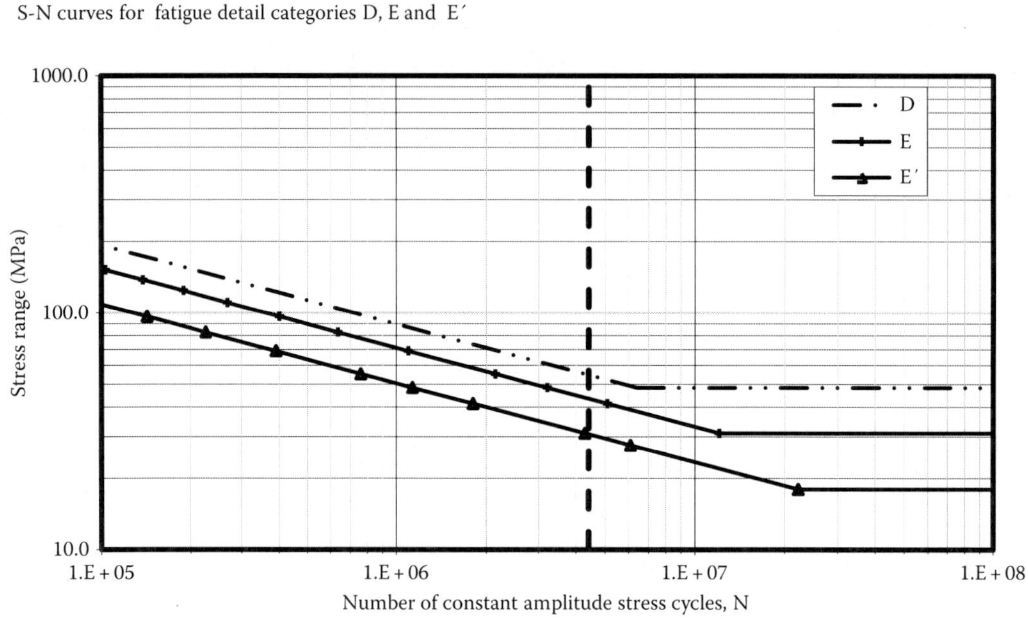

(b)

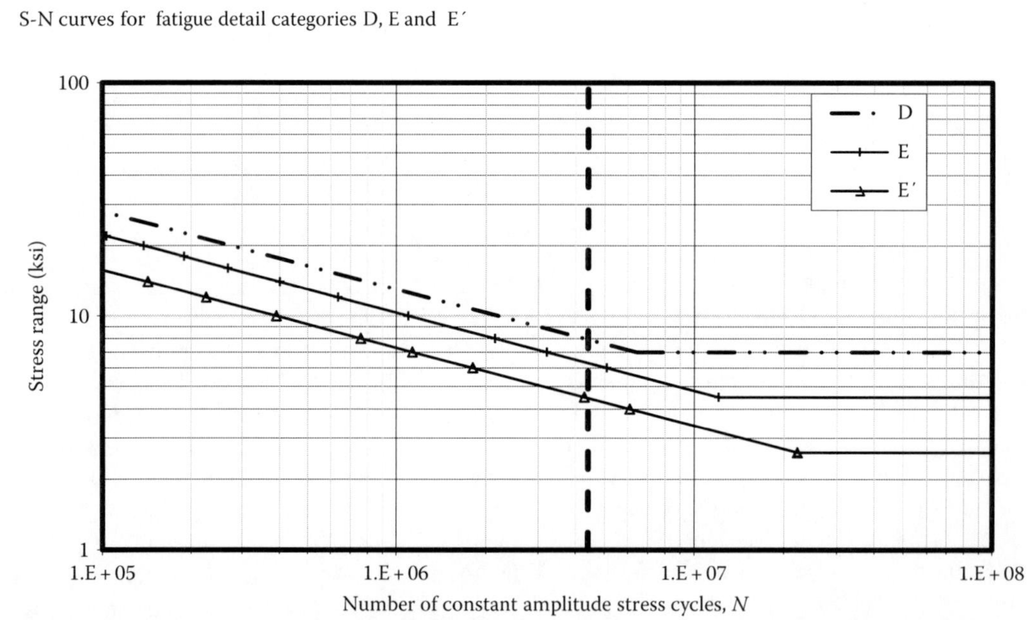

FIGURE 5.34 Allowable fatigue stress range at 4.4×10^6 cycles for category D, E, and E′ details.

AREMA (2015) also recommends that Fatigue Detail Category E and E′ details not be used in Fracture Critical Members (FCM). Caution regarding the use of Category D details is also expressed. It is generally good practice that designers avoid any poor fatigue details in all main carrying members and, particularly, in nonredundant or FCM members.

TABLE 5.6
Maximum Number of Variable-Amplitude Cycles per Train at the CAFL for Various Fatigue Detail Categories

	Maximum Number of Cycles per Train at CAFL (10^6)		
Fatigue Detail Category	EM270 (E60) Characteristic Loading	EM245 (E55) Characteristic Loading	EM220 (E50) Characteristic Loading
A	3	3	4
B	4	5	7
B′	5	6	8
C	6	8	11
C′	4	5	6
D	9	11	15
E	16	21	29
E′	30	39	53

TABLE 5.7
Allowable Fatigue Stress Range at 2,000,000 Cycles [used for loaded lengths or spans greater than 30.5 m (100 ft) long]

Fatigue Detail Category	Allowable Fatigue Stress Range (MPa) at 2,000,000 Cycles	Allowable Fatigue Stress Range (ksi) at 2,000,000 Cycles	S_{Rfat} (CAFL) (MPa)	S_{Rfat} (CAFL) (ksi)
A	159 (AREMA uses 165, which is CAFL)	23 (AREMA uses 24, which is CAFL)	165	24
B	124	18	110	16
B′	100	14.5	83	12
C	90	13	69	10
C′	90	13	83	12
D	69	10	48	7
E	55	8	31	4.5
E′	40	5.8	18	2.6

Table 5.3 outlines the fatigue limit state criteria for typical members and corresponding load combinations (see Table 4.10).

5.3.2.3 Other Design Criteria for Steel Railway Bridges

There are other specific design criteria relating to both the strength and serviceability design of steel railway bridges that require consideration by the design engineer.

5.3.2.3.1 Secondary Stresses in Truss Members and Girders

For steel railway superstructures, the members most likely to be subjected to combined stresses are as follows:

- Truss chord and web members. Where secondary stresses exceed 27.5 MPa (4 ksi) in tension members and 20.7 MPa (3 ksi) in compression members the truss members must be designed as combined axial-flexural members (see Chapters 4 and 8).

- Truss end posts which typically resist bending and axial forces due to portal bracing effects superimposed on the axial compression as a member in the main truss (see Chapters 5 and 8).
- Truss hangers* which are stressed primarily in axial tension, but the effects of out-of-plane bending must also be investigated in their design. Hanger allowable fatigue stress range is of critical importance due to the cyclical tensile live load regime and relatively short influence line.
- Girders and trusses with external steel prestressing cables (Dunker et al., 1985; Troitsky, 1990).

5.3.2.3.2 Minimum Thickness of Material

Material thickness is related to strength and serviceability. AREMA (2015) recommends that steel members should not have any components less than 10 mm (3/8 in.) thick (with exception of fillers), but some design engineers specify a greater minimum material thickness [typically 12 mm (1/2 in.)]. Gusset plates used to connect chord and web members in trusses should be proportioned for the force transmitted, but should not be less than 12 mm (1/2 in.) thick in any case (see Chapter 9).

Where components are subject to corrosive conditions, they should be made thicker than otherwise required (as determined by judgment of the design engineer) or protected against corrosion by painting or metallic coating (usually hot-dipped or spray applied zinc or aluminum). Atmospheric corrosion resistant (weathering) steel (see Chapter 2) does not protect against corrosion by standing water, and/or consistently wet or corrosive environments. Therefore, the design engineer should also carefully consider drainage holes in thin horizontal elements and deck drainage in the design of a bridge.

5.3.2.3.3 Camber

Camber is a serviceability related criterion. AREMA (2015) recommends that plate girder spans in excess of 27.5 m (90 ft) long be cambered for dead load deflection.† Trusses are recommended for greater camber based on dead load deflection plus the deflection from a uniform live load of 45 kN per track m (3000 lb per track ft) at each panel point‡.

5.3.2.3.4 Web Members in Trusses

AREMA (2015) recommends that truss web members and their connections be designed for the live load that increases the total stress by 33% over the design stress in the most highly stressed chord member of the truss. For this load case, allowable stresses in the truss web members are increased by 133% (see Chapter 4).

REFERENCES

American Railway Engineering and Maintenance-of-Way Association (AREMA), 2015, Chapter 15—Steel structures, in *Manual for Railway Engineering*, Lanham, MD.

Anderson T.L., 2005, *Fracture Mechanics*, 3rd Edition, CRC Press, Boca Raton, FL.

Armenakas, A.E., 1988, *Classical Structural Analysis*, McGraw-Hill, New York.

Armenakas, A.E., 2006, *Advanced Mechanics of Materials and Applied Elasticity*, CRC Press, Boca Raton, FL.

Bakht, B. and Jaeger, L., 1985, *Bridge Analysis Simplified*, McGraw-Hill, New York.

Bleich, F., 1952, *Buckling Strength of Metal Structures*, 4th ed. McGraw-Hill, New York.

Christy, C.T., 2006, *Engineering with the Spreadsheet*, ASCE Press, Reston, VA.

Cook, R.D., 1981, *Concepts and Applications of Finite Element Analysis*, 2nd ed., John Wiley and Sons, New York.

---

* Truss vertical members without diagonals at bottom chord panel point.

† Girder camber is typically cut into the web plate at Fabrication (see Chapter 10).

‡ Truss camber can be accomplished during fabrication by vertically offsetting truss joints through changing the length of the truss members (see Chapter 10).

Davison, B. and Owens, G.W., 2003, *Steel Designer's Manual*, The Steel Construction Institute, Blackwell Publishing, Oxford, UK.

Dexter, R. J., 2005, Fatigue and fracture, Chapter 34, in *Handbook of Structural Engineering*, Chen, W.F. and Lui, E.M., Eds, CRC Press, Boca Raton, FL.

Dick, S.M., 2002, Bending moment approximation analysis for use in fatigue life evaluation of steel railway girder bridges, PhD Thesis, University of Kansas, Lawrence, KS.

Dunker, K.F., Klaiber, F.W., and Sanders, W.W., 1985, Design manual for strengthening single-span composite bridges by post-tensioning, Iowa State University Engineering Research Report, Ames, IA.

Fisher, J.W., 1984, *Fatigue and Fracture in Steel Bridges*, John Wiley & Sons, New York.

FRA Track Safety Standards, 2014, Track geometry, FRA 49 CFR 213, Subpart C, USDOT, Washington, DC.

Fu, C.C. and Wang, S., 2015, *Computational Analysis and Design of Bridge Structures*, CRC Press, Taylor & Francis, Boca Raton, FL.

Galambos, T.V., Ed., 1988, Chapter 3 Centrally loaded columns, in *Guide to Stability Design Criteria for Metal Structures*, McGraw-Hill, New York.

Grinter, L.E., 1942, *Theory of Modern Steel Structures*, Volume 1, Macmillan, New York.

Hertzberg, R.W., 1989, *Deformation and Fracture Mechanics of Engineering Materials*, 3rd ed., John Wiley & Sons, New York.

Holt, E.C., 1952, *Buckling of a Pony Truss Bridge*, Column Research Council, Report No. 2.

Holt, E.C., 1956, *The Analysis and Design of Single Span Pony Truss Bridges*, Column Research Council, Report No. 3.

Jaeger, L. and Bakht, B., 1989, *Bridge Analysis by Micro-Computer*, McGraw-Hill, New York.

Ketchum, M.S., 1924, *Structural Engineer's Handbook*, Part 1, McGraw-Hill, New York.

Kulak, G.L. and Smith, I.F.C., 1995, Analysis and design of fabricated steel structures for fatigue, University of Alberta Structural Engineering Report No. 190, Edmonton, AB.

Martin, H.C. and Carey, G.F., 1973, *Introduction to Finite Element Analysis*, McGraw-Hill, New York.

Paterson, D., Rakoczy, A., and Dick, S., 2016, Rail through-plate girder 3-D analysis for fundamental evaluation of knee brace behavior, *Proceedings of the NASCC Steel Conference*, American Institute of Steel Construction (AISC), Orlando, FL.

Petroski, H., 1995, *Engineers of Dreams*, Random House, New York.

Salmon, C.G. and Johnson, J.E., 1980, *Steel Structures Design and Behavior*, 2nd Edition, Harper and Row, New York.

Steinman, D.B., 1953, *Suspension Bridges*, John Wiley & Sons, New York.

Taly, N., 1998, *Design of Modern Highway Bridges*, McGraw-Hill, New York.

Troitsky, M.S., 1990, *Prestressed Steel Bridges Theory and Design*, Van Nostrand Reinhold, New York.

Waddell, J.A.L., 1916, *Bridge Engineering*, Volume 1, John Wiley & Sons, New York.

Weaver, W. and Johnston, P.R., 1984, *Finite Elements for Structural Analysis*, Prentice-Hall Inc., Englewood Cliffs, NJ.

Wilson, E.L., 2004, *Static and Dynamic Analysis of Structures*, Computers and Structures, Berkeley, CA.

Zienkiewicz, O.C., 1983, *The Finite Element Method*, 3rd ed., McGraw-Hill, New York.

6 Design of Axial Force Steel Members

6.1 INTRODUCTION

Members designed to carry primarily axial forces are found in steel railway bridges as main truss members (e.g., chords, hangers, posts, diagonal web members, and end posts), span bracing members, steel tower columns and bracing, and spandrel columns in arches. Truss and bracing members may be in axial tension, compression, or both (due to stress reversal from moving train and wind loads). Axial members must be designed considering the yield and instability criteria of the strength limit state and the serviceability limit state. Members in axial tension must also be designed for the fatigue limit state (if subjected to cyclical stresses) and the fracture criteria of the strength limit state. Furthermore, some axial tension and compression members are subjected to additional stresses due to flexure* and must be designed for these combined stresses (see Chapter 8).

6.2 AXIAL TENSION MEMBERS

Axial tension main members in steel railway superstructures are often fracture critical and nonredundant. Therefore, the strength (yielding and ultimate) and fatigue limit states require careful consideration during design. Brittle fracture is mitigated by appropriate material selection, design detailing, and fabrication quality control (see Chapters 2, 5, and 10).

6.2.1 Strength of Axial Tension Members

The strength of a tension member is contingent upon yielding of the gross area, A_g, occurring prior to failure (defined at ultimate strength) of the effective net area,† A_e, or

$$F_y A_g \le \phi F_u A_e, \tag{6.1}$$

where

F_y = tensile yield stress of the steel
F_u = ultimate tensile stress of the steel
$\phi = 0.85$ = connection strength capacity reduction factor (Salmon and Johnson, 1980)
AREMA (2015) uses a safety factor of 9/5 which, when substituted into Equation 6.1, provides

$$0.56 F_y A_g \le 0.47 F_u A_e. \tag{6.2}$$

The net area, A_n, is determined from the gross area, A_g, with connection holes removed, and may require further reduction to an effective net area, A_e, to account for the effects of stress concentrations and eccentricities at connections. Therefore, the allowable strength of the tension member, T_{all}, is

---

* Flexural stresses in axial force members are typically due to bending forces created by end conditions (frame action, connection fixity, and eccentricity), the presence of transverse loads and/or load eccentricities.

† The effective net area is the net area reduced to account for tensile stresses not uniformly distributed across the net area.

$$T_{all} = 0.56 F_y A_g \tag{6.3a}$$

or

$$T_{all} = 0.47 F_u A_e. \tag{6.3b}$$

Based on Equation 6.3a, AREMA (2015) recommends $T_{all} = 0.55 F_y A_g$ as the resistance to tensile yielding of the gross area. The design of the tension member should be established based on the lesser T_{all} given by Equations 6.3a or b.

6.2.1.1 Net Area, A_n, of Tension Members

The net area, A_n, is determined at the cross section of the member with the greatest area removed for perforations or other openings in the member.* The gross area, A_g, across a bolted tension member connection is reduced by the area of the holes. The net area at the connection, A_{nc}, is at the potential tensile failure line, w_{nc}, of least length. The length of potential failure lines at connections is (Cochrane, 1922).

$$w_{nc} = w_g - \sum_{i=1}^{N_b} d_b + \sum_{j=1}^{N_b-1} \frac{s_j^2}{4g_j}, \tag{6.4}$$

where

w_g = the gross length across the connection (gross width of the axial member)
N_b = the number of bolt holes in the failure line
d_b = the effective diameter of the bolt holes = bolt diameter + 3 mm (+ 1/8 in.)
s = the hole stagger or pitch (the hole spacing in the direction parallel to the load)
g = the hole gage (the hole spacing in the direction perpendicular to the load).
The net area is

$$A_n = w_{nc}(t_m) = A_g - \left(\sum_{i=1}^{N_b} d_b - \sum_{j=1}^{N_b-1} \frac{s_j^2}{4g_j} \right)(t_m), \tag{6.5}$$

where t_m = the thickness of the member.
The calculation of net area is shown in Example 6.1.

Example 6.1a (SI Units)

Member U1–L1 is connected with gusset plates to the bottom chord of the truss in Figure E6.1 by M22 (22 mm diameter.) ASTM F3125 Grade A325 bolts as shown in Figure E6.2. Determine the net area of the member if it comprises two laced C 310 × 45 channels.

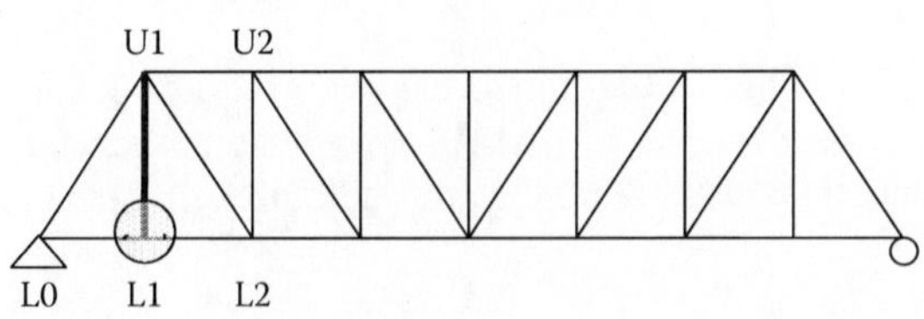

FIGURE E6.1 Elevation of truss.

* Perforations and opening are stress raisers and also require consideration in fatigue design (see Chapter 5).

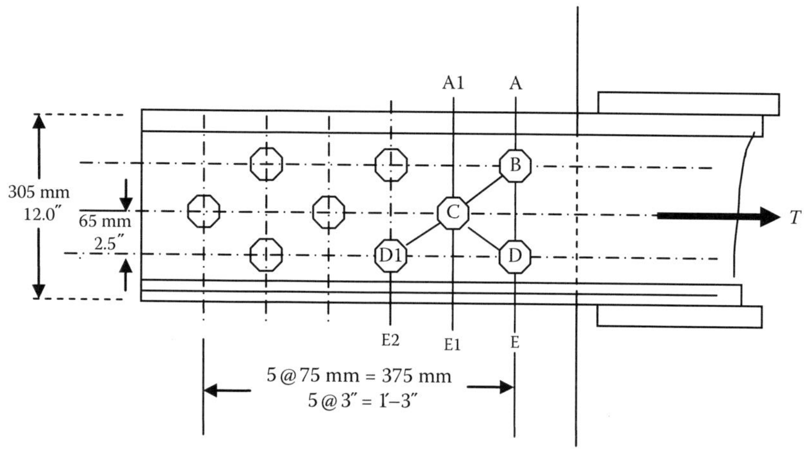

FIGURE E6.2 Member U1-L1 connection at bottom chord.

For a C 310 × 45 channel:

$A = 5690\ mm^2$
$t_w = 9.8\,mm$
$b_f = 77\,mm$
$t_f = 12.7\,mm$
Path A–B–D–E: $A_n = 5690 - 2(25)(9.8) = 5200\ mm^2$
Path A–B–C–D–E: $A_n = 5690 - \{3(25) - 2[(75^2/(4(65)))]\}(9.8) = 5379\ mm^2$
Path A–B–C–E1: $A_n = 5690 - \{2(25) - 1[((75^2/(4(65)))]\}(9.8) = 5412\ mm^2$
Path A–B–C–D1–E2: $A_n = 5690 - \{3(25) - 2[(75^2/(4(65)))]\}(9.8) = 5379\ mm^2$
Path A1–C–E1: $A_n = 5690 - 1(25)(9.8) = 5445\ mm^2$
Path A1–C–D1–E2: $A_n = 5690 - \{2(25) - 1[(75^2/(4(65)))]\}(9.8) = 5412\ mm^2$

Therefore, for the member U1–L1: $A_n = 2[5200] = 10{,}400\ mm^2$ (91.4% of A_g).

Example 6.1b (US Customary and Imperial Units)

Member U1–L1 is connected with gusset plates to the bottom chord of the truss in Figure E6.1 by 7/8″ diameter. ASTM A325 bolts as shown in Figure E6.2. Determine the net area of the member if it comprises two laced C 12 × 30 channels.

For a C 12 × 30 channel:

$A = 8.82\ in.^2$
$t_w = 0.375\,in.$
$b_f = 3.17\,in.$
$t_f = 0.50\,in.$
Path A–B–D–E: $A_n = 8.82 - 2(1)(0.375) = 8.07\ in.^2$
Path A–B–C–D–E: $A_n = 8.82 - \{3(1) - 2[(3^2/(4(2.5)))]\}(0.375) = 8.37\ in.^2$
Path A–B–C–E1: $A_n = 8.82 - \{2(1) - 1[(3^2/(4(2.5)))]\}(0.375) = 8.41\ in.^2$
Path A–B–C–D1–E2: $A_n = 8.82 - \{3(1) - 2[(3^2/(4(2.5)))]\}(0.375) = 8.37\ in.^2$
Path A1–C–E1: $A_n = 8.82 - 1(1)(0.375) = 8.45\ in.^2$
Path A1–C–D1–E2: $A_n = 8.82 - \{2(1) - 1[(3^2/(4(2.5)))]\}(0.375) = 8.41\ in.^2$

Therefore, for the member U1–L1: $A_n = 2[8.07] = 16.14\ in.^2$ (91.5% of A_g).

6.2.1.2 Effective Net Area, A_e, of Tension Members

Shear lag occurs at connections when the tension load is not transmitted by all of the member elements in the connection. Therefore, at tension member connections with elements in different planes (e.g., splices, flanges of channels with the web only connected, angles with only one leg

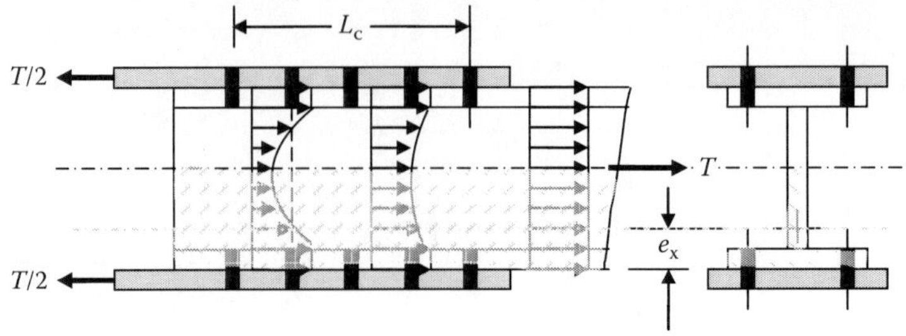

FIGURE 6.1 Shear lag at tension connection.

connected, webs of I-sections with only flanges connected), an effective area is determined to reflect that the tensile force is not uniformly distributed across the net area at the connection (Figure 6.1). Shear lag effects are related to the length of the connection and the efficacy of the tension member with respect to the transfer of forces on the shear plane between the member and the connection plate (Munse and Chesson, 1963).

The connection efficiency, U_c, is described by the ratio of the eccentricity, e_x, (the distance between the center of gravity of connected member elements and the shear plane) and connection length, L_c, as

$$U_c = \left(1 - \frac{e_x}{L_c}\right). \tag{6.6}$$

Therefore, where a joint is arranged such that not all of the member elements in the connection are fastened with bolts or with a combination of longitudinal and transverse welds, the effective net area, A_e, is

$$A_e = U_c A_n, \tag{6.7}$$

where

A_n = net area (see Section 6.2.1.1) (gross area for welded connections)

U_c = connection efficiency or shear lag reduction factor ≤0.90.

AREMA (2015) recommends shear lag reduction factors, U_c, between 0.75 and 1.00, depending on weld length for connections between members and plates that use only longitudinal welds.

Angle members connected by only one leg are particularly susceptible to shear lag effects. AREMA (2015) recommends shear lag reduction factors, U_c, between 0.60 and 0.80, depending on the number of bolts per fastener line in the connection.*

For members that are continuous through a joint (such as truss chord members continuous across several panels) or connections where load transfer between the chord segments is efficient, AREMA (2015) indicates shear lag may not be of concern and U_c may be effectively taken as 1.00. However, in these circumstances, engineering judgment may indicate that consideration of $A_e = 0.90A_n$ is appropriate for design (Bowles, 1980).

The calculation of effective net area is shown in Example 6.2.

* The larger value of 0.80 is used for angles when there are four or more bolts per fastener line in the connection (i.e., a relatively long connection).

Example 6.2a (SI Units)

Member U1–L1 is 8.5 m long and connected to the bottom chord of the truss in Figure E6.1 with M22 (22 mm diameter) ASTM F3125 Grade A325M bolts in gusset plates as shown in Figure E6.3a and b. Determine the effective net area for strength design of the member if it comprises a W310 × 118 rolled section.

For a W310 × 118 section:

$A = 15{,}000\ \text{mm}^2$
$t_w = 12\ \text{mm}$
$b_f = 307\ \text{mm}$
$t_f = 19\ \text{mm}$
$h_w = 242\ \text{mm}$
$d = 314\ \text{mm}.$

With elastic properties:

$I_x = 275 \times 106\ \text{mm}^4$
$S_x = 1750 \times 103\ \text{mm}^3$
$r_x = 135\ \text{mm}$
$I_y = 90.2 \times 106\ \text{mm}^4$
$S_y = 588 \times 103\ \text{mm}^3$
$r_y = 77.5\ \text{mm} = r_{min.}$

Net area:

Path A–B–C: $A_n = 15{,}000 - 4(25)(19) = 13{,}100\ \text{mm}^2$
Path A–B–D–E: $A_n = 15{,}000 - 2\{2(25) - [(100^2/(4(230)))]\}(19) = 13{,}513\ \text{mm}^2$

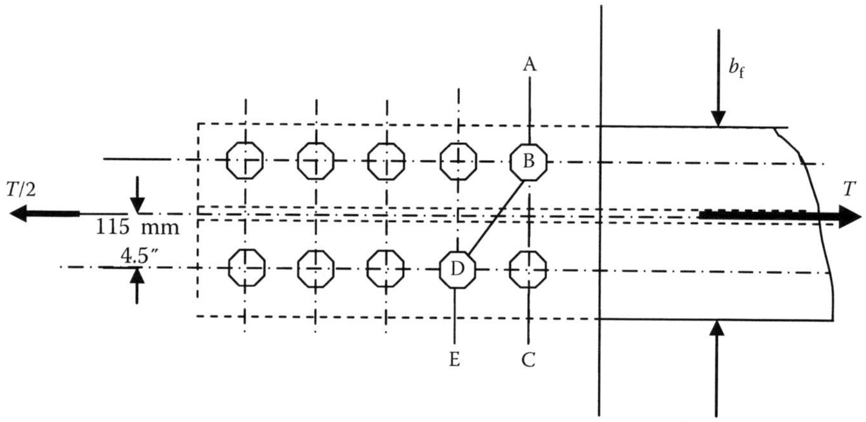

FIGURE E6.3a Member U1-L1 connection at bottom chord.

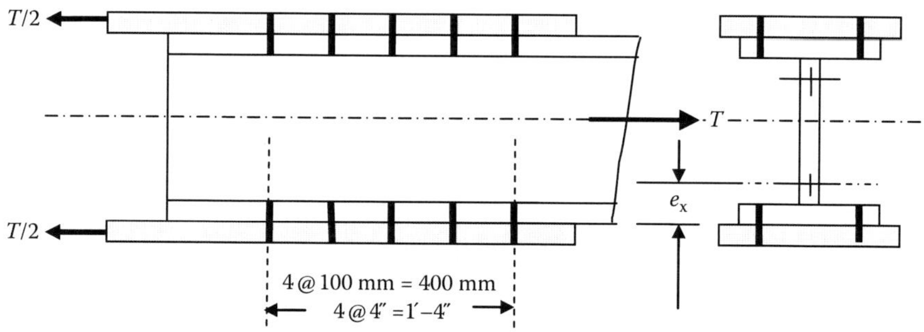

FIGURE E6.3b Member U1-L1 connection at bottom chord.

Effective net area:

$e_{x\sim}$ {(19)(307)(19/2) + (12)((314/2) − 19)[((314/2) − 19)/2 + 19]}/{(19)(307) + (12)((314/2) − 19)}
= 27 mm
L_c = 400 mm
U_c = 1 − (27/407) = 0.93, use minimum 0.90
A_e = 0.90(13,100) = 11,790 mm^2
L/r_{min} = (8500)/77.5 = 110.

Example 6.2b (US Customary and Imperial Units)

Member U1–L1 is 27.25 ft long and connected to the bottom chord of the truss in Figure E6.1 with 7/8″ diameter ASTM A325 bolts in gusset plates as shown in Figure E6.3a and b. Determine the effective net area for strength design of the member if it comprises a W12 × 79 rolled section.

For a W12 × 79 section:

A = 23.2 in.^{2}
t_w = 0.47 in.
b_f = 12.08 in.
t_f = 0.735 in.
h_w = 10.61 in.
d = 12.38″.

With elastic properties:

I_x = 662 in.^{4}
S_x = 107 in.^{3}
r_x = 5.34 in.
I_y = 216 in.^{4}
S_y = 35.8 in.^{3}
r_y = 3.05 in. = r_{min}.

Net area:

Path A–B–C: A_n = 23.2 − 4(1)(0.735) = 20.26 in.^{2}
Path A–B–D–E: A_n = 23.2 − 2{2(1) − [4^2/4(9)]}(0.735) = 20.91 in.^{2}

Effective net area:

$e_{x\sim}$ {(0.735)(12.08)(0.735/2) + (0.47)(12.38/2 − 0.735)[(12.38/2 − 0.735)/2 + 0.735]}/{(0.735)(12.08) + (0.47)(12.38/2 − 0.735)} = 1.06″
L_c = 16″
U_c = 1 − (1.06/16) = 0.93, use minimum 0.90
A_e = 0.90(20.26) = 18.23 in.^{2}
L/r_{min} = (27.25)(12)/3.05 = 107

6.2.2 Fatigue Strength of Axial Tension Members

The fatigue strength of an axial tension member is

$$T_{fat} = S_{rfat} A_{efat}, \tag{6.8}$$

where

A_{efat} = the effective gross or net area of only the member elements that are directly connected (e.g., the flange elements in Figure 6.1). This reduction accounts for shear lag effects for fatigue design, which occur at stress levels below fracture.* For slip-critical (or friction-type) bolted connections,†

* Shear lag for strength design is evaluated at stress levels near fracture.

† Slip-critical connections are recommended for main member connections and all connections subject to stress reversal and/or cyclical loading (Chapter 9).

the effective net area for fatigue design, $A_{efat,}$ is taken as the gross area of only the member elements that are directly connected.

S_{rfat} = the allowable fatigue stress range depending on number of design cycles, and connection and fabrication details of the tension member. For design, the number of cycles is generally assumed to be >2,000,000 for single track bridges with relatively short influence lines (for spans ≤30.5 m (100 ft) as recommended by AREMA (2015)). Since fatigue design is based on nominal stresses, S_{rfat} is recommended for various fatigue detail categories (A, B, B′, C, C′, D, E, E′, and F) depending on connection or detail geometry (see Chapter 5). Table 6.1 indicates the allowable fatigue stress ranges used for the design of tension members at connections for >2,000,000 stress cycles.*

The allowable fatigue stress ranges for detail categories consider stress concentrations related to member discontinuities (such as change in section) (Figure 6.2a) or apertures in the member (such as bearing connection holes, access or drainage openings) (Figure 6.2b). The magnitude of the stress

TABLE 6.1
Allowable Fatigue Stress Range for Number of Design Stress Range Cycles >2,000,000

Member or Connection Condition	S_{rfat} (MPa)	S_{rfat} (ksi)
Plain member	165	24
Bolted slip-resistant connection	110	16
Partial penetration groove and fillet welded connection	18–69	2.6–10
Full penetration weld connection	18–110	2.6–16

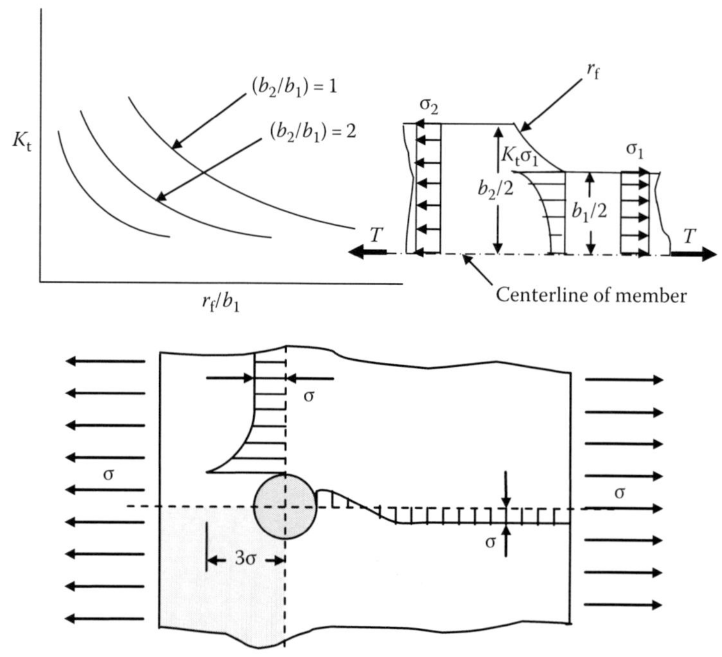

FIGURE 6.2 (a) Stress concentration factors for a flat bar with transition fillets in axial tension. (b) Stress concentration factors for a flat bar with round holes in axial tension.

* For welded connections, the allowable fatigue stress range depends on type of stress, direction of stress, direction of weld, weld continuity, and transition details. AREMA (2015) makes recommendations considering these factors.

TABLE 6.2
Stress Concentration Factors for AREMA, 2015 Fatigue Detail Categories

Fatigue Detail Category	Stress Concentration Factor, K_t
A	1.00
B	1.15
C	2.35
C′	2.75
D	3.6
E	4.8

concentrations may be determined by elasticity theory and fracture mechanics methods or through use of published stress concentration factors,* K_t, (Pilkey, 1997; Anderson, 2005; Armenakas, 2006).

Member transition fillets of usual dimensions that might be considered for steel bridge tension members may result in stress concentration factors in the order of, $K_t = 1.5$–2.0. Figure 6.2b illustrates that, for the simple case of uniaxial tension in a flat plate with a single circular hole, stress concentration factors may reach, $K_t = 3.0$. Table 6.2 indicates AREMA (2015) fatigue design detail categories and corresponding stress concentration factors (Sweeney, 2006).

However, since the allowable fatigue stress ranges recommended by AREMA (2015) are based on nominal stress test results, it is usually not necessary to explicitly consider stress concentration factors in the design of axial tension members unless the design details are particularly severe. Therefore, the designer must carefully consider the use of transitions, apertures or other discontinuities in the detailing of axial tension members subjected to fluctuating stresses and ensure that an allowable fatigue stress range based on the appropriate detail category is considered.

6.2.3 Serviceability of Axial Tension Members

In order to preclude excessive deflection (e.g., sag of long members under self-weight) or vibration (e.g., from wind loads on bracing members), the flexibility of axial tension members must be limited. The maximum slenderness between points of support, L_u, is recommended by AREMA (2015) as

$$\frac{L_u}{r_{min}} \leq 200, \tag{6.9}$$

where

$$r_{min} = \sqrt{\frac{I_{min}}{A_g}} \text{ and}$$

I_{min} = the minimum moment of inertia of the tension member in bending about an axis between supports, L_u.

* In some cases obtained by photoelastic testing.

Example 6.3a (SI Units)

Determine the design criteria for Load Case D1 (see Table 4.10) for member U1–L1 in Figure E6.1 for Cooper's EM360 load. Use Grade 350 ($F_y = 350$ MPa) steel with ultimate stress, F_U, of 490 MPa. Assume tangent track over the bridge (CF = 0) and a connection as shown in Example 6.2a.

The forces in member U1–L1 are as follows:

Dead load force = DL = +48 kN
Maximum live load force = LL1 = +579 kN (see Chapter 5)
Minimum live load force = LL2 = 0
Maximum live load impact = 39.28% (L=6 m see Chapter 4)
Mean live load impact = 0.40(39.28)% = 15.71% (see Chapter 4)
Range of live load force = LL_{range} = LL1 + LL2 = +579 kN.

Load combinations for Load Case D1:

$P_{rangeLL+I} = 1.157(579) = 670$ kN
$P_{max} = +48 + 1.393(579) = +855$ kN.

Strength considerations:

$$A_g \geq \frac{855(1000)}{0.55(350)} \geq 4440 \text{ mm}^2 \left(\text{tensile yielding}\right)$$

$$A_n \geq \frac{855(1000)}{0.47(490)} \geq 3713 \text{ mm}^2 \left(\text{fracture}\right).$$

Due to strength-related shear lag effects at the connection,

$$A_e \geq \frac{3713}{U_c} \geq \frac{3713}{0.90} \geq 4126 \text{ mm}^2 \left(\text{fracture}\right).$$

Fatigue considerations (including fatigue shear lag):

$A_{efat} \geq \dfrac{670(1000)}{110} \geq 6091 \text{ mm}^2$ for the gross area of connected elements of the member (Table 6.1)

(AREMA (2015) recommends slip-resistant connections for main members such as U1–L1)

$A_e \geq \dfrac{670(1000)}{165} \geq 4061 \text{ mm}^2$ for the net area net of the member away from connection (Table 6.1).

Stiffness considerations:

$$\frac{L}{r_{min}} \leq 200$$

$$r_{min} \geq \frac{(8500)}{200} \geq 42.5 \text{ mm}.$$

Select a member with following section properties:

Minimum gross area of the member = $A_g \geq 4400$ mm^2 (tensile yielding)
Minimum gross area for the portions of the member connected = $A_g' \geq 6091$ mm^2 (fatigue)
Minimum effective net area of the member at the connection = $A_e \geq 4126$ mm^2 (fracture)

Minimum net area of the member away from the connection = $A_n \geq 4061$ in.^{2} (fatigue)
Minimum radius of gyration of member = $r_{min} \geq 42.5$ mm.

Example 6.3b (US Customary and Imperial Units)

Determine the design criteria for Load Case D1 (see Table 4.10) for member U1–L1 in Figure E6.1 for Cooper's E80 load. Use Grade 50 ($F_y = 50$ ksi) steel with ultimate stress, F_U, of 70 ksi. Assume tangent track over the bridge (CF = 0) and a connection as shown in Example 6.2b.

The forces in member U1–L1 are as follows:

Dead load force = DL = +10.82 kips
Maximum live load force = LL1 = +128.8 kips (see Chapter 5)
Minimum live load force = LL2 = 0
Maximum live load impact = 39.28% (L = 19.55′ see Chapter 4)
Mean live load impact = 0.40(39.28)% = 15.71% (see Chapter 4)
Range of live load force = LL_{range} = LL1 + LL2 = +128.80 + (0.00) = 128.80 kips.

Load combinations for Load Case D1:

$P_{rangeLL+I} = 1.157(128.8) = 149.0$ kips
$P_{max} = +10.82 + 1.393(128.80) = +190.2$ kips.

Strength considerations:

$$A_g \geq \frac{190.2}{0.55(50)} \geq 6.92 \text{ in.}^2 (\text{tensile yielding})$$

$$A_n \geq \frac{190.2}{0.47(70)} \geq 5.78 \text{ in.}^2 (\text{fracture}).$$

Due to strength-related shear lag effects at the connection,

$$A_e \geq \frac{5.78}{U_c} \geq 6.42 \text{ in.}^2 (\text{fracture}).$$

Fatigue considerations (including fatigue shear lag):

$A_{efat} \geq \dfrac{149.0}{16} \geq 9.31$ in.^{2} for the gross area of connected elements of the member (Table 6.1)

(AREMA, 2015 recommends slip-resistant connections for main members such as U1–L1)

$A_e \geq \dfrac{149.0}{24} \geq 6.21$ in.^{2} for the net area net of the member away from connection (Table 6.1).

Stiffness considerations:

$$\frac{L}{r_{min}} \leq 200$$

$$r_{min} \geq \frac{(27.25)(12)}{200} \geq 1.64 \text{ in.}$$

Select a member with following section properties:

Minimum gross area of the member = $A_g \geq 6.92$ in.^{2} (tensile yielding)
Minimum gross area for the portions of the member connected = $A_g' \geq 9.31$ in.^{2} (fatigue)
Minimum effective net area of the member at the connection = $A_e \geq 6.42$ in.^{2} (fracture)

Minimum net area of the member away from the connection = $A_n \geq 6.21$ in.^{2} (fatigue)
Minimum radius of gyration of member = $r_{min} \geq 1.64$ in.

Example 6.4a (SI Units)

Determine the design criteria for Load Case D1 (see Table 4.10) for member U1–L2 in the 8-panel 48 m long Pratt truss shown in Figure E6.1 for Cooper's EM360 load. Use Grade 350 ($F_y = 350$ MPa) steel with ultimate stress, F_U, of 490 MPa.

Forces in member U1–L2 are as follows:

Dead load force = DL = +243 kN
Maximum live load force = LL1 = +1495 kN
Minimum live load force = LL2 = −33.3 kN
Maximum live load impact = 20.75% ($L = 48$ m see Chapter 4)
Mean live load impact = 0.65(20.758)% = 13.49% (see Chapter 4)

Range of live load force = $LL_{range} = +1495 - (-33.3) = 1528$ kN (AREMA (2015) recommends that all stress ranges be considered tensile stress ranges, due to the potential for pre-existing mean tensile stresses—see Chapters 5, 9, and 10).

Load combinations for Load Case D1:

$P_{rangeLL+I} = 1.135(1528) = 1735$ kN
$P_{max} = +243 + 1.208(+1495) = 2049$ kN

Strength considerations:

$$A_g \geq \frac{2049(1000)}{0.55(350)} \geq 10,644 \text{ mm}^2 (\text{tensile yielding})$$

$$A_e \geq \frac{2049(1000)}{0.47(490)} \geq 8897 \text{ mm}^2 (\text{fracture}).$$

Due to strength-related shear lag effects at the connection (assuming that $U_c = 0.90$),

$$A_e \geq \frac{8897}{U_c} = 9886 \text{ mm}^2 (\text{fracture}).$$

Fatigue considerations:

$$A_{efat} \geq \frac{1735(1000)}{110} \geq 15,773 \text{ mm}^2 \text{ for the gross area of connected elements of the member}$$

$$A_e \geq \frac{1735(1000)}{165} \geq 10,515 \text{ mm}^2 \text{ for the net area net of the member away from connection.}$$

Stiffness considerations:

$$\frac{L}{r_{min}} \leq 200$$

$$r_{min} \geq \frac{\left(\sqrt{(8.5)^2 + (8.0)^2}\right)(1000)}{200} \geq 58.4 \text{ mm}.$$

Select a member with following section properties:

Minimum gross area of the member = $A_g \geq 10,644$ mm^2 (tensile yielding)
Minimum gross area for the portions of the member connected = $A_g' \geq 15,773$ mm^2 (fatigue)

Minimum net effective area of the member at the connection = $A_e \geq 9886$ mm^2 (fracture)
Minimum net area of the member away from the connection = $A_e \geq 10,515$ mm^2 (fatigue)
Minimum radius of gyration = $r_{min} \geq 58.4$ mm.

Example 6.4b (US Customary and Imperial Units)

Determine the design criteria for Load Case D1 (see Table 4.10) for member U1–L2 in the 8-panel 156.40 ft long Pratt truss shown in Figure E6.1 for Cooper's E80 load. Use Grade 50 (F_y = 50 ksi) steel with ultimate stress, F_U, of 70 ksi.

Forces in member U1–L2 are as follows:

Dead load force = DL = +54.56 kips
Maximum live load force = LL1 = +332.4 kips
Minimum live load force = LL2 = –7.4 kips
Maximum live load impact = 20.75% (L = 156.4′ see Chapter 4)
Mean live load impact = 0.65(20.758)% = 13.49% (see Chapter 4)

Range of live load force = LL_{range} = +332.4 – (–7.4) = 339.8 kips (AREMA (2015) recommends that all stress ranges be considered tensile stress ranges, due to the potential for pre-existing mean tensile stresses—see Chapters 5 and 9).

Load combinations for Load Case D1:

$P_{rangeLL+I} = 1.135(339.8) = 385.7$ kips
$P_{max} = +54.56 + 1.208(+332.4) = 456.1$ kips

Strength considerations:

$$A_g \geq \frac{456.1}{0.55(50)} \geq 16.6 \text{ in.}^2 (\text{tensile yielding})$$

$$A_e \geq \frac{456.1}{0.47(70)} \geq 13.9 \text{ in.}^2 (\text{fracture}).$$

Due to strength-related shear lag effects at the connection (assuming that $U_c = 0.90$),

$$A_e \geq \frac{13.9}{U_c} = 15.4 \text{ in.}^2 (\text{fracture}).$$

Fatigue considerations:

$$A_{efat} \geq \frac{385.7}{16} \geq 24.1 \text{ in.}^2 \text{ for the gross area of connected elements of the member}$$

$$A_e \geq \frac{385.7}{24} \geq 16.1 \text{ in.}^2 \text{ for the net area net of the member away from connection.}$$

Stiffness considerations:

$$\frac{L}{r_{min}} \leq 200$$

$$r_{min} \geq \frac{\left(\sqrt{(27.25)^2 + (19.55)^2}\right)(12)}{200} \geq 2.01 \text{ in.}$$

Select a member with following section properties:

Minimum gross area of the member = $A_g \geq 16.6$ in.^{2} (tensile yielding)

Minimum gross area for the portions of the member connected = $A_g' \geq 24.1$ in.^{2} (fatigue)
Minimum net effective area of the member at the connection = $A_e \geq 15.4$ in.^{2} (fracture)
Minimum net area of the member away from the connection = $A_e \geq 16.1$ in.^{2} (fatigue)
Minimum radius of gyration = $r_{min} \geq 2.01$ in.

6.2.4 Design of Axial Tension Members for Steel Railway Bridges

Tension members in steel railway bridges may comprise eyebars, cables, structural shapes, and built-up sections. Eyebars are not often used in modern bridge superstructure fabrication, and suspension or cable-stayed bridges are unusual for freight railway structures due to flexibility concerns (see Chapter 1). Structural shapes such as W, WT, C, and angles are frequently used for steel railway bridge tension members.

It is often necessary to fabricate railway superstructure tension members of several structural shapes due to the large magnitude live loads and tension members that undergo stress reversals. Bending effects (see Chapter 8) and connection geometry may also dictate the use of built-up tension members. The components must be adequately fastened together to ensure integral behavior of the tension member. In cases where a box-type member is undesirable, such as where the ingress of water is difficult to preclude, open tension members are used. Built-up open tension members are often fabricated with lacing bars and stay (tie or batten) plates or perforated cover plates.* Shear deformation in tension members, which is primarily due to self-weight and wind loads, is relatively small and AREMA (2015) recognizes this by providing nominal recommendations for lacing bars and stay plates in axial tension members.

Lacing bar width should be a minimum of three times the fastener diameter to provide adequate edge distance and the thickness for single flat bar lacing bars should be at least 1/40 of the length† for main structural members and 1/50 of the length for bracing members. Stay plates should be used at the ends of built-up tension members and at intermediate locations where lacing bar continuity is interrupted for the connection of other members.‡ The length of the stay plates at the ends of laced bar built-up tension members must be at least 85% of the distance between connection lines across the member. The length of the stay plates at intermediate locations of laced bar built-up tension members must be at least 50% of the distance between connection lines across the member.

The thickness of perforated cover plates should be at least 1/50 of the length between closet adjacent fastener lines. Perforated cover plate thickness is based on transverse shear, V, at the centerline of the cover plate. The maximum transverse shear stress, τ_V, at the center of the cover plate is

$$\tau_V = \frac{3V}{2bt_{pc}}, \tag{6.10}$$

where

b = the width of the perforated cover plate
t_{pc} = the thickness of the perforated cover plate.

Therefore, the longitudinal shear force, V', over the distance between the centers of perforations or apertures, l_p, is

$$V' = \tau_V(l_p t_{pc}) = \frac{3Vl_p}{2b} \tag{6.11}$$

* Perforated cover plates are most commonly used for modern built-up truss members.
† Lacing bar length is the distance between fastener centers.
‡ For example, stay plates are used each side of members that interrupt lacing bars.

and the shear stress over the net area of the plate between the centers of perforations, τ_{pc}, is

$$\tau_{pc} = \frac{V'}{(l_p - c)t_{pc}} = \frac{3Vl_p}{2bt_{pc}(l_p - c)}, \tag{6.12}$$

where

c = the length of the perforation.

Rearrangement of Equation 6.12 yields

$$t_{pc} = \frac{3Vl_p}{2\tau_{all}b(l_p - c)}, \tag{6.13}$$

where

τ_{all} = allowable shear stress (0.35 F_y recommended by AREMA (2015)).

The transverse shear force, V, is generally small in tension members and perforated cover plate thickness is primarily dependent on the requirements for axial tension and recommended minimum material thickness (see Chapter 5). The net section through the perforation of the plate is included in the member net area.

Example 6.5

Use the AREMA (2015) recommendations to select bolted lacing bars and stay plates for tension member U1–L1 of Example 6.1. The member cross section comprises two C 310 × 45 (C 12 × 30) channels 305 mm (12″) apart back-to-back (Figure E6.4).

Lacing bars:

Use ~ 60° angle between lacing bar and longitudinal axis of the hanger.
Minimum width for single flat lacing bars = 90 mm (3.5 in.)
Minimum thickness of bar = (1/40)(380/sin 60°) = 11 mm [(1/40)(15/sin 60°) = 0.43 in.]
For 22 mm (7/8 in.) diameter bolts, the minimum bar width = 3(22) = 66 mm [3(7/8) = 2.63 in.].

Stay plates:

Use at ends and intermediate locations where lacing not present.
Minimum intermediate plates length = (3/4)(317) = 238 mm [(3/4)(12.5) = 9.38 in.]
Minimum thickness of stay plate = (1/50)(380) = 7.6 mm [(1/50)(15)=0.30 in.]
For 22 mm (7/8″) diameter bolts maximum fastener spacing = 4(22) = 88 mm [4(7/8) = 3.5 in.]
Minimum stay plate length = 250 mm [10 in.] (minimum of 3 bolts per side of stay plate)

The bridge designer should use these minimum requirements to select practical design dimensions of lacing bars and stay plates for the axial tension member.

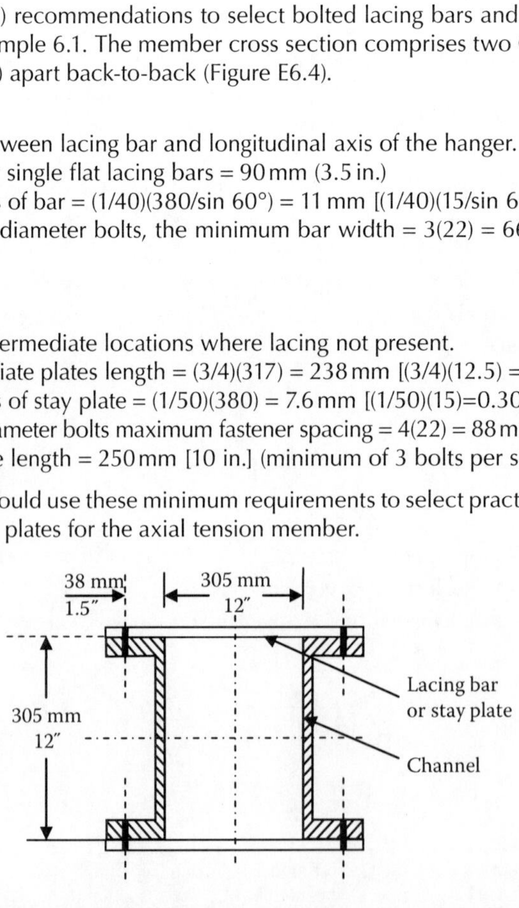

FIGURE E6.4 Member U1-L1 cross section.

6.3 AXIAL COMPRESSION MEMBERS

Axial compression main members in steel railway superstructures are often nonredundant. Therefore, the strength (yielding and stability) limit state requires careful consideration during design.

6.3.1 Strength of Axial Compression Members

The strength of a steel compression member is contingent upon its susceptibility to instability or buckling. For very short members, failure is governed by yield stress, F_y.* However, for members with greater slenderness, inelastic or elastic instability, depending on the degree of member slenderness, will control failure at a critical buckling force, P_{cr}.

6.3.1.1 Elastic Compression Members

Long and slender members will buckle at loads generating compressive stresses below the proportional limit, F_p. (Figure 6.3). The magnitude of this elastic critical buckling force depends on the stiffness, length, and end conditions of the compression member, as well as imperfections in loading and geometry.

6.3.1.1.1 Elastic Buckling with Load, P, Applied along Centroidal Axis of Member

Assuming that:

- The member has no geometric imperfections (perfectly straight)
- Plane sections remain plane after deformation
- Flexural deflection is small and is only deformation considered (shear deflection is neglected)
- Hooke's law is applied

The differential equation of the deflection curve is

$$\frac{d^2y(x)}{dx^2}+k^2y(x)=U, \tag{6.14}$$

where

$y(x)$ = the lateral deflection of the compression member

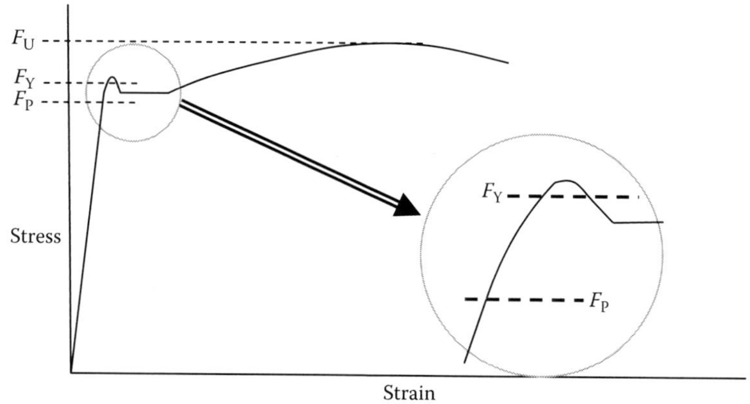

FIGURE 6.3 Typical stress–strain curve for structural steel.

* Compressive yield stress is almost equal to tensile yield stress (see Chapter 2).

U depends on the effects on load, P, of the compression member end conditions

$$k^2 = \frac{P}{EI}$$

P = the load applied at the end and along the centroidal axis of the compression member.

The solution of Equation 6.14 for the elastic critical buckling force, P_{cr}, is readily accomplished by consideration of the appropriate boundary conditions (Wang et al., 2005). The elastic critical buckling force for various member end conditions is shown in Table 6.3. Figures 6.4 and 6.5 illustrate the various compression member end conditions shown in Table 6.3.

The critical buckling force can be expressed as

$$P_{cr} = \frac{\pi^2 EI}{(KL)^2}. \tag{6.15}$$

Considering $I = A_g r^2$, the critical buckling stress may be obtained from Equation 6.15 as

$$F_{cr} = \frac{P_{cr}}{A_g} = \frac{\pi^2 E}{(KL/r)^2}. \tag{6.16}$$

Rearrangement of Equation 6.15 yields

$$K = \text{effective length factor} = \frac{\pi}{L}\sqrt{\frac{EI}{P_{cr}}}. \tag{6.17}$$

TABLE 6.3
Elastic Critical Buckling Force for Concentrically Loaded Members with Various End Conditions

End Condition	U	P_{cr}
Both ends pinned (Figure 6.4a)	0	$\frac{\pi^2 EI}{L^2}$
Both ends fixed (Figure 6.4b)	$-\left(\frac{V}{EI}\right)x + \frac{M}{EI}$	$4\frac{\pi^2 EI}{L^2}$
One end fixed and other end free (Figure 6.4c)	$k^2\Delta$	$\frac{\pi^2 EI}{4L^2}$
One end hinged and other end fixed (Figure 6.4d)	$\left(\frac{M}{EIL}\right)x$	$2.046\frac{\pi^2 EI}{L^2}$
One end guided and other end fixed (Figure 6.5a)	$\frac{P\Delta}{2EI}$	$\frac{\pi^2 EI}{L^2}$
One end hinged and other end guided (Figure 6.5b)	0	$\frac{\pi^2 EI}{4L^2}$

L, length of member between end supports; V = shear force in member; M, bending moment at end of member; Δ, lateral deflection at free or guided end of fmember with other end fixed

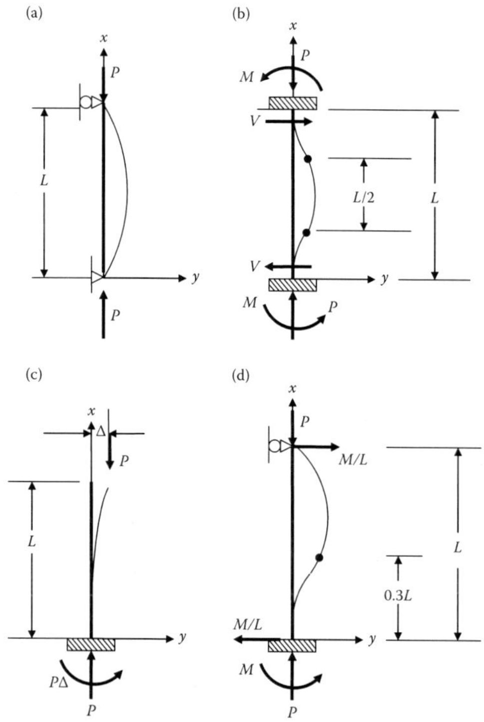

FIGURE 6.4 Compression member end conditions. (a) pinned-pinned, (b) fixed-fixed, (c) fixed-free, and (d) fixed-hinged.

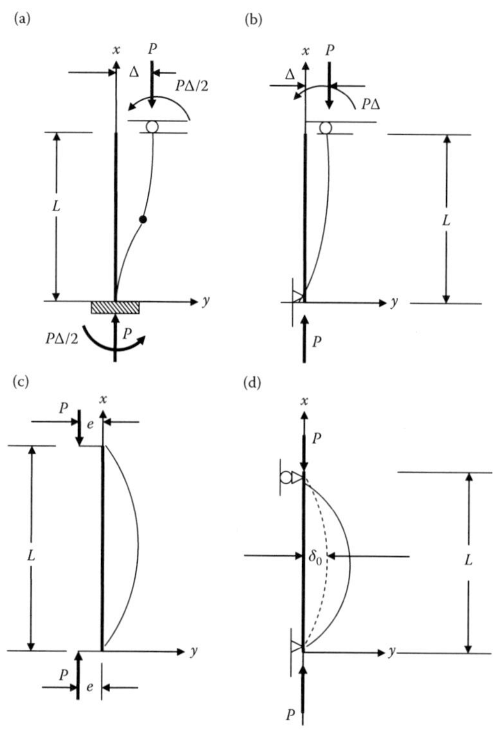

FIGURE 6.5 Compression member end and initial conditions. (a) fixed-guided, (b) hinged-guided, (c) eccentric load, and (d) geometric imperfection.

TABLE 6.4
Effective Length Factors for Various Compression Member End Conditions

End Condition	K
Both ends pinned (Figure 6.4a)	1.00
Both ends fixed (Figure 6.4b)	0.50
One end fixed and other end free (Figure 6.4c)	2.00
One end hinged and other end fixed (Figure 6.4d)	0.70
One end guided and other end fixed (Figure 6.5a)	1.00
One end hinged and other end guided (Figure 6.5b)	2.00

Values of K from Equation 6.17 for the various end conditions and corresponding critical buckling force, P_{cr}, in Table 6.3 are shown in Table 6.4.

However, due to the ideal conditions of the mathematical model, which may not be representative of actual end conditions, the effective length factors in Table 6.4 are generally reduced for the ASD of compression members. For railway trusses with moving live loads, the forces in members framing into the end of a member under consideration will be less than maximum when the member under consideration is subject to maximum force. Therefore, the ideally pinned end condition is not established because this force arrangement imposes rotational restraints at the end of the member under consideration. For members with equal rotational restraint at each end, an approximate effective length factor has been developed as (Newmark, 1949).

$$K = \frac{\bar{C}+2}{\bar{C}+4}, \tag{6.18}$$

where

$$\bar{C} = \frac{\pi^2 EI}{LR_k}$$

R_k = the equivalent rotational spring constant.

For members and end conditions typically used in steel railway trusses, Equation 6.18 provides $K = 0.75–0.90$. Furthermore, theoretical solutions for truss members indicate that, for constant cross section chord members, the effective length factor, K, can be estimated as

$$K = \sqrt{1 - \frac{5}{4n}}, \tag{6.19}$$

where

n = number of truss panels (typically, $K = 0.85–0.95$).

The same studies also indicated that, for web members typically used in steel railway trusses, the effective length factor, K, is generally between 0.70 and 0.90 (Bleich, 1952).

AREMA (2015) recommends two effective length factors, K, to represent actual steel railway bridge compression member end conditions. For true pin-end connections, $K = 0.875$ is recommended. For all other end conditions (with bolted or welded end connections), AREMA (2015) recommends $K = 0.75$ for design purposes.

A safety factor must be applied to Equations 6.15 and 6.16 to account for small load eccentricities and geometric imperfections. AREMA (2015) uses a factor of safety, FS, of 1.95 to arrive at the allowable compressive strength, C_{all}, of

$$C_{all} = \frac{P_{cr}}{1.95} = \frac{0.514\pi^2 EI}{(KL)^2} \tag{6.20}$$

or

$$F_{all} = \frac{C_{all}}{A_g} = \frac{F_{cr}}{1.95} = \frac{0.514\pi^2 E}{(KL/r)^2}. \tag{6.21}$$

Elastic buckling, described by Equation 6.21, will occur at values of $KL/r \geq C_c$. C_c is defined by the intersection of the Euler buckling curve (Figure 6.6) with a transition curve from compressive yielding, F_y, as shown by the vertical lines in Figure 6.6 for members with $K = 0.5$, 1.0, and 2.0. The transition curve represents the effects of eccentricities, initial imperfections, and residual stresses introduced during fabrication and erection of steel railway bridge compression members.* For elastic buckling (at large KL/r) the degree of member end restraint, expressed in terms of the effective length factor, K, greatly affects the allowable compressive stress, F_{all}, as shown within the shaded area of Figure 6.6.

If they exist, explicit consideration of relatively large load eccentricities and/or geometric imperfections must be made for long and slender compression members.

6.3.1.1.2 *Elastic Buckling with Load Applied Eccentric to the Centroidal Axis of Member*

Assuming that:

- The member has no geometric imperfections (perfectly straight)
- Plane sections remain plane after deformation
- Flexural deflection is small and is only deformation considered (shear deflection is neglected)
- Hooke's law is applied

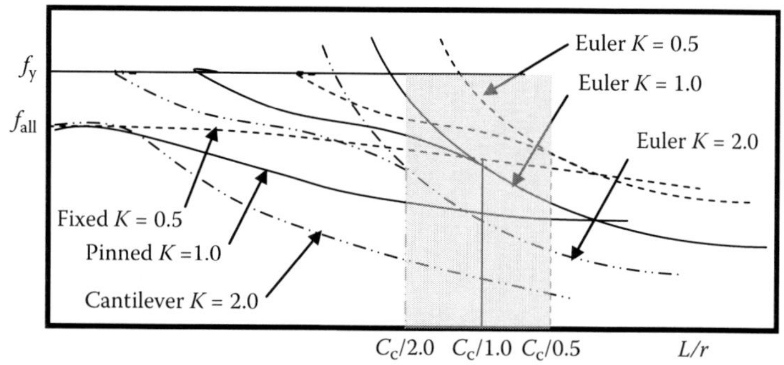

FIGURE 6.6 Effect of end restraint on allowable stresses and slenderness values at elastic (Euler) buckling.

* The transition curve describes the inelastic buckling of members with KL/r less than that for elastic buckling but greater than the maximum KL/r value for compressive yielding.

The solution of the differential equation of the deflection curve (Equation 6.14), where $U = -k^2 e$ and e = eccentricity of load (Figure 6.5c), is the secant formula (Chen and Lui, 1987)

$$P_{cr} = \frac{P_y}{1 + \dfrac{ec}{r^2} \sec \dfrac{\pi}{2} \sqrt{\dfrac{P_{cr} L^2}{\pi^2 EI}}}, \tag{6.22}$$

where

$$P_y = A_g F_y$$

c = the distance from the neutral axis to the extreme fiber of the member cross section
r = the radius of gyration of the member cross section.

The secant formula was considered appropriate for inelastic buckling of members from initial curvature and load eccentricity. However, it does not include consideration of residual stresses, which are of considerable importance in modern steel structures. Therefore, Equation 6.22 is no longer used to determine the critical buckling force of compression members. The equation was used in the AREMA recommended practice prior to 1969, but was discontinued as a basis for compression member design because of the difficulty associated with its use and indications that Euler type formulas are more appropriate for eccentrically loaded compression members (AREMA, 2015).

6.3.1.1.3 Elastic Buckling of Members with Geometric Imperfections (Initial Out-of-Straightness)

Assuming that:

- The member is concentrically loaded
- Plane sections remain plane after deformation
- Flexural deflection is small and is only deformation considered (shear deflection is neglected)
- Hooke's law is applied

The solution of the differential equation of the deflection curve (Equation 6.14), where $U = -k^2 \delta_0 \sin \dfrac{\pi x}{L}$ (Figure 6.5d) is the Perry-Robertson formula (Chen and Lui, 1987)

$$P_{cr} = \frac{P_y}{1 + \delta_0 c / r^2 \left(1/1 - \dfrac{P_{cr} L^2}{\pi^2 EI} \right)}, \tag{6.23}$$

where

δ_0 = out-of-straightness at middle of member (Figure 6.5d).

For low values of L/r (generally less than about 60) out-of-straightness geometric imperfections are usually not an important design consideration.

6.3.1.2 Inelastic Compression Members

Steel railway superstructure members of typical length and slenderness will buckle at loads above the proportional limit, F_p, (Figure 6.3) when some cross section fibers have already yielded before the initiation of instability. Therefore, the effective modulus of elasticity is less than the initial value.

This nonlinear behavior occurs primarily as a result of residual stresses* but may also be a result of initial curvature and force eccentricity.

These material and/or geometric imperfections, or nonlinearities, are considered by replacing the elastic modulus, E, with an effective modulus, E_{eff}. Therefore, inelastic critical buckling force solutions are analogous to those shown in Table 6.3 with elastic modulus, E, replaced with the effective modulus, E_{eff}, so that

$$P_{cr} = \frac{\pi^2 E_{eff} I}{(KL)^2}. \tag{6.24}$$

Engesser (see Chapter 1) proposed both the tangent modulus, E_t, (Equation 6.25) and the reduced modulus, E_r, (Equation 6.26) for the effective modulus. The tangent modulus is

$$E_t = \frac{d\sigma}{d\varepsilon} = E\left(\frac{F_y - \sigma}{F_y - c\sigma}\right). \tag{6.25}$$

The reduced modulus, for symmetric I-sections (and neglecting web area), is (Timoshenko and Gere, 1961)

$$E_r = \frac{2EE_t}{E - E_t}, \tag{6.26}$$

where

$d\sigma$ = the change in stress
$d\varepsilon$ = the change in strain
σ = the applied stress = P/A
c = 0.96–0.99 for structural steel.

The reduced modulus, E_r, is less than the tangent modulus, E_t, as shown in Figure 6.7.

An inelastic compression member theory was also proposed by Shanley (Tall, 1974). The theory indicates that actual inelastic compression member behavior lies between that of the tangent and reduced modulus load curves. However, because test results are closer to the tangent modulus curve values (Chen and Lui, 1987), the tangent modulus is often used in the development of modern inelastic compression curves and equations.

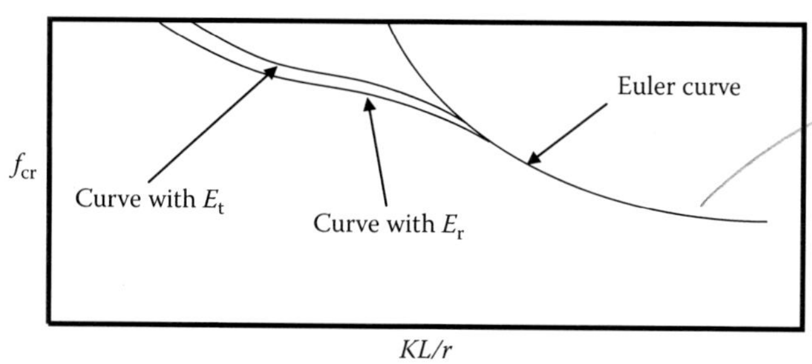

FIGURE 6.7 Typical compression member curves.

* Since residual stresses are most affected by size, the use of high-strength steel can make their effect relatively smaller. Annealing to reduce residual stresses (heat treatment) may increase the strength of a compression member.

The tangent modulus theory (Equation 6.25) includes material imperfection considerations, but it does not explicitly consider the effects of geometric imperfections (member out-of-straightness) and residual stresses in compression members.

Geometric imperfections (unintentional member out-of-straightness and eccentricity) have a detrimental effect on the inelastic critical buckling force of compression members of relatively large slenderness. The American Institute of Steel Construction (AISC, 1980) Allowable Stress Design (ASD) provisions recognized this by increasing the factor of safety, FS, to 115% of 5/3 for compression members with an effective slenderness ratio at the value for Euler elastic buckling, $C_c = KL/r$. This variable FS is

$$\mathrm{FS} = \frac{5}{3} + \frac{3}{8}\frac{(KL/r)}{C_c} - \frac{1}{8}\left(\frac{(KL/r)}{C_c}\right)^3 \quad \text{for } KL/r \le C_c \tag{6.27}$$

Geometric imperfections are also implicitly recognized in the AREMA (2015) recommendations through the use of a higher, although constant, FS for axial compression (FS = 1.95) than that used for axial tension (FS = 1.82). A similar cubic polynomial as Equation 6.27 was used in Chapter 5 to investigate a variable FS using the AREMA (2015) criteria. However, due to the potential for geometric imperfections to exacerbate instability for members loaded with relatively large magnitude live loads, the higher factor of safety is likely appropriate for even less slender compression members in railway superstructures.

The rolling of structural steel plates and shapes, and fabrication bending, cutting, and/or welding procedures may create residual stresses (see Chapter 10) that affect the inelastic critical buckling stress in a compression member. The pattern of compressive and tensile residual stresses is very dependent on member cross section and dimensions. The presence of varying residual stresses will affect the material compressive stress-strain curve (Figure 6.8) and establish a different effective modulus of elasticity in each direction across a compression member cross section. If the tangent modulus is taken as the effective modulus of elasticity, it will differ depending on the direction of buckling and will underestimate compression member strength. Therefore, buckling direction (weak or strong axis) must be considered independently to determine compression member strength when allowing for residual stresses.

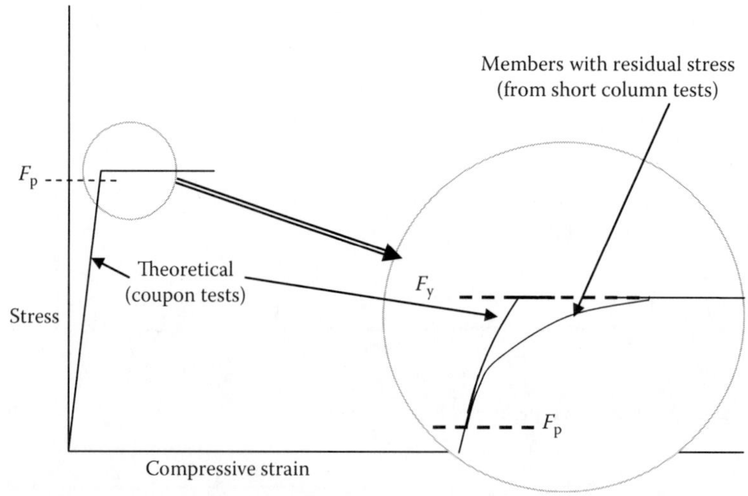

FIGURE 6.8 Typical compressive stress–strain curve for structural steel.

The Column Research Council (CRC) conducted tests and performed analytical studies of weak and strong axis inelastic buckling with linear and parabolic residual stress distributions across the compression member cross section. These studies revealed that, within the inelastic range, the compression member curves were parabolic (Figure 6.9).

The residual stresses used in the CRC studies were about $0.3F_y$. However, the value of $0.5F_y$ is used in Figure 6.9 in order to conservatively represent the residual stresses and provide a smooth transition to the Euler elastic buckling curve at $KL/r = C_c$ (elastic behavior and buckling below the proportional limit of $0.5F_y$). Therefore, for $KL/r < C_c$, the Johnson parabola (Equation 6.29) may be used to represent inelastic behavior (Tall, 1974). The Johnson parabola is

$$F_{cr} = F_y - B\left(\frac{KL}{r}\right)^2. \tag{6.28}$$

The value of the constant B with $F_p = 0.5F_y$ and $F_r = F_y - F_p$ (Figure 6.8) is (Bleich, 1952):

$$B = \frac{F_r}{F_y \pi^2 E}(F_y - F_r) = \frac{1}{4\pi^2 E} \tag{6.29}$$

and the inelastic critical buckling stress is

$$F_{cr} = F_y\left(1 - \frac{F_y}{4\pi^2 E}\left(\frac{KL}{r}\right)^2\right) \tag{6.30}$$

or

$$\frac{F_{cr}}{F_y} = 1 - 0.25\lambda_c^2, \tag{6.31}$$

where

$$\lambda_c = \left(\frac{KL}{r}\right)\sqrt{\frac{F_y}{\pi^2 E}}.$$

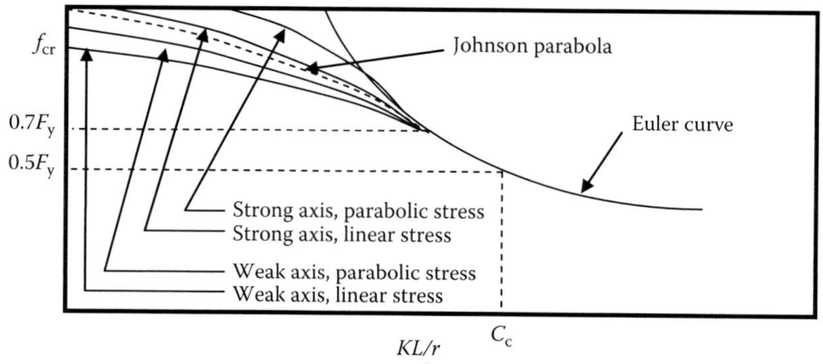

FIGURE 6.9 Weak and strong axis compression member curves using linear and parabolic residual stress distributions.

The uncertainties involved in the determination of K will result in overestimates for F_{cr}, particularly when KL/r is between 40 and 100 (AREMA, 2015). To mitigate this, AREMA (2015) adopts the conservative straight line approximation of

$$F_{cr} = F_y - B\left(\frac{KL}{r}\right). \tag{6.32}$$

The value of the constant, B, (the slope of the line) can be established through the development of compression member curves using variable and constant safety factors with Equation 6.30. In this manner, using an appropriate safety factor, AREMA (2015) recommends an allowable compression stress of

$$F_{all} = \frac{F_{cr}}{FS} = 0.60F_y - \left(635\frac{F_y}{E}\right)^{3/2}\left(\frac{KL}{r}\right), \text{ in SI units} \tag{6.33a}$$

or

$$F_{all} = \frac{F_{cr}}{FS} = 0.60F_y - \left(17500\frac{F_y}{E}\right)^{3/2}\left(\frac{KL}{r}\right), \text{ in US Customary or Imperial units.} \tag{6.33b}$$

This curve is made to intersect the Euler elastic curve at $0.20F_y$ (Figures 6.10 and 6.11*) in order to conservatively represent the effects of eccentricities, initial imperfections, and residual stresses introduced during fabrication and erection of steel railway superstructure compression members.

Inserting $F_{cr} = 0.20F_y$ into Equations 6.33a and b yields

$$\frac{KL}{r} = 5.034\sqrt{\frac{E}{F_y}}. \tag{6.34}$$

Equation 6.34 is the limiting slenderness ratio or critical buckling coefficient, C_c, to preclude elastic buckling of compression members with the AREMA (2015) straight line approximation. Values of C_c for various grades of steel are given in Table 6.5.

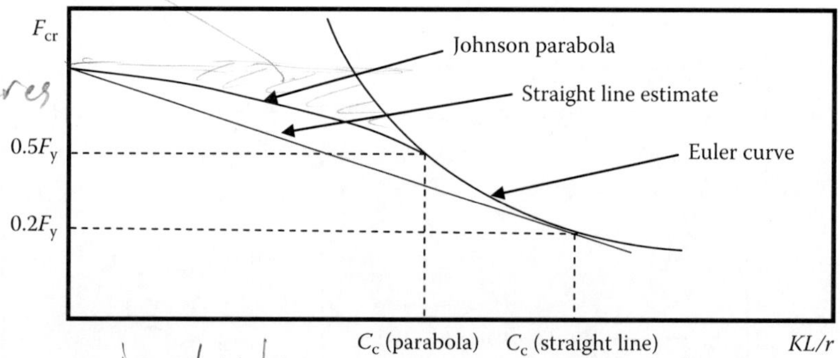

FIGURE 6.10 Parabolic and linear compression member curves.

* In Figure 6.11 the curves intersect at $0.2F_y$ = 70 MPa (10 ksi).

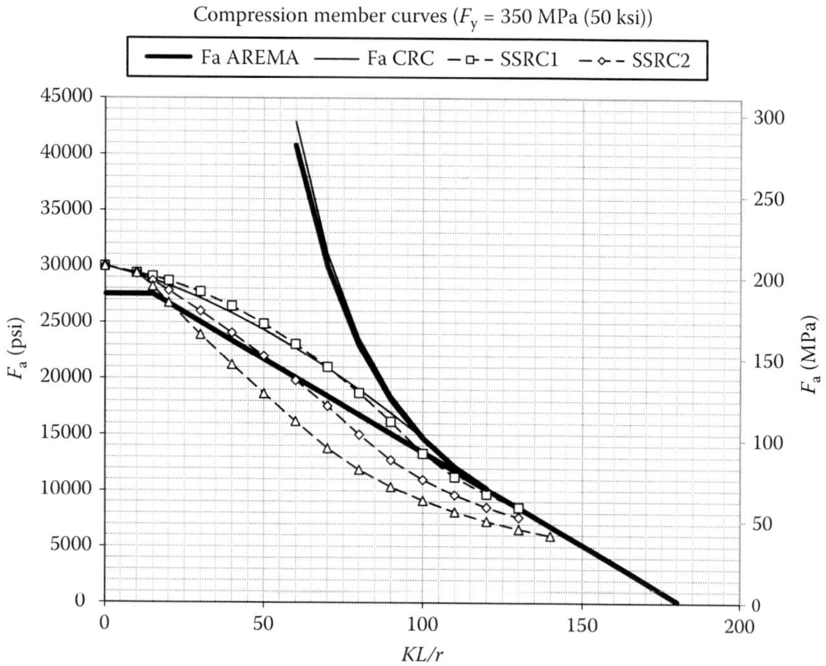

FIGURE 6.11 CRC parabolic and AREMA linear compression member strength curves (F_y = 350 MPa (50 ksi)).

TABLE 6.5
Critical Buckling Coefficients

Steel Yield Stress (F_y) (MPa)	Steel Yield Stress (F_y) (ksi)	Critical Buckling Coefficient (C_c)
260	36	143
300	44	129
350	50	121

The allowable compressive force, C_{all}, is

$$C_{all} = F_{all} A_g. \tag{6.35}$$

The CRC curve (Equation 6.30) with the variable factor of safety used by AISC ASD (Equation 6.27) and the AREMA (2015) curve (Equation 6.33) are shown in Figure 6.11 for steel with F_y = 350 MPa (50 ksi).

6.3.1.3 Yielding of Compression Members

When $F_{all} = 0.55F_y$, the slenderness ratio, KL/r, from Equation 6.33 is

$$\frac{KL}{r} \le 0.629\sqrt{\frac{E}{F_y}}. \tag{6.36}$$

Compression members are very short and without the potential for instability for slenderness below that given by Equation 6.36. Values of KL/r from Equation 6.36 for various grades of steel are given in Table 6.6.

TABLE 6.6
Short Compression Member Buckling Coefficient

Steel Yield Stress (F_y) (MPa)	Steel Yield Stress (F_y) (ksi)	KL/r from Equation 6.36
260	36	18
300	44	16
350	50	15

Therefore, the allowable compressive force, C_{all}, based on yielding is

$$C_{all} = 0.55 F_y A_g. \tag{6.37}$$

6.3.1.4 Compression Member Design for Steel Railway Superstructures

Equations 6.21, 6.33, and 6.37 are the AREMA (2015) recommendations for allowable compressive stress considering elastic stability, inelastic stability, and compressive yielding, respectively. Equation 6.34 provides the value of KL/r that delineates elastic and inelastic stability and Equation 6.36 provides the value of KL/r that delineates inelastic and yielding behavior. The AREMA (2015) strength criteria for axial compression members with steel yield strength of 350 MPa (50 ksi) is shown in Figure 6.11 with some other compression member design criteria.

6.3.2 Serviceability of Axial Compression Members

Limiting the compression member slenderness ratio based on effective length, KL, to that of Equation 6.34 precludes the possibility of elastic buckling in order to avoid sudden stability failures in steel railway superstructures. However, slenderness ratio, L/r, must also be limited to values that will preclude excessive vibration or deflection, which is of particular concern for compression member stability (e.g., to avoid excessive secondary flexural compressive stress due to member curvature). AREMA (2015) recommends that

$$\frac{L}{r_{min}} \leq 100 \text{ for main compression members,} \tag{6.38}$$

$$\frac{L}{r_{min}} \leq 120 \text{ for wind and sway bracing compression members.} \tag{6.39}$$

Example 6.6a (SI Units)

Determine the design criteria for Load Case D1 (see Table 4.10) for member U3–U4 in the 8-panel 48 m long Pratt truss shown in Figure E6.1 for Cooper's EM360 load. Use Grade 350 ($F_y = 350$ MPa) steel with ultimate stress, F_U, of 490 MPa.

The forces in member U3–U4 are as follows:

Dead load force = DL = −441 kN
Maximum live load force = LL1 = 0 kN
Minimum live load force = LL2 = −2506 kN
Maximum live load impact = 20.75% ($L = 48$ m see Chapter 4)

Load combinations for Load Case D1:

$$P_{max} = -441 + 1.21(-2506) = -3473 \text{ kN}.$$

Stiffness considerations:

$$\frac{L}{r_{min}} \leq 100$$

$$r_{min} \geq \frac{(48.0/8)(1000)}{100} \geq 60 \text{ mm } \left(\text{main member of truss}\right)$$

$$\frac{KL}{r_{min}} \leq 75 \ \left(\text{bolted end connection}\right).$$

Strength considerations:

$$F_{cr(min)} = 0.60(350) - \left(\frac{635(350)}{200,000}\right)^{\frac{3}{2}}(75) = 210 - 88 = 122 \text{ MPa at } KL/r = 75 \ \left(\text{Figure 6.11}\right)$$

and

$$A_g = \frac{3473(1000)}{122} = 28,470 \text{ mm}^2 \left(\text{minimum gross area required}\right).$$

Example 6.6b (US Customary and Imperial Units)

Determine the design criteria for Load Case D1 (see Table 4.10) for member U3–U4 in the 8-panel 156.40 ft long Pratt truss shown in Figure E6.1 for Cooper's E80 load. Use Grade 50 (F_y = 50 ksi) steel with ultimate stress, F_U, of 70 ksi.

The forces in member U3–U4 are as follows:

Dead load force = DL = −98.10 kips
Maximum live load force = LL1 = 0 kN
Minimum live load force = LL2 = −557.2 kips
Maximum live load impact = 20.75% (L = 156.38′ see Chapter 4)

Load combinations for Load Case D1:

$$P_{max} = -98.1 + 1.21\left(-557.2\right) = -771.2 \text{ kips}.$$

Stiffness considerations:

$$\frac{L}{r_{min}} \leq 100$$

$$r_{min} \geq \frac{(19.55)(12)}{100} \geq 2.35 \text{ in. } \left(\text{main member of truss}\right)$$

$$\frac{KL}{r_{min}} \leq 75 \ \left(\text{bolted end connection}\right).$$

Strength considerations:

$$F_{cr(min)} = 0.60(50,000) - \left(\frac{17,500(50)}{29x10^3}\right)^{\frac{3}{2}}(75) = 30,000 - 12,430 = 17,570 \text{ psi at } KL/r = 75 \text{ (Figure 6.11)}$$

and

$$A_g = \frac{771,200}{17,570} = 43.9 \text{ in.}^2 \text{(minimum gross area required)}.$$

6.3.3 Axial Compression Members in Steel Railway Superstructures

It is often necessary to fabricate railway superstructure compression members of several structural shapes due to large magnitude live loads and the potential for instability. Bending effects and connection geometry may also dictate the use of built-up compression members. The components must be adequately fastened to ensure integral behavior of the compression member. In cases where a box-type member is undesirable, such as where the ingress of water is difficult to preclude, open compression members are used.* Built-up open compression members are often fabricated with lacing bars and stay (tie or batten) plates or perforated cover plates.

6.3.3.1 Buckling Strength of Built-Up Compression Members

Only bending deformations were considered in the development of Equation 6.24 for buckling strength, P_{cr}. The effect of shear forces was neglected. For solid section compression members, this is appropriate. However, for built-up compression members, the shear forces may create deformations of the open section that reduce the overall stiffness and, thereby, reduce the buckling strength. Therefore, the curvature of the compression member due to shear must be included in Equation 6.14 to determine the critical buckling load, $\overline{P}_{cr}$, for built-up compression members. The shear force is (Figure 6.12)

$$V(x) = P \sin \varphi(x) = P \frac{dy(x)}{dx} \tag{6.40}$$

with curvature, γ_v, given as

$$\gamma_v = \frac{\beta}{A_g G_{eff}} \frac{dV(x)}{dx} = \frac{\beta P}{A_g G_{eff}} \frac{d^2 y(x)}{dx^2}, \tag{6.41}$$

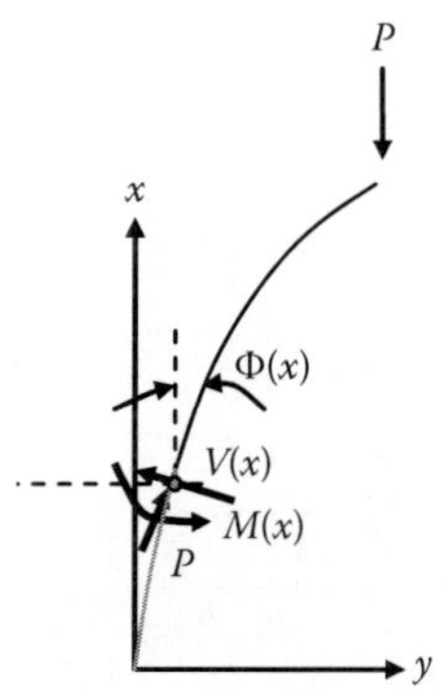

FIGURE 6.12 Bending and shear forces at compression member cross section.

* This is generally the case for railway compression members exposed to rain, ice, and snow.

where

β = a numerical factor to correct for nonuniform stress distribution across the cross section of the compression member

A_g = the gross cross-sectional area of the compression member

G_{eff} = the effective shear modulus = $E_{eff}/(2(1 + \upsilon))$

υ = Poisson's ratio = 0.3 for steel.

The substitution of Equation 6.41 into 6.14 with $E = E_{eff}$ yields the differential equation

$$\frac{d^2y(x)}{dx^2}+\frac{k^2y(x)}{\left(1+\frac{\beta P}{A_g G_{eff}}\right)}=\frac{U}{\left(1+\frac{\beta P}{A_g G_{eff}}\right)} \tag{6.42}$$

with solution analogous to Equation 6.15 of

$$\overline{P}_{cr}=\frac{\pi^2 E_{eff} I}{(\alpha KL)^2}=\frac{P_{cr}}{\alpha^2}, \tag{6.43}$$

where

P_{cr} = the critical buckling load for the compression member with gross cross-sectional area, A_g, moment of inertia, I, and length, L (Equation 6.15), and

$$\alpha=\sqrt{\left(1+\frac{\beta P_{cr}}{A_g G_{eff}}\right)}. \tag{6.44}$$

Equation 6.43 may be written as

$$\overline{P}_{cr}=\frac{P_{cr}}{\left(1+\frac{\beta P_{cr}}{A_g G_{eff}}\right)}. \tag{6.45}$$

Equation 6.45 illustrates that the critical buckling load, $\overline{P}_{cr}$, for built-up compression members can be determined based on the critical buckling load, P_{cr}, for solid (non–built-up) members of the same cross-sectional area, A_g.

Most modern steel railway superstructure compression members are assumed to be pin connected at each end (K = 0.75). Therefore, the critical buckling strength of built-up compression members of various configurations (using lacing and batten bars, and perforated cover plates) with pinned ends will be considered further.

Equation 6.44 may be written as

$$\alpha=\sqrt{\left(1+\frac{\beta P_{cr}}{A_g G_{eff}}\right)}=\sqrt{1+\Omega P_{cr}}, \tag{6.46}$$

where

$$\Omega=\frac{\beta}{A_g G_{eff}}=\frac{2\beta(1+\nu)}{A_g E_{eff}}=\frac{1}{P_\Omega}. \tag{6.47}$$

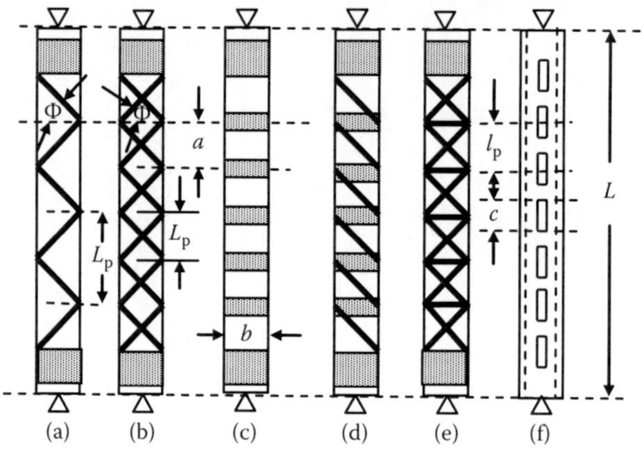

FIGURE 6.13 Various built-up compression members comprising lacing bars with and without batten plates; and perforated cover plates.

The value of Ω is determined through investigation of the deformations of the lacing bars, batten plates and/or perforated cover plates caused by lateral displacements from the shear force, *V*. The results of such investigations for various built-up compression members (Figure 6.13) are presented in the next sections.

6.3.3.1.1 *Critical Buckling Strength of Laced Bar Built-Up Compression Members without Shear Transfer Batten Plates (Pinned at Each End) (Figures 6.13a, b, and e)*

Shear is resisted by pin-connected truss behavior of lacing bars (for double lacing, consider tension resistance only).

$$\Omega = \frac{1}{A_{lb} E_{eff} \sin\phi \cos^2\phi}, \tag{6.48a}$$

$$\alpha = \sqrt{1+\Omega P_{cr}} = \sqrt{1+\left(\frac{13.2}{(L/r)^2}\right)\left(\frac{A_g}{A_{plb}}\right)\frac{1}{\sin\phi\cos^2\phi}}, \tag{6.48b}$$

where

A_{plb} = the cross-sectional area of the diagonal lacing bars in each panel of the member.
For single lacing $A_{plb} = A_{lb} = t_{lb}w_{lb}$
For double lacing $A_{plb} = 2A_{lb} = 2t_{lb}w_{lb}$
t_{lb} = thickness of lacing bar
w_{lb} = width of lacing bar
r = radius of gyration of the compression member = $\sqrt{I/A_g}$
Φ = the angle of the lacing bar from the line perpendicular to the member axis (should be about 30° for single lacing and 45° for double lacing).

6.3.3.1.2 *Critical Buckling Strength of Built-Up Compression Members with Batten Plates Only (Pinned at Each End) (Figure 6.13c)*

Shear is resisted by flexure of batten plates and main member elements.

$$\Omega = \frac{ab}{12E_{eff}I_{bb}} + \frac{a^2}{24E_{eff}I}, \tag{6.49a}$$

$$\alpha = \sqrt{1+\Omega P_{cr}} = \sqrt{1+\left(\frac{1.10}{(L/r)^2}\right)\left(\left(\frac{A_g}{A_{bb}}\right)\frac{ab}{r_{bb}^2}+\frac{a^2}{2r^2}\right)}, \tag{6.49b}$$

where

a = the distance between the centroid of batten plates
b = the distance between the centroid of the main compression elements of the member (effective batten plate length)
I_{bb} = the moment of inertia of the batten plate = $t_{bb}(w_{bb})^3/12$
t_{bb} = the batten plate thickness
w_{bb} = the batten plate width
A_{bb} = the batten plate cross-sectional area = $t_{bb}w_{bb}$, and

$$r_{bb} = \sqrt{\frac{I_{bb}}{A_{bb}}}.$$

If the shear rigidity of the batten plates is small, reduction of the built-up compression member critical buckling force will result. Inclusion of the batten plate shearing strain into Equation 6.49a yields

$$\Omega = \frac{ab}{12E_{eff}I_{bb}} + \frac{a^2}{24E_{eff}I} + \frac{\beta a}{A_{bb}G_{eff}b}, \tag{6.49c}$$

$$\alpha = \sqrt{1+\Omega P_{cr}} = \sqrt{1+\left(\frac{1.10}{(L/r)^2}\right)\left(\left(\frac{A_g}{A_{bb}}\right)\left(\frac{ab}{r_{bb}^2}+\frac{31.2a}{b}\right)+\frac{a^2}{2r^2}\right)}. \tag{6.49d}$$

6.3.3.1.3 *Critical Buckling Strength of Laced Bar Built-Up Compression Members with Shear Transfer Batten Plates (Pinned at Each End)* (Figure 6.13d)*

Shear is resisted by pin-connected truss behavior of lacing bars (for double lacing, consider tension resistance only):

$$\Omega = \frac{1}{A_{lb}E_{eff}\sin\phi\cos^2\phi} + \frac{b}{A_{bb}E_{eff}a}, \tag{6.50a}$$

$$\alpha = \sqrt{1+\Omega P_{cr}} = \sqrt{1+\left(\frac{13.2}{(L/r)^2}\right)\left(\left(\frac{A_g}{A_{plb}}\right)\left(\frac{1}{\sin\phi\cos^2\varphi}\right)+\left(\frac{A_g}{A_{plb}}\right)\left(\frac{1}{\tan\phi}\right)\right)}. \tag{6.50b}$$

6.3.3.1.4 *Critical Buckling Strength of Built-Up Compression Members with Perforated Cover Plates (Pinned at Each End) (Figure 6.13f)*

Most built-up compression members in modern steel railway superstructures are fabricated with the main elements connected by perforated cover plates. Shear is resisted by flexure of the main member elements because the perforated cover plates act as rigid batten plates between the perforations:

$$\Omega = \frac{9c^3}{32l_pE_{eff}I}, \tag{6.51a}$$

---

* This equation was developed and then later used by Engesser in connection with investigations into the collapse of the Quebec Bridge (see Chapter 1).

$$\alpha = \sqrt{1+\Omega P_{cr}} = \sqrt{1+\left(\frac{3.71}{(L/r)^2}\right)\left(\frac{c^3}{l_p r^2}\right)} \tag{6.51b}$$

where

l_p = the distance between the center of perforations

c = the length of the perforation.

The perforation length, c, can be expressed in terms of the distance between the center of perforations, l_p, so that

$$\alpha = \sqrt{1+\left(\frac{3.71\gamma^3}{(L/r)^2}\right)\left(\frac{l_p}{r}\right)^2}. \tag{6.51c}$$

where

$$\gamma = \frac{c}{l_p}.$$

6.3.3.1.5 *Design of Built-Up Compression Members*

The overall critical buckling strength of the built-up compression member, $\overline{P_{cr}}$, (Equation 6.45) is contingent upon the main element web plates, cover plates, lacing bars, and/or batten plates being of adequate strength and stability as individual components (local buckling).

In order to ensure that the webs of main elements of built-up compression members do not buckle prior to $\overline{P_{cr}}$, a minimum web plate thickness, t_w, is recommended by AREMA (2015)* as

$$t_w \geq \frac{0.90b\sqrt{F_y/E}}{\sqrt{F_{all}/f}}, \tag{6.52}$$

where $F_{all}/f \leq 4$

F_{all} = allowable compressive stress (Equations 6.21, 6.33, and $0.55F_y$ (see Equation 6.37) which consider elastic stability, inelastic stability and compressive yielding, respectively)

f = the calculated compressive stress in the member

Also, in order to ensure that cover plates for main elements of built-up compression members do not buckle locally prior to $\overline{P_{cr}}$, a minimum cover plate thickness, t_{cp}, is recommended by AREMA (2015) as

$$t_{cp} \geq \frac{0.72b\sqrt{F_y/E}}{\sqrt{F_{all}/f}}. \tag{6.53}$$

The thickness of lacing bars, stay plates, batten plates, and perforated cover plates must also be considered to complete the design of built-up compression members.

6.3.3.1.6 *Design of Lacing Bars, Stay Plates, Batten Plates, and Perforated Plates*

AREMA (2015) recommends that lacing bars, batten plates, and perforated plates be designed for a total shear force normal to the member in the plane of the lacing bars, batten plates, or cover plates consisting of self-weight, wind, and 2.5% of the axial compressive force in the member. The shear

* Based on uniform compression of elastic plates theory (Timoshenko and Gere, 1961).

forces related to self-weight and wind are generally small and may be neglected in many cases. The shear force normal to the member in the plane of the lacing bars, batten plates, or cover plates related to the axial compressive force can be estimated by considering the conditions of Figure 6.5c. The solution of the differential equation of the deflection curve (Equation 6.14), where $U = Pe$ is (Bowles, 1980)

$$y(x) = e\left(\tan\frac{k^2L}{2}\sin k^2y(x) + \cos k^2y(x) - 1\right). \tag{6.54}$$

Differentiating Equation 6.54 yields

$$\frac{\mathrm{d}y(x)}{\mathrm{d}x} = k^2e\tan\frac{k^2L}{2} \tag{6.55}$$

and substitution of Equation 6.55 into 6.40 yields

$$V = Pk^2e\tan\frac{k^2L}{2}. \tag{6.56}$$

AREMA (2015) recommends

$$k^2e\tan\frac{k^2L}{2} = 0.025 \tag{6.57}$$

so that

$$V = 0.025P, \tag{6.58}$$

where

$$V \geq \frac{A_rF_y}{150} \geq \frac{PF_y}{150F_{all}} \tag{6.59}$$

and

$$A_r = P/F_{all}.$$

Equation 6.59 was developed considering force eccentricity, initial curvature, and flexure of the compression member (Hardesty, 1935). Figure 6.14 illustrates that the AREMA (2015) recommendation for minimum shear force (Equation 6.59) ensures that shear forces with relatively greater proportion to the axial compressive force are used for the design of weaker compression members (more slender members that approach the Euler buckling behavior) when $F_{all}/F_y < 0.26$.

The shear force, V, forms the basis of lacing bar, batten plate, and perforated cover plate design for built-up compression members.

6.3.3.1.6.1 Lacing Bars for Compression Members The spacing of lacing bars along the main member must be designed to preclude buckling of portions of the main member elements between the lacing bar connections. AREMA (2015) limits the slenderness ratio, L_p/r_p of elements between lacing bar connections to 2/3 of the member slenderness ratio, L/r. This is appropriate in order to consider not only the local buckling effects over the length,

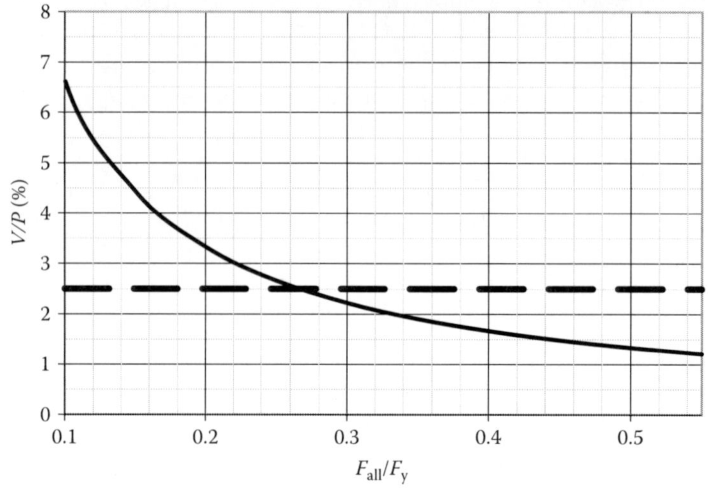

FIGURE 6.14 AREMA minimum shear force for built-up compression member design.

L_p, but also the interaction between global and local buckling (or compound buckling) (Duan et al., 2002). AREMA (2015) recommends that the lacing bar spacing be such that

$$\frac{L_p}{r_p} \leq 40 \leq \frac{2L}{3r}, \tag{6.60}$$

where

L_p = the length of the main member element between lacing bar connections (see Figures 6.13a and b)
= $2a$ (for single lacing)
= a (for double lacing)

r_p = the minimum radius of gyration of the main member element.

The lacing bars on each side of the main member must be designed to resist the shear force $V/2$ in the plane of the lacing bars. Therefore, the force in each lacing bar is

$$P_{lb} = \frac{V}{2\cos\phi}. \tag{6.61}$$

The critical buckling stress and the minimum cross-sectional area of the bar, A_{lb}, can be determined, using Equation 6.33 with $K = 1.0$, as

$$F_{all} = 0.60F_y - \left(635\frac{F_y}{E}\right)^{\frac{3}{2}}\left(\frac{L_{lb}}{r_{lb}}\right), \quad \text{in SI Units} \tag{6.62}$$

or

$$F_{all} = 0.60F_y - \left(17500\frac{F_y}{E}\right)^{\frac{3}{2}}\left(\frac{L_{lb}}{r_{lb}}\right), \quad \text{in US Customary or Imperial units} \tag{6.63}$$

where

$L_{lb} = \sqrt{a^2 + b^2}$ for single lacing systems (Figures 6.13a and d) (length of the lacing bar between connections at the main member)

$L_{lb} = 0.70\sqrt{a^2 + b^2}$ for double lacing systems (Figures 6.13b and e) (equivalent length of the lacing bar between connections at the main member)

$r_{lb} = \sqrt{I_{lb} / A_{lb}} = 0.29 t_{lb}$, for flat lacing bars.

Double lacing systems are used to reduce lacing bar slenderness and thickness. Therefore, AREMA (2015) recommends that double lacing connected at the center be used when $b > 380$ mm (15 in.) and the lacing bar width is <90 mm (3.5 in.).

The minimum thickness of lacing bars, based on the slenderness recommendations of AREMA (2015), is indicated in Table 6.7. In order to ensure adequate edge distance for connections and to accommodate the AREMA (2015) recommendation that the lacing bar connection bolt diameter not exceed 1/3 of the lacing bar width, w_{lb}, the minimum lacing bar width should be 70 mm (2–5/8 in.) for 22 mm (7/8 in.) diameter bolts.

6.3.3.1.6.2 Stay Plates for Built-Up Compression Members Stay plates must be used at the ends of laced bar built-up compression members and at locations where the lacing bars are interrupted (e.g., at a connection with another member).

The length of the stay plates at the ends of laced bar built-up compression members must be at least 25% greater than the distance between connection lines across the member (distance b in Figure 6.13). The length of the stay plates at intermediate locations of laced bar built-up compression members must be at least 75% of the distance between connection lines across the member. The minimum thickness of stay plates, t_{sp}, recommended by AREMA (2015), is shown in Table 6.8.

The center-to-center spacing of bolts must not exceed 4 bolt diameters and no less than 3 bolts should be used to connect stay plates to main elements of the built-up compression member. Welded stay plates shall utilize a minimum 8 mm (5/16″) continuous fillet weld along the stay plate longitudinal edges.

Example 6.7 outlines the design of a laced built-up compression member.

6.3.3.1.6.3 Batten Plates for Compression Members Built-Up compression members using only batten plates (Figure 6.13c) are not generally used* in steel railway superstructure compression elements due to strength and stability requirements. When batten plates are used in conjunction with laced bar compression members (Figure 6.13d) they may be designed for minimum slenderness considerations since the lacing bars are assumed to resist the applied shear forces.

TABLE 6.7
Minimum Lacing Bar Thickness

Member	t_{lbmin} Single Lacing	t_{lbmin} Double Lacing
Main	$L_{lb}/40$	$L_{lb}/60$
Bracing	$L_{lb}/50$	$L_{lb}/75$

TABLE 6.8
Minimum Stay Plate Thickness

Member	Minimum Stay Plate Thickness, t_{sp}
Main	$b/50$
Bracing	$b/60$

* For this reason, AREMA (2015) does not contain any specific recommendations relating to the use of batten plates.

However, if used without lacing bars, the spacing of batten plates, a, along the main member must also be designed to preclude buckling of portions of the main member elements between the batten plate connections. Based on the AREMA (2015) recommendations for lacing bars, it appears reasonable that batten plate spacing, a, be such that

$$\frac{a_{bp}}{r_p} \leq 40 \leq \frac{2L}{3r}, \tag{6.64}$$

where

$a_{bp} = a - w_{bp} + 2e_{bp}$

e_{bp} = the edge distance to the first fastener in the batten plate

w_{bp} = the width of the batten plate

Furthermore, when batten plates are used without lacing bars, the batten plates and main member elements* on each side of the main member must be designed to resist bending created by the shear force $V/2$ in each panel between the batten plates.

The force creating bending in each batten plate, P_{bp}, is

$$P_{bp} = \frac{1}{2}\frac{V}{2}\left(\frac{a}{b}\right) = \frac{Va}{4b}. \tag{6.65}$$

Therefore, the bending moment, M_{bp}, in each batten plate is

$$M_{bp} = P_{bp}\left(\frac{b}{2}\right) = \frac{Va}{8} \tag{6.66}$$

such that, based on an allowable stress of $0.55F_y$, the minimum thickness of the batten plate, t_{bp}, is

$$t_{bp} \geq \frac{1.36Va}{F_y(w_{bp})^2} \geq \frac{Pa}{29.3F_y(w_{bp})^2}, \tag{6.67}$$

where

w_{bp} = the width of the batten plate

P = the compressive force in the member = fA_g

f = the calculated compressive stress in the member

The bending moment in the main member element, M_p, is

$$M_p = \frac{1}{2}\frac{V}{2}\left(\frac{a}{2}\right) = \frac{Va}{8}. \tag{6.68}$$

The bending stress in each main member element from the shear forces applied on each side of the member is then calculated to ensure it does not exceed $0.55\ F_y$.†

Also, when batten plates are used without lacing bars, the batten plate spacing, a, is critical in regard to overall stability of the built-up compression member as indicated by Equations 6.49b and d.

* This criterion for main member elements will usually govern the design of built-up compression members using only batten plates.

† The main member elements between batten plates can be considered as flexural compression members with $L/r = 0$ (Tall, 1974).

The minimum thickness recommended in Table 6.8 for stay plates for main and bracing members also appears appropriate for batten plates. Also, if used, the minimum width and connection geometry of batten plates should consider the criteria outlined for stay plates.

Example 6.7 outlines the design of a batten plate built-up compression member.

6.3.3.1.6.4 Perforated Cover Plates for Compression Members In order to avoid the fabrication cost of laced bar built-up compression members and the bending inefficiencies of using batten plates, perforated cover plates are often used for built-up compression members in modern railway steel superstructures.

AREMA (2015) recommends that, to ensure stability of the main member elements at perforations, the effective slenderness ratio, C_{pc}, about the member axis (Figure 6.15) not exceed 20% or 33% of the member slenderness ratio about an axis perpendicular to the plane of the perforation.

The effective slenderness ratio is

$$C_{pc} = \frac{c}{r_{pc}} \le 20 \le \frac{L}{3r}, \tag{6.69}$$

where

c = the length of the perforation

$$r_{pc} = \sqrt{I_{pc}/A_{pc}}$$

I_{pc} = the moment of inertia of half of the member (one "flange") about the member axis at the center of the perforation

A_{pc} = the area of half of the member (one "flange").

AREMA (2015) also presents other recommendations related to perforated cover plates for built-up compression member design as

$$c \le 2w_{perf} \tag{6.70}$$

$$c \le l_p - b', \tag{6.71}$$

where

w_{perf} = the width of the perforation

l_p = the distance between the center of perforations

b' = the width of the perforated plate between the inside lines of fasteners.

The thickness of perforated cover plates should be governed by the largest of the following expressions relating to local plate stability criteria:

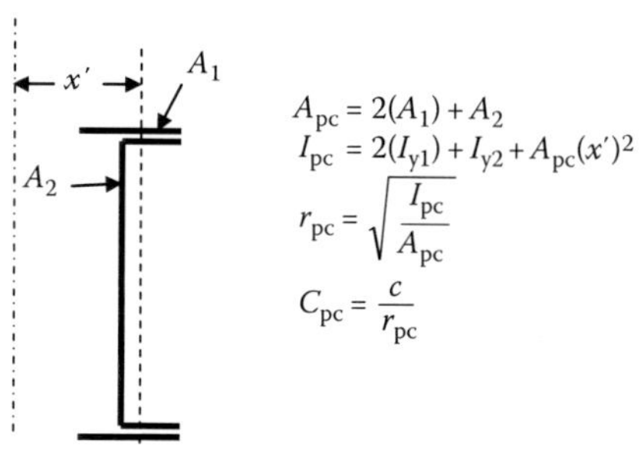

FIGURE 6.15 Element of compression member at a perforation.

$$t_{pc} \geq \frac{b'}{50}, \tag{6.72}$$

$$t_{pc} \geq 1.17(b' - w_{perf})\sqrt{F_y/E}, \tag{6.73}$$

$$t_{pc} \geq \frac{0.90b'\sqrt{F_y/E}}{\sqrt{F_{all}/f}}, \tag{6.74}$$

where $F_{all}/f \leq 4$.

Perforated cover plate thickness is also based on the transverse shear, V, at the centerline of the cover plate. The minimum perforated cover plate thickness (Equation 6.13) is

$$t_{pc} \geq \frac{3Vl_p}{2\tau_{all}b(l_p - c)}, \tag{6.75}$$

where

τ_{all} = the allowable shear stress ($0.35F_y$ is recommended by AREMA (2015)).

The gross section through the perforation (with only the perforated area removed) of the plate is included in the member gross cross-sectional area for compression design.

Example 6.7 outlines the design of a perforated cover plate built-up compression member.

Example 6.7a (SI Units)

Design a 6.0 m long compression member to resist a 2450 kN load as a solid, built-up laced bar, built-up batten plate, and built-up perforated cover plate member.

a. Design of solid section:
   Compression member: W360 × 134 rolled section.
   $A = 17{,}100$ mm^2
   $I_x = 415 \times 10^6$ mm^4
   $S_x = 2330 \times 10^3$ mm^3
   $r_x = 156$ mm
   $I_y = 151 \times 10^6$ mm^4
   $S_y = 817 \times 10^3$ mm^3
   $r_y = 94.0$ mm

$$\frac{L}{r_{min}} = \frac{6000}{94.0} = 63.8 \leq 100, \text{ OK}$$

$$\frac{KL}{r_{min}} = \frac{0.75(6000)}{94.0} = 47.9 \leq 5.034\sqrt{\frac{E}{F_y}} \leq 120$$

$$\frac{KL}{r_{min}} = 47.6 \geq 0.629\sqrt{\frac{E}{F_y}} \geq 15$$

$$F_{all} = 0.60F_y - \left(635\frac{F_y}{E}\right)^{3/2}\left(\frac{KL}{r_{min}}\right) = 210 - 1.17\left(\frac{KL}{r_{min}}\right) = 210 - 56 = 154 \text{ MPa}$$

$$P_{all} = F_{all}(A) = 154(17{,}100)/1000 = 2633 \geq 2450 \text{ kN, OK.}$$

b. Design of laced section:
   Compression member: 2 – C380 × 74 channels laced 165 mm back-to-back with 100 mm × 14 mm lacing bars as shown in Figure E6.5a.

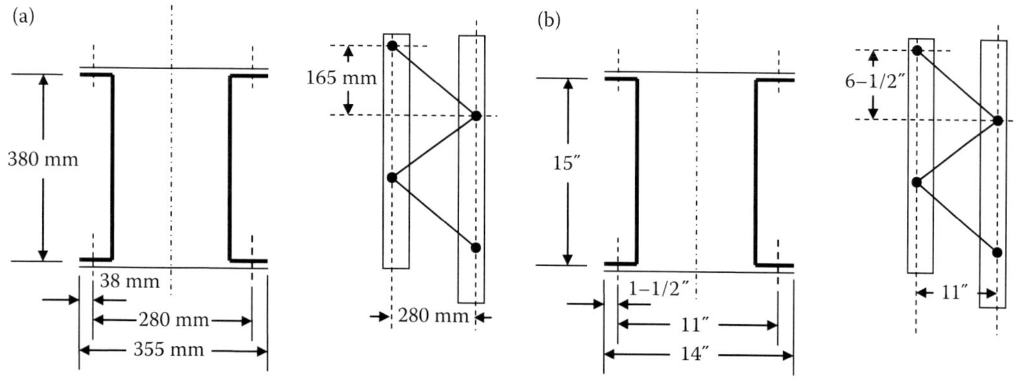

FIGURE E6.5 Laced section column.

For C380 × 74 channel:

$A = 9480\ \text{mm}^2$
$I_x = 168 \times 10^6\ \text{mm}^4$
$S_x = 881 \times 10^3\ \text{mm}^3$
$r_x = 133\ \text{mm}$
$I_y = 4.60 \times 10^6\ \text{mm}^4$
$S_y = 62.4 \times 10^3\ \text{mm}^3$
$r_y = 22.0\ \text{mm}$
$\bar{x} = 20.3\ \text{mm}.$

For laced member:

$A = 2(9480) = 18{,}960\ \text{mm}^2$
$I_x = 2(168 \times 10^6) = 336 \times 10^6\ \text{mm}^4$
$I_y = 2(4.60 \times 10^6) + 2(9480)(82.5+20.3)^2 = 210 \times 10^6\ \text{mm}^4$

$$r_y = \sqrt{\frac{210 \times 10^6}{18{,}960}} = 105\ \text{mm}$$

$$\frac{L}{r_{min}} = \frac{(6000)}{105} = 57.0 \le 100,\ \text{OK}$$

$$\frac{KL}{r_{min}} = \frac{0.75(6000)}{105} = 42.8 \le 5.034\sqrt{\frac{E}{F_y}} \le 120$$

$$\frac{KL}{r_{min}} = 42.8 \ge 0.629\sqrt{\frac{E}{F_y}} \ge 15.$$

Therefore,

$$F_{all} = 0.60F_y - \left(635\frac{F_y}{E}\right)^{3/2}\left(\frac{KL}{r_{min}}\right) = 210 - 1.17\left(\frac{KL}{r_{min}}\right) = 210 - 50.0 = 160\ \text{MPa}$$

$$P_{all} = F_{all}(A) = 160(18{,}960)/1000 = 3034 \ge 2450\ \text{kN, OK.}$$

If the effects of shear deformation are included (Equations 6.43 and 6.48b):

$$\alpha = \sqrt{1+\left(\frac{13.2}{(L/r)^2}\right)\left(\frac{A}{A_{\text{plb}}\sin\phi\cos^2\phi}\right)} = \sqrt{1+\left(\frac{13.2}{(57.0)^2}\right)\left(\frac{18,960}{2(100)(14)\sin 30.6^\circ\cos^2 30.6^\circ}\right)} = 1.04$$

$$\bar{P}_{\text{all}} = \frac{P_{\text{all}}}{\alpha^2} = \frac{3034}{1.07} = 2828 \geq 2450 \text{ kN, OK.}$$

Check the design of 100 mm. × 14 mm. lacing bars at 30.6° to horizontal:

$$t_{\text{lb}} \geq \frac{\sqrt{165^2+280^2}}{40} = \frac{325}{40} = 8.1 \text{ mm, OK}$$

$$V = 0.025P = 0.025(2450) = 61.3 \geq \frac{PF_y}{150F_{\text{all}}} \geq \frac{2450(350)}{150(160)} \geq 35.7 \text{ kN, OK}$$

$$P_{\text{lb}} = \frac{V}{2\cos 30.6^\circ} = \frac{61.3}{2\cos 30.6^\circ} = 35.7 \text{ KN}$$

$$F_{\text{all}} = 30-1.17\left(\frac{(K)L_{\text{lb}}}{r_{\text{lb}}}\right) = 210-1.17\left(\frac{(1.0)325}{0.29t_{\text{lb}}}\right) = 210-1.17\left(\frac{(1.0)325}{0.29(14)}\right) = 210-93.7 = 116 \text{ MPa}$$

$$P_{\text{all}} = (116)(100)(14)/1000 = 162 \text{ kN} > 35.7 \text{ kN, OK.}$$

Check the design of main member with lacing bars at 30.6° to horizontal:

$\frac{L_p}{r_p} = \frac{2(165)}{22.0} = 15 \leq 40$, OK (length of main member element (channel) between lacing bar connections)

$$\frac{L_p}{r_p} = 15 \leq \frac{2}{3}(57.0) \leq 38.0, \text{ OK.}$$

c. Design of battened section:
   Compression member: 2 – C380 × 74 channels battened 165 mm back-to-back with 150 mm × 14 mm batten plates as shown in Figure E6.6a.

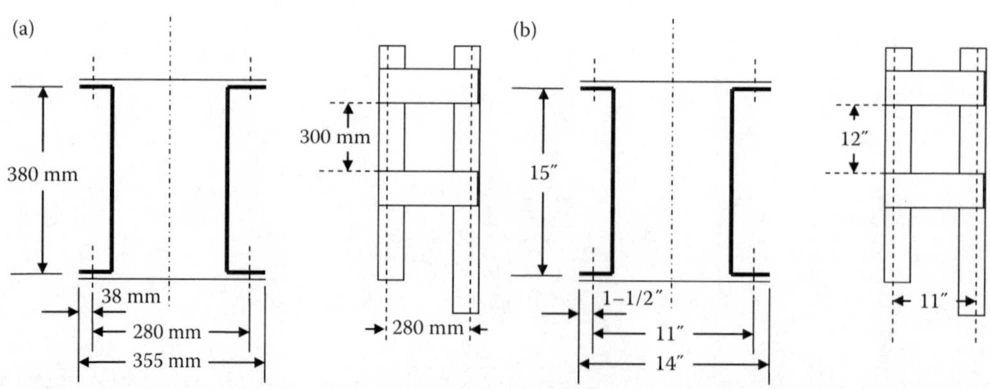

FIGURE E6.6 Battened section column.

C380 × 74 channel battened member section properties and allowable compression force (3034 kN) are same as for the laced section design.
If the effects of shear deformation are included (Equations 6.43 and 6.49d):

$$\alpha = \sqrt{1+\left(\frac{1.10}{(L/r)^2}\right)\left(\left(\frac{A}{A_{bb}}\right)\left(\frac{ab}{r_{bb}^2}+\frac{31.2a}{b}\right)+\frac{a^2}{2r^2}\right)}$$

$$= \sqrt{1+\left(\frac{1.10}{(57.0)^2}\right)\left(\left(\frac{18,960}{2(150)(14)}\right)\left(\frac{(450)(280)}{165^2}+\frac{31.2(450)}{280}\right)+\frac{(450)^2}{2(105)^2}\right)} = 1.04$$

$$\bar{P}_{all} = \frac{P_{all}}{\alpha^2} = \frac{3034}{1.09} = 2792 \geq 2450 \text{ kN, OK.}$$

Check the design of 150 mm. × 14 mm batten plates at 450 mm. center-to-center spacing:

$$t_{bp} \geq \frac{b}{40} = \frac{280}{40} = 7.0 \text{ mm, OK}$$

$$t_{bp} \geq \frac{Pa}{29.3F_y(w_{bp})^2} \geq \frac{2450(450)(1000)^2}{18,960(350)(150)^2} = 7.4 \text{ mm, OK.}$$

Check the design of main member with 150 mm batten plates at 450 mm center-to-center spacing:
$\frac{L_p}{r_p} = \frac{(300+76)}{22.0} = 17.1 \leq 40$, OK (length of main member element (channel) between closest connections of adjacent batten plates)

$$\frac{L_p}{r_p} = \frac{(300+76)}{22.0} = 17.1 \leq \frac{2}{3}(57.0) \leq 38.0, \text{ OK}$$

$$M_p = \frac{Va}{8} = \frac{61.3(450)}{8} = 3448 \text{ kNmm}$$

$$f_p = \frac{3448(1000)}{62.4 \times 10^3} = 55.3 \text{ MPa}$$

$$F_{all} = 210 - 1.17\left(\frac{KL}{r_{min}}\right) = 210 - 1.17\left(\frac{(1.0)450}{22.0}\right) = 210 - 23.9 = 186 \geq 55.3 \text{ MPa, OK.}$$

d. Design of perforated cover plated section (100 mm × 200 mm perforations at 450 mm center-to-center)
Compression member: 2 – C310 × 45 channels 150 mm back-to-back connected with 14 mm thick perforated cover plates as shown in Figure E6.7a.
For C310 × 45 channel:

$A = 5690$ mm^2
$I_x = 67.3 \times 10^6$ mm^4
$S_x = 442 \times 10^3$ mm^3
$r_x = 109$ mm
$I_y = 2.12 \times 10^6$ mm^4
$S_y = 33.6 \times 10^3$ mm^3
$r_y = 19.3$ mm
$\bar{x} = 17.0$ mm.

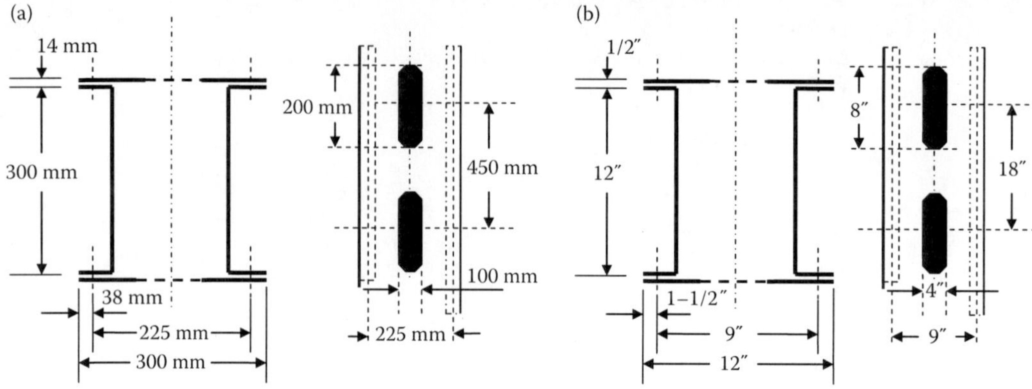

FIGURE E6.7 Perforated plate section column.

For cover plated member:

$$A = 2(5690)+2(300)(14) - 2(100)(14) = 16{,}980 \text{ mm}^2$$

$$I_x = 2(67.3\times10^6) + 2(200)(14)(157)^2 + 2(200)(14)^3/12 = 272.7\times10^6 \text{ mm}^4$$

$$r_x = \sqrt{272.7\times\frac{10^6}{16{,}980}} = 127 \text{ mm}$$

$$I_y = 2(2.12\times10^6) + 2(5690)(75+17.0)^2 + 2(2)(100)^3(14)/12 + 2(2)(100)(14)(50+50)^2$$
$$= 161.2 \times 10^6 \text{ mm}^4$$

$$r_y = \sqrt{161.2\times\frac{10^6}{16{,}980}} = 97.4 \text{ mm}$$

$$\frac{L}{r_{\min}} = \frac{(6000)}{97.4} = 61.6 \le 100, \text{ OK}$$

$$\frac{KL}{r_{\min}} = \frac{0.75(6000)}{97.4} = 46.2 \le 5.034\sqrt{\frac{E}{F_y}} \le 120$$

$$\frac{KL}{r_{\min}} = 46.2 \ge 0.629\sqrt{\frac{E}{F_y}} \ge 15$$

$$F_{\text{all}} = 0.60F_y - \left(635\frac{F_y}{E}\right)^{3/2}\left(\frac{KL}{r_{\min}}\right) = 210 - 1.17\left(\frac{KL}{r_{\min}}\right) = 210 - 54.1 = 156 \text{ MPa}$$

$$P_{\text{all}} = F_{\text{all}}(A) = \frac{156(16{,}980)}{1000} = 2649 \ge 2450 \text{ kN, OK.}$$

If the effects of shear deformation are included (Equations 6.43 and 6.51b):

$$\alpha = \sqrt{1+\left(\frac{3.71}{\left(\frac{L}{r}\right)^2}\right)\left(\frac{c^3}{I_p r^2}\right)} = \sqrt{1+\left(\frac{3.71}{(61.6)^2}\right)\left(\frac{200^3}{450(97.4)^2}\right)} = 1.00$$

$$\bar{P}_{\text{all}} = \frac{P_{\text{all}}}{\alpha^2} = \frac{2649}{1.00} = 2649 \geq 2450 \text{ kN, OK.}$$

Check the design of 14 mm cover plates with 100 mm × 200 mm perforations at 450 mm center-to-center spacing as shown in Figure E6.7a:

The properties of half the member at the center of perforation about its own axis are as follows (Figure E6.8):

$$z' = \frac{2(100)(14)(100)+5690(17.0+75)}{2(100)(14)+5690} = 94.6 \text{ mm}$$

$$A_{pc} = 2(100)(14)+5690 = 8490 \text{ mm}^2$$

$$I_{pc} = 2.12 \times 10^6 + 5690(94.6-92)^2 + 2(14)(100)^3/12 + 2(100)(14)(100-94.6)^2$$
$$= 4573 \times 10^3 \text{ mm}^4$$

$$r_{pc} = \sqrt{\frac{4573\times 10^3}{8490}} = 23.2 \text{ mm}$$

$$C_{pc} = \frac{200}{23.2} = 8.6 \leq 20, \text{ OK}$$

$$C_{pc} = 8.6 \leq \frac{L}{3r} \leq \frac{1}{3}(61.6) \leq 20.5, \text{ OK}$$

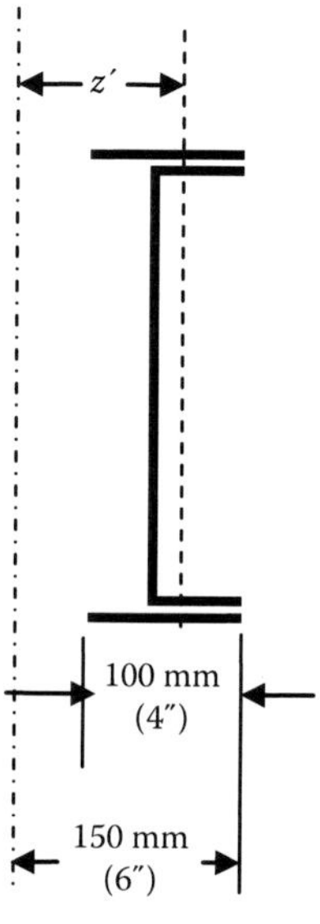

FIGURE E6.8 Perforated plate column half section.

TABLE E6.1a
Comparison of Compression Member Designs

Compression Member	Figure	Allowable Force (kN)	Shear Deformation Effect (%)	Gross Area of Member (Through Perforation for Perforated Plates) (mm^2)	Description
Solid	n/a	2,633	0	17,100	*W* shape
Laced	6.13a, b, and e	2,828	7	18,960	2 – *C* shapes (A = 18,960 mm^2); laced only; no shear transfer batten plates
Battened	6.13c	2,792	9	18,960	2 – *C* shapes (A = 18,960 mm^2); no lacing; with shear transfer batten plates
Perforated Plated	6.13f	2,649	~0	16,980	2 – *C* shape (A = 11,380 mm^2); cover plates

$$c = 200 \le 2w_{\text{perf}} \le 2(100) \le 200 \text{ in., OK}$$

$$c = 200 \le l_{\text{p}} - b' \le 450 - 225 \le 225 \text{ mm } \big(b' = \text{width between inside lines of fasteners}\big), \text{ OK}$$

$$t_{\text{pc}} \ge \frac{b}{50} \ge \frac{225}{50} \ge 4.5 \text{ mm, OK}$$

$$t_{\text{pc}} \ge 1.17(b - w_{\text{perf}})\sqrt{\frac{F_{\text{y}}}{E}} \ge 1.17(225 - 100)\sqrt{\frac{F_{\text{y}}}{E}} \ge 6.1 \text{ mm, OK}$$

$$t_{\text{pc}} \ge \frac{0.90b\sqrt{\frac{F_{\text{y}}}{E}}}{\sqrt{\frac{F_{\text{all}}}{f}}} \ge \frac{0.90(225)\sqrt{\frac{F_{\text{y}}}{E}}}{\sqrt{\frac{156}{(2450(1000)/16980)}}} \ge 8.2 \text{ mm, OK}$$

$$\sqrt{\frac{F_{\text{all}}}{f}} = 1.04 \le 2, \text{ OK}$$

$$t_{\text{pc}} \ge \frac{3Vl_{\text{p}}}{2\tau_{\text{all}}b(l_{\text{p}} - c)} \ge \frac{3(61.3)(450)(1000)}{2(0.35)(350)(225)(450 - 200)} = 6.0 \text{ mm, OK.}$$

A summary of the compression member designs is shown in Table E6.1a.

The benefits of the perforated plate contribution to the compression member area and moment of inertia, and the relative weakness of battened only compression members are apparent in Table E6.1a.

Example 6.7b (US Customary and Imperial Units)

Design a 19.55 ft long compression member to resist a 550 kip load as a solid, built-up laced bar, built-up batten plate, and built-up perforated cover plate member.

a. Design of solid section:

Compression member: W14 × 90 rolled section.

$A = 26.5$ in.^{2}

$I_x = 999$ in.^{4}

$S_x = 143$ in.^{3}
$R_x = 6.14$ in.
$I_y = 362$ in.^{4}
$S_y = 49.9$ in.^{3}
$R_y = 3.70$ in.

$$\frac{L}{r_{min}} = \frac{(19.55)(12)}{3.70} = 63.4 \leq 100, \text{ OK}$$

$$\frac{KL}{r_{min}} = \frac{0.75(19.55)(12)}{3.70} = 47.6 \leq 5.034\sqrt{\frac{E}{F_y}} \leq 121$$

$$\frac{KL}{r_{min}} = 47.6 \geq 0.629\sqrt{\frac{E}{F_y}} \geq 15$$

$$F_{all} = 0.60F_y - \left(17{,}500\frac{F_y}{E}\right)^{3/2}\left(\frac{KL}{r_{min}}\right) = 30 - 0.166\left(\frac{KL}{r_{min}}\right) = 30 - 7.9 = 22.1 \text{ ksi}$$

$$P_{all} = F_{all}(A) = 22.1(26.5) = 586 \geq 550 \text{ kips, OK.}$$

b. Design of laced section:
Compression member: 2 – C 15 × 50 channels laced 6–1/2 in. back–to-back with 4 in. × 1/2 in. lacing bars as shown in Figure E6.5b.
For C15 × 50 channel:

$A = 14.7$ in.^{2}
$I_x = 404$ in.^{4}
$S_x = 53.8$ in.^{3}
$r_x = 5.24$ in
$I_y = 11.0$ in.^{4}
$S_y = 3.78$ in.^{3}
$r_y = 0.867$ in.
$\bar{x} = 0.80$ in.

For laced member:

$A = 2(14.7) = 29.4$ in.^{2}
$I_x = 2(404) = 808$ in.^{4}
$I_y = 2(11.0) + 2(14.7)(3.25+0.80)^2 = 504$ in.^{4}

$$r_y = \sqrt{\frac{504}{29.4}} = 4.14 \text{ in.}$$

$$\frac{L}{r_{min}} = \frac{(19.55)(12)}{4.14} = 56.7 \leq 100, \text{ OK}$$

$$\frac{KL}{r_{min}} = \frac{0.75(19.55)(12)}{4.14} = 42.5 \leq 5.034\sqrt{\frac{E}{F_y}} \leq 121$$

$$\frac{KL}{r_{min}} = 42.5 \geq 0.629\sqrt{\frac{E}{F_y}} \geq 15.$$

Therefore,

$$F_{all} = 0.60F_y - \left(17{,}500\frac{F_y}{E}\right)^{3/2}\left(\frac{KL}{r_{min}}\right) = 30 - 0.166\left(\frac{KL}{r_{min}}\right) = 30 - 7.1 = 22.9 \text{ ksi}$$

$$P_{all} = F_{all}(A) = 22.9(29.4) = 675 \geq 550 \text{ kips, OK.}$$

If the effects of shear deformation are included (Equations 6.43 and 6.48b):

$$\alpha = \sqrt{1 + \left(\frac{13.2}{(L/r)^2}\right)\left(\frac{A}{A_{plb}\sin\phi\cos^2\phi}\right)} = \sqrt{1 + \left(\frac{13.2}{(56.7)^2}\right)\left(\frac{29.4}{2(4)(0.5)\sin 30.6^\circ\cos^2 30.6^\circ}\right)} = 1.04$$

$$\bar{P}_{all} = \frac{P_{all}}{\alpha^2} = \frac{675}{1.08} = 625 \geq 550 \text{ kips, OK.}$$

Check the design of 4 in. × 1/2 in. lacing bars at 30.6° to horizontal:

$$t_{lb} \geq \sqrt{6.5^2 + \frac{11^2}{40}} = \frac{12.8}{40} = 0.32 \text{ in., OK}$$

$$V = 0.025P = 0.025(550) = 13.8 \geq \frac{PF_y}{150F_{all}} \geq \frac{550(50)}{150(22.9)} \geq 8.0 \text{ kips, OK}$$

$$P_{lb} = \frac{V}{2\cos 30.6^\circ} = \frac{13.8}{2\cos 30.6^\circ} = 8.0 \text{ kips}$$

$$F_{all} = 30 - 0.166\left(\frac{(K)L_{lb}}{r_{lb}}\right) = 30 - 0.166\left(\frac{(1.0)12.8}{0.29t_{lb}}\right)$$

$$= 30 - 0.166\left(\frac{(1.0)12.8}{0.29(0.5)}\right) = 30 - 14.7 = 15.3 \text{ ksi.}$$

P_{all} = (15.3)(4)(0.5) = 30.6 kips > 8.0 kips, OK.
Check the design of main member with lacing bars at 30.6° to horizontal:

$\frac{L_p}{r_p} = \frac{2(6.5)}{0.867} = 15 \leq 40$, OK (length of main member element (channel) between lacing bar connections)

$$\frac{L_p}{r_p} = 15 \leq \frac{2}{3}(56.7) \leq 37.8, \text{ OK.}$$

c. Design of battened section:
Compression member: 2 – C 15 × 50 channels battened 6–1/2 in. back-to-back with 6 in. × 1/2 in. batten plates as shown in Figure E6.6b.
C 15 × 50 channel battened member section properties and allowable compression force (675 kips) are same as for the laced section design.

If the effects of shear deformation are included (Equations 6.43 and 6.49d):

$$\alpha = \sqrt{1+\left(\frac{1.10}{(L/r)^2}\right)\left(\left(\frac{A}{A_{bb}}\right)\left(\frac{ab}{r_{bb}^2}+\frac{31.2a}{b}\right)+\frac{a^2}{2r^2}\right)}$$

$$= \sqrt{1+\left(\frac{1.10}{(56.7)^2}\right)\left(\left(\frac{29.4}{2(6)(0.5)}\right)\left(\frac{(18)(11)}{(6^2)}+\frac{31.2(18)}{11}\right)+\frac{(18)^2}{2(4.14)^2}\right)} = 1.05$$

$$\bar{P}_{all} = \frac{P_{all}}{\alpha^2} = \frac{675}{1.10} = 615 \geq 550 \text{ kips, OK.}$$

Check the design of 6 in. × 1/2 in. batten plates at 18 in. center-to-center spacing:

$$t_{bp} \geq \frac{b}{40} = \frac{11}{40} = 0.28 \text{ in., OK}$$

$$t_{bp} \geq \frac{Pa}{29.3F_y(w_{bp})^2} \geq \frac{550(18)}{29.3(50)(6)^2} = 0.19 \text{ in., OK.}$$

Check the design of main member with 6 in. batten plates at 18 in. center-to-center spacing: $\frac{L_p}{r_p} = \frac{(12+3)}{0.867} = 17.3 \leq 40$, OK (length of main member element (channel) between closest connections of adjacent batten plates)

$$\frac{L_p}{r_p} = \frac{(12+3)}{0.867} = 17.3 \leq \frac{2}{3}(56.7) \leq 37.8, \text{ OK}$$

$$M_p = \frac{Va}{8} = \frac{15(18)}{8} = 33.8 \text{ in-kips}$$

$$f_p = \frac{33.8}{3.78} = 8.9 \text{ ksi}$$

$$F_{all} = 30 - 0.166\left(\frac{KL}{r_{min}}\right) = 30 - 0.166\left(\frac{(1.0)18}{0.867}\right) = 30 - 3.5 = 26.5 \geq 8.9 \text{ ksi, OK.}$$

d. Design of perforated cover plated section (4 in. × 8 in. perforations at 18 in. center-to-center)
Compression member: 2 – C 12 × 30 channels 6 in. back-to-back connected with ½ in. thick perforated cover plates as shown in Figure E6.7b.
For C 12 × 30 channel:

$A = 8.82$ in.^{2}
$I_x = 162$ in.^{4}
$S_x = 27.0$ in.^{3}
$r_x = 4.29$ in.
$I_y = 5.14$ in.^{4}
$S_y = 2.06$ in.^{3}
$r_y = 0.763$ in.
$\bar{x} = 0.67$ in.

For cover plated member:

$A = 2(8.82) + 2(12)(0.5) - 2(4)(1/2) = 25.6$ in.^{2}
$I_x = 2(162) + 2(8)(0.5)(6.25)^2 + 2(8)(0.5)^3/12 = 637$ in.^{4}

$$r_x = \sqrt{\frac{637}{25.6}} = 5.00 \text{ in.}$$

$$I_y = 2(5.14) + 2(8.82)(3+0.67)^2 + 2(2)(4)^3(0.5)/12 + 2(2)(4)(0.5)(2+2)^2 = 387 \text{ in.}^4$$

$$r_y = \sqrt{\frac{387}{25.6}} = 3.89 \text{ in.}$$

$$\frac{L}{r_{\min}} = \frac{(19.55)(12)}{3.89} = 60.4 \le 100, \text{ OK}$$

$$\frac{KL}{r_{\min}} = \frac{0.75(19.55)(12)}{3.89} = 45.3 \le 5.034\sqrt{\frac{E}{F_y}} \le 121$$

$$\frac{KL}{r_{\min}} = 45.3 \ge 0.629\sqrt{\frac{E}{F_y}} \ge 15$$

$$F_{\text{all}} = 0.60F_y - \left(17{,}500\frac{F_y}{E}\right)^{3/2}\left(\frac{KL}{r_{\min}}\right) = 30 - 0.166\left(\frac{KL}{r_{\min}}\right) = 30 - 7.5 = 22.5 \text{ ksi}$$

$$P_{\text{all}} = F_{\text{all}}(A) = 22.5(25.6) = 576 \ge 550 \text{ kips, OK.}$$

If the effects of shear deformation are included (Equations 6.44 and 6.52b):

$$\alpha = \sqrt{1+\left(\frac{3.71}{(L/r)^2}\right)\left(\frac{c^3}{I_p r^2}\right)} = \sqrt{1+\left(\frac{3.71}{(60.4)^2}\right)\left(\frac{8^3}{18(3.89)^2}\right)} = 1.00$$

$$\bar{P}_{\text{all}} = \frac{P_{\text{all}}}{\alpha^2} = \frac{576}{1.00} = 576 \ge 550 \text{ kips, OK.}$$

Check the design of 1/2 in. cover plates with 4 in. × 8 in. perforations at 18 in. center-to-center spacing as shown in Figure E6.7b:

The properties of half the member at the center of perforation about its own axis are (see Figure E6.8):

$$z' = \frac{2(4)(0.5)(4) + 8.82(3.67)}{2(4)(0.5) + 8.82} = 3.77 \text{ in.}$$

$$A_{pc} = 2(4)(0.5) + 8.82 = 12.82 \text{ in.}^2$$

$$I_{pc} = 5.14 + 8.82(3.77-3.67)^2 + 2(0.5)(4)^3/12 + 2(4)(0.5)(4-3.77)^2 = 10.77 \text{ in.}^4$$

$$r_{pc} = \sqrt{\frac{10.77}{12.82}} = 0.92 \text{ in.}$$

$$C_{pc} = \frac{8}{0.92} = 8.7 \le 20, \text{ OK}$$

$$C_{pc} = 8.7 \le \frac{L}{3r} \le \frac{1}{3}(60.4) \le 20.1, \text{ OK}$$

TABLE E6.1b
Comparison of Compression Member Designs

Compression Member	Figure	Allowable Force (kips)	Shear Deformation Effect (%)	Gross Area of Member (Through Perforation for Perforated Plates) (in.^{2})	Description
Solid	n/a	586	0	26.5	*W* shape
Laced	6.13a, b, and e	625	8	29.4	2 – *C* shapes ($A = 29.4$ in.^{2}); laced only; no shear transfer batten plates
Battened	6.13c	615	10	29.4	2 – *C* shapes ($A = 29.4$ in.^{2}); No lacing; with shear transfer batten plates
Perforated plated	6.13f	576	~0	25.6	2 – *C* shape ($A = 17.64$ in.^{2}); cover plates

$$c = 8 \le 2w_{\text{perf}} \le 2(4) \le 8 \text{ in., OK}$$

$$c = 8 \le l_p - b' \le 18 - 9 \le 9 \text{ in. } (b' = \text{width between inside lines of fasteners}), \text{ OK}$$

$$t_{pc} \ge \frac{b}{50} \ge \frac{9}{50} \ge 0.18 \text{ in., OK}$$

$$t_{pc} \ge 1.17(b - w_{\text{perf}})\sqrt{\frac{F_y}{E}} \ge 1.17(9-4)\sqrt{\frac{F_y}{E}} \ge 0.24 \text{ in., OK}$$

$$t_{pc} \ge \frac{0.90b\sqrt{F_y/E}}{\sqrt{F_{\text{all}}/f}} \ge \frac{0.90(9)\sqrt{F_y/E}}{\sqrt{22.5/(550/25.6)}} \ge 0.33 \text{ in., OK}$$

$$\sqrt{\frac{F_{\text{all}}}{f}} = 1.02 \le 2, \text{ OK}$$

$$t_{pc} \ge \frac{3Vl_p}{2\tau_{\text{all}}b(l_p - c)} \ge \frac{3(13.8)(18)}{2(0.35)(50)(9)(18-8)} = 0.24 \text{ in., OK.}$$

A summary of the compression member designs is shown in Table E6.1b.

The benefits of the perforated plate contribution to the compression member area and moment of inertia, and the relative weakness of battened only compression members are apparent in Table E6.1b.

REFERENCES

American Institute of Steel Construction (AISC), 1980, *Manual of Steel Construction*, 8th ed., AISC, Chicago, IL.

American Railway Engineering and Maintenance-of-Way Association (AREMA), 2015, Chapter 15—Steel structures, *Manual for Railway Engineering*, Lanham, MD.

Anderson, T.L., 2005, *Fracture Mechanics*, CRC Press, Boca Raton, FL.

Armenakas, A.E., *Advanced Mechanics of Materials and Applied Elasticity*, CRC Press, Boca Raton, FL.

Bleich, F., 1952, *Buckling Strength of Metal Structures*; 1st ed., McGraw-Hill, New York.

Bowles, J.E., 1980, *Structural Steel Design*, McGraw-Hill, New York.

Chen, W.F. and Lui, E.M., 1987, *Structural Stability*, Elsevier, New York.

Cochrane, V.H., 1922, Rules for rivet hole deductions in tension members, *Engineering News-Record*, Vol. 89, No. 16, 847–848.

Duan, L., Reno, M., and Uang, C., 2002, Effect of compound buckling on compression strength of built-up members, *Engineering Journal*, Vol. 39, No. 1, 30–37.

Hardesty, S., 1935, Live loads and unit stresses, American Railway Engineering Associaton *AREA Proceedings,* Vol. 36., 770–773. Chicago, IL.

Munse, W.H., and Chesson, E., 1963, Riveted and bolted joints: Net section design, *Journal of the Structural Division*, Vol. 89, No. 1, 107–126.

Newmark, N.M., 1949, A simple approximate formula for effective end-fixity of columns, *Journal of Aeronautical Sciences*, Vol. 16, No. 2, 116.

Pilkey, W.D., 1997, *Peterson's Stress Concentration Factors*, John Wiley & Sons, New York.

Salmon, C.G. and Johnson, J.E., 1980, *Steel Structures Design and Behavior*, Harper & Row, New York.

Sweeney, R.A.P., 2006, What's important in railroad bridge fatigue life evaluation, *Bridge Structures*, Vol. 2, No. 4, 191–198.

Tall, L. (editors), 1974, *Structural Steel Design*, 2nd ed., John Wiley & Sons, New York.

Timoshenko, S.P. and Gere, J.M., 1961, *Theory of Elastic Stability*, McGraw-Hill, New York.

Wang, C.M., Wang, C.Y. and Reddy, J.N., 2005, *Exact Solutions for Buckling of Structural Members*, CRC Press, Boca Raton, FL.

7 Design of Flexural Steel Members

7.1 INTRODUCTION

Members designed to primarily carry bending or flexural forces are typically found in steel railway superstructures such as girders, beams, floor beams, and stringers. These beams and girder members experience normal tensile, normal compressive and shear stresses, and are designed considering strength, serviceability, and fatigue limit states.

Flexural members must be designed in accordance with the American Railway Engineering and Maintenance-of-Way Association (AREMA, 2015) allowable stress design (ASD) method to resist normal tensile stresses based on yield strength at the net section. Flexural members subjected to cyclical or fluctuating normal tensile stress ranges must be designed with due consideration of stress concentration effects and metal fatigue. The members must also be designed for strength to resist shear and compressive normal stresses with due attention to stability.

Flexural member serviceability design for stiffness is achieved by respecting live load vertical deflection limits for simple spans used in freight rail operations. Adequate lateral rigidity to resist vibrations and deflections from wind and live load effects (e.g., lateral loads from track misalignments, rail wear, track fastener deterioration, and track curvature) is provided by considering the lateral deflection criteria with the design lateral forces recommended by AREMA (2015) (see Chapter 5).

Steel beams and girders for railway superstructures can be of noncomposite (steel, timber, or independent concrete deck) or of composite (integral concrete deck) material design. Noncomposite material beams and girders are used in the design of both open and ballasted deck spans. Composite material design is efficient and typically provides a ballasted deck span. The relative merits of each system with respect to steel railway superstructure design, fabrication, and erection are discussed in Chapter 3.

7.2 STRENGTH DESIGN OF NONCOMPOSITE FLEXURAL MEMBERS

Steel girders, beams, floor beams, and stringers are designed as noncomposite flexural members. They must be designed for the internal normal flexural and shear stresses caused by combinations of external actions (loads or forces) (see Chapter 4).

7.2.1 Bending of Laterally Supported Beams and Girders

Elastic strains in beams and girders with at least one axis of symmetry that undergoes bending with small deformations, and where plane sections through the beam longitudinal axis remain plane, will have a linear distribution. Furthermore, it is assumed that Poisson's effect and shear deformations can be neglected due to practical beam member geometry (Wang et al., 2000). Therefore, for an elastic design where stress is proportional to strain (Hooke's Law), the distribution of stress is shown in Figure 7.1. It should be noted that no instability or stress concentration effects are considered.

Equilibrium of moments (ignoring signs because tension and compression are easily located by inspection) results in

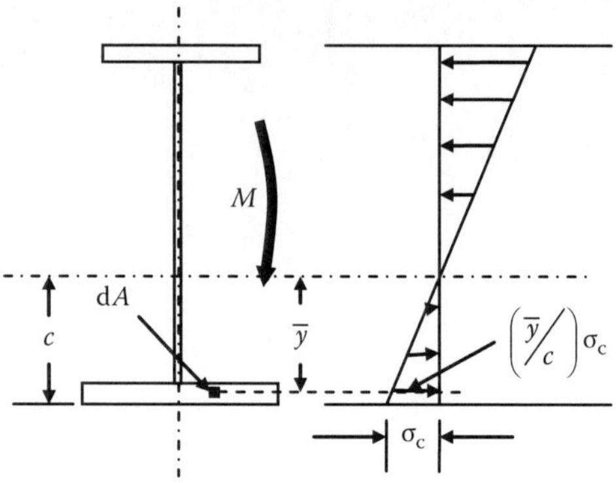

FIGURE 7.1 Bending of a beam.

$$M = 0 = \int\left(\frac{\bar{y}}{c}\sigma_c\right)dA(\bar{y}) = \frac{\sigma_c}{c}\int \bar{y}^2\,dA = \frac{\sigma_c}{c}I_x. \tag{7.1}$$

Therefore, the maximum bending stress is

$$\sigma_{max} = \frac{Mc}{I_x} = \frac{M}{S_x}, \tag{7.2}$$

where

$$S_x = \frac{I_x}{c}.$$

The AREMA (2015) ASD recommendations use a factor of safety against tensile yield stress of 1.82. The required section modulus of the beam or girder is then

$$S_x \geq \frac{M}{0.55F_y}, \tag{7.3}$$

where

M = the externally applied bending moment
$\bar{y}$ = the distance from the neutral axis to the area under consideration
c = the distance from the neutral axis to the extreme fiber of the beam or girder
dA = the infinitesimal area under consideration
σ_c = the normal stress at the extreme fiber of the beam or girder
σ_{max} = the maximum normal stress (at the top or bottom extreme fiber)
F_y = the specified steel yield stress
$I_x = \int y^2\,dA$ = the vertical bending moment of inertia about the beam or girder neutral axis
$S_x = I_x/c$ = the vertical bending section modulus about the beam or girder neutral axis Equation 7.3 enables the determination of the section modulus based on the allowable tensile stress. However, the equation does not address instability in the compression region of the beam or girder. If compression region stability (usually by lateral support of the compression flange) is sustained, Equation 7.3 may be used to determine both required net and gross-sectional properties of the beam or girder. Lateral support of the compression flange may be provided by a connected steel or concrete deck,

and/or either diaphragms, cross bracing frames, or struts at appropriate intervals. However, if the compression flange is laterally unsupported, instability must be considered as it may reduce the beam or girder strength (by reducing the allowable compressive stress).

7.2.2 Bending of Laterally Unsupported Beams and Girders

If the compression flange of a beam or girder is not supported at sufficiently close intervals, it is susceptible to lateral-torsional instability prior to yielding and may not be able to fully participate in resisting bending moment applied to the beam or girder.

In addition to the vertical translation or deflection, y, simply supported doubly symmetric elastic beams subjected to uniform bending will buckle with lateral translation, w, and torsional translation or twist, ϕ, as shown in Figure 7.2.

It is assumed that $I_x >> I_y$ so that vertical deformation effects may be neglected with respect to the lateral deformation. It is also assumed that vertical deformation has no effect on torsional twist and the effect of prebuckling on in-plane deflections may be ignored because $EI_x >> EI_y >> GJ >> EI_w/L^2$. The equilibrium equation for out-of-plane bending (in terms of flexural resistance) is

$$M\frac{\mathrm{d}^2\phi(x)}{\mathrm{d}x^2} = -EI_y\frac{\mathrm{d}^4 w(x)}{\mathrm{d}x^4} \tag{7.4}$$

and the equation of equilibrium for torsion (in terms of warping and twisting resistance) is

$$M\frac{\mathrm{d}^2 w(x)}{\mathrm{d}x^2} = -EI_w\frac{\mathrm{d}^4\phi(x)}{\mathrm{d}x^4} + GJ\frac{\mathrm{d}^2\phi(x)}{\mathrm{d}x^2} \tag{7.5}$$

with boundary conditions

$$w(0) = w(L) = \frac{\mathrm{d}^2 w(0)}{\mathrm{d}x^2} = \frac{\mathrm{d}^2 w(L)}{\mathrm{d}x^2} = \phi(0) = \phi(L) = \frac{\mathrm{d}^2\phi(0)}{\mathrm{d}x^2} = \frac{\mathrm{d}^2\phi(L)}{\mathrm{d}x^2} = 0, \tag{7.6}$$

where L is the length of the beam or girder between lateral supports ($L = 0$ when members are continuously laterally supported at compression flanges).

Equations 7.4 and 7.5 are satisfied when (Trahair, 1993)

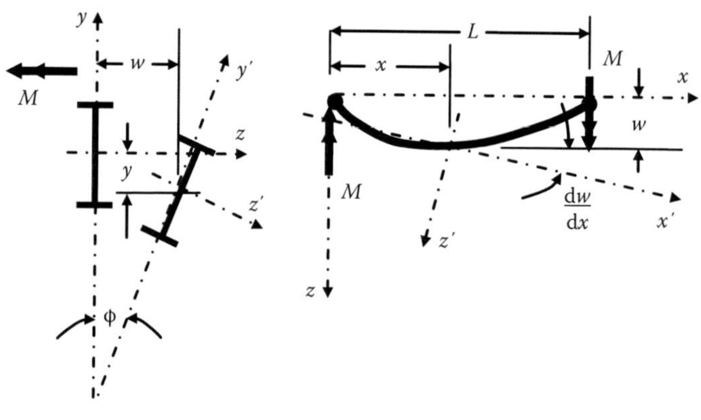

FIGURE 7.2 Bending of beam in buckled position.

$$\frac{w(L/2)}{w(x)} = \frac{\phi(L/2)}{\phi(x)} = \frac{1}{\sin \pi x/L}, \tag{7.7}$$

and

$$M_{cr} = \pi \sqrt{\frac{\pi^2 E^2 I_w I_y}{L^4} + \frac{EI_y GJ}{L^2}}, \tag{7.8}$$

where

$\phi(x)$ = the angle of twist about the shear center axis

E = the tensile modulus of elasticity [~200,000 MPa (29,000 ksi) for steel] (see Chapter 2)

G = the shear modulus of elasticity = $E/2(1+\upsilon)$

υ = Poisson's ratio (0.3 for steel)

J = the torsional constant, which depends on element dimensions. Equations for some common cross sections are given in Table 7.1.

$w(x)$ = the lateral deflection along the x axis

$I_y = \int x^2 \, dA$ = the lateral bending moment of inertia about the beam or girder vertical axis of symmetry

$I_w = \int \omega^2 \, dA = C_w$ = torsional moment of inertia or warping constant (ω is defined in terms of the position of the shear center and the thickness of the member). Warping constant values are available in many references (Roark and Young, 1982; Seaburg and Carter, 1997). Equations for some common cross sections are given in Table 7.1.

M_{cr} = critical lateral-torsional buckling moment

For an I-section with equal flanges and $t_w = t_f$ (see Table 7.1)

$$I_y = Ar_y^2 \approx 2\frac{t_f b^3}{12}, \tag{7.9}$$

$$I_w = C_w = \frac{h^2 t_f b^3}{24} = \frac{h^2 I_y}{4}, \tag{7.10}$$

$$J = 0.3At_f^2, \tag{7.11}$$

$$S_x = \frac{2Ar_x^2}{d}, \tag{7.12}$$

$$r_x \approx 0.4d \text{ (radius of gyration in the vertical direction)}, \tag{7.13}$$

$$r_y \approx 0.2b \text{ (radius of gyration in the lateral direction on the compression side of the neutral axis)}, \tag{7.14}$$

$$h \approx d, \tag{7.15}$$

$$G = \frac{E}{2(1+\upsilon)} = 0.38E. \tag{7.16}$$

Substitution of Equations 7.9 through 7.16 into 7.8 provides

TABLE 7.1
Torsional Warping Constants for Common Cross Sections

Cross Section ◆ Denotes position of shear center	Warping Constant, C_w Torsion Stiffness Constant, J
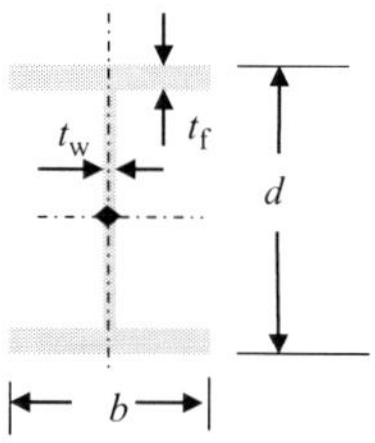	$h=d-t_f$ $C_w=\dfrac{h^2 t_f b^3}{24}$ $J=\dfrac{1}{3}\left(2bt_f^3+ht_w^3\right)$
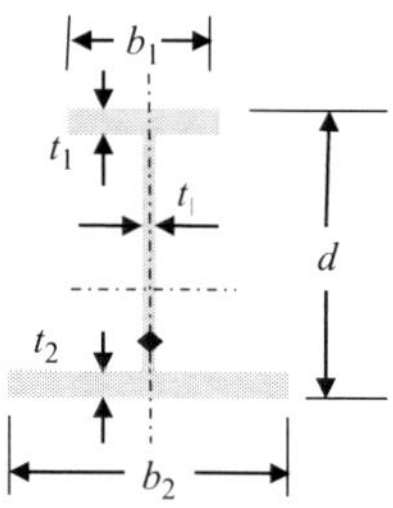	$h=d-\left(\dfrac{t_1}{2}\right)-\left(\dfrac{t_2}{2}\right)$ $C_w=\dfrac{h^2 t_1 t_2 b_1^3 b_2^3}{12(t_1 b_1^3+t_2 b_2^3)}$ $J=\dfrac{1}{3}\left(t_1^3 b_1+t_2^3 b_2+t_w^3 h\right)$
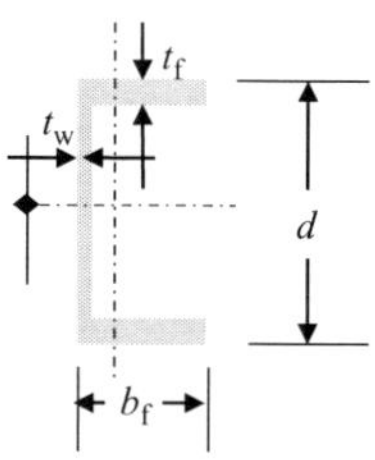	$h=d-t_f$ $b=b_f-t_w/2$ $C_w=\dfrac{t_f b^3 h^2}{12}\left(\dfrac{3bt_f+2ht_w}{6bt_f+ht_w}\right)$ $J=\dfrac{1}{3}\left(2bt^3+ht_w^3\right)$
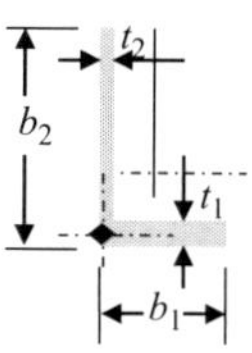	$b=b_1-\dfrac{t_2}{2}$ $h=b_2-\dfrac{t_1}{2}$ $C_w=\dfrac{1}{36}\left(b^3t_1^3+h^3t_2^3\right)$ (for small t_1 and $t_2 C_w \sim 0$) $J=\dfrac{1}{3}\left(bt_1^3+ht_2^3\right)$

$$f_{cr}=\sqrt{\left(\frac{15.4E}{(L/r_y)^2}\right)^2+\left(\frac{0.67E}{Ld/bt_f}\right)^2}, \tag{7.17}$$

where

$f_{cr}=M_{cr}/S_x=$ the critical lateral-torsional buckling stress

The first term in Equation 7.17 represents the warping torsion effects and the second term describes pure torsion effects. For torsionally strong sections (shallow sections with thick flanges), the warping effects are negligible and

$$f'_{cr}=\left(\frac{0.67E}{Ld/bt_f}\right)=\left(\frac{0.21\pi E}{Ld/bt_f}\right)=\left(\frac{0.24\pi E}{Ld\left(\sqrt{1+\upsilon}\right)/bt_f}\right). \tag{7.18}$$

For torsionally weak sections (deep sections with thin flanges and web, typical of railway plate girders), the pure torsion effects are negligible and

$$f''_{cr} = \left(\frac{15.4E}{(L/r_y)^2}\right) = \left(\frac{1.56\pi^2 E}{(L/r_y)^2}\right) \tag{7.19}$$

which is analogous to determining the elastic (Euler) column strength of the flange.

Using a factor of safety of 9/5 = 1.80, Equations 7.18 and 7.19 for torsionally strong and weak sections, respectively, are

$$F'_{cr} = \frac{f'_{cr}}{1.80} = \left(\frac{0.13\pi E}{Ld\left(\sqrt{1+\upsilon}\right)/bt_f}\right). \tag{7.20}$$

and

$$F''_{cr} = \frac{f''_{cr}}{1.80} = \left(\frac{0.87\pi^2 E}{(L/r_y)^2}\right). \tag{7.21}$$

Due to residual stresses, unintended load eccentricities, and fabrication imperfections, axial compression member strength is based on an inelastic buckling strength parabola when $F''_{cr} \geq F_y/2$ and a Euler (elastic) buckling curve for $F''_{cr} \leq F_y/2$ (see Chapter 6 on axial compression member behavior). Since the critical lateral-torsional buckling condition for torsionally weak sections corresponds to the elastic (Euler) column strength of the flange, a similar stability condition is assumed as shown in Figure 7.3.

Where $F''_{cr} = (0.55F_y)/2$, Equation 7.21 (elastic buckling curve) is equal and tangent to the inelastic buckling strength parabola (transition curve). From Equation 7.21, the slenderness, L/r_y at $(0.55F_y)/2$ is

$$\frac{L}{r_y} = 5.55\sqrt{\frac{E}{F_y}}. \tag{7.22}$$

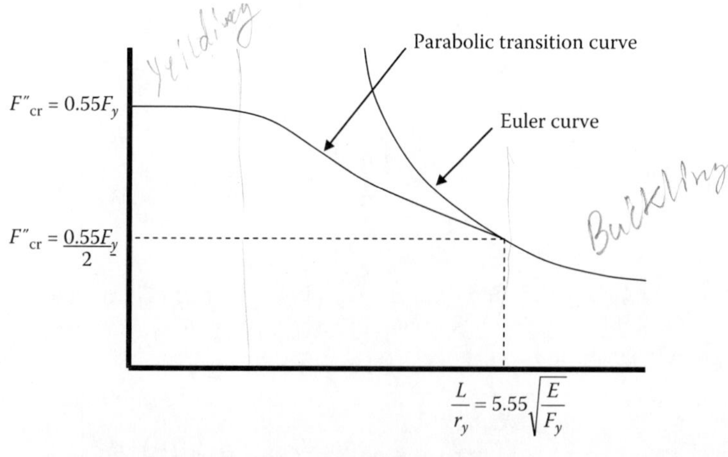

FIGURE 7.3 Lateral-torsional buckling curve for flexural compression.

The parabolic transition equation is

$$F''_{cr} = A - B\left(\frac{L}{r_y}\right)^2 \tag{7.23}$$

which becomes

$$F''_{cr} = 0.55F_y - \frac{0.55F_y^2}{6.3\pi^2 E}\left(\frac{L}{r_y}\right)^2 \tag{7.24}$$

with

$A = F''_{cr} = 0.55F_y$ (when $L/r_y = 0$)

$B = \dfrac{0.55F_y^2}{6.3\pi^2 E}$ (when $F''_{cr} = 0.55F_y/2$ and L/r_y given by Equation 7.22).

AREMA (2015) recommends a conservative approach using Equations 7.20 and 7.24 independently and adopting the larger of the two buckling stresses, F'_{cr} or F''_{cr}, for the design of flexural members. AREMA (2015) also restricts beam and girder slenderness to $L/r_y \le 5.55\sqrt{E/F_y}$ in order to preclude elastic buckling.

It should be noted that Equations 7.20 and 7.24 are developed based on the assumption of a uniform moment (no shear forces). Moment gradients related to concentrated or moving load effects on simply supported beams and girders can be considered through the use of modification factors (Salmon and Johnson, 1980). Modification factors, C_b, based on loading and support conditions are available in the literature of structural stability. The equations for pure and warping effects are then

$$F'_{cr} = \left(\frac{0.13\pi E C_b}{Ld\left(\sqrt{1+\upsilon}\right)/bt}\right), \tag{7.25}$$

$$F''_{cr} = 0.55F_y - \frac{0.55F_y^2}{6.3\pi^2 E C_b}\left(\frac{L}{r_y}\right)^2. \tag{7.26}$$

AREMA (2015) conservatively neglects this effect (C_b=1) and uses Equations 7.20 and 7.24 as the basis for steel beam and girder flexural design because the actual moment gradient along the unbraced length of a beam or girder is difficult to assess for moving train live loads.

7.2.3 Shearing of Beams and Girders

Shear stresses exist in beams and girders due to the change in bending stresses at adjacent sections. The distribution of shear stress through the cross section of a beam or girder can be complex. Simplifying assumptions are made to approximate the theoretical solution for use in routine design.

7.2.3.1 Shearing of Rectangular Beams

Solid rectangular sections are typically not used for steel superstructure members.* Nevertheless, the investigation of shear stresses across a rectangular section provides a basis for simplification of

* However, secondary elements of built-up members are often of rectangular section (see Chapter 6).

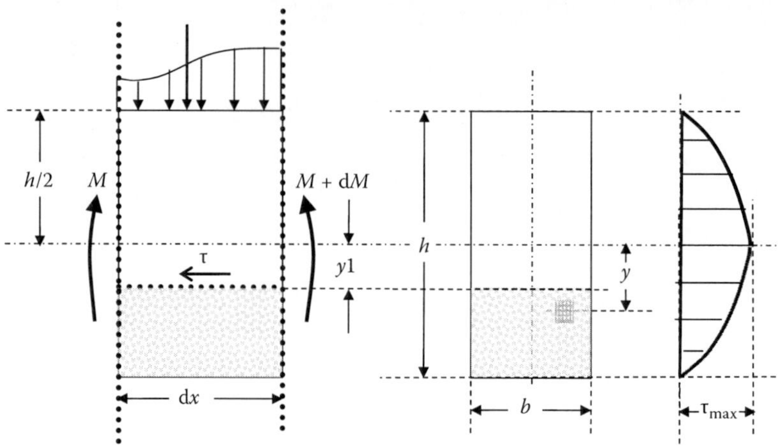

FIGURE 7.4 Shearing of a rectangular beam.

the theoretical solution for shear stresses across an I-shaped section. Shear stresses, τ, in rectangular sections can be assumed to act parallel to the vertical axis and be uniform across the thickness, b, of the beam (Figure 7.4).

Equilibrium of the normal forces across an infinitesimal distance, dx, provides

$$\tau b\mathrm{d}x = \int \frac{(M+\mathrm{d}M)y}{I}\,\mathrm{d}A - \int \frac{My}{I}\,\mathrm{d}A, \tag{7.27a}$$

$$\tau = \frac{\mathrm{d}M}{(Ib)\mathrm{d}x}\int y\,\mathrm{d}A = \frac{V}{Ib}\int y\,\mathrm{d}A. \tag{7.27b}$$

For a rectangular section,

$$\int y\,\mathrm{d}A = \int_{y1}^{h/2} yb\,\mathrm{d}y = b\left(\frac{y^2}{2}\right)_{y1}^{h/2} = \frac{b}{2}\left(\frac{h^2}{4} - y1^2\right). \tag{7.27c}$$

At the neutral axis, $y1 = 0$ and

$$\int y\,\mathrm{d}A = \frac{bh^2}{8}. \tag{7.27d}$$

Substitution of Equation 7.27d into 7.27b yields

$$\tau_{\max} = \frac{V}{I}\left(\frac{h^2}{8}\right) = \frac{VQ}{Ib}, \tag{7.27e}$$

where $Q = \dfrac{bh^2}{8}$.

Substitution of $I = bh^3/12$ into Equation 7.27e yields $\tau_{\max} = \dfrac{3V}{2A}$.

Therefore, the allowable shear force with $A = bh$ and $\tau_{\mathrm{all}} = 0.35F_y$ is

$$V_{\mathrm{all}} = 0.23F_y bh. \tag{7.27f}$$

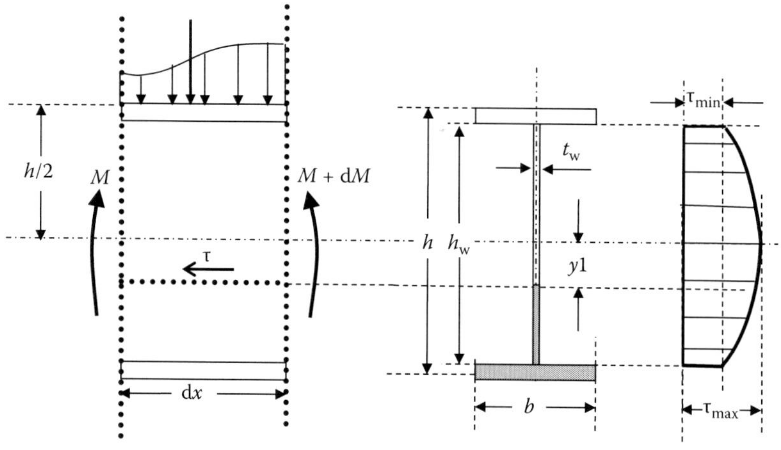

FIGURE 7.5 Shearing of shapes and plate girders.

7.2.3.2 Shearing of I-Shaped Sections

We allow the simplifying assumptions that shear stresses, τ, in I-shaped sections can also be assumed to act parallel to the vertical axis and be uniform across the thickness, t_w, of the web (Figure 7.5).*

$$\tau = \frac{dM}{(It_w)dx} \int y dA = \frac{V}{It_w} \int y dA. \tag{7.28a}$$

For an I-shaped section,

$$\int y dA = \frac{b(h^2 - h_w^2) + t_w(h_w^2 - 4y1^2)}{8}. \tag{7.28b}$$

At the neutral axis, $y1 = 0$, and

$$\int y dA = \frac{b(h^2 - h_w^2) + t_w(h_w^2)}{8}. \tag{7.28c}$$

Substitution of Equation 7.28c into 7.28a yields

$$\tau_{max} = \frac{V}{It_w}\left(\frac{b(h^2 - h_w^2) + t_w(h_w^2)}{8}\right) = \frac{VQ}{It_w}, \tag{7.28d}$$

where

$$Q = \frac{b(h^2 - h_w^2) + t_w(h_w^2)}{8}.$$

* This assumption is acceptable for I-sections, but not theoretically correct. Since shear stresses act parallel to the surfaces of a member, the flange shear should be determined on vertical sections through the flange rather than horizontal sections. Shear stresses evaluated on horizontal sections through the flange will provide a distribution typically referred to as the "top hat" distribution across the cross section. This distribution is incorrect, but sufficiently accurate for practical I-sections.

At the flange-to-web interface, $y1 = h_w/2$, and

$$\int y\,dA = \frac{b(h^2 - h_w^2)}{8}. \tag{7.28e}$$

Substitution of Equation 7.28e into 7.28a yields

$$\tau_{min} = \frac{V}{It_w}\left(\frac{b(h^2 - h_w^2)}{8}\right) = \frac{VQ}{It_w}, \tag{7.28f}$$

where

$$Q = \frac{b(h^2 - h_w^2)}{8}.$$

7.2.3.3 Design for Shearing of Shapes and Plate Girders

Considering that typical values of $(h - h_w)$ will be small, Equation 7.28d yields

$$\tau_{max} = \frac{V}{It_w}\left(\frac{b(h^2 - h_w^2) + t_w(h_w^2)}{8}\right) \approx \frac{V}{It_w}\left(\frac{t_w(h_w^2)}{8}\right) \tag{7.29}$$

and $Q \approx \left(\frac{t_w(h_w^2)}{8}\right)$ = the statical moment of area about the neutral axis for the web plate only. Equation 7.29 considering the web plate only with $I_{web} = t_w h_w^3/12$ may be expressed as $\tau_{max} \approx 3V/(2t_w h_w)$, which is the shear stress on a rectangular section, $h_w t_w$. Therefore, considering the shear force acting over the web plate area only is acceptable for routine superstructure design and the average shear stress is

$$\tau_{avg} \approx \frac{V}{t_w h_w} \tag{7.30}$$

which is typically within 10% of the maximum shear stress (Gere and Timoshenko, 1984).*

AREMA (2015) ASD uses an allowable shear stress based on tensile yield stress $\left(\tau_y = F_y/\sqrt{3}\right.$, see Chapter 2). The allowable shear stress is $\tau_{all} = 0.55F_y/\sqrt{3} = 0.32F_y$, and AREMA (2015) uses $0.35F_y$. Therefore, the allowable shear force is

$$V_{all} = 0.35F_y t_w h_w. \tag{7.31}$$

7.2.4 Biaxial Bending of Beams and Girders

Biaxial bending is not generally a concern for ordinary steel railway longitudinal beams and girders. However, in some cases, biaxial bending of stringers and floor beams or unsymmetrical bending of floor beams† may warrant consideration (see Chapter 8).

Stresses in perpendicular principal directions may be superimposed at critical symmetric sections, as shown in Equations 7.32 and 7.33.

* This is typically an accurate though slightly nonconservative approach to shear design.

† May occur at transverse floor beams with horizontal longitudinal live load forces applied due to traction and/or braking (see Chapter 4).

$$1.0 \le \pm \frac{M_x}{F_{bx}S_x} \pm \frac{M_y}{F_{by}S_y}, \tag{7.32}$$

$$F_v \le \pm \frac{V_x Q_x}{I_x t_x} \pm \frac{V_y Q_y}{I_y t_y}, \tag{7.33}$$

where

F_{bx}, F_{by} = the allowable bending stress in the x and y directions, respectively
F_v = the allowable shear stress

7.2.5 Preliminary Design of Beams and Girders

For planning purposes, the preliminary proportioning of plate girders may be necessary before detailed design. Preliminary dimensions may be needed for estimating weight and cost, and assessing site geometry constraints for erection. Preliminary proportions may also be required to estimate splice requirements, and for developing fabrication (see Chapter 10), shipping, and erection methodologies (see Chapter 11).

There are various techniques used by experienced bridge engineers to assess the preliminary dimensions of beams and girders. One method is by simplification of the bending resistance from equal flanges and webs independently as (Figure 7.6)

$$M = M_f + M_w = bt_f(d - t_f)(f_{bf}) + \left(\frac{t_w(h)^2}{6}\right)(f_{bw}), \tag{7.34}$$

where

M_f = the moment carried by flanges
M_w = the moment carried by webs
f_{bf} = the allowable flange bending stress
f_{bw} = the allowable web bending stress

Assuming that $f_{bf} \sim f_{bw} \sim f_b$ and for usual railway beams and girders $h \sim (d - t_f)$, Equation 7.34 yields

$$bt_f = A_f = \frac{M}{f_b h} - \frac{A_w}{6}. \tag{7.35}$$

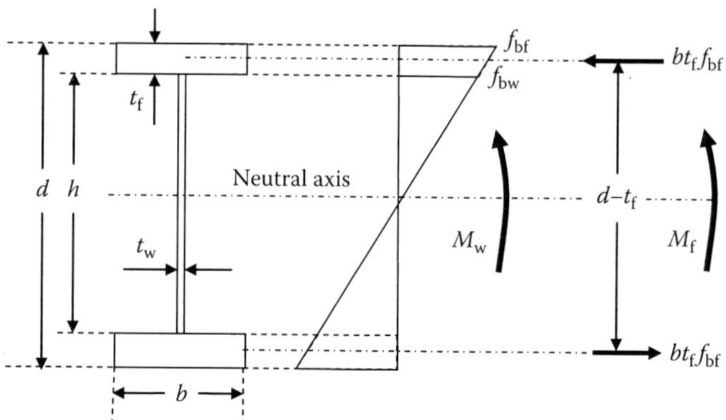

FIGURE 7.6 Preliminary proportioning of girder flanges.

An estimate of flange size can be made based on preliminary web height and web thickness, $t_w \geq V/0.35F_y h$, where V is the maximum applied shear force, as

$$A_f = \frac{M}{f_b h} - \frac{V}{0.35F_y(6)} = \frac{M}{f_b h} - \frac{V}{2.10F_y}. \tag{7.36}$$

The preliminary web height is typically estimated from typical *L*/*d* ratios for economic design (see Chapter 3), available plate sizes, plate slenderness, underclearance requirements, existing substructure geometry, and/or aesthetic considerations.

Steel freight railway girder spans can be economically designed with a minimum depth to span ratio of about 1/15. Typically, web height is constrained by factors such as, available plate sizes or site geometry. In such cases, plate girders may still be reasonably economically designed with depth-to-span ratios in the range of 1/10–1/12, and in some cases as deep as about 1/8.

If web height is unconstrained by factors such as, available plate sizes or site geometry, an optimum web height can be estimated using Equation 7.35 to determine the beam or girder cross-sectional area, A, as

$$A = 2A_f + A_w = 2\left(\frac{M}{f_b h} + \frac{t_w h}{3}\right). \tag{7.37}$$

Minimizing the cross-sectional area, A, in terms of web height, h, with t_w constant (assuming shear strength governs) provides an estimate of optimum web height, h_{opt}, as

$$\frac{\partial A}{\partial h} = 0 = \frac{-M}{f_b h^2} + \frac{t_w}{3}, \tag{7.38a}$$

$$h_{opt} = \sqrt{\frac{3M}{f_b t_w}}. \tag{7.38b}$$

Minimizing the cross-sectional area, A, in terms of web height, h, with (h/t_w) constant (assuming stability governs) provides an estimate of optimum web height, h_{opt}, as

$$\frac{\partial A}{\partial h} = 0 = \frac{-M}{f_b h^2} + \frac{2h}{3(h/t_w)} \tag{7.39a}$$

$$h_{opt} = \sqrt[3]{\frac{3M}{2f_b}\left(\frac{h}{t_w}\right)}. \tag{7.39b}$$

For typical steel railway plate girders, stability governs web plate design and the approximate optimum web height is best represented by Equation 7.39b, where M = maximum bending moment and f_b = allowable bending stress = $0.55F_y$. The optimum web height can be expected to be within about 10% of this estimated value. Once the estimated optimum web height has been established, flange size and web thickness can be estimated based on experience and approximations that flange width, b, should typically be between $d/4$ and $d/3$. Flange plate thickness, t_f, may be estimated as between $b/12$ and $b/5$ with due consideration given to compression flange buckling and practical thickness for fabrication. Web plate thickness, based on stability considerations, may be estimated as a minimum of $h/160$ for F_y = 250 MPa (36 ksi) steel, $h/135$ for F_y = 350 MPa (50 ksi) steel and $h/115$ for F_y = 490 MPa (70 ksi) steel (see Equation 7.60b).

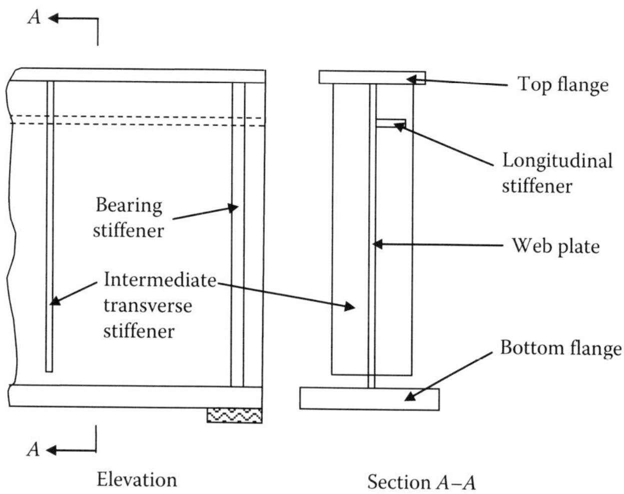

FIGURE 7.7 Cross section of a modern plate girder.

7.2.6 Plate Girder Design

Modern plate girders typically consist of welded flange and stiffened web plates (Figure 7.7). Railway bridge girders are generally of large size and ensure that the stability of plate elements in compression zones and consideration of stress concentration effects in tension zones are critical components of the design.

Simple span plate girders of about 45 m (150 ft)* long or less are generally economical for railway superstructure construction (fabrication and erection). Longer spans are feasible using continuous construction. However, continuous spans are less frequently used due to uplift considerations (related to the large live load to dead load ratio of steel railway superstructures) and stresses imposed by foundation movements that may not be able to be considered during design.

Box girders are also relatively rare in railway superstructure construction due to welded stiffener fatigue concerns but may be used where large torsional stiffness is required (for example, curved bridges).

Hybrid girders using both high-performance steel (HPS) and high-strength low-alloy (HSLA) steels (see Chapter 2) may be economical for some long span applications (particularly over supports of continuous spans where flexural and shear stresses may be relatively high). However, many ordinary steel railway superstructure designs will be governed by deflection (which is stiffness related) and fatigue requirements (see Chapter 5). The serviceability and fatigue limit states are essentially material independent for structural steel. These criteria and fabrication issues must be carefully considered when assessing the use of hybrid girders with plate elements of differing steel type and grade.

The main elements of a plate girder (flange plates, web plates, main element splices, bearing stiffeners, and their respective connections) are designed to resist tensile and compressive normal axial and bending stresses, and shear stresses in the cross section. In ASD, secondary elements (stiffener plates) are designed to provide stability to the main elements of the girder cross section.

7.2.6.1 Main Girder Elements

Flanges of modern welded plate girders are made using a single plate in the cross section. The required thickness and width of the flange plates will be governed by strength, stability, fatigue, and serviceability criteria. Cover plates should not be used as flange plate dimensions may be varied

* These limits typically apply to two-girder superstructures. Four-girder superstructures up to about 60 m (200 ft) in length are also reasonably economical.

along the length of the girder as required.* The thickness of flange plates may be limited by issues related to steelmaking, cost, design, and fabrication.

Nonuniform cooling of thick plates during the steelmaking process detrimentally affects fracture toughness. Modern steelmaking processes such as thermo mechanical control process (see Chapter 2) may alleviate many metallurgical concerns related to thick plates, but designers should carefully review other issues related to cost, design, and fabrication.

For steel plate girder flange design, costs are related to

- Plate length availability and requirements for flange splices,
- The raw material of thicker plate,
- Fabrication handling, cutting, and welding of thicker plates, and
- The increase of cross-sectional area for typical girders with thick flange plates require careful consideration during preliminary design.

The design of flange plates to minimize typical plate girder cross-sectional area should consider thin flange plates that

- Reduce the shear flow and weld size at the flange-to-web connection, and
- Utilize wider flanges† that resist lateral-torsional buckling of compression flanges.

The fabrication of steel plate girder flanges must consider

- The welding details, materials, and procedures required for thick plates, and
- The effects of welding on lamellar tearing‡ and toughness.§

Modern railway girder superstructures with spans less than about 50 m (160 ft) long can be economically designed with flanges less than 70 mm (2–3/4 in.) thick. Plate girder designs with a maximum flange thickness of about 65 mm (2–1/2 in.) will typically preclude concerns with respect to steelmaking, cost, design, and welded fabrication.

Webs of plate girders are generally relatively thin plates when designed in accordance with shear strength criteria. However, because of slenderness, the required thickness of the web plate also depends on flexural and shear stability considerations. Therefore, in order to avoid thick plates, the web plates are often stiffened longitudinally (to resist flexural buckling) and/or transversely (to resist shear buckling). The height of web plates in long plate girders may be quite large and designers should carefully review available plate sizes from steelmaking, fabrication, and shipping perspectives in order to avoid costly longitudinal splices and limit, to within practical requirements, transverse splices.

Welded or bolted splices may be used when required due to available plate length, shipping, and/or erection limitations. If shipping and/or erection considerations do not govern, welded shop splices are typically used.¶ Bolted splices are generally used to accommodate shipment or erection constraints and may be used, if approved by the design engineer, for splices required due to limitations regarding plate size availability. These splices are generally designed for the shear and bending moments at the spliced section and/or specific strength criteria recommended by AREMA (2015).

* Designers should note that it is often less costly to fabricate flanges without changes in dimension due to butt welding and transitioning requirements. Designs that utilize varying flange plate widths and lengths may be desirable and economical for long span fabrication, but consultation with experienced bridge fabricators is often warranted.

† For through plate girders, minimum required lateral clearances must be considered.

‡ Lamellar tearing occurs in thick plates due to large through-thickness strains produced by fabrication process effects such as weld metal shrinkage at highly restrained locations (i.e., joints and connections).

§ Ductility is affected by the triaxial strains created by weld shrinkage and restraint in thick plates (see Chapters 2 and 10).

¶ Typically, there are more stringent fabrication QC and QA requirements for welded splices in tension flanges (see Chapter 10).

The connection of web and flange plates in modern plate girder fabrication is generally performed by high quality automatic welding (see Chapter 10). These welds must be designed to resist the total longitudinal shear at the connection as well as other loads directly applied to the flanges and/or web (see Chapter 9).

7.2.6.1.1 Girder Tension Flanges and Splices

7.2.6.1.1.1 Tension Flanges Overall, girder bending capacity at service loads is not reduced by the usual number and pattern of holes in a girder cross section (the same capacity as for the gross section but stresses are distributed differently because of localized stress concentrations). Nevertheless, AREMA (2015) recommends that girder tension flanges can be designed based on the moment of inertia of the entire net section, I_{xn}, (using the neutral axis determined from the gross section*) and the tensile yield stress. However, the design based on the moment of inertia of the entire net section is appropriate for tension flange splices and as protection from the effects of occasional tensile overload stresses. Many designers' proportion tension flanges are based on net section properties, being not greater than 85% of the gross-sectional properties.

The plate girder net section modulus, S_{xn}, is determined as

$$S_{xn} = \frac{I_{xn}}{c_t} \geq \frac{M_{t\max}}{F_{all}} \geq \frac{M_{t\max}}{0.55F_y}, \tag{7.40}$$

or

$$S_{xn} \geq \frac{\Delta M}{F_{fat}}, \tag{7.41}$$

where

c_t = the distance from the neutral axis to the extreme fiber in tension

M_{tmax} = the maximum tensile bending moment due to all load effects and combinations (see Chapter 4)

ΔM = the maximum bending moment range due to fatigue load (see Chapter 4)

F_{fat} = the allowable fatigue stress range for the appropriate fatigue detail category (see Chapter 5).

Residual stresses, which must be considered in the design of dynamically loaded axial and flexural tensile members, and all axial compression members, are of negligible concern in the design of statically loaded bending members. This is because the presence of residual stresses may cause an initial inelastic behavior but subsequent statically applied loads of the same or smaller magnitude will result in elastic behavior (Brockenbrough, 2011). Residual stresses are not explicitly considered in fatigue design because tensile fatigue strength is based on nominal stress tests on elements and members containing residual stresses from manufacture or fabrication. Therefore, residual stresses are generally not explicitly considered in the design of bending members such as plate girders.

7.2.6.1.1.2 Tension Flange Splices AREMA (2015) recommends that splices in main members have strength not less than that of the member being spliced, regardless of the actual forces at the splice location. It is also recommended that splices in girder flanges can be comprised of

* The neutral axis must consist of a smooth line because fiber stresses cannot suddenly change and will vary only slightly at the typically few cross sections with holes. Therefore, the neutral axis will be essential at the location of the neutral axis of the gross section along the girder length.

elements which are not lesser in section than the flange element being spliced. Two elements in the same flange cannot be spliced at the same location.*

Therefore, bolted† splice elements in girder flanges should

- Have a cross-sectional area that is at least equal to that of the flange element being spliced, and
- Be constructed of splice elements of sufficient cross section and location such that the moment of inertia of the spliced member is no less than that of the member alone at the splice location.

Splice elements may be single or double plates. Single plate splices are generally used on the exterior surfaces of flange plates to ensure a greater moment of inertia at the splice.‡ Two plates§ are often used for larger girder splices where single-shear bolted connections are too long and a double-shear connection is required.

The splice fasteners (see Chapter 9) should be designed to transfer the force in the element being spliced to the splice material. Welded splices are usually made with complete joint penetration (CJP) (full penetration) groove welds with strength at least equal to the base material being spliced.

7.2.6.1.2 Girder Compression Flanges and Splices

7.2.6.1.2.1 Compression Flanges AREMA (2015) recommends that girder compression flanges can be designed based on the moment of inertia of the entire gross section, I_{xg}, and the tensile¶ yield stress.

Therefore, the plate girder gross-sectional modulus, S_{xg}, is

$$S_{xg} = \frac{I_{xg}}{c_c} \geq \frac{M_{cmax}}{F_{call}}, \tag{7.42}$$

where

c_c = the distance from the neutral axis to the extreme fiber in compression

M_{cmax} = the maximum compressive bending moment due to all load effects and combinations (see Chapter 4)

F_{call} = the allowable compressive stress, which is based on stability considerations as the girder compression flange is susceptible to lateral-torsional instability prior to yielding

In addition to lateral-torsional buckling effects on the allowable compressive stress, vertical and torsional buckling effects must be considered to ensure compression flange stability.

7.2.6.1.2.1.1 Lateral-Torsional Buckling Compression flange lateral-torsional instability is controlled by limiting allowable stresses to those given by Equations 7.20 and 7.24. However, Equation 7.20 was developed assuming an I-section with equal flanges. Therefore, the smallest flange area, $A_f = bt_f$, should be used in Equation 7.20 when establishing the critical buckling stress. It should also be noted that Equation 7.24 precludes sinelastic buckling by ensuring that L/r_y of the girder compression zone does not exceed the value of Equation 7.22, which is presented again as Equation 7.43. The length, L, in Equation 7.22 is the largest distance between compression flange lateral supports, L_p, and r_{cy} is determined as the

* This is applicable to built-up section flanges, which are not often used for modern plate girder fabrication.

† Shop or field splices.

‡ Single plate splices are typically restricted to small span beams.

§ It is a good practice that the centroid of the splice plates on each side of the flange plate can be coincident with the centroid of the flange being spliced.

¶ Tensile yield stress is almost equal to compressive yield stress (Chapter 2).

minimum radius of gyration of the compression flange and portion of the web in compression (from neutral axis to edge of web plate).

The larger of either Equations 7.20 and 7.24, presented again as Equations 7.44 and 7.45, respectively, is adopted to determine the allowable compressive bending stress, F_{call}, for design of the compression flange. Therefore, the compression flange design requirements are

$$\frac{L_p}{r_{cy}} \leq 5.55\sqrt{\frac{E}{F_y}} \tag{7.43}$$

and the larger of

$$F_{call} = \left(\frac{0.13\pi E}{L_p d\left(\sqrt{1+\upsilon}\right)/A_f}\right) \tag{7.44}$$

or

$$F_{call} = 0.55F_y - \frac{0.55F_y^2}{6.3\pi^2 E}\left(\frac{L_p}{r_{cy}}\right)^2, \tag{7.45}$$

where

L_p = the largest distance between compression flange lateral supports

$A_f = bt_f$ = the area of the smallest flange in the girder (even if tension flange)

r_{cy} = the minimum radius of gyration of the compression flange and that portion of the web in compression. However, as shown in Figure 7.3, F_{call} cannot exceed $0.55F_y$

7.2.6.1.2.1.2 Vertical Flexural Buckling If the web plate buckled due to bending in the compression zone, it would be unable to provide support for the attached compression flange and the compression flange could then buckle vertically as shown in Figure 7.8. To avoid compression flange vertical buckling, flexural buckling of the web plate is precluded by limiting the web height, h, to thickness, t_w, ratio or by including a longitudinal stiffener.

The critical elastic buckling stress of a rectangular plate is

$$F_{cr} = \frac{k\pi^2 E t_w^2}{12(1-\upsilon^2)h^2}, \tag{7.46}$$

where

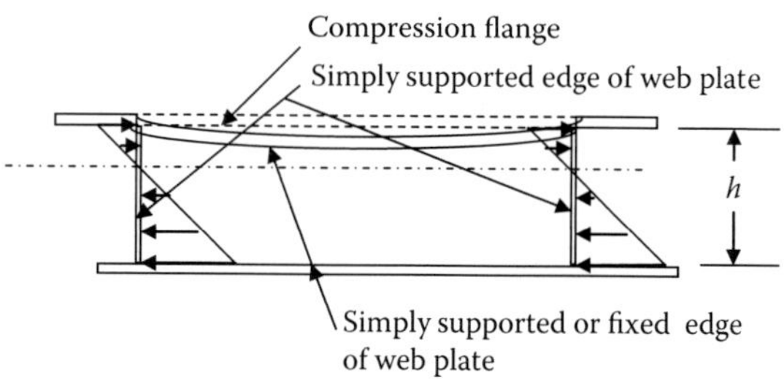

FIGURE 7.8 Pure flexural buckling of the web plate causing compression flange buckling.

F_{cr} = the critical buckling stress
k = a buckling coefficient depending on loading and plate edge conditions
υ = Poisson's ratio (0.3 for steel)

k ranges from 23.9 for simply supported edge conditions to 39.6 for fixed edge conditions assumed at the two edges (at the flanges) of a long plate in pure bending (Timoshenko and Woinowsky-Kreiger, 1959). AREMA (2015) conservatively uses $k = 23.9$ and reduces the web height to thickness ratio to 90% of the theoretical value to account for web geometry imperfections. Rearrangement and substitution of $k = 23.9$ into Equation 7.46 yield

$$\frac{h}{t_w} \leq 0.90\sqrt{\frac{23.9\pi^2 E}{12(1-0.3^2)(F_{cr})}} \leq 4.18\sqrt{\frac{E}{F_{cr}}} \tag{7.47}$$

which will preclude elastic buckling due to pure bending (Figure 7.9).

Rearrangement of Equation 7.47 provides

$$t_w \geq 0.24h\sqrt{\frac{F_{cr}}{E}}. \tag{7.48}$$

and if $F_{cr} = 0.55F_y$,

$$t_w \geq 0.18h\sqrt{\frac{F_y}{E}}. \tag{7.49}$$

The allowable compressive bending stress, F_{cr}, is given by Equations 7.44 and 7.45. Therefore, when the actual calculated flexural stress at the compression flange, f_c, is less than F_{cr}

$$t_w \geq 0.18h\sqrt{\frac{F_y}{E}}\sqrt{\frac{f_c}{F_{cr}}} \tag{7.50}$$

and longitudinal stiffeners are not required for web flexural buckling stability.

However, in cases where longitudinal stiffeners are provided, the minimum web thickness criteria to avoid flexural buckling (and thereby prevent vertical buckling of the compression flange) of Equation 7.50 are reduced. The optimum location for longitudinal web plate stiffeners is at $0.22h$ from the compression flange (Rockey and Leggett, 1962). When a longitudinal stiffener is placed

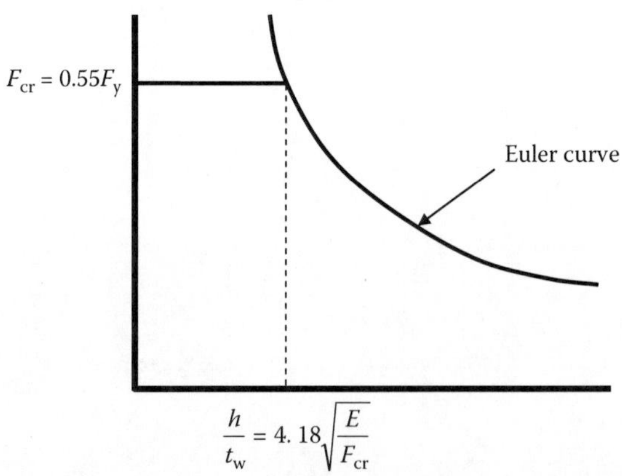

FIGURE 7.9 Elastic buckling curve for rectangular plate buckling under pure bending.

at $h/5$ from the inside surface of the compression flange, as recommended by AREMA (2015), the critical elastic buckling stress of a rectangular plate (Equation 7.46) is

$$F_{cr} = \frac{129\pi^2 E t_w^2}{12(1-\upsilon^2)h^2} \tag{7.51}$$

with theoretical $k = 129$ (Galambos, 1988). Equation 7.51 with $F_{cr} = 0.55F_y$ can be expressed as

$$t_w \geq 0.08h\sqrt{\frac{F_y}{E}}. \tag{7.52}$$

Equation 7.52 indicates that the web thickness to preclude elastic critical flexural buckling of the web with a longitudinal stiffener can be 43% ($\sqrt{23.9/129}$) of that required without a longitudinal stiffener (Equation 7.49). AREMA (2015) recommends that the web thickness with a longitudinal stiffener be no less than 50% of that required without a longitudinal stiffener.

7.2.6.1.2.1.3 Torsional Buckling Torsional buckling of the compression flange is essentially the buckling problem of uniform compression on a plate free at one side and partially restrained at the other (Salmon and Johnson, 1980). The critical elastic plate buckling stress is

$$F_{cr} = \frac{k\pi^2 E t_f^2}{12(1-\upsilon^2)b^2} \tag{7.53}$$

and the limiting width-to-thickness ratio* at $F_{cr} = F_y$ is

$$\frac{b}{2t_f} \leq \sqrt{\frac{k\pi^2 E}{12(1-0.3^2)F_y}} = 0.95\sqrt{\frac{kE}{F_y}}. \tag{7.54}$$

However, this is an elastic buckling curve, and at $F_{cr} = F_y$, the plate axial strength is overestimated (above the transition curve as shown in Figure 7.10).

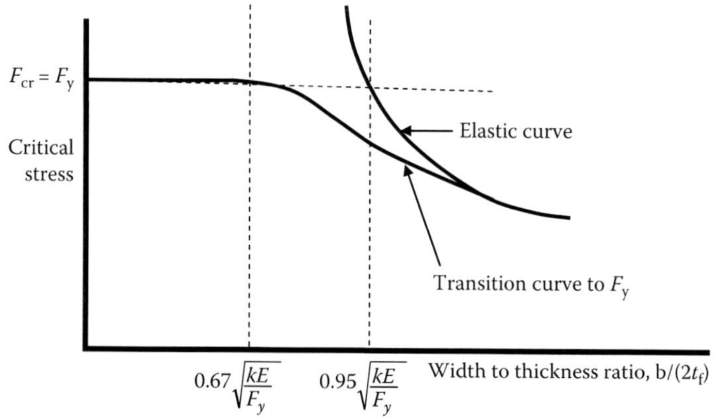

FIGURE 7.10 Plate buckling curve for uniform compression.

* Based on the yield strength of the plate.

To mitigate this, it is customary to use a limiting width-to-thickness ratio of

$$\frac{b}{2t_f} \le 0.67\sqrt{\frac{kE}{F_y}} \tag{7.55}$$

which is the approximate value corresponding to the transition curve at $F_{cr}=F_y$. For plates with a free edge, the buckling coefficient, k, is 0.425 with one edge considered as simply supported and 1.277 with another edge considered as fixed (Bleich, 1952). Tests have indicated that the lowest value of buckling coefficient, k, for partially restrained elements is about 0.70 (typical of a girder flange) (Tall, 1974). Therefore, substitution of $k=0.7$ into Equation 7.55 yields

$$\frac{b}{2t_f} \le 0.56\sqrt{\frac{E}{F_y}}. \tag{7.56}$$

AREMA (2015) recommends that this width-to-thickness ratio for local flange buckling can be decreased further based on practical experience with local compression forces from the deck, ties,* fabrication tolerances, and other unaccounted effects. The recommended compression flange width-to-thickness ratio (with FS = 1.30) is

$$\frac{b}{2t_f} \le 0.43\sqrt{\frac{E}{F_y}}, \tag{7.57}$$

where no ties bear directly on the compression flanges and (with FS = 1.60)

$$\frac{b}{2t_f} \le 0.35\sqrt{\frac{E}{F_y}}, \tag{7.58}$$

where ties bear directly on the compression flanges.

7.2.6.1.2.2 Compression Flange Splices Splices in girder compression flanges are treated in a similar manner to those in tension flanges. The requirements are outlined in the section on girder tension flange splice design.

7.2.6.1.3 Girder Web Plates and Splices

7.2.6.1.3.1 Web Plates Economical railway girders have relatively thin web plates. Therefore, in addition to designing the web plate to carry shear forces (Equation 7.31), it is also necessary to ensure stability of the web plates in girders. Figure 7.11 indicates the forces on the web plate that may create instability.

The stability criteria for shear, bending, and compression forces are developed separately and combined to investigate web plate stability. Inelastic buckling, due to residual stresses, load eccentricities, and geometric tolerances, is modeled by a buckling strength transition parabola formulation consistent with other structural stability criteria (e.g., axial and flexural member compression).

7.2.6.1.3.1.1 Elastic Buckling under Pure Bending Flexural buckling of the web plate is precluded by limiting the web height, h, to thickness, t_w, ratio or including a longitudinal stiffener.

* Particularly, if poorly framed.

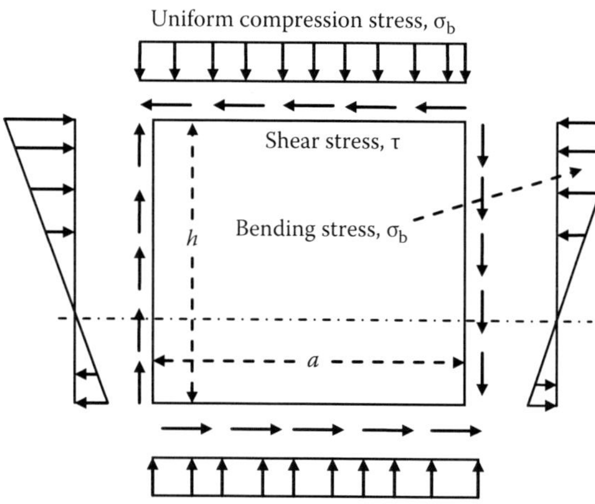

FIGURE 7.11 Stresses on girder web plates.

The elastic buckling of the web plate under bending was considered above in conjunction with the investigation of vertical buckling of the compression flange in Section 7.2.6.1.2.1.2.

Web plate design without longitudinal stiffeners considering Equation 7.50, presented again as Equation 7.59, requires a minimum web plate thickness to preclude elastic flexural buckling of

$$t_w \geq 0.18h\sqrt{\frac{F_y}{E}}\sqrt{\frac{f_c}{F_{cr}}} \geq 0.24h\sqrt{\frac{f_c}{E}}. \tag{7.59}$$

Rearrangement of Equation 7.59 and substitution of $F_{cr}=0.55F_y$ (to preclude elastic buckling per Figure 7.9) yield the criteria that

$$\frac{h}{t_w} \leq 4.18\sqrt{\frac{E}{f_c}}. \tag{7.60a}$$

Otherwise, longitudinal stiffeners are required for web flexural buckling stability.

The limit for h/t_w with $f_c=0.55F_y$ in Equation 7.60a is

$$\frac{h}{t_w} \leq 5.64\sqrt{\frac{E}{F_y}} \tag{7.60b}$$

which is 160 for $F_y=250$ MPa (36 ksi) steel, 135 for $F_y=350$ MPa (50 ksi) steel, and 115 for $F_y=490$ MPa (70 ksi) steel.

Web plate design with a longitudinal stiffener at $0.20h$ from the compression flange considering Equation 7.52, presented again as Equation 7.61, requires a minimum web plate thickness to preclude elastic flexural buckling of

$$t_w \geq 0.08h\sqrt{\frac{F_y}{E}}\sqrt{\frac{f_c}{F_{cr}}} \geq 0.10h\sqrt{\frac{f_c}{E}}. \tag{7.61}$$

AREMA (2015) recommends a minimum web plate thickness of

$$t_w \geq 0.12h\sqrt{\frac{f_c}{E}}. \tag{7.62}$$

7.2.6.1.3.1.2 Elastic Buckling under Pure Shear The critical elastic plate buckling shear stress is

$$\tau_{cr} = \frac{k\pi^2 E t_w^2}{12(1-\upsilon^2)h^2}, \tag{7.63}$$

where

k = 4.0 + 5.34/(a/h)2 for a/h ≤ 1

$k = \frac{4.0}{(a/h)^2} + 5.34 \quad \text{for } \frac{a}{h} > 1$ (However, AREMA (2015) does not permit $a/h > 1$.)

$k = 5.34$ for infinitely long simply supported plate under pure shear (Timoshenko and Gere, 1961) (see Figure 7.12)

Shear yield stress, τ_y, is related to tensile yield stress, F_y, as (see Chapter 2)

$$\tau_{cr} = \tau_y = \frac{F_y}{\sqrt{3}}. \tag{7.64}$$

Therefore, from Equation 7.63

$$\frac{h}{t_w} \le 2.89\sqrt{\frac{E}{F_y}}. \tag{7.65}$$

7.2.6.1.3.1.3 Inelastic Buckling under Pure Shear In order to account for residual stresses (which, however, are generally not large in girder webs) and geometric eccentricities (such as out-of-flatness, which is typical in girder webs), Equation 7.65 is reduced to

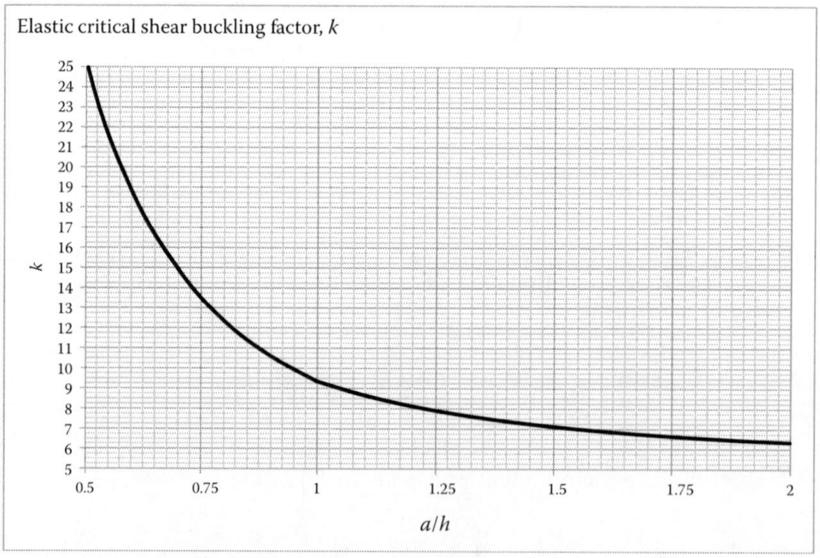

FIGURE 7.12 Elastic buckling under pure shear.

$$\frac{h}{t_w} \le 2.12\sqrt{\frac{E}{F_y}} \tag{7.66}$$

for design purposes. Therefore, if $h \ge 2.12 t_w \sqrt{E/F_y}$, transverse web stiffeners are required.

It is well known that if shear stresses are increased beyond the elastic critical buckling stress (Equation 7.63), plate girder webs do not collapse but exhibit postbuckling strength through tension field action in the web.

7.2.6.1.3.1.4 Postbuckling Shear Strength of the Web The postbuckling shear strength, V_{tf}, arises from tension field action in the web following elastic buckling. Horizontal force equilibrium and taking moments about point O in Figure 7.13 yields

$$\Delta F_f = (\sigma_t t_w a \sin\theta)\cos\theta = \frac{\sigma_t t_w a \sin 2\theta}{2}, \tag{7.67}$$

$$\Delta F_f\left(\frac{h}{2}\right) - \frac{V_{tf} a}{2} = 0, \tag{7.68}$$

where σ_t is the tensile membrane stress in the web that develops following elastic buckling and is equal to $F_y - \sqrt{3}\tau_{cr}$ from consideration of the von Mises failure criterion (see Chapter 2) for combined shear and inclined tensions (at an angle θ in Figure 7.13).

Substitution of Equation 7.67 into 7.68 yields

$$V_{tf} = \frac{\sigma_t h t_w \sin 2\theta}{2} = \frac{\sigma_t h t_w \left(\dfrac{1}{\sqrt{1+(a/h)^2}}\right)}{2} = \frac{\sigma_t h t_w}{2\sqrt{1+(a/h)^2}}. \tag{7.69}$$

7.2.6.1.3.1.5 Ultimate Shear Buckling Strength of the Web The ultimate shear buckling strength is the elastic critical shear buckling strength plus the postbuckled shear strength of the girder web plate. The elastic critical shear buckling strength (from Equation 7.63) is

$$V_{cr} = \frac{\left(4.0 + \dfrac{5.34}{(a/h)^2}\right)\pi^2 E t_w^3}{12(1-\upsilon^2)h}, \tag{7.70}$$

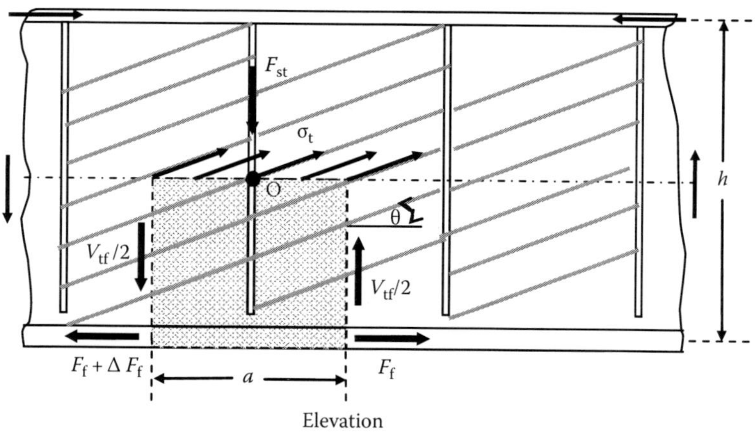

FIGURE 7.13 Tension field in the web plate.

and the ultimate shear buckling strength [for $(a/h) \le 1$ as required by AREMA (2015)] is

$$V_u = \frac{\left(4.0 + \frac{5.34}{(a/h)^2}\right)\pi^2 E t_w^3}{12(1-\upsilon^2)h} + \frac{\sigma_t h t_w}{2\sqrt{1+(a/h)^2}}. \tag{7.71}$$

The ratios of τ_{tf}/τ_{cr} for values of a/h between 0 and 2 for various values of F_y/τ_{cr} are shown in Figure 7.14. The figure illustrates that τ_{tf}/τ_{cr} is low (about 0.2 at $a/h = 1$) for τ_{cr} greater than about $F_y/2$, which is typically the case in plate girder design. The τ_{tf}/τ_{cr} ratios are only considerable when the elastic critical shear buckling stress is very small. Tension field behavior is not used in ASD, but the designer should be aware that approximately 20%–30% increase in shear buckling strength exists due to tension field action for typical plate girder designs with τ_{cr} greater than about $F_y/2$.

However, to develop the ultimate shear buckling strength, the intermediate transverse stiffeners must be designed for the forces required to develop the tension field in the web plate. Vertical force equilibrium in Figure 7.13 yields

$$F_{st} = (\sigma_t t_w a \sin\theta)\sin\theta = \sigma_t t_w a \sin^2\theta = \frac{\sigma_t t_w a}{2}\left(1 - \frac{(a/h)}{\sqrt{1+(a/h)^2}}\right). \tag{7.72}$$

7.2.6.1.3.1.6 Combined Elastic Bending and Shear *Strength Criteria*: The web plate is subjected to a combination of shear forces and bending moment depending on location in the span. Since shear stress is greatest at the neutral axis where normal flexural stresses are zero and normal bending stress is greatest at the flange where shear stresses are less than average, it is generally sufficient to design for shear and flexural allowable stresses independently. Also, in ordinary steel railway girder design, the bending moment carried by the web plate is relatively small (see Equation 7.35). However, the design engineer may need to review shear and flexure interaction at locations where

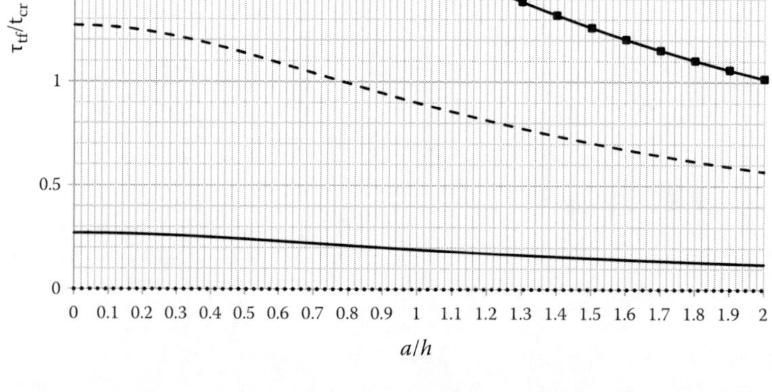

FIGURE 7.14 Ratio of shear buckling stress from tension field action to critical elastic shear buckling stress.

- Flexural stress is at maximum allowable and shear stress is greater than 55% of allowable shear stress, or
- Shear stress is at maximum allowable and bending stress exceeds 70% of allowable flexural stress

These interaction criteria are plotted in Figure 7.15.

An interaction equation can be developed, using an FS = 1.82, as

$$f_b \le \left(0.75 - 1.05\frac{f_v}{F_y}\right)F_y \le 0.55F_y, \tag{7.73}$$

where

f_v = the shear stress in the web
$F_v = 0.35F_y$ = the allowable web shear stress
f_b = the flexural stress in the web
$F_b = 0.55F_y$ = the allowable web flexural stress

Stability Criteria: Shear and flexural buckling may have to be considered together when f_v/τ exceeds 0.40* (Timoshenko and Gere, 1961). For a simply supported plate, a simple circular interaction formula (Equation 7.74) has been found to closely represent experimental data (Galambos, 1988).

$$\left(\frac{f_b}{F_{cr}}\right)^2 + \left(\frac{f_v}{\tau_{cr}}\right)^2 = 1. \tag{7.74}$$

7.2.6.1.3.2 Web Plate Splices AREMA (2015) recommends that splices in the web plates of girders can be designed with

- A plate each side of the web with each plate designed for half the shear strength of the gross section of the web plate and having a minimum net moment of inertia of half the net moment of inertia of the web plate.
- The combined forces of the flexural strength of the net section of the web with the maximum shear force at the splice.

The web splice fasteners (see Chapter 9) should be designed to transfer the shear force, *V*, and moment, *Ve*, due to eccentricity, *e*, of the centroid of the bolt group from the location of the shear

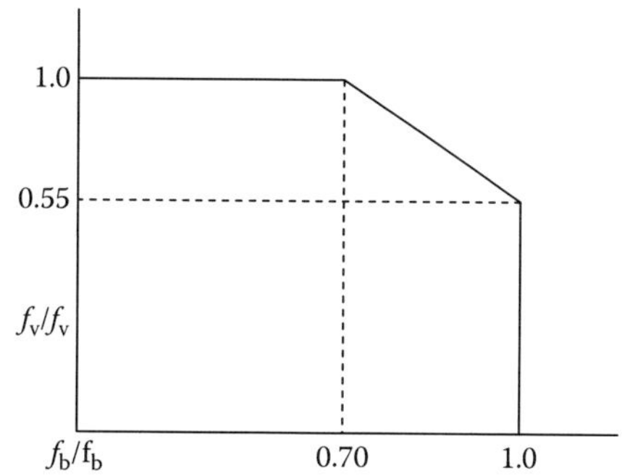

FIGURE 7.15 Web plate combined bending and shear.

* When $f_v/\tau_{cr} < 0.4$, the critical bending stress is negligibly affected by the presence of shear stresses.

force. Welded splices are usually made with CJP groove welds with strength at least equal to the base material being spliced. The entire cross section should be welded.

7.2.6.1.4 Girder Flange-to-Web Plate Connection

In modern steel plate girder superstructures, the flange-to-web plate connection is made with welds. AREMA (2015) indicates that CJP, partial joint penetration (PJP), or fillet welds may be used for the flange-to-web connection.

PJP and fillet welds in deck plate girders (DPGs) with open decks or noncomposite concrete decks must be designed such that fatigue strength is controlled by weld toe cracking (to preclude cracking in the weld throat). Therefore, some design engineers specify CJP flange-to-web welds for open DPG and noncomposite ballasted deck plate girder (BDPG) spans to ensure that vertically applied wheel loads can be safely resisted by the top flange-to-web weld.

7.2.6.1.4.1 Top Flange-to-Web Connection (Simply Supported Girder Spans) In addition to vertical wheel loads, the top flange-to-web weld connection must transmit horizontal shear caused by the varying flange bending moment, dM, along the girder length. The change in flange force, dP_f, due to bending along a length of girder, dx, (Figure 7.16) is

$$\mathrm{d}P_\mathrm{f} = \frac{\mathrm{d}M}{I}\bar{y}(A_\mathrm{f}) = \frac{(V\mathrm{d}x)}{I}\bar{y}(A_\mathrm{f}) = \frac{VQ_\mathrm{f}}{I}\mathrm{d}x. \tag{7.75}$$

The horizontal shear flow, $q_f = \mathrm{d}P_f/\mathrm{d}x$, for which the top (compression) flange weld is designed, is

$$q_\mathrm{f} = \frac{\mathrm{d}P_\mathrm{f}}{\mathrm{d}x} = \frac{VQ_\mathrm{f}}{I}, \tag{7.76}$$

where

V = the shear force

$\bar{y}$ = the distance from top flange-to-web connection to the neutral axis

$Q_\mathrm{f} = A_\mathrm{f}\bar{y}$ (statical = moment of top flange area about the neutral axis)

The shear force from wheel live load, W, with 80% impact (AREMA, 2015) acting in a vertical direction along the top flange-to-web connection of DPG and noncomposite BDPG spans is

$$w = \frac{1.80(W)}{S_\mathrm{w}}, \tag{7.77}$$

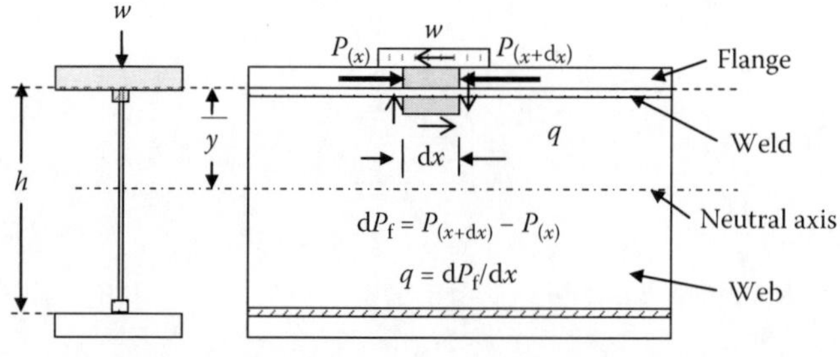

FIGURE 7.16 Forces transferred between flange and web.

TABLE 7.2
Allowable Weld Stresses

Weld Type	Allowable Shear Stress (MPa)	Allowable Shear Stress (ksi)
CJP or PJP	$0.35F_y$	$0.35F_y$
Fillet (415 MPa (60 ksi) electrode)	115 but $<0.35F_y$ on base metal	16.5 but $<0.35F_y$ on base metal
Fillet (480 MPa (70 ksi) electrode)	130 but $<0.35F_y$ on base metal	19.0 but $<0.35F_y$ on base metal
Fillet (550 MPa (80 ksi) electrode)	150 but $<0.35F_y$ on base metal	22.0 but $<0.35F_y$ on base metal

where

S_w = the wheel load longitudinal distribution [S_w = 915 mm (3 ft) for open deck girders or S_w = 1525 mm (5 ft) for ballasted deck girders].

The resultant force per unit length of weld is

$$q = \sqrt{q_f^2 + w^2}. \tag{7.78}$$

The required effective area of weld can then be established based on the allowable weld stresses recommended by AREMA (2015), as shown in Table 7.2 (see also Chapter 9).

7.2.6.1.4.2 Bottom Flange-to-Web Connection (Simply Supported Girder Spans) The horizontal shear flow for which the bottom (tension) flange weld is to be designed is

$$q_f = \frac{VQ_f}{I} \tag{7.79a}$$

or

$$\Delta q_f = \frac{\Delta VQ_f}{I}, \tag{7.79b}$$

where

ΔV = the shear force range from live load plus impact

$\bar{y}$ = the distance from bottom flange-to-web connection to the neutral axis

$Q_f = A_f\bar{y}$ (statical moment of bottom flange area about the neutral axis)

The required effective area of weld can then be established based on the allowable weld shear stress for maximum shear flow and allowable fatigue stress (typically Category B or B′ depending on weld backing bar usage).

7.2.6.1.5 Girder Bearing Stiffeners

Concentrated loads (e.g., reactions at the ends of girders) create localized compressive stresses that may exceed yield stress. The localized yielding, or web crippling, may be resisted by web plates of sufficient thickness or by pairs of stiffeners. Web crippling can be conservatively analyzed as shown in Figure 7.17.

The minimum web plate thickness is

$$t_w \geq \frac{R}{0.75F_y(L_B + k)} \text{ for the end reaction, } R \tag{7.80a}$$

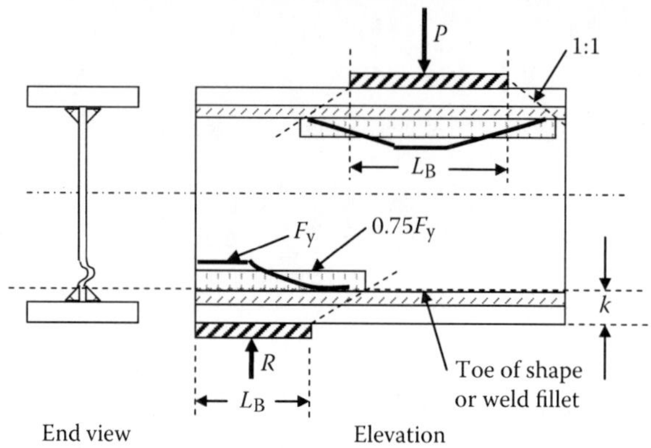

FIGURE 7.17 Web crippling (stress distribution at toe of fillet shown).

$$t_w \geq \frac{P}{0.75F_y(L_B + 2k)} \quad \text{for an interior concentrated load, } P, \tag{7.80b}$$

where

L_B = the length of bearing

Steel railway girders will generally require bearing stiffeners due to the high magnitude loads. However, in situations where concentrated loads may not cause web crippling in accordance with Equations 7.80a or 7.80b, it is often advisable to install at least nominal stiffeners, in any case (an example illustrating the benefit of nominal stiffeners at locations of concentrated loads is given in Akesson (2008)). Bearing stiffeners must be connected to both flanges and extend to near the edge of the flange. Bearing stiffeners are designed for the following criteria:

- Compression member behavior (yield and stability)
- Bearing stress
- Local plate buckling

7.2.6.1.5.1 Compression Member Behavior of Bearing Stiffeners The bearing stiffener is designed as a compression member with an effective cross section comprised of the area of the stiffener elements, A_{bs}, and a portion of the web, A_{wbs}, related to the web thickness. The effective area, A_{ebs}, and the effective moment of inertia, I_{ebs}, of the bearing stiffener cross sections shown in Figure 7.18 are

$$A_{ebs} = 2A_{bs} + A_{wbs} = 2A_{bs} + 12(t_w)^2 \quad \text{for the end reaction, } R \tag{7.81a}$$

$$I_{ebs} = 2I_{bs} + 2A_{bs}\bar{y}^2 + t_w^4 \quad \text{for the end reaction, } R \tag{7.81b}$$

$$A_{ebs} = 2A_{bs} + 25(t_w)^2 \quad \text{for an interior concentrated load, } P \tag{7.82a}$$

$$I_{ebs} = 2I_{bs} + 2A_{bs}\bar{y}^2 + 2.08t_w^4 \quad \text{for an interior concentrated load, } P. \tag{7.82b}$$

The bearing stiffener may be designed as a compression member using $K = 0.75$ (AREMA, 2015) (also see Chapter 6) for an allowable compressive stress, F_{call}, of

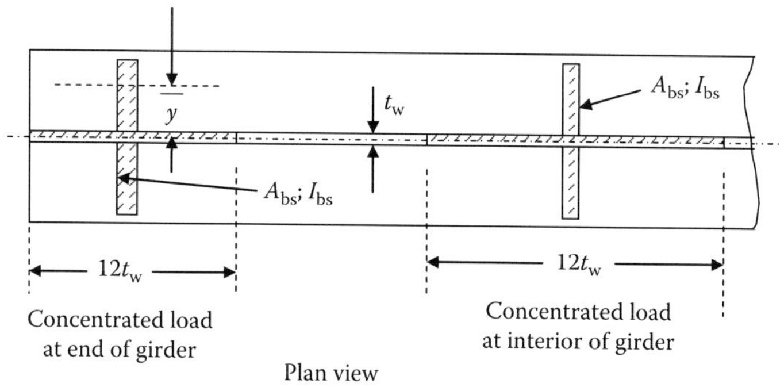

FIGURE 7.18 Bearing stiffener effective cross section.

$$F_{call} = 0.55F_y; \text{ when } \frac{h}{r_{ebs}} \prec 0.839\sqrt{\frac{E}{F_y}}, \tag{7.83}$$

$$F_{call} = 0.60F_y - \left(625\frac{F_y}{E}\right)^{\frac{3}{2}}\left(\frac{0.75h}{r_{ebs}}\right) \text{ when } 0.839\sqrt{\frac{E}{F_y}} \le \frac{h}{r_{ebs}} \le 6.712\sqrt{\frac{E}{F_y}} \text{ and}$$
$$F_y \text{ and } E \text{ in MPa (SI units)}, \tag{7.84a}$$

$$F_{call} = 0.60F_y - \left(17500\frac{F_y}{E}\right)^{\frac{3}{2}}\left(\frac{0.75h}{r_{ebs}}\right); \text{ when } 0.839\sqrt{\frac{E}{F_y}} \le \frac{h}{r_{ebs}} \le 6.712\sqrt{\frac{E}{F_y}} \text{ and}$$
$$F_y \text{ and } E \text{ in psi (US Customary and Imperial units)}, \tag{7.84b}$$

$$F_{call} = \frac{0.685\pi^2 E}{(h/r_{ebs})^2}; \quad \text{when } \frac{h}{r_{ebs}} \succ 6.712\sqrt{\frac{E}{F_y}}, \tag{7.85}$$

where

h = the height of the bearing stiffener (the clear distance between the girder top and bottom flanges) and $r_{ebs} = \sqrt{I_{ebs}/A_{ebs}}$.

The allowable force on the bearing stiffener is

$$P_{call} = F_{call}\left(A_{ebs}\right) \tag{7.86}$$

which should not exceed the maximum reaction, R, or concentrated load, P.

7.2.6.1.5.2 Bearing Stresses Since a part of the bearing stiffener area, A_{bs}, is removed from the top and bottom to clear the shape or girder weld fillets, the reduced bearing area, A'_{bs}, must be considered. AREMA (2015) recommends an allowable bearing stress for milled stiffeners and parts in contact of $0.83F_y$. Based on this, the reduced bearing stiffener area (area of bearing stiffener in contact with flange plate), A'_{bs}, is

$$A'_{bs} \ge \frac{R}{0.83F_y}. \tag{7.87}$$

7.2.6.1.5.3 Local Plate Buckling Local buckling of bearing stiffeners is essentially the problem of uniform compression on a plate free at one side and partially restrained at the other. The maximum permissible width-to-thickness ratio is, therefore, the same as that established previously for local buckling of a girder compression flange plate (Equation 7.57)

$$\frac{b_{\mathrm{bs}}}{t_{\mathrm{bs}}} \leq 0.43\sqrt{\frac{E}{F_{\mathrm{y}}}}, \tag{7.88}$$

where

b_{bs} = the width of the outstanding leg of the bearing stiffener
t_{bs} = the thickness of the outstanding leg of the bearing stiffener

7.2.6.2 Secondary Girder Elements

Stiffeners are secondary elements, but of paramount importance to ensure the stability of some main load carrying elements of plate girders. Specifically, the web plate is usually stiffened by transverse, and sometimes, longitudinal stiffeners. The web stiffeners generally consist of welded plates or bolted angles. Welded web stiffeners may have an effect on the main member allowable fatigue stress range if the stiffener attachments are within tensile regions of the plate girder* (see Chapter 5). Bolted stiffeners may reduce the net section of the girder and require greater fabrication effort,† but welded attachment fatigue concerns are eliminated.

7.2.6.2.1 Longitudinal Web Plate Stiffeners

Equation 7.60 indicates that if $\frac{h}{t_{\mathrm{w}}} \succ 4.18\sqrt{\frac{E}{f_{\mathrm{c}}}}$ longitudinal stiffeners are required to preclude web flexural buckling instability. The minimum recommended (AREMA, 2015) web plate thickness with longitudinal stiffeners is 50% of Equation 7.59 or

$$t_{\mathrm{w}} \geq 0.09h\sqrt{\frac{F_{\mathrm{y}}}{E}}\sqrt{\frac{f_{\mathrm{c}}}{F_{\mathrm{cr}}}} \geq 0.12h\sqrt{\frac{f_{\mathrm{c}}}{E}}. \tag{7.89}$$

The longitudinal stiffeners should be proportioned such that they have a flexural rigidity, EI_{ls}, which creates straight nodes in the buckled plate. The plate buckling coefficient, k, for critical buckling stress with a longitudinal stiffener at 25% of web depth is $k = 101$. Using energy methods, it can be shown that (Bleich, 1952)

$$I_{\mathrm{ls}} = \frac{ht_{\mathrm{w}}^3}{12(1-\upsilon^2)}\left[\left(12.6+50\left(\frac{A_{\mathrm{ls}}}{ht_{\mathrm{w}}}\right)\right)\left(\frac{a}{h}\right)^2 - 3.4\left(\frac{a}{h}\right)^3\right]. \tag{7.90}$$

AREMA (2015)‡ uses a similar but simpler formula

$$I_{\mathrm{ls}} = ht_{\mathrm{w}}^3\left(2.4\left(\frac{a}{h}\right)^2 - 0.13\right), \tag{7.91}$$

---

* For example, the bottom of transverse web stiffeners is attached near the tension flange in simply supported plate girders. Such attachments, particularly if welded, should be reviewed with respect to their effect on the allowable fatigue stress range.

† However, modern steel fabrication CNC equipment (see Chapter 10) has substantially increased quality and reduced the cost of bolt hole drilling.

‡ Many other ASD design codes, recommendations, guidelines, and manuals use the same, or similar, equation.

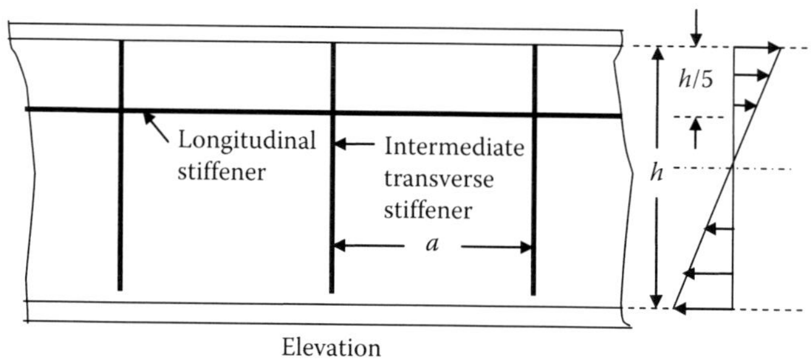

FIGURE 7.19 Web plate under pure bending.

where

I_{ls} = the moment of inertia for a single longitudinal stiffener about the face of the web plate (if longitudinal stiffeners are used on both sides of the web, the moment of inertia is taken about the centerline of the web*)

$A_{ls} = b_{ls}t_{ls}$ = the cross-sectional area of the longitudinal stiffener

a = the distance between intermediate transverse stiffeners (Figure 7.19)

Equation 7.91 also fits experimental data for pure bending with longitudinal stiffeners at $h/5$ and small values of A_{ls}/ht_w (0.05–0.25) (Salmon and Johnson, 1980).

The thickness of the longitudinal stiffener, t_{ls}, to avoid local buckling is essentially the buckling problem of uniform compression on a plate free at one side and partially restrained at the other. From Equation 7.55, the width-to-thickness ratio is

$$\frac{b_{ls}}{t_{ls}} \leq 0.67\sqrt{\frac{kE}{F_y}} \tag{7.92}$$

which is the approximate value corresponding to the transition curve at $F_{cr} = F_y$ (Figure 7.10). The buckling coefficient, k, is 1.277 for plates with one edge fixed and the other edge free (typical of a longitudinal stiffener) (Bleich, 1952). Substitution of $k = 1.277$ into Equation 7.92 yields

$$\frac{b_{ls}}{t_{ls}} \leq 0.76\sqrt{\frac{E}{F_y}}. \tag{7.93}$$

Assuming the actual calculated stress $f = F_y$ and applying a safety factor of 1.82

$$\frac{b_{ls}}{t_{ls}} \leq 0.42\sqrt{\frac{E}{f}} \tag{7.94}$$

$$t_{ls} \geq 2.39 b_{ls}\sqrt{\frac{f}{E}} \tag{7.95}$$

from which the minimum longitudinal stiffener dimensions for stability can be determined.

7.2.6.2.2 Transverse Web Plate Stiffeners

Recent investigations have determined that it is appropriate to consider the design of transverse web stiffeners as flexural members resisting bending forces created by the restraint that the transverse

* It is usually not necessary and, therefore, unusual to use longitudinal stiffeners on both sides of the web.

stiffener imposes on lateral deflections of the web plate at the shear strength limit state (Kim et al., 2007). Therefore, if transverse stiffeners are required, the necessary spacing, a, to provide adequate rigidity through creation of nodal lines is determined by considering that AREMA (2015) restricts $a/h \le 1$ so that the shear buckling coefficient is (Equation 7.63)

$$k = 4.0 + \frac{5.34}{(a/h)^2}. \tag{7.96}$$

Therefore, the critical shear buckling stress is

$$\tau_{cr} = \frac{\left(4.0 + \frac{5.34}{(a/h)^2}\right)\pi^2 E t_w^2}{12(1-\upsilon^2)h^2} \tag{7.97}$$

which may be rearranged to provide

$$\frac{a}{t_w} = \sqrt{\frac{5.34\pi^2}{12(1-\upsilon^2)(\tau_{cr}/E) - 4\pi^2(t_w/h)^2}}. \tag{7.98}$$

Using a factor of safety, FS = 1.5, Equation 7.98 is*

$$\frac{a}{t_w} = \sqrt{\frac{1.336}{1.5(0.277)(\tau_{cr}/E) - (t_w/h)^2}}. \tag{7.99}$$

Flexural buckling is precluded (Equation 7.47) where

$$\frac{h}{t_w} \le 4.18\sqrt{\frac{E}{F_{cr}}} \le 5.64\sqrt{\frac{E}{F_y}} \tag{7.100}$$

and considering $\tau_{cr} \le 0.35F_y$ Equation 7.99 becomes

$$\frac{a}{t_w} = 3.42\sqrt{\frac{E}{F_y}} \tag{7.101},$$

or (with $\tau = 0.35F_y$)

$$\frac{a}{t_w} = 2.02\sqrt{\frac{E}{\tau}} \tag{7.102}$$

and AREMA (2015) recommends

$$a \le 1.95 t_w \sqrt{\frac{E}{\tau}} \tag{7.103}$$

* AREMA (2015) uses the relatively lower factor of safety of 1.5 for shear buckling in recognition of the postbuckling strength of web plates in shear.

to establish transverse stiffener spacing. Based on web plate imperfection tolerances, AREMA (2015) also provides a practical recommendation for maximum stiffener spacing of 2450 mm (96 in.).

An equation for the required moment of inertia of a transverse web stiffener was developed from analytical and experimental tests as (Bleich, 1952)

$$I_{ts} = \frac{4at_w^3}{12(1-\upsilon^2)}\left(7\left(\frac{h}{a}\right)^2 - 5\right), \tag{7.104}$$

which may be simplified to

$$I_{ts} = 2.5a_0t_w^3\left(\left(\frac{h}{a}\right)^2 - 0.7\right), \tag{7.105}$$

where

a_o = the actual stiffener spacing used in the design, which must be: $<h$ (see development of Equation 7.96); <2450 mm (96 in.)

a = the required stiffener spacing from Equation 7.103

Equation 7.104 is valid for $0.2 \le a/h \le 1$ (based on the limits of the testing). These limits are within practical steel railway girder web plate dimensions. Therefore, Equation 7.105 is recommended by AREMA (2015) to determine the required moment of inertia of transverse web stiffeners.

In elastic design, the stiffeners are not required to carry force[*] and, therefore, there is no need to design them, nor their connections, for strength. The dimensions of the transverse stiffener are determined from Equation 7.105, which is based solely on rigidity considerations, and only nominal welded or bolted connection to the web is required.[†] However, for superstructures with relatively large skew, curvature, or track eccentricity, the stiffener connection to the web must also be designed to resist the forces from out-of-plane bending of the beams or girders they are connected to and forces due to lateral distribution of the live load. Wrap around fillet welds must not be used for welding transverse stiffeners to either the web or the compression flange.

Furthermore, welds connecting transverse stiffeners to web plates should not be made close to the tension flanges because of stress concentration effects. Extensive testing and analytical work have established that the stiffener weld should be minimum four to six times the web plate thickness from the near toe of the tension flange-to-web weld (Basler and Thurlimann, 1959). AREMA (2015) recommends this distance as $6t_w$. Careful consideration of details is required (such as provision of bolted angles at the bottom of the stiffeners (D'Andrea et al., 2001) or peening pretreatments[‡]) where brace frames are attached to transverse stiffeners that may precipitate out-of-plane distortion in the web gap. Even though fabrication cost may be increased, some design engineers will provide bolted transverse stiffener connections, particularly when they serve as bracing connection plates, in order to preclude detrimental out-of-plane web gap weld fatigue effects.

AREMA (2015) recommends connection of intermediate transverse stiffeners to the compression flange to provide additional stability to the stiffener itself and against torsional buckling of the compression flange. Intermediate transverse stiffeners connected to only one side of the web plate are recommended to be connected to the compression flange. This stiffener to compression flange

---

[*] Such as the compressive force, if tension field action is assumed in the web (stiffener is analogous to truss post with web behavior like a truss tension diagonal).

[†] Bolt spacing in intermediate transverse stiffener connections to the web is typically specified to not exceed the recommended "sealing" spacing (see Chapter 9).

[‡] Ultrasonic impact treatment is a modern pre-treatment to improve these poor fatigue details at the base of welded transverse stiffeners (Roy and Fisher, 2006).

connection may be accomplished with bolts, fillet welds, or by careful grinding to ensure a uniform and tight fit against the flange.

AREMA (2015) also recommends that intermediate transverse stiffeners connected to the interior of through plate girder (TPG) and ballasted through plate girder (BTPG) web plates can be connected to the compression flange to minimize out-of-plane deformations of the web from end rotation of the floor beams. This stiffener to compression flange connection may be made with bolts or fillet welds.

Furthermore, AREMA (2015) recommends that intermediate transverse stiffeners in TPG or BTPG spans within a distance from the end of the girder equal to the depth of the girder are connected to the tension flange. The web plate stiffeners must not be welded to the tension flanges, as such a transverse weld is a very poor fatigue detail.

If lateral bracing is attached to an intermediate transverse stiffener, the connection at the top flange must be designed to transmit 2.5% of compression flange force and other lateral forces from wind, centrifugal, or nosing (see Chapters 4 and 5). From a lateral bracing perspective, the connection at the bottom flange is less important as it resists only forces from wind. AREMA (2015) recommends that transverse web stiffeners can be adequately attached to both top and bottom flanges when bracing is connected to the stiffeners (although not required for rolled beams on single track spans without skew or curvature). The stiffener to flange connections may be made with bolts or, for compression flanges, fillet welds.

7.2.7 Box Girder Design

Box girders have a high flexural capacity and torsional rigidity. The design of box girders is generally analogous to the design of plate girders. However, the large compression flange makes it necessary to utilize stiffened steel plates or concrete slabs. Steel plate decks are often used when span lengths are large enough such that the dead load from a concrete slab deck becomes a disproportionate portion of the total load on the span. The compression flange typically also serves as the ballasted deck.

7.2.7.1 Steel Box Girders

Steel box girders typically employ an orthotropic steel deck plate. The strength (yield and stability), fatigue, and serviceability design of orthotropic deck plates require careful consideration of fabrication details. The design of orthotropic plate deck bridges is beyond the scope of this book and the reader is referred to books by Wolchuk (1963), Lekhnitski (1968), Cusens and Pama (1979), Troitsky (1987), Szilard (2004), and others which provide definitive information regarding the analysis and design of orthotropic steel deck plate superstructures.

7.2.7.2 Steel–Concrete Composite Box Girders

The design of steel–concrete composite box girder spans is also generally analogous to that for steel–concrete composite plate girder spans. The latter are discussed in greater detail in this chapter.

7.3 SERVICEABILITY DESIGN OF NONCOMPOSITE FLEXURAL MEMBERS

AREMA (2015) recommends that the midspan deflection of simply supported spans due to live load plus impact, LL+I, does not exceed $L/640$, where L = span length. Some engineers or bridge owners recommend even more stringent live load plus impact deflection criteria to attain stiffer spans which offer improved performance from a vehicle–bridge interaction dynamics perspective (see Chapter 4).

It is also recommended that camber can be provided for dead load deflections in spans exceeding 27.5 m (90 ft). Camber of truss spans is recommended to account for deflections from dead load plus a live load of 45 kN/track meter (3000 lb/track foot).

The serviceability criteria for steel railway spans are also discussed in Chapter 5.

TABLE E7.1a

Design Force (Load Case D1 in Table 4.10)	Shear Force, V (kN)	Bending Moment, M (kN m)
Dead load, DL	550	4060
Live load + 33% impact (max.)	1770	11,440
Maximum (DL + LL + I)	2320	15,500
Live load +12% impact (fatigue)	1490s	9630

Example 7.1a (SI Units)

A 30 m simple span steel DPG is to be designed for the forces shown in Table E7.1a using ASTM A709M Grade 350WT3 steel.

Preliminary Girder Design

The girder height, based on typical span-to-depth ratios (see Chapter 3), is between 30(1000)/15 = 2000 mm and 30(1000)/10 = 3000 mm.

For $F_y = 350$ MPa steel, and considering a maximum h/t_w of 134 from Equation 7.49, the optimum girder height is (Equation 7.39b)

$h_{opt} = \sqrt[3]{\dfrac{3(15,500)(10^6)}{2(0.55)(350)}}(134) \approx 2530$ mm, this would typically correspond to a girder depth of about 2700 mm, assuming 65 mm thick flange plates.

Try $d = 2500$ mm to minimize web plate height.

Web thickness may be estimated based on experience, strength, and stability considerations. Based on strength, web plate thickness is

$t_w \geq \dfrac{2320(1000)}{0.35(350)(2500-130)} \geq 8.0$ mm, try 10 mm [9.5 mm is minimum thickness of material as per AREMA (2015)]. However, based on stability, where a minimum web slenderness of 135 is recommended for $F_y = 350$ MPa steel (Equation 7.60b), web plate thickness is

$t_w \geq \dfrac{2500-130}{135} \geq 17.6$ mm, without longitudinal stiffeners (8.8 mm (use minimum t_w of 10 mm) with longitudinal stiffeners). Therefore, use an 18 mm thick web without longitudinal stiffeners. The designer should confirm that 18 mm thick Grade 350WT3 plate is available with a minimum width of 2400 mm (allowing for trimming) in sufficient lengths to avoid excessive or poorly located vertical web plate splices.

Flange size may be estimated based on experience and approximations such that flange width, b, can be between $d/4$ and $d/3$ and thickness, t_f, can be between $b/12$ and $b/5$ with due consideration of compression flange buckling and maximum practical thickness (also see Chapters 2 and 10). Flange width, b, should be between about 625 and 825 mm, and flange thickness between about 50 and 165 mm. Flange plates thicker than about 65 mm are not economical or practical (also see Chapter 10) and, therefore, a flange thickness of 65 mm is tried. The flange area can be estimated based on strength using Equation 7.35 (with $A_w = (18)(2500-130) = 42,660$ mm²) as

$A_f = \dfrac{9630(10^6)}{110(2500-130)} - \dfrac{42660}{6} = 29830$ mm^2 (f_b = 110 MPa = allowable stress for fatigue Category B with no welded attachments, which means that if transverse web stiffeners are required, a bolted connection to the web will be necessary). Try 65 mm × 525 mm (A_f = 34,125 mm²) Grade 350WT3 bottom flange plate.

$A_f = \dfrac{15,500(10^6)}{0.55(350)(2500-130)} - \dfrac{42,660}{6} = 26,864$ mm^2 ($f_b = 192.5$ MPa = maximum allowable stress). Try 65 mm × 425 mm (A_f = 27,625 mm²) Grade 350WT3 top flange plate.

The designer should confirm that 65 mm thick Grade 350WT3 plate is available in sufficient lengths to avoid excessive or poorly located transverse flange plate splices (see Chapter 9).

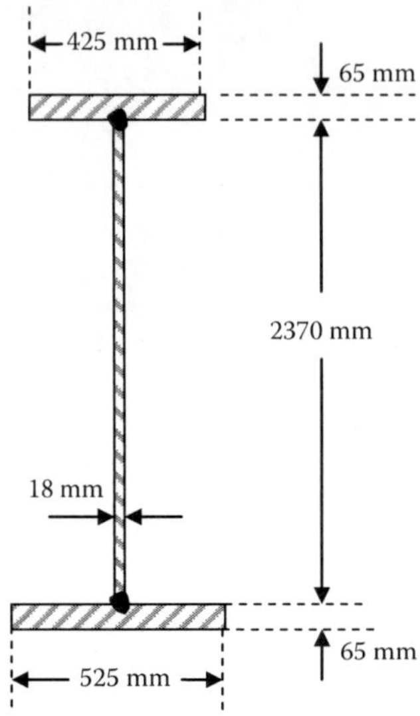

FIGURE E7.1a Girder cross section.

TABLE E7.2a

Element	A (mm²)	y_b (mm)	$Ay_b \times 10^3$ (mm³)	$y_b - y$ (mm)	$A(y_b - y)^2 \times 10^6$ (mm⁴)	$I_o \times 10^6$ (mm⁴)
Top flange	27,625	2467.5	68,165	1217.5	40,949	9.7
Web	42,660	1250	53,325	0	0	19,968
Bottom flange	34,125	32.5	1109	−1217.5	50,584	12.0
	104,410		122,599		91,532	19,990

The girder section properties are shown in Figure E7.1a and Table E7.2a.

$$\frac{\sum Ay_b}{\sum A} = \frac{122,599\times10^6}{104,410\times10^3} = 1174 \text{ mm (neutral axis to underside of the bottom flange plate)}$$

$$I_g = \sum A(y_b - y)^2 + \sum I_o = (91,532+19,990)\times10^6 = 111,522\times10^6 \text{ mm}^4$$

$$r_{cy} = \sqrt{\frac{I_{cy}}{A}} = \sqrt{\frac{[(1250-65)(18)^3 + 65(425)^3]}{12[425(65)+(1250-65)(18)]}} = 92.2 \text{ mm}$$

$$S_g = \frac{111,522\times10^6}{(2500-1174)} = 84,104\times10^3 \text{ mm}^3$$

$$S_n = \frac{n_{ng}(111,522\times10^6)}{1174} = (94,994\times10^6)n_{ng} \text{ mm}^3$$

$$n_{ng} = \frac{I_n}{I_g}$$

Assume that $n_{ng} = 0.90$ (a subsequent check is required after detailed design of bolted connections)

$I_n \sim 0.90(111,522\times10^6) \sim 100,370\times10^6 \text{ mm}^4$

$$S_n \approx \frac{100,370\times10^6}{1174} \sim 85,494 \times 10^3 \text{mm}^3$$

$\sigma_{tmax} \sim 15{,}500(10^6)/84{,}104(10^3) = 184$ MPa $< F_{tall} < 0.55(350) < 192.5$ MPa OK (a subsequent check of assumed dead load is required after final proportioning and detailing).

For 30 m long plate girder; design for >2,000,000 fatigue cycles with mean impact percentage 35% of maximum impact

$\Delta\sigma_{max} \sim 9630(10^6)/85{,}494(10^3) = 113$ MPa. Therefore, allowing for a small 2.5% MPa overstress, details with fatigue detail less than Category B (allowable fatigue stress range = 110 MPa) should not be used near the bottom flange area of the span.

$\tau_{max} = 2320(1000)/(2500 - 130)(18) = 54.4$ MPa $< F_{vall} < 0.35(350) < 122.5$ MPa OK.

AREMA (2015) recommends maximum of 3660 mm between points of top flange lateral support.

$$F_{call} = 0.55(350) - \frac{0.55(350)^2}{6.3\pi^2(200{,}000)}\left(\frac{3660}{92.2}\right)^2 = 192.5 - 8.5 = 184 \text{ MPa}$$

$$> \sigma_{cmax} > 15{,}500(10^6)/84{,}104(10^3) = 184 \text{ MPa OK.}$$

Weight of girder = $104{,}410 \times 10^3(7850)/(1000)^3$ = 820 kg/m (24,588 kg total per girder without stiffeners).

The girder will be laterally supported (after erection) by brace frames at maximum spacing of 3660 mm.

Girder Design (for Fabrication and Erection Loads)

During fabrication and erection (see Chapters 10 and 11), the entire girder, or a portion of its length, may be laterally unsupported. It is typical that the girder will be supported at its bearing locations during fabrication, erection staging, and in final position. It is also typical that the girder will be lifted at two locations, *a*, from each end of the girder and erection lifts will not be done coincidently with windy conditions. Girder self-weight ~ 1.15(820) ~ 950kg/m (with 15% contingency load)

$w_{girder} = (950)(9.81)/1000 = 9.32$ kN/m.

For typical railway girder lifting arrangements, the maximum bending stress will occur between the lifting locations, $b = L - 2a$. However, since the girder may be laterally unsupported for its entire length, the maximum bending stress is*

$M_{girder} = (wL/2)[(L/4) - a] = (9.32(30)/2)[(30/4) - a] = 139.8(7.50 - a)$

assuming support or lifting of the girder at the ends ($a = 0$)

$\sigma_{girder} = [950(9.81)(1000)(30)^2/(8(84{,}104(10^3)))] = 12.5$ MPa

using an allowable bending stress of $1.25(0.55)F_y = 0.69F_y$ (see Chapter 4)

$$\left(\frac{L}{r_y}\right)_{girder} = \sqrt{\frac{0.69(350) - 12.5}{0.69(350)^2/6.3\pi^2(200{,}000)}} = 184$$

and the maximum unsupported length for fabrication and erection is $L_u = 92.2(184)/1000 = 16.9$ m ($L_u/L = 0.56$).

Detailed Girder Design

Detailed Design of the Tension Flange

$$S_{xn} = \frac{I_{xn}}{c_t} \geq \frac{M_{tmax}}{F_{all}} \geq \frac{M_{tmax}}{0.55F_y} \geq \frac{15{,}500(10^6)}{0.55(350)} \geq 80{,}519 \times 10^3 \text{ mm}^3 \left(\text{OK, } 85{,}494 \times 10^3 \text{mm}^3 \text{ provided}\right)$$

for Category B weld

* See Chapter 11 for a more detailed analysis of lifting girders at location, a, from each end of the girder.

$$S_{xn} \geq \frac{M_{max-range}}{F_{fat}} \geq \frac{(9630)(10^6)}{110} \geq 87{,}545 \times 10^3 \text{ mm}^3 \text{(OK, with 2.5\%}$$

overstress with $85{,}494 \times 10^3 \text{ mm}^3$ provided).

A subsequent check on the net section is required after detailed design is completed.

Detailed Design of the Compression Flange

Brace frames will be placed at equal intervals of 3.33 m,

$$\frac{L_p}{r_{cy}} = \frac{3.33(1000)}{92.2} = 36.1 \leq 5.55\sqrt{\frac{E}{F_y}} \leq 134 \text{ OK}$$

The allowable compressive stress is

$$F_{call} = 0.55(350) - \frac{0.55(350)^2}{6.3\pi^2(200{,}000)}\left(\frac{(3330)}{92.2}\right)^2 = 192.5 - 7.1 = 185 \text{ MPa},$$

or

$$F_{call} = \left(\frac{0.13\pi(200{,}000)(425)(65)}{(3330)(2500)(\sqrt{1+0.3})}\right) = 238 \text{ MPa}$$

Since $F_{call} \geq 0.55F_y = 192.5$; $F_{call} = 192.5$ MPa

$$S_{xg} = \frac{I_{xg}}{c_c} \geq \frac{M_{cmax}}{F_{call}} \geq \frac{15{,}500(10^6)}{192.5} \geq 80{,}519 \times 10^3 \text{ in.}^3 \left(\text{OK, } 84{,}104 \times 10^6 \text{ mm}^3 \text{ provided}\right)$$

Vertical buckling of the compression flange is avoided by precluding flexural buckling of the web when

$$t_w \geq 0.18h\sqrt{\frac{F_y}{E}}\sqrt{\frac{f_c}{F_{cr}}} \geq 0.18(2500 - 2(65))\sqrt{\frac{350}{200{,}000}}\sqrt{\frac{15{,}500(10^6)/84{,}104 \times 10^3}{192.5}}$$

$$\geq 17.85\sqrt{\frac{184.3}{192.5}} \geq 17.5 \text{ mm} \left(\text{OK, 18 mm provided}\right)$$

The plate girder will have ties directly supported on the top (compression) flange, and consideration of local buckling provides

$$\frac{b}{2t_f} = \frac{425}{2(65)} = 3.3 \leq 0.35\sqrt{\frac{E}{F_y}} \leq 8.4 \text{ OK.}$$

Detailed Design of the Web Plate

$$A_w \geq \frac{V}{0.35F_y} \geq \frac{2320(1000)}{0.35(350)} \geq 18{,}939 \text{ mm}^2 \left(\text{OK, 42,660 mm}^2 \text{ provided}\right)$$

Web plate flexural buckling is considered as

$$\frac{h}{t_w} = \frac{2500-2(65)}{18} = 132 \leq 4.18\sqrt{\frac{E}{f_c}} \leq 4.18\sqrt{\frac{200{,}000}{184.3}} \leq 138.$$

Therefore, no longitudinal stiffeners are required for web flexural stability. This was also shown in the calculation related to compression flange vertical buckling above.

Web plate shear buckling is considered as

$$h = 2370 \geq 2.12t_w\sqrt{\frac{E}{F_y}} \geq 2.12(18)\sqrt{\frac{200{,}000}{350}} \geq 912 \text{ mm.}$$

Therefore, transverse web stiffeners are required.

Consideration of combined bending and shear yields

$$f_b = \frac{15{,}500(10^6)}{84{,}104(10^3)} = 184 \leq \left(0.75-1.05\frac{f_v}{F_y}\right)F_y \leq \left(0.75-1.05\frac{2320(1000)/42{,}660}{350}\right)350 = 205 \text{ MPa.}$$

However, $f_b \times 0.55F_y = 192.5$ MPa OK.

This interaction criterion does not generally require checking and, in the case where $f_v/(0.35F_y) = 0.44 < 0.55$ (see Figure 7.15) and the moment carried by the web plate is approximately $\left(\frac{42{,}660/6}{42{,}660/6+((425+525)/2)(65)}\right)100 = 20.0$ % of the total moment, combined bending and shear need not be considered.

Flange-to-Web Connection

For the top flange-to-web weld,

$$q = \sqrt{\left(\frac{VQ_f}{I}\right)^2 + \left(\frac{1.80W}{S_w}\right)^2} = \sqrt{\left(\frac{2320(1000)((425)(65))(1250-32.5)}{111{,}522(10^6)}\right)^2 + \left(\frac{1.80(180)(1000)}{915}\right)^2}$$

$$= \sqrt{(700.0)^2 + (354.1)^2} = 784.5 \text{ N/mm.}$$

For a CJP weld, the weld size must be $\geq \frac{0.5\sqrt{2}(784.5)}{0.35(350)} \geq 4.5$ mm OK since web thickness is 18 mm.

For the bottom flange-to-web weld,

$$q = \frac{\Delta VQ_f}{I} = \frac{(1490)(1000)((525)(65))(1250-32.5)}{100{,}370(10^6)} = 616.8 \text{ N/mm}$$

For a CJP weld with backing bar removed, the weld size must be $\geq \frac{0.5\sqrt{2}(616.8)}{110} \geq 4.0$ in. OK since web thickness is 18 mm.

In general, CJP weld design does not need to be considered.

Design of Web Plate Stiffeners

Use angles bolted to the web in order to preclude fatigue issues related to welding at the base of the intermediate transverse stiffeners.

The spacing of the intermediate transverse stiffeners is

$a_0 < h = 2370\,\text{mm}$

$< 2440\,\text{mm}$

$$< a \le 1.95 t_w \sqrt{\frac{E}{\tau}} \le 1.95(18)\sqrt{\frac{200{,}000}{2320(1000)/42{,}660}} \le 2129\,\text{mm}$$

Use a stiffener spacing of (3330)(12)/2 = 1665 (1/2 the distance between brace frames)

$$I_{ts} = 2.5 a_0 t_w^3 \left(\left(\frac{h}{a}\right)^2 - 0.7\right) = 2.5(1665)(18)^3\left(\left(\frac{2370}{2129}\right)^2 - 0.7\right) = 13{,}090\times 10^3\,\text{mm}^4$$

As shown in Figure E7.2aa, a single 150 × 100 × 13 angle on one side of the web plate provides $I_{ts} = 7.03\times10^6 + 3080(49.9)^2 = 14{,}699\times10^3\ \text{mm}^4$ OK.

Design of Bearing Stiffeners

As shown in Figure E7.2ab, the bearing stiffeners consist of 4 – *L*200 × 100 × 16.

$$A_{ebs} = 2A_{bs} + A_{wbs} = 2A_{bs} + 12(t_w)^2 = 2(2)(4540) + 12(18)^2 = 22{,}048\ \text{mm}^2$$

$$I_{ebs} = 2I_{bs} + 2A_{bs}\bar{y}^2 + t_w^4 = 2(2)(18.7\times10^6) + 2(2)(4540)(72.8)^2 + (18)^4 = 171.2\times10^6\ \text{mm}^4$$

$$r_{ebs} = \sqrt{\frac{171.2\times10^6}{22{,}048}} = 88.1\ \text{mm}$$

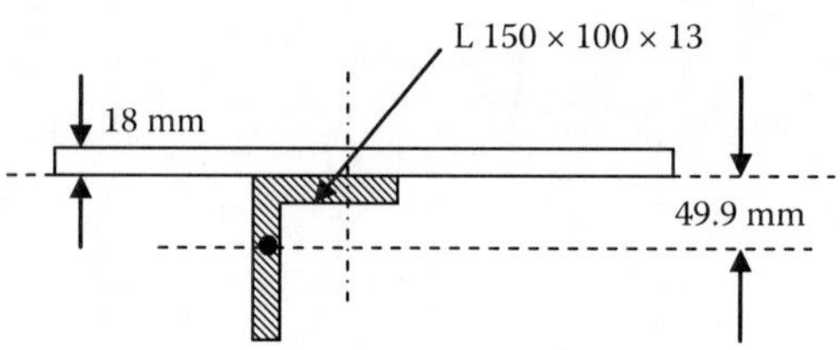

FIGURE E7.2aa Web plate stiffener.

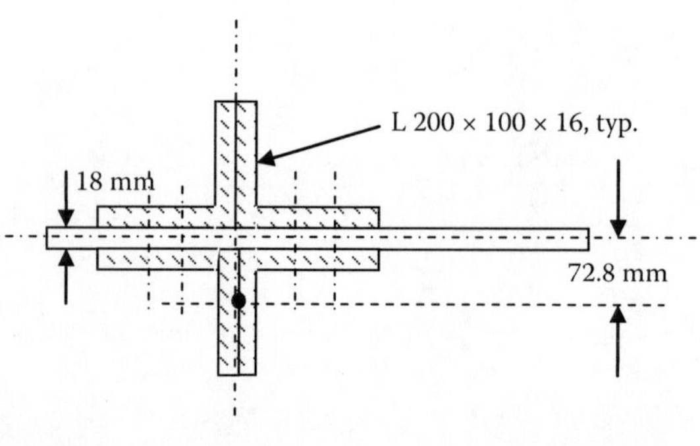

FIGURE E7.2ab Bearing stiffener.

$$\frac{h}{r_{ebs}} = \frac{2370}{88.1} = 26.9$$

$0.839\sqrt{E/F_y} = 20.1$ and, therefore, the governing expression for allowable compressive stress is (Equation 7.84)

$$F_{call} = 0.60F_y - \left(625\frac{F_y}{E}\right)^{3/2}\left(\frac{0.75h}{r_{ebs}}\right) = 210 - 23.1 = 187 \text{ MPa}$$

$P_{call} = (187)(22{,}048)/1000 = 4121 \text{ kN} > 2320 \text{ kN}$ OK.

The reduced bearing stiffener area assuming a 13 mm clip (made to clear the fillet at the web-to-flange junction from double bevel CJP groove welds),

$$A'_{bs} = 2(2)(200-13)(16) = 11{,}968\,\text{mm}^2$$

$$R = 2320 \le A'_{bs}(0.83F_y) \le (11{,}968)(0.83)(350)/1000 \le 3477 \text{ kN OK}$$

Local buckling of outstanding compression elements:

$$\frac{b_{bs}}{t_{bs}} = \frac{200-16}{16} = 11.5 \ge 0.43\sqrt{\frac{E}{F_y}} \ge 10.3 \text{ NG, try } 4 - L\ 200 \times 100 \times 20$$

$$\frac{b_{bs}}{t_{bs}} = \frac{200-20}{20} = 9.0 \le 0.43\sqrt{\frac{E}{F_y}} \le 10.3 \text{ OK}$$

Serviceability Design Deflection Criteria

The deflections are estimated as (see Chapter 5)

$$\Delta_{LL+I} = \frac{0.104M_{LL+I}L^2}{EI} = \frac{0.104(11{,}440)(10^6)[30{,}000]^2}{200{,}000(111{,}522 \times 10^6)} = 48.0 \text{ mm.}$$

The section $I_g = 111{,}522 \times 10^6$ mm^4 provides a deflection that is 2.5% greater than the AREMA (2015) criterion of $L/640 = 30{,}000/640 = 46.9$ mm, OK.

$$\Delta_{DL} = \frac{0.104M_{DL}L^2}{EI} = \frac{0.104(4060)(10^6)[30{,}000]^2}{200{,}000(111{,}522 \times 10^6)} = 17.0 \text{ mm; therefore, consider a camber of 18 mm.}$$

In practice, dimensions of the various main and secondary elements may be revised further to attain greater economy of material.

Example 7.1b (US Customary and Imperial Units)

A 90 ft simple span steel DPG is to be designed for the forces shown in Table E7.1b using ASTM A709 Grade 50WT2 steel.

Preliminary girder design:

The girder height, based on typical span-to-depth ratios (see Chapter 3), is between 90(12)/15 = 72″ and 90(12)/10 = 108″.

For $F_y = 50$ ksi steel, and considering a maximum h/t_w of 134 from Equation 7.49, the optimum girder height is (Equation 7.39b)

TABLE E7.1B

Design Force (Load Case D1 in Table 4.10)	Shear Force, V (kips)	Bending Moment, M (kip ft)
Dead load, DL	110	2500
Live load + 37% impact (max.)	376	7314
Maximum (DL + LL + I)	486	9814
Live load +13% impact (fatigue)	310	6032

$h_{opt} = \sqrt[3]{\frac{3(9814)(12)}{2(0.55)(50)}}(134) = 95''$, which would typically correspond to a girder depth of about 100″, assuming 2–1/2″ thick flange plates.

Try $d = 90''$ which is used to minimize web plate height.

Web thickness may be estimated based on experience, strength, and stability considerations. Based on strength, web plate thickness is

$t_w \geq \frac{486}{0.35(50)(90-5)} \geq 0.33''$, try 3/8″ [minimum thickness of material as per AREMA (2015)].

However, based on stability, where a minimum web slenderness of 135 is recommended for 50 ksi steel (Equation 7.60b), web plate thickness is

$t_w \geq \frac{90-5}{135} \geq 0.63''$, without longitudinal stiffeners (0.32″ (use minimum t_w of 3/8″) with longitudinal stiffeners). Therefore, use a 5/8″ thick web without longitudinal stiffeners. The designer should confirm that 5/8″ thick Grade 50WT2 plate is available with a minimum width of 86″ (allowing for trimming) in sufficient lengths to avoid excessive or poorly located vertical web plate splices.

Flange size may be estimated based on experience and approximations such that flange width, b, can be between $d/4$ and $d/3$ and thickness, t_f, can be between $b/12$ and $b/5$ with due consideration of compression flange buckling and maximum practical thickness (also see Chapters 2 and 10). Flange width, b, should be between about 22″ and 30″, and flange thickness between about 2″ and 6″. Flange plates thicker than about 2–1/2″ are not economical or practical (also see Chapter 10) and, therefore, a flange thickness of 2–1/2″ is tried. The flange area can be estimated based on strength using Equation 7.35 (with $A_w = (5/8)(90-5) = 53.12$ in.^{2}) as

$A_f = \frac{6032(12)}{16(90-5)} - \frac{53.1}{6} = 44.4$ in.^{2} ($f_b = 16$ ksi = allowable stress for fatigue Category B with no welded attachments, which means that if transverse web stiffeners are required, a bolted connection to the web will be necessary). Try 2–1/2″ × 20″ ($A_f = 50$ in.^{2}) Grade 50WT2 for bottom flange plate.

$$A_f = \frac{9814(12)}{0.55(50)(90-5)} - \frac{53.1}{6} = 41.5 \text{ in.}^2 \ (f_b = 27.5 \text{ ksi} = \text{allowable bending stress})$$

Try 2–1/2″×20″ ($A_f = 50.0$ in.^{2}) also for top flange plate. The designer should confirm that 2–1/2″ thick Grade 50WT2 plate is available in sufficient lengths to avoid excessive or poorly located transverse flange plate splices.

The girder section properties are shown in Figure E7.1b and Table E7.2b.

$$\frac{\sum Ay_b}{\sum A} = \frac{6890.6}{153.13} = 45.00 \text{ in. (neutral axis to underside of the bottom flange),}$$

$$I_g = \sum A(y_b - y)^2 + \sum I_o = 191{,}406 + 32{,}038 = 223{,}444 \text{ in.}^4$$

$$r_{cy} = \sqrt{\frac{I_{cy}}{A}} = \sqrt{\frac{[(45-2.5)(5/8)^3 + 2.5(20)^3]}{12[20(2.50)+(45-2.5)(5/8)]}} = 4.67 \text{ in.}$$

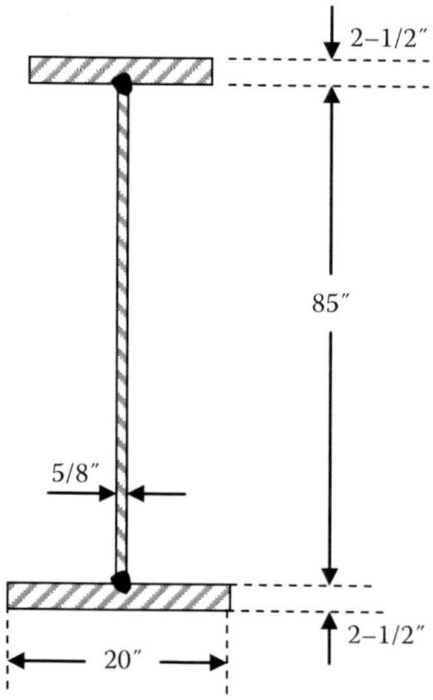

FIGURE E7.1b Girder cross section.

TABLE E7.2b

Element	A (in.^{2})	y_b (in.)	Ay_b (in.^{3})	$y_b - y$ (in)	$A(y_b - y)^2$ (in.^{4})	I_o (in.^{4})
Top flange	50.00	88.75	4437.5	43.75	95,703	26.0
Web	53.13	45.00	2390.6	0	0	31,986
Bottom flange	50.00	1.25	62.5	−43.75	95,703	26.0
	153.13		6890.6		191,406	32,038

$$S_g = \frac{223,444}{45.0} = 4965 \text{ in.}^3$$

$$S_n = \frac{n_{ng}(223,444)}{45.0} = 4965 n_{ng} \text{in.}^3$$

$$n_{ng} = \frac{I_n}{I_g}.$$

Assume that $n_{ng} = 0.90$ (a subsequent check is required after detailed design of bolted connections)

$$I_n \sim 0.90(223,444) \sim 201,100 \text{ in.}^4$$

$$S_n \approx \frac{201,100}{45} \sim 4469 \text{ in.}^3$$

$\sigma_{tmax} \sim 9814(12)/4469 = 26.4$ ksi $< F_{tall} < 0.55(50) < 27.5$ ksi OK (a subsequent check of assumed dead load is required after final proportioning and detailing).

For 90 ft long plate girder; design for >2,000,000 fatigue cycles with mean impact percentage 35% of maximum impact

$\Delta\sigma_{max} \sim 6032(12)/4469 = 16.2$ ksi. Therefore, allowing for a very small overstress, details with fatigue detail less that Category B (allowable fatigue stress range = 16 ksi) should not be used near the bottom flange area of the span.

$\tau_{max} = 486/(85)(5/8) = 9.15$ ksi $< F_{vall} < 0.35(50) < 17.5$ ksi OK.

AREMA (2015) recommends maximum of 12 ft between points of top flange lateral support.

$$F_{\text{call}} = 0.55(50) - \frac{0.55(50)^2}{6.3\pi^2(29{,}000)}\left(\frac{12(12)}{4.67}\right)^2 = 27.5 - 0.73 = 26.8 \text{ ksi}$$

$$> \sigma_{\text{cmax}} > 9814(12)/4965 = 23.7 \text{ ksi OK.}$$

Weight of girder = 153.1(490/144) = 521 lb/ft (46,900 lb total per girder).

The girder will be laterally supported (after erection) by brace frames at maximum spacing of 12 ft.

Girder Design (for Fabrication and Erection Loads)

During fabrication and erection (see Chapters 10 and 11), the entire girder, or a portion of its length, may be laterally unsupported. It is typical that the girder will be supported at its bearing locations during fabrication, erection staging, and in final position. It is also typical that the girder will be lifted at two locations, *a*, from each end of the girder and erection lifts will not be done coincidently with windy conditions.

Girder self-weight ~ 1.15(521) ~ 600 lb/ft (with 15% contingency load)

For typical railway girder lifting arrangements, the maximum bending stress will occur between the lifting locations, $b = L - 2a$. However, since the girder may be laterally unsupported for its entire length, the maximum bending stress is*

$\sigma_{\text{girder}} = [0.6(90)^2/8](12)/4469 = 1.6$ ksi

using an allowable bending stress of $1.25(0.55)F_y = 0.69F_y$ (see Chapter 4)

$$\left(\frac{L}{r_y}\right)_{\text{girder}} = \sqrt{\frac{0.69(50) - 1.6}{0.69(50)^2/6.3\pi^2(29{,}000)}} = 185$$

and the maximum unsupported length for fabrication and erection is $L_u = 4.67(185)/12 = 72.1$ ft ($L_u/L = 0.80$). The wide top flange benefits girder stability during fabrication and erection (see Example 7.1a for girder with narrow top flange and allowable unsupported length of only 56% of girder length).

Detailed Girder Design

Detailed Design of the Tension Flange

$$S_{xn} = \frac{I_{xn}}{c_t} \geq \frac{M_{t\max}}{F_{\text{all}}} \geq \frac{M_{t\max}}{0.55F_y} \geq \frac{9814(12)}{0.55(50)} \geq 4282 \text{ in.}^3 \left(\text{OK, } 4469 \text{ in.}^3 \text{ provided}\right)$$

for Category B weld,

$$S_{xn} \geq \frac{M_{\text{max-range}}}{F_{\text{fat}}} \geq \frac{(6032)(12)}{16} \geq 4524 \text{ in.}^3 \text{ (OK, ~1\% overstress with 4469 in.}^3 \text{ provided).}$$

A subsequent check on the net section is required after detailed design is completed.

Detailed Design of the Compression Flange

Brace frames will be placed at equal intervals of 11.25 ft.

* See Chapter 11 for a more detailed analysis of lifting girders at location, a, from each end of the girder.

$$\frac{L_p}{r_{cy}} = \frac{11.25(12)}{4.67} = 28.9 \le 5.55\sqrt{\frac{E}{F_y}} \le 134 \text{ OK}$$

The allowable compressive stress is

$$F_{call} = 0.55(50) - \frac{0.55(50)^2}{6.3\pi^2(29{,}000)}\left(\frac{(11.25)(12)}{4.67}\right)^2 = 27.5 - 0.64 = 26.9 \text{ ksi}$$

or

$$F_{call} = \left(\frac{0.13\pi(29{,}000)(50.0)}{(11.25)(12)(90)(\sqrt{1+0.3})}\right) = 42.8 \text{ ksi}$$

$$\text{Since } F_{call} \ge 0.55F_y = 27.5;\ F_{call} = 27.5 \text{ ksi}$$

$$S_{xg} = \frac{I_{xg}}{c_c} \ge \frac{M_{cmax}}{F_{call}} \ge \frac{9814(12)}{27.5} \ge 4282 \text{ in.}^3 \left(\text{OK, } 4965 \text{ in.}^3 \text{ provided}\right)$$

Vertical buckling of the compression flange is avoided by precluding flexural buckling of the web when

$$t_w \ge 0.18h\sqrt{\frac{F_y}{E}}\sqrt{\frac{f_c}{F_{cr}}} \ge 0.18(90 - 2(2.5))\sqrt{\frac{50}{29{,}000}}\sqrt{\frac{9814(12)/4965}{27.5}}$$

$$\ge 0.64\sqrt{\frac{23.7}{27.5}} \ge 0.59 \text{ in } \left(\text{OK, } 0.625 \text{ in provided}\right)$$

The plate girder will have ties directly supported on the top (compression) flange, and consideration of local buckling provides

$$\frac{b}{2t_f} = \frac{20}{2(2.50)} = 4.0 \le 0.35\sqrt{\frac{E}{F_y}} \le 8.4 \text{ OK}$$

Detailed Design of the Web Plate

$$A_w \ge \frac{V}{0.35F_y} \ge \frac{486}{0.35(50)} \ge 27.8 \text{ in.}^2 \left(\text{OK, } 53.13 \text{ in.}^2 \text{ provided}\right)$$

Web plate flexural buckling is considered as

$$\frac{h}{t_w} = \frac{85}{0.625} = 136 \le 4.18\sqrt{\frac{E}{f_c}} \le 4.18\sqrt{\frac{29{,}000}{23.7}} \le 146$$

Therefore, no longitudinal stiffeners are required for web flexural stability. This was also shown in the calculation related to compression flange vertical buckling above.

Web plate shear buckling is considered as

$$h = 85 \ge 2.12t_w\sqrt{\frac{E}{F_y}} \ge 2.12(0.625)\sqrt{\frac{29{,}000}{50}} \ge 31.9 \text{ in.}$$

Therefore, transverse web stiffeners are required.

Consideration of combined bending and shear yields

$$f_b = \frac{9814(12)}{4469} = 26.3 \le \left(0.75 - 1.05\frac{f_v}{F_y}\right)F_y \le \left(0.75 - 1.05\frac{486/53.13}{50}\right)50 = 27.9 \text{ ksi.}$$

$$\text{However, } f_b \times 0.55F_y = 27.5 \text{ ksi OK}$$

This interaction criterion does not generally require checking and, in the case where $f_v/(0.35F_y) =$ $0.52 < 0.55$ (see Figure 7.13) and the moment carried by the web plate is approximately $\left(\frac{53.1/6}{53.1/6+(20)(2.5)}\right)100 = 15.0\%$ of the total moment, combined bending and shear need not be considered.

Flange-to-Web Connection

For the top flange-to-web weld,

$$q = \sqrt{\left(\frac{VQ_f}{I}\right)^2 + \left(\frac{1.80W}{S_w}\right)^2} = \sqrt{\left(\frac{486(50.0)(45-1.25)}{223,444}\right)^2 + \left(\frac{1.80(40)}{3(12)}\right)^2} = \sqrt{4.62^2 + 2.00^2} = 5.04 \text{ k/in.}$$

For a CJP weld, the weld size must be $\ge \frac{0.5\sqrt{2}(5.04)}{0.35(50)} \ge 0.20$ in. OK since web thickness is 0.625 in.

For the bottom flange-to-web weld,

$$q = \frac{VQ_f}{I} = \frac{(310)(50.0)(45-1.25)}{201,100} = 3.20 \text{ k/in.}$$

For a CJP weld with backing bar removed, the weld size must be $\ge \frac{0.5\sqrt{2}(3.20)}{16} \ge 0.15$ in OK since web thickness is 0.625 in.

In general, CJP weld design does not need to be considered.

Design of Web Plate Stiffeners

Use angles bolted to the web in order to preclude fatigue issues related to welding at the base of the intermediate transverse stiffeners.

The spacing of the intermediate transverse stiffeners is

$a_0 < h = 85$ in.

< 96 in.

$$< a \le 1.95t_w\sqrt{\frac{E}{\tau}} \le 1.95(0.625)\sqrt{\frac{29,000}{486/53.13}} \le 68.6 \text{ in.}$$

Use a stiffener spacing of (11.25)(12)/2 = 67.5 in (1/2 the distance between brace frames)

$$I_{ts} = 2.5a_0t_w^3\left(\left(\frac{h}{a}\right)^2 - 0.7\right) = 2.5(67.5)(0.625)^3\left(\left(\frac{85}{68.6}\right)^2 - 0.7\right) = 34.4 \text{ in.}^4$$

As shown in Figure E7.2ba, a single 6 × 4 × 1/2 angle on one side of the web plate provides I_{ts} = 17.4 + 4.75(1.99)^2 = 36.2 in.^{4} OK.

Design of Bearing Stiffeners

As shown in Figure E7.2bb, the bearing stiffeners consist of 4 – *L*8 × 4 × 1/2.

$$A_{ebs} = 2A_{bs} + A_{wbs} = 2A_{bs} + 12(t_w)^2 = 2(2)(5.75) + 12(0.625)^2 = 27.69 \text{ in.}^2$$

$$I_{ebs} = 2I_{bs} + 2A_{bs}\bar{y}^2 + t_w^4 = 2(2)(38.5) + 2(2)(5.75)(3.17)^2 + (0.625)^4 = 385.6 \text{ in.}^4$$

$$r_{ebs} = \sqrt{\frac{385.6}{27.69}} = 3.73 \text{ in.}$$

$$\frac{h}{r_{ebs}} = \frac{85}{3.73} = 22.8$$

$0.839\sqrt{E/F_y} = 20.2$ and, therefore, the governing expression for allowable compressive stress is (Equation 7.84)

$$F_{call} = 0.60F_y - \left(17,500\frac{F_y}{E}\right)^{3/2}\left(\frac{0.75h}{r_{ebs}}\right) = 30,000 - 165.7(17.1) = 27,169 \text{ psi}$$

$$P_{call} = (27.17)(27.69) = 752 \text{ kips} > 486 \text{ kips OK.}$$

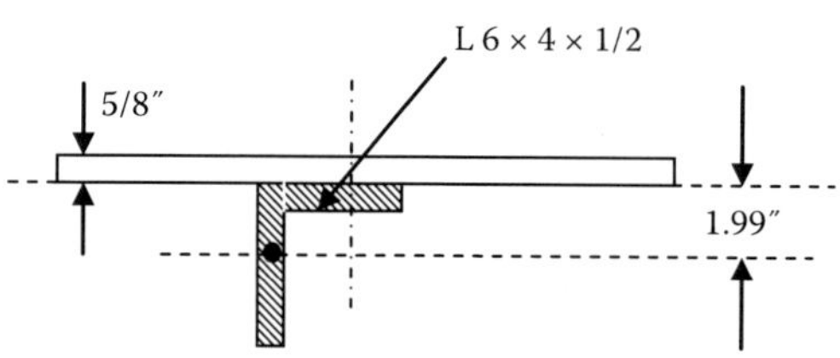

FIGURE E7.2ba Web plate stiffener.

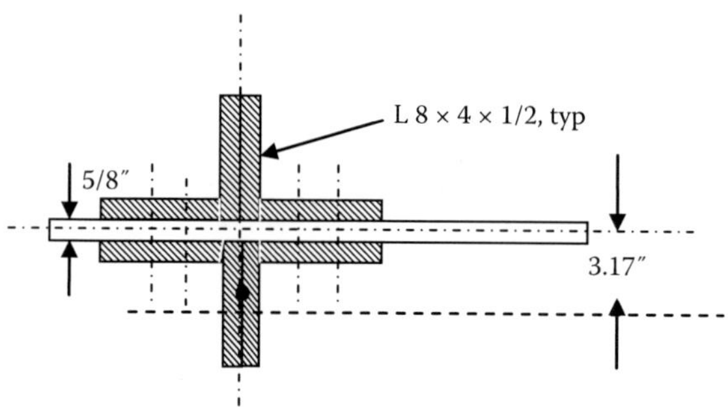

FIGURE E7.2bb Bearing stiffener.

TABLE E7.3

Deflection Criteria, L/Δ_{LL+I}	Required Gross Moment of Inertia, I_g, (in.^{4})
500	170,180
640 (AREMA, 2015)	217,832
800	272,290
1000	340,362

The reduced bearing stiffener area assuming a 1/2″ clip (made to clear the fillet at the web-to-flange junction from double bevel CJP groove welds), $A'_{bs} = 2(2)(8-0.50)(0.5) = 15.0$ in.^{2}

$$R = 486 \leq A'_{bs}(0.83F_y) \leq (15.0)(0.83)(50) \leq 622.5 \text{ kips OK}$$

Local buckling of outstanding compression elements:

$$\frac{b_{bs}}{t_{bs}} = \frac{8-0.5}{0.5} = 15.0 \leq 0.43\sqrt{\frac{E}{F_y}} \leq 10.4 \text{ NG, try } L\ 8\times4\ \times3/4.$$

$$\frac{b_{bs}}{t_{bs}} = \frac{8-0.75}{0.75} = 9.7 \leq 0.43\sqrt{\frac{E}{F_y}} \leq 10.4 \text{ OK}$$

Serviceability Design Deflection Criteria

The deflections are estimated as (see Chapter 5)

$$\Delta_{LL+I} = \frac{0.104M_{LL+I}L^2}{EI} = \frac{0.104(7314)(12)[90(12)]^2}{29{,}000(223{,}444)} = 1.64 \text{ in.}$$

The section $I_g = 223{,}444$ in.^{4} provides a deflection that is 2.5% less than the AREMA (2015) criterion of $L/640 = 90(12)/640 = 1.69$ in.

$$\Delta_{DL} = \frac{0.104M_{DL}L^2}{EI} = \frac{0.104(2500)(12)[90(12)]^2}{29{,}000(223{,}444)} = 0.56 \text{ in.; therefore, consider a camber of 1/2″.}$$

The required gross moment of inertia for LL + I deflection criteria of $L/f\Delta$ is (see Chapter 5)

$$I \geq \frac{M_{LL+I}Lf_\Delta}{1934} \geq \frac{7314(90)f_\Delta}{1934} \geq 340.4f_\Delta \text{ in.}^4$$

The required section gross moment of inertia for various deflection criteria, $f\Delta$, is shown in Table E7.3.

In practice, dimensions of various main and secondary elements may be revised further to attain greater economy of material.

7.4 STRENGTH DESIGN OF STEEL AND CONCRETE COMPOSITE FLEXURAL MEMBERS

Railway bridges with ballasted decks are beneficial from operational, structural, and maintenance perspectives (see Chapter 3). The ballast may be placed on timber, steel, or concrete decks.

Timber decks are generally not effective from structural and maintenance standpoints. Steel plate decks have only the strength or stiffness to span small lengths under railway live loads. Therefore, steel plate decks are generally not feasible for deck type bridges with girders or trusses spaced at wide distances unless supplemental support to the steel deck plate is provided. Floor systems are used to support stiffened or unstiffened steel plate decks unless the longitudinal members are closely spaced.* The use of steel plate decks† is often appropriate for long span construction to reduce the superstructure dead load stresses. Steel plate decks are often used and supported by closely spaced transverse floor beams between the girders of BTPG superstructures (see Appendix A). Deck plates may be fabricated in panels for shipment and erection in the field.‡ Steel deck plates supported on transverse floor beams provide an effective ballast and track support system for BTPG spans. However, for BDPG spans with widely spaced girders, a concrete deck with greater stiffness is required to support the ballast and track (see Appendix B).

Reinforced and/or prestressed concrete decks have the strength and stiffness required for use as ballasted decks in ordinary steel railway bridge construction (Figure 7.20).

Concrete decks may be noncomposite (not positively connected to the steel bridge span) or made composite. Relative slip between concrete deck and steel span will occur with noncomposite construction and, even with the substantial dead load, the deck may translate under the action of modern train braking and locomotive traction forces (see Chapter 4). To resist slippage at the noncomposite deck to superstructure interface, bridge designers typically provide for connection between concrete deck and steel superstructure for even noncomposite designs. Therefore, consideration of the benefits of composite steel and concrete behavior is warranted for design. Composite steel and concrete construction has the following benefits for steel railway bridge superstructures:

- Ease of site access for the materials used in railway bridge construction§ (provided that concrete transport or site batching is available, there is reduced shipping and erection of large steel sections and/or plate decks requiring field bolting).
- Improved train ride and reduced track and deck maintenance (given that adequate deck drainage and waterproofing are provided).
- Reduction in the weight of fabricated steel (typically between 10% and 20%)
- Reduction of superstructure depth (may be required for clearances, see Chapter 3).
- Increased superstructure stiffness (improved performance under live load).
- Increased capacity against overloads.
- Deck acts to resist lateral forces at top flange of girders (no need for horizontal bracing and reduced requirements related to vertical bracing, see Chapter 5). AREMA (2015) also

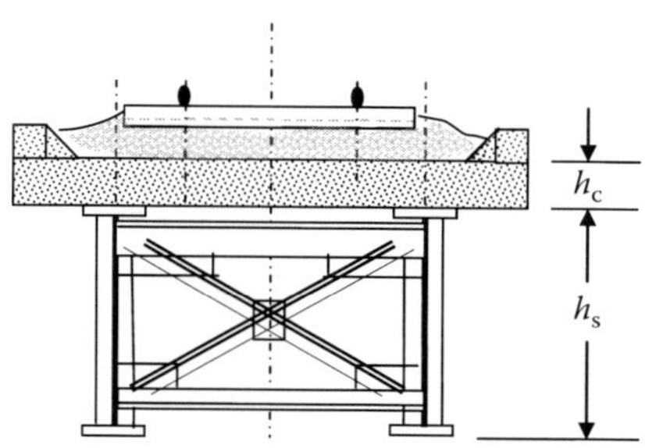

FIGURE 7.20 Cross section of a typical composite steel and concrete BDPG span.

* The fabrication and erection difficulties associated with steel decks on closely spaced longitudinal members may make them impractical except for use in short spans.

† In particular, orthotropic steel plate decks.

‡ Typically, 3 m (10 ft) to 4 m (13 ft) panels comprised of a deck plate welded to four to six floor beams are field fastened to adjacent transverse floor beams by bolting or welding. Bolted connections are often more appropriate for erection purposes.

§ For cast-in-place concrete decks. Composite precast concrete decks are feasible using grouted recesses (or pockets) for shear transfer devices (often shear studs or channels welded to the top flange of the girder). However, maintenance issues with grout longevity and at connections among precast concrete deck panels are typical with precast decks supporting large magnitude railway axle loads.

recommends that noncomposite decks have at least 25 mm (1 in.) of steel flange embedded in the concrete to be considered as sufficient lateral resistance to preclude the need for top lateral bracing.

Therefore, while composite construction is often utilized, its effectiveness depends on the mechanical interaction between concrete deck and steel superstructure. This mechanical connection between steel and concrete is usually accomplished with proprietary shear studs developed specifically for this purpose. However, other mechanical shear transfer connectors have been used to attain composite action. AREMA (2015) recommends the use of shear studs of 19 mm (3/4 in.) or 22 mm (7/8 in.) diameter, *d*, with a minimum length of 4*d* or channels with a minimum height of 75 mm (3 in.). Mechanical connectors are typically welded to the steel superstructure and embedded in the concrete deck when cast (for cast-in-place construction) or within grouted recesses (for precast concrete construction*).

Steel girders, beams, floor beams, and stringers are built as composite flexural members in railway bridge spans and must be designed to resist the internal normal and shear stresses created by combinations of external actions at various limit states (see Chapter 4). However, in addition to the usual strength (yielding and stability), fatigue, fracture, and serviceability design criteria, structures of composite materials require consideration of stiffness and strain compatibility at the interface between steel and concrete materials.

7.4.1 Flexure in Composite Steel and Concrete Spans

When a composite steel and concrete span bends, the horizontal shear at the steel to concrete interface must be resisted in order that the materials act integrally. The number, spacing, and size of mechanical connectors govern the interface behavior in terms of shear strength and stiffness of the connection. The ASD method of AREMA (2015) indicates a linear elastic analysis at service loads. Linear elastic analysis may be used provided complete (or near complete) interaction occurs at the interface. The flexural stress is then

$$\sigma = \frac{Mc}{I}, \tag{7.106}$$

where

M = the bending moment
σ = the normal (flexural) stress
c = the distance from the neutral axis to the extreme fiber
I = the moment of inertia

The degree of interaction at the steel–concrete interface is dependent on the horizontal stiffness of the interface connection. The horizontal stiffness of the interface connection, k_i, is determined from the horizontal shear load vs. interface slip relationship (Figure 7.21). All connections between concrete decks and steel superstructures will exhibit some degree of slip† or partial interaction. Partial interaction analysis requires consideration of nonlinear behavior, although "equivalent" linear elastic analyses of partial interaction have been developed (Newmark et al., 1951). Complete interaction enables a linear elastic analysis to be performed with a connection stiffness, $k_i = \infty$.

The relationship between interface connection stiffness, slip, and slip strain in a simply supported composite steel and concrete span is shown in Figure 7.22. With $k_i = 0$, no interaction occurs and with $k_i = \infty$, complete interaction occurs. As indicated above, practical structures will behave

* Care must be exercised in precast concrete deck slab construction to ensure that grouted and mortared recesses, joints, and slab bedding are properly designed and constructed to establish and maintain composite action of the superstructure.

† Slippage must occur in order to mobilize shear connector resistance.

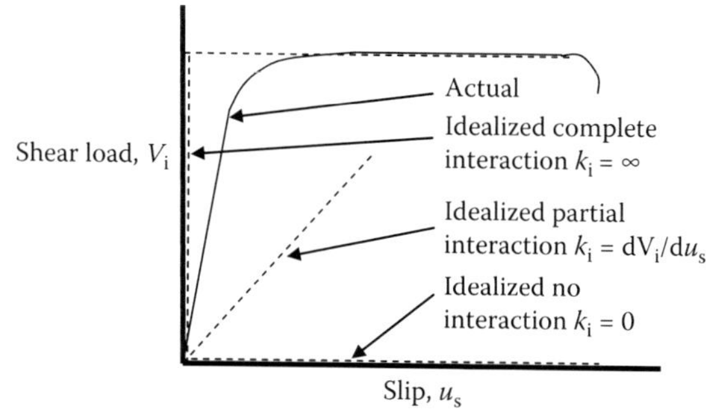

FIGURE 7.21 Stiffness of the interface connection.

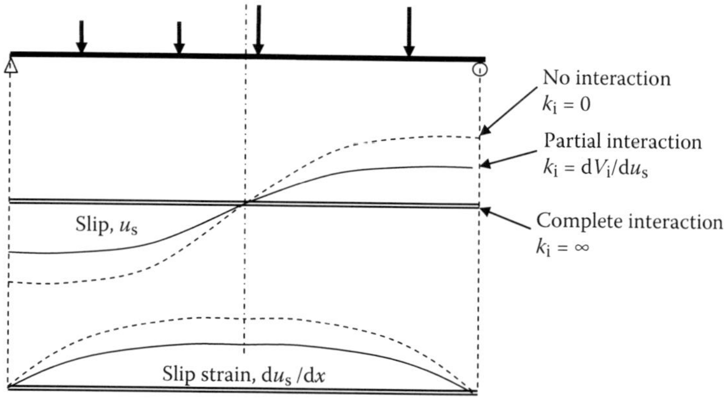

FIGURE 7.22 Slip and slip strain distribution in a simply supported composite beam.

between these extremes and exhibit partial interaction. However, because of the relatively large number of shear connectors required for strength in railway girders, the connection stiffness will be very large and may be idealized as infinitely stiff with complete interaction. Complete interaction strain compatibility indicates that, slip, $u_s = 0$, and slip strain, $du_s/dx = \varepsilon_c - \varepsilon_s = 0$, need not be considered in the flexural analysis. The strain distribution through a composite steel and concrete beam is shown in Figure 7.23 for no, partial, and complete interactions.

Transformed section methods may be used to determine cross-sectional stresses at service load levels since complete interaction allows for a linear elastic analysis (elastic E_s and E_c), with the same stress and strain profile (Gere and Timoshenko, 1984). The modular ratio, $n = E_s/E_c$, can be established and used as the transformation ratio for the steel and concrete elements. However, long-term dead load stresses do not remain constant due to creep and shrinkage of the concrete deck. The dead load stresses will increase in the steel elements. Long-term effects are considered by a simplified approach to shrinkage and creep that uses a plastic modulus, $n_{cr} = 3n$ (Viest et al., 1958).

The elastic stress distribution will also depend on how load is transferred to the composite steel–concrete span (i.e., dependent on construction scheme). If the steel beams are supported by shoring or falsework until after the concrete deck hardens,* the entire composite section resists load. If, however, the construction is unshored (typical of railway bridges constructed over waterways, highways, or other railways), the composite section does not resist load until after the concrete deck hardens

* AREMA (2015) indicates that this occurs when the concrete has attained 75% of its specified 28 days compressive strength.

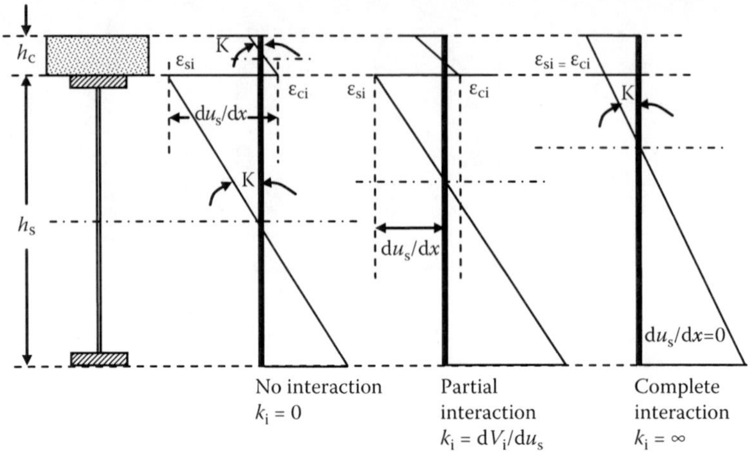

FIGURE 7.23 Strain profile through composite beam at various connection stiffness (no, partial and complete interaction) [κ=curvature of beam (typically greatest at or near center of span)].

and the steel beams or girders must resist a portion of the dead load (concrete deck, steel beams, and construction equipment). The stress distributions for complete interaction through the composite section for these two construction methods are shown in Figures 7.24a and b.

7.4.2 Shearing of Composite Beams and Girders

The linear elastic shear stress on the composite section is

$$\tau = \frac{VQ}{It} \tag{7.107}$$

and the shear flow at any section is

$$q = t\tau = \frac{VQ}{I}, \tag{7.108}$$

where

- V = the shear force
- τ = the shear stress
- q = the shear flow (along the length of the girder)
- I = the moment of inertia
- $Q = Ay_{na}$ = the statical moment of area about the neutral axis
- t = the thickness of the element

7.4.2.1 Web Plate Shear

AREMA (2015) recommends that shear is resisted by the steel girder web only. With maximum shear occurring at the neutral axis of the web plate, the minimum gross cross-sectional area of the steel girder web plate, A_w, is (Equation 7.31)*

* This is based on an "average" shear stress instead of the calculation of the shear stress through the cross section using Equation 7.107. For some wide flange (I-beam) sections, the "average" shear stress may be about 75% of the maximum shear stress through the cross section calculated using Equation 7.107.

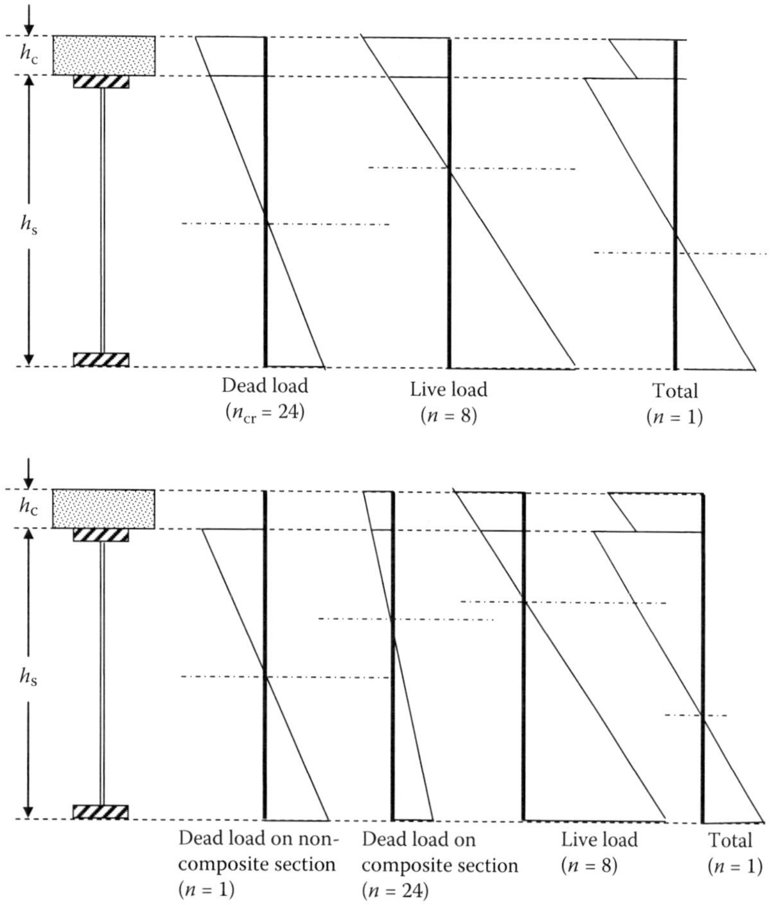

FIGURE 7.24 (a) Stress profile through a composite beam using shored or falsework supported construction. (b) Stress profile through a composite beam using unshored or unsupported construction.

$$A_w \geq \frac{V}{0.35F_y}. \tag{7.109}$$

Pure flexural, pure shear, and combined flexural and shear buckling of the web plate must also be considered in the same manner required for noncomposite girders in Section 7.2.6.1.3.

7.4.2.2 Shear Connection between Steel and Concrete

The shear connection strength is also affected by the method of construction. Shored or supported construction requires that the shear flow at the steel to concrete interface can be determined based on composite section properties for short- and long-term dead load, and short-term live load effects. In unshored or unsupported construction, shear flow at the steel to concrete interface must be determined based on composite section properties for the long-term effects of dead load and short-term live load effects. The shear flow, q_i, at the steel to concrete interface is

$$q_i = \tau b_f = \frac{VQ_c}{I_{cp}}, \tag{7.110}$$

where

V = the shear force

$Q_c = A_c y_{na}$ = the statical moment of the concrete deck about the neutral axis

A_c = the transformed area of the concrete deck

b_f = the width of the girder flange

y_{na} = the distance from the centroid of the concrete slab to the neutral axis

I_{cp} = the moment of inertia of the composite section

Shear transfer connectors must be designed to resist the shear flow at the steel to concrete interface. Shear studs with 19 mm (3/4″) or 22 mm (7/8″) diameters are typically used. AREMA (2015) recommends that shear studs can be at least 75 mm (3″) long and 100 mm (4″) long shear studs are commonly used. The allowable strength of shear studs is

$$S_r = \frac{C_{sr}\pi(d_s)^2}{4} \tag{7.111a}$$

$$S_m = \frac{C_{sm}\pi(d_s)^2}{4}, \tag{7.111b}$$

where

S_r = the allowable horizontal design force for fatigue per connector

S_m = the allowable maximum horizontal design force per connector

d_s = the diameter of shear stud

For shear studs using SI units,

$$C_{sr} = 50 \text{ MPa for fatigue design cycles}, N \geq 2{,}000{,}000 \text{ cycles}$$

$$= 70 \text{ MPa for fatigue design cycles}, N = 2{,}000{,}000 \text{ cycles}$$

$$C_{sm} = 140 \text{ MPa}$$

For shear studs using US Customary and Imperial units

$$C_{sr} = 7000 \text{ psi for fatigue design cycles}, N \geq 2{,}000{,}000 \text{ cycles}$$

$$= 10{,}000 \text{ psi for fatigue design cycles}, N = 2{,}000{,}000 \text{ cycles}$$

$$C_{sm} = 20{,}000 \text{ psi}$$

For channels, the recommended strength is

$$S_r = D_{sr} w_c, \tag{7.112a}$$

$$S_m = D_{sm} w_c, \tag{7.112b}$$

where

w_c = the length of channel perpendicular to shear flow (transverse to flange).

For channels using SI units

$$D_{sr} = 370 \text{ N/mm for fatigue design cycles}, N \geq 2{,}000{,}000 \text{ cycles}$$

$$= 430 \text{ N/mm for fatigue design cycles}, N = 2{,}000{,}000 \text{ cycles}$$

$$D_{sm} = 630 \text{ N/mm}$$

For channels using US Customary and Imperial units,

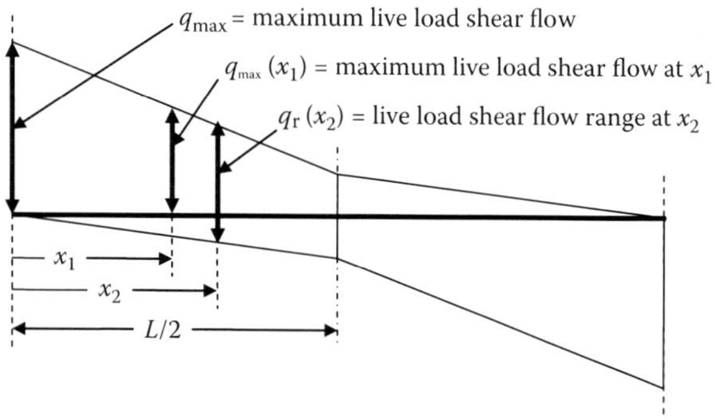

FIGURE 7.25 Distribution of shear flow along span length.

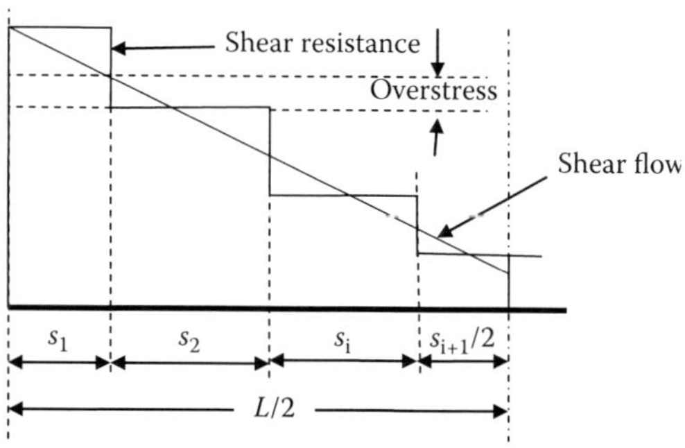

FIGURE 7.26 Distribution of shear resistance (studs or channels) along span length.

$$D_{sr} = 2100 \text{ lb/in. for fatigue design cycles, } N \geq 2{,}000{,}000 \text{ cycles}$$

$$= 2400 \text{ lb/in. for fatigue design cycles, } N = 2{,}000{,}000 \text{ cycles}$$

$$D_{sm} = 3600 \text{ lb/in}$$

The distribution of shear connectors along the span is made based on the magnitude and variation of the shear flow along the span length. Since live load shear flow varies along the span length, L, the shear connector spacing may also differ. The form of a typical shear flow influence line for the determination of live load maximum and range of shear flow is shown in Figure 7.25. Based on the maximum shear flow and live load shear flow range, a practical spacing over a length, s_i, with some acceptable overstress (usually about 10%) can be made as illustrated in Figure 7.26.

7.5 SERVICEABILITY DESIGN OF COMPOSITE FLEXURAL MEMBERS

AREMA (2015) recommends that midspan deflection of simply supported spans due to live load plus impact, ΔLL+I, should not exceed $L/640$, where L=span length. Some engineers or bridge owners recommend even more stringent live load plus impact deflection criteria to attain stiffer spans which offer improved performance from a structural* and train-superstructure dynamics perspective (see Chapter 4).

It is also recommended that camber can be provided for dead load deflections exceeding 25 mm (1 in.). For composite spans, the dead load deflections depend on the construction method employed (shored or unshored).

* Concrete bridge decks generally exhibit better behavior and less cracking on stiffer spans.

The serviceability criteria for steel railway spans are also discussed in Chapter 5. Example 7.2 outlines the design of a composite steel–concrete span for Cooper's EM360 and Cooper's E80 live load considering both shored and unshored construction for the flexural design. Appendix B outlines the design of a composite steel–concrete BDPG span considering unshored construction.

Example 7.2a (SI Units)

A 30 m simple span steel (F_y = 350 MPa) BDPG is to be designed for the forces shown in Table E7.4a.

Section Properties of the Span

The steel girder section properties are shown in Figure E7.1a and Table E7.2a (see Example 7.1a). The composite steel and concrete girder section properties are shown in Figure E7.3a and Tables E7.5a and E7.6a for short-term loads, and Tables E7.7a and E7.8a for long-term loads.

TABLE E7.4a

Design Force (Load Case D1 in Table 4.10)	Shear Force, V (kN)	Bending Moment (kN m)
Dead load on unshored steel section, DL1	350	2300
Dead load on composite section (unshored), DL2	200	1760
Total DL (DL1 + DL2)	550	4060
Live load + 33% impact (maximum)	1770	11,440
Maximum	2320	15,500
Live load +12% impact (fatigue)	1490	9630

TABLE E7.5a

Element	A (mm^2)	y_b (mm)	$Ay_b \times 10^3$ (mm^3)	$y_b - y$ (mm)	$A(y_b - y)^2 \times 10^6$ (mm^4)	$I_o \times 10^6$ (mm^4)
Steel section	104,410	1174	122,599	−584	35,610	111,522
Concrete slab (n = 8)	70,313	2625	184,570	867	52,853	366
Σ	174,723		307,169		88,463	111,888

TABLE E7.6a

Location (Figure E7.3a)	n	c (mm)	I (Gross or Net Depending on Location of NA) $\times 10^6$ (mm^4)	nS (gross or net) $\times 10^6$ (mm^3)
Top concrete	8	992	200,351	1616
Bottom concrete	8	742	200,351	2160
Top steel	1	742	200,351	270
Bottom steel	1	1758	189,162	108

TABLE E7.7a

Element	A (mm^2)	y_b (mm)	$Ay_b \times 10^3$ (mm^3)	$y_b - y$ (mm)	$A(y_b-y)^2 \times 10^6$ (mm^4)	$I_o \times 10^6$ (mm^4)
Steel section	104,410	1174	122,599	−266	7388	111,522
Concrete slab (n = 24)	23,438	2625	61,523	1185	32,912	122
Σ	127,848		184,122		40,300	111,644

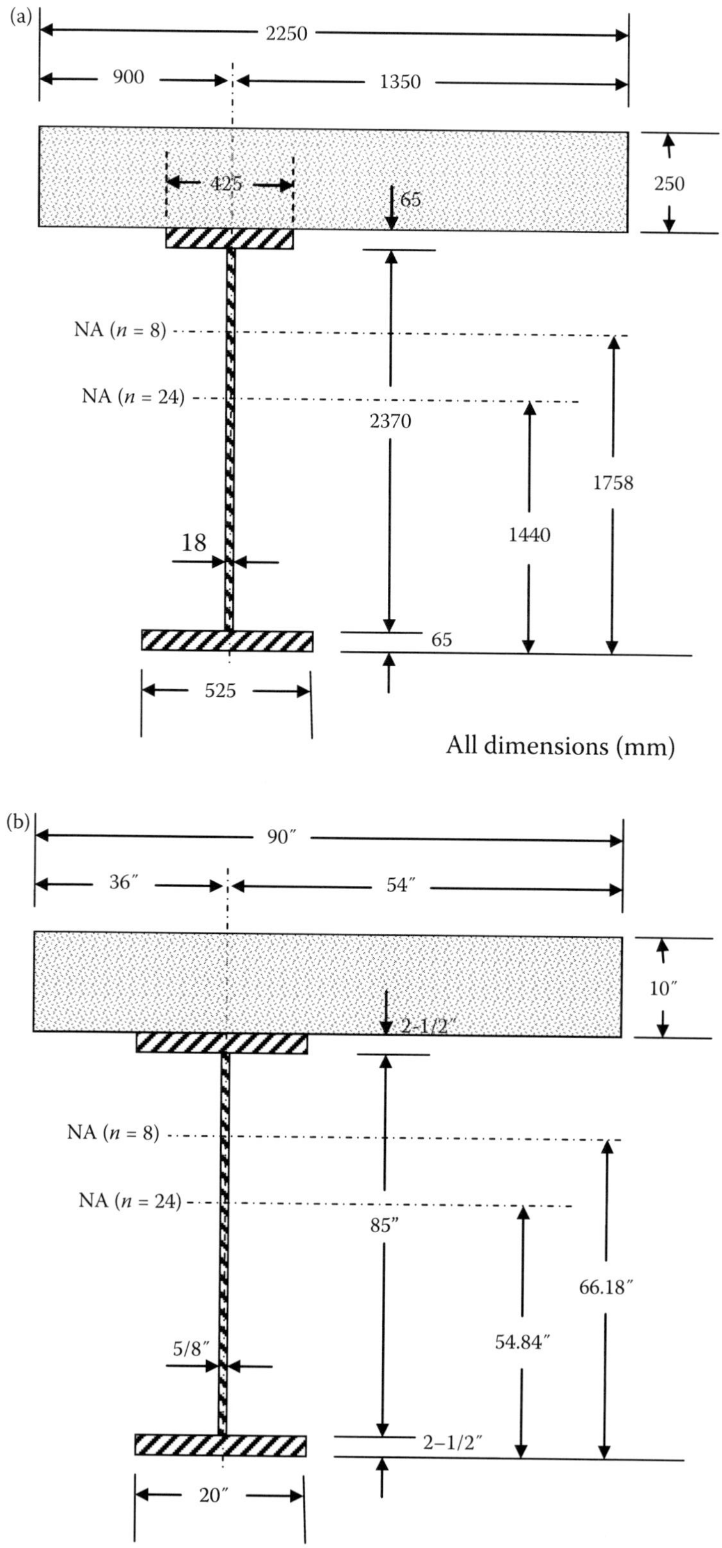

FIGURE E7.3 (a) Composite section with $n=8$ and $n=24$ (elastic and plastic modular ratios). (b) Composite section with $n=8$ and $n=24$ (elastic and plastic modular ratios).

TABLE E7.8a

Location (Figure E7.3a)	n	c (mm)	I (Gross or Net Depending on Location of NA) $\times 10^6$ (mm^4)	$nS \times 10^6$ (Gross or Net) (mm^3)
Top concrete	24	1310	151,944	2784
Bottom concrete	24	1060	151,944	3440
Top steel	1	1060	151,944	143
Bottom steel	1	1440	140,780	98

Composite steel–concrete section properties:

Short-Term Loads

$$\frac{\sum Ay_b}{\sum A} = \frac{307,169(1000)}{174,723} = 1758 \text{ mm}$$

$$I_g = \sum A(y_b - y)^2 + \sum I_o = (88,463 + 111,888) \times 10^6 = 200,351 \times 10^6 \text{ mm}^4$$

Assuming I_n steel section $= 0.90 I_o = 0.90(111,888 \times 10^6) = 100,699 \times 10^6$ mm^4
I_n composite section $= (88,463 + 100,699) \times 10^6 = 189,162 \times 10^6$ mm^4.

Long-Term Loads

$$\frac{\sum Ay_b}{\sum A} = \frac{184,122(1000)}{127,848} = 1440 \text{ mm}$$

$$I_g = \sum A(y_b - y)^2 + \sum I_o = (40,300 + 111,644) \times 10^6 = 151,944 \times 10^6 \text{ mm}^4.$$

Assuming I_n steel section $= 0.90 I_o = 0.90(111,644) \times 10^6 = 100,480 \times 10^6$ mm^4
I_n composite section $= (40,300 + 100,480) \times 10^6 = 140,780 \times 10^6$ mm^4.

Flexure and Shear Design

Flexural stresses are summarized in Figure E7.4a and Table E7.9a or Figure E7.5a and Table E7.10a for unshored and supported deck construction, respectively.

In this example, there is not a great difference in unshored and shored flexural stresses due to the relatively small dead load stress on the noncomposite (steel only) section during unshored construction.

Shear Stresses

$f_v = (2320)(1000)/[(2370)(18)] = 54.4$ MPa (shear resisted entirely by steel girder web).

Allowable Stresses (30 MPa 28 Day Minimum Compressive Strength Concrete and Grade 350 Steel)

$$F_{call} \text{ concrete} = 0.40 f_c' = 0.40(30) = 12 \text{ MPa} > 8.6 \text{ MPa OK}$$

$$F_{tall} \text{ steel} = F_{call} \text{ steel} = 0.55 F_y = 0.55(350) = 192.5 \text{ MPa} > 150.8 \text{ MPa OK}$$

(a)

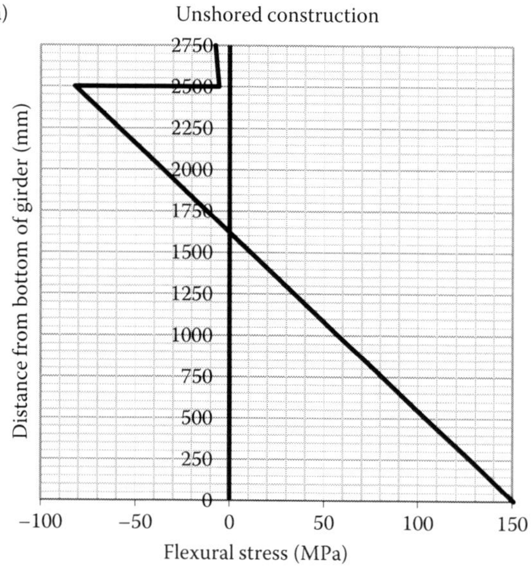

(b)

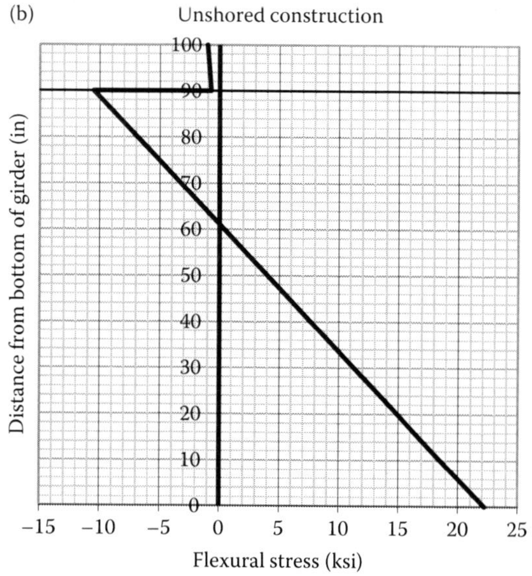

FIGURE E7.4 (a) Composite steel and concrete section flexural stresses—unshored construction. (b) Composite steel and concrete section flexural stresses—unshored construction.

TABLE E7.9a

Location (Figure E7.3a)	DL1 Flexural Stress on Noncomposite Section ($n=1$) (MPa)	DL2 Flexural Stress on Composite Section ($n=24$) (MPa)	Maximum LL+I Flexural Stress on Composite Section ($n=8$) (MPa)	Range of LL+I Flexural Stress on Composite Section ($n=8$) (MPa)	Maximum Flexural Stress (MPa)
Top concrete	—	0.6	7.1	6.0	7.7
Bottom concrete	—	0.5	5.3	4.5	5.8
Top steel	27.3	12.3	42.4	35.7	82.0
Bottom steel	26.9	18.0	105.9	89.2	150.8

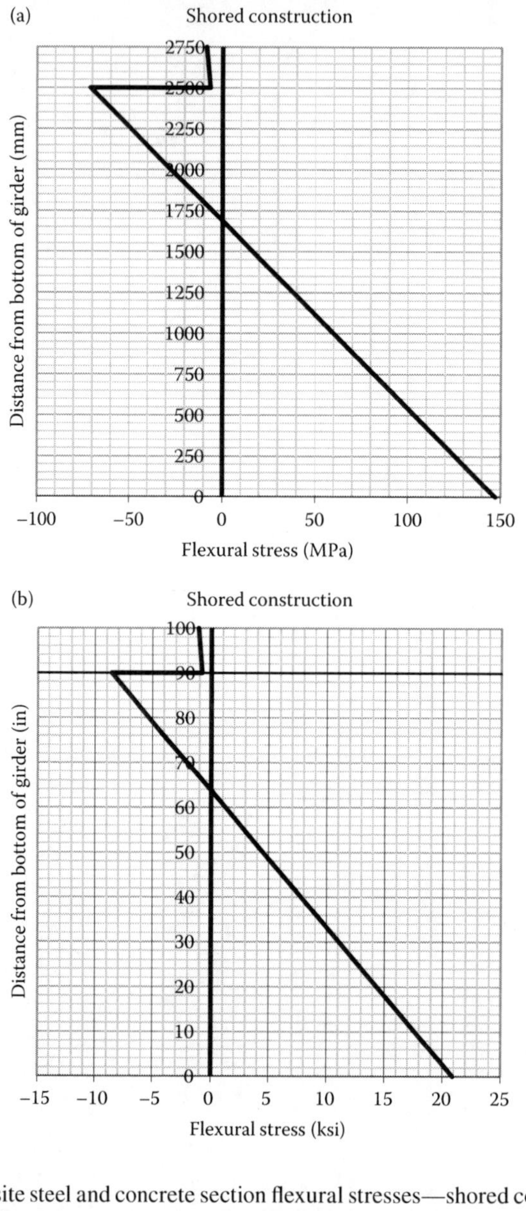

FIGURE E7.5 (a) Composite steel and concrete section flexural stresses—shored construction. (b) Composite steel and concrete section flexural stresses—shored construction.

TABLE E7.10a

Location (Figure E7.3a)	DLFlexural Stress on Composite section ($n = 24$) (MPa)	Maximum LL + I Flexural Stress on Composite Section ($n = 8$) (MPa)	Range of LL + I Flexural Stress on Composite Section ($n = 8$) (MPa)	Maximum Flexural Stress (MPa)
Top concrete	1.5	7.1	6.0	8.6
Bottom concrete	1.2	5.3	4.5	6.5
Top steel	28.4	42.4	35.7	70.8
Bottom steel	41.4	105.9	89.2	147.3

$$F_{\text{fat}} = 110 \text{ MPa (Category B with loaded length of 30 m)} > 89.2 \text{ MPa, OK}$$

$$F_{\text{vall}} = 0.35(350) = 122.5 \text{ MPa} > 54.4 \text{ MPa, OK}$$

Girder Design for Fabrication and Erection Loads

Assuming concrete deck will be cast on site (typically on falsework and slid into place once deck hardened, see Chapter 10), unsupported length will be as determined in Example 7.1a

Detailed Design of the Girder

Detailed Design of Web Plate

Flexural buckling:

$$\frac{h}{t_w} = \frac{2370}{18} = 131.7 \le 4.18\sqrt{\frac{E}{f_c}} \le 4.18\sqrt{\frac{200{,}000}{150.8}} \le 152.2.$$

Therefore, no longitudinal stiffeners are required for web flexural buckling stability.

Shear buckling:

$$h = 2370 \ge 2.12 t_w\sqrt{\frac{E}{F_y}} \ge 2.12(18)\sqrt{\frac{200{,}000}{350}} \ge 912 \text{ mm.}$$

Therefore, transverse web stiffeners are required.

Combined bending and shear:

$$f_b = 150.8 \le \left(0.75 - 1.05\frac{f_v}{F_y}\right)F_y \le \left(0.75 - 1.05\frac{54.4}{350}\right)350 \le 205.4 \text{ MPa,}$$

but no greater than 192.5 MPa, OK.

Flange-to-web connection:

Unshored construction only is considered in the flange-to-web connection design for brevity. Similar calculations may be performed if a shored construction method is utilized.

The section properties at the top and bottom welds are shown in Tables E7.11a and E7.12a, respectively. Shear flow at the weld, calculated based on these section properties, is shown in Table E7.13a.

Maximum shear flow = 898 N/mm

TABLE E7.11a

Section	A (mm^2)	y_{na} (mm)	$Ay_{na} \times 10^3$ (mm^3)	$I_g \times 10^6$ (mm^4)	Ay_{na}/I_g (mm^{-1})
Noncomposite (steel only)	27,625	1291	35,664	111,522	3.20×10^{-4}
Composite-long term ($n = 24$)	23,438	1185	27,774	151,944	3.70×10^{-4}
	27,625	1028	28,399		
Composite-short term ($n = 8$)	70,313	867	60,961	200,351	4.02×10^{-4}
	27,625	710	19,614		

TABLE E7.12a

Section	A (mm²)	y_{na} (mm)	$Ay_{na} \times 10^3$ (mm³)	$I_n \times 10^6$ (mm⁴)	Ay_{na}/I_n (mm⁻¹)
Noncomposite (steel only)	34,125	1142	38,971	100,370	3.88×10^{-4}
Composite-long term ($n = 24$)	34,125	1408	48,048	140,780	3.41×10^{-4}
Composite-short term ($n = 8$)	34,125	1726	58,900	189,162	3.11×10^{-4}

TABLE E7.13a

Location (Figure E7.3a)	DL1 Shear Flow on Noncomposite Section ($n = 1$) (N/mm)	DL2 Shear Flow on Composite Section ($n = 24$) (N/mm)	Maximum LL + I Shear Flow on Composite Section ($n = 8$) (N/mm)	Range of LL + I Shear Flow on Composite Section ($n = 8$) (N/mm)	Maximum Shear Flow (N/mm)
Top weld	112	74	712	599	898
Bottom weld	136	68	550	463	754

Maximum LL + I range shear flow = 599 N/mm [AREMA recommends that even stress ranges in welds in compression zones can be considered for fatigue due to the relatively high "effective mean stress" created by residual stresses from welding processes (see Chapter 10)]

Allowable weld shear stress = 120 MPa (fillet)

Allowable fatigue stress range = 110 MPa (Fatigue Detail Category B)

Weld size = 898/[2(0.71)120] = 5.3 mm for maximum shear flow

Weld size = 599/[2(0.71)110] = 3.8 mm for cyclical shear flow

Use min 8 mm fillet welds or CJP welds (some designers will specify CJP welds because of the vertical load transmitted directly to the top flange-to-web weld).

Shear Stud design:

Unshored construction only is considered for shear stud design for brevity. Similar calculations may be performed if a shored construction method is utilized.

Also for brevity, shear stud spacing will be calculated at only three locations on the girder as indicated in Table E7.14a (in the design of practical girders, a smaller interval is recommended*).

The negative live load shear flow (the shear reverses when a wheel passes over a stud) is estimated as a linear interpolation of the center span live load shear as shown in Figure E7.6a.

Shear flow at the steel–concrete interface (Table E7.15a) is

$$q_{LL+I} = \frac{V_{LL+I}Q_{cp(n=8)}}{I_{cp(n=8)}} = \frac{60,961(V_{LL+I})}{200,351(1000)} = \frac{V_{LL+I}}{3287}$$

$$q_{max} = q_{LL+I} + q_{DL2} = q_{LL+I} + \frac{V_{DL2}Q_{cp(n=24)}}{I_{cp(n=24)}} = \frac{V_{LL+I}}{3287} + \frac{27,774V_{DL2}}{191,544(1000)} = \frac{V_{LL+I}}{3287} + \frac{V_{DL2}}{6897}$$

TABLE E7.14a

Location (Figure E7.6a)	Distance, x, from End a (mm)	$+V_{LL+I}$ (kN)	$-V_{LL+I}$ (kN) (Linear Interpolation)	V_{DL2} (kN)	V_r (kN)	V_{max} (kN)
a	0	1301	0	200	1490	1969
b	7500	828	178	100	1152	1468
c	15,000 (center)	355	355	0	813	966

* Equivalent uniform load charts (such as Steinman charts) are useful in determining live load shear forces at various locations along the span (see Chapter 5).

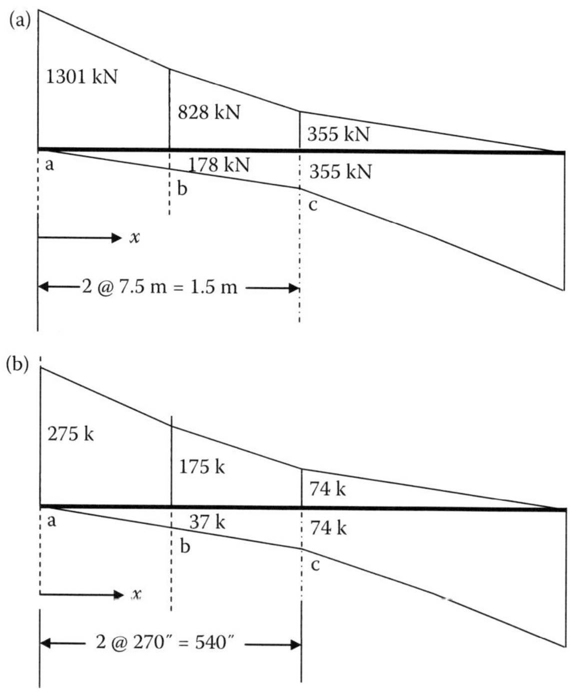

FIGURE E7.6 (a) Shear forces along the span. (b) Shear forces along the span.

TABLE E7.15a

Location (Figure E7.6)	Distance, x, from End a (mm)	V_r (kN)	V_{DL2} (kN)	q_r (N/mm)	q_{DL2} (N/mm)	q_{max} (N/mm)
a	0	1490	200	453	29	567
b	7500	1152	100	350	14	431
c	15,000	813	0	247	0	294

The shear flow due to dead load on the composite section is small and ignored in the practical design of composite steel and concrete girders. This is also evident from comparison of the allowable design load for shear stress range, S_r, and the allowable design load for maximum shear stress, S_m:

$$S_r = \frac{50\pi(22)^2}{4(1000)} = 19.0 \text{ kN}$$

$$S_m = \frac{140\pi(22)^2}{4(1000)} = 53.2 \text{ kN}$$

For three shear studs across the flange width

s = spacing required (Table E7.16a) = $3(19.0)(1000)/q_r = 57 \times 10^3/q_r$.

The actual shear stud spacing can be arranged in order that the maximum overstress is, for example, 10% as shown in Figure E7.7a.

Design of web plate stiffeners:

Use angles bolted to the web in order to preclude issues related to welding at the base of transverse stiffeners. Use a single 150 × 100 × 13 angle at 1665 mm centers as shown in Example 7.1a.

TABLE E7.16a

Location	q_r (N/mm)	s (mm)
a	453	126
b	350	163
c	247	231

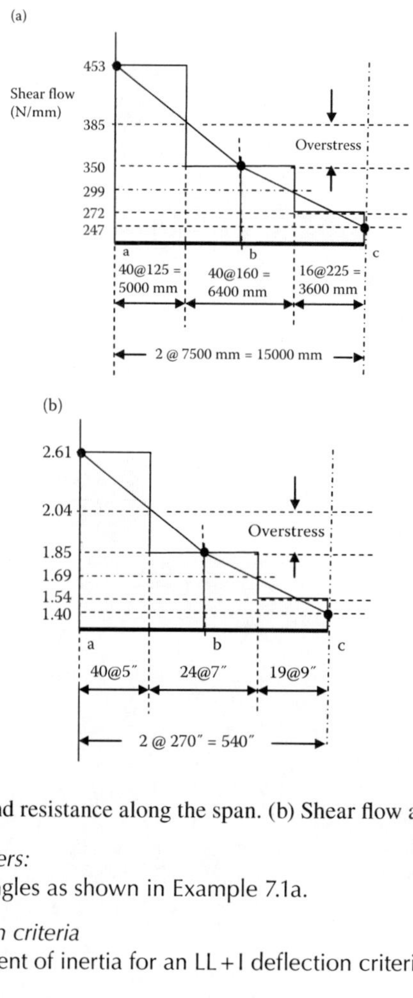

FIGURE E7.7 (a) Shear flow and resistance along the span. (b) Shear flow and resistance along the span.

Design of bearing stiffeners:
Use four 200×100×16 angles as shown in Example 7.1a.

Serviceability—deflection criteria
The required gross moment of inertia for an LL+I deflection criteria of $L/f\Delta$ is (see Chapter 5)

$$I \geq 520M_{LL+I}Lf_{\Delta} \geq 520(11,440)(30)f_{\Delta} \geq 178.5\times10^{6} f_{\Delta} \text{ mm}^4.$$

The required section gross moment of inertia for various deflection criteria, f_D, is shown in Table E7.17a.

TABLE E7.17a

Deflection Criteria, L/Δ_{LL+I}	Required Gross Moment of Inertia, I_g, $\times 10^6$ (mm^4)
500	89,232
640 (AREMA, 2015)	114,217
800	142,771
1000	178,464

The section $I_g = 200{,}351$ in.[4] provides a very stiff structure.
The deflections are estimated as (see Chapter 5)

$$\Delta_{LL+I} = \frac{0.104 M_{LL+I} L^2}{EI} = \frac{0.104(11,440)[30,000]^2}{200,000(200,351)} = 26.7 \text{ mm}$$

$$\Delta_{DL} = \frac{0.104 M_{DL1} L^2}{EI_{g1}} = \frac{0.104 M_{DL2} L^2}{EI_{g2}} = \frac{0.104(2300)[30000]^2}{200,000(111522)} + \frac{0.104(1760)[30,000)]^2}{200,000(151,944)}$$

$$= 9.7 \text{ mm} + 5.4 \text{ mm} = 15.1 \text{ mm}.$$

No camber is required.

Summary of the design

Examples 7.1a and 7.2a are not intended to be examples of optimum design but to provide numerical examples of noncomposite and composite steel and concrete girder designs. The summary of stresses and deflections of the two span designs [noncomposite (Example 7.1a) and composite (Example 7.2a)] shown in Table E7.18a reveals that reductions in element sizes can be made and the advantages of composite girder design may be exploited.

Example 7.2b (US Customary and Imperial Units)

A 90 ft simple span steel ($F_y = 50$ ksi) BDPG is to be designed for the forces as shown in Table E7.4b.

Section properties of the span:

The steel girder section properties are shown in Figure E7.1b and Table E7.2b (see Example 7.1b). The composite steel and concrete girder section properties are shown in Figure E7.3b and Tables E7.5b and E7.6b for short-term loads, and Tables E7.7b and E7.8b for long-term loads.

TABLE E7.18a

Location (Figures E7.1a and E7.3a)	Noncomposite Section Stress (MPa) (Figure E7.1a)	Unshored Composite Section Stress (MPa) (Figure E7.3a)	Shored Composite Section Stress (MPa) (Figure E7.3a)
Top concrete (maximum flexure)	—	7.7	8.6
Bottom concrete (maximum flexure)	—	5.8	6.5
Top steel (maximum flexure)	184	82	71
Web (maximum shear)	54	54	54
Bottom steel (maximum flexure)	181	151	147
Bottom steel (LL+I flexure)	113	89	89

TABLE E7.4b

Design Force (Load Case D1 in Table 4.10)	Shear Force, V (kips)	Bending Moment (kips-ft)
Dead load on unshored steel section, DL1	70	1430
Dead load on composite section (unshored), DL2	40	1070
Total DL (DL1+DL2)	110	2500
Live load+37% impact (maximum)	376	7314
Maximum	486	9814
Live load +13% impact (fatigue)	310	6032

TABLE E7.5b

Element	A (in.^{2})	y_b (in.)	Ay_b (in.^{3})	y_b-y (in.)	$A(y_b-y)^2$ (in.^{4})	I_o (in.^{4})
Steel section	153.13	45.00	6891	−21.18	68,667	223,444
Concrete slab ($n=8$)	112.50	95.00	10,688	28.82	93,467	938
Σ	265.6		17,578		162,134	224,382

TABLE E7.6b

Location (Figure E7.3b)	n	c (in.)	I (Gross or Net Depending on Location of NA) (in.^{4})	nS (Gross or Net) (in.^{3})
Top concrete	8	33.82	386,516	91,429
Bottom concrete	8	23.82	386,516	129,812
Top steel	1	23.82	386,516	16,227
Bottom steel	1	66.18	363,234	5489

TABLE E7.7b

Element	A (in.^{2})	y_b (in.)	Ay_b (in.^{3})	y_b-y (in.)	$A(y_b-y)^2$ (in.^{4})	I_o (in.^{4})
Steel section	153.13	45.00	6891	−9.84	14,825	223,444
Concrete slab ($n=24$)	37.50	95.00	3563	40.16	60,483	313
Σ	190.63		10,454		75,308	223,757

TABLE E7.8b

Location (Figure E7.3b)	n	c (in.)	I (Gross or Net Depending on Location of NA) (in.^{4})	nS (Gross or Net) (in.^{3})
Top Concrete	24	45.16	299,065	158,936
Bottom Concrete	24	35.16	299,065	204,140
Top Steel	1	35.16	299,065	8506
Bottom Steel	1	54.84	276,408	5040

Composite Steel–Concrete Section Properties

Short-Term Loads

$$\frac{\sum Ay_b}{\sum A} = \frac{17{,}578}{265.6} = 66.18 \text{ in.}$$

$$I_g = \sum A(y_b - y)^2 + \sum I_o = 162{,}134 + 224{,}382 = 386{,}516 \text{ in.}^4$$

Assuming I_n steel section $= 0.90 I_o = 0.90(223{,}444) = 201{,}100$ in.^{4}

I_n composite section = 162,134 + 201,100 = 363,234 in.

Long-Term Loads

$$\frac{\sum Ay_b}{\sum A} = \frac{10,454}{190.63} = 54.84 \text{ in.}$$

$$I_g = \sum A(y_b - y)^2 + \sum I_o = 75,308 + 223,757 = 299,065 \text{ in.}^4$$

Assuming I_n steel section = $0.90I_o$ = 0.90(223,444) = 201,100 in.^{4}
I_n composite section = 75,308 + 201,100 = 276,408 in.^{4}.

Flexure and Shear Design

Flexural stresses are summarized in Figure E7.4b and Table E7.9b or Figure E7.5b and Table E7.10b for unshored and supported deck construction, respectively.

In this example, there is not a great difference in unshored and shored flexural stresses due to the relatively small dead load stress on the non- composite (steel only) section during unshored construction.

Shear Stresses

f_v = (486)/[(85)(0.625)] = 9.15 ksi (shear resisted entirely by steel girder web).

TABLE E7.9b

Location (Figure E7.3b)	DL1 Flexural Stress on Noncomposite Section (n = 1) (ksi)	DL2 Flexural Stress on Composite Section (n = 24) (ksi)	Maximum LL + I Flexural Stress on Composite Section (n = 8) (ksi)	Range of LL + I Flexural Stress on Composite Section (n = 8) (ksi)	Maximum Flexural Stress (ksi)
Top concrete	—	0.08	0.96	0.79	1.04
Bottom concrete	—	0.06	0.68	0.56	0.74
Top steel	3.46	1.51	5.41	4.47	10.38
Bottom steel	3.84	2.55	15.99	13.19	22.38

TABLE E7.10b

Location (Figure E7.3)	DL Flexural Stress on Composite Section (n = 24) (ksi)	Maximum LL + I Flexural Stress on Composite Section (n = 8) (ksi)	Range of LL + I Flexural Stress on Composite Section (n = 8) (ksi)	Maximum Flexural Stress (ksi)
Top concrete	0.19	0.96	0.79	1.15
Bottom concrete	0.15	0.68	0.56	0.83
Top steel	3.53	5.41	4.47	8.94
Bottom steel	5.95	15.99	13.19	21.94

Allowable Stresses

F_{call} concrete $= 0.40f_c' = 0.40(3)$

$= 1.2$ ksi (minimum 28 day concrete compressive strength of 3000 psi and Grade 50 steel) > 1.15 ksi, OK

F_{tall} steel $= F_{call}$ steel $= 0.55F_y = 0.55(50) = 27.5$ ksi > 22.4 ksi, OK
$F_{fat} = 16$ ksi (Category B with loaded length of 90 ft) > 13.2 ksi, OK
$F_{vall} = 0.35(50) = 17.50$ ksi > 9.15 ksi, OK

Girder Design for Fabrication and Erection Loads

Assuming concrete deck will be cast on site (typically on falsework and slid into place once deck hardened, see Chapter 10), unsupported length will be as determined in Example 7.1b.

Detailed Design of the Girder

Detailed design of web plate

Flexural buckling:

$$\frac{h}{t_w} = \frac{85}{0.625} = 136 \le 4.18\sqrt{\frac{E}{f_c}} \le 4.18\sqrt{\frac{29{,}000}{22.6}} \le 150.$$

Therefore, no longitudinal stiffeners are required for web flexural buckling stability.

Shear Buckling

$$h = 85 \ge 2.12t_w\sqrt{\frac{E}{F_y}} \ge 2.12(0.625)\sqrt{\frac{29{,}000}{50}} \ge 31.9 \text{ in}$$

Therefore, transverse web stiffeners are required.

Combined bending and shear:

$f_b = 22.38 \le \left(0.75 - 1.05\dfrac{f_v}{F_y}\right)F_y \le \left(0.75 - 1.05\dfrac{486/53.13}{50}\right)50 \le 27.9$ ksi, but no greater than 27.5 ksi OK.

Flange-to-web connection:

Unshored construction only is considered in the flange-to-web connection design for brevity. Similar calculations may be performed if a shored construction method is utilized.

The section properties at the top and bottom welds are shown in Tables E7.11b and E7.12b, respectively. Shear flow at the weld, calculated based on these section properties, is shown in Table E7.13b.

Maximum shear flow = 5.37 k/in.

Maximum LL + I range shear flow = 3.51 k/in. [AREMA recommends that even stress ranges in welds in compression zones can be considered due to the relatively high "effective mean stress" created by residual stresses from welding processes (see Chapter 10)]

Allowable weld shear stress = 17.5 ksi (fillet)
Allowable fatigue stress range = 16 ksi (Fatigue Detail Category B)
Weld size = 5.37/[2(0.71)17.5] = 0.21 in for maximum shear flow

TABLE E7.11b

Section	A (in.^{2})	y_{na} (in.)	Ay_{na} (in.^{3})	I_g (in.^{4})	Ay_{na}/I_g (in.$^{-1}$)
Noncomposite (steel only)	50.00	43.75	2188	223,444	9.79×10^{-3}
Composite-long term ($n=24$)	37.50	40.16	1506	299,065	10.71×10^{-3}
	50.00	33.91	1696		
Composite-short term ($n=8$)	112.50	28.82	3242	386,516	11.31×10^{-3}
	50.00	22.57	1129		

TABLE E7.12b

Section	A (in.^{2})	y_{na} (in.)	Ay_{na} (in.^{3})	I_n (in.^{4})	Ay_{na}/I_n (in.$^{-1}$)
Noncomposite (steel only)	50.00	43.75	2188	201,100	10.88×10^{-3}
Composite-long term ($n=24$)	50.00	53.59	2680	276,408	9.70×10^{-3}
Composite-short term ($n=8$)	50.00	64.93	3247	363,234	8.94×10^{-3}

TABLE E7.13b

Location (Figure E7.3)	DL1 shear Flow on Noncomposite Section ($n=1$) (k/in.)	DL2 Shear Flow on Composite Section ($n=24$) (k/in.)	Maximum LL + I Shear Flow on Composite Section ($n=8$) (k/in.)	Range of LL + I Shear Flow on Composite Section ($n=8$) (k/in.)	Maximum Shear Flow (k/in.)
Top weld	0.69	0.43	4.25	3.51	5.37
Bottom weld	0.76	0.39	3.37	2.77	4.52

Weld size = 3.52/[2(0.71)16] = 0.16 in for cyclical shear flow

Use min 5/16″ fillet welds or CJP welds (some designers will specify CJP welds because of the vertical load transmitted directly to the top flange-to-web weld).

Shear Stud Design

Unshored construction only is considered in shear stud for brevity. Similar calculations may be performed if a shored construction method is utilized.

Also for brevity, shear stud spacing will be calculated at only three locations on the girder as indicated in Table E7.14b (in the design of practical girders, a smaller interval is recommended*).

TABLE E7.14b

Location (Figure E7.6b)	Distance, x, from End a (in.)	$+V_{LL+I}$ (kips)	$-V_{LL+I}$ (kips) (Linear Interpolation)	V_{DL2} (kips)	V_r (kips)	V_{max} (kips)
a	0	275	0	40	310	417
b	270	175	37	20	220	287
c	540 (center)	74	74	0	167	203

* Equivalent uniform load charts (such as Steinman charts) are useful in determining live load shear forces at various locations along the span.

The negative live load shear flow (the shear reverses when a wheel passes over a stud) is estimated as a linear interpolation of center span live load shear as shown in Figure E7.6b.

Shear flow at the steel–concrete interface (Table E7.15b) is

$$q_{LL+I} = \frac{V_{LL+I}Q_{cp(n=8)}}{I_{cp(n=8)}} = \frac{3242(V_{LL+I})}{386,516} = \frac{V_{LL+I}}{119}$$

$$q_{max} = q_{LL+I} + q_{DL2} = q_{LL+I} + \frac{V_{DL2}Q_{cp(n=24)}}{I_{cp(n=24)}} = \frac{V_{LL+I}}{119} + \frac{1506V_{DL2}}{299,065} = \frac{V_{LL+I}}{119} + \frac{V_{DL2}}{199}$$

The shear flow due to dead load on the composite section is small and ignored in the practical design of composite steel and concrete girders. This is also evident from comparison of the allowable design load for shear stress range, S_r, and the allowable design load for maximum shear stress, S_m:

$$S_r = \frac{7.0\pi(0.875)^2}{4} = 4.21 \text{ kips}$$

$$S_m = \frac{20.0\pi(0.875)^2}{4} = 12.0 \text{ kips}$$

For three shear studs across the flange width,

s = spacing required (Table E7.16b) = $3(4.21)/q_{LL+I} = 12.63/q_{LL+I}$.

The actual shear stud spacing can be arranged in order that the maximum overstress is, for example, 10% as shown in Figure E7.7b.

Design of web plate stiffeners:

Use angles bolted to the web in order to preclude issues related to welding at the base of transverse stiffeners. Use a single 6×4×1/2 angle at 67.5 in. centers as shown in Example 7.1b.

Design of bearing stiffeners: Use four 8×4×1/2 angles as shown in Example 7.1b.

Serviceability—deflection criteria

The required gross moment of inertia for an LL + I deflection criteria of L/f_D is (see Chapter 5)

$$I \geq \frac{M_{LL+I}Lf_{\Delta}}{1934} \geq \frac{7314(90)f_{\Delta}}{1934} \geq 340.4f_{\Delta} \text{ in.}^4$$

The required section gross moment of inertia for various deflection criteria, f_D, is shown in Table E7.17b.

TABLE E7.15b

Location (Figure E7.6)	Distance, x, from end a (in.)	V_r (kips)	V_{DL2} (kips)	q_r (k/in.)	q_{DL2} (k/in.)	q_{max} (k/in.)
a	0	310	40	2.61	0.20	3.37
b	270	220	20	1.85	0.10	2.39
c	540	167	0	1.40	0	1.71

TABLE E7.16b

Location	q_r (k/in.)	s (in.)
a	2.61	4.9
b	1.85	6.8
c	1.40	9.0

TABLE E7.17b

Deflection Criteria, L/Δ_{LL+I}	Required Gross Moment of Inertia, I_g, (in.^{4})
500	170,180
640 (AREMA, 2008)	217,832
800	272,290
1000	340,362

TABLE E7.18b

Location (Figures E7.1b and E7.3b)	Noncomposite Section Stress (ksi) (Figure E7.1b)	Composite Section Stress (ksi) (Figure E7.3b)
Top concrete (maximum flexure)	—	1.04
Bottom concrete (maximum flexure)	—	0.74
Top steel (maximum flexure)	23.7	10.38
Web (maximum shear)	9.15	9.15
Bottom steel (maximum flexure)	26.4	22.4
Bottom steel (LL + I flexure)	16.2	13.2

The section $I_g = 386{,}516$ in.^{4} provides a very stiff structure.
The deflections are estimated as (see Chapter 5).

$$\Delta_{LL+I} = \frac{0.104 M_{LL+I} L^2}{EI} = \frac{0.104(7314)(12)[90(12)]^2}{29{,}000(386{,}516)} = 0.95 \text{ in.}$$

$$\Delta_{DL} = \frac{0.104 M_{DL1} L^2}{EI_{g1}} = \frac{0.104 M_{DL2} L^2}{EI_{g2}} = \frac{0.104(1430)(12)[90(12)]^2}{29{,}000(223{,}444)} + \frac{0.104(1070)(12)[90(12)]^2}{29{,}000(299{,}065)}$$

$$= 0.32 + 0.18 = 0.50''.$$

No camber is required.

Summary of the design

Examples 7.1b and 7.2b are not intended to be examples of optimum design but to provide numerical examples of noncomposite and composite steel and concrete girder designs. The summary of stresses and deflections of the two span designs [noncomposite (Example 7.1b) and composite (Example 7.2b)] shown in Table E7.18b reveals that reductions in element sizes can be made and the advantages of composite girder design may be exploited.

REFERENCES

Akesson, B., 2008, *Understanding Bridge Collapses*, Taylor & Francis, London, UK.

American Railway Engineering and Maintenance-of-Way Association (AREMA), 2015, Steel structures, in *Manual for Railway Engineering*, Chapter 15, Lanham, MD.

Basler, K. and Thurlimann, B., 1959, *Plate Girder Research*, National Engineering Conference Proceedings, AISC, Chicago, IL.

Bleich, F., 1952, *Buckling Strength of Metal Structures*, McGraw-Hill, New York.

Brockenbrough, R.L., 2011, Properties of structural steels and effects of steelmaking and fabrication, in *Structural Steel Designer's Handbook*, Chapter 1, 5th edition, Brockenbrough, R.L., Ed., McGraw-Hill, New York.

Cusens, A.R. and Pama, R.P., 1979, *Bridge Deck Analysis*, John Wiley & Sons, New York.

D'Andrea, M., Grondin, G.Y., and Kulak, G.L., 2001, Behaviour and rehabilitation of distortion-induced fatigue cracks in bridge girders, University of Alberta Structural Engineering Report No. 240, Edmonton, Canada.

Galambos, T.V. Ed., 1988, *Guide to Stability Design Criteria for Metal Structures*, John Wiley & Sons, New York.

Gere, J.M. and Timoshenko, S.P., 1984, *Mechanics of Materials*, Wadsworth, Belmont, CA.

Kim, Y.D., Jung, S.-K., and White, D.W., 2007, Transverse stiffener requirements in straight and horizontally curved steel I-girders, *Journal of Bridge Engineering,* Vol. 12, No. 2, 174–183.

Lekhnitskii, S. G., 1968, *Anisotropic Plates*, 2nd edition, Gordon and Breach, New York.

Newmark, N.M., Seiss, C.P. and Viest, I.M., 1951, Tests and analysis of composite beams with incomplete interaction, *Proceedings of the Society for Experimental Stress Analysis*, Vol. 9, No. 1, 75–92.

Roark, R.J. and Young, W.C., 1982, *Formulas for Stress and Strain*, McGraw-Hill, New York.

Rockey, K.C. and Leggett, D.M.A., 1962, The buckling of a plate girder web under pure bending when reinforced by a single longitudinal stiffener, *Proceedings of the Institute of Civil Engineers*, Vol. 21, No. 1, 161–188.

Roy, S. and Fisher, J.W., 2006, Modified AASHTO design S–N curves for post-weld treated welded details, *Journal of Bridge Engineering*, Vol. 2, No. 4, 207–222.

Salmon, C.G. and Johnson, J.E., 1980, *Steel Structures Design and Behavior*, Harper and Row, New York.

Seaburg, P. A. and Carter, C. J., 1997, *Torsional Analysis of Structural Steel Members*, AISC, Chicago, IL.

Szilard, R., 2004, *Theories and Applications of Plate Analysis*, John Wiley & Sons, New York.

Tall, L., Ed., 1974, *Structural Steel Design*, John Wiley & Sons, New York.

Timoshenko, S.P. and Gere, J.M., *Theory of Elastic Stability*, 2nd edition, McGraw-Hill, New York.

Timoshenko, S. and Woinowsky-Krieger, S., 1959, *Theory of Plates and Shells*, McGraw-Hill, New York.

Trahair, N.S., 1993, *Flexural-Torsional Buckling of Structures*, CRC Press, Boca Raton, FL.

Troitsky, M.S., 1987, *Orthotropic Bridges: Theory and Design*, 2nd edition, James F. Lincoln Arc Welding Foundation, Cleveland, OH.

Viest, I.M., Fountain, R.S., and Singleton, R.C., 1958, *Composite Construction in Steel and Concrete*, McGraw-Hill, New York.

Wang, C.M., Reddy, J.N., and Lee, K.H., 2000, *Shear Deformable Beams and Plates*, Elsevier, Kidlington, Oxford, UK.

Wolchuk, R., 1963, *Design Manual for Orthotropic Steel Plate Deck Bridges*, AISC, Chicago, IL.

8 Design of Steel Members for Combined Forces

8.1 INTRODUCTION

Structural steel members in railway superstructures are usually designed to resist only axial or transverse loads as outlined in Chapters 6 and 7, respectively. These external loads create internal normal and shear stresses in members of the superstructure. However, in some situations, it is necessary to consider members subjected to combinations of stresses.

Combined stresses in railway bridges typically arise from biaxial bending of unsymmetrical cross sections, unsymmetrical bending from transverse force eccentricities, and combined axial and bending forces caused by eccentricities, member out of straightness, self-weight,* and applied lateral loads such as wind. For linear elastic materials and small deformations, superposition of combined stresses is appropriate.

8.2 BIAXIAL BENDING

If bending moments, M_x and M_y, are applied at the centroid of an unsymmetrical section as shown in Figure 8.1, bending will occur in both the yz and xz planes.

However, in unsymmetrical cross sections these planes are not principal planes, and each moment contributes to a portion of the total bending about each axis. The flexural stress, σ_p, at a location, p, with coordinates x and y is

$$\sigma_\text{p} = E\varepsilon_x + E\varepsilon_y = -E(\kappa_x x + \kappa_y y), \tag{8.1}$$

where

ε_x = the strain on the xz plane
ε_y = the strain on the yz plane
κ_x = the curvature on the xz plane
κ_y = the curvature on the yz plane
E = the modulus of elasticity of steel = 200,000 MPa (29,000 ksi).

The bending moments, M_x and M_y, can then be written as

$$M_y = \int \sigma x \,\text{d}A = -E\left(\kappa_y \int xy\,\text{d}A + k_x \int x^2\,\text{d}A\right) = -E\left(\kappa_y I_{xy} + \kappa_x I_y\right), \tag{8.2}$$

$$M_x = \int \sigma y \,\text{d}A = -E\left(\kappa_y \int y^2\,\text{d}A + k_x \int xy\,\text{d}A\right) = -E\left(\kappa_y I_x + \kappa_x I_{xy}\right), \tag{8.3}$$

where

$I_{x=}$ the moment of inertia about the x axis
$I_{y=}$ the moment of inertia about the y axis.

Equations 8.2 and 8.3 may be solved simultaneously for κ_x and κ_y and substituted into Equation 8.1 to obtain

* This is the case for members that are not in a vertical plane.

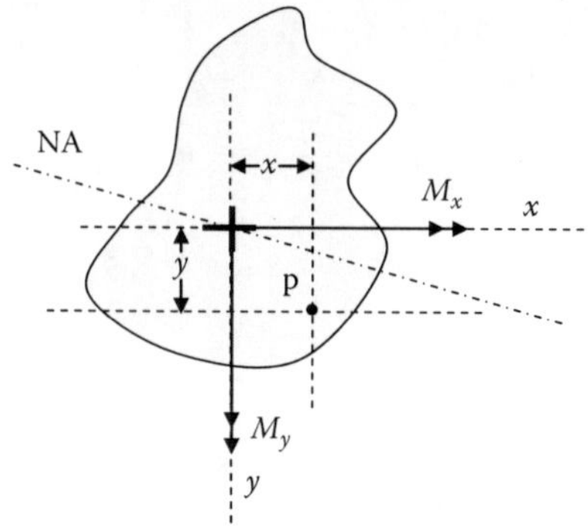

FIGURE 8.1 Biaxial bending of unsymmetrical cross section.

$$\sigma = \frac{M_x I_y - M_y I_{xy}}{I_x I_y - I_{xy}^2} y + \frac{M_y I_x - M_x I_{xy}}{I_x I_y - I_{xy}^2} x. \tag{8.4}$$

Steel members in railway superstructures subjected to biaxial bending usually have two axes of symmetry. Therefore,

$$I_{xy} = 0 \tag{8.5}$$

and Equation 8.4 becomes

$$\sigma = \frac{M_x}{I_x} y + \frac{M_y}{I_y} x = f_{bx} + f_{by} \le F_b, \tag{8.6}$$

where

f_{bx} = the normal stress from bending moment, M_x, about the x axis
f_{by} = the normal stress from bending moment, M_y, about the y axis
F_b = the allowable bending stress

However, since the allowable bending stress may not be the same in each plane of bending, Equation 8.6 may be expressed as

$$\frac{f_{bx}}{F_{bx}} + \frac{f_{by}}{F_{by}} = \frac{(M_x/I_x)y}{F_{bx}} + \frac{(M_x/I_y)x}{F_{by}} \le 1, \tag{8.7}$$

where

F_{bx} = the allowable bending stress in the direction of x axis
F_{by} = the allowable bending stress in the direction of y axis.

The interaction Equation 8.7 may be used for design considering both tensile and compressive flexural stresses by using the appropriate allowable bending stress for flexural tension (F_{bx}, F_{by} = $F_{tall} = 0.55F_y$) or compression (F_{bx}, $F_{by} = F_{call}$).

8.3 UNSYMMETRICAL BENDING (COMBINED BENDING AND TORSION)

The best design strategy is to avoid torsion. However, in some cases, it is unavoidable. Torsion is combined with bending when transverse loads are not applied through the shear center of the member.

When a torsional moment is applied, pure torsion always exists. Pure torsion creates shearing stresses in the flanges and webs of structural shapes such as, channels and I-shaped beams. However, warping torsion also exists when cross sections do not remain plane due to some form of restraint. Warping torsion creates shearing and normal stresses in the flanges of I shapes and normal stresses in the flanges and web of channels. These torsional shear and normal stresses must be superimposed on the shear and normal stresses in flanges and webs due to flexure.

The torsional moment resistance, T, of a cross section to a constant torsional moment is

$$T = T_t + T_w, \tag{8.8}$$

where

$$T_t = \text{the pure torsional (or St. Venant) moment resistance} = GJ\frac{d\theta}{dz} \tag{8.9}$$

$$T_w = \text{the warping torsional moment resistance} = -EC_w\frac{d^3\theta}{dz^3}. \tag{8.10}$$

Here,

z = the longitudinal axis of the beam or girder
G = the shear modulus of elasticity of steel [~77,000 MPa (11,200 ksi)]
E = the tensile modulus of elasticity of steel [~200,000 MPa (29,000 ksi)]
J = the torsional constant of the cross section
C_w = the warping constant of the cross section.

The shear stresses from pure torsion effects of the applied torsional moments are

$$\tau_t = \frac{T_t t}{J} \tag{8.11}$$

and the substitution of Equation 8.9 into 8.11 yields

$$\tau_t = \frac{T_t t}{J} = Gt\left(\frac{d\theta}{dz}\right), \tag{8.12}$$

where

t = the thickness of the element.

The shear stresses from warping effects of the applied torsional moments on an I-shaped section are

$$\tau_w = \frac{3}{2}\frac{T_w}{A_f h} \tag{8.13}$$

and the substitution of Equation 8.10 into 8.13 yields

$$\tau_w = -\frac{3}{2}\frac{EC_w}{A_f h}\left(\frac{d^3\theta}{dz^3}\right) = -\frac{Eb_f^2 h}{16}\left(\frac{d^3\theta}{dz^3}\right), \tag{8.14}$$

where

A_f = the area of the flange = $b_f t_f$

$$C_w = \frac{I_y h^2}{4} = \frac{(I_f)h^2}{2} = \frac{t_f b_f^3 h^2}{24}$$

h = the distance between the centroids of the flanges for I-shaped members.

The normal stresses from warping effects of the applied torsional moments on an I-shaped section are determined by considering the normal stress from the lateral bending of the flanges

$$\sigma_{\mathrm{w}} = \frac{M_{\mathrm{l}}x}{I_{\mathrm{f}}} = \frac{EI_{\mathrm{f}}h}{2}\left(\frac{\mathrm{d}^2\theta}{\mathrm{d}z^2}\right) = \frac{EC_{\mathrm{w}}}{h}\left(\frac{\mathrm{d}^2\theta}{\mathrm{d}z^2}\right), \tag{8.15}$$

where

x = the distance on the flange from the neutral axis of the flexural stress distribution in the flange (maximum at $x = b/2$)

M_{l} = the lateral bending moment on one flange

I_{f} = the moment of inertia of the flange.

The differential equation of torsion is (from Equations 8.8 through 8.10)

$$T = GJ\frac{\mathrm{d}\theta}{\mathrm{d}z} - EC_{\mathrm{w}}\frac{\mathrm{d}^3\theta}{\mathrm{d}z^3}. \tag{8.16}$$

For torsional moments that vary uniformly along the length (z axis) of a member, $\mathrm{d}T/\mathrm{d}z$, is

$$\frac{\mathrm{d}T}{\mathrm{d}z} = t' = GJ\frac{\mathrm{d}^2\theta}{\mathrm{d}z^2} - EC_{\mathrm{w}}\frac{\mathrm{d}^4\theta}{\mathrm{d}z^4}. \tag{8.17}$$

For torsional moments that vary linearly along the length (z axis) of a member, $\mathrm{d}T/\mathrm{d}z$ is

$$\frac{\mathrm{d}T}{\mathrm{d}z} = \frac{t'z'}{L} = GJ\frac{\mathrm{d}^2\theta}{\mathrm{d}z^2} - EC_{\mathrm{w}}\frac{\mathrm{d}^4\theta}{\mathrm{d}z^4}, \tag{8.18}$$

where

t' = the maximum torsional moment applied at the end support

$L - z'$ = the distance from the end support with maximum torsional moment

L = the length of the span.

The angle of twist, θ, is provided by the solution of Equations 8.16, 8.17, or 8.18 and depends on both loading and boundary conditions. The general form of the angle of twist, θ, can be expressed as (Kuzmanovic and Willems, 1983)

$$\theta = A\sinh\left(\frac{z}{a}\right) + B\cosh\left(\frac{z}{a}\right) + C + D(z), \tag{8.19}$$

where

A, B, C, and D are constants depending on the boundary conditions and loading

$D(z)$ = an expression in terms of z, depending on loading

$a = \sqrt{\frac{EC_{\mathrm{w}}}{GJ}}$ (a characteristic length).

The equations for the angle of twist and its derivatives for a concentrated torsional moment, T', applied at the center of a simply supported span are (Salmon and Johnson, 1980)

$$\theta = A\sinh\left(\frac{z}{a}\right) + B\cosh\left(\frac{z}{a}\right) + C + \frac{T'z}{2GJ} = \frac{T'a}{2GJ}\left(\frac{z}{a} - \frac{\sinh(z/a)}{\cosh(L/2a)}\right), \tag{8.20a}$$

$$\frac{d\theta}{dz} = \frac{T'}{2GJ}\left(1 - \frac{\cosh(z/a)}{\cosh(L/2a)}\right), \tag{8.20b}$$

$$\frac{d^2\theta}{dz^2} = \frac{T'}{2GJa}\left(-\frac{\sinh(z/a)}{\cosh(L/2a)}\right), \tag{8.20c}$$

$$\frac{d^3\theta}{dz^3} = \frac{T'}{2GJa^2}\left(-\frac{\cosh(z/a)}{\cosh(L/2a)}\right). \tag{8.20d}$$

Example 8.1 illustrates the use of Equations 8.20a through d for the determination of combined stresses due to torsion and flexure.

The equations for the angle of twist and its derivatives for a uniformly distributed torsional moment, t', are (Kuzmanovic and Willems, 1983)

$$\theta = \frac{t'a^2}{GJ}\left(-\tanh\left(\frac{L}{2a}\right)\sinh\left(\frac{z}{a}\right) + \cosh\left(\frac{z}{a}\right) - \frac{z^2}{2a^2} + \frac{zL}{2a^2} - 1\right), \tag{8.21a}$$

$$\frac{d\theta}{dz} = \frac{t'a}{GJ}\left(-\tanh\left(\frac{L}{2a}\right)\cosh\left(\frac{z}{a}\right) + \sinh\left(\frac{z}{a}\right) - \frac{z}{a} + \frac{L}{2a}\right), \tag{8.21b}$$

$$\frac{d^2\theta}{dz^2} = \frac{t'}{GJ}\left(-\tanh\left(\frac{L}{2a}\right)\sinh\left(\frac{z}{a}\right) + \cosh\left(\frac{z}{a}\right) - 1\right), \tag{8.21c}$$

$$\frac{d^3\theta}{dz^3} = \frac{t'}{GJa}\left(-\tanh\left(\frac{L}{2a}\right)\cosh\left(\frac{z}{a}\right) + \sinh\left(\frac{z}{a}\right)\right). \tag{8.21d}$$

Example 8.3 illustrates the use of Equations 8.21a through d for the determination of combined stresses due to torsion and flexure.

Equation 8.19 and its derivatives have also been solved for other typical boundary and loading conditions and are provided for design use in equations and charts (Seaburg and Carter, 1997). However, even with such design aids, the solution of Equations 8.12, 8.14, and 8.15 is generally too cumbersome for routine design work and an approximate method, based on a flexure analogy, is often employed.

In this method, it is assumed that the torsional moment acts as a horizontal force couple in the plane of the flanges. The horizontal forces create bending moments in the flanges and the problem is solved using a simplified analysis involving flexure only. If the torsion is created by eccentric vertical loads, an equivalent static system consisting of a vertical load applied at the shear center and horizontal forces applied at each flange is appropriate (Figure 8.2). Examples 8.2 and 8.4 illustrate the use of the flexure analogy for torsional stresses created by an eccentric vertical load. If the torsion is created by an applied horizontal force, an equivalent static system consisting of vertical and horizontal loads applied at the shear center (creating biaxial bending) and horizontal forces applied at each flange is appropriate (Figure 8.3a). For the latter case, an equivalent static system as shown in Figure 8.3b may also be used. The method is conservative as it ignores pure torsion and assumes that torsional moment is resisted entirely by warping torsion. Therefore, normal stresses due to warping are overestimated. Modification factors that reduce the normal lateral bending stress have been developed to moderate this conservative approach.

However, designers should use the flexure analogy with caution, particularly in cases where torsional effects are relatively large or for members with unusual loading or support conditions.

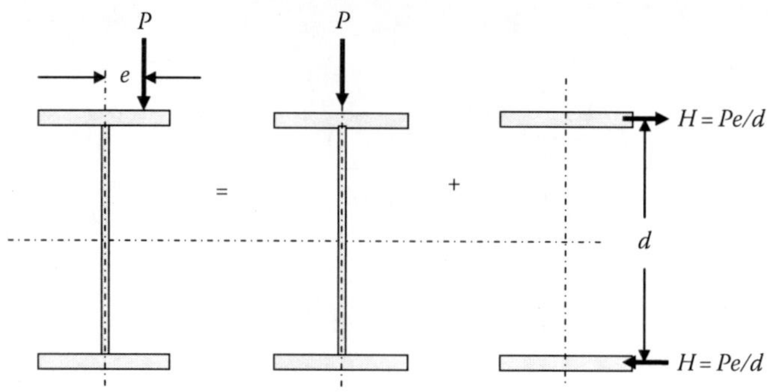

FIGURE 8.2 Equivalent static system for eccentrically applied vertical load.

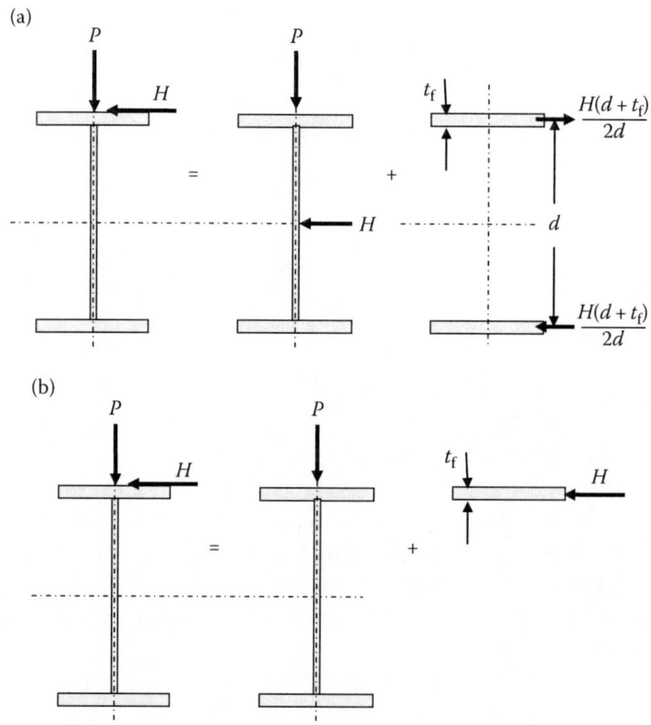

FIGURE 8.3 (a) Equivalent static system for applied vertical and horizontal loads. (b) Alternative equivalent static system for applied vertical and horizontal loads.

Examples 8.1 (concentrated torsional moment) and 8.3 (uniformly distributed torsional moment) outline solutions developed from Equation 8.16. Examples 8.2 and 8.4 outline the solution of the same problems using the flexure analogy.

Example 8.1 (US Customary and Imperial Units)

A machinery girder in a movable bridge is to support a concentrated load from equipment with an eccentricity of 6″ as shown in Figure E8.1.

Note that only the stresses from the 35 kip equipment load are considered in this example. Dead loads and other loads on the machinery girder are not considered.

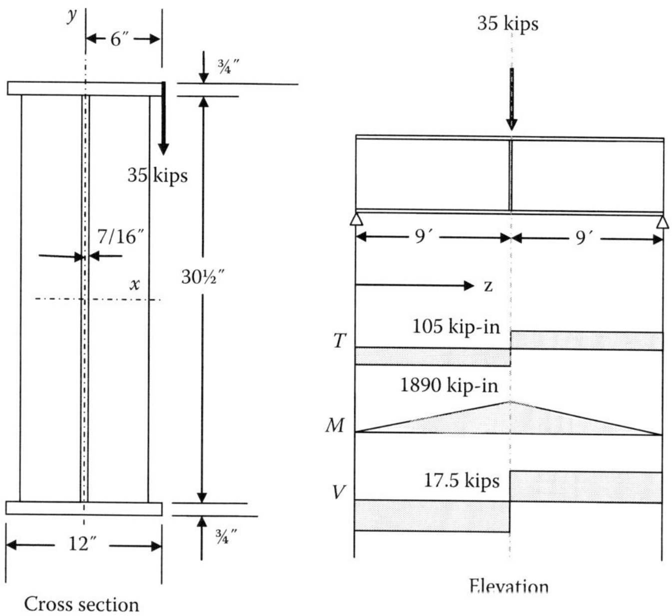

FIGURE E8.1 Girder cross section and forces.

Section properties are

$I_x = 5430$ in.^{4}
$S_x = 339$ in.^{3}
$I_y = 216$ in.^{3}

$$J = \sum \frac{bt^3}{3} = \frac{2(12(0.75)^3) + (30.5(0.44)^3)}{3} = 4.23 \text{ in.}^4$$

$$C_w = \frac{I_y h^2}{4} = \frac{216(31.25)^2}{4} = 52{,}734 \text{ in.}^6$$

$$a = \sqrt{\frac{EC_w}{GJ}} = 180 \text{ in.}$$

$$\frac{L}{a} = 1.20$$

$$\frac{T'}{2GJ} = \frac{Pe}{2GJ} = \frac{35(6)}{2(11200)(4.23)} = 2.22 \times 10^{-3} \text{ in.}^{-1}.$$

TABLE E8.1
Pure (St. Venant) Torsion Shear Stresses

	Pure Torsion Shear Stress, t_t (ksi)	
Element of Girder	**$z = 0$ (end)**	**$z = L/2 = 108''$ (center)**
Flange, $t = 0.75''$	2.92	0
Web, $t = 0.44''$	1.70	0

The solution of the differential Equation 8.16 is

$$\theta = A\sinh\left(\frac{z}{a}\right) + B\cosh\left(\frac{z}{a}\right) + C + \frac{T}{2GJ}z.$$

For boundary conditions $\theta = \dfrac{d^2\theta}{dz^2} = 0$ at $z = 0$ and $z = L$

$$\theta = \frac{T'a}{2GJ}\left(\frac{z}{a} - \frac{\sinh(z/a)}{\cosh(L/2a)}\right) = 0.40\left(\frac{z}{180} - \frac{\sinh(z/180)}{1.186}\right)$$

$$\frac{d\theta}{dz} = \frac{T'}{2GJ}\left(1 - \frac{\cosh(z/a)}{\cosh(L/2a)}\right) = 2.22\times10^{-3}\left(1 - \frac{\cosh(z/180)}{1.186}\right)$$

$$\frac{d^2\theta}{dz^2} = \frac{T'}{2GJa}\left(-\frac{\sinh(z/a)}{\cosh(L/2a)}\right) = 1.23\times10^{-5}\left(-\frac{\sinh(z/180)}{1.186}\right)$$

$$\frac{d^3\theta}{dz^3} = \frac{T'}{2GJa^2}\left(-\frac{\cosh(z/a)}{\cosh(L/2a)}\right) = 6.87\times10^{-8}\left(-\frac{\cosh(z/180)}{1.186}\right).$$

Pure (St. Venant) torsion (Table E8.1):

$$\tau_t = Gt\left(\frac{d\theta}{dz}\right) = 11200(t)2.22\times10^{-3}\left(1 - \frac{\cosh(z/180)}{1.186}\right) = 24.82t\left(1 - \frac{\cosh(z/180)}{1.186}\right).$$

Warping torsion (lateral bending of flanges) (Tables E8.2 and E8.3):

$$\tau_w = \frac{-Eb_f^2h}{16}\left(\frac{d^3\theta}{dz^3}\right) = \frac{-29000(12)^2(31.25)}{16}6.87\times10^{-8}\left(-\frac{\cosh(z/180)}{1.186}\right) = 0.472\cosh(z/180)$$

$$\sigma_w = \frac{Eb_fh}{4}\left(\frac{d^2\theta}{dz^2}\right) = \frac{29000(12)(31.25)}{4}1.23\times10^{-5}\left(-\frac{\sinh(z/180)}{1.186}\right) = -28.20\sinh(z/180).$$

TABLE E8.2
Warping Torsion Shear Stresses

	Warping Torsion Shear Stress, τ_w (ksi)	
Element of Girder	**z = 0 (end)**	**z = L/2 = 108″ (center)**
Flange, t = 0.75″	0.47	0.56

TABLE E8.3
Warping Torsion Normal Stresses

	Warping Torsion Normal Stress, σ_w (ksi)	
Element of Girder	**z = 0 (end)**	**z = L/2 = 108″ (center)**
Flanges, t = 0.75″	0	−17.95

TABLE E8.4
Flexural Normal Stresses

Element of Girder	Flexural Normal Stress, σ_b (ksi)	
	$z = 0$ (end)	$z = L/2 = 108''$ (center)
Top flange, $t = 0.75''$	0	−5.58

TABLE E8.5
Flexural Shear Stresses

Element of Girder	Flexural Shear Stress, τ_b (ksi)	
	$z = 0$ (end)	$z = L/2 = 108''$ (center)
Flange, t = 0.75″	0.58	0.58
Web, $t = 0.44''$	1.40	1.40

Flexure of girder (Tables E8.4 and E8.5):

$$M = \frac{35(216)}{4} = 1890 \text{ kip-in.}$$

$$\sigma_b(0) = 0$$

$$\sigma_b\left(\frac{L}{2}\right) = \pm\frac{1890}{339} = \pm 5.58 \text{ ksi}$$

$$V = 35/2 = 17.5 \text{ kips}$$

$$Q_{\text{flange}} = ((12 - 0.44)(0.75))\left(\frac{32 - 0.75}{2}\right) = 135.5 \text{ in.}^3$$

$$Q_{\text{web}} = ((12)(0.75))\left(\frac{32 - 0.75}{2}\right) + \frac{30.5}{2}(0.44)\frac{30.5}{4} = 191.8 \text{ in.}^3$$

$$\tau_{b\,\text{flange}}(0) = \frac{17.5(135.5)}{5430(0.75)} = 0.58 \text{ ksi}$$

$$\tau_{b\,\text{web}}(0) = \frac{17.5(191.8)}{5430(0.44)} = 1.40 \text{ ksi.}$$

Combined stresses (Tables E8.6 and E8.7):

Both pure, t_t, and warping, t_w, shear stresses due to torsion are relatively small but warping normal stress is large. The warping normal stress is (17.95/23.53)100 = 76% of the flange normal stresses.

TABLE E8.6
Combined Shear Stresses

Element of Girder	Shear Stress (ksi)	
	$z = 0$ (end)	$z = L/2 = 108''$ (center)
Flange, $t = 0.75''$	3.97	1.14
Web, $t = 0.44''$	3.10	1.40

TABLE E8.7
Combined Normal Stresses

	Normal Stress (ksi)	
Element of Girder	**$z = 0$ (end)**	**$z = L/2 = 108''$ (center)**
Flange, $t = 0.75''$	0	−23.5

Example 8.2 (US Customary and Imperial Units)

Use the flexure analogy (Figure E8.2) for torsion to find the shear and normal stresses due to combined flexure and torsion of Example 8.1.

$$H = \frac{35(6)}{32 - 0.75} = 6.72 \text{ kips}$$

$$M_H = \frac{6.72(216)}{4} = 362.9 \text{ kip-in.}$$

$$\sigma_{bH} = \frac{362.9}{\left(\frac{0.75(12)^2}{6}\right)} = 20.2 \text{ ksi.}$$

Modification factors, *b*, that reduce the normal lateral bending stress have been developed as a corrective measure since the flexure analogy overestimates normal flange stresses due to warping. For the case of $L/a = 1.20$ and torsional moment, *T*, applied at center of span, $\beta \sim 0.94$ (Salmon and Johnson, 1980) (Kulak and Grondin, 2002).

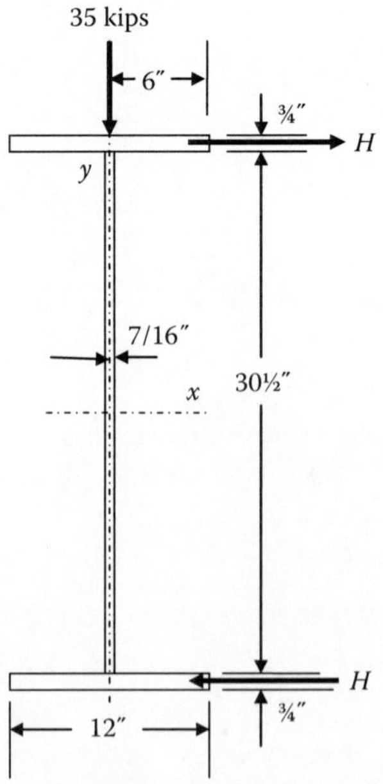

FIGURE E8.2 Girder cross section and forces.

$$\sigma_{bH} = \frac{362.9\beta}{\left(\frac{0.75(12)^2}{6}\right)} = 20.2\beta = 20.2(0.94) = 19.0 \text{ ksi}$$

which is close to the value of 18.0 ksi obtained in Example 8.1. The flexure analogy models flange normal warping stresses well for many typical steel railway bridge elements.

$$V_H = \frac{6.72}{2} = 3.36 \text{ kips}$$

$$\tau_{bH} = \frac{3(3.36)}{2(12(0.75))} = 0.56 \text{ ksi.}$$

Combined stresses:

$$\sigma_{flange} = \sigma_b(L/2) + \sigma_w(L/2) = 5.58 + 19.0 = 24.6 \text{ ksi}$$

which is close to the value of 23.5 ksi obtained in Example 8.1.

$$\tau_{web} = \tau_b = 1.40 \text{ ksi}$$

$$\tau_{flange} = \tau_b + \tau_w = 0.58 + 0.56 = 1.14 \text{ ksi.}$$

The shear stress in the flange and web is correct at $z = L/2$, where pure torsional shear stresses, t_t, are zero but underestimated at $z = 0$ because pure torsion is not considered in the flexure analogy. However, for many typical steel railway bridge elements, torsional shear stresses due to torsion are of much less concern than flange normal stresses due to warping torsion. In such cases, the flexure analogy is appropriate for ordinary torsion design problems.

Example 8.3 (SI Units)

The end floorbeam in a ballasted through plate girder bridge is subjected to a uniformly distributed load, w, at an eccentricity of 65 mm as shown in Figure E8.3.

The solution of the differential equation is (Equation 8.21a)

$$\theta = 0.082\left(-0.926\sinh\left(\frac{z}{1685}\right) + \cosh\left(\frac{z}{1685}\right) - \frac{z^2}{5678x10^3} + \frac{z}{1032} - 1\right)$$

and its differentials are

$$\frac{d\theta}{dz} = 48.66\times10^{-6}\left(-0.926\cosh\left(\frac{z}{1685}\right) + \sinh\left(\frac{z}{1685}\right) - \frac{z}{1685} + 1.633\right)$$

$$\frac{d^2\theta}{dz^2} = 28.88\times10^{-9}\left(-0.926\sinh\left(\frac{z}{1685}\right) + \cosh\left(\frac{z}{1685}\right) - 1\right)$$

$$\frac{d^3\theta}{dz^3} = 17.14\times10^{-12}\left(-0.926\cosh\left(\frac{z}{1685}\right) + \sinh\left(\frac{z}{1685}\right)\right).$$

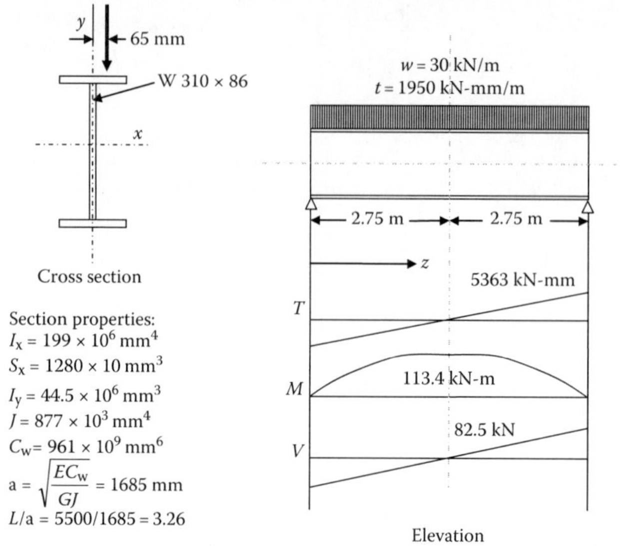

FIGURE E8.3 Girder cross section and forces.

Pure (St. Venant) torsion (Table E8.8):

$$\tau_t = Gt\left(\frac{d\theta}{dz}\right) = 3.75(t)\left(-0.926\cosh\left(\frac{z}{1685}\right) + \sinh\left(\frac{z}{1685}\right) - \frac{z}{1685} + 1.633\right).$$

Warping torsion (lateral bending of flanges) (Tables E8.9 and E8.10):

$$\tau_w = \frac{-Eb_f^2 h}{16}\left(\frac{d^3\theta}{dz^3}\right) = \frac{-200000(254)^2(310)}{16}(17.14\times10^{-12})\left(-0.926\cosh\left(\frac{z}{1685}\right) + \sinh\left(\frac{z}{1685}\right)\right)$$

$$\tau_w = 4.28\left(-0.926\cosh\left(\frac{z}{1685}\right) + \sinh\left(\frac{z}{1685}\right)\right)$$

TABLE E8.8
Pure (St. Venant) Torsion Shear Stresses

	Pure Torsion Shear Stress, τ_t (MPa)	
Element of Girder	$z = 0$ **(end)**	$z = L/2 = 2750$ **mm (center)**
Flange, $t = 16$ mm	42.4	0
Web, $t = 9$ mm	23.9	0

TABLE E8.9
Warping Torsion Shear Stresses

	Warping Torsion Shear Stress, τ_w (MPa)	
Element of Girder	$z = 0$ **(end)**	$z = L/2 = 2750$ **mm (center)**
Flange, $t = 16$ mm	−3.96	0

TABLE E8.10
Warping Torsion Normal Stresses

	Warping Torsion Normal Stress, σ_w (MPa)	
Element of Girder	$z = 0$ **(end)**	$z = L/2 = 2750$ **mm (center)**
Flanges, $t = 16$ mm	0	−70.8

$$\sigma_w = \frac{Eb_fh}{4}\left(\frac{d^2\theta}{dz^2}\right) = \frac{200000(254)(310)}{4}(28.88\times10^{-9})\left(-0.926\sinh\left(\frac{z}{1685}\right)+\cosh\left(\frac{z}{1685}\right)-1\right)$$

$$\sigma_w = 113.7\left(-0.926\sinh\left(\frac{z}{1685}\right)+\cosh\left(\frac{z}{1685}\right)-1\right).$$

Flexure of girder (Tables E8.11 and E8.12):

$$M = 113.4 \text{ kN-m}$$

$$\sigma_b(0) = 0:$$

$$\sigma_b(L/2) = \pm\frac{113.4(103)}{1280} = \pm 88.6 \text{ MPa}$$

$$V = 82.5 \text{ kN}$$

$$Q_{\text{flange}} = 298\times10^3 \text{ mm}^3$$

$$Q_{\text{web}} = 708\times10^3 \text{ in}^3$$

$$\tau_{b\text{ flange}}(0) = \frac{82.5(1000)(298\times10^3)}{199\times10^6(16)} = 7.72 \text{ MPa}$$

$$\tau_{b\text{ web}}(0) = \frac{82.5(708)}{199(9)} = 32.6 \text{ MPa}.$$

TABLE E8.11
Flexural Normal Stresses

	Flexural Normal Stress, σ_b (MPa)	
Element of Girder	$z = 0$ **(end)**	$z = L/2 = 2750$ **mm (center)**
Top flange, $t = 16$ mm	0	−88.6

TABLE E8.12
Flexural Shear Stresses

	Flexural Shear Stress, τ_b (MPa)	
Element of Girder	$z = 0$ **(end)**	$z = L/2 = 2750$ **mm (center)**
Flange, $t = 16$ mm	7.72	0
Web, $t = 9$ mm	32.6	0

TABLE E8.13
Combined Shear Stresses

	Shear Stress (MPa)	
Element of Girder	**$z = 0$ (end)**	**$z = L/2 = 2750\,\text{mm}$ (center)**
Flange, $t = 16\,\text{mm}$	$42.4 + 3.96 + 7.72 = 54.1$	0
Web, $t = 9\,\text{mm}$	$23.9 + 32.6 = 56.5$	0

TABLE E8.14
Combined Normal Stresses

	Normal Stress (MPa)	
Element of Girder	**$z = 0$ (end)**	**$z = L/2 = 2750\,\text{mm}$ (center)**
Flange, $t = 16\,\text{mm}$	0	$70.8 + 88.6 = 159.4$

Combined stresses (Tables E8.13 and E8.14):

In this case, where there are no bearing stiffeners to transfer loads between flanges, local flexural normal stresses in the flange from the eccentric load must also be considered and superimposed on top flange stresses. The local flexural stresses are

$$\sigma_{\text{b flange}}(z) = \frac{1950}{(16)^2/6} = 45.7\,\text{MPa}.$$

Therefore, the total flexural stress = 70.8 + 88.6 + 45.7 = 205 MPa.

The normal stress due the torsional load is ((70.8 + 45.7)/205)100 = 57% of the flange normal stresses (warping stresses less severe for distributed torsional load).

Example 8.4 (SI Units)

Use the flexure analogy for torsion to find the shear and normal stresses due to combined flexure and torsion of Example 8.3.

$$H = \frac{30(65)}{310 - 16} = 6.63\ \text{kN/m}$$

$$M_{\text{H}} = \frac{6.63(5.5)^2}{8} = 25.07\ \text{kN-m}$$

$$\sigma_{\text{bH}} = \frac{25.07 \times 10^6}{\left(\dfrac{16(254)^2}{6}\right)} = 145.7\ \text{MPa}.$$

Modification factors, b, that reduce the normal lateral bending stress have been developed as a corrective measure since the flexure analogy overestimates normal flange stresses due to warping. For the case of $L/a = 3.26$ and uniform torsional moment, t', $\beta = 0.48$ (Salmon and Johnson, 1980).

$$\sigma_{\text{bH}} = 20.1\beta = 145.7(0.48) = 69.9\ \text{MPa}$$

which is very close to the value of 70.8 MPa obtained in Example 8.3.

$$V_H = \frac{6.63(5.5)}{2} = 18.23 \text{ kN}$$

$$\tau_{bH} = \frac{3(18.23)(1000)}{2(254(16))} = 6.73 \text{ MPa}.$$

Combined stresses:

$$\sigma_{flange} = \sigma_b(L/2) + \sigma_w(L/2) = 88.62 + 69.9 = 158.5 \text{ MPa}$$

which is very close to the value of 70.8 + 88.6 = 159.4 MPa obtained in Example 8.3.

$$\tau_{web} = \tau_b = 30 \text{ MPa}$$

$$\tau_{flange} = \tau_b + \tau_w = 7.65 + 6.73 = 14.4 \text{ MPa}.$$

The shear stress in the flange and web are underestimated at $z = 0$ because pure torsion is not considered in the flexural analogy.

8.4 COMBINED AXIAL FORCES AND BENDING OF MEMBERS

Members are subjected to axial forces and bending moments due to axial force eccentricities (often unintentional and related to connection eccentricities, member out-of-straightness, and/or secondary deflection effects) and when axial members are laterally loaded (typically by self-weight and/or wind). These normal axial and flexural stresses must be combined. Axial tension combined with bending is generally of lesser concern than axial compression combined with bending due to the potential for instability of slender compression members.

8.4.1 Axial Tension and Uniaxial Bending

The tensile load reduces the bending effects on the member when tensile axial loads act simultaneously with bending. For the beam shown in Figure 8.4

$$M = \frac{wL^2}{8} - T\Delta. \tag{8.22}$$

However, the deflection, Δ, is dependent on the bending moment, M, which is itself dependent on the deflection Δ. The deflection

$$\Delta = \frac{5wL^4}{384EI} - \frac{T\Delta L^2}{8EI} \tag{8.23}$$

may be solved iteratively.* The bending moment, M, is

$$M = \frac{wL^2}{8} - \frac{T}{EI}\left(\frac{5wL^4}{384} - \frac{T\Delta L^2}{8}\right). \tag{8.24}$$

* A digital computer algorithm is generally required.

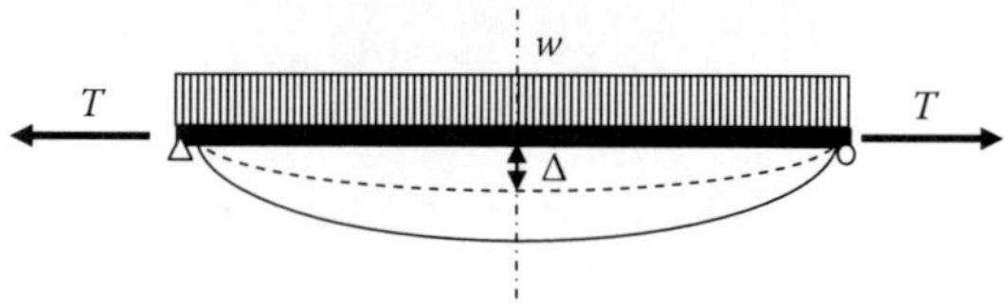

FIGURE 8.4 Member subjected to combined tensile axial force and bending.

However, since the effect of the tensile force on the deflection, Δ, can conservatively be neglected in the analysis* of linear elastic members, the principle of superposition may be applied to ensure that failure by yielding does not occur. American Railway Engineering and Maintenance-of-Way Association (AREMA, 2015) recommends that if bending (even with superimposed axial tensile stresses) causes compression in some part of the cross section, the flexural compressive stress and stability criteria should be considered. Therefore, the allowable flexural stress may differ from the allowable axial tensile stress, which provides the interaction equation

$$\pm\frac{\sigma_b}{F_b}+\frac{\sigma_t}{F_t}\le 1, \tag{8.25}$$

where

σ_b = the maximum tensile or compressive bending stress
σ_t = the maximum tensile axial stress
F_b = the allowable tensile or compressive bending stress
F_t = the allowable axial tensile stress on gross section = $0.55F_y$.

When flexure with axial tension results in only tensile stresses, Equation 8.25 is

$$\sigma_b+\sigma_t\le 0.55F_y \tag{8.26}$$

which is the AREMA (2015) recommendation. When flexure with axial tension results in compressive stresses, AREMA (2015) recommends

$$-\sigma_b+\sigma_t\le F_{call}. \tag{8.27}$$

where

F_{call} = the allowable compressive bending stress.

8.4.2 Axial Compression and Uniaxial Bending

The compressive load increases the bending effects on the member when compressive axial loads act simultaneously with bending (Figure 8.5). For the beam shown in Figure 8.5

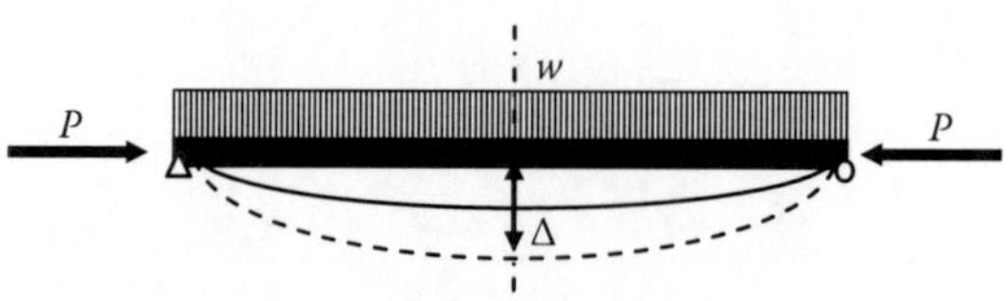

FIGURE 8.5 Member subjected to combined compressive axial force and bending.

* In usual structures the effect is small (Bresler et al., 1968).

$$M = \frac{wL^2}{8} + P\Delta. \tag{8.28}$$

Again, the deflection, Δ, is dependent on the bending moment, M, which is dependent on the deflection Δ. The deflection is

$$\Delta = \frac{5wL^4}{384EI} + \frac{P\Delta L^2}{8EI} \tag{8.29}$$

and the bending moment, M, is determined as

$$M = \frac{wL^2}{8} - \frac{P}{EI}\left(\frac{5wL^4}{384} + \frac{P\Delta L^2}{8}\right). \tag{8.30}$$

Equation 8.29 indicates that the deflection builds upon itself (a deflection causes more deflection) and instability may occur due to this P–Δ effect. Therefore, an iterative solution to Equation 8.30 is required, which is not efficient for routine design work. Alternatively, for some boundary conditions and loads, the differential equation for axial compression and flexure may be solved. However, for routine design work limitations on combined stresses or semi-empirical interaction equations have been developed. AREMA (2015) uses interaction equations for both the yielding and stability criteria.

8.4.2.1 Differential Equation for Axial Compression and Bending in a Simply Supported Beam

Consider a member loaded with a general uniform lateral load, $w(z)$, a concentrated load, Q, at location, a, end moments, M_A and M_B, and compressive axial force, P, as shown in Figure 8.6 .

The bending moments at z due to loads M_A, M_B, Q, and $w(z)$ are combined into a collective bending moment, M_p, such that

$$M_p(z) = M_{w(z)} + M_{MA} + M_{MB} + M_{Q(z)}, \tag{8.31}$$

where

$M_{w(z)}$ = the bending moment at z due to $w(z)$
M_{MA}, M_{MB} = the bending moment due to M_A and M_B
$M_{Q(z)}$ = the bending moment at z due to $Q(z)$.
The moment, M_z, at z is then

$$M_Z = M_p(z) + Py. \tag{8.32}$$

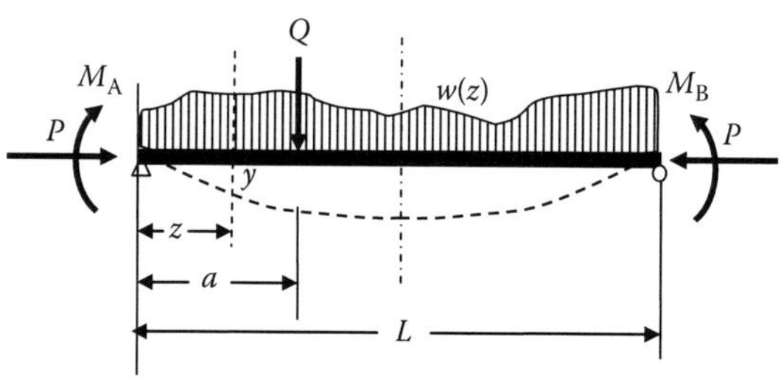

FIGURE 8.6 General loading of combined axial compression and bending member.

The substitution of $M_Z = -EI\dfrac{d^2y}{dz^2}$ into Equation 8.32 yields

$$\frac{d^2y}{dz^2} + k^2 y = -\frac{M_p(z)}{EI}, \tag{8.33}$$

where

$$k^2 = \frac{P}{EI}.$$

Differentiating twice yields

$$\frac{d^4y}{dz^4} + k^2\frac{d^2y}{dz^2} = -\frac{1}{EI}\left(\frac{d^2M_p(z)}{dz^2}\right). \tag{8.34}$$

Equation 8.34 is the differential equation for axial compression and bending.

8.4.2.1.1 Axial Compression and Bending from a Uniformly Distributed Transverse Load

$$M_p(z) = M_{w(z)} = \frac{wz(L-z)}{2} \tag{8.35}$$

$$\frac{d^2M_p(z)}{dz^2} = -w \tag{8.36}$$

and the differential equation for axial compression and flexure (Equation 8.34) is

$$\frac{d^4y}{dz^4} + k^2\frac{d^2y}{dz^2} = \frac{w}{EI}. \tag{8.37}$$

The solution of Equation 8.37 is (Chen and Lui, 1987)

$$y(z) = \frac{w}{EIk^4}\left(\tan\frac{kL}{2}\sin kz + \cos kz - 1\right) - \frac{w}{2EIk^2}z(L-z), \tag{8.38}$$

$$\frac{d^2y(z)}{dz^2} = -\frac{w}{EIk^2}\left(\tan\frac{kL}{2}\sin kz + \cos kz - 1\right), \tag{8.39}$$

$$M_Z = -EI\frac{d^2y}{dz^2} = \frac{w}{EIk^2}\left(\tan\frac{kL}{2}\sin kz + \cos kz - 1\right). \tag{8.40}$$

The maximum moment at the center of the span is

$$M_{z=L/2} = \frac{w}{k^2}\left(\sec\frac{kL}{2} - 1\right) = \frac{wL^2}{8}\left(\frac{8\left(\sec\frac{kL}{2} - 1\right)}{k^2L^2}\right), \tag{8.41}$$

where $\left(\dfrac{8\left(\sec\dfrac{kL}{2}-1\right)}{k^2L^2}\right)$ is a moment magnification factor accounting for the effects of the axial compressive force, P. The secant function can be expanded in a power series as (Beyer, 1984)

$$\sec\frac{kL}{2}=1+\frac{1}{2}\left(\frac{kL}{2}\right)^2+\frac{5}{24}\left(\frac{kL}{2}\right)^4+\frac{61}{720}\left(\frac{kL}{2}\right)^6+\cdots. \tag{8.42}$$

Since $P_e=\pi^2EI/L^2$ (Euler buckling load) and $k=\sqrt{P/EL}$,

$$\frac{kL}{2}=\frac{\pi}{2}\sqrt{\frac{P}{P_e}}. \tag{8.43}$$

The substitution of Equation 8.43 into Equations 8.41 and 8.42 provides

$$M_{z=L/2}=\frac{wL^2}{8}\left(1+1.028\left(\frac{P}{P_e}\right)+1.031\left(\frac{P}{P_e}\right)^2+1.032\left(\frac{P}{P_e}\right)^3+\cdots\right) \tag{8.44a}$$

$$=\frac{wL^2}{8}\left(1+1.028\left(\frac{P}{P_e}\right)\left(1+1.003\left(\frac{P}{P_e}\right)+1.004\left(\frac{P}{P_e}\right)^2+\cdots\right)\right). \tag{8.44b}$$

Equation 8.44b may be approximated as

$$M_{z=L/2}\approx\frac{wL^2}{8}\left(1+1.028\left(\frac{P}{P_e}\right)\left(1+\left(\frac{P}{P_e}\right)+\left(\frac{P}{P_e}\right)^2+\left(\frac{P}{P_e}\right)^3+\cdots\right)\right), \tag{8.45}$$

$$\approx\frac{wL^2}{8}\left(1+1.028\left(\frac{P}{P_e}\right)\left(\frac{1}{1-\left(\frac{P}{P_e}\right)}\right)\right), \tag{8.46}$$

$$\approx\frac{wL^2}{8}\left(\frac{1}{1-P/P_e}\right), \tag{8.47}$$

where $(1/1-(P/P_e))$ is an approximate moment magnification factor appropriate for use in design.

8.4.2.1.2 Axial Compression and Bending from a Concentrated Transverse Load

$$M_p(z)=M_{Q(z)}=\frac{Qz(L-a)}{L}\quad\text{for } 0\le z\le a, \tag{8.48a}$$

$$M_p(z)=M_{Q(z)}=\frac{Qa(L-z)}{L}\quad\text{for } a\le z\le L. \tag{8.48b}$$

The substitution of Equations 8.48a and b into 8.33 yields

$$\frac{d^2y}{dz^2}+k^2y=-\frac{Qz(L-a)}{EIL} \quad \text{for } 0 \le z \le a, \tag{8.49a}$$

$$\frac{d^2y}{dz^2}+k^2y=-\frac{Qa(L-z)}{EIL} \quad \text{for } a \le z \le L. \tag{8.49b}$$

The general solutions to Equations 8.49a and b are

$$y=A\sin kz+B\cos kz-\frac{Qz(L-a)}{EILk^2} \quad \text{for } 0 \le z \le a, \tag{8.50a}$$

$$y=C\sin kz+D\cos kz-\frac{Qa(L-z)}{EILk^2} \quad \text{for } a \le z \le L. \tag{8.50b}$$

Differentiating Equations 8.50a and 8.50b with boundary conditions of $y(0) = y(L) = 0$ and noting that displacement, $y(a)$, and slope, $dy(a)/dz$, are continuous at $z = a$ provides

$$M_Z=-EI\frac{d^2y}{dz^2}=-\frac{Q}{k}\frac{\sin k(L-a)}{\sin kL}\sin kz, \quad \text{for } 0 \le z \le a, \tag{8.51a}$$

$$M_Z=-EI\frac{d^2y}{dz^2}=\frac{Q\sin ka}{k}\left(\frac{\sin kz}{\tan kL}-\cos kz\right), \quad \text{for } a \le z \le L. \tag{8.51b}$$

The maximum moment at the center of the span (obtained by the substitution of $z = L/2$ into Equations 8.51a or b) is

$$M_{z=L/2}=\frac{QL}{4}\left(\frac{2\tan kL/2}{kL}\right). \tag{8.52}$$

The tangent function in Equation 8.52 can be expanded in a power series as (Beyer, 1984)

$$\tan\frac{kL}{2}=\frac{kL}{2}+\frac{1}{3}\left(\frac{kL}{2}\right)^3+\frac{2}{15}\left(\frac{kL}{2}\right)^5+\frac{17}{315}\left(\frac{kL}{2}\right)^7+\cdots \tag{8.53}$$

which, when substituted into Equation 8.52, and after making further simplifications similar to those outlined in Section 8.4.2.1.1, yields

$$M_{z=L/2}\approx\frac{QL}{4}\left(\frac{1-0.2(P/P_e)}{1-(P/P_e)}\right), \tag{8.54}$$

where $\left(\frac{1-0.2(P/P_e)}{1-(P/P_e)}\right)$ is an approximate moment magnification factor appropriate for use in design.

8.4.2.1.3 Axial Compression and Bending from End Moments

$$M_p(z) = M_{MA} + M_{MB} = M_A - \left(\frac{M_A + M_B}{L}\right)z. \quad (8.55)$$

The substitution of Equation 8.55 into Equation 8.33 yields

$$\frac{d^2y}{dz^2} + k^2 y = -\frac{M_A}{EI} + \left(\frac{M_A + M_B}{EIL}\right)z. \quad (8.56)$$

The general solution to Equation 8.56 is

$$y = A\sin kz + B\cos kz + \frac{M_A + M_B}{EILk^2}z - \frac{M_A}{EIk^2}. \quad (8.57)$$

Considering boundary conditions of $y(0) = y(L) = 0$

$$y = -\frac{(M_A \cos kL + M_B)}{EIk^2 \sin kL}\sin kz + \frac{M_A}{EIk^2}\cos kz + \frac{M_A + M_B}{EILk^2}z - \frac{M_A}{EIk^2}. \quad (8.58)$$

Differentiation of Equation 8.58 yields

$$\frac{dy}{dz} = -\frac{(M_A \cos kL + M_B)}{EIk \sin kL}\cos kz - \frac{M_A}{EIk}\sin kz + \frac{M_A + M_B}{EILk^2}, \quad (8.59)$$

$$\frac{d^2y}{dz^2} = \frac{(M_A \cos kL + M_B)}{EI \sin kL}\sin kz - \frac{M_A}{EI}\cos kz, \quad (8.60)$$

$$\frac{d^3y}{dz^3} = \frac{k(M_A \cos kL + M_B)}{EI \sin kL}\cos kz + \frac{kM_A}{EI}\sin kz. \quad (8.61)$$

The bending moment and shear forces are

$$M_Z = -EI\frac{d^2y}{dz^2} = -\frac{(M_A \cos kL + M_B)}{\sin kL}\sin kz + M_A \cos kz, \quad (8.62)$$

$$V_z = -EI\frac{d^3y}{dz^3} = -\frac{k(M_A \cos kL + M_B)}{\sin kL}\cos kz - kM_A \sin kz. \quad (8.63)$$

The maximum moment occurs where the shear force is zero. If Equation 8.63 is equated to zero we obtain

$$\tan kz_M = -\frac{M_A \cos kL + M_B}{M_A \sin kL}, \quad (8.64)$$

where

z_M = the location of maximum moment along the z axis.

Expressions for $\sin(kz_M)$ and $\cos(kz_M)$ may be obtained from Equation 8.64 for the substitution into Equation 8.62 to obtain the maximum bending moment, M_{max}, as

$$M_{max} = -M_B\left(\sqrt{\frac{(M_A/M_B)^2 - 2(M_A/M_B)\cos kL + 1}{\sin^2 kL}}\right). \tag{8.65}$$

If the end moments are equal and opposite in direction (i.e., beam bent in single curvature),

$$M = M_A = -M_B. \tag{8.66}$$

Equation 8.65 is

$$M_{max} = M\sqrt{\frac{2(1-\cos kL)}{\sin^2 kl}} = M\sec\frac{kL}{2} \tag{8.67}$$

which is the secant formula. In this case, the maximum moment occurs at $z_M = L/2$.

8.4.2.1.4 Combined Axial Compression and Flexural Loading

The bending moments from combined transverse uniformly distributed and axial compression loads, transverse concentrated and axial compression loads, and end moment and axial compression loads have been developed from the differential Equations 8.33 and 8.34. For combined loads, such as those shown in Figure 8.6, the bending moments from each load case may be superimposed, provided that the axial compression force is the same for each load case.

A general way of superimposing the effects from each load case is to superpose the deflected shapes by summation of Equations 8.38, 8.50, and/or 8.58, depending on the applicable load cases. The location, z_M, of maximum bending moment, M_{max}, may be obtained by setting $d^3y/dz^3 = 0$ and the maximum bending moment obtained by the substitution of z_M into the equation for bending moment, $-EI(d^2y/dz^2)$.

However, it is evident that the design of members subjected to simultaneous bending and axial compression by methods involving the solution of differential Equations 8.33 and 8.34 is relatively complex and not well suited to routine design work. Because of this, interaction equations have been developed based on bending moment–curvature–axial compression relationships.

8.4.2.2 Interaction Equations for Axial Compression and Uniaxial Bending

Interaction equations used in allowable stress design (ASD) are determined from interaction equations developed for ultimate loads. When members remain elastic, the axial compressive force, P, has no effect on the moment–curvature–axial compression relationship. However, in the inelastic range, where partial yielding has occurred, the moment–curvature–axial compression relationship is dependent on the axial compression. The nonlinear relationship between the bending moment, M, and the axial compressive force, P, depends on the curvature, ϕ, of the member. The ultimate axial compressive load is attained when yielding under combined bending and axial compression reduces the member stiffness, due to partial yielding of the cross section, and creates instability. Also, in the range of inelastic behavior, the superposition of effects due to bending, M, and axial compression, P, is not applicable.

The moment–curvature–axial compression relationships for members are determined analytically for various degrees of member yielding.* A typical moment–curvature–axial compression relationship is shown in Figure 8.7.

* Indicated by depth of yielded material from the fibers with the largest compressive strain.

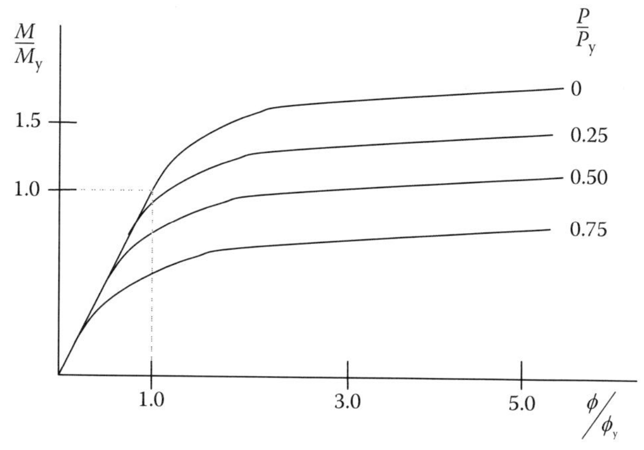

FIGURE 8.7 Typical plot of moment–curvature relationship for a member subjected to bending and axial compression.

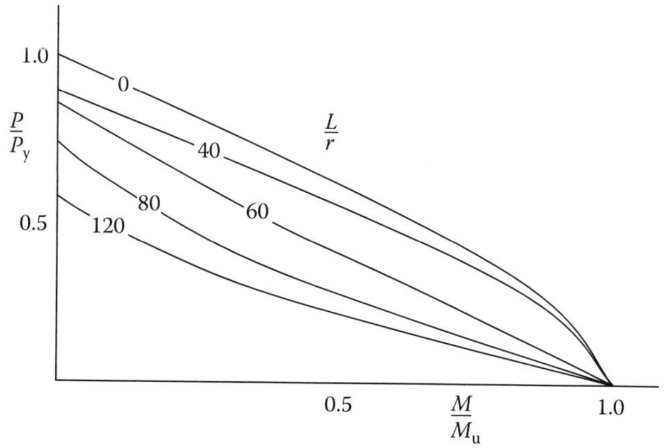

FIGURE 8.8 Typical interaction curves for a member subjected to bending and axial compression.

Once the moment–curvature–axial compression relationship is established, each value of P/P_y with a slenderness ratio, KL/r (where r is the radius of gyration in the plane of bending), is combined with various values of M/M_y until instability occurs (at M_u). Since M/M_y creates deflection, Δ, which creates an additional bending moment $(P/P_y)\Delta$, an iterative analysis, such as numerical integration by Newmark's method or another step-by-step numerical integration technique, is often used. Once the appropriate values of P/P_y and M/M_u are determined (when deflections calculated in two successive iterations are in sufficient agreement) for various values of KL/r, interaction diagrams, such as that shown in Figure 8.8, may be produced. These interaction curves may be approximated by interaction equations for various values of L/r.

The curve for $L/r = 0$ may be approximated as

$$\frac{P}{P_y} + \frac{M}{1.18M_u} = 1.0 \tag{8.68}$$

or

$$\frac{\sigma_a}{0.55F_y} + \frac{\sigma_b}{1.18F_u} = 1.0. \tag{8.69}$$

For service load design, Equation 8.69 may be conservatively expressed as

$$\frac{\sigma_a}{0.55F_y} + \frac{\sigma_b}{F_b} = 1.0 \tag{8.70}$$

where

σ_a = the normal stress due to applied axial compression
σ_b = the normal stress due to applied bending moment
F_u = the ultimate bending stress
F_b = the allowable compressive stress for bending alone.

This interaction equation is applicable to members with low slenderness, such as locations that are braced in the plane of bending, where yielding will be the failure criteria. However, for members with larger slenderness, the stability criterion must also be investigated in addition to the yielding criterion.

The yield criterion for members of larger slenderness is established as Equation 8.70 but considering F_a instead of F_y due to the potential for allowable axial compressive stresses to be controlled by instability when $\frac{KL}{r} \geq 0.629\sqrt{\frac{E}{F_y}}$ (see Chapter 6). Equation 8.70 is then

$$\frac{\sigma_a}{F_a} + \frac{\sigma_b}{F_b} = 1.0. \tag{8.71}$$

The interaction curves may be approximated for various values of slenderness, L/r, by interaction equations as

$$\frac{P}{P_{cr}} + \frac{M}{M_u(1 - P/P_e)} = 1.0 \tag{8.72}$$

or

$$\frac{\sigma_a}{F_{cr}} + \frac{\sigma_b}{F_u(1 - \sigma_a/\sigma_e)} = 1.0, \tag{8.73}$$

where

P_{cr} = the critical axial buckling load
P_e = the Euler buckling load = $\frac{\pi^2 EI}{L^2}$
F_{cr} = the critical axial buckling stress
σ_e = the Euler buckling stress = $\frac{\pi^2 E}{(KL/r)^2}$.

Equation 8.73, using FS = 1.95 for service load design for axial buckling, may be conservatively expressed as

$$\frac{\sigma_a}{F_a} + \frac{\sigma_b}{F_b(1 - \sigma_a/(\sigma_e/1.95))} = 1.0 \tag{8.74}$$

or

$$\frac{\sigma_a}{F_a} + \frac{\sigma_b}{F_b\left(1 - \frac{\sigma_a}{0.514\pi^2 E}\left(\frac{KL}{r}\right)^2\right)} = 1.0, \tag{8.75}$$

where

F_a = the allowable stress for axial compression alone

K = the effective length factor $= \frac{\pi}{L}\sqrt{\frac{EI}{P_{cr}}}$ (see Chapter 6).

The yield criterion (Equation 8.71) and the stability criterion (Equation 8.75) should be investigated for all members subject to simultaneous bending and axial compression.

8.4.3 Axial Compression and Biaxial Bending

The strength of members subjected to axial compression and biaxial bending is complex. Theoretical procedures have been developed for short members and longer members to produce interaction curves (Culver, 1966; Chen and Astuta, 1977) and confirmed as reasonable by experiment and in computer studies for typical members by Birnstiel (1968), Pillai (1980), and others (Galambos, 1988).

Since a design methodology for axial compression and biaxial bending must include the case of axial compression and uniaxial bending, it would appear reasonable to extend the interaction (Equations 8.71 and 8.75) to the case of biaxial bending with axial compression. Therefore, the interaction formula relating to the stability criterion is

$$\frac{\sigma_a}{F_a} + \frac{\sigma_{bx}}{F_{bx}\left(1-\sigma_a/0.514\pi^2 E(K_xL_x/r_x)^2\right)} + \frac{\sigma_{by}}{F_{by}\left(1-\sigma_a/0.514\pi^2 E(K_yL_y/r_y)^2\right)} = 1.0. \quad (8.76)$$

For members with low slenderness (where yielding controls), at locations of supports or where braced in the plane of bending, Equation 8.76 is

$$\frac{\sigma_a}{F_a} + \frac{\sigma_{bx}}{F_{bx}} + \frac{\sigma_{bx}}{F_{bx}} = 1.0, \quad (8.77)$$

where

$F_a = 0.55F_y$ when $L/r = 0$ (at locations of bracing and at supports)

σ_{bx} = the normal bending stress about the x axis

σ_{by} = the normal bending stress about the y axis

F_{bx} = the allowable bending stress about the x axis

F_{by} = the allowable bending stress about the y axis

$\frac{K_xL_x}{r_x}, \frac{K_yL_y}{r_y}$ = the effective slenderness ratio of the member about x and y axes, respectively (see Chapter 6).

8.4.4 AREMA Recommendations for Combined Axial Compression and Biaxial Bending

AREMA (2015) recommends that members subjected to axial compression and biaxial bending be designed in accordance with Equations 8.76, 8.77, and 8.70 extended for biaxial bending.

However, AREMA (2015) recognizes that, for members with relatively small axial compressive forces, the secondary effects are negligible. Therefore, when $\frac{\sigma_a}{F_a} \le 0.15$, Equation 8.76 may be expressed as

$$\frac{\sigma_a}{F_a} + \frac{\sigma_{bx}}{F_{bx}} + \frac{\sigma_{by}}{F_{by}} \le 1.0. \quad (8.78)$$

Both yielding and stability effects must be considered when $\frac{\sigma_a}{F_a} \succ 0.15$. AREMA (2015) recommends that when $\frac{\sigma_a}{F_a} \succ 0.15$, Equation 8.70 (the yield criterion), extended for biaxial bending, and Equation 8.76 (the stability criterion) be used for design as

$$\frac{\sigma_a}{0.55F_y} + \frac{\sigma_{bx}}{F_{bx}} + \frac{\sigma_{by}}{F_{by}} \le 1.0 \quad (8.79)$$

and

$$\frac{\sigma_a}{F_a} + \frac{\sigma_{bx}}{F_{bx}\left(1 - \frac{\sigma_a}{0.514\pi^2 E}\left(\frac{K_x L_x}{r_x}\right)^2\right)} + \frac{\sigma_{by}}{F_{by}\left(1 - \frac{\sigma_a}{0.514\pi^2 E}\left(\frac{K_y L_y}{r_y}\right)^2\right)} \le 1.0, \quad (8.80)$$

where

σ_a = the normal axial compressive stress
σ_{bx} = the normal flexural stress about the x axis
σ_{by} = the normal flexural stress about the y axis
F_a = the allowable axial compressive stress for axial compression only (Chapter 6)
F_{bx} = the allowable flexural compressive stress for bending only (Chapter 7) about the x axis
F_{by} = the allowable flexural compressive stress for bending only (Chapter 7) about the y axis.

Equation 8.79 relates to the yield criterion that is appropriate to consider at support locations, for members with very low slenderness ratios and at locations braced in the planes of bending. The design of typical members subjected to axial compression and bending is usually governed by the stability criterion of Equation 8.80.

8.5 COMBINED BENDING AND SHEAR OF PLATES

Combined bending and shear may be significant in the webs of plate girders as outlined in Chapter 7 concerning plate girder design.

REFERENCES

American Railway Engineering and Maintenance-of-Way Association (AREMA), 2015, Chapter 15—Steel structures, *Manual for Railway Engineering*, Lanham, MD.

Beyer, W.H. (ed.), 1984, *Standard Mathematical Tables*, 27th ed., CRC Press, Boca Raton, FL.

Birnstiel, C., 1968, Experiments on H-columns under biaxial bending, *Journal of Structural Engineering*, Vol. 94, No. 10, 2429–2450.

Bresler, B., Lin, T.Y., and Scalzi, J.B., 1968, *Design of Steel Structures*, 2nd ed., John Wiley & Sons, New York.

Chen, W.F. and Astuta, T., 1977, *Theory of Beam Columns*, Vols. 1 and 2, McGraw-Hill, New York.

Chen, W.F. and Lui, E.M., 1987, *Structural Stability*, Elsevier, New York.

Culver, C.G., 1966, Exact solution of the biaxial bending equations, *Journal of Structural Engineering*, Vol. 92, No. 2, 63–84.

Galambos, T.V., 1988, *Guide to Stability Design Criteria for Metal Structures*, John Wiley & Sons, New York.

Kulak, G.L. and Grondin, G.Y., 2002, *Limit States Design in Structural Steel*, 7th ed., CISC, Toronto, ON.

Kuzmanovic, B.O. and Willems, N., 1983, *Steel Design for Structural Engineers*, 2nd ed., Prentice-Hall, Englewood Cliffs, NJ.

Pillai, U.S., 1980, Comparison of test results with design equations for biaxially loaded steel beam columns, Civil Engineering Research Report No. 80-2, Royal Military College of Canada, Kingston, Canada.

Salmon, C.G. and Johnson, J.E., 1980, *Steel Structures Design and Behavior*, 2nd ed., Harper & Row, New York.

Seaburg, P.A. and Carter, C.J., 1997, *Torsional Analysis of Structural Steel Members*, AISC, Chicago, IL.

9 Design of Connections for Steel Members

9.1 INTRODUCTION

The design of connections is as important to the safety and reliability of steel railway superstructures as the design of the axial and flexural members that are connected to form the superstructure (see Chapters 6 and 7, respectively). Connections in modern steel railway superstructures are made with welds, bolts*, and/or pins†. Typically, these connections transmit axial shear (e.g., truss member to gusset plate connections and beam flange splices), combined axial tension, and shear (e.g., semirigid and rigid beam framing connections that transmit shear and moment) or eccentric shear (e.g., welded flexible beam framing connections and web plate splices). Connection behavior is often complex but may be modeled for routine design with relatively simple mathematical models. AREMA (2015) recommends design criteria for axial and flexural member connection design.

Truss member end connections at the top chord of deck trusses or the bottom chord of through trusses should be designed for the allowable strength of the member. Vertical post end connections at the top chord of deck trusses without diagonals in adjacent panels and at hangers in through trusses should be designed for 125% of the calculated maximum force in the member. Connections for members subjected to tensile cyclical stress ranges from live load (see Chapter 4) must also be designed considering allowable fatigue stress ranges‡.

Beam framing connections generally behave as rigid (fixed or with substantial rotational restraint), semirigid (intermediate level of rotational restraint), or flexible (little or no rotational restraint) at service loads. The connections transfer only shear forces if considered as flexible (i.e., as simply supported beam end connections). AREMA (2015) recommends that flexible beam framing end connections in beams (typically assumed in the design of stringers and floor beams) and girders be designed for 125% of the calculated shear force. Connections considered as semirigid or rigid may be designed for the combined bending moment and shear force applied at the joint. Rotational end restraint may be modeled for analysis using rotational springs with spring stiffness, $k\upsilon = M_r/\upsilon$, at the ends of the beam. Knowledge of $k\upsilon$, from analytical and experimental research, enables the determination of the end bending moment, M_r, for connection design.

Secondary and bracing member connections must be designed for the lesser of the allowable strength of the member or 150% of the calculated maximum force in the member. The connections in members used as struts and ties to reduce the unsupported length of other members should be designed for 2.5% of the force in the member being braced.

9.2 WELDED CONNECTIONS

Welding is the metallurgical fusion of steel components or members through an atomic bond. The steel must be melted to create the coalescence and, therefore, requires a relatively large quantity

* Rivets may be used in the design of some historical structures. However, riveting is not a modern or often used fastening method and the engineer should confirm that expertise in riveting is available for fabrication and erection.

† Pins are generally only used in special circumstances in modern steel railway superstructures such as at suspended spans of cantilever structures (Chapter 5) or as components of support bearings in long spans.

‡ For example, for slip-resistant bolted connections without the presence of stresses from out-of-plane bending, the allowable fatigue stress range is 124 MPa (18 ksi) for less than 2 million stress range cycles and 110 MPa (16 ksi) for greater than 2 million stress range cycles (Fatigue Detail Category B).

and concentration of heat energy. The heat energy is usually supplied by an electric arc created between a metal electrode and the base metal during the welding of structural steel components and members.

There are many electric arc welding processes in use. Shielded metal arc welding (SMAW), submerged arc welding (SAW), and flux cored arc welding (FCAW) are the most commonly used processes for railway superstructure fabrication (see Chapter 10). Several passes are often required during the welding process to ensure fine-grain metallurgy of the deposited weld metal. The maximum size of weld placed in a single pass depends on the position of the weld and is specified in the American Welding Society Bridge Welding Code (ANSI/AASHTO/AWS D1.5M/D1.5, 2010).

Residual stresses (already introduced into rolled plates and members at steel mills) combined with welding heat cycles can cause distortions to occur. Residual (or locked-in) stresses may be increased if the distortions are restrained. The distortions and/or residual stresses occur when welds contract more than the base metal along the weld longitudinal axis and/or due to the transverse contraction of weld metal (which tends to pull plates together and may involve a transverse angular distortion). Weld balancing and multipass welding procedures can often mitigate distortion or excessive residual stress. However, when residual stresses are inevitable, such as in thick butt welded plates, supplementary heat treatments are often required to "stress relieve" the weld and base metal adjacent to the weld.

Lamellar tearing can occur in thick plates with welded joints where tensile stresses are directed through the plate thickness by weld shrinkage [usually not of concern for plates less than about 40 mm (1 1/2 in.) thick]. Joint preparation and welding sequence is important to mitigate lamellar tearing.

Therefore, it is of critical importance that welding processes and procedures produce quality welds with proper profiles, good penetration, complete contact with the base metal at all surfaces, and no cracks, porosity, and/or inclusions.

There are many weld and joint types used for welded connections in railway superstructure steel fabrication. Fillet and groove welds are the most prevalent weld types for steel railway superstructure fabrication (slot and plug welds are generally not used). Stud welding is used in composite span fabrication to attach studs to the top flange of beam and girder spans (see Chapter 7). Butt and "T" joints are the most common welded joints used in steel railway superstructures. However, corner, lap, and edge joints might be used in secondary and nonstructural steel elements.

Welded connection design must consider the force path and strain compatibility between weld and base material. Important aspects of welded connection design are base metal weldability (see Chapter 2), deposited weld metal quality, element thickness, restraint conditions, and other details such as preparation and weld quantity. In general, the engineer should design the smallest welds possible to mitigate distortion and residual stresses in welded joints. However, when welding does cause slight distortion, it may be corrected by mechanical and heat straightening of the component or member. Steel fabricators have procedures for mechanical and heat straightening. Many of these procedures are based on Federal Highway Administration guidelines (FHWA, 1998).

Welds are continuous and rigid. Therefore, they facilitate crack propagation. As a consequence, welds must not be susceptible to fatigue crack initiation and weld designs must avoid conditions that create stress concentrations (such as excessive weld reinforcement, concentrations at intersecting welds*, highly constrained joints, discontinuous backing bars, plug and slot welds). Weld fracture in nonredundant fracture critical members (FCM) must be carefully considered in weld design. The requirements for base metal materials, welding process, consumables (including weld metal toughness), joint preparation, and preheat and interpass temperatures recommended by AREMA (2015) and AWS (2010) will provide for welds which are not susceptible to fracture.

* This might occur, for example, where horizontal gusset plates and vertical transverse or intermediate stiffeners are connected near the tension flange of a girder.

Welding inspection is critical for quality control and assurance, and is an economical method of ensuring that welding is properly performed to produce connections in conformance with the design requirements. Welds are typically inspected by magnetic particle, ultrasonic, and radiographic methods.

9.2.1 Welding Processes for Steel Railway Bridges

The arc welding processes most commonly used for railway superstructure steel fabrication are SMAW, SAW, and FCAW* (see Chapter 10).

9.2.1.1 Shielded Metal Arc Welding

The shielded metal arc welding (SMAW) process is a manual welding process where the consumable electrode (the electrode metal is transferred to the base metal) is coated with powdered materials in a silicate binder. The heated coating converts to a shielding gas to ensure no oxidation and atmospheric contamination† of the weld material in the arc stream and pool. The coating residue forms as a slag on the weld surface (provided enough time exists for the slag to float to the weld surface) and some is absorbed. The coating also assists with stabilizing the electric arc during the welding process.

9.2.1.2 Submerged Arc Welding

The submerged arc welding (SAW) process is an automatic welding process where the bare metal consumable electrode is covered with a granular fusible flux material. This flux shields the arc stream and pool. Economical and uniform welds with good mechanical properties, ductility, and corrosion resistance are produced by the modern SAW process.

9.2.1.3 Flux Cored Arc Welding

The flux cored arc welding (FCAW) process is often semiautomatic and uses a continuous wire as the consumable electrode. The wire is annular with the core filled with the flux material‡. In this manner, the flux material at the core behaves similarly to the coating in the SMAW or granular flux in the SAW process.

9.2.1.4 Stud Welding

Stud welding is essentially an SMAW process made automatic. The stud acts as the electrode as it is driven into a molten pool of metal. A ceramic ferrule placed at the base of the stud provides the molten pool of metal and also serves as the electrode coating for protection of the weld. A complete penetration weld (with a small exterior annular fillet weld) is achieved across the entire stud shank or body. This welding process is used extensively in the fabrication of composite steel and concrete beams and girders (Chapter 7).

9.2.1.5 Welding Electrodes

The electrodes for structural steel welding are specified for the base metal and welding process to be used in the qualified welding procedure specifications (WPS) (see Chapter 10). The electrode designation in the WPS refers to requirements related to use, such as strength, toughness, welding position, power supply, coating type, and arc type/characteristics. Low hydrogen electrodes are often specified to provide superior weld properties and preclude possible embrittlement of the weld from absorbed hydrogen.

---

* For nonredundant fracture critical members (FCM) only SMAW, SAW, or FCAW processes are generally permitted. Therefore, other arc welding processes such as gas metal arc welding (GMAW) and electroslag welding (ESW) are not commonly used in steel railway superstructure fabrication.

† Generally in the form of nitrides and oxides which promote brittle weld behavior.

‡ Exterior coatings on the continuous wire would be removed while being fed through the electrode holder to reach the arc.

9.2.2 Weld Types

Welds are groove, fillet, slot, or plug welds. Slot and plug welds exhibit poor fatigue behavior and are not recommended for use in steel railway superstructure fabrication.

9.2.2.1 Groove Welds

Groove welds generally require joint preparation and may be complete joint penetration (CJP) or partial joint penetration (PJP) welds. CJP welds are made through the thickness of the pieces being joined. PJP welds are made without weld penetration through the thickness of the pieces being joined. Groove welds are single bevel or double bevel* square, V, U, or J welds. Descriptions of these welds and their prequalification requirements† for both CJP and PJP welds are shown in AWS (2010). The size, *a*, of a CJP groove weld is the thinner of the plates joined (Figure 9.1). The size, *a*, of a PJP weld is usually the depth of the preparation chamfer‡ (Figure 9.2).

Minimum PJP groove weld sizes that ensure fusion are recommended in AWS (2010). These minimum recommended PJP groove weld sizes depend on the thickness of the thickest plate or element in the joint. The minimum PJP groove weld size is recommended as 6 mm (1/4 in.), except for joints with plates or elements of base metal thickness greater than 19 mm (3/4 in.) where the

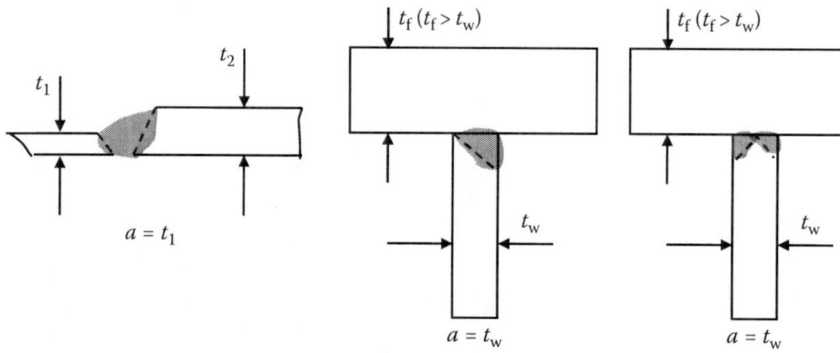

FIGURE 9.1 Size of CJP groove welds.

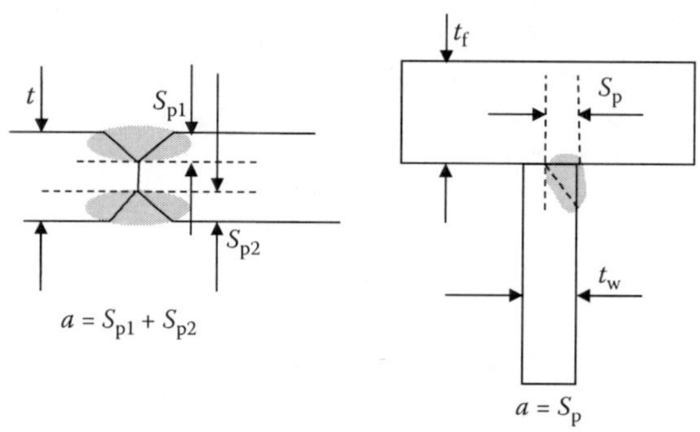

FIGURE 9.2 Size of PJP groove welds.

---

* Double bevel, V, U, or J welds are generally required for plates greater than about 16 mm (5/8″) thick to avoid excessive weld material consumption, distortion, and/or residual stresses.

† Prequalification requirements relate to the welding process, base metal thickness, groove preparation, welding position, and supplemental gas shielding if FCAW is used.

‡ For some welds, the entire depth of the preparation cannot be used and the PJP weld size may be taken as 3 mm (1/8″) or 6 mm (1/4″) less than the preparation depth. In the case of PJP square butt joints, the weld size should not exceed 75% of the plate thickness.

minimum recommended weld size is 8 mm (5/16 in.). However, the weld size need not exceed the thickness of the thinnest part in the joint. PJP welds should not be used for members loaded such that there is tensile stress normal to the effective throat of the PJP weld.

AREMA (2015) recommends that only CJP groove welds be used for connections, with the exception that PJP groove welds may be used to connect plate girder flange and web plates (T-joint in Figure 9.2).

9.2.2.2 Fillet Welds

Fillet welds do not require any joint preparation and are readily made by the SMAW, SAW and FCAW processes. The size of a fillet weld, *a* or *b*, is determined based on the thickness of the weld throat, t_e, required to resist shear (Figure 9.3). The throat depth is 0.707*a* for fillet welds with equal leg length, a = b (the usual case). Minimum fillet weld sizes that ensure fusion are recommended in AWS (2010). Minimum fillet weld size, a_{min}, is a function of the thickness of the thickest plate or element in the joint. Minimum size for single pass fillet welds is generally recommended as 6 mm (1/4 in.), except for plates or elements with base metal thickness greater than 19 mm (3/4 in.) where minimum fillet weld size is 8 mm (5/16 in.). However, the weld size need not exceed the thickness of the thinnest part in the joint, in which case care must be taken to provide adequate preheating of the weld area.

The minimum connection plate thickness recommendations of AREMA (2015) should preclude cutting of an element, plate, or component by fillet weld penetration. Maximum fillet weld size, a_{max}, is recommended in AWS (2010) in order to avoid excessive base metal melting and the creation of potential stress concentrations. The maximum fillet weld size is the thickness of the thinnest of the plates or elements in a joint for elements or plates with thickness less than 6 mm (1/4 in.). The maximum fillet weld size is the thickness of the thinnest plate or element less 2 mm (5/64 in.) for joints with thickness of the thinnest plates or elements greater than 6 mm (1/4 in.).

The minimum effective length of fillet welds is generally recommended as 4a. AREMA (2015) recommends that fillet welds used to resist axial tension that is eccentric to the weld line or cyclical tensile stresses must be returned continuously around any corner for a minimum of 2*a*. AREMA (2015) also recommends that wrap-around fillet welds not be used when welding intermediate transverse stiffeners to girder webs.

9.2.3 Joint Types

Welds are used in lap, edge, "T," corner, and butt joints. Welded lap joints are generally used only in secondary members and edge joints are used only in nonstructural members. However, "T," corner, and butt joints are commonly used for main girder fabrication and splicing of steel railway superstructure elements.

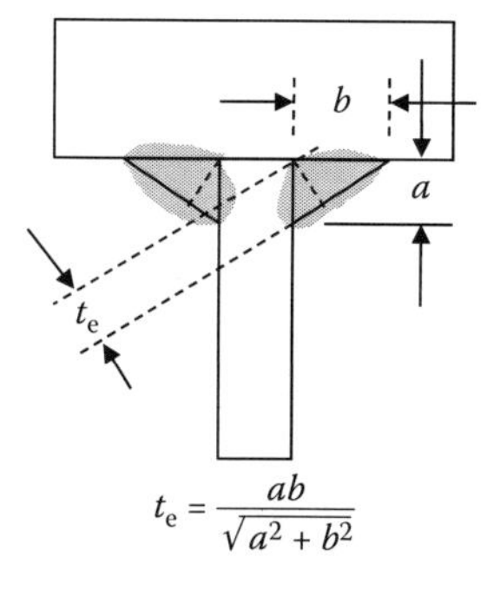

FIGURE 9.3 Size of fillet welds.

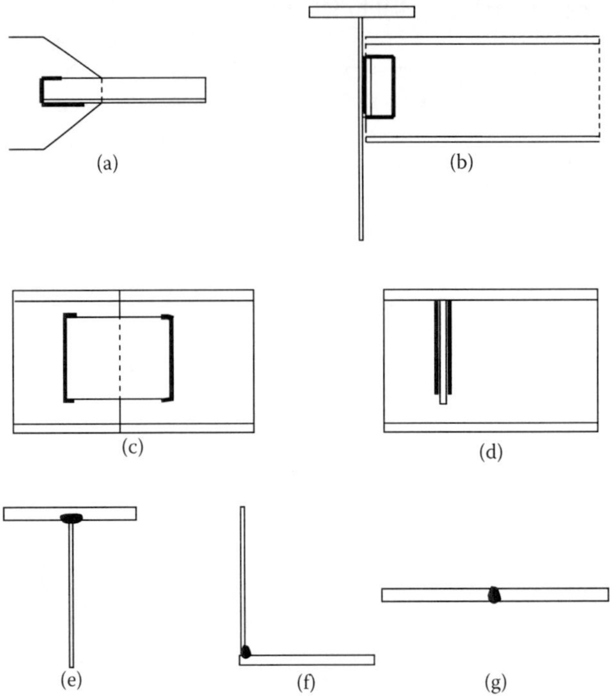

FIGURE 9.4 Typical welded joints in steel railway superstructures.

Welded lap joints (Figures 9.4a through d) are simple joints sometimes used in secondary members of steel railway superstructures. The joints in Figure 9.4a through c are also typically subjected to eccentric loads. Figure 9.4d shows a type of lap joint used to connect attachments, such as stiffeners, to girder web plates. Lap joints typically use fillet welds.

Welded "T" and corner joints (Figures 9.4e and f) are typically used to connect web plates and flange plates of plate and box girder spans, respectively. "T" and corner joints may use fillet or groove welds and are typically subjected to horizontal shear from girder bending along the longitudinal weld axis.

Welded butt joints (Figure 9.4g) often join plate ends (such as at girder flange and web plate splices) with CJP groove welds. Butt joints are also used in welded splices of entire elements or sections. There is no force eccentricity in typical butt joints, but, particularly in tension zones, butt joints require careful consideration of residual stresses*. Weld and connection element transitions for butt welded plates of different thickness and/or width are recommended by AREMA (2015). Butt joints should not be used to join plates with a difference in both thickness and width unless the element resists only axial compression†. Edge preparation and careful alignment during welding are critical for good quality butt joints.

9.2.4 Welded Joint Design

9.2.4.1 Allowable Weld Stresses

Fillet welds transmit forces by shear stress in the weld throat and groove welds transmit loads in the same manner as the elements that are joined (e.g., by shear, axial, and/or bending stresses).

Therefore, for fillet welds, the allowable shear stress is the smaller of $0.28F_u$ through the weld throat based on electrode strength or $0.35F_y$ at the weld leg based on base metal strength.

* Stress relieving is usually required.

† Stress concentrations may be large for butt welded connections subject to tension.

For groove welds, the allowable shear stress is $0.35F_y$ and the allowable tension or compression stress is $0.55F_y$ based only on base metal strength. This is because CJP welds using matching electrodes (as specified by AWS (2010)) are at least as strong as the base metal under static load conditions. In addition, CJP welds made in accordance with the AREMA (2010) Fracture Control Plan (FCP) will be of equal or greater fatigue strength than the base metal.

However, for PJP groove or fillet welds subjected to shear, axial, and/or flexural tensile stresses due to live load, the allowable fatigue stress range for the appropriate Fatigue Detail Category and equivalent number of constant stress cycles (Chapter 5) must be considered. The allowable fatigue stress ranges may be small and govern the required weld size for weak fatigue details such as transversely loaded fillet or PJP welds, and fillet or groove welds used on attachments with poor transition details*.

9.2.4.2 Fatigue Strength of Welds

Stress concentrations are often created by welding. The welding process may introduce discontinuities within the weld, distortion of members, residual stresses, and stress raisers due to poor weld profiles. Stress concentration factors typically range from 1.0 to 1.6 for butt joint welds and from 1.0 and 2.8 (or more) for other joints (Kuzmanovic and Willems, 1983). These stress concentration effects are included in the nominal stress range fatigue testing of many different weld types, joint configurations, and loading directions. This provides the design criteria, in terms of the allowable fatigue stress ranges, for the various Fatigue Detail Categories recommended in AREMA (2015). Further discussion of allowable fatigue stresses for design is contained in Chapter 5.

9.2.4.3 Weld Line Properties

It is intuitive and convenient to design welds as line elements. The effective weld area, A_e, (on which allowable stresses are assumed to act) is

$$A_e = t_e L_w, \tag{9.1}$$

where

t_e = the thickness of the thinner element for CJP groove welds
= the depth of preparation chamfer—3 mm (−1/8″) for PJP groove weld root angles between 45° and 60° (for SMAW and SAW welds)
= the depth of preparation chamfer for PJP groove weld root angles greater than or equal to 60° (for SMAW and SAW welds)
= the throat length = 0.707a (for fillet welds with equal legs, a)
L_w = the length of the weld.

If the weld is considered as a line, the allowable force per unit length, F_w, on the weld is

$$F_w = t_e(f_{all}), \tag{9.2}$$

where

f_{all} = the allowable weld stress
and, from Equation 9.2,

$$t_e = \frac{F_w}{f_{all}}. \tag{9.3}$$

* Generally such details should be avoided to preclude low allowable fatigue stress ranges, which may render the superstructure design uneconomical.

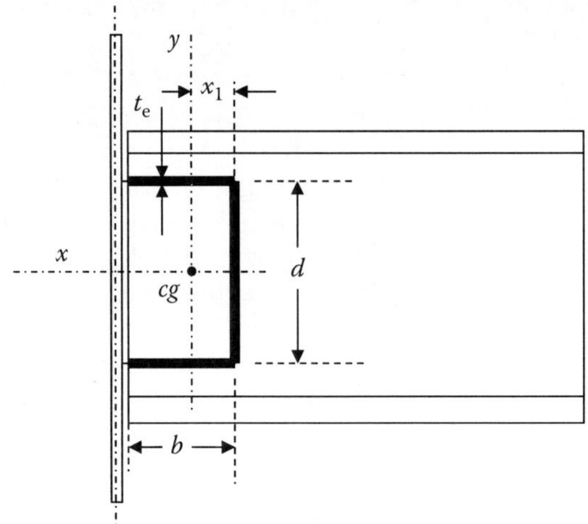

FIGURE E9.1

Therefore, considering the weld as a line provides a direct method of designing welds not subjected to eccentricities that cause bending and/or torsion of the weld line. The moment of inertia in the direction parallel to and perpendicular to the longitudinal weld axis is required for situations where welds are subjected to bending and torsion. Example 9.1 illustrates the calculation of weld line properties for a particular weld configuration.

Example 9.1

Determine the weld line properties for the weld configuration shown in Figure E9.1.

$$x_1 = \frac{2t_e b\left(b/2\right)}{2bt_e + dt_e} = \frac{b^2}{2b+d}$$

$$I_x = t_e\left(\frac{d^3}{12} + 2b\left(d/2\right)^2\right)$$

$$I_y = t_e\left(\frac{2b^3}{12} + 2b\left(\frac{b}{2} - x_1\right)^2 + dx_1^2\right)$$

$$I_p = I_x + I_y = t_e\left(\frac{d^3}{12} + 2b\left(d/2\right)^2 + \frac{2b^3}{12} + 2b\left(\frac{b}{2} - x_1\right)^2 + dx_1^2\right) = t_e\left(\frac{8b^3 + 6bd^2 + d^3}{12}\right)$$

$$S_x = \frac{I_x}{d/2} = t_e\left(\frac{d^2}{6} + bd\right).$$

Weld line properties for any weld configuration may be determined as shown in Example 9.1. Table 9.1 provides weld line properties for some commonly used weld configurations.

9.2.4.4 Direct Axial Loads on Welded Connections

Axial weld connections should have at least the strength of the members being connected and be designed to avoid large eccentricities.

TABLE 9.1
Properties of Weld Lines

Weld Configuration	Location of cg (x_1 and y_1)	Section Modulus/Weld Size	Polar Moment of Inertia about cg/Weld Size
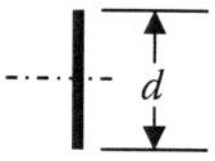	$x_1 = 0$ $y_1 = \frac{d}{2}$	$\frac{S_x}{t_e} = \frac{d^2}{6}$	$\frac{I_p}{t_e} = \frac{d^3}{12}$
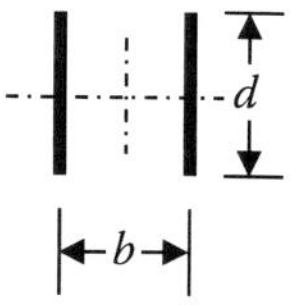	$x_1 = \frac{b}{2}$ $y_1 = \frac{d}{2}$	$\frac{S_x}{t_e} = \frac{d^2}{3}$	$\frac{I_p}{t_e} = \frac{d(3b^2 + d^2)}{6}$
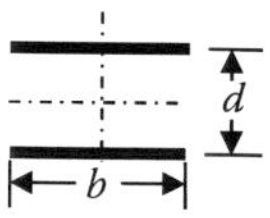	$x_1 = \frac{b}{2}$ $y_1 = \frac{d}{2}$	$\frac{S_x}{t_e} = bd$	$\frac{I_p}{t_e} = \frac{d(3d^2 + b^2)}{6}$
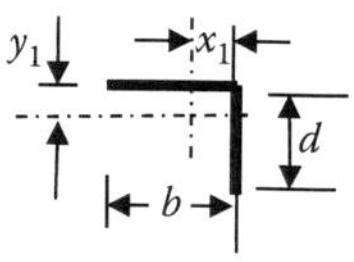	$x_1 = \frac{b^2}{2(b+d)}$ $y_1 = \frac{d^2}{2(b+d)}$	$\frac{S_x}{t_e} = \frac{d^2 + 4bd}{6}$	$\frac{I_p}{t_e} = \frac{(b+d)^4 - 6b^2d^2}{12(b+d)}$
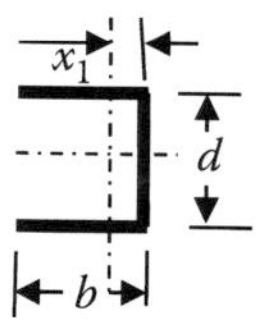 (see Example 9.1)	$x_1 = \frac{b^2}{2b+d}$ $y_1 = \frac{d}{2}$	$\frac{S_x}{t_e} = bd + \frac{d^2}{6}$	$\frac{I_p}{t_e} = \frac{8b^3 + 6bd^2 + d^3}{12}$
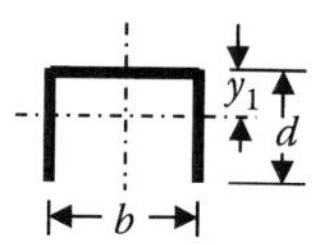	$x_1 = \frac{b}{2}$ $y_1 = \frac{d^2}{2d+b}$	$\frac{S_x}{t_e} = \frac{d^2 + 2bd}{3}$	$\frac{I_p}{t_e} = \frac{8d^3 + 6db^2 + b^3}{12}$
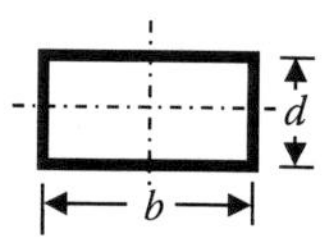	$x_1 = \frac{b}{2}$ $y_1 = \frac{d}{2}$	$\frac{S_x}{t_e} = bd + \frac{d^2}{3}$	$\frac{I_p}{t_e} = \frac{(b+d)^3}{6}$

Groove welds are often used for butt welds in axial tension or compression members (Figure 9.4g). Eccentricities are avoided and, with electrodes properly chosen to the match the base metal [see AWS (2010)], CJP groove welds are designed in accordance with the base metal strength and thickness.

Fillet welds are designed to resist shear stress on the effective area, A_e. The size of fillets welds is often governed by the thickness of the elements being joined and it is necessary to determine the length of fillet weld required to transmit the axial force without eccentricity at the connection.

Example 9.2 outlines the design of an axially loaded full strength fillet weld connection that eliminates eccentricities.

Example 9.2a (SI Units)

Design the welded connection for some secondary wind bracing shown in Figure E9.2a. The steel has $F_y = 290$ MPa and $F_u = 435$ MPa, and E60XX ($F_u = 414$ MPa) electrodes are used for the SMAW fillet weld.

Considering an estimated shear lag coefficient of 0.90 (see Chapter 6), the member strength is

$$T = 0.55(290)(3740)/1000 = 596.5 \text{ kN}$$

or

$$T = (0.90)0.47(435)(3740)/1000 = 688.2 \text{ kN}.$$

The minimum fillet weld size is 6 mm and maximum fillet weld size is 14 mm. Try a 10 mm fillet weld.

$$t_e = (10)(0.707) = 7.1 \text{ mm}.$$

From Equation 9.2, the allowable strength of the weld line is

$$F_w = t_e(f_{all}) = 7.1(130) = 919 \text{ N/mm}$$

$$F_E = 919(100)/1000 = 91.9 \text{ kN}.$$

Taking moments about F_{L2} yields

$$F_{L1} = \frac{T(50.9) - F_E(50)}{100} = \frac{596.5(25.9) - 91.9(50)}{100} = 108.5 \text{ kN}$$

$$F_{L2} = T - F_{L1} - F_E = 596.5 - 108.5 - 91.9 = 396.1 \text{ kN}.$$

Length of weld L1 = 108.5(1000)/919 = 118 mm, use 120 mm
Length of weld L2 = 396.1(1000)/919 = 431 mm, say 440 mm.

The effect of the force eccentricity is to require that the fillet welds are balanced such that weld L2 is 320 mm longer than weld L1. The length of weld L2 may be reduced by reducing the eccentricity.

Shear stress on the base metal $= \dfrac{596.5(1000)}{10(120+440)} = 106.5$ MPa. This is 5% over the allowable stress of 0.35(290) × 101.5 MPa, OK since design is based on the strength of the member.

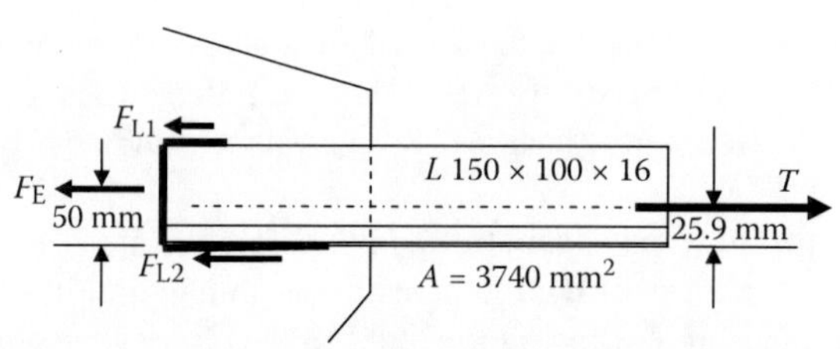

FIGURE E9.2a

Shear stress on the fillet weld throat $= \dfrac{596.5(1000)}{0.707(10)(120+440)} = 150.7$ MPa, which is much greater than the allowable stress of 0.28(414) = 116 MPa. Therefore, larger weld size and/or electrode strength is required. Try using 12 mm fillet welds made with E70XX electrodes ($F_u = 483$ MPa) $t_e = (12)(0.707) = 8.5$ mm.

From Equation 9.2, the allowable strength of the weld line is

$$F_w = t_e(f_{all}) = 8.5(130) = 1103 \text{ N/mm}$$

$$F_E = 1103(100)/1000 = 110.3 \text{ kN}.$$

Taking moments about F_{L2} yields

$$F_{L1} = \frac{T(50.9) - F_E(50)}{100} = \frac{596.5(25.9) - 110.3(50)}{100} = 99.3 \text{ kN}$$

$$F_{L2} = T - F_{L1} - F_E = 596.5 - 99.3 - 100.3 = 396.9 \text{ kN}.$$

Length of weld L1 = 99.3(1000)/1103 = 90 mm
Length of weld L2 = 396.9(1000)/1103 = 360 mm.

Shear stress on the base metal $= \dfrac{596.5(1000)}{12(90+360)} = 110.5$ MPa. This is 9% over the allowable stress of 0.35(290) = 101.5 MPa; therefore, increase the weld length to $\dfrac{596.5(1000)}{12(101.5)} = 490$ mm (say 500 mm). Length of weld L1 = (90/450)(500) = 100 mm and length of weld L2 = (360/450)(500) = 400 mm.

Shear stress on the fillet weld throat $= \dfrac{596.5(1000)}{0.707(12)(100+400)} = 140.6$ MPa, which is 4% greater than the allowable stress of 0.28(483) = 135 MPa, OK since design is based on the strength of the member.

If the connection length is assumed to be (100 + 400)/2 = 250 mm, the shear lag coefficient, U, is $U = (1 - x/L) = (1 - 25.9/250) = 0.90$, which is equal to the $U = 0.90$ assumed.

Example 9.2b (US Customary and Imperial Units)

Design the welded connection for some secondary wind bracing shown in Figure E9.2b. The steel is Grade 50 (F_y = 50 ksi) and E70XX electrodes are used for the SMAW fillet weld.

Considering an estimated shear lag coefficient of 0.90 (see Chapter 6), the member strength is

$$T = 0.55(50)(5.75) = 158.1 \text{ kips}$$

or

$$T = (0.90)0.47(70)(5.75) = 170.3 \text{ kips}.$$

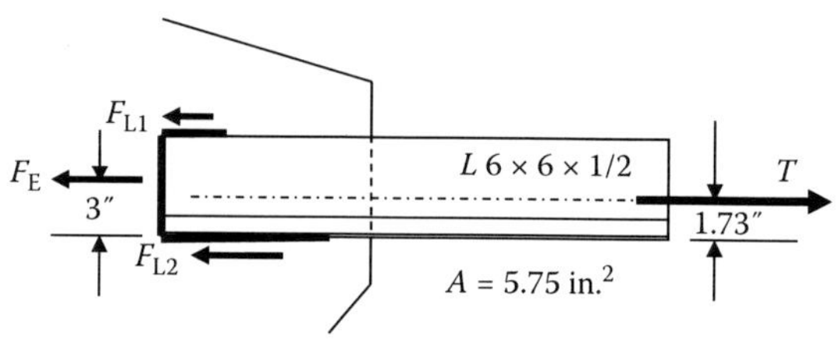

FIGURE E9.2b

The minimum fillet weld size is 1/4 in. and maximum fillet weld size is 7/16 in. Try a 5/16 in. fillet weld.

$t_e = (5/16)(0.707) = 0.22$ in.

From Equation 9.2, the allowable strength of the weld line is

$$F_w = t_e(f_{all}) = 0.22(19) = 4.18 \text{ kips/in.}$$

$$F_E = 4.18(6) = 25.1 \text{ kips.}$$

Taking moments about F_{L2} yields

$$F_{L1} = \frac{T(1.73) - F_E(3)}{6} = \frac{158.1(1.73) - 25.1(3)}{6} = 33.0 \text{ kips}$$

$$F_{L2} = T - F_{L1} - F_E = 158.1 - 33.0 - 25.1 = 100.0 \text{ kips.}$$

Length of weld L1 = 33.0/4.18 = 7.9 in., say 8 in.

Length of weld L2 = 100.0/4.18 = 23.9 in., say 24 in.

The effect of the force eccentricity is to require that the fillet welds are balanced such that weld L2 is 16 in. longer than weld L1. The length of weld L2 may be reduced by reducing the eccentricity.

Shear stress on base metal $= \dfrac{158.1}{(5/16)(8+24)} = 15.8 \text{ ksi} \geq 0.35(50) \geq 17.5$ ksi, OK.

Shear stress on fillet weld throat $= \dfrac{158.1}{0.707(5/16)(8+24)} = 22.4$ ksi, which is much greater than the allowable stress of 0.28(70) = 19.6 ksi. Increase the weld size to 3/8″ and the shear stress on fillet weld throat is $\dfrac{158.1}{0.707(3/8)(32)} = 18.6$ ksi, which is less than the allowable stress of 0.28(70) = 19.6 ksi.

If the connection length is assumed to be (8 + 24)/2 = 16 in., the shear lag coefficient, U, is

$U = (1 - x/L) = (1 - 1.73/16) = 0.89$, which is sufficiently close to the $U = 0.90$ assumed.

9.2.4.5 Eccentrically Loaded Welded Connections

Even small load eccentricities must be considered in design since welded connections have no initial pretension (such as that achieved by the application of torque to bolts). Many welded connections are loaded eccentrically (e.g., the connections shown in Figure 9.4b through d). Eccentric loads will result in combined shear and torsional moments or combined shear and bending moments, depending on the direction of loading with respect to weld orientation in the connection.

9.2.4.5.1 Connections Subjected to Shear Forces and Bending Moments

A connection such as that of Figure 9.4d is shown in greater detail in Figure 9.5. The fillet welds each side of the stiffener resist both shear forces and bending moments.

The shear stress in the welds is

$$\tau = \frac{P}{A} = \frac{P}{2t_e d} \tag{9.4}$$

and the flexural stress (using S_x from Table 9.1) is

$$\sigma_b = \frac{M}{S_x} = \frac{3Pe}{t_e d^2}. \tag{9.5}$$

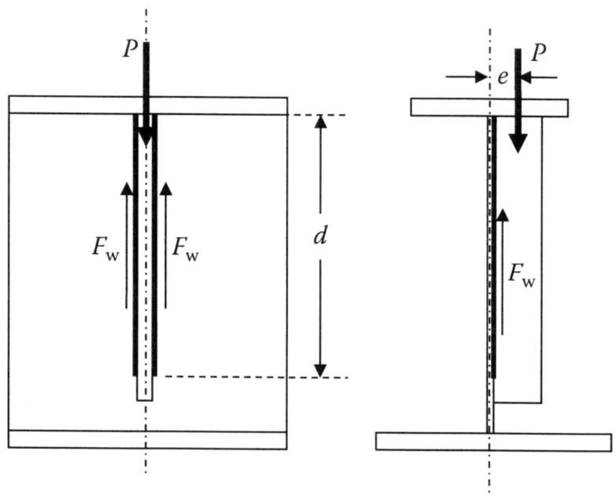

FIGURE 9.5 Bending and shear forces on fillet welds.

The stress resultant is

$$f = \sqrt{\tau^2 + \sigma_b^2} = \frac{P}{2t_e d}\sqrt{1 + \left(\frac{6e}{d}\right)^2}. \tag{9.6}$$

9.2.4.5.2 Connections Subjected to Shear Forces and Torsional Moments

A connection such as that of Figure 9.4b is shown in greater detail in Figure 9.6. The fillet welds each side of the leg of the connection angle (or plate) against the beam web resist both shear forces and torsional moments.

The shear stress in the welds is

$$\tau = \frac{P}{A} = \frac{P}{2t_e(2b + d)} \tag{9.7}$$

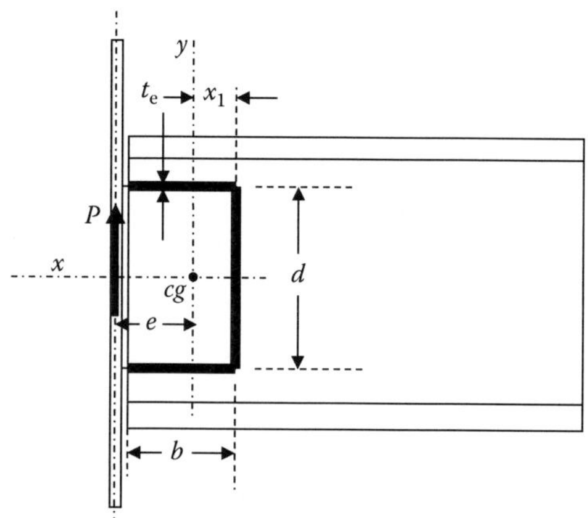

FIGURE 9.6 Torsional and shear forces on fillet welds.

and the torsional stress (using I_p from Table 9.1) is

$$\sigma_{tx} = \frac{Ty}{I_p} = \frac{6(Pe)y}{t_e\left(8b^3 + 6bd^2 + d^3\right)} \tag{9.8a}$$

$$\sigma_{ty} = \frac{Tx}{I_p} = \frac{6(Pe)x}{t_e\left(8b^3 + 6bd^2 + d^3\right)}, \tag{9.8b}$$

where

x = the distance from the centroid to the point of interest on the weld in the x-direction

y = the distance from the centroid to the point of interest on the weld in the y-direction.

The stress resultant at any location on the weld described by locations x and y is

$$f = \sqrt{(\tau + \sigma_{ty})^2 + \sigma_{tx}^2}. \tag{9.9}$$

9.2.4.5.3 *Beam Framing Connections*

Welded beam framing connections are not used in main members of steel railway superstructures due to the cyclical load regime (see Example 9.3). Nevertheless, when used on secondary members, such as walkway supports, welded beam framing connections are subject to the shear force, P, and an end bending moment, M_e, on the welds on the outstanding legs of the connection angles. The legs of the connection angles fastened to the web of the beam are also subject to an eccentric shear force, which creates a torsional moment, Pe (Figure 9.7).

In usual design practice, beam framing connections are often assumed to transfer shear only (i.e., it is assumed the beam is simply supported and $M_e = 0$). However, in reality, due to end restraint, some proportion of the fixed end moment, δM_f, typically exists ($\delta = M_e/M_f$ and M_f = fixed end beam moment). Welded connection behavior in structures is often semirigid with a resulting end moment (Blodgett, 2002). The magnitude of the end moment depends on the rigidity of the support. For example, a beam end connection to a heavy column flange may be quite rigid ($\delta \rightarrow 1$) while a beam end connection framing into the thin web of a girder or column may be quite flexible ($\delta \rightarrow 0$) (Figure 9.8).

A rigid connection may be designed for the end moment due to full fixity, M_f, and corresponding shear force, P. A semirigid connection will require an understanding of the end moment (M_e)—end rotation (ϕ_e) relationship (often nonlinear) to determine rotational stiffness and the end moment to be used in conjunction with the shear force for design. Moment–rotation curves, developed from

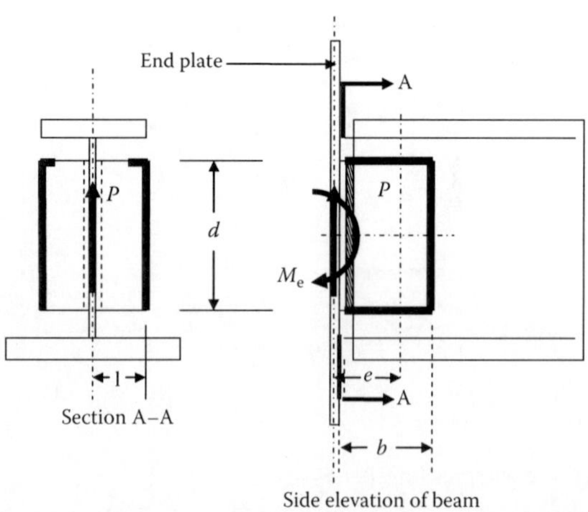

FIGURE 9.7 Simple welded beam framing connection.

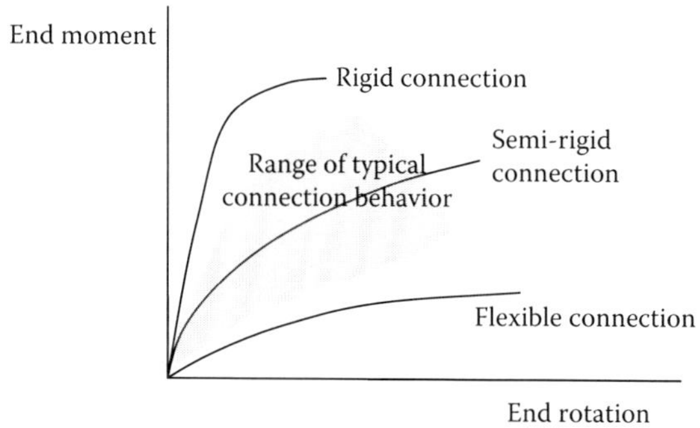

FIGURE 9.8 Typical moment–rotation curves for welded beam end connections.

theory and experiment, for welded joint configurations* are available in the technical literature (e.g., Faella et al., 2000).

A flexible end connection will deform and resist very little bending moment. Simple beam framing connections (Figure 9.7) that exhibit the characteristics of a flexible connection may be designed for shear force, P, only ($M_e = 0$). AREMA (2015) recognizes that most connections actually exhibit some degree of semirigid behavior and allows flexible connection design (with an angle thickness that allows for deformation†) provided the design shear force is increased by 25%. Therefore, flexible welded beam framing connections may be designed considering shear on the outstanding legs of the connection and, due to the eccentricity of the shear force, combined shear, and torsion on the legs of the connection angles fastened to the web of the beam. Otherwise, a semirigid connection design considering both the beam end moment and the shear force is required.

The outstanding legs of the angles in a simple beam framing connection (often referred to as clip angles) must deform sufficiently to allow for flexible connection behavior. An approximate solution for the maximum thickness of an angle to allow sufficient deformation in a welded beam framing connection, over a depth, d, can be developed assuming the shear force, P, is applied at a distance, e (see Figures 9.6 and 9.7). The bending stress, f_{wa}, in the leg of the angle connected to the end plate (e.g., a beam, girder, or column web plate) from a load, P, applied at an eccentricity, e, is (from Equation 9.5)

$$f_{wa} = \frac{M}{S} = \frac{3Pe}{t_a d^2}, \tag{9.10}$$

where

P = the shear force applied at an eccentricity, e
t_a = the thickness of the angle
d = the depth of the connected angle.

The tensile force, T_P, on the connection angle (pulling the angle away from the end plate connection) is

$$T_P = f_{wa}(2)t_a = \frac{6Pe}{d^2}. \tag{9.11a}$$

This tensile force, T_P, creates a bending moment in the angle legs (assuming the legs behave as simply supported beams of length $2l$) of

$$M_{wa} = \frac{T_P(2l)}{4} \tag{9.11b}$$

* Mainly for beam to column flange connections.
† In particular, the outstanding legs must be sufficiently flexible to disregard any beam bending effects.

and the stress in the angle leg is

$$f_{wa}=\frac{6M_{wa}}{t^2}=\frac{3T_{\mathrm{P}}l}{t^2}=\frac{18Pel}{t^2d^2}. \tag{9.12}$$

The deformation of the connection angles is

$$\Delta=\frac{T_{\mathrm{P}}(2l)^3}{48EI}=\frac{2T_{\mathrm{P}}l^3}{Et^3}=\frac{f_{wa}l^2}{1.5Et}. \tag{9.13}$$

Example 9.3 illustrates the design of a welded beam framing connection.

Example 9.3a (SI Units)

Design the welded simple beam framing connection shown in Figure E9.3a for a shear force of P = 250 kN. The uniformly loaded beam is 6.0 m long, has a strong axis moment of inertia of 500 × 10^6 mm^4, and frames into the web of a plate girder. The allowable shear stress on the fillet welds is 120 MPa. Electrodes are E70XX (F_u = 490 MPa). The return weld in section X–X will have a minimum length of twice the weld size and may be neglected in the design.

P' = 1.25(250) = 312.5 kN (AREMA recommendation for flexible connection design)

From Table 9.1:

$$e=b-x_1=150-\frac{150^2}{2(150)+300}=150-\frac{22,500}{600}=112.5\text{ mm}.$$

Shear and bending on welds in section X–X:

Bending stress in the welds:

From Table 9.1, the section modulus of the welds, S_w, is

$$S_w=\frac{t_e d^2}{3}=30,000t_e\text{ mm}^3$$

$$\sigma_w=\pm\frac{M_e}{S_w}=\frac{3M_e}{t_e d^2}=\pm\frac{M_e}{30\times10^3 t_e}.$$

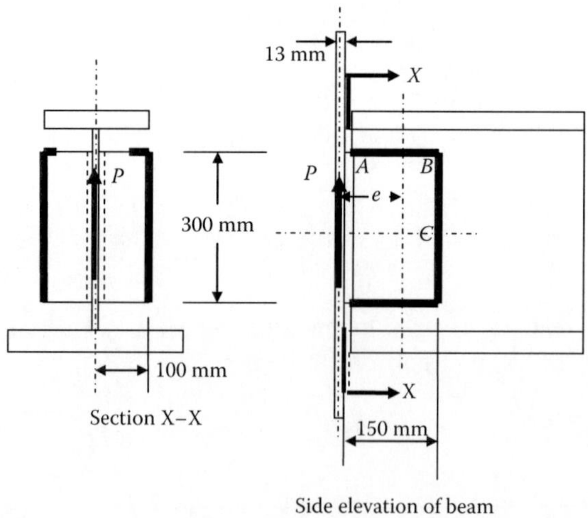

FIGURE E9.3a

Since, for a flexible connection, $M_e = 0$, there is no flexural stress in the weld.
Shear stress in the weld:

$$\tau = \frac{P'}{2t_e d} = \frac{520.8}{t_e} \text{ MPa.}$$

The stress resultant is

$$f_1 = \sqrt{\tau^2 + \sigma_b^2} = \frac{1}{t_e}\sqrt{520.8^2 + 0^2} = \frac{520.8}{t_e} \text{ MPa.}$$

Therefore, the required weld thickness is

$$t_e \geq \frac{520.8}{0.28(490)} \geq 3.8 \text{ mm for shear in the throat of welds}$$

or

$$t_e \geq \frac{520.8}{(120)(0.707)} \geq 6.1 \text{ mm for shear of the equal leg welds on the end plate.}$$

Angle thickness to allow deformation:
From Equations 9.12 and 9.13, the deformation of the connection angles is

$$\Delta = \frac{12Pel^3}{d^2Et^3} = \frac{12(250)(1000)(112.5)(100)^3}{(300)^2(200{,}000)t^3} = \frac{18{,}750}{t^3} \text{ mm.}$$

If the beam is uniformly loaded with distributed load, w,

$$w = \frac{2P}{L} = \frac{2(250)(1000)}{6000} = 83.33 \text{ N/mm,}$$

the end rotation, Y_b, is

$$\theta_b = \frac{wL^3}{24EI} = \frac{83.33(6000)^3}{24(200{,}000)I} = \frac{3750\times 10^3}{I} \text{ rad.}$$

The rotation occurs about the bottom the angle so that the deformation at the top of the angle, D, is

$$\Delta = 300\theta_b = \frac{1125\times 10^6}{I} \text{ mm}$$

so that

$$t \leq 0.0255(I)^{1/3}.$$

For a beam with $I = 500 \times 10^6 \text{ mm}^4$,

$$t \leq 0.0255(500\times 10^6)^{1/3} \leq 20.3 \text{ mm.}$$

The angle thickness should be based on the requirement for transmitting shear or the minimum element thickness recommended by AREMA (2015), but should not be greater than about 20 mm thick (note the calculation of 20.3 mm is approximate) to ensure adequate flexibility for consideration as a flexible beam framing connection.

TABLE E9.1a

Location	x (mm)	y (imm)	$t_e f_2$ (kN/mm)
A	112.5	150	0.495
B	37.5	150	0.394
C	37.5	0	0.318

Shear and torsion on welds in side elevation:

The shear stress in the welds is

$$\tau = \frac{P'}{2t_e(2b+d)} = \frac{260.4}{t_e} \text{ MPa.}$$

The torsional stress (using I_p from Table 9.1) is

$$\sigma_{tx} = \frac{6(P'e)y}{t_e\left(8b^3+6bd^2+d^3\right)} = \frac{1.563y}{t_e} \text{ MPa}$$

$$\sigma_{ty} = \frac{1.563x}{t_e} \text{ MPa.}$$

The stress resultant, f_2, at any location on the weld described by locations x and y is

$$f_2 = \sqrt{(\tau+\sigma_{ty})^2+\sigma_{tx}^2} = \frac{1}{t_e}\sqrt{(260.4+1.563x)^2+(1.563y)^2}.$$

The weld stresses are computed in Table E9.1a for various locations on the welds on the beam web.

Therefore, the required weld thickness is

$t_e \geq \dfrac{0.495(1000)}{0.28(490)} \geq 3.6$ mm for shear and torsion in throat of welds on the beam web.

$t_e \geq \dfrac{495}{(120)(0.707)} \geq 5.8$ mm for shear and torsion of equal leg welds on the beam web.

Fatigue must be considered if the applied load is cyclical. A connection such as that shown in Figure E9.3a is a very poor connection from a fatigue perspective with an allowable fatigue shear stress range of 55 MPa on the fillet weld throat. In that case, the required weld thickness for shear and bending of the welds on the end plate will greatly exceed the maximum allowable fillet weld size based on the thickness of the thinnest plate or element in the connection.

Example 9.3b (US Customary and Imperial Units)

Design the welded simple beam framing connection shown in Figure E9.3b for a shear force of $P = 52$ kips. The uniformly loaded beam is 20 ft long, has a strong axis moment of inertia of 1200 in.^{4}, and frames into the web of a plate girder. The allowable shear stress on the fillet welds is 17.5 ksi. Electrodes are E70XX. The return weld in section X–X will have a minimum length of twice the weld size and may be neglected in the design.

$P' = 1.25(52) = 65$ kips (AREMA recommendation for flexible connection design)

From Table 9.1:

$$e = b - x_1 = 6 - \frac{b^2}{2b+d} = 6 - \frac{36}{24} = 4.5 \text{ in.}$$

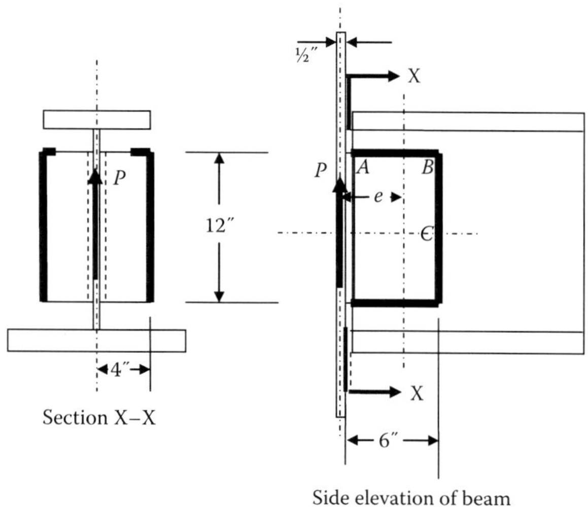

FIGURE E9.3b

Shear and bending on welds in section X–X:

Bending stress in the welds:

From Table 9.1, the section modulus of the welds, S_w, is

$$S_w = \frac{t_e d^2}{3} = 48t_e \text{ in.}^3$$

$$\sigma_w = \pm\frac{M_e}{S_w} = \frac{3M_e}{t_e d^2} = \pm\frac{M_e}{48t_e}.$$

Since, for a flexible connection, $M_e = 0$, there is no flexural stress in the weld.

Shear stress in the weld:

$$\tau = \frac{P'}{2t_e d} = \frac{2.71}{t_e} \text{ ksi.}$$

The stress resultant is

$$f_1 = \sqrt{\tau^2 + \sigma_b^2} = \frac{1}{t_e}\sqrt{2.71^2 + 0^2} = \frac{2.71}{t_e} \text{ ksi.}$$

Therefore, the required weld thickness is

$t_e \geq \dfrac{2.71}{0.28(70)} \geq 0.14$ in. for shear in the throat of welds

or

$t_e \geq \dfrac{2.71}{(17.5)(0.707)} \geq 0.22$ in. for shear of the equal leg welds on the end plate.

Angle thickness to allow deformation:

From Equations 9.12 and 9.13 the deformation of the connection angles is

$$\Delta = \frac{12Pel^3}{d^2Et^3} = \frac{12(52)(4.5)(4)^3}{(12)^2(29{,}000)t^3} = \frac{0.0430}{t^3} \text{ in.}$$

If the beam is uniformly loaded with distributed load, w,

$$w = \frac{2P}{L} = \frac{2(52)}{20(12)} = 0.433 \text{ k/in.}$$

the end rotation, Y_b, is

$$\theta_b = \frac{wL^3}{24EI} = \frac{0.433(240)^3}{24(29{,}000)I} = \frac{8.60}{I} \text{ rad.}$$

The rotation occurs about the bottom the angle so that the deformation at the top of the angle, D, is

$$\Delta = 12\theta_b = \frac{103.2}{I}$$

so that

$$t \leq 0.0747(I)^{1/3}.$$

For a beam with $I = 1200$ in.^{4},

$$t \leq 0.0747(1200)^{1/3} \leq 0.79 \text{ in.}$$

The angle thickness should be based on the requirement for transmitting shear or the minimum element thickness recommended by AREMA (2015), but should not be greater than about 3/4″ (note the calculation of 0.79″ is approximate) to ensure adequate flexibility for consideration as a flexible beam framing connection.

Shear and torsion on welds in side elevation:

The shear stress in the welds is

$$\tau = \frac{P'}{2t_e(2b+d)} = \frac{1.35}{t_e} \text{ ksi.}$$

The torsional stress (using I_p from Table 9.1) is

$$\sigma_{tx} = \frac{6(P'e)y}{t_e\left(8b^3 + 6bd^2 + d^3\right)} = \frac{0.203y}{t_e} \text{ ksi}$$

$$\sigma_{ty} = \frac{0.203x}{t_e} \text{ ksi.}$$

The stress resultant, f_2, at any location on the weld described by locations x and y is

$$f_2 = \sqrt{(\tau + \sigma_{ty})^2 + \sigma_{tx}^2} = \frac{1}{t_e}\sqrt{(1.35 + 0.203x)^2 + (0.203y)^2}.$$

The weld stresses are computed in Table E9.1b for various locations on the welds on the beam web.

Therefore, the required weld thickness is

$t_e \geq \dfrac{2.57}{0.28(70)} \geq 0.13$ in. for shear and torsion in throat of welds on the beam web.

TABLE E9.1b

Location	x (in)	y (in)	$t_e f_2$ (kips/in.)
A	4.5	6	2.57
B	1.5	6	2.06
C	1.5	0	1.66

or $t_e \geq \dfrac{2.57}{(17.5)(0.707)} \geq 0.21$ in. for shear and torsion of equal leg welds on the beam web.

Fatigue must be considered if the applied load is cyclical. A connection such as that shown in Figure E9.3b is a very poor connection from a fatigue perspective with an allowable fatigue shear stress range of 8 ksi on the fillet weld throat. In that case, the required weld thickness for shear and bending of the welds on the end plate will greatly exceed the maximum allowable fillet weld size based on the thickness of the thinnest plate or element in the connection.

Eccentrically loaded welded connections should not be used to join members subjected to cyclical live loads. The very low fatigue strength of these joints (e.g., transversely loaded fillet welds subject to tension or stress reversal) makes them unacceptable for joints in main carrying members. For such joints, bolted connections are more appropriate.

9.2.4.6 Girder Flange to Web "T" Joints

Connection such as those in Figure 9.4e and f transmit horizontal shear forces from bending and, if present, vertical shear forces due to direct transverse loads. This connection between girder flanges and web plates may be made with CJP groove welds, PJP groove welds, or fillet welds using the SAW process. Some engineers specify fillet or PJP groove welds when the connection is subjected to only horizontal shear from bending. CJP groove welds may be required when weld shear due to direct loading is combined with horizontal shear from bending.

The horizontal shear flow, q_f, for which the flange-to-web weld is designed, is (see Chapter 7)

$$q_f = \frac{dP_f}{dx} = \frac{VQ_f}{I}, \tag{9.14}$$

where

V = the maximum shear force

$Q_f = A_f \bar{y}$ (the statical moment of the flange area about the neutral axis)

$\bar{y}$ = the distance from the flange-to-web connection to the neutral axis.

If present*, the shear force, acting in a vertical direction is (see Chapter 7)

$$w = \frac{1.80(W)}{S_w}, \tag{9.15}$$

where

W = the wheel live load, W [with 80% impact (AREMA (2015)]

S_w = wheel load longitudinal distribution (S_w = 915 mm (3 ft) for open deck girders or S_w = 1525 mm (5 ft) for ballasted deck girders).

The resulting force per unit length of weld (from Equations 9.14 and 9.15) is

$$q = \sqrt{q_f^2 + w^2}. \tag{9.16}$$

* For example, vertical loads are transferred through the flange-to-web weld in open and noncomposite deck plate girder spans.

TABLE 9.2
Allowable Weld Stresses

Weld Type	Stress State	Allowable Stress MPa (ksi)
CJP or PJP	Tension or Compression	$0.55F_y$
CJP or PJP	Shear	$0.35F_y$
Fillet (E60XX electrode)	Shear	115 (16.5) but $< 0.35F_y$ on base metal
Fillet (E70XX electrode)	Shear	130 (19.0) but $< 0.35F_y$ on base metal
Fillet (E80XX electrode)	Shear	155 (22.0) but $< 0.35F_y$ on base metal

The required effective area of weld can then be established based on the allowable weld stresses recommended by AREMA (2015) as shown in Table 9.2.

The required effective area of welds subject to horizontal shear from tensile flexure ranges must also be established based on the allowable weld fatigue stresses recommended by AREMA (2015) (typically Category B or B′ depending on weld backing bar usage).

9.3 BOLTED CONNECTIONS

Bolting is the connection of steel components or members by mechanical means. Bolted connections are relatively easy to make and inspect (in contrast to the equipment and skills required for welding and the inspection of welds). The strength of the mechanical connection is significantly affected by the bolt installation process.

9.3.1 Bolting Processes for Steel Railway Superstructures

9.3.1.1 Snug-Tight Bolt Installation

Connection strength depends on the bearing, shear, and tensile strength of the bolts and connected material for bolts installed without pretension (snug-tight). The bolts are made snug-tight (ST) manually with a wrench or power tool applied to the nut. The full effort of a person installing a bolt with a wrench or a nut installed with a power tool until wrench impact will generally provide the small pretension required to retain the nut on the bolt in statically loaded structures. These are bearing-type connections. Bearing-type connections are generally not used in steel railway superstructures due to live load stress reversals and cyclical stresses in main members, and vibration in both main and secondary members.

9.3.1.2 Pretensioned Bolt Installation

Pretensioned (PT) joints are typically required if the connection is subjected to load reversals, fatigue loading without reversal, tensile fatigue loading on ASTM F-3125 Grade A325M (A325) bolts, or tension loading on Grade A490M (A490) bolts. Connections made with pretensioned bolts rely on friction between plates or element surfaces (faying surfaces) for strength. PT bolted joints are made by ST bolt installation followed by increasing the torque applied to the bolt. The applied torque creates tension in the bolt (and corresponding compression of, and friction between, the connection elements).

9.3.1.3 Slip-Critical Bolt Installation

Slip-critical (SC) joints are generally required for shear resistance with fatigue load reversal, where oversize or slotted holes are used in the connection, or where no slip of faying surfaces is tolerable in the superstructure. Therefore, SC connections are used extensively in modern steel railway

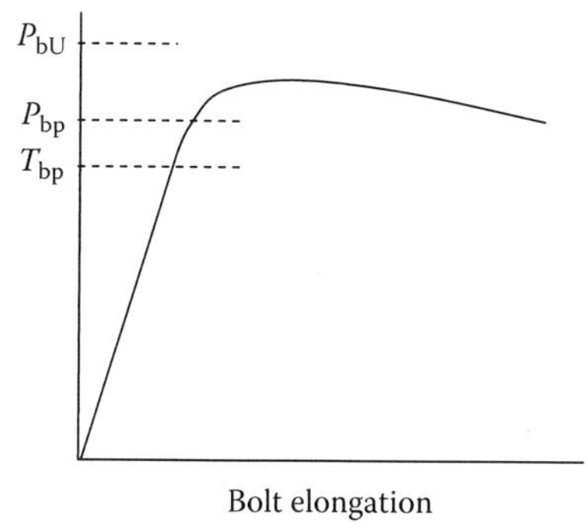

FIGURE 9.9 Bolt tension forces and elongation during application of bolt torque.

superstructures. SC connections also rely on friction between plates or element surfaces (faying surfaces*) for strength. SC bolted joints are made by snug-tight bolt installation followed by increasing the torque applied to the bolt. The applied torque creates tension in the bolt (and corresponding compression of, and friction between, the connection elements). The minimum required bolt pretension, T_{bP}, is

$$T_{bP} \geq 0.70 P_{bU} \geq 0.70 F_{bU} A_{st}, \tag{9.17}$$

where
P_{bU} = the minimum specified tensile strength of the bolt
F_{bU} = the minimum specified tensile stress of the bolt material
A_{st} = the tensile stress area of the bolt = the cross-sectional area through the threaded portion of the bolt.

To attain this, minimum bolt tension AREMA (2015) recommends that nuts be rotated between 1/3 and 1 turn from the ST condition, depending on bolt length and angle of connection plates with respect to the bolt axis†. This will establish a pretension in the bolt, P_{bP}, which is greater than the minimum required bolt pretension, T_{bP}, for the bolt, as shown in Figure 9.9. Alternatively, SC bolted joints may be made using specialized twist-off type bolts or direct tension indicators.

AREMA (2015) provides recommendations for the allowable shear stress‡ in ASTM F3125 Grades A325M (A325) and A490M (A490) bolts in SC connections.

9.3.2 Bolt Types

Fasteners used in modern steel structures are either common or high-strength bolts.

9.3.2.1 Common Steel Bolts

Common§ bolts are specified by ASTM Standard A307. A307 bolts are generally not used in applications involving live load stress reversals, cyclical stresses, and/or vibration. A307 bolts are also not used in steel railway superstructure fabrication due to their low strength.

* Allowable shear stress for bolts depends on the Class of faying surface. Class A, B, C, and D surface slip coefficients for SC connections are described in AREMA (2015).

† Bolts will generally not fail until nut rotation exceeds about 1.75 times from the snug tight condition (Kulak, 2002).

‡ AREMA (2015) recommendations for allowable shear stress on Grade A325M/A325 and A490M/A490 bolts are based on Class A surface with a slip coefficient of 0.33.

§ Also called machine, ordinary, unfinished, or rough bolts.

TABLE 9.3
Minimum Tensile Strength of High-Strength Steel Bolts

Bolt Grade	F_{bU} (MPa)	F_{bU} (ksi)
A325M/A325	830	120
A490M/A490	1040	150
F1852	n/a	120
F2280	n/a	150

9.3.2.2 High-Strength Steel Bolts

High-strength steel bolts are specified by ASTM Standard F3125. This standard includes Grades A325M (A325) and A490M (A490) heavy hex head bolts, and Grades F1852 and F2280 twist-off bolts. The minimum tensile strength, F_{bU}, for Grade A325M (A325), A490M (A490), F1852, and F2280 bolts is shown in Table 9.3. ASTM F3125 type 1 high-strength steel bolts are produced from carbon and alloy steels*. ASTM F3125 type 3 high-strength steel bolts are produced from atmospheric corrosion resistant steel (see Chapter 2).

Equation 9.17 can be used to establish the minimum required bolt pretension, T_{bP}, as shown in Tables 9.4a and b. In order to account for the threaded portion of bolts, an effective bolt area, $A_{st} = 0.75(A_b)$, is used, such that

$$T_{bP} \geq 0.70 F_{bU} A_{st} \geq 0.53 F_{bU} A_b, \tag{9.18}$$

where
A_b = the cross-sectional area of the bolt based on the nominal bolt diameter.

9.3.3 Joint Types

Bolts are used in lap, "T," corner, and butt joints. Bolted lap joints (Figures 9.10a, 9.10b, 9.10c, and 9.10d) are often used in members of steel railway bridges. The joints in Figures 9.10b and 9.10c may be subjected to eccentric loads. Figure 9.10c shows a beam splice arrangement using bolted lap joints. Figure 9.10d shows a type of lap joint used to connect attachments such as stiffeners to girder web plates.

Bolted "T" and corner joints (Figures 9.10e and 9.10f) are rarely used to connect web plates and flange plates of plate and box girder bending members in modern steel railway superstructures. However, "T" and corner joints may be used in the fabrication of built-up axial members.

Bolted butt joints (Figure 9.10g) are typically used join plate ends in a similar fashion to the flange splice joints shown in Figure 9.10c.

9.3.4 Bolted Joint Design

9.3.4.1 Allowable Bolt Stresses

Forces in a connection are transmitted through the effective shear, bearing, and tensile strength of the bolts. Bearing-type (ST) and slip-resistant (PT or SC) connections exhibit different behavior in effective shear, but similar bolt bearing† and tensile behavior‡.

* In some cases with boron added.
† PT or SC connections rely on bolt bearing strength following joint failure.
‡ Applied tension forces readily dissipate pretension forces.

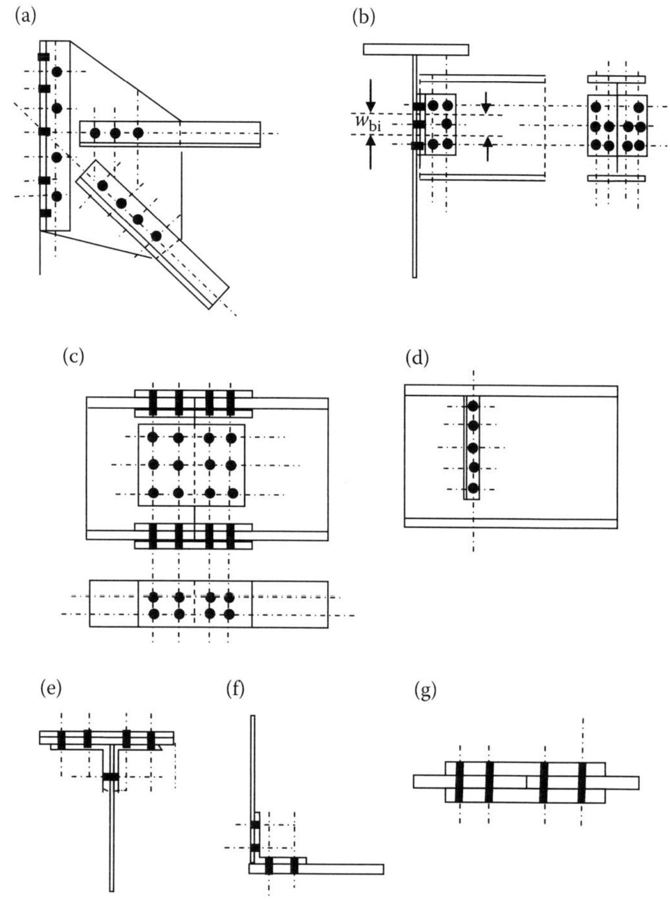

FIGURE 9.10 Typical bolted joints in steel railway superstructures.

9.3.4.1.1 *Allowable Effective Shear Stress*

The allowable effective shear force on bearing-type or ST connections is based on the allowable shear strength of the bolt shanks in the joint. Slip-resistant PT or SC connections have an effective shear strength based on the magnitude of the prestress force and the shear slip coefficient of the steel connection elements. Following the failure of slip-resistant connections, the connection will behave as a bearing-type connection.

9.3.4.1.1.1 Allowable Effective Shear Stress in Bearing-Type Connections Figure 9.11 illustrates that the behavior of a bolt under shear load is inelastic and without a well-defined yield stress. Therefore, bolt strength is determined based on its ultimate shear strength.

Experimentation has shown that the ultimate shear strength, F_{bv}, is in direct proportion to the ultimate tensile strength, P_{bU}, and is not affected by bolt prestress (Kulak et al., 1987). Therefore, the allowable shear stress of a bolt (including a reduction of 0.80 due to the approximation) is

$$f''_{bv} \approx \frac{(0.80)0.62F_{bU}}{FS} \approx \frac{0.50F_{bU}}{FS}. \tag{9.19}$$

The allowable shear stress, f''_{bv}, for Grade A325M (A325) bolts is 210 MPa (30 ksi) (considering the nominal bolt diameter), if it is assumed that FS = 2.0 and F_{bU} = 830 MPa (120 ksi). The allowable

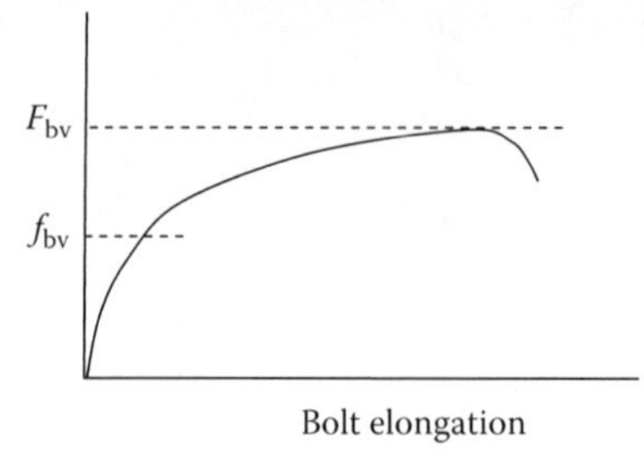

FIGURE 9.11 Bolt shear stress and elongation.

shear stress, f''_{bv}, for Grade A325M (A325) bolts is 0.70(210) = 147 MPa (21 ksi), if shear is assumed through the threaded portion of the bolt.

AREMA (2015) recommends only slip-resistant SC connections and, therefore, does not provide an allowable stress, f''_{bv}, based on shearing of the bolt shank. In slip-resistant connections, service loads are transmitted by friction and bolt shank shearing will not govern the design.

9.3.4.1.1.2 Allowable Effective Shear Stress in Slip-Resistant Connections The shear slip force, P_{bv}, is

$$P_{bv} = mnf'_{bv}(A_b) = k_s m\alpha \sum_{i=1}^{n} T_{bPi}, \tag{9.20}$$

where

f'_{bv} = the effective allowable bolt shear stress
k_s = the shear slip coefficient of the steel connection
m = the number of slip planes (faying surfaces)
n = the number of bolts in the connection
T_{bPi} = the specified pretension in bolt i (Table 9.4)
$\alpha = T_{bi}/T_{bPi}$
T_{bi} = actual pretension in bolt i.

Therefore, the effective allowable shear stress (which is based on the magnitude of the prestress force and the shear slip coefficient) is

TABLE 9.4a
Minimum Required Bolt Pretension for High-Strength Steel Bolts in Slip-Resistant Connections

	Minimum Required Bolt Pretension, T_{bP} (kN)	
Bolt Diameter, d_b (mm)	**Grade A325M Bolts**	**Grade A490M Bolts**
12	50	62
16	88	111
20	138	173
22	167	210
24	199	249
36	448	561

TABLE 9.4b
Minimum Required Bolt Pretension for High-Strength Steel Bolts in Slip-Resistant Connections

Bolt Diameter, d_b (in.)	Minimum Required Bolt Pretension, T_{bP} (kips)	
	Grade A325 Bolts	Grade A490 Bolts
1/2	12	16 (AREMA uses 15)
5/8	19	24
3/4	28	35
7/8	38 (AREMA uses 39)	48 (AREMA uses 49)
1	50 (AREMA uses 51)	62 (AREMA uses 64)
1½	98 (AREMA uses 103)	140 (AREMA uses 148)

$$f'_{bv} = \frac{k_s \alpha \sum_{n=1}^{n} T_{bPi}}{n A_b}. \tag{9.21a}$$

Using FS = 2.0 and A_{st} (0.75(A_b)), Equation 9.21a for Grade A325M (A325) bolts, when the specified pretension in each bolt, T_{bPi}, is equal, is

$$f'_{bv} = \frac{k_s \alpha T_{bPi}}{A_b} = \frac{k_s \alpha \left(0.70 F_{bU} A_{st}\right)}{A_b} = 63\alpha k_s. \tag{9.21b}$$

In tests done to establish an empirical relationship for the effective allowable bolt shear stress, the slip probability level, mean slip coefficient, k_{sm}, and bolt pretension are not explicitly determined. They are combined into a slip factor, D, that incorporates the k_s and k_{sm} relationship, and α of Equation 9.21b as

$$f'_{bv} = \frac{k_s \alpha T_{bPi}}{A_b} = (0.53) D k_{sm} F_{bU} = 63 D k_{sm}. \tag{9.22}$$

AREMA (2015) outlines three SC connection faying surface conditions for design (Table 9.5). Tests done with turn-of-nut and calibrated wrench bolt installations will yield different results for the

TABLE 9.5
Mean Slip Coefficients for Steel Faying Surfaces

Class	Surface Description	Mean Slip Coefficient, k_{sm}
A	Clean mill scale and blast cleaned surface before coating	0.33
B	Blast cleaned surface with or without coating	0.50
C	Galvanized and roughened surfaces	0.40

TABLE 9.6
Slip Coefficient for A325 Bolts with 5% Slip Probability

	Slip Coefficient, D	
Mean Slip Coefficient, k_{sm}	**Turn-of-Nut Installation**	**Calibrated Wrench Installation**
0.33	0.82	0.72
0.50	0.90	0.79
0.40	0.90	0.78

slip factor, D. The "Specification for Structural Joints Using ASTM A325 or A490 Bolts" (RCSC, 2000) provides values of slip factor, D, based on a 5% slip probability* and method of installation as shown in Table 9.6.†

Substitution of the mean slip coefficient k_{sm} (Table 9.5) and slip factor D (Table 9.6) into Equation 9.22 provides the effective allowable shear stress for a 5% slip probability as shown in Tables 9.7a and b. It is usual practice to specify turn-of-nut bolt installation and use $f'_{bv} = 115$ MPa (17.0 ksi) for design of SC connections. This provides an allowable shear force per bolt of 45.4 kN (10.2 kips) for a 22 mm (7/8 in.) diameter bolt with a nominal cross-sectional area of 380 mm^2 (0.60 in.^{2}). The "Specification for Structural Joints Using High Strength Bolts" (RCSC, 2014) incorporates Class C surfaces into Class A and specifies slip coefficients of 0.30 for Class A and 0.50 for Class B surfaces. Therefore, the effective allowable shear stress, f'_{bv}, in the AREMA (2015) recommendations and RCSC (2014) specifications differ as shown in Table 9.8a and b. The RCSC (2014) specifications are based on recent research concerning bolted connection strength and modern shop

TABLE 9.7a
Effective Allowable Shear Stress for A325 Bolts Based on 5% Slip Probability

	Effective Allowable Shear Stress, f'_{bv} (MPa)	
Mean Slip Coefficient, k_{sm}	**Turn-of-Nut Installation**	**Calibrated Wrench Installation**
0.33	117.2	103.4
0.50	195.8 (AREMA uses 193.1)	171.7
0.40	156.5 (AREMA uses 151.7)	135.8

TABLE 9.7b
Effective Allowable Shear Stress for A325 Bolts Based on 5% Slip Probability

	Effective Allowable Shear Stress, f'_{bv} (ksi)	
Mean Slip Coefficient, k_{sm}	**Turn-of-Nut Installation**	**Calibrated Wrench Installation**
0.33	17.0	15.0
0.50	28.4 (AREMA uses 28.0)	24.9
0.40	22.7 (AREMA uses 22.0)	19.7

* A slip probability of 5% (corresponds to a 95% confidence level for the test data) is appropriate for usual steel railway superstructure design.

† RCSC (2000) also provides slip factors for 1% and 10% slip probability.

TABLE 9.8a
Effective Allowable SC Shear Stress for A325 Bolts (AREMA and AISC/RCSC)

SC Bolted Connection Faying Surface Class		Surface Description		Minimum Slip Coefficient		Allowable Shear Stress for Grade A325 Bolts in Standard Holes by Turn-of-Nut Method (MPa)	
AREMA	AISC/RCSC	AREMA	AISC/RCSC	AREMA	AISC/RCSC	AREMA	AISC/RCSC
A	A & C	Clean mill scale and blast cleaned		0.33	0.30	115	90
B	B	Blast cleaned surfaces		0.50	0.50	190	150
C	D	Hot-dip galvanized	Blast cleaned	0.40	0.45	150	135

TABLE 9.8b
Effective Allowable SC Shear Stress for A325 Bolts (AREMA and AISC/RCSC)

SC Bolted Connection Faying Surface Class		Surface Description		Minimum Slip Coefficient		Allowable Shear Stress for Grade A325 Bolts in Standard Holes by Turn-of-Nut Method (ksi)	
AREMA	AISC/RCSC	AREMA	AISC/RCSC	AREMA	AISC/RCSC	AREMA	AISC/RCSC
A	A & C	Clean mill scale and blast cleaned		0.33	0.30	17.0	12.9
B	B	Blast cleaned surfaces		0.50	0.50	28.0	21.5
C	D	Hot-dip galvanized	Blast cleaned	0.40	0.45	22.0	19.3

practice considerations, and it is expected that AREMA may incorporate applicable results of this research into recommended practices. Most modern steel railway superstructure fabrication processes include bolted joint preparation of surfaces with slip coefficients of about 0.50.

9.3.4.1.1.3 Allowable Bearing Stress in Connections Bearing failures are manifested as either yielding due to bearing of the connection elements against the bolt and/or block shearing of the connection elements near an edge.

The ultimate bearing strength, F_B, of the connection element material bearing on the bolt shank is related to the ultimate tensile strength, F_u, by the following linear relationship (Kulak et al., 1987):

$$F_B = \left(\frac{l_e}{d_b}\right) F_U, \tag{9.23}$$

where

l_e = the distance from the centerline of the bolt to the nearest edge in the direction of the force
d_b = the diameter of the bolt.

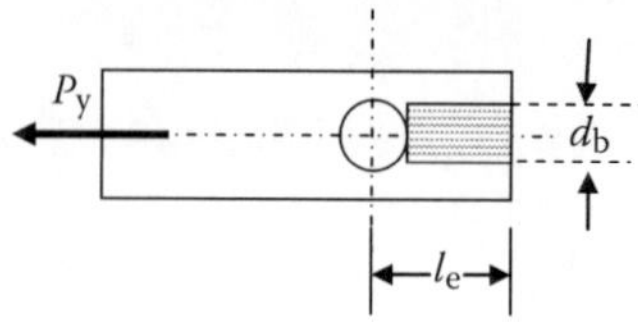

FIGURE 9.12 Shear block failure due to bolt bearing.

Using an FS = 2.5 against bearing on the plate material and the AREMA (2015) recommendation that $l_e \geq 3d_b$ results in an allowable bolt bearing stress, f_B, of

$$f_B = \frac{F_B}{FS} = 1.2F_u. \tag{9.24}$$

The yield strength in pure shear F_v (see Chapter 2) is

$$F_v = \frac{F_y}{\sqrt{3}} \tag{9.25}$$

and the yield strength, P_y, of the shear block failure shown in Figure 9.12 is

$$P_y = 2t_p\left(l_e - \frac{d_b}{2}\right)\frac{F_y}{\sqrt{3}} = 1.15t_p\left(l_e - \frac{d_b}{2}\right)F_y, \tag{9.26}$$

where

t_p = the plate thickness.

The bearing strength of the bolt, F_{bB}, is

$$F_{bB} = f_B d_b t_p, \tag{9.27}$$

which must not exceed the yield strength of the shear block given by Equation 9.26. Therefore,

$$P_y = 1.15t_p\left(l_e - \frac{d_b}{2}\right)F_y \geq F_{bB} = f_B d_b t_p. \tag{9.28a}$$

Rearrangement of Equation 9.28a yields

$$\frac{l_e}{d_b} \geq \frac{0.87f_B}{F_y} + 0.5, \tag{9.28b}$$

which may be conservatively simplified (for $l_e/d_b \geq 1.4$*) to

$$\frac{l_e}{d_b} \geq \frac{f_B}{F_y}. \tag{9.29}$$

Rearrangement of Equation 9.29 and using FS = 2.0† provides the allowable bearing stress as

$$f_B \leq \frac{l_e F_y}{d_b} \leq \frac{l_e F_u}{d_b(FS)} = \frac{l_e F_u}{2d_b}. \tag{9.30}$$

Equations 9.24 and 9.30 are the allowable bearing stresses on bolts recommended by AREMA (2015).

* This is the case in practical structures.

† A lower FS is used due to the conservative nature of the assumptions made to develop Equation 9.29.

9.3.4.1.1.4 Allowable Tension Stress in Connections Figure 9.13 illustrates that the behavior of a bolt under tensile load is elastic for small elongations. The strength of a bolt loaded in direct tension is not affected by pretension stresses from installation by a method that applies the pretension to the bolt by torqueing (Kulak, 2002). This is because the pretension load is readily dissipated as a direct tensile load is applied to a connection. Therefore, bolt strength is determined based on ultimate tensile strength.

The allowable tensile stress, f_{bt}, in a Grade A325M (A325) bolt, using an FS = 2.0 and A_{st}) $0.75(A_b)$, is

$$f_{bt} = \frac{0.75F_{bU}}{FS} = 310\text{ MPa }(45\text{ ksi}). \tag{9.31}$$

However, as shown in Figure 9.14, the bolts in tension connections are subjected to additional tensile forces, T_Q, created by the prying action resulting from flexibility of the connection leg.

From Figure 9.14, the bending moment at the bolt line, M_b, is

$$T(b) = M_f + \left({}^{A_{nb}}\!\!\diagup_{A_{gf}}\right) M_b = M_f + \left({}^{A_{nb}}\!\!\diagup_{A_{gf}}\right) \alpha M_f = M_f\,(1+\alpha\eta) \tag{9.32}$$

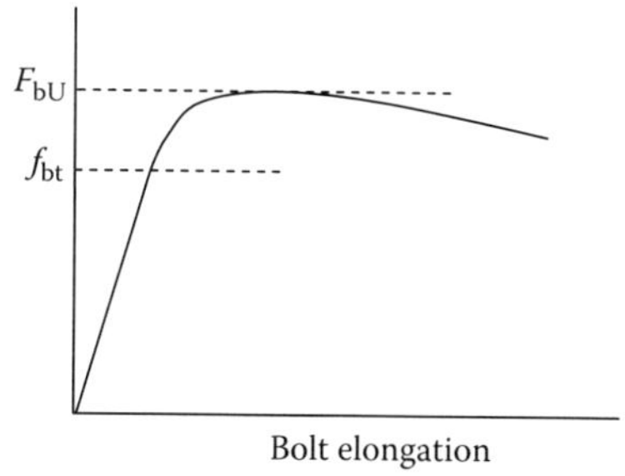

FIGURE 9.13 Bolt direct tension forces and elongation.

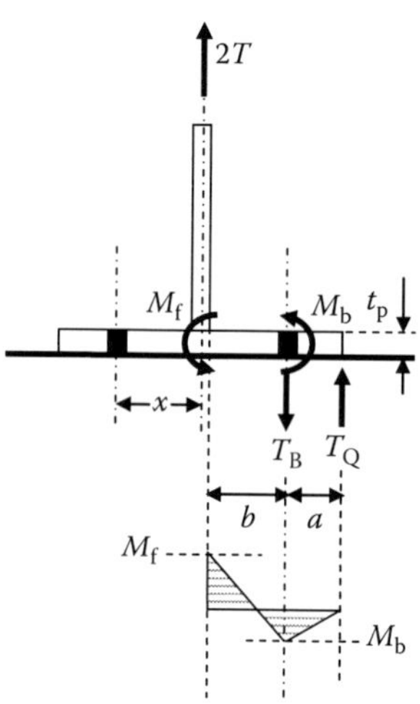

FIGURE 9.14 Prying action on bolted tension joint.

and with $M_f = \frac{w_b t_p^2}{4} F_y$

$$T_Q(a) = M_b = \alpha\eta M_f = \frac{\alpha\eta w_b t_p^2 F_y}{4}, \tag{9.33}$$

where

T = the applied tensile force per bolt
T_Q = the prying tensile force per bolt
$T_B = T + T_Q$ = the total tensile force per bolt
$\eta = A_{nb}/A_{gf}$
A_{nb} = the net area of the flange at the bolt line
A_{gf} = the gross area of the flange at the intersection with the web plate
$\alpha = M_b/M_f$ and depends on T_Q/T
w_{bi} = tributary area for prying of each bolt i (see Figure 9.10b).

The bolt tension, $T_B = T + T_Q$, with substitution of Equations 9.32 and 9.33, is

$$T_B = T + T_Q = \frac{M_f(1+\eta\alpha)}{b} + \frac{\alpha\eta w_b t_p^2 F_y}{4a} = \frac{M_f(1+\eta\alpha)}{b} + \frac{\alpha\eta M_f}{a}$$

$$= T\left(1 + \frac{\alpha\eta b}{(1+\alpha\eta)a}\right). \tag{9.34}$$

Therefore,

$$T_Q = T\left(\frac{\alpha\eta b}{(1+\alpha\eta)a}\right). \tag{9.35}$$

Further manipulation of these equations provides the thickness, t_p, as

$$t_p \le \sqrt{\frac{4T_B ab}{w_b F_y (a + \alpha\eta(a+b))}}. \tag{9.36}$$

However, Equations 9.34, 9.35, and 9.36 are difficult to use in routine design. Analytical and experimental studies have provided a semiempirical equation (in US Customary and Imperial units) as (Douty and McGuire, 1965)

$$T_Q = T\left(\frac{0.50 - \left(w_b t_p^4 / 30ab^2 A_b\right)}{a/b((a/3b)+1) + w_b t_p^4 / 6ab^2 A_b}\right). \tag{9.37}$$

Equation 9.37 may be further simplified as (Kulak, Fisher and Struik, 1987)

$$\text{in SI units (mm)}: T_Q = T\left(\frac{3b}{8a} - \frac{t_p^3}{328 \times 10^3}\right) \tag{9.38a}$$

$$\text{and in US Customary and Imperial units (in.)}: T_Q = T\left(\frac{3b}{8a} - \frac{t_p^3}{20}\right). \tag{9.38b}$$

Further analytical and empirical studies (Nair et al., 1974) have provided other empirical equations for the prying force, but Equation 9.38 is simple and conservative for use in routine design of bolted connections subjected to tension.

Connections with bolts subjected to direct tension should generally be avoided in the main members of steel railway superstructures. Bolt tension and prying may occur combined with shear in connections such as those shown in Figures 9.10a and 9.10b. AREMA (2015) recommends the allowable tensile stress on fasteners, including the effects of prying, as 300 MPa (44 ksi) and 370 MPa (54 ksi) for Grades A325M (A325) and A490M (A490) bolts, respectively.

9.3.4.1.1.5 Allowable Combined Tension and Shear Stress in Connections Connections in steel railway superstructures may be subjected to combined shear and tension forces (e.g., the beam connection of Figure 9.10b). An ultimate strength interaction equation developed from tests (Chesson, Faustino, and Munse, 1965) is shown in Figure 9.15 (solid line) and Equation 9.39

$$\left(\frac{F_{bt}}{F_{bU}}\right)^2 + \left(\frac{F_{bv}}{F_{vU}}\right)^2 = 1.0 \tag{9.39}$$

where

F_{bt} = the ultimate tensile stress under combined shear and tension
F_{bU} = the ultimate tensile stress under tension only
F_{bv} = the ultimate shear stress under combined shear and tension
F_{vU} = the ultimate shear stress under shear only = 0.62 F_{bU} (see Equation 9.19).

Therefore, Equation 9.39 may be expressed as

$$\left(\frac{F_{bt}}{F_{bU}}\right)^2 + 2.60\left(\frac{F_{bv}}{F_{bU}}\right)^2 = 1.0 \tag{9.40}$$

Equation 9.39, in terms of allowable stresses, is

$$\left(\frac{\sigma_{bt}}{f_{bt}}\right)^2 + \left(\frac{\tau_{bv}}{f_{bv}}\right)^2 = 1.0, \tag{9.41}$$

where

σ_{bt} = the tensile stress in the bolt (including prying action effects)
f_{bt} = the allowable bolt tensile stress for bearing-type connections, or

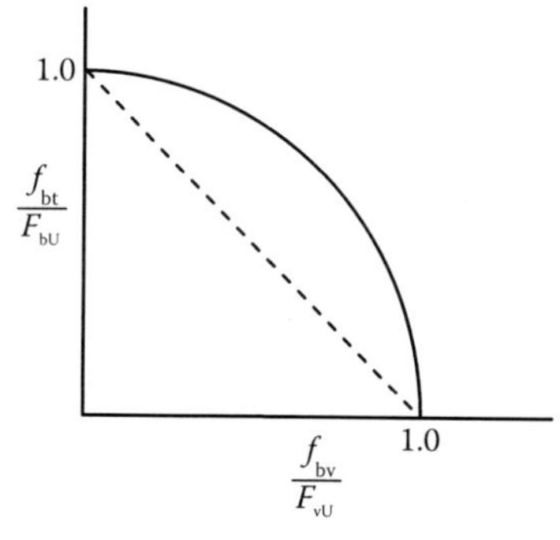

FIGURE 9.15 Bolt shear and tension interaction.

= the nominal tensile stress from pretension for SC connections = T_{bP}/A_b

τ_{bv} = the shear stress in bolt

f_{bv} = the allowable effective bolt shear stress = f'_{bv} for SC connections (Equation 9.22), or = f''_{bv} for bearing-type connections (Equation 9.19)

T_{bP} = the bolt pretension (see Equation 9.17 and Table 9.4)

A_b = the nominal area of the bolt ~ 1.33 A_{st}

A_{st} = the effective bolt area through the threaded portion of the bolt.

Therefore, for slip-resistant connections

$$\left(\frac{\sigma_{bt}}{(T_{bP}/A_b)}\right)^2 + \left(\frac{\tau_{bv}}{f'_{bv}}\right)^2 = 1.0 \tag{9.42}$$

The elliptical Equation 9.42 may be simplified by a straight-line approximation (dashed line for ultimate stress values in Figure 9.15) as

$$\left(\frac{\sigma_{bt}}{(T_{bP}/A_b)}\right) + \left(\frac{\tau_{bv}}{f'_{bv}}\right) = 1.0 \tag{9.43}$$

and may be rearranged as

$$\tau_{bv} \le f'_{bv}\left(1 - \frac{\sigma_{bt}}{(T_{bP}/A_b)}\right). \tag{9.44}$$

If τ_{bv} is taken as the allowable shear stress, f_{bv}, when combined with tension, Equation 9.44 becomes

$$f_{bv} \le f'_{bv}\left(1 - \frac{\sigma_{bt}}{(T_{bP}/A_b)}\right), \tag{9.45}$$

which is the allowable shear stress for combined shear and tension recommended by AREMA (2015).

9.3.4.1.1.6 Allowable Fatigue Stresses in Bolted Connections The allowable fatigue stress of bolted joints depends on whether the bolts are loaded primarily in shear, such as in lap and butt joints (Figures 9.10a, 9.10b, 9.10c, and 9.10g) or tension. AREMA (2015) recommends that all joints subject to fatigue by cyclical stresses must be slip-resistant SC connections.

9.3.4.1.1.6.1 Allowable Shear Fatigue Stress in Bolted Connections Bearing-type connections are subject to fatigue damage accumulation and crack initiation at the edge of, or within, holes due to localized tensile stress concentrations. Slip-resistant connections are subject to fretting fatigue.

AREMA (2015) recommends the allowable stress range, based on Fatigue Detail Category B, of 125 MPa (18 ksi) for the base metal of slip-critical connections subjected to 2 million cycles or less; or 110 MPa (16 ksi) for the base metal of slip-critical connections subjected to greater than 2 million stress range cycles (see Chapter 5). Bolts will generally not experience fatigue failure prior to the base metal and, therefore, AREMA (2015) includes no recommendations concerning allowable shear stress ranges for bolts.

9.3.4.1.1.6.2 Allowable Tensile Fatigue Stress in Bolted Connections The stress range in a bolt of a SC connection is affected by the pretension applied to the bolt and the rigidity of the connection joint.

The stress range is typically considerably less than the nominal tensile stress in the bolt in relatively rigid SC connections with small prying forces. The prying force should be limited to a maximum of 20% of the external load on the bolt for connections subjected to cyclical stresses in steel railway superstructures. AREMA (2015) recommends an allowable tensile stress range of 215 MPa (31.0 ksi) for Grade A325M (A325)* bolts where prying forces do not exceed 20% of the external load on the bolt.

The AREMA (2015) criteria are appropriate given the need to discourage the use of bolted connections subject to direct cyclical tensile stresses in structures such as steel railway superstructures.

9.3.4.2 Axially Loaded Members with Bolts in Shear

These are typically truss member end connections with forces transferred from the axial members through connections consisting of bolts in shear and gusset plates (Figure 9.10a). Axial tension member end connections and gusset plates must be designed considering yield, fracture, and block shear (tear-out) failure modes. Axial compression member end connection gusset plates must be designed considering buckling and block shear.

9.3.4.2.1 Axial Member End Connection

The number of bolts required in the axial force connection may be determined by considering the allowable bolt shear stress, f'_{bv}, and, if cyclically loaded, the allowable bolt shear fatigue stress [110 MPa (16 ksi) or 125 MPa (18 ksi) for SC connections depending on the number of design stress range cycles]. The allowable bearing stress, f_B, should be used for bearing-type connections. Bearing stress may also be considered for SC connections as a precaution following slip failure. Determination of the number of bolts required to transfer forces from the member in the connection will determine the net section area required for the member and the basic dimensions of the connection gusset plates.

The design of axial members for strength (yield and fracture), fatigue, stability, and serviceability is discussed in Chapter 6. The effects of the connection in terms of net area and shear lag effects were considered in the design criteria for axial tension members. However, the design of axial member end connections requires attention to the localized effects of bolt bearing stresses (in regard to member element thickness) and fracture or rupture by block shear.

Axial member bolted connections should conform to the recommended minimum bolt spacing and edge distance criteria of AREMA (2015). The minimum bolt spacing (center-to-center bolts) is three times the bolt diameter. The minimum bolt spacing, s_b, along a line of force, based on bearing considerations, from Equation 9.30 is

$$s_b \geq \frac{2d_b\sigma_{bc}}{F_u} + \frac{d_b}{2}, \tag{9.46}$$

where

σ_{bc} = the bearing stress on the connection element of area $d_b t_p$

t_p = the thickness of connection element.

The recommended minimum edge distance, e_b, is $40 + 4t_p \times 150$ mm ($1.5 + 4t_p \times 6$ in.) and, based on bearing stress considerations,

$$e_b \geq \frac{2d_b\sigma_{bc}}{F_u}. \tag{9.47}$$

* AREMA (2015) recommends an allowable tensile stress range of 260 MPa (38.0 ksi) for A490M (A490) bolts.

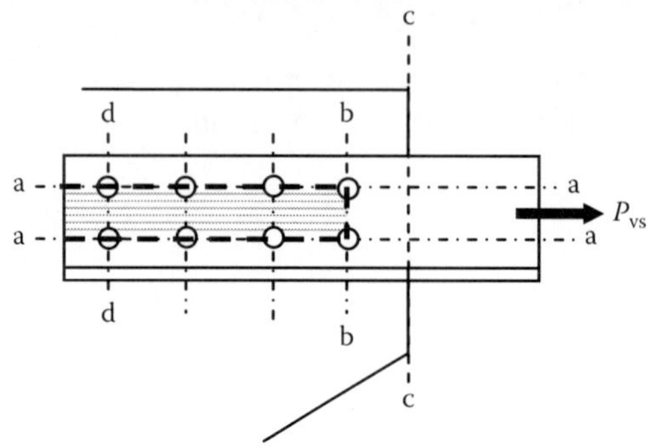

FIGURE 9.16 Block shear failure in member at axial tension connection.

Block shear failure occurs in members at ultimate shear stress on planes along the bolt lines (lines a–a in Figure 9.16) and ultimate tensile stress on planes between bolt lines (line b–b in Figure 9.16). The failure ultimately results in the tear-out of a section (shaded area of the member in Figure 9.16).

If the allowable shear stress is limited to F_{vU}) $0.60F_u$/FS (see Equation 9.19) and the allowable tensile stresses is F_u/FS, the allowable block shear strength, P_{vs}, using FS = 2.0, is

$$P_{vs} = 0.30F_UA_{nv} + 0.50F_uA_{nt}. \tag{9.48}$$

However, a combination of yielding on one plane and fracture on the other plane is likely depending on the connection configuration. When the net fracture strength in tension is greater than the net fracture strength in shear, $F_uA_{nt} \geq F_{vU}A_{nv} \geq 0.60F_uA_{nv}$, yielding will occur on the gross shear plane and the allowable block shear strength, P_{vs}, is

$$P_{vs} = 0.35F_yA_{gv} + 0.50F_uA_{nt}. \tag{9.49a}$$

Conversely, when the net fracture strength in tension is less than the net fracture strength in shear, $F_uA_{nt} < F_{vU}A_{nv} < 0.60F_uA_{nv}$, yielding will occur on the gross tension plane and the allowable block shear strength, P_{vs}, is

$$P_{vs} = 0.30F_UA_{nv} + 0.55F_yA_{gt}, \tag{9.49b}$$

where

F_u = the ultimate tensile stress of a connection element
F_y = the tensile yield stress of a connection element
A_{gv} = gross area subject to shear stress (thickness times gross length along lines a–a in Figure 9.16)
A_{nv} = the net area subject to shear stress (the thickness times the net length along lines a–a in Figure 9.16)
A_{gt} = the gross area subject to tension stress (the thickness times the gross length along lines b–b in Figure 9.16)
A_{nt} = the net area subject to tension stress (the thickness times the net length along line b–b in Figure 9.16).

AREMA (2015) recommends determination of the allowable block shear strength using Equations 9.48 and either Equations 9.49a or 9.49b, depending on whether the net fracture strength in tension is greater or less than the net fracture strength in shear.

Connection shear lag effects that require consideration for axial tension member design are considered in Chapter 6. Shear lag is taken into account for design by determination of a reduced cross-sectional area or effective net area, A_e, which is based on the connection efficiency.

9.3.4.2.2 Gusset Plates

In general, gusset plates should be designed to be as compact as possible. This not only reduces material consumption, but also reduces slenderness ratios and free edge distances for greater buckling strength. Gusset plates have been traditionally designed using beam theory to determine axial, bending, and shear stresses at various critical sections of the gusset plate. However, the slender beam model is not an accurate model* and considering the limit states of block shear (tear-out) and axial stress, based on an appropriate area, it is used for the design of ordinary gusset plates.

Block shear in a gusset plate is analogous to the situation shown in Figure 9.16 but with the tear-out section extending from the edge of the gusset plate (line c–c in Figure 9.16) to the furthest line of bolts (line d–d in Figure 9.16). Equation 9.48 and Equations 9.49a or b are also used to determine the allowable block shear strength of the gusset plate at each member end connection.

The axial stress in the gusset plate is required for comparison to allowable tensile and compressive axial stresses. Testing has shown that an effective length, w_e, perpendicular to the last bolt line (line b–b in Figure 9.17), on which axial stresses act, may bc bascd on lines 30° to the bolt row lines (lines a–a in Figure 9.17) from the first perpendicular bolt line (line c–c in Figure 9.17) (Whitmore, 1952).

The Whitmore effective length, w_e, is

$$w_e = 2l_c \tan(30^\circ) + s_r = 1.15l_c + s_r. \tag{9.50}$$

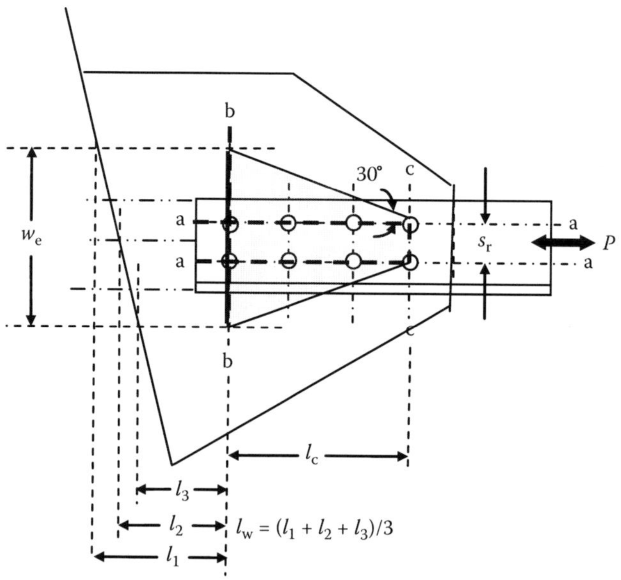

FIGURE 9.17 Whitmore stress block in gusset plate at axial member connection.

* For example, tests show that shear stresses in gusset plates are closer to V/A than 1.5 V/A as predicted by beam theory.

The effective length, w_e, must be reduced if it intersects other members or contains elements with different strengths (e.g., the gusset plate of Figure 9.18a with $\bar{w}_e$ as shown in Figure 9.18b).

The axial tensile design of the gusset plate is then based on

$$\sigma_{at} = \frac{P}{w_e t_p} \leq 0.55 F_y \tag{9.51a}$$

or

$$\sigma_{at} = \frac{P}{w_{ne} t_p} \leq 0.47 F_U. \tag{9.51b}$$

The limit state of yielding on the gross section, $w_e t_p$, immediately below the last line of bolts is represented by Equation 9.51a* and that of tensile failure on the net section, $w_{ne} t_p$, through the last line of bolts by Equation 9.51b. Compressive design (considering stability) is based on

$$\sigma_{ac} = \frac{P}{w_e t_p} \leq F_{call}, \tag{9.52}$$

where

w_{ne} = the net effective length

σ_{at} = the axial tension stress on the effective area, $w_e t_p$ or $w_{ne} t_p$

σ_{ac} = the axial compression stress on the effective area, $w_e t_p$

F_{call} = the allowable axial compression stress based on the effective slenderness ratio, Kl_w/r_w (many engineers restrict $Kl_w/r_w \leq 100$ to 120)

Kl_w = the effective buckling length of the gusset plate

l_w = the average distance, $(l_1 + l_2 + l_3)/3$, from the last line of bolts (line b–b in Figure 9.17 and lines d–d in Figure 9.18b) to edge of the gusset plate. l_w may extend to the first row of bolts at an interface member such as a bottom chord element (line f–f in Figure 9.18b).

$$r_w = t_p/\sqrt{12}$$

t_p = the thickness of the gusset plate

K = the effective length factor, typically taken as between 0.50 and 0.65 for properly braced gusset plates (Thornton and Kane, 1999)†.

The use of block shear rupture and the Whitmore section analysis may be sufficient for the design of ordinary gusset plates (Figure 9.18b). However, for heavily loaded railway truss members it is often appropriate to also check beam theory‡ shear forces, bending moments and axial forces at critical sections (e.g., lines f–f, g–g, h–h, and k–k in Figure 9.18c).

The critical sections, such as lines g–g and h–h in Figure 9.18c, should be reviewed for combined stresses (see Chapter 8) from the following:

- Shear yielding on the gross section of the gusset plate from the resultant horizontal V (on section g–g) and vertical P (on section h–h) forces in the members

* Equation 9.51a is slightly conservative as the Whitmore section is taken through the center of the last line of bolts.

† If gusset plates are not braced against lateral movement, K may be greater than 1 (see Chapter 6).

‡ An alternative to slender beam theory, the uniform force method, which is strongly dependent on connection geometry, has been used for building design (Thornton and Kane, 1999).

- Axial tension or compression on the gross section of the gusset plate from the resultant horizontal V (on section h–h) and vertical P (on section g–g) forces in the members
- Bending moments, for example, $M = \pm V(d_e) \pm P(e_c)$ at section g–g in Figure 9.18c.

Critical sections such as lines f–f and k–k in Figure 9.18c should be reviewed for combined stresses from:

- Shear fracture on the net section of the gusset plate from the resultant horizontal V (on section f–f) and vertical P (on section k–k) forces in the members

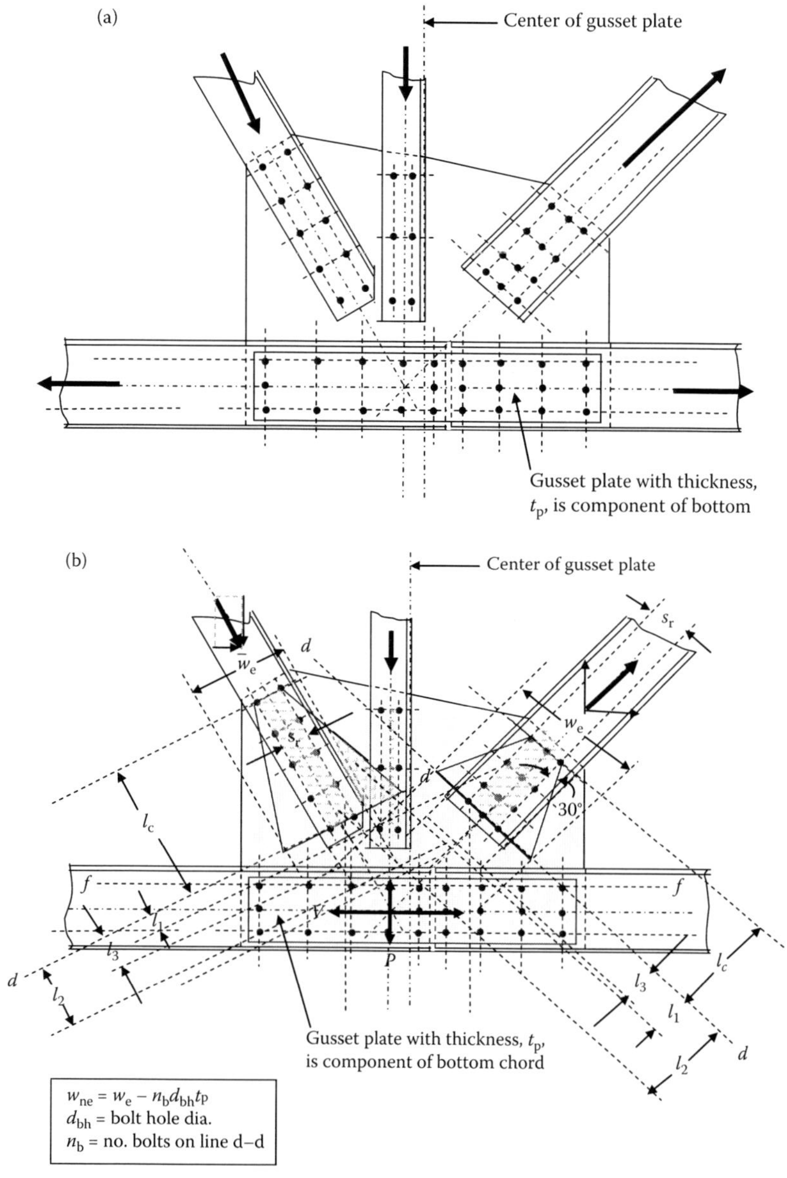

FIGURE 9.18 (a) Typical truss gusset plate connection arrangement, (b) typical truss gusset plate connection block shear and axial stress (on Whitmore section).

(Continued)

(c)

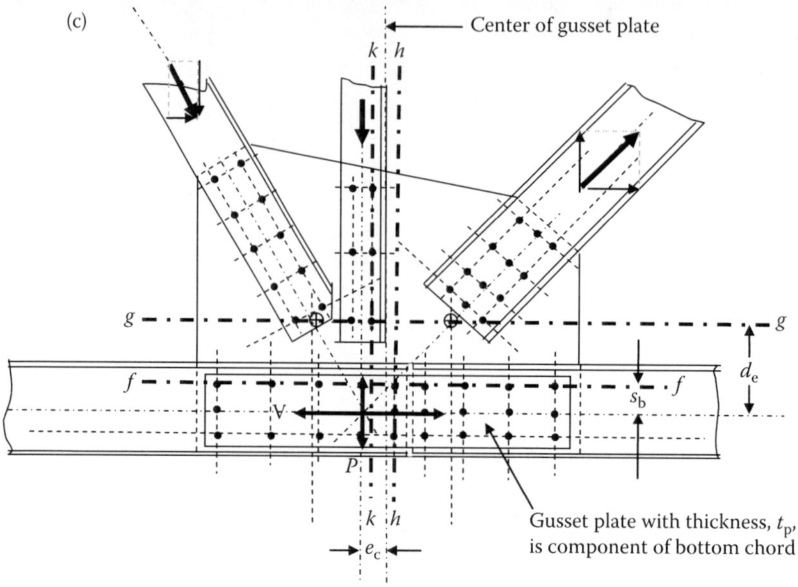

(d)

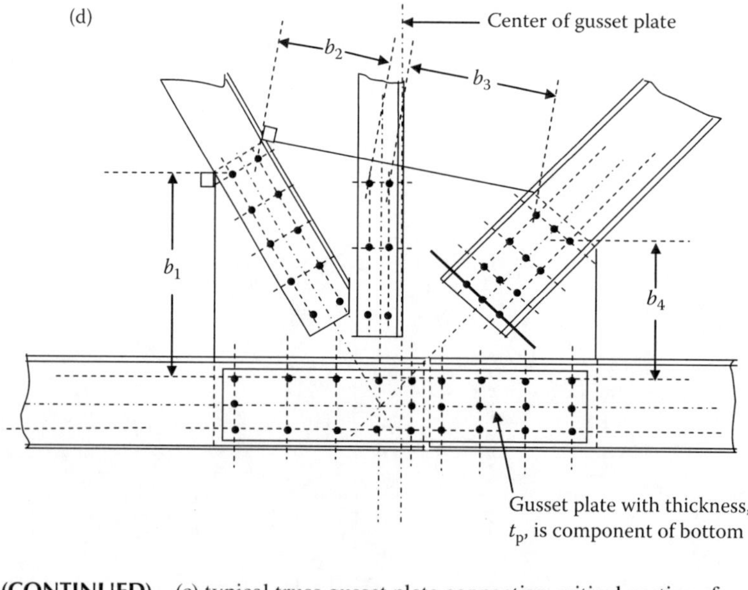

FIGURE 9.18 (CONTINUED) (c) typical truss gusset plate connection critical sections for axial, shear and bending (beam theory) and (d) typical truss gusset plate connection free edge lengths for local buckling.

- Axial tension on the net section or compression on the gross section of the gusset plate from the resultant horizontal V (on section k–k) and vertical P (on section f–f) forces in the members
- Bending moments on sections f–f and k–k.

Depending on connection geometry, other critical sections and interfaces may require review for combined stresses due to slender beam theory forces.

For gusset plates in very complex connections or in long span trusses, the detailed analysis of gusset plate connections by finite element analysis is often warranted.

In addition, free edge lengths on the gusset plate should be minimized to preclude localized buckling effects. Many engineers restrict b_i/t_p ratios ($i = 1, 2, 3, 4$ in Figure 9.18d) to less than

$2.06\sqrt{E/F_y}$. Edge stiffening angles should be used when the free edge distance is large and should be proportioned such that $b_i/r_w \leq 120$.

Example 9.4 outlines the design of an axially loaded bolted connection using block shear and Whitmore stress block analyses.

Example 9.4a (SI Units)

Design the slip-resistant bolted connection for the wind bracing member shown in Figure E9.4a. The steel is Grade 350 (F_y = 350 MPa and F_U = 450 MPa) and 22 mm diameter A325M high-strength steel bolts are used in the connection for this 3 m long member.

$$T = C = 200 \text{ kN}$$

f_{bv} = 115 MPa (slip-resistant connection allowable bolt shear).

Secondary and bracing member connections must be designed for the lesser of the allowable strength of the member or 150% of the calculated maximum forces in the member.

T' = 1.5(200) = 300 kN.

Allowable tensile strength of member (see Chapter 6):

Shear lag effect, $U = 1 - \left(\frac{\bar{x}}{L}\right) = 1 - \left(\frac{43.4}{225}\right) = 0.81$. However, AREMA recommends use of 0.60 for single angle connections.

Effective net area = A_{ne} = 0.60(4540–2(25)(16)) = 2244 mm²

Allowable strength = 0.55(350)(2244)/1000 = 432 kN

Design connection for 300 kN axial tension.

Member:

Number of bolts in single shear (single shear plane or faying surface):

$$n = \frac{300(1000)}{115(\pi d_b^2/4)} = \frac{300(1000)}{115(380)} = \frac{300(1000)}{43715} = 6.9 \text{ use min. 8 bolts.}$$

Check bearing stress

$$\sigma_{bc} = \frac{300(1000)}{8(12)(25)} = 125 \text{ MPa for min. 12 mm thick gusset plate}$$

$$f_B \leq \frac{l_e F_u}{2d_b} \leq \frac{(90)(450)}{2(22)} \leq 920 \text{ MPa}$$

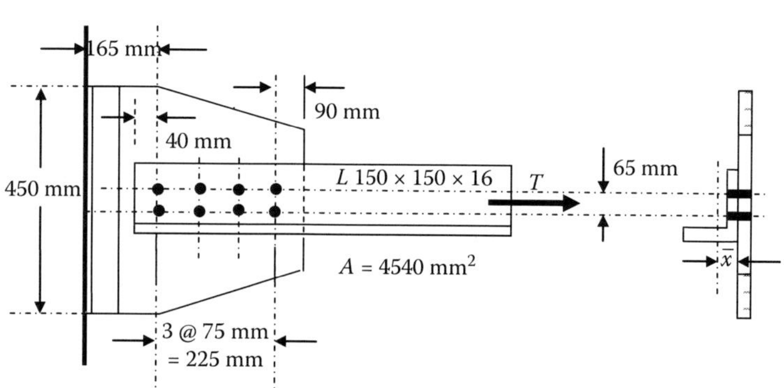

FIGURE E9.4a

or

$$f_B \le 1.2F_u \le 1.2(450) \le 540 \text{ MPa, OK.}$$

Tensile stress in member:

$$\bar{x} = 43.4 \text{ mm}$$

$$A_e = 2244 \text{ mm}^2.$$

$$\text{Tensile axial stress in the angle net section} = \sigma_a = \frac{300(1000)}{2244}$$

$$= 134 \text{ MPa} \le 0.47(450) \le 212 \text{ MPa, OK}$$

$$\text{Tensile axial stress in the angle gross section} = \sigma_a = \frac{300(1000)}{4540}$$

$$= 66.1 \text{ MPa} \le 0.55(350) \le 192.5 \text{ MPa, OK.}$$

Block shear failure in angle:

$A_{gt} = 65(16) = 1040 \text{ mm}^2$
$A_{nt} = 1040 - 16(25) = 640 \text{ mm}^2$
$A_{gv} = 2(225 + 40)(16) = 8480 \text{ in.}^2$
$A_{nv} = 8480 - 8(16)(25) = 5280 \text{ mm}^2$

$$P_{vs} = 0.30F_U A_{nv} + 0.50F_U A_{nt} = 0.30(450)(5280)/1000 + 0.50(450)(640)/1000$$

$$= 712.8 + 144.0 = 856.8 kN$$

Tensile ultimate strength $= F_u A_{nt} = (450)(640)/1000 = 288.0$ kN
Shear ultimate strength $= 0.60F_u A_{nv} = 0.60(450)(5280)/1000 = 1426$ kN.

Therefore, tensile yielding on the gross section and shear fracture on the net section is appropriate to consider.

$$P_{vs} = 0.30F_U A_{nv} + 0.55F_y A_{gt} = 0.30(450)(5280)/1000 + 0.55(350)(1040)/1000$$

$$= 712.8 + 200.2 = 913.0 \text{ kN.}$$

The allowable block shear is 856.8 kN ≥ 300 kN, OK.

The member design is governed by the tensile fracture criterion due to the considerable shear lag effect associated with the single angle connection used for this secondary member.

Gusset plate:

Block shear failure in gusset plate:

$A_{gt} = 65(12) = 780 \text{ mm}^2$
$A_{nt} = 780 - 12(25) = 480 \text{ mm}^2$
$A_{gv} = 2(225 + 90)(12) = 7560 \text{ mm}^2$
$A_{nv} = 7560 - 8(12)(25) = 5160 \text{ mm}^2$

$$P_{vs} = 0.30F_U A_{nv} + 0.50F_U A_{nt} = 0.30(450)(5160)/1000 + 0.50(450)(480)/1000$$

$$= 696.6 + 108.0 = 804.6 \text{kN}$$

Tensile ultimate strength $= F_u A_{nt} = (450)(480)/1000 = 216.0$ kN

Shear ultimate strength $= 0.60 F_u A_{nv} = 0.60(450)(5160)/1000 = 1393$ kN.

Therefore, tensile yielding on the gross section and shear fracture on the net section is appropriate to consider.

$$P_{vs} = 0.30 F_U A_{nv} + 0.55 F_y A_{gt} = 0.30(450)(5160)/1000 + 0.55(350)(780)/1000$$

$$= 696.6 + 150.2 = 846.8\,\text{kN}.$$

The allowable block shear is 804.6 kN ≥ 300 kN, OK.

Axial tension in gusset plate

$$w_e = 2l_c \tan(30^\circ) + s_r = 1.15(225) + 65 = 324.8 \text{ mm}$$

$$\sigma_{at} = \frac{300(1000)}{(324.8)(12)} = 77.0 \le 0.55 F_y \le 192.5 \text{ MPa, OK}$$

or

$$\sigma_{at} = \frac{300(1000)}{(324.8 - 2(25))(12)} = 91.0 \le 0.47 F_U \le 212 \text{ MPa, OK.}$$

Axial compression in gusset plate:

$$C = -200 \text{ kN}$$

$f_{bv} = 115$ MPa (slip-resistant connection allowable bolt shear)

$C' = 1.5(-200) = -300$ kN.

Allowable compressive strength of member (see Chapter 6):

$r_{min} = r_{xy} = 29.4$ mm

$$\frac{kL}{r_{min}} = \frac{0.75(3000)}{29.4} = 76.5$$

$$F_{call} = 0.60(350) - 1.171(76.5) = 120.4 \text{ MPa.}$$

Allowable strength = (120.4)(4540)/1000 = 546.5 kN compression

Design connection for 300 kN axial compression.

Compressive stress in gusset plate:

$$\sigma_{ac} = \frac{C'}{w_e t_p} = \frac{300(1000)}{324.8(12)} = 77.0 \text{ MPa}$$

$$\frac{Kl_w}{r_w} = \frac{0.65(165)\sqrt{12}}{12} = 31.0 \ge 0.629\sqrt{\frac{E}{F_y}} \ge 17.8$$

$$F_c = 0.60 F_y - \left(\frac{635 F_Y}{E}\right)^{3/2}\left(\frac{Kl_w}{r_w}\right) = 0.60 F_y - 1.171\left(\frac{Kl_w}{r_w}\right) = 210 - 1.171(31.0)$$

$$= 173.7 \text{ MPa} \ \ge\ 77.0 \text{ MPa, OK.}$$

If the connection shown in Figure E9.4a has a compression diagonal creating a vertical force of 150 kN and horizontal force of 125 kN in addition to the 200 kN tensile force:

$P = T = 200 - 125 = 75$ kN

$V = 150$ kN

$M = 150(165) = 24{,}750$ kNmm

$A_g = 450(12) = 5400\,\text{mm}^2$

$A_n = 5400 - 2(25)(12) = 4800\,\text{mm}^2$

$S_n = 12(450)^2/6 - 2(25)(12)(32.5) = 385.5 \times 10^3\,\text{mm}^3$

$t_v = 150(1000)/5400 = 27.8$ MPa (1.5V/A not used since slender beam theory not theoretically valid)

$\sigma_a = 125(1000)/4800 = 26.0$ MPa

$\sigma_b = 24{,}750/385.5 = 64.2$ MPa.

Use a linear interaction formula to examine combined stress effects;

$$\frac{27.8}{0.35(350)} + \frac{26.0}{0.55(350)} + \frac{64.2}{0.55(350)} = 0.23 + 0.14 + 0.33 = 0.70 \leq 1.00 \text{ OK}$$

Example 9.4b (US Customary and Imperial Units)

Design the slip-resistant bolted connection for the wind bracing member shown in Figure E9.4b. The steel is Grade 50 ($F_y = 50$ ksi and $F_U = 65$ ksi) and 7/8 in. diameter A325 high-strength steel bolts are used in the connection for this 7 ft long member.

$$T = C = 50 \text{ kips}$$

$f_{bv} = 17$ ksi (slip-resistant connection allowable bolt shear).

Secondary and bracing member connections must be designed for the lesser of the allowable strength of the member or 150% of the calculated maximum forces in the member.

$T' = 1.5(50) = 75$ kips.

Allowable tensile strength of member (see Chapter 6):

Shear lag effect, $U = 1 - \left(\frac{\bar{x}}{L}\right) = 1 - \left(\frac{1.68}{9}\right) = 0.81$. However, AREMA recommends use of 0.60 for single angle connections.

Effective net area $= A_{ne} = 0.60(5.75 - 2(1)(0.5)) = 2.85$ in.^{2}

Allowable strength $= 0.55(50)(2.85) = 78.4$ kips

Design connection for 75 kips axial tension.

Member:

Number of bolts in single shear (single shear plane or faying surface):

$$n = \frac{75}{17\left(\pi d_b^2/4\right)} = \frac{75}{17(0.60)} = \frac{75}{10.2} = 7.4 \text{ use min. 8 bolts.}$$

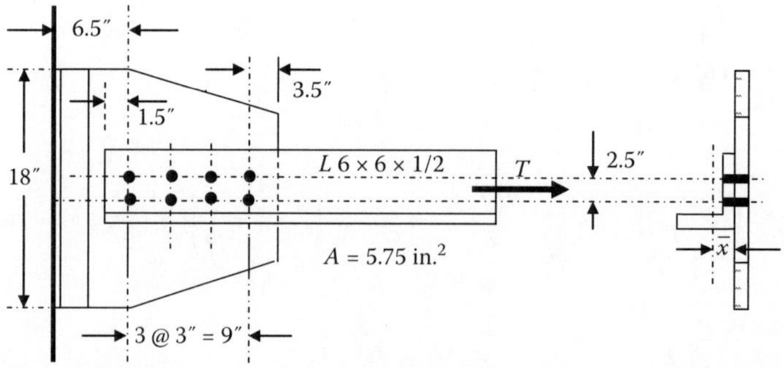

FIGURE E9.4b

Check bearing stress

$$\sigma_{bc} = \frac{75}{8(0.375)(1.00)} = 25.0 \text{ ksi for min. } 3/8'' \text{ thick gusset plate}$$

$$f_B \leq \frac{l_e F_u}{2d_b} \leq \frac{(3.5)(65)}{2(7/8)} \leq 130 \text{ ksi}$$

or

$$f_B \leq 1.2F_u \leq 1.2(65) \leq 78 \text{ ksi, OK.}$$

Tensile stress in member:

$$\bar{x} = 1.68 \text{ in}$$

$$A_e = 2.85 \text{ in.}^2$$

$$\text{Tensile axial stress in the angle net section} = \sigma_a = \frac{75}{2.85} = 26.3 \text{ ksi} \leq 0.47(65) \leq 30.6 \text{ ksi, OK}$$

$$\text{Tensile axial stress in the angle gross section} = \sigma_a = \frac{75}{5.75} = 13.0 \text{ ksi} \leq 0.55(50) \leq 27.5 \text{ ksi, OK.}$$

Block shear failure in angle:

$A_{gt} = 2.5(0.5) = 1.25$ in.^{2}
$A_{nt} = 1.25 - 0.5(1) = 0.75$ in.^{2}
$A_{gv} = 2(9 + 1.5)(0.5) = 10.50$ in.^{2}
$A_{nv} = 10.50 - 8(0.5)(1.0) = 6.50$ in.^{2}

$$P_{vs} = 0.30F_U A_{nv} + 0.50F_U A_{nt} = 0.30(65)(6.50) + 0.50(65)(0.75) = 126.8 + 24.4 = 151.2 \text{ kips}$$

$$\text{Tensile ultimate strength} = F_u A_{nt} = (65)(0.75) = 48.8 \text{ kips}$$

$$\text{Shear ultimate strength} = 0.60F_u A_{nv} = 0.60(65)(6.50) = 253.4 \text{ kips.}$$

Therefore, tensile yielding on the gross section and shear fracture on the net section is appropriate to consider.

$$P_{vs} = 0.30F_U A_{nv} + 0.55F_y A_{gt} = 0.30(65)(6.50) + 0.55(50)(1.25) = 126.8 + 34.4 = 161.2 \text{ kips.}$$

The allowable block shear stress is 151.2 kips ≥ 75 kips, OK.

The member design is governed by the tensile fracture criterion due to the considerable shear lag effect associated with the single angle connection used for this secondary member.

Gusset plate:

Block shear failure in gusset plate:

$A_{gt} = 2.5(0.375) = 0.94$ in.^{2}
$A_{nt} = 0.94 - 0.375(1) = 0.56$ in.^{2}
$A_{gv} = 2(9 + 3.5)(0.375) = 9.38$ in.^{2}
$A_{nv} = 9.38 - 8(0.375)(1.0) = 6.38$ in.^{2}

$$P_{vs} = 0.30F_U A_{nv} + 0.50F_U A_{nt} = 0.30(65)(6.38) + 0.50(65)(0.56) = 124.3 + 18.2 = 142.5 \text{ kips}$$

$$\text{Tensile ultimate strength} = F_u A_{nt} = (65)(0.56) = 36.4 \text{ kips}$$

$$\text{Shear ultimate strength} = 0.60F_u A_{nv} = 0.60(65)(6.38) = 248.6 \text{ kips.}$$

Therefore, tensile yielding on the gross section and shear fracture on the net section is appropriate to consider.

$$P_{vs} = 0.30F_U A_{nv} + 0.55F_y A_{gt} = 0.30(65)(6.38) + 0.55(50)(0.94) = 124.3 + 25.9 = 150.2 \text{ kips.}$$

The allowable block shear stress is 142.5 kips ≥ 75 kips, OK.

Axial tension in gusset plate

$$w_e = 2l_c \tan(30^\circ) + s_r = 1.15(9) + 2.5 = 12.85''$$

$$\sigma_{at} = \frac{75}{(12.85)(0.375)} = 15.6 \le 0.55F_y \le 27.5 \text{ ksi, OK}$$

or

$$\sigma_{at} = \frac{75}{(12.85 - 2(1))(0.375)} = 18.4 \le 0.47F_U \le 30.6 \text{ ksi, OK.}$$

Axial compression in gusset plate:

$$C = -50 \text{ kips}$$

$f_{bv\,=}$ 17 ksi (slip-resistant connection allowable bolt shear)

$C' = 1.5(-50) = -75$ kips.

Allowable compressive strength of member (see Chapter 6):

$r_{min} = r_{xy} = 1.18$ in.

$$\frac{kL}{r_{min}} = \frac{0.75(7)(12)}{1.18} = 53.4$$

$$F_{call} = 0.60(50) - 0.165(53.4) = 21.2 \text{ ksi}$$

Allowable strength = (21.2)(5.75) = 121.6 kips compression.

Design connection for 75 kips axial compression.

Compressive stress in gusset plate:

$$\sigma_{ac} = \frac{C'}{w_e t_p} = \frac{75}{12.85(0.375)} = 15.6 \text{ ksi}$$

$$\frac{Kl_w}{r_w} = \frac{0.65(6.5)\sqrt{12}}{0.375} = 39.0 \ge 0.629\sqrt{\frac{E}{F_y}} \ge 15.2$$

$$F_c = 0.60F_y - \left(\frac{17{,}500F_Y}{E}\right)^{3/2}\left(\frac{Kl_w}{r_w}\right) = 0.60F_y - 165.7\left(\frac{Kl_w}{r_w}\right) = 30{,}000 - 165.7(39.0)$$

$$= 23{,}533 \text{ psi} = 23.5 \text{ ksi} \ge 15.6 \text{ ksi, OK.}$$

If the connection shown in Figure E9.4b has a compression diagonal creating a vertical force of 35 kips and horizontal force of 25 kips in addition to the 50 kip tensile force:

$P = T = 50 - 25 = 25$ kips

$V = 35$ kips

$M = 35(6.5) = 227.5$ kips-in.

$A_g = 18(0.375) = 6.75$ in.^{2}

$A_n = 6.75 - 2(1)(0.375) = 6.00$ in.^{2}

$S_n = 0.375(18)^2/6 - 2(1)(0.375)(1.25) = 19.3$ in.^{3}

$t_v = 35/6.75 = 5.2$ ksi (1.5V/A not used since slender beam theory not theoretically valid)
$\sigma_a = 25/6.00 = 4.2$ ksi
$\sigma_b = 227.5/19.3 = 11.8$ ksi.
Use a linear interaction formula to examine combined stress effects:

$$\frac{5.3}{0.35(50)} + \frac{4.2}{0.55(50)} + \frac{11.8}{0.55(50)} = 0.30 + 0.15 + 0.43 = 0.88 \leq 1.00, \text{ OK.}$$

9.3.4.3 Eccentrically Loaded Connections with Bolts in Shear and Tension

Small load eccentricities may often be ignored in slip-resistant bolted connections, but larger eccentricities should be considered in the design. Many bolted connections are loaded eccentrically (e.g., the connections shown in Figures 9.10b and 9.10c). Eccentric loads will result in combined shear and torsional moments or combined shear and bending moments, depending on the direction of loading with respect to the bolts in the connection.

9.3.4.3.1 Connections Subjected to Shear Forces and Bending Moments

A connection similar to that shown in Figure 9.10d is shown in Figure 9.19. The bolts on each side of the bracket resist both shear forces and bending moments.

9.3.4.3.1.1 Bolt Shear Stress The shear stress on the bolts is

$$\tau_b = \frac{P}{n_s n_b A_b}, \tag{9.53}$$

where
n_b = the number of bolts
n_s = the number of shear planes
A_b = the nominal cross-sectional area of the bolt.

9.3.4.3.1.2 Bolt Tensile Stress The tension, σ_{ti}, on bolt i from bending moment, $M = Pe$ is

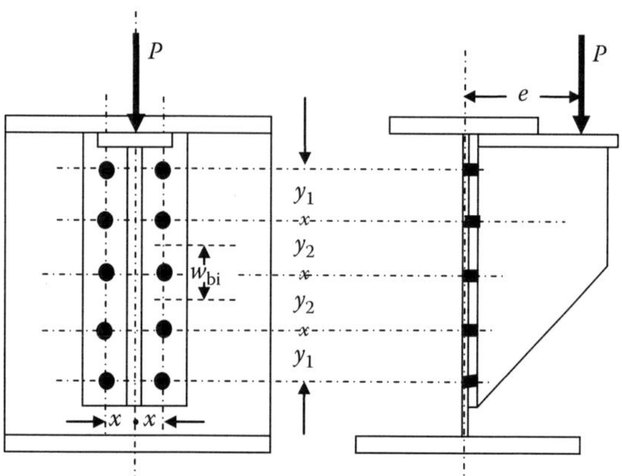

FIGURE 9.19 Bending and shear forces on bolted connection.

$$\sigma_{ti} = \frac{M}{A_b S_{bi}} = \frac{Pe}{A_b S_{bi}} = \frac{Peh_{bi}}{A_b I_b}, \quad (9.54)$$

where

A_b = the nominal cross-sectional area of bolt i

S_{bi} = the effective "section modulus" of bolt $i = I_b/h_{bi}$

I_b = the effective "moment of inertia" of the bolt group

h_{bi} = the distance from bolt i to the neutral axis of the bolt group.

For the connection in Figure 9.19, $I_b = 4(y_1 + y_2)^2 + 4(y_2)^2$ and the bolt tension on the most highly stressed bolt, σ_t (bolt farthest from neutral axis of the bolt group), is $\sigma_t = \frac{Peh_{bi}}{A_b I_b} = \frac{Pe(y_1 + y_2)}{4\left((y_1 + y_2)^2 + (y_2)^2\right)A_b}$. If $y = y_1 = y_2$, $\sigma_t = \frac{Pe}{10yA_b}$. Prying action effects, which will increase the bolt tension, must also be considered in the connection design (e.g., by using Equation 9.38).

9.3.4.3.1.3 Combined Shear and Tension The bolts in Figure 9.19 are subject to shear force, $F_{bv} = \tau_b A_b$, and tensile force, $T_B = T + T_Q$, which must be combined to determine the allowable stress in the bolts of the connection. The allowable shear stress for combined shear and tension in a SC connection is (from Equation 9.45)

$$f_{bv} = \frac{F_{bv}}{A_b} \le f'_{bv}\left(1 - \frac{(T_B)}{(T_{bP})}\right), \quad (9.55)$$

where

f'_{bv} = the allowable bolt shear stress for SC connections

$T_B = T + T_Q$ = the total bolt tensile force

T_{bP} = the bolt pretension (see Table 9.4)

A_b = the nominal area of the bolt.

Example 9.5a (SI Units)

Review the design of the SC single shear plane connection shown in Figure 9.19 using 22 mm diameter A325M bolts for a load P = 225 kN with eccentricity, e = 150 mm. The steel is ASTM A709M Grade 350 with F_y = 350 MPa and F_u = 450 MPa. The connection geometry is similar to Figure 9.19 with the following:

$y = y_1 = y_2$ = 100 mm

x = 125 mm

w_{bi} = 100 mm

a = 40 mm (see Figure 9.14)

b = 125 − 6 = 119 mm (see Figure 9.14 with 12 mm web plate)

t_p = 12 mm (see Figure 9.14)

n_b = 10 (number of bolts)

n_s = 1.

Shear

$$\tau'_{bv} = \frac{P}{n_s n_b A_b} = \frac{225(1000)}{(1)10(380)} = 59.2 \text{ MPa}.$$

Tension

$$T = \frac{Pe}{10y} = \frac{225(150)}{10(100)} = 33.8 \text{ kN}$$

$$T_Q = T\left(\frac{3b}{8a} - \frac{t_p^3}{328\times10^3}\right) = 33.8\left(\frac{3(120)}{8(40)} - \frac{(12)^3}{328\times10^3}\right) = 37.8 \text{ kN}$$

$$\sigma_{bt} = \frac{(33.8+37.8)(1000)}{380} = 188.4 \text{ MPa} \le f_{bt} \le 300 \text{ MPa, OK.}$$

Combined shear and tension

$$f_{bv} = f'_{bv}\left(1 - \frac{(T)}{(T_{bP})}\right) = 115\left(1 - \frac{(33.8+37.8)}{167}\right) = 65.7 \text{ MPa} \ge \tau_{bv} \ge 59.2 \text{ MPa, OK}$$

(T_{bP} = 167 kN from Table 9.4a).

Check bearing stress

$\sigma_{bc} = \dfrac{225(1000)}{10(12)(25)} = 75.0$ MPa, assuming that the minimum thickness of the connection plate and the web is 12 mm.

$f_B \le \dfrac{l_e F_u}{2d_b} \le \dfrac{(40)(450)}{2(22)} \le 409.1$ MPa, assuming that the minimum edge distance is 40 mm, OK.

or

$$f_B \le 1.2F_u \le 1.2(450) \le 540 \text{ MPa.}$$

Example 9.5b (US Customary and Imperial Units)

Review the design of the SC single shear plane connection shown in Figure 9.19 using 7/8 in. diameter A325 bolts for a load P = 55 kips with eccentricity, e = 6 in. The steel is ASTM A709 Grade 50 with F_y = 50 ksi and F_u = 65 ksi. The connection geometry is similar to Figure 9.19 with:

$y = y_1 = y_2 = 4.0''$
$x = 4.5$ in.
$w_{bi} = 4.0$ in.
$a = 1.5$ in. (see Figure 9.14)
$b = 4.25$ in. [see Figure 9.14 with 1/2″ web plate]
$t_p = 0.5$ in.
$n_b = 10$ (number of bolts)
$n_s = 1$.

Shear

$$\tau'_{bv} = \frac{P}{n_s n_b A_b} = \frac{55}{(1)10(0.60)} = 9.2 \text{ ksi.}$$

Tension

$$T = \frac{Pe}{10y} = \frac{55(6)}{10(4.0)} = 8.3 \text{ kips}$$

$$T_Q = T\left(\frac{3b}{8a} - \frac{t_p^3}{20}\right) = 8.3\left(\frac{3(4.25)}{8(1.5)} - \frac{(0.5)^2}{20}\right) = 8.7 \text{ kips}$$

$$\sigma_{bt} = \frac{(8.3+8.7)}{0.60} = 28.3 \text{ ksi} \times f_{bt} \times 44.0 \text{ ksi, OK.}$$

Combined shear and tension

$$f_{bv} = f'_{bv}\left(1 - \frac{(T)}{(T_{bP})}\right) = 17\left(1 - \frac{(8.3+8.7)}{39}\right) = 9.6 \text{ ksi} \geq \tau_{bv} \geq 9.2 \text{ ksi, OK}$$

(T_{bP} = 39 kips from Table 9.4b).

Check bearing stress

$$\sigma_{bc} = \frac{55}{10(0.50)(1.00)} = 11.0 \text{ ksi, assuming that the minimum thickness of the connection plate}$$

and the web is 0.5″.

$$f_B \leq \frac{l_e F_u}{2d_b} \leq \frac{(1.5)(65)}{2(7/8)} \leq 55.7 \text{ ksi, assuming that the minimum edge distance is 1.5″, OK}$$

or

$$f_B \leq 1.2F_u \leq 1.2(65) \leq 78 \text{ ksi.}$$

9.3.4.3.2 Connections Subjected to Eccentric Shear Forces (Combined Shear and Torsion)

A connection subjected to eccentric shear is shown in Figure 9.20. The bolts in the connection resist direct shear forces from, P, and torsional shear forces from the moment, Pe.

The direct shear stress, τ, on the bolts (all bolts with same A_b), is

$$\tau = \frac{P}{n_s \sum A_b} = \frac{P}{n_s n_b A_b} \tag{9.56}$$

and the torsional shear stress, τ_T, on the bolts is

$$\tau_T = \frac{Per_T}{n_s J_b} = \frac{Per_T}{n_s \sum n_b A_b r_T^2} = \frac{Per_T}{n_s A_b \sum n_b (x_T^2 + y_T^2)}. \tag{9.57}$$

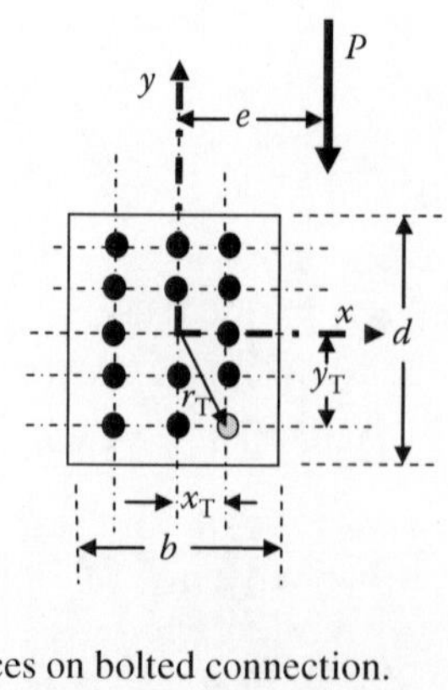

FIGURE 9.20 Eccentric shear forces on bolted connection.

Equation 9.57 can be developed in the x- and y-directions as

$$\tau_{Tx} = \frac{Pey_T}{n_s A_b \sum n_b (x_T^2 + y_T^2)} \tag{9.58a}$$

$$\tau_{Ty} = \frac{Pex_T}{n_s A_b \sum n_b (x_T^2 + y_T^2)}, \tag{9.58b}$$

where

n_b = the total number of bolts in the connection
n_s = the number of shear planes
$J_b = \sum n_b A_b r_T^2$ = the polar moment of inertia of the connection
$r_T = \sqrt{x_T^2 + y_T^2}$ = is the distance from the bolt to the centroid of the bolt group
x_T = the distance from the centroid to the bolt in the x-direction
y_T = the distance from the centroid to the bolt in the y-direction
The resultant shear stress on any bolt described by locations x and y is

$$f = \sqrt{(\tau + \tau_{Ty})^2 + \tau_{Tx}^2}. \tag{9.59}$$

9.3.4.3.3 *Beam Framing Connections*

Bolted beam framing connections are often used in the main members of steel railway superstructures (Figure 9.21a)*. These framing connections are subject to shear forces, P, and member end bending moments, M_e. Furthermore, the legs of the connection angles fastened to the web of the beam (a double-shear connection) may also be subject to a torsional moment, Pe, due to the eccentric application of shear force†.

Beam framing connections are often assumed to transfer shear only (i.e., it is assumed the beam is simply supported and $M_e = 0$), provided that adequate connection flexibility exists. However, due to some degree of end rotational restraint, an end moment, M_e, typically exists. The magnitude of the end moment depends on the rigidity of the support and can be of considerable magnitude (Al-Emrani, 2005).

A rigid connection may be designed for the shear force, P, and the corresponding end moment due to full fixity, M_f. A semirigid connection will require consideration of the end moment, M_e,—end rotation, ϕ_e, relationship (often nonlinear) to determine rotational stiffness and the end moment for design. Moment–rotation curves, developed from theory and experiment, for bolted joint configurations‡ are available in the technical literature on connection design (e.g., Faella et al., 2000; Leon, 1999). A flexible end connection will deform and resist very little bending moment. Therefore, simple beam framing connections that exhibit the characteristics of a flexible connection (see Figure 9.8) may be designed for shear force, P, only ($M_e = 0$). AREMA (2015) recognizes that all connections actually exhibit some degree of semirigid behavior, and allows flexible connection design (with a bolt configuration that allows for adequate deformation and flexibility) provided the design shear force is increased by 25%. Otherwise, a semirigid connection design considering both beam end moment and shear is required§.

* These connections can be single-shear or double-shear connections depending on configuration. For example, in the floor systems of many steel railway superstructures, a double-shear connection exists at interior floor beams and, typically, a single shear connection at end floor beams.

† Depending on whether these effects are accounted for in the structural analysis.

‡ Typically for beam to column flange connections.

§ Finite element analysis models using rotational spring boundary conditions for beam end connections can realistically model semirigid connection behavior.

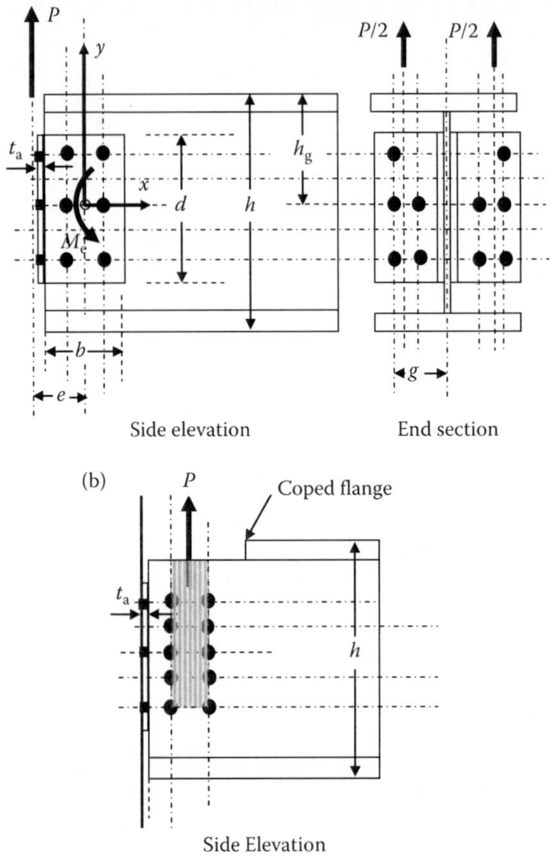

FIGURE 9.21 (a) Bolted beam framing connection and (b) bolted beam framing connection subject to block shear.

Therefore, flexible bolted beam framing connections must be designed considering 125% of the shear on the outstanding legs of the connection and, due to the eccentricity of the shear force, combined shear and torsion on the leg of the connection angles fastened to the web of the beam. However, for flexible bolted connections*, it is usual practice to disregard the moment, Pe, due to the typically relatively small eccentricity, e.

The angles in the simple beam framing connection (often referred to as clip angles) must deform in order to allow an adequate degree of flexible connection behavior. Bolted connections are often made more flexible, which is particularly necessary for stringers, by providing a minimum gage distance, g, over a distance, h_g, from the top of the beam. AREMA (2015) recommends $h_g \geq h/3$ and

$$g \geq \sqrt{\frac{Lt_a}{8}}, \tag{9.60}$$

where

L = length of stringer span
t_a = thickness of angle.

* In contrast to more rigid welded beam end connections where the moment, Pe, due to eccentricity, e, should be considered in the design.

Equation 9.60 (using consistent units for L and t_a) is based on analytical and experimental work regarding the fatigue strength of typical stringer to floor beam connections (Yen et al., 1991).

It is important to design beam framing connections for the appropriate flexibility, shear and moment as they are often critical members of through span floor systems*.

If the beam flanges are coped at the connection, the design must also consider block shear (the combination of shear or tension yielding on one plane and tension or shear fracture on the other that may cause tear-out of the shaded area shown in Figure 9.21b). AREMA (2015) recommends the determination of allowable block shear strength using Equations 9.48 and either 9.49a or 9.49b, depending on whether the net fracture strength in tension is greater or less than the net fracture strength in shear.

Examples 9.6 and 9.7 illustrate bolted beam end framing connection design assuming no beam end moment (flexible connection) and with a beam end moment (semirigid or rigid)†, respectively.

Example 9.6 (SI Units)

Design the bolted simple beam framing connection using 150 × 100 × 12 angles as shown in Figure E9.5 for a shear force of $P = 350$ kN. The beam is 6 m long and frames into the web of a plate girder with a single shear connection. The allowable shear stress on the high-strength bolts is 115 MPa. The shear force is developed from a routine analysis considering a simply supported beam with complete connection flexibility.

$P' = 1.25(350) = 437.5$ kN (AREMA recommendation for flexible connection design).

The angle thickness should be based on the requirement for transmitting shear or minimum element thickness recommended by AREMA (2015). In this example the angle thickness is 12 mm.

Bolt configuration to allow deformation:

$$g \geq \sqrt{\frac{Lt_a}{8}} \geq \sqrt{\frac{(6000)(12)}{8}} = 95 \text{ mm} \leq 115, \text{OK}.$$

$h_g = 230\,\text{mm} \geq h/3 \geq 535/3 \geq 178\,\text{mm}$, OK,

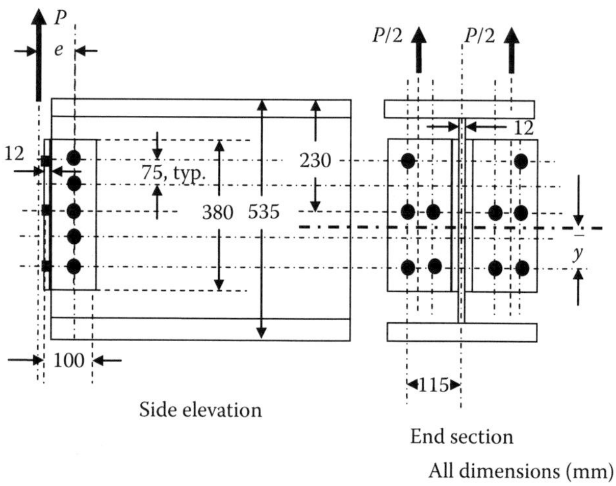

FIGURE E9.5

* For example, in a reliability analysis of a through span bridge, the stringer-to-floor beam connections had a reliability index considerably lower than other members of the superstructure (Rakoczy and Nowak, 2013).

† The beam end moments are generally determined by relatively sophisticated structural analyses that, for example, consider flexural members with equivalent rotational spring stiffness.

therefore flexible connection (neglect end bending effects).
Shear stress on the bolts:

$\tau = \dfrac{P'}{n_s n_b A_b} = \dfrac{437.5(1000)}{10(380)(1)} = 115\,\text{MPa} \le 115\ \text{MPa, OK}$ (single shear connection, use 5 bolts in double shear connection).

Check bearing stress

$$\sigma_{bc} = \frac{437.5(1000)}{10(12)(24)} = 152 \le f_B\ \text{MPa (both angles and beam web are 12 mm thick)}$$

$$f_B = \frac{l_e F_u}{2d_b} = \frac{(40)(450)}{2(22)} = 409\ \text{MPa}, \text{ assuming a minimum 40 mm. loaded edge distance, OK}$$

or

$$f_B = 1.2F_u = 1.2(450) = 540\ \text{MPa}.$$

If the applied load is cyclical, fatigue must be considered. A connection such as that shown in Figure E9.5 has an allowable fatigue shear stress range of 110 MPa for connections subjected to greater than 2 million stress range cycles.

Example 9.7 (US Customary and Imperial Units)

Design the bolted beam framing connection using 6 × 6 × 1/2 in. angles as shown in Figure E9.6 for a shear force, $P = 70$ kips and end moment, $M_e = 25$ ft-kip. The beam is 20 ft long and frames into the web of a plate girder with single shear connections. The allowable shear stress on the high-strength bolts is 17.0 ksi. The shear force and bending moment were developed from an analysis considering partial rigidity of the connection, so the AREMA recommendation of a 25% increase in shear force is not used.

$P' = 70$ kips
$M_e = 25$ ft-kips = 300 in.-kips in direction creating tension at top of beam end
Shear and bending on bolts in the end section:
Bolt forces due to bending:
The centroid of the connection is

$$\bar{y} = \frac{2(12+9+6+3)}{10} = 6.0\ \text{in.}$$

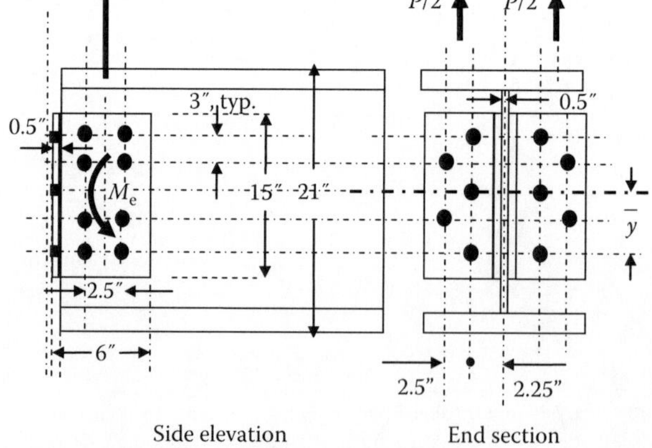

FIGURE E9.6

The section modulus of the top row of bolts in connection is

$$S_{b1} = \frac{2((12-6.0)^2+(9-6.0)^2+(6-6.0)^2+(3-6.0)^2+(0-6.0)^2)}{(12-6.0)} = 30.0 \text{ bolt-in (in tension in}$$

simple beam connection).

The section modulus of the second row of bolts in connection is

$$S_{b2} = \frac{2((12-6.0)^2+(9-6.0)^2+(6-6.0)^2+(3-6.0)^2+(0-6.0)^2)}{(9-6.0)} = 60.0 \text{ bolt-in}$$

$$\sigma_{b1} = \frac{M_e}{S_{b1}A_b} = +\frac{300}{30.0(0.6)} = 16.67 \text{ ksi fortop row bolts} \leq 44.0 \text{ ksi, OK}$$

$$\sigma_{b2} = \frac{M_e}{S_{b2}A_b} = +\frac{300}{60.0(0.6)} = 8.33 \text{ ksi for second row bolts}$$

$$T_{b1} = \sigma_{b1}A_b = 16.67(0.6) = 10.0 \text{ kips for top row bolts}$$

$$T_{b2} = \sigma_{b2}A_b = 8.33(0.6) = 5.0 \text{ kips for second row bolts.}$$

Prying action (Equation 9.38):

$$T_{Q1} = 9.4\left(\frac{3b}{8a} - \frac{(t_p)^3}{20}\right) = 10.0\left(\frac{3(2.25-0.25-.05)}{8(3.75)} - \frac{(0.5)^3}{20}\right) = 0.14(10.0) = 1.4 \text{ kips for top row bolts}$$

$$T_{Q2} = 9.4\left(\frac{3b}{8a} - \frac{(t_p)^3}{20}\right) = 5.0\left(\frac{3(4.00)}{8(1.25)} - \frac{(0.5)^3}{20}\right) = 1.19(5.0) = 6.0 \text{ kips for second row bolts}$$

Tension in top row bolts = 10.0 + 1.4 = 11.4 kips
Tension in second row bolts = 5.0 + 6.0 = 11.0 kips.
Shear stress on the bolts:

$$\tau = \frac{P'}{n_s n_b A_b} = \frac{70}{10(0.60)(1)} = 11.67 \text{ ksi.}$$

Combined shear and tension

$$f_{bv} = f'_{bv}\left(1 - \frac{(T_b)}{(T_{bP})}\right) = 17\left(1 - \frac{11.4}{39}\right) = 12.0 \text{ ksi} \geq 11.67 \text{ ksi, OK.}$$

Check bearing stress

$$\sigma_{bc} = \frac{70}{10(0.50)(1.00)} = 14.0 \leq f_B \text{ ksi}$$

$$f_B = \frac{l_e F_u}{2d_b} = \frac{(1.5)(65)}{2(7/8)} \leq 55.7 \text{ ksi, assuming a minimum 1.5 in. loaded direction edge distance, OK}$$

or

$$f_B = 1.2F_u = 1.2(65) \leq 78 \text{ ksi.}$$

Shear and torsion on bolts in side elevation:
The direct shear stress, t, on the bolts is

$$\tau = \frac{P'}{n_s n_b A_b} = \frac{70}{(2)8(0.60)} = 7.3 \text{ ksi}$$

and the torsional shear stress, t_T, on the highest stressed bolts is

$$\tau_{Tx} = \frac{M_e y_T}{n_s A_b \sum n_b (x_T^2 + y_T^2)} = \frac{300(1.25)}{2(0.6)\left(4(6^2 + 1.25^2) + 4(3^2 + 1.25^2)\right)}$$

$$= \frac{300(1.25)}{2(0.6)(192.5)} = 1.6 \text{ ksi}$$

$$\tau_{Ty} = \frac{M_e x_T}{n_s A_b \sum n_b (x_T^2 + y_T^2)} = \frac{300(6)}{2(0.6)(192.5)} = 7.8 \text{ ksi.}$$

The maximum resultant shear stress on the bolts is

$$f = \sqrt{(\tau + \tau_{Ty})^2 + \tau_{Tx}^2} = \sqrt{(7.3 + 7.8)^2 + 1.6^2} = 15.2 \text{ ksi} \times 17.0 \text{ ksi, OK.}$$

Check bearing stress

$$\sigma_{bc} = \frac{(15.2)(0.60)}{1(0.50)(1.00)} = 18.2 \le f_B \text{ ksi}$$

$$f_B = \frac{l_e F_u}{2 d_b} = \frac{(1.5)(65)}{2(7/8)} = 55.7 \text{ ksi, assuming a minimum 1.5 in. loaded edge distance, OK}$$

or

$$f_B = 1.2 F_u = 1.2(65) = 78 \text{ ksi.}$$

9.3.4.4 Axially Loaded Connections with Bolts in Direct Tension

Connections with bolts subject to direct tension should generally be avoided in the main members of steel railway superstructures. However, when bolts are subjected to direct tension, the additional bolt forces created by prying action of the connection leg must also be considered (e.g., by using Equation 9.38). The effects of the prying action on the allowable fatigue design stresses must also be considered as shown in Example 9.8

Example 9.8a (SI Units)

Design the bolted hanger-type connection shown in Figure E9.7a for an axial force consisting of
$P_{DL} = 80$ kN (dead load)
$P_{LL+I} = 270$ kN (live load plus impact)
Use 22 mm diameter A325M bolts.
$b = 50–12 = 38$ mm
$a = 75$ mm
$t_p = 12$ mm.

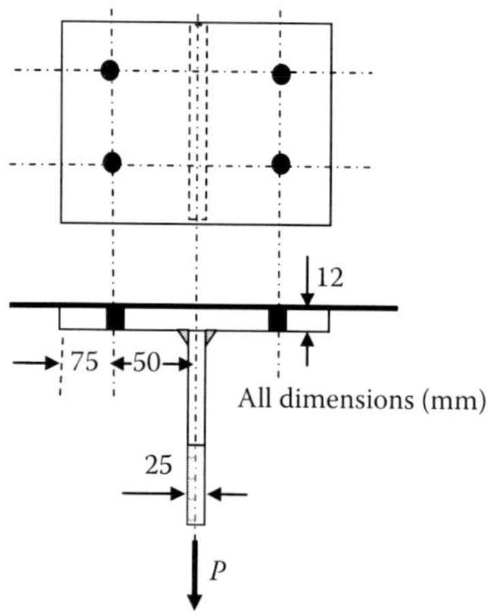

FIGURE E9.7a

$$T_Q = T\left(\frac{3b}{8a} - \frac{t_p^3}{328\times10^3}\right) = \frac{(80+270)}{4}\left(\frac{3(38)}{8(75)} - \frac{(12)^3}{328\times10^3}\right) = 87.5(0.18) = 15.8 \text{ kN}$$

$$T_B = T + T_Q = 87.5 + 15.8 = 103.3 \text{ kN}$$

$$\sigma_{bt} = \frac{T_B}{A_b} = \frac{103.3(1000)}{(380)} = 272 \le 300 \text{ MPa, OK.}$$

The tensile stress range, including prying stress, is

$$\Delta\sigma_{bt} = \frac{270(1.18)(1000)}{(4)380} = \frac{79.7(1000)}{380} = 210 \text{ MPa.}$$

$$\text{Since } \frac{T_Q}{T}\% = \frac{15.8}{87.5}(100) = 18\% \le 20\%,$$

the allowable tensile stress range is 215 MPa ≥ 210 MPa, OK.

Example 9.8b (US Customary and Imperial Units)

Design the bolted hanger-type connection shown in Figure E9.7b for an axial force consisting of
P_{DL} = 20 kips (dead load)
P_{LL+I} = 60 kips. (live load plus impact)
Use 7/8 in. diameter A325 bolts.
b = 2.0–0.5 = 1.5 in.
a = 3 in.
t_p = 0.5 in.

$$T_Q = T\left(\frac{3b}{8a} - \frac{t_p^3}{20}\right) = \frac{(20+60)}{4}\left(\frac{3(1.5)}{8(3)} - \frac{(0.5)^3}{20}\right) = 20(0.18) = 3.60 \text{ kips}$$

$$T_B = T + T_Q = 20 + 3.60 = 23.6 \text{ kips}$$

$$\sigma_{bt} = \frac{T_B}{A_b} = \frac{23.6}{(0.6)} = 39.3 \le 44.0 \text{ ksi, OK.}$$

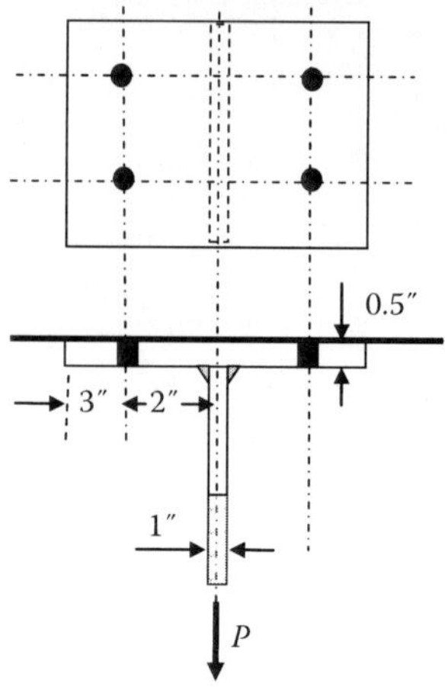

FIGURE E9.7b

The tensile stress range, including prying stress, is

$$\Delta\sigma_{bt} = \frac{60(1.18)}{(4)0.6} = \frac{17.7}{0.6} = 29.5 \text{ ksi}.$$

$$\text{Since } \frac{T_Q}{T}\% = \frac{3.60}{20}(100) = 18\% \le 20\%,$$

the allowable tensile stress range is 31.0 ksi ≥ 29.5 ksi, OK.

9.3.4.5 Axial Member Splices

A common axial member bolted splice involves the use of lap joints on the member elements as shown in Figure 9.10g. The bolted connection is designed as a SC connection, and it is recommended to include a review of bearing stresses in case of joint failure by slippage. Splices located in the center of truss members must also be sufficiently rigid to resist bending from self weight and other lateral forces. AREMA (2015) also recommends that truss web member axial splice connections be designed for 133% of allowable stress using the live load that will increase the maximum chord stress in highest stresses chord by 33% (see Chapter 4).

9.3.4.5.1 Axial Tension Member Splices

Main member axial tension splices should be designed for the strength of member. For secondary members, AREMA (2015) recommends that the splice be designed for the lesser of the strength of the member or 150% of the maximum calculated tension.

Steel rods or bars may be spliced by turnbuckles and sleeve nuts. Rolled or built-up tension members are spliced by bolted plates and, therefore, designed as net area tension members (see Chapter 6) with due consideration of block shear at the connection. Generally, all elements of tension members are spliced on each side of the element to avoid eccentricities and shear lag effects. The connection bolts are designed for direct shear and bearing strength.

9.3.4.5.2 Axial Compression Member Splices

Splice plates and bolts must transmit 50% of the force and be placed on four sides of the member in a manner that provides for the accurate and firm fit of the abutting elements in rolled or built-up compression members that are faced or finished to bear. This may result in compression members with only nominal splice plates and the designer may wish to ensure adequate bending rigidity by designing the splice for bending and shear from a minimum transverse force of 2.5% of the member axial compression. The connection bolts are designed for direct shear and bearing strength.

9.3.4.6 Beam and Girder Splices

Conventional beam and girder bolted splices involve the use of lap joints as shown in Figure 9.10c. The plates used in these splices are designed in accordance with AREMA (2015) as outlined in Chapter 7 concerning plate girder design. The bolted flange and web splices are designed as SC connections, with a bearing check in case of joint failure by slippage.

9.3.4.6.1 Beam and Girder Flange Splices

Beam and girder flange splices should be designed for the strength of member being spliced. Also, as outlined in Chapter 7, bolted splice elements in girder flanges should:

- Have a cross-sectional area that is at least equal to that of the flange element being spliced
- Comprise splice elements of sufficient cross section and location such that the moment of inertia of the member at the splice is no less than that of the member adjacent to the splice location.

The bolts in tension and compression flange lap joint splices are subjected to direct shear in transferring flange forces from bending, F_f, between the girder flange and splice plates. Depending on whether one or two plate splices are used, the bolts will be in single or double shear, respectively.

9.3.4.6.2 Beam and Girder Web Splices

As outlined in Chapter 7, bolted splices in girder web plates should be designed

- To transfer the shear force, V, including moment, Ve, due to eccentricity, e, of centroid of bolt group
- For the gross shear and net flexural strength of the web plate
- For the combined forces of the flexural strength of the net section of the web with the maximum shear force at the splice.

The bolts in girder web lap joint splices are subjected to direct and torsional shear in transferring web plate forces from shear and bending between the girder web and splice plates. The web splice plates must be designed for the gross shear strength of the web plate, have a net moment of inertia not less than that of the web plate, and resist the combined web plate net flexural strength and maximum shear force at the splice. Two web splice plates must always be used so the bolts are in double shear.

The design of girder flange and web splice connection plates and bolts is outlined in Example 9.9.

Example 9.9a (SI Units)

Design the bolted flange and web splices for the girder shown in Figure E9.8a with flange splices located where $M = 12{,}000$ kNm and web splices where $M_V = 8000$ kNm and $V = 2000$ kN.

The girder section has the following properties (see Example 7.1a):

$I_g = 111{,}522 \times 10^6\,\text{mm}^4$

$S_g = 84{,}104 \times 10^3\,\text{mm}^3$

$S_n = 85{,}494 \times 10^3\,\text{mm}^3$

$A_w = 42{,}660\,\text{mm}^2$

Allowable girder compressive bending capacity, $M_{call} = 0.55(F_y)(S_g) = 0.55(350)(84{,}104)/1000 =$ 16,190 kNm (assuming fully laterally supported).

Allowable girder tensile bending capacity, $M_{tall} = 0.55(F_y)(S_n) = 0.55(350)(85{,}494)/1000 =$ 16,458 kNm.

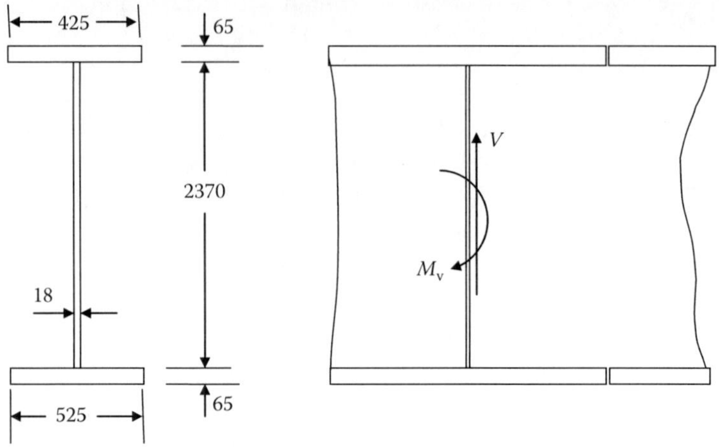

FIGURE E9.8a

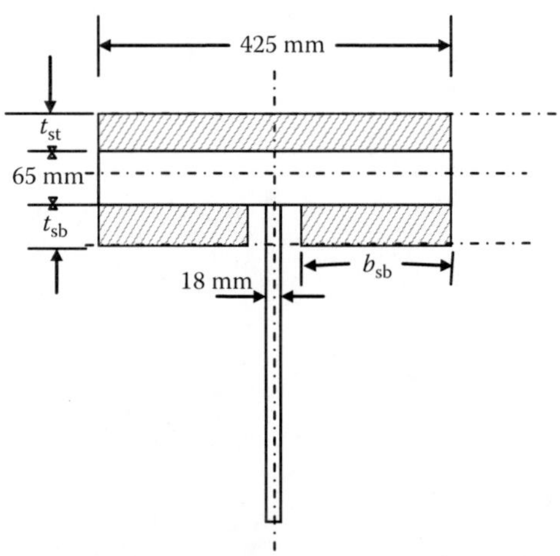

FIGURE E9.9a

Allowable girder shear capacity, $V_{all} = 0.35(F_y)(A_w) = 0.35(350)(42{,}660)/1000 = 5226$ kN.

Compression flange splices (Figure E9.9a):

There must be enough bolts each side of the splice to develop the flange strength.

The maximum force in the compression flange plate at the splice location (average at the centroid) is

$$F_f = \frac{12{,}000}{84{,}104}\frac{(1+1185/1250)}{2}(65)(425) = 142.7(0.97)(27{,}625)/1000 = 3839 \text{ kN}.$$

The allowable force in the compression flange plate at the splice location (average at the centroid) is

$$F_f = \frac{16{,}190}{84{,}104}\frac{(1+1185/1250)}{2}(65)(425) = 192.5(0.97)(27{,}625)/1000 = 5158 \text{ kN}.$$

Transfer of this force from girder flange to splice plate is by a SC lap joint using A325M bolts with $f'_{bv} = 120$ MPa. The number of double shear bolts required is

$$n_{fs} = \frac{5158(1000)}{(2)(120)(380)} = 57\,\text{bolts}$$ each side of the splice. Use 14 rows of 4 bolts (56 bolts OK for force of 5106 kN).

For the splice cross section to be at least equal to the flange cross section ($65 \times 425 = 27{,}625\ \text{mm}^2$);

$t_{st} = 50\,\text{mm}$

$t_{sb} = 25\,\text{mm}$

$b_{sb} = 150\,\text{mm}$.

Flange thickness at splice = 50 + 65 + 25 = 140 mm.

The area of the splice is $50(425) + 2(25)(150) = 21{,}250 + 7500 = 28{,}750\,\text{mm}^2 \geq 27{,}625\,\text{mm}^2$, OK.

$$y_{\text{splice}} = 140 - \frac{(21,250)(115) + (7500)(12.5)}{28,750} = 140 - 88 = 52 \text{ mm from top of splice}$$

The top flange splice centroid is $52 - (50 + 65/2) = -30.5$ mm from the top flange centroid (30.5 mm farther from the girder neutral axis) and the moment of inertia of the splice will be greater than the moment of inertia of the flange plate being spliced.

The spliced flange eccentricity = 30.5 mm. This eccentricity creates a maximum force bending moment of 3587(30.5)/1000 = 109 kNm which is small (less than 1% of the maximum moment at the splice location) and, therefore, acceptable without further consideration.

Check bearing stress (on thinnest plate)

$$\sigma_{bc} = \frac{5158(1000)}{56(25)(25)} = 147 \leq f_B \text{ MPa}$$

$$f_B = \frac{l_e F_u}{2d_b} = \frac{(40)(450)}{2(22)} = 409 \text{ MPa, assuming a minimum 40 mm loaded edge distance, OK}$$

or

$$f_B = 1.2F_u = 1.2(450) = 540 \text{ MPa}.$$

Tension flange splices (Figure E9.10a):

There must be enough bolts each side of the splice to develop the flange strength.

The maximum force in the tension flange plate at the splice location (average at the centroid) is

$$F_f = \frac{12,000}{85,494}\frac{(1+1185/1250)}{2}((65)(525) - 4(25)(65)) = 140.4(0.97)(34,125 - 6500)/1000 = 3762 \text{ kN}.$$

The allowable force in the tension flange plate at the splice location (average at the centroid) is

$$F_f = \frac{16,458}{85,494}\frac{(1+1185/1250)}{2}((65)(525) - 4(25)(65)) = 192.5(0.97)(27,625)/1000 = 5158 \text{ kN}.$$

Transfer of this force from girder flange to splice plate is by a SC lap joint using A325M bolts $f'_{bv} = 120\,\text{MPa}$. The number of double shear bolts required is

$$n_{fs} = \frac{5158(1000)}{(2)(120)(380)} = 57\,\text{bolts}$$ each side of the splice. Use 14 rows of 4 bolts (56 bolts OK for force of 5106 kN).

For the splice cross section to be at least equal to flange cross section $(525(65)) - 4(25)(65) = 27{,}625\,\text{mm}^2$ try

$t_{st} = 50\,\text{mm}$

$t_{sb} = 38\,\text{mm}$

$b_{sb} = 150\,\text{mm}$.

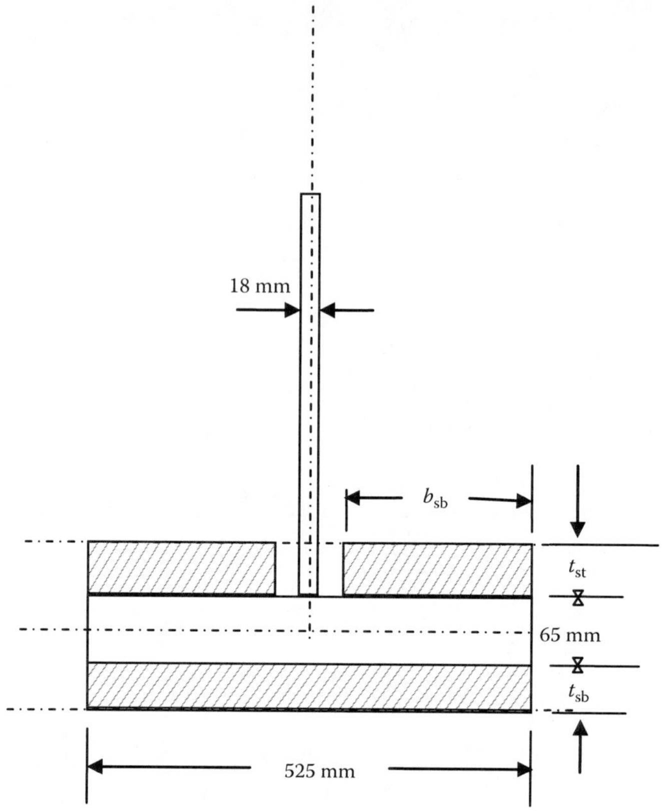

FIGURE E9.10a

Flange thickness at splice = 50 + 65 + 38 = 153 mm

The gross area of the splice is 50(525) + (2(38)(150))= 26,250 + 11,400 = 37,650 mm^2

The net area of the splice is (50(525) – 4(25)(50)) + (2(38)(150) – 4(25)(38))= 21,250 + 7600 = 28,850 mm^2 ≥ 27,625 mm^2, OK

$$y_{\text{splice}} = \frac{(26,250)(25)+11,400(134)}{37,650} = 58 \text{ mm.}$$

The bottom flange splice centroid is 58 – (50 + 65/2) = –24.5 mm from the bottom flange centroid (24.5 mm farther from the girder neutral axis).

The girder gross moment of inertia = $I_g = 111,522 \times 10^6$ mm^4.

Girder strength at the flange splices:

The girder neutral axis is at $\dfrac{27,625(2467.5)+42,660(1250)+34,125(32.5)}{(27,625+42,660+34,125)} = 1174$ mm. from the underside of the bottom flange and the girder net section moment of inertia is

$$I_n = 111,522 \times 10^6 - \left(4(25)(65)(1174-(65/2))^2\right) = (111,522-8470)\times 10^6 = 103,052 \times 10^6 \text{ mm}^4$$

(3% greater than assumed in Example 7.1a).

The girder allowable tensile bending capacity, M_{tall} = 0.55(350)(103,052/1174) = 192.5(87.78) = 16,897 kNm

The gross and net moment of inertia of the spliced section with both top and bottom flange splices in the same section are

$$I_g = 28{,}750(1293.5+30.5)^2 + 37{,}650(1141.5+24.5)^2 + (4427+2(195))(1000) +$$

$$(5469+2(686))(1000) + \frac{18(2370)^3}{12} = 121{,}560\times 10^6\ \text{mm}^4.$$

$$I_n = 121{,}560\times 10^6 - \left(4(25)(153)(1097.5)^2\right) = (121{,}560 - 18{,}429)\times 10^6$$

$$= 103{,}132\text{x}10^6 \geq 103{,}052\times 10^6\ \text{mm}^4$$

OK, the girder moment of inertia at the splice is greater than moment of inertia of the cross section being spliced.

Check bearing stress

$$\sigma_{bc} = \frac{5158(1000)}{56(25)(35)} = 105 \leq f_B\ \text{MPa}$$

$f_B = \dfrac{l_e F_u}{2d_b} = \dfrac{(40)(450)}{2(22)} = 409$ MPa, assuming a minimum 40 mm loaded edge distance, OK

or

$$f_B = 1.2F_u = 1.2(450) = 540\ \text{MPa}.$$

Top and bottom flange splices will consist of 56 bolts each side of the double shear splice. Long joints, particularly after slippage, do not provide for an equal distribution of bolt shear stress at gross section yielding. Therefore, the average bolt shear strength will be decreased in longer joints and, effectively, the joint has a lower factor of safety against yielding than bolts in shorter joints. However, theoretical and experimental investigations have shown that for joints less than about 1250 mm long, the factor of safety (FS) remains at least 2.0, which is acceptable (Kulak et al., 1987). For long joints, it is often recommended to consider reducing the allowable bolt shear stress by 20% to ensure FS ≥ 2.0.

Web plate splices:

Try a web splice using 14 mm plates with 48 bolts each side of the splice in the 2370 mm web plate shown in Figure E9.11a.

The gross section shear strength of the girder web plate is

V_{all} = (18)(2370)(0.35)(350)/1000 = 5226 kN.

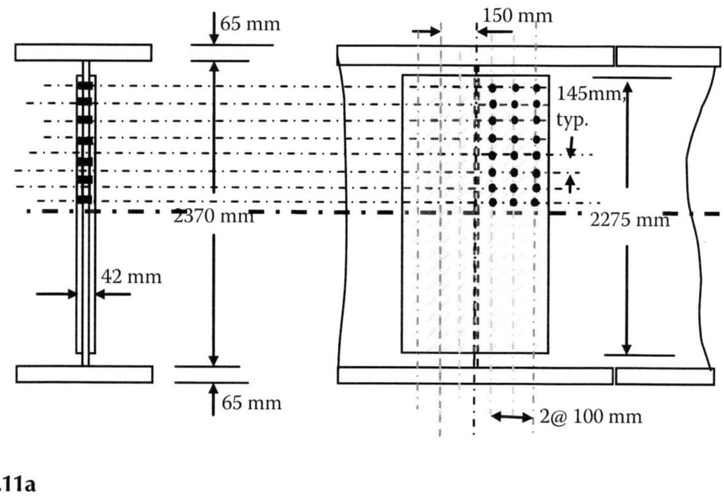

FIGURE E9.11a

The gross section shear strength of the web splice is

$$V_{sw} = \frac{2((2275)(14)(0.35)(350))}{1.5} = 5202 \text{ kN (less than 1\% less than web plate strength), OK}$$

$$f_{vs} = \frac{1.5(5226)(1000)}{2(2275)(14)} = 123 \text{ MPa} \leq 0.35\ (350) \leq 123 \text{ MPa, OK.}$$

The net section moment of inertia of the web plate is

$$= \left(\frac{18(2370)^3}{12} - (25)(18)(2)[1087.5^2 + 942.5^2 + 797.5^2 + 652.5^2 + 507.5^2 + 362.5^2 + 217.5^2 + 72.5^2] \right)$$

$$= (19,968 - (25)(18)(2)[3.574]) \times 10^6 = (19,968 - 3217) \times 10^6 = 16,751 \times 10^6 \text{ mm}^4.$$

The net section flexural strength of the girder web is

$$M_w = 0.55(350)\frac{(16,751)}{1185} = 2721 \text{ kNm} \text{ (17\% of girder minimum (compressive bending capacity).}$$

The net section moment of inertia of the web splice is

$$= 2\left(\frac{14(2275)^3}{12} - (25)(14)(2)[3574(1000)] \right) = 22,470 \times 10^6 \text{ mm}^4.$$

The net section flexural strength of the web splice is

$$M_{sw} = 0.55(350)\frac{(22,470)}{1185} = 3650 \text{ kNm (23\% of girder minimum (compressive) bending capacity).}$$

The web splice plates have a net moment of inertia and net section bending strength greater than that of the web plate.

Design for the combined forces of the flexural strength of the net section of the web and the maximum shear force at the splice:

$M_c = M_w = 2721$ kNm

$V_c = 2000$ kN

$\tau_c = 2000(1000)/(2(14)(2275)) = 31.4$ MPa

$\sigma_c = 2721 \times 10^6/(2(14)(2275)^2/6) = 112.7$ MPa.

The interaction between flexure and shear in the web splice plates is 31.4/(0.35(350)) + (112.7/(0.55(350)) = 0.26 + 0.59 = 0.84 ≤ 1.0, OK.

The web splice flexural stress at the net section flexural strength of the girder web is

$$= \frac{2721(1137.5)}{22,470} = 137.8 \text{ MPa.}$$

The interaction between flexure and shear in the web splice plates is (see Chapter 7)

$$f_b = \left(0.75 - 1.05\frac{f_v}{F_y}\right)F_y = \left(0.75 - 1.05\frac{31.4}{350}\right)350 = 0.66(350) = 229.5 \text{ MPa, use } 0.55(350) = 192.5$$

MPa ≥ 137.8 MPa, OK.

The maximum direct shear stress, t, on the bolts is

$$\tau = \frac{V}{n_s n_b A_b} = \frac{2000(1000)}{2(48)(380)} = 54.8 \text{ MPa (all bolts with same Ab)}$$

and the maximum torsional shear stress in the *x*- and *y*–directions is

$$\tau_{Tx} = \frac{(Ve + M_{sw})y_T}{n_s A_b J_s}$$

$$\tau_{Ty} = \frac{(Ve + M_{sw})x_T}{n_s A_b J_s}$$

$$Ve = 2000(150)/1000 = 300 \text{ kNm}$$

$$M_{sw} = 2721 \text{ kNm}$$
$$V_e + M_w = 300 + 2721 = 3021 \text{ kNm}$$

$$J_s = \sum n_b(x_T^2 + y_T^2) = 4\left[3\left(3574(1000)\right) + 8\left(250^2 + 150^2 + 50^2\right)\right] = 4(10{,}722 + 700)(1000)$$

$$= 45{,}688 \times 10^3 \text{bolt-mm}^2$$

$$\tau_{Tx} = \frac{3021(1087.5)(1000)}{2(380)(45{,}688)} = 94.6 \text{ MPa}$$

$$\tau_{Ty} = \frac{3021(250)(1000)}{2(380)(45{,}688)} = 21.8 \text{ MPa.}$$

The resultant shear stress on the most highly stressed bolt is

$f = \sqrt{(\tau + \tau_{Ty})^2 + \tau_{Tx}^2} = \sqrt{(54.8 + 21.8)^2 + 94.6^2} = 121.7$ MPa, which exceeds the allowable shear stress of 120 MPa by less than 1.5%, OK.

Check bearing stress

$$\sigma_{bc} = \frac{121.7(380)}{1(14)(25)} = 132 \le f_B \text{ MPa}$$

$f_B = \dfrac{l_e F_u}{2d_b} = \dfrac{(40)(450)}{2(22)} = 409$ MPa, assuming a minimum 40 mm loaded edge distance, OK

or

$$f_B = 1.2F_u = 1.2(450) = 540 \text{ MPa.}$$

Example 9.9b (US Customary and Imperial Units)

Design the bolted flange and web splices for the girder shown in Figure E9.8b with flange splices located where $M = 8800$ kips-ft and web splices where $M_V = 6000$ kips-ft and $V = 450$ kips.

The girder has the following properties (see Example 7.1b):

$I_g = 223{,}444$ in.^{4}
$S_g = 4965$ in.^{3}
$S_n = 4469$ in.^{3}
$A_w = 53.13$ in.^{2}

Allowable girder compressive bending capacity, $M_{call} = 0.55(F_y)(S_g) = 0.55(50)(4965)/12 = 11{,}378$ kips-ft (assuming fully laterally supported).

Allowable girder tensile bending capacity, $M_{tall} = 0.55(F_y)(S_n) = 0.55(50)(4469)/12 = 10{,}241$ kips-ft.

Allowable girder shear capacity, $V_{all} = 0.35(F_y)(A_w) = 0.35(50)(53.13) = 929.8$ kips.

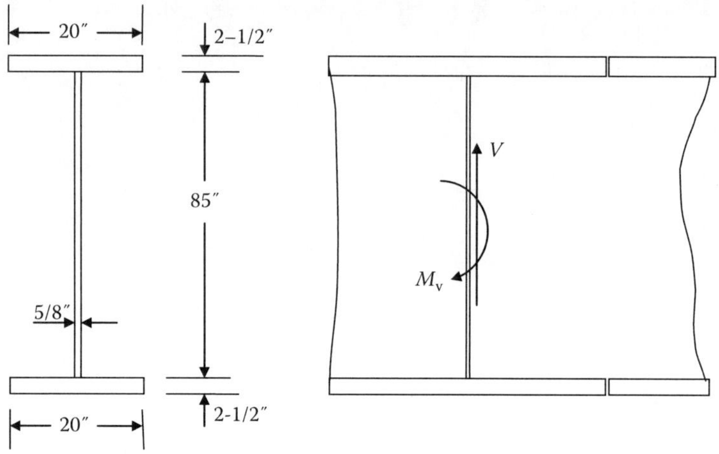

FIGURE E9.8b

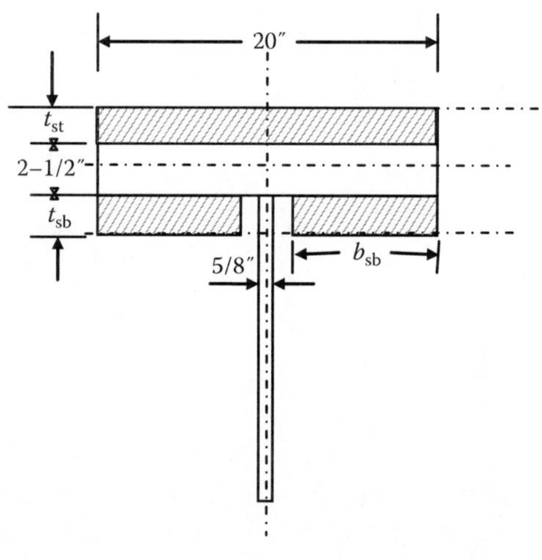

FIGURE E9.9b

Compression flange splices (Figure E9.9b):

There must be enough bolts each side of the splice to develop the flange strength.

The maximum force in the compression flange plate at the splice location (average at the centroid) is

$$F_f = \frac{8800(12)}{4965}\frac{(1+42.5/45)}{2}(2.5)(20) = 21.3(0.97)(50) = 1031.5 \text{ kips}.$$

The allowable force in the compression flange plate at the splice location (average at the centroid) is

$$F_f = \frac{11,378(12)}{4965}\frac{(1+42.5/45)}{2}(2.5)(20) = 27.5(0.97)(50) = 1333.7 \text{ kips}.$$

Transfer of this force from girder flange to splice plate is by a slip-resistant lap joint using A325 bolts $f'_{bv} = 17.0$ ksi. The number of single shear bolts required is

$$n_{fs} = \frac{1333.7}{(2)(17.0)(0.6)} = 65$$ bolts each side of the splice. Use 16 rows of 4 bolts (64 bolts OK for force of 1314 kips).

For the splice cross section to be at least equal to flange cross section (50 in.^{2})

$t_{st} = 1.75$ in.

$t_{sb} = 1.00$ in.

$b_{sb} = 8$ in.

Thickness of splice = 1.75 + 1.00 + 2.5 = 5.25″.

The area of the splice is 1.75(20) + 2(1.00)(8) = 35.0 + 16.0 = 51.0 in.^{2} ≥ 50 in.^{2}, OK

$$y_{splice} = 5 - \frac{(35.0)(4.375) + (16.0)(0.5)}{51.0} = 5.25 - 3.16 = 2.09'' \text{ from top of the splice.}$$

The top flange splice centroid is 2.09 − (1.75 + 2.5/2) = −0.91 in. from the top flange centroid (0.91 in. farther from the girder neutral axis).

The spliced flange eccentricity = 0.91 in. This eccentricity creates a maximum force bending moment of 1031.5(0.91)/12 = 78.2 kips-ft which is small (less than 1% of the maximum moment at the splice location) and, therefore, is acceptable without further consideration.

Check bearing stress

$$\sigma_{bc} = \frac{1333.7}{64(1.00)(1.00)} = 20.8 \le f_B \text{ ksi}$$

$$f_B = \frac{l_e F_u}{2d_b} = \frac{(1.5)(65)}{2(7/8)} = 55.7 \text{ ksi, assuming a minimum 1.5 in. loaded edge distance, OK}$$

or

$$f_B = 1.2F_u = 1.2(65) = 78 \text{ ksi.}$$

Tension flange splices (Figure E9.10b)

There must be enough bolts each side of the splice to develop the flange strength.

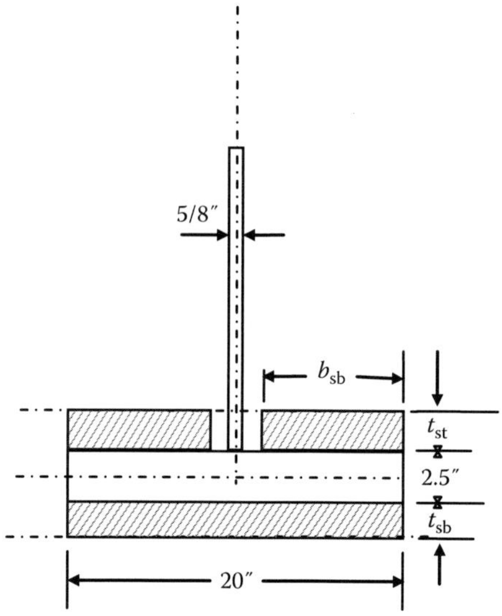

FIGURE E9.10b

The maximum force in the tension flange plate at the splice location (average at the centroid) is

$$F_f = \frac{8800(12)}{4965}\frac{(1+42.5/45)}{2}(50-4(1.0)(2.5)) = 21.3(0.97)(40) = 825.2 \text{ kips.}$$

The allowable force in the tension flange plate at the splice location (average at the centroid) is

$$F_f = \frac{10241(12)}{4469}\frac{(1+42.5/45)}{2}(40) = 27.5(0.97)(40) = 1067.0 \text{ kips.}$$

Transfer of this force from girder flange to splice plate is by a slip-resistant lap joint using A325 bolts f'_{bv} = 17.0 ksi. The number of double shear bolts required is

$n_{fs} = \dfrac{1067}{(2)(17.0)(0.60)}$ = 52 bolts each side of the splice. Use 13 rows of 4 bolts.

For the splice cross section to be at least equal to flange cross section (50 – 4(1.00)(2.5) = 40 in.^{2}) try

$t_{st} = 2.0$ in.

$t_{sb} = 1.25$ in.

$b_{sb} = 8$ in.

Flange thickness at splice = 2.0 + 2.5 + 1.25 = 5.75″.

The gross area of the splice is 2.0(20) + (2(1.25)(8.0))= 40.0 + 20.0 = 60.0 in.^{2}

The net area of the splice is [(2.0(20) – 4(1.0)(2.0)) + (2(1.25)(8.0) – 4(1.25)(1.0))= 32.0 + 15.0 = 47.0 in.^{2} ≥ 40.0 in.^{2}, OK

$$y_{splice} = \frac{(40.0)(1.0)+20.0(5.13)}{60.0} = 2.38 \text{ in.}$$

The bottom flange splice centroid is (2.0 + 2.5/2) – 2.38 = 0.88 in. from the bottom flange centroid (0.88″ farther from the girder neutral axis).

Girder strength at compression and tension flange splices:

The girder gross moment of inertia = $I_g = 223,444$ in.^{4}

The girder neutral axis is at $\dfrac{50.0(88.75)+53.13(45)+50.0(1.25)}{(153.13)} = 45.0$ in. from the underside of the bottom flange and the girder net section moment of inertia is

$I_n = 223,444 - \left(4(1.0)(2.5)(45-(2.5/2))^2\right) = (223444 - 19141) = 204,303$ in.^{4} (1.6% greater than assumed in Example 7.1b).

The girder allowable tensile bending capacity, M_{tall} = 0.55(50)(204,303/45) = 27.5(4540)/12 = 10,404 kips-ft.

The gross and net moment of inertia of the spliced section with both top and bottom flange splices in the same section is

$$I_g = 51.0(43.75+0.91)^2 + 60.0(43.75+0.88)^2 + (8.93+2(0.67)) + (13.33+2(1.30)) + \frac{0.625(85)^3}{12}$$

$$= 101720 + 119,510 + 1.3 + 15.9 + 31986 = 253,233 \text{ in.}^4$$

$I_n = 253,233 - \left(4(1.0)(5.75)(44.13)^2\right) = (253,233 - 44,781) = 208,452 \geq 204,303$ in.^{4}, OK, the moment of inertia at the splice is greater than moment of inertia of the cross section being spliced.

Check bearing stress

$$\sigma_{bc} = \frac{1066.0}{52(1.00)(1.00)} = 20.5 \le f_B \text{ ksi}$$

$$f_B = \frac{l_e F_u}{2d_b} = \frac{(1.5)(65)}{2(7/8)} = 55.7 \text{ ksi, assuming a minimum 1.5 in. loaded edge distance, OK}$$

or

$$f_B = 1.2F_u = 1.2(65) = 78 \text{ ksi.}$$

Since the weaker tension flange splice will govern at ultimate conditions, a stronger compression flange splice is not required. Therefore, both top and bottom flange splices will consist of 52 bolts each side of the double shear splice, provided the splice is strong enough to transmit the actual compression flange force of 1031.5 kips.

$$n_{fs} = \frac{1031.5}{(2)(17.0)(0.6)} = 51 \le 52 \text{ provided, OK.}$$

Long joints, particularly after slippage, do not provide for an equal distribution of bolt shear stress at gross section yielding. Therefore, the average bolt shear strength will be decreased in longer joints and, effectively, the joint has a lower factor of safety against yielding than bolts in shorter joints. However, theoretical and experimental investigations have shown that for joints less than about 50 in. long, the factor of safety (FS) remains at least 2.0, which is acceptable (Kulak et al., 1987). For long joints, it is often recommended to consider reducing the allowable bolt shear stress by 20% to ensure FS ≥ 2.0.

Web plate splices:

Try a web splice using 1/2″ plates with 42 bolts each side of the splice in the 85″ web plate shown in Figure E9.11b.

The gross section shear strength of the girder web plate is $V_{all} = (0.625)(85)(0.35)(50) = 929.8$ kips.

The gross section shear strength of the web splice is

$$V_{sw} = \frac{2((81)(0.50)(0.35)(50))}{1.5} = 945 \text{ kips} \ge 929.8 \text{ kips, OK}$$

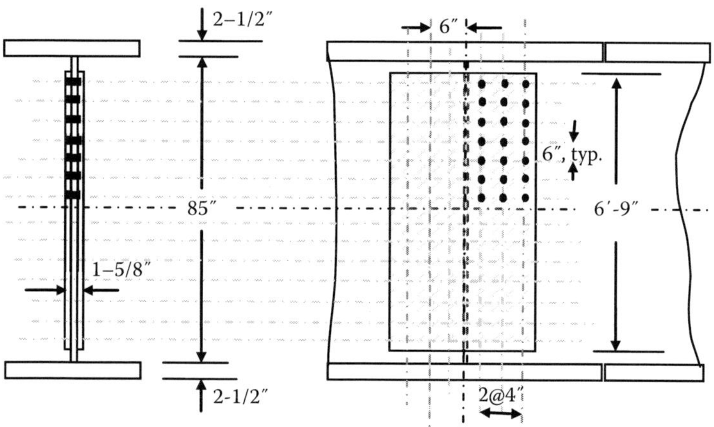

FIGURE E9.11b

$$f_{vs} = \frac{1.5(929.8)}{2(81)(0.5)} = 17.2 \le 17.5 \text{ ksi, OK.}$$

The net section moment of inertia of the girder web plate is

$$= \left(\frac{0.625(85)^3}{12} - (1.00)(0.625)(2)(39^2+33^2+27^2+21^2+15^2+9^2+3^2)\right) = 31{,}986 - 5{,}119 = 26{,}867 \text{ in.}^4.$$

The net section flexural strength of the girder web is

$$M_{sw} = 0.55(50)\frac{(31{,}986-5{,}119)}{42.5(12)} = 1449 \text{ ft-kips.}$$

The net section moment of inertia of the web splice is

$$= 2\left(\frac{0.5(81)^3}{12} - (1.00)(0.5)(2)(39^2+33^2+27^2+21^2+15^2+9^2+3^2)\right) = 36{,}097 \text{ in.}^4$$

The web splice plates have a net moment of inertia greater than that of the web plate.

Design for the combined forces of the flexural strength of the net section of the web and the maximum shear force at the splice:

$M_c = M_w = 1449$ ft-kips
$V_c = 450$ kips
$\tau_c = 450/(2(0.5)(81)) = 5.56$ ksi
$\sigma_c = 1449(12)/(2(0.5)(81)^2/6) = 15.9$ ksi.

The interaction between flexure and shear in the web splice plates is $5.56/(0.35(50)) + (15.9/(0.55(50)) = 0.32 + 0.58 = 0.90 \le 1.0$, OK.

The web splice flexural stress at the net section flexural strength of the girder web is

$$= \frac{1449(12)(40.5)}{36{,}097} = 19.5 \text{ ksi.}$$

The interaction between flexure and shear in the web splice plates is (see Chapter 7)

$$f_b = \left(0.75 - 1.05\frac{f_v}{F_y}\right)F_y = \left(0.75 - 1.05\frac{5.56}{50}\right)350 = 0.63(50) \text{ ksi, use } 0.55(50) = 27.50 \text{ ksi} \ge 19.5 \text{ ksi, OK.}$$

The maximum direct shear stress, t, on the bolts is

$$\tau = \frac{V}{n_s n_b A_b} = \frac{450}{2(42)(0.60)} = 8.9 \text{ ksi } (\text{all bolts with same } A_b)$$

and the maximum torsional shear stress in the x- and y-directions is

$$\tau_{Tx} = \frac{(Ve + M_{sw})y_T}{n_s A_b J_s}$$

$$\tau_{Ty} = \frac{(Ve + M_{sw})x_T}{n_s A_b J_s}$$

$$Ve = 450(0.5) = 225.0 \text{ ft-kips}$$

$$M_{sw} = 1449 \text{ ft-kips}$$

$$J_s = \sum n_b(x_T + y_T)^2 = 4[(39^2+10^2)+(33^2+10^2)+(27^2+10^2)+(21^2+10^2)+(15^2+10^2)+$$

$$+(9^2+10^2)+(3^2+10^2)+(39^2+6^2)+(33^2+6^2)+(27^2+6^2)+(21^2+6^2)+(15^2+6^2)+$$

$$(9^2+6^2)+(3^2+6^2)+(39^2+2^2)+(33^2+2^2)+(27^2+2^2)+(21^2+2^2)+(15^2+2^2)+$$

$$+(9^2+2^2)+(3^2+2^2)] = 4(13{,}265) = 53{,}060 \text{ bolt-in.}^2$$

$$\tau_{Tx} = \frac{1674(12)(39)}{2(0.60)(53{,}060)} = 12.3 \text{ ksi}$$

$$\tau_{Ty} = \frac{1674(12)(10)}{2(0.60)(53{,}060)} = 3.2 \text{ ksi.}$$

The resultant shear stress on the most highly stressed bolt is

$f = \sqrt{(\tau + \tau_{Ty})^2 + \tau_{Tx}^2} = \sqrt{(8.9+3.2)^2 + 12.3^2} = 17.2$ ksi, which is considered acceptable at 1.3% overstress, OK.

Check bearing stress

$$\sigma_{bc} = \frac{17.2(0.60)}{1(0.5)(1.00)} = 20.6 \le f_B \text{ ksi}$$

$$f_B = \frac{l_e F_u}{2d_b} = \frac{(1.5)(65)}{2(7/8)} = 55.7 \text{ ksi, assuming a minimum 1.5 in. loaded edge distance, OK}$$

or

$$f_B = 1.2F_u = 1.2(65) = 78 \text{ ksi.}$$

REFERENCES

Al-Emrani, M., 2005, Fatigue performance of stringer-to-floor-beam connections in riveted railway bridges, *Journal of Bridge Engineering*, Vol. 10, No. 2, 179–185.

American Railway Engineering and Maintenance-of-Way Association (AREMA), 2015, Steel structures Chapter 15, in *Manual for Railway Engineering*, Lanham, MD.

American Welding Society (AWS) AASHTO/AWS D1.5M/D1.5, 2010, *Bridge Welding Code*, Miami, FL.

Blodgett, O.W., 2002, *Design of Welded Structures*, Lincoln Arc Welding Foundation, Cleveland, OH.

Chesson, E., Faustino, N.L., and Munse, W.H., 1965, High-strength bolts subjected to tension and shear, *Journal of the Structural Division*, Vol. 91, No. 5, 155–180.

Douty, R.T. and McGuire, W., 1965, High strength bolted moment connections, *Journal of the Structural Division*, Vol. 91, No. 101–128.

Faella, C., Piluso, V., and Rizzano, G., 2000, *Structural Steel Semirigid Connections*, CRC Press, Boca Raton, FL.

Federal Highway Administration (FHWA), 1998, *Heat-Straightening Repairs of Damaged Steel Bridges*, FHWA-IF-99-004, Washington, DC.

Kulak, G., 2002, *High Strength Bolts*, AISC Steel Design Guide 17, Chicago, IL.

Kulak, G., Fisher, J.W., and Struik, J.H.A, 1987, *Guide to Design Criteria for Bolted and Riveted Joints*, 2nd ed., John Wiley & Sons, New York.

Kuzmanovic, B.O. and Willems, N., *Steel Design for Structural Engineers*, 2nd ed., Prentice-Hall, Englewood Cliffs, NJ.

Leon, R.T., 1999, Partially restrained connections. Chapter 4, in *Handbook of Structural Steel Connection Design and Details* (Tamboli, A.R., ed.), McGraw-Hill, New York.

Nair, R.S., Birkemoe, P.C., and Munse, W.H., 1974, High strength bolts subjected to tension and prying, *Journal of the Structural Division*, Vol. 100, ST2, 351–372.

Rakoczy, A.M. and Nowak, A.S., 2013, Reliability-based strength limit state for steel railway bridges, *Structure and Infrastructure Engineering*, Vol. 10, No. 9, 1248–1261.

Research Council on Structural Connections (RCSC), 2000, Specification for Structural Joints Using ASTM A325 or A490 Bolts, AISC, Chicago, IL.

Research Council on Structural Connections (RCSC), 2014, Specification for Structural Joints Using High-Strength Bolts, AISC, Chicago, IL.

Thornton, W.A. and Kane, T., 1999, Design of connections for axial, moment and shear forces. Chapter 2, in *Handbook of Structural Steel Connection Design and Details* (Tamboli, A.R., ed.), McGraw-Hill, New York.

Whitmore, R.E., 1952, *Experimental Investigation of Stresses in Gusset Plates*, University of Tennessee Experiment Station Bulletin 16.

Yen, B.T, Zhou, Y., Wang, D., and Fisher, J.W., 1991, Fatigue behavior of stringer-floorbeam connections, Lehigh University Report 91-07, Bethlehem, PA.

10 Construction of Steel Railway Bridges

Superstructure Fabrication

10.1 INTRODUCTION

Designers[*] of steel railway bridge superstructures must consider fabrication constructability in order to initiate practical and cost-effective bridge designs. Effective fabrication[†] is contingent upon design drawings and technical specifications[‡] that clearly communicate the design engineer's intent. The design documentation must be accurate and complete in order that the fabricator can provide the railroad company with relevant fabrication costs and create precise shop (or detailed) drawings for review by the designer and subsequent fabrication. A plethora of requests for information (RFI) related to the design drawings and/or specifications during cost estimating and/or shop drawing production may be an indication that this was not realized.

The fabrication process is complex and generally involves material preparation, punching/drilling, assembly (bolting and/or welding), coatings application, finishing, and quality control (QC) inspection. However, prior to engaging in these shop activities, shop drawings (produced from the design drawings and specifications) and material procurement (based on the shop drawings and specifications) are required. These fabrication planning functions are typically on the steel superstructure fabrication critical path, reiterating the need for clear and, where possible, standardized design drawings and technical specifications. Standardized design details, specifications, and inspection requirements contribute significantly to the achievement of high quality, low cost, and expeditious steel superstructure fabrication.

The superstructure designer must consider the availability and procurement lead time for the required material grade and member dimensions (typically plates and rolled shapes) early in the design process in order that material procurement is not delayed by shop drawing bills of materials (BOM) including unavailable or not readily obtainable shapes or plates.[§] In addition, as the superstructure design process proceeds from conceptual to detailed, the designer must continually consider typical shop practices relating to material preparation, fastening, and assembly to realize a successful design. When final design drawings and technical specifications are completed, the fabricator must promptly produce shop drawings and associated BOM for cost estimating, material procurement, production sequence, and shop scheduling purposes. The BOM should clearly indicate the number, overall dimensions, and weight of plates and shapes required. The BOM may also be used by superstructure erectors for the determination of required construction site storage and crane lift capacity requirements[¶] (see Chapter 11).

---

[*] Designers are typically professional engineers employed by the railroad company and/or consulting engineering firms acting on the railroad's behalf or, in the case of design-build projects, the contractor's behalf.

[†] From quality, schedule, and cost perspectives.

[‡] Design drawings, models, and specifications are critical design documentation in the contract documents for the fabrication and/or erection of the steel superstructure. Discrepancies between design documents may occur and, in such cases, documents other than the specifications are considered to govern.

[§] The designer must consider material procurement early in the design stage and the fabricator should verify availability at the time of shop drawing production.

[¶] In such cases, the erection and fabrication contractors may need to clearly communicate concerning any constraints to the size and weight of assemblies for erection.

Shop drawing reviews, RFI responses (typically for design clarifications or confirmations), and approvals must be promptly managed by design engineers in order that material procurement and fabrication processes may proceed without delay. Shop drawing reviews by design engineers are typically performed for general conformance to the design drawings and technical specifications. Nevertheless, the responsibility for the accuracy of shop drawings rests with the fabricator. The design engineer must also, on the bridge owner's behalf, arrange for any required fabrication quality assurance (QA) inspections.

10.2 FABRICATION PLANNING

10.2.1 Project Cost Estimating

In some cases, fabricators may be able to estimate costs for initial bids based on typical or similar fabrication projects to those shown on preliminary* design drawings provided, if practicable, by the design engineer. Alternatively, the fabricator may be able to estimate costs for preliminary bids based on initial quantity take-offs from preliminary design drawings.† However, typically, the fabricator will estimate costs for bids based on the technical specifications and quantity take-offs from issued or released design drawings. Bid estimates must also include costs related to transportation (of raw and finished materials), storage, handling, and profit. Detailed cost estimates, made after bid estimates, are based on material, labor, and machine effort costs developed from the quantity take-offs from the shop drawings and BOM, and, if necessary, updated costs of transportation, storage, handling, and profit.

10.2.2 Shop Drawings for Steel Fabrication

Shop drawings serve as clearly written communication between owner (typically a railroad company), design engineer, fabricator, and QC/QA inspectors. Shop drawings are prepared by the fabricator's own, and/or contracted, steel detailers working from approved (issued or released) design drawings and technical specifications.

The design drawings must clearly outline steel grades, member attributes,‡ member positions, working points, coatings, and, typically for railway superstructures, main member connection details. Bolted connection details should include bolt sizes, locations, quantities, and material grade. Some connections, such as light bracing connections, not detailed on design drawings, may be designed by the fabricator§ based on connection forces¶ shown on the design drawings. Welded connections shown on design drawings or in technical specifications must indicate the type, size, length, and weld metal strength required.

Shop drawings prepared from approved design drawings must include material requirements (quantity, grades, dimensions, and weights of plates and shapes), and detail the size of components, location of holes, and weld types. Detailers must be aware of any adjustments to the dimensions necessary to accommodate camber, weld shrinkage, edge finishing, or other fabrication process requirements. Shop drawings must also indicate tolerances** for cutting, drilling, and assembly,

* Railroad companies and bridge owners must be aware of the potential for change orders that may increase cost and/or schedule due to substantial revisions to the preliminary design documents on which the bid was based. However, such changes in scope must be demonstrated to be substantive enough to generate changes in cost and/or schedule of the superstructure fabrication.

† This may involve the necessity of scaling some dimensions not shown on preliminary drawings, which should be done by only experienced estimators.

‡ Plate and shape sizes, thicknesses, camber, and controlling dimensions, and if a fracture critical member (FCM).

§ Connections in steel railway superstructures that are designed by the fabricator should be performed by, or under the direction of, a professional engineer and not a steel detailer. The railroad company's design engineer will typically review the fabricator's connection design documentation prior to fabrication.

¶ For the allowable stress design method of American Railway Engineering and Maintenance-of-Way Association (AREMA) Chapter 15, the forces should be provided at the service load level.

** Fabrication tolerances for steel railway superstructures are recommended in AREMA Chapter 15, Part 3. The tolerances apply to members fabricated by welding but are also considered as reasonable tolerances for rolled shapes [except as specified by the American Society for Testing Materials (ASTM) A6] and bolted fabrication.

and designate fracture critical members (FCMs) and welds requiring specific minimum material toughness requirements.* The fabricator will typically indicate the use of prequalified welds on the shop drawings. The prequalified welds are qualified in accordance with an appropriate bridge welding code.† FCM welds also require specific processes and procedures to attain the required toughness for fracture resistance.‡ Welding procedure specifications (WPS) for the welds must also be indicated on shop drawings. Mechanical connections requiring slip-critical (SC) fasteners (typically, bolted connections in steel railway superstructures are specified to be SC as outlined in Chapter 9) must also be clearly indicated on shop drawings.

For small projects, shop drawings and BOM may be produced manually or with relatively simple computer aided design (CAD) applications. However, detailers for larger projects in modern fabrication shops may work from computer models to create digital information and working drawings for computer numerically controlled (CNC) operation of shop processes involving cutting, punching, drilling, and welding. Modern digital shop drawing files may be efficiently developed from collaborative digital design drawing files, and subsequently used to create machine program codes for CNC operations. If well managed, this integrated approach to design and fabrication can reduce the schedule and cost of fabrication.

Integrated digital technology is currently used for many large construction projects§ and may become the norm for even typical fabrication projects in the future. In addition to decreasing schedule and cost, CNC operations can also enhance fabrication quality. Also, digital drawing files may be used by large erection contractors for field erection operations. Synergistic digital design, shop, and erection drawing files can also improve erection safety, schedule, and cost.

Detailed cost estimating, production sequence, and schedule planning may proceed following approval of shop drawings and associated BOM. Shop fabrication can commence after material procurement, verification, and production planning review.

10.2.3 Fabrication Shop Production Scheduling and Detailed Cost Estimating

Final fabrication cost estimates are typically based on a shop production schedule¶ to carefully determine the material and labor quantities required to complete the fabrication in accordance with the approved shop drawings. The shop production schedule must be monitored and updated to remain current during the fabrication process. Monitoring and updating the shop production schedule will enable the fabricator to control the cost and schedule of the production.

Shape and plate requirements (in terms of dimensions and grades) are based on quantity take-offs from the shop drawings and BOM. The estimated cost and the availability of the specified shape and plate materials will depend on steel production mill rolling schedules and the accessibility of materials from steel service centers.** In addition to raw materials costs, the price of shop supplies, consumables,†† bolts, and other related materials‡‡ is necessary for detailed cost estimates. Estimated labor and machine effort costs to perform the various shop activities relating to material preparation,

* The design engineer must review shop drawings to ensure that all FCMs are identified for fabrication in accordance with the specified fracture control plan (FCP).

† For example, American Welding Society (AWS) AASHTO/AWS D1.5M/D1.5—Bridge Welding Code in the United States and Canadian Standards Association (CSA) W59—Welded Steel Construction (Metal Arc Welding) in Canada.

‡ The requirements are outlined in AREMA Chapter 15, Section 1.14, which references the AASHTO/AWS D1.5M/D1.5 FCP.

§ For example, building information modeling (BIM) is used in the design, construction, and management of many large scope civil engineering projects.

¶ The shop production schedule must consider the fabrication production sequence and resources (labor and equipment) available for the fabrication work.

** Typically, these companies stock more commonly used steel grades, shapes, and plate sizes.

†† Typically, welding electrodes.

‡‡ For example, bearings, walkways, coatings, and drainage systems.

punching/drilling, bolting, welding, coatings, finishing, assembly, and QC* inspection must also be skillfully estimated. The total estimated cost of fabrication must also include the cost of office† and shop overheads, inventory, raw materials transportation and storage, materials handling and layout, transportation of fabricated superstructure assemblies to the erection site, and profit.

The dimensions of raw materials and fabricated assemblies required for longer span steel railway superstructures may be large and require special consideration of costs for transportation,‡ storage, and materials handling. Shipment of large dimensional raw materials and most fabricated assemblies is often feasible and economic where fabricators have access to a rail line. This is particularly the case for the transportation of fabricated steel railway superstructure assemblies to be erected on a railway line.

10.2.4 Material Procurement for Fabrication

Since the superstructure designer must determine the availability and consider the procurement lead time of various grades and sizes of steel plates and shapes early in the design process, design philosophies associated with effective material procurement need to be deliberated. General design principles associated with material procurement are the use of, if possible, readily available plate and shape sizes and grades and the use of thin plates.§ Thinner plates are economical to purchase, handle, and fabricate. The procurement lead time required for thinner plates is less than that for thicker heat-treated or quenched and tempered (Q&T) low-alloy steel plates (see Chapter 2). Thinner plates, with good strength and toughness properties, may be more readily produced by the thermomechanical control process (TMCP) than thicker plates requiring a longer production cycle and more costly Q&T operations. Thin plates are produced with larger width and length dimensions than thick plates, enabling, in many cases, fabrication without splicing.¶ In addition, optimum girder sections use relatively thin flange plates (see Chapter 7). Nevertheless, the large plate sizes and thickness required for long-span railway girder webs and flanges may require special order from either steel service centers or steel production mills. Successful material specification is initiated with a review of current steel mill and service center plate and rolled shape inventories.** Plates and shapes used for FCM must be ordered as killed fine-grain practice steel (see Chapter 2). Mill orders must also indicate any supplementary requirements for fracture toughness, chemistry,†† and/or heat treatment.‡‡ Welded repairs to raw materials are not acceptable for FCM plates and shapes. To expedite fabrication, some raw materials, such as large plates and consumables, should be procured, based on the project design drawings and technical specifications, early in the fabrication planning phase.

The major raw materials (plates and rolled shapes) may be procured from the fabricators' inventory, steel production mills, or steel service centers. Some fabricators may purchase and inventory commonly specified plate and quantities of commonly rolled shapes obtained from steel production mills. Purchasing directly from mills is economical, but quantity and grade availability, particularly for shapes, depend on the mill production schedule, which is not likely coincident with

---

* QC inspection is typically performed by the fabricator's qualified inspectors. Specialty QC inspectors (QCIs) may be necessary for special inspections such as radiographic testing or stress relieving.

† Including the cost of shop drawing preparation.

‡ Including loading and support equipment on special rail cars (dimensions for transportation are typically governed by railway clearances at bridges and tunnels) or trucks and highway permits.

§ With due consideration, from a design perspective, of the instability (buckling) of compression members or elements of members in compression, and shear flow in wide tension flanges.

¶ This is beneficial for girder web plates with due consideration of shear strength, shear buckling, and flexural buckling considerations (see Chapter 7).

** Updated shape and plate inventories are often available at steel mill or service center websites on the internet.

†† When there is significant tensile stress in the through-thickness direction of plates, design engineers may specify low sulfur steel plate production with a maximum of 0.010% sulfur.

‡‡ Supplementary requirements for fracture toughness, chemistry, and/or heat treatment will increase the cost of raw materials so the design engineer must carefully consider the necessity of specification of supplementary mill order requirements.

the project fabrication schedule. In such cases, purchase of materials from steel service centers is often required. It is more costly than purchasing from steel mills, but steel service centers typically carry inventories of small plates and rolled shapes purchased from various steel production mills. Nevertheless, smaller quantities of rolled shapes are often procured from steel service centers. Mill tolerances for plates and shapes are required in order to enable effective fabrication of superstructure assemblies and are specified in the appropriate standard or specification.*

Hot rolled shapes and plates have residual stresses created by an unequal cooling rate across the section from the steel production mill shape and plate rolling processes. These residual stresses are generally uniform along the length of the shape or plate. Figure 10.1 shows the residual stress pattern in hot rolled plates with rolled edges† typically supplied by universal plate mills. Figure 10.2 illustrates the residual stress pattern created across a rolled shape as the center of the flange cools

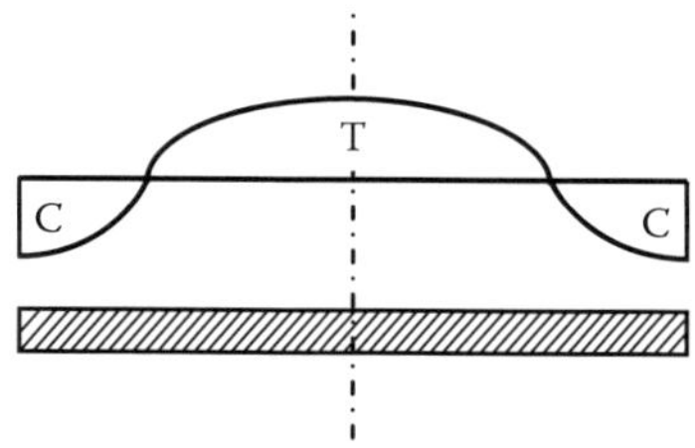

FIGURE 10.1 Residual stresses in hot rolled plates.

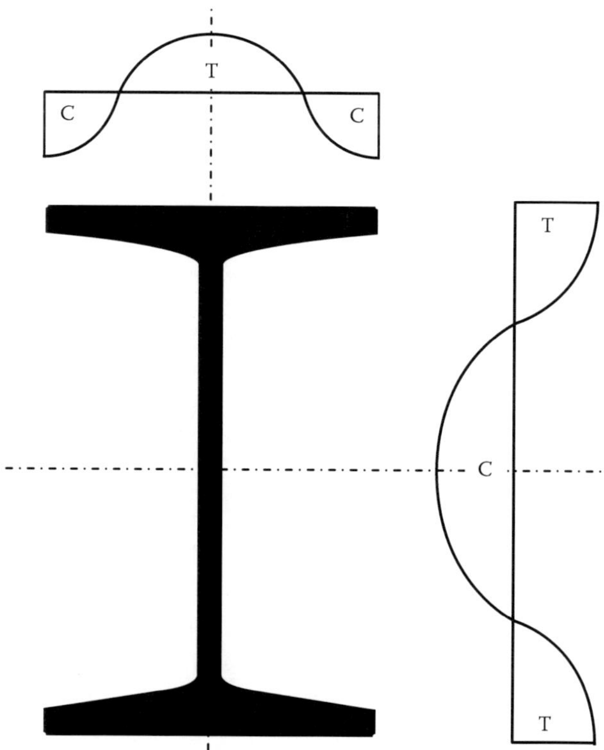

FIGURE 10.2 Residual stresses in hot rolled I beam shapes.

* In North America, ASTM A6/A6M generally applies.
† Often referred to as universal mill plate steel.

more slowly developing residual tensile stresses, which are balanced for equilibrium by compressive residual stresses at the edges of flanges. These residual stresses are typically not of concern for routine superstructure design, but may have to be considered in some specific design cases.*

Other raw materials required for steel superstructure fabrication, such as fasteners, consumables, bearing materials, coatings, walkway components, drainage devices, and/or other appurtenances, are often obtained from specialty suppliers or manufacturers.

During, and at the completion of material procurement, the fabricator must verify mill certificates and arrange to conduct or document supplemental tests† for submission to the owner and/or design engineer. Once shop drawings are approved and materials procured, the fabrication process may commence.

10.3 STEEL FABRICATION PROCESSES

The fabrication process includes many interrelated and complex shop activities. The fabrication process generally involves material preparation, cutting, punching/drilling, fit-up assembly, bolting, welding, coatings application, finishing, and QC inspection. Cutting, straightening, bending, curving, cambering, welding, and heat treatment processes also have specific requirements for nonredundant FCM fabrication.

10.3.1 Material Preparation

The material quality and dimensions of plates and shapes received from steel service centers or steel production mills must be in accordance with the requirements of ASTM A6 or other applicable specification.‡ Nevertheless, raw materials may arrive at the fabrication shop in need of preparation for the fabrication process. Material preparation involves layout, marking, and cutting and may include straightening, bending, curving, cambering, surface preparation, and heat treatment. Heat and/or mechanical straightening may be used to adjust cross section, flatness, straightness, camber, and/or sweep to within specified tolerances. If required, residual stresses in plates and shapes may be relieved by heat treatment.§

10.3.1.1 Layout and Marking of Plates and Shapes

Materials must be prepared so they may be placed and stored on the shop floor at a location and position, and in time for pieces to be readily accessed during the relevant fabrication sequence. Plates and shapes may require straightening (to within specified tolerances for subsequent fabrication) before layout (Figure 10.3). Piece marking should be done with crayons, tags, or low-stress steel die stamps specified as Fatigue Category B or better (see Chapter 5). Nevertheless, stamped piece marks must not be located near welds, holes, edges, or other discontinuities. FCM materials must be piece marked in accordance with the approved shop drawings.

10.3.1.2 Cutting of Plates and Shapes

Plate and shape cutting methods used in preparation for steel shop fabrication¶ depend on edge preparation specification, dimensional tolerances, and material thickness. A common steel cutting method is thermal flame cutting with oxy-acetylene or pure oxygen gases. Metals may also be cut by thermal plasma gas cutting. Plasma cutting is fast and provides smooth and clean cut surfaces but is typically less effective than flame cutting for steel plates in excess of about 25 mm (1 in.)

* For example, residual stresses are considered for compression member buckling (see Chapter 6).
† For example, supplemental Charpy V-Notch (CVN) toughness testing of ASTM A588 steel to determine suitability for FCM use.
‡ Dimensional tolerances are typically specified in the appropriate standard or specification.
§ However, this is typically not required at the material preparation stage.
¶ Specifically, assembly and welding.

FIGURE 10.3 Layout of a long plate girder. (Courtesy of Dr. K. Frank, Hirschfeld Industries, Austin, TX, USA. With Permission.)

thick and effectively limited to cutting plates less than about 65 mm (2.5 in.) thick. Thermally cut surface roughness in the form of small notches (potentially detrimental stress raisers depending on the service load stress state of the component) may be precluded with speed- and direction-controlled automatic cutting (Figure 10.4). Otherwise, supplemental trimming of rough cut surfaces by grinding or machine planing may be required. Automated or mechanized thermal cutting also increases shop productivity [e.g., girder flanges can be economically flame cut at same time using guides (Figure 10.5)].

Steel railway superstructure fabrication may require the cutting of relatively thick plates, which is usually accomplished with mechanically guided thermal flame cutting equipment. Thermal cutting should be done from both sides of plates to be used as girder flanges to limit distortions. FCM plates are typically specified to be thermally cut to size for fabrication. Sheared edge and universal

FIGURE 10.4 Controlled automatic cutting. (Courtesy of Y. Martin, Canam Bridges Inc., St. Foy, Qc, Canada. With Permission.)

FIGURE 10.5 Flame cutting girder flange plates. (Courtesy of Dr. K. Frank, Hirschfeld Industries, Austin, TX, USA. With permission.)

mill plate* for FCM must be trimmed at least 5 mm (3/16 in.) by thermal cutting. However, the rapid cooling of plates near thermal cuts, particularity cuts in high-strength steel and thick plates, may create heat-affected zones (HAZ) with reduced toughness. It is costly, but to preclude potential detrimental edge effects† in thick plates, a slow cutting rate combined with preheat and/or postheat treatment may be necessary. Thermally cut holes in Q&T steel should have a minimum of 1.5 mm (1/16 in.) of cut surface material removed by grinding or another appropriate process. Thermal cutting of thick plates may create residual stresses as shown in Figure 10.6.

Plates of moderate thickness may be sheared and trimmed.‡ Thinner plates may also be sheared or, in some modern shops, cut by laser and water cutting systems. Shapes may be cut with saws or thermal cutting equipment. Girder web and flange plates are generally trimmed about 6 mm (1/4 in.) from sheared edges.

Plates and shapes may require finishing following cutting. Finishing by edge planing or end milling is often required for compression members (e.g., the milled ends of truss compression members should be in full contact for uniform bearing during service).

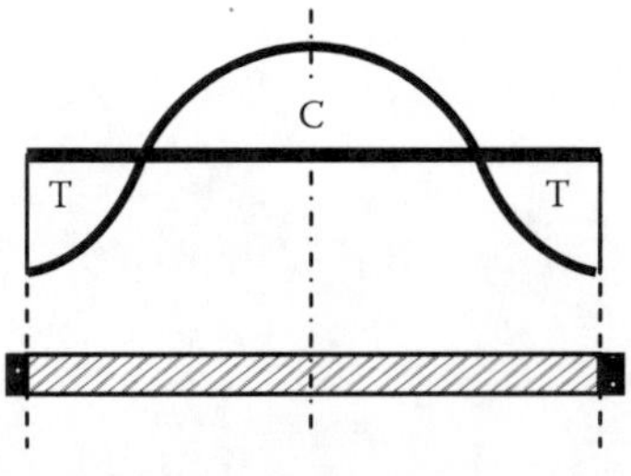

FIGURE 10.6 Residual stresses in thermally cut thick plates.

* Plate rolled to width and thickness by vertical and horizontal rolls, respectively.

† Thermally cut edge surfaces must not contain any gouges or other discontinuities.

‡ Typically trimmed about 1.5 mm (1/16 in.).

10.3.1.3 Straightening, Bending, Curving, and Cambering of Plates and Shapes

Raw materials may arrive at the fabrication shop in need of straightening and, in preparation for fabrication, some plates and shapes may require localized curving (weak direction bending) or cambering (strong direction bending).

Shapes of relatively small section and thin plates that are supplied cambered or with sweep may be straightened in cold bending machines. However, plates thicker than about 50 mm (2 in.) with sweep or camber generally require heat or flame straightening. Heat straightening must be performed below phase shift temperatures and is generally specified to be accomplished at a maximum temperature of 650°C (1200 F). In some cases, the aid of applied forces from mechanical or hydraulic jacks is required.* Heat straightening requires considerable experience and skill. Detailed guidelines concerning the heat straightening of plates and shapes and the effects of heating on steel material properties are provided by the US Department of Transportation Federal Highway Administration (FHWA, 1998).

Locally bent plates required for fabrication may be bent cold or with heat. Cold bending is an intentional inelastic deformation. The inelastic deformation is created by the application of localized stresses into the plastic deformation or strain hardening regions of material behavior (see Chapter 2). Subsequent unloading is elastic and, therefore, generates localized residual strain. Cold work may detrimentally affect strength and ductility, which may be restored by stress-relieving processes such as normalizing or annealing. The plate must not crack from localized loss of strength and/or ductility resulting from the cold bending process; and the minimum cold bend radius is controlled by rolling direction, steel grade, and plate thickness. To preclude cracking at plate edges, the edges must be smooth and chamfered in the area of the cold bend. High-strength high-performance steel, Q&T, and TMCP steel plates are most effectively bent by uniformly heating the area to be bent to a maximum of 600°C (1100 F)† followed by crack testing.‡

Curved plates may be achieved by cold bending or through applied heat using the same principles that were used for the heat straightening of plates with a weak direction bend or sweep (Figure 10.7).

Lightly cambered plates may also be straightened by cold bending. Alternatively, small camber may be removed by plate trimming. However, heat application using hydraulic jacks is typically required for larger camber.

FCM must not be cold bent. Corrective or intentional bending, curving, or cambering of FCM must be performed by heat application.

10.3.1.4 Surface Preparation

The surfaces of steel plates and shapes may require preparation for appearance, coatings, and connection faying surfaces, or to meet surface tolerances specified by the design engineer.§ If required, atmospheric corrosion-resistant steel may be blast cleaned to the Steel Structures Painting Council (SSPC) specification SP6 to enhance appearance.¶ Surface preparation to SSPC specification SP6 is shown in Figure 10.8. Surfaces with deposits of materials detrimental to atmospheric corrosion resistance, coatings adherence, or appearance may be removed in accordance with the appropriate SSPC specification SP1, SP2, SP3, or SP7.**

Bolted SC connection faying surface requirements are recommended by AREMA (2015) and are summarized in Tables 10.1 and 10.2. It should be noted that the Research Council on Structural

---

* In such cases, the magnitude of the applied force must not create stresses in excess of 50% of the specified material yield stress, F_y.

† To avoid detrimental changes in material physical properties.

‡ It is often specified to dye-penetrant test (DPT) or magnetic particle test (MPT) the surface and edges of plate bends made by hot bending.

§ Typically, ASTM A6.

¶ May be specified for exterior beam, girder, and truss surfaces at specific sites.

** SSPC specifications for solvent, hand tool, power tool, and brush-off blast cleaning, respectively.

FIGURE 10.7 Weak direction curvature of a plate girder. (Courtesy of Dr. K. Frank, Hirschfeld Industries, Austin, TX, USA. With permission.)

FIGURE 10.8 Surface preparation to SSPC specification SP6. (Courtesy of Y. Martin, Canam Bridges Inc., St. Foy, QC, Canada. With permission.)

Connections (RCSC) and the American Institute of Steel Construction (AISC) have recently revised the slip coefficients and allowable SC connection shear stresses as shown in Tables 10.3 and 10.4 and many engineers currently use these requirements for the design and fabrication of SC connections.

10.3.1.5 Heat Treatment

Shop normalizing and annealing may be required.* Supplementary normalizing or quench and tempering heat treatments may also be specified for specific applications of FCM. Congruently, FCM

* For example, normalizing moderately thick plates over 40 mm (1–1/2 in.) thick that have not been TMCP produced may be specified to ensure good metallurgical and physical properties.

TABLE 10.1
SC Connection Faying Surface Requirements (AREMA)

SC Bolted Connection Faying Surface Class	Description	Slip Coefficient (Coating for Each Class Must Meet or Exceed)	Allowable Shear Stress for Grade A325 Bolts in Standard Holes (MPa)
A	Clean mill scale and blast cleaned	0.33	115
B	Blast cleaned surfaces	0.50	190
C	Hot-dip galvanized	0.40	150

TABLE 10.2
SC Connection Faying Surface Requirements (AREMA)

SC Bolted Connection Faying Surface Class	Description	Slip Coefficient (Coating for Each Class Must Meet or Exceed)	Allowable Shear Stress for Grade A325 Bolts in Standard Holes (ksi)
A	Clean mill scale and blast cleaned	0.33	17.0
B	Blast cleaned surfaces	0.50	28.0
C	Hot-dip galvanized	0.40	22.0

TABLE 10.3
SC Connection Slip Coefficients and Allowable Shear Stresses (RCSC and AISC)

SC Bolted Connection Faying Surface Class	Surface Description	Slip Coefficient (Coating for Each Class Must Meet or Exceed)	Allowable Shear Stress for Grade A325 Bolts in Standard Holes (MPa)
A and C	Clean mill scale and blast cleaned	0.30	90
B	Blast cleaned	0.50	150
D	Blast cleaned	0.45	135

TABLE 10.4
SC Connection Slip Coefficients and Allowable Shear Stresses (RCSC and AISC)

SC Bolted Connection Faying Surface Class	Surface Description	Slip Coefficient (Coating for Each Class Must Meet or Exceed)	Allowable Shear Stress for Grade A325 Bolts in Standard Holes (ksi)
A and C	Clean mill scale and blast cleaned	0.30	12.9
B	Blast cleaned	0.50	21.5
D	Blast cleaned	0.45	19.3

welds may require postwelding heat treatments* for hydrogen diffusion.† Heat treatment techniques are outlined in AWS D1.5M/D1.5 and other industry standards and guidelines.

10.3.2 Punching and Drilling of Plates and Shapes

Holes may be punched, drilled, subdrilled and reamed, or subpunched and reamed for mechanical connection bolts. Holes in beams, girders, trusses, and FCM carrying vertical live load should be either drilled full size; or subpunched or subdrilled and reamed to full size. Holes for connections in cross frame, diaphragm, and bracing members may be punched full size, unless otherwise specified by the design engineer. In some cases, design drawings and specifications may specify that punching full size is not permitted and holes must be reamed or drilled full size. Punching of holes is limited to material 19 mm (3/4 in.) thick [22 mm (7/8 in.) for mild carbon steel‡]. Subdrilling or subpunching is done 3 or 6 mm (1/8 or 1/4 in.) smaller than the finished hole diameter, depending on the thickness of the material and number of plies being reamed. In any case, cost- and schedule-effective punching, reaming, and drilling are achieved during fabrication from design drawings showing standard size holes of the same diameter§.

Nevertheless, shop assembly of connections stressed by vertical live load, or other connections designated by the design engineer, may be required for drilling and reaming of fastener holes.

Shop and field connections in plate girders, truss chords, and beams stressed by vertical live load must be drilled full size; or subdrilled or subpunched and reamed to full size with pieces assembled¶. Shop and field connections not stressed by vertical live load (typically bracing, intermediate stiffeners, walkway components, etc.) may have holes punched (unless specifically prohibited by the design engineer), drilled, subdrilled and reamed, or subpunched and reamed, with or without assembly of parts. Field splices in beams, girders, and truss chords must be reamed or drilled full size with members assembled. In many cases, assembly of connections in the shop is difficult, and drilling and reaming to final hole diameter may be accomplished using fabricated connection plates or templates. In many cases, assembly of connections in the shop is difficult, and drilling and reaming to final hole diameter may be accomplished using fabricated connection plates or templates. However, drilling full size without assembly is feasible with CNC process drilling. Steel railway girders are typically simply supported and often less than 45 m (150 ft.) long, which will typically allow shop assembly of the girders in order to drill and ream connection holes.

The punching, drilling, and reaming requirements vary for the different assembly requirements associated with shop and field connection fasteners. Nevertheless, the primary concerns for both punched and drilled hole connections are hole surface quality and dimensional accuracy (to ensure true connection fit).

10.3.2.1 Hole Quality

Punched, drilled, or reamed holes must be made 2 mm (1/16 in.) larger than the nominal diameter of the fastener. Mechanical fasteners used in steel railway superstructures are typically 22 mm (7/8 in.) diameter high strength steel bolts** in 24 mm (15/16 in.) diameter holes.

Fastener holes may behave as local stress concentrations and, particularly in cyclically stressed tension zones, holes with poor edge conditions may precipitate cracking. Finished holes must be clean and devoid of cracks, notched edges, or edge burrs. It is good practice to chamfer the edges of punched, drilled, or reamed holes 1.5 mm (1/16 in.) by light grinding. Punched holes, particularly in

* Postweld heat treatments are any thermal applications greater than 480°C (900 F) other than short period heating temperatures from welding and/or heat straightening, curving, or cambering.

† May be specified to prevent cracking and lamellar tearing.

‡ Such as ASTM A36 or A709 Grade 36 steel.

§ Typically, 24 mm (15/16 in.) diameter holes.

¶ This is particularly important for field connections to ensure proper field fit to preclude erection delays.

** In accordance with ASTM F3125-15 Grade A325M (A325) specifications.

thicker plates, can result in localized embrittlement at the hole surface that may create short cracks radiating from the edge of the hole. Such cracking can lead to fatigue crack initiation and propagation or brittle fracture (if temperature and material toughness are low). Drilling avoids such cracking, and reaming removes any embrittled or cracked edge material.

10.3.2.2 Punching and Drilling Accuracy for Shop and Field Fasteners

To ensure dimensional accuracy, holes for field connections that are drilled, subdrilled and reamed, or subpunched and reamed without assembly of pieces must be made using steel templates with hardened steel bushings. The templates must be precisely positioned and secured to the work to ensure dimensional accuracy of the connection. If CNC full size drilling without assembly is used, it is often necessary to prepare representative check shop assemblies for periodic verification that the CNC equipment is providing accurate connection dimensions.

10.3.3 Shop Assembly for Fit-Up of Steel Plates and Shapes

Members and connections must fit to within specified tolerances* for welded and bolted shop assembly and subsequent field erection.† The fabrication process includes shop assembly following material preparation, punching, drilling (or subpunching, subdrilling for subsequent reaming), and cutting. Shop assembly involves bringing together and alignment of parts, components, and/or segments of the superstructure using fit-up bolts, clamps, wedges, pins, guy lines, struts, and tack welds (transverse tack welds on tension flanges are prohibited by AREMA 2015). In some cases, the temperature differential between shop fabricated members or assemblies and planned field erection temperature of members or assemblies may need careful consideration when determining fabricated member dimensions.

Longitudinal beams, girders, and trusses are shop assembled for overall geometry, dimensional accuracy,‡ connection alignment, hole and edge conditions, gap tolerances, and interaction with other accessories§ of the superstructure. Relatively complex spans are often completely shop assembled (assembly of both longitudinal and transverse members) for fit prior to shipment¶ (Figures 10.9a and b).

10.3.3.1 Fabrication of Cambered Superstructure Assemblies

Shop assembly of longitudinal beams, girders, and trusses may require consideration of specified camber** requirements. Longitudinal beams and girders for railway spans must be assembled to reflect the AREMA (2015) recommendation that girders be cambered for dead load when greater than 27.5 m (90 ft) long. Trusses must also be assembled to accommodate the AREMA 2015 recommendation that trusses be cambered for a load comprising the dead load plus a uniform live load of 45 kN/track meter (3000 lbs per track foot).††

Longitudinal beams, girders, and trusses are typically fabricated in the no-load‡‡ (or cambered) position at shop temperature. No-load assembly may occur in the horizontal or vertical position

* For example, fabrication and erection tolerances in AREMA Chapter 15, AWS D1.5M/D1.5, AISC Code of Standard Practice for Steel Buildings and Bridges and/or other appropriate specification for fabrication and erection tolerances.

† This is of critical importance for railway superstructure where erection must typically occur in an accelerated manner with minimum interruption to railway traffic. For example, North American Class 1 railroads typically may provide only 4–8 h/day (sometimes at night) a few days a week with only a few longer planned track outages.

‡ Length, height, center-to-center distances, squareness, verticality (plumbness), camber, and other dimensions critical for further shop assembly and/or field erection.

§ For example, bearings, shear connectors for composite construction, handrails, and drainage devices.

¶ Depending on handling and shipping constraints, the span may be shipped assembled or may be disassembled and shipped in smaller assemblies.

** Strong direction curvature of beams, girders, or trusses fabricated as the difference between the geometric shape (shape under specified camber load at the normal ambient temperature of bridge site) and the no-load shape at shop temperature.

†† Representing light railway cars.

‡‡ Members fabricated for field erection fit-up as though no self-weight or other dead load is present.

FIGURE 10.9 (a) Complete shop assembly of a BTPG span (one girder). (b) Complete shop assembly of a BTPG span (two girders). ((a) Courtesy of F. Mathieu, SGS Inc., Quebec City, QC, Canada; (b) Courtesy of the author, Calgary, AB, Canada.)

if properly sized and placed supports are provided.* Laydown assembly (horizontal assembly of girders and trusses) (Figure 10.3) will match and connect segments or members with geometry that avoids self-weight deflections.†

Rolled shapes may be cambered by heat application. However, it is recommended that the specified camber for girders be cut into the web plate rather than being made by heat application. The specified camber for trusses can be achieved by varying the length of chord members. Tension members are fabricated shorter than, and compression members longer than, the member's geometric

* For truss assembly, such support must be installed at truss panel points.

† Self-weight deflections can be avoided for assembly in the vertical position if supported in the no-load (or cambered) positions.

length[*] by an amount equal to the axial deformation from the specified camber load.[†] In some cases, the variation in member length related to the difference between shop member fabrication temperature and ambient temperature at erection may also need to be considered (Durkee, 2014). Furthermore, if the gusset plates at chord joints are fabricated for connecting members at geometric angles, secondary bending stresses may occur in members as they are erected. The secondary bending moments are created in members at the truss joints as members are erected and joints are fitted up.[‡] In many typical simply supported trusses, the secondary steel erection stresses may be relieved upon application of the dead load to be carried by the trusses. Secondary bending stresses are also established in truss members when the truss is subjected to live loads creating a deflected shape different from the geometric shape.[§] Therefore, the assembly of members of trusses requires that designers are aware of truss camber fabrication methods in order to avoid, or account for, secondary bending stresses in calculations for truss member design.

10.3.3.2 Shop Assembly of Longitudinal Beams, Girders, and Trusses

Full[¶] or complete[**] girder and truss shop assembly may be required to ensure members and connections fit within specified tolerances for field erection.

Longitudinal beams, girders, and some trusses less than about 45 m (150 ft) long can usually be fully shop assembled by drilling or reaming connection holes to final size. For typical steel railway superstructures, accurate secondary member[††] erection fit-up is readily achieved where fabrication drilling is conducted by CNC or by using templates with hardened bushings in holes. In addition, longitudinal beams, girders, and trusses less than about 45 m (150 ft) may be generally shipped fully[‡‡] assembled.

In some cases, complete assembly (Figure 10.9) is required[§§] to ensure accelerated erection of the span or due to proposed erection procedures.[¶¶] Some short plate girder spans may be shipped completely assembled (fully assembled girders including transverse members[***]). Longer through plate girder spans are typically completely shop assembled but may be disassembled and shipped with girders and floor system segments as separate subassemblies.[†††] Assembly of the field connections of primary members[‡‡‡] in the shop and reaming, as necessary, is also typically required.

In many cases, full or complete girder and truss assembly to ensure members and connections fit within specified tolerances for field erection is practical in the fabrication shop. However,

---

[*] The geometric length of a member is the length shown in its final configuration, as shown on the design drawings. Final configuration occurs when the member is subjected to the specified camber load.

[†] Designers must also consider any member strengthening required for erection and account for the section change when calculating axial deformation from the specified camber load.

[‡] Truss members will require forced fitting when erected in final supported condition. Forced fitting will induce the secondary stresses in the truss members.

[§] The specification of a camber load exceeding the dead load (dead load plus a uniform live load of 45 kN/track meter (3000 lbs per track foot) reduces secondary bending stresses in members when the truss is subjected to live loads creating a deflected shape different from the geometric shape.

[¶] Longitudinal beam, girder, or truss assembly without transverse member assembly.

[**] Longitudinal beam, girder, or truss assembly with transverse member assembly.

[††] For example, diaphragm, cross frame, and lateral bracing members.

[‡‡] Long girders and trusses up to about 45 m (150 ft) long may be shipped by rail on bolsters (allowing for the "string-lining" effect at track curves) on two adjacent flat deck freight cars.

[§§] Many railway superstructures are less than about 45 m (150 ft) in length but may require complete shop assembly to ensure that field erection can occur without interruption to railway traffic. This is typically the case for ballasted through plate girder (BTPG) spans, movable bridge spans, viaduct structures, spans with skewed or otherwise complex connections of transverse members, and where floor system stringers are continuous.

[¶¶] Such as fit-up assurances for incremental erection, falsework support, staged construction, or complex field connections.

[***] Typically, bracing for deck plate girder spans and floor systems for through plate girder spans.

[†††] Due to shipping or handling constraints or erection contractor preference (e.g., when cranes with adequate lifting capacity for completely assembled spans are not available or able to access the site).

[‡‡‡] For example, field connections of girders, trusses, floorbeams, and stringers.

handling of large* or complex segments is typically difficult within the fabrication shop environment and sequential or progressive assembly may be required. The design engineer, owner, fabricator, and erector† must carefully consider the appropriate progressive shop assembly method for the superstructure being fabricated. For example, progressive assembly of cross frames and diaphragms for skewed spans is difficult and requires careful planning with the design engineer.

10.3.3.3 Progressive Shop Assembly of Longitudinal Beams, Girders, and Trusses

Progressive shop assembly of long girders and trusses may be required to ensure members and connections fit within specified tolerances for subsequent shop assembly operations and field erection.

Full assembly (assembly without transverse members) of longitudinal beams, girders, and trusses greater than about 45 m (150 ft) long may be accomplished by drilling or reaming connection holes to final size with the main members progressively assembled. Accurate secondary member erection fit-up is achieved where fabrication drilling is conducted by CNC or by using templates with hardened bushings in holes. Progressive full assembly involves the fitting-up of geometrically controlled superstructure segments.‡ In addition, progressive full assembly of longitudinal segments requires consideration of camber requirements for each segment or panel. Also, relative rotations and deflections associated with the shop assembly method need to be considered. The minimum full assembly length is typically specified as three§ adjacent girder segments, truss panels, or chord lengths (between field splices) not less than about 45 m (150 ft) long. The progressive fully assembled length will not be less than the minimum full assembly length¶ by adding advancing sections or panels before the removal of rear segments.

Unless the superstructure is highly skewed or has complex connections,** progressive complete assembly (progressive assembly with transverse members) is generally not necessary. Nevertheless, in some cases, progressive complete assembly is required to ensure accelerated erection of the span or due to proposed erection procedures. Through plate girder spans longer than about 45 m (150 ft) may be progressively completely assembled and shipped with girder segments and floor system segments†† as separate subassemblies. Assembly of the field connections of primary members in the shop and reaming, as necessary, is also usually required.

Some special assemblies may require that the fabricator engage in large section lay downs or vertical 3D assemblies. Such assembly requirements are not within usual shop practice and should be specified by the design engineer in consultation with structural steel fabrication engineers.

10.3.3.4 Shop Assembly of Bolted Splices and Connections

Splices and bolted connections are shop assembled to ensure that they are within fabrication and erection tolerances (Figures 10.10a and b). Splices and bolted connections in primary members should be fully assembled‡‡ before reaming or drilling to full size. Assembled splice plates should

---

* Segment or piece size that can be handled depends on shop size and crane capacities. Nevertheless, members or segments of longitudinal beams, girders, and trusses about 45 m (150 ft) long are typically the maximum length for practical shop handling.

† If the erector can be brought into the process prior to fabrication, which is particularly beneficial for complex bridges or site conditions. Otherwise, experienced erection engineers may be able to provide useful guidance on typical field erection procedures appropriate for the superstructure to be fabricated.

‡ Girder or beam segments, arch rib sections or truss panels.

§ In some cases, progressive shop assembly using only two girder segments or truss panels is used.

¶ Typically two or three girder segments or truss panels depending on the superstructure owner's requirements.

** Or if directed by the owner or design engineer.

†† Typically, there are multiple floor system segments per girder segment, which should be standardized as much as practicable to reduce fabrication cost, schedule, and errors. TPG and BTPG spans longer than about 45 m (150 ft) are not common for heavy axle load freight railway superstructures.

‡‡ Components assembled with drift pins and fit-up bolts to bring pieces into position.

FIGURE 10.10 (a) Girder web plate splice. (Courtesy of Dr. K. Frank, Hirschfeld Industries. With permission.) (b) Girder flange plate splice. (Courtesy of Dr. K. Frank, Hirschfeld Industries.) ((a) Courtesy of Dr. K. Frank, Hirschfeld Industries, Austin, TX, USA; (b) Courtesy of Dr. K. Frank, Hirschfeld Industries, Austin, TX, USA.)

be match marked and disassembled to clean solvents, oil, burrs, and mill scale* from the connection faying surfaces.

Splice assembly may be precluded when full size CNC hole drilling with fabricator QC and segment verification procedures,† acceptable to the design engineer, are used. Alternatively, assembly with the member flange or web ends and splices individually drilled to templates may be acceptable.

Connection assembly may also be precluded with CNC drilling or use of hardened bushing templates. Connection angles in floor systems‡ are not milled or otherwise finished unless required

* Typically removed by sand blasting.

† In many cases, partial or segmental assembly of longitudinal superstructures is used to ensure tolerances for subsequent field erection. Verification assemblies, in which at least the first and last segments are tested for accuracy, are often a key component of the procedure.

‡ Typically used in stringer-to-floorbeam, floorbeam-to-girder, and floorbeam-to-truss chord connections.

during assembly. Specifications typically stipulate the permissible connection thickness reduction allowed by milling.*

10.3.3.5 Fit-Up for Shop Welded Splices and Connections†

Shop welded connections are made with fillet or groove welds. Shop welded splices are typically made with complete joint penetration (CJP) groove welds. Shop splices of flange plates and web plates in girder spans should be a minimum of about 150 mm (6 in.) apart. Shop welded splices in shapes are made through the entire section.

Welding codes specify fillet welded joint fit-up gaps.‡ Fillet welds made within specified gap tolerances will not require consideration of increasing the weld size or other measures to ensure adequate throat thickness is achieved. If used, fillet welds for the attachment of stiffeners to girder web plates should proceed from the tension flange toward the compression flange (Figure 10.11). Transverse stiffener welds must not intersect flange-to-web or longitudinal stiffener welds.§

Fit-up tolerances for welding are indicated in the WPS for prequalified groove welds.¶ Maximum plate separation for groove welds for butt joints (Figure 10.12) is also specified in welding codes.**

10.3.3.6 Fabrication and Erection Tolerances

Mill tolerances for plates and shapes are required in order to enable effective fabrication of superstructure assemblies. Fabrication tolerances†† for plates and shapes are required in order to ensure intended performance and enable the effective erection of the superstructure assemblies.

FIGURE 10.11 Fillet welding of transverse stiffener to girder web plate. (Courtesy of Dr. K. Frank, Hirschfeld Industries, Austin, TX, USA. With permission.)

* Often reduction of thickness is limited to about 1.5 mm (1/16 in.). However, in some cases, the design engineer may authorize thickness reductions by milling of up to 3 mm (1/8 in.) for thicker connection angles.

† AREMA (2015) does not permit field welding of main members.

‡ Typically less than 3 mm (1/8 in.) depending on elements being connected. The fit-up gap tolerances may be more stringent for girder flange-to-web fillet welds than those used to connect stiffeners and connection plates.

§ This is a very poor fatigue detail (see Chapter 5). AREMA (2015) outlines acceptable and unacceptable details for welded stiffeners and attachments.

¶ AWS D1.5M/D1.5, 2010 Sections 2 and 3.

** Typically specified separation of less than 2 mm (1/16 in.) and misalignment of less than 3 mm (1/8 in.).

†† Final member tolerances (e.g., length tolerances) are typically more stringent for fabricated members than mill tolerances.

FIGURE 10.12 Plate preparation for groove welding. (Courtesy of Dr. K. Frank, Hirschfeld Industries, Austin, TX, USA. With permission.)

Fabrication tolerances are typically specified in the appropriate standard or specification.* AREMA (2015) provides recommended tolerances for welded girders and trusses, which are also considered as good practice for bolted assemblies. Tolerances for member dimensions, straightness, curvature, sweep, camber, warp and tilt, connections, joints, splices, and stiffener location and fit are recommended. In situations where tolerances are not provided by the pertinent specification or standard, it is not acceptable to assume a default value of zero. In such instances, the designer and fabricator must communicate and agree upon an appropriate tolerance for the member or connection being fabricated.

AREMA (2015) recommends explicit values for the allowable deviations in length for members with faced† end connection angles (e.g., stringers and floorbeams), members with milled ends (e.g., truss compression members), framed members not milled or faced (e.g., bracing and diaphragms), and other superstructure members. The recommended tolerances for compression member ends are also good practice for tension member fabrication.

Allowable deviations in the depth or width of girders are recommended in terms of the design depth or width. These should also apply to girder and truss member sections at bolted splices and to welded built-up members.

AREMA (2015) recommends allowable deviations in straightness or curvature (sweep), and camber in terms of length, and with explicit values. Except for members that are very stiff, the straightness tolerances are acceptable for field erection where member flexibility will allow for connection fit-up. However, it may be necessary to shop assemble very stiff members to ensure field connection accuracy. Girders fabricated within the sweep tolerances will generally enable effective erection of diaphragms, vertical brace frames, and/or lateral bracing due to the lateral flexibility of the girders.

Web plate flatness is not typically a problem for web plates designed in accordance with the AREMA (2015) recommended minimum thickness. Web plate distortions from shrinkage of web-to-flange and stiffener welds can be controlled by use of the appropriate welding procedures (Figure 10.13). Girder web plate flatness tolerances are provided in AREMA (2015) in terms of offsets from a template and plate thickness. The maximum deviation from flatness should not

* AREMA, AWS, AISC, or other approved fabrication and erection tolerances.

† Facing requirements are also outlined.

FIGURE 10.13 Balanced SAW welding of girder flange to web. (Courtesy of Dr. Frank, Hirschfeld Industries, Austin, TX, USA. With permission.)

exceed 75% of the web plate thickness. Camber ordinate tolerances* are typically specified in terms of member length.†

Tolerances for the twist, warp, and tilt of member elements and cross sections along the length of the member are important for erection. AREMA (2015) recommends allowable deviation in twist or parallelism of corresponding elements in a member (e.g., truss elements and girder flanges) in terms of bevel (or angle) over a given length of the member. It is important to consider these tolerances for the "squaring" of diaphragms during fabrication. Tolerances relating to the combined warp and tilt of fabricated beam and girder flanges are also recommended by AREMA (2015). Allowable deviations are maximum measured offsets in terms of flange width and fixed values.

AREMA (2015) recommends allowable deviation between the centerline of member flanges at splices and connections in terms of member depth. AWS D1.5M/D1.5 indicates this tolerance with a fixed value of 6 mm (1/4 in.). The AREMA recommendations result in larger tolerances for web plates exceeding about 1600 mm (63 in.) in depth as shown in Figure 10.14.

AREMA (2015) recommends specific allowable deviations in stiffener position or location. A smaller allowable tolerance is recommended if members are connected to the stiffeners.‡ Tolerances for bearing and intermediate transverse stiffener fit are also recommended. Bearing stiffeners are typically milled or ground to bear against flanges and/or connected to the flanges with full penetration groove welds. AREMA (2015) provides recommended fabrication tolerances for stiffener warp, tilt, and surface contact. Tolerances for the length of fill plates under stiffeners are also recommended.

AREMA (2015) also provides recommendations for the allowable deviations regarding bearing surface contact flatness and the tolerances for gaps. Allowable tolerances for surface contact deviations for column base plates and cap plates§ are provided with recommended end treatments. Seating and base tolerances for erection purposes should be considered.¶

---

* Typically, camber is cut into the web plate of long girders.

† For short spans, the camber tolerances may be near or exceed the specified camber. AREMA (2015) requires spans longer than 27.4 m (90 ft) be cambered.

‡ For example, cross bracing members in deck plate girder spans.

§ Used in steel viaduct towers.

¶ These may be stringent for some erection methods.

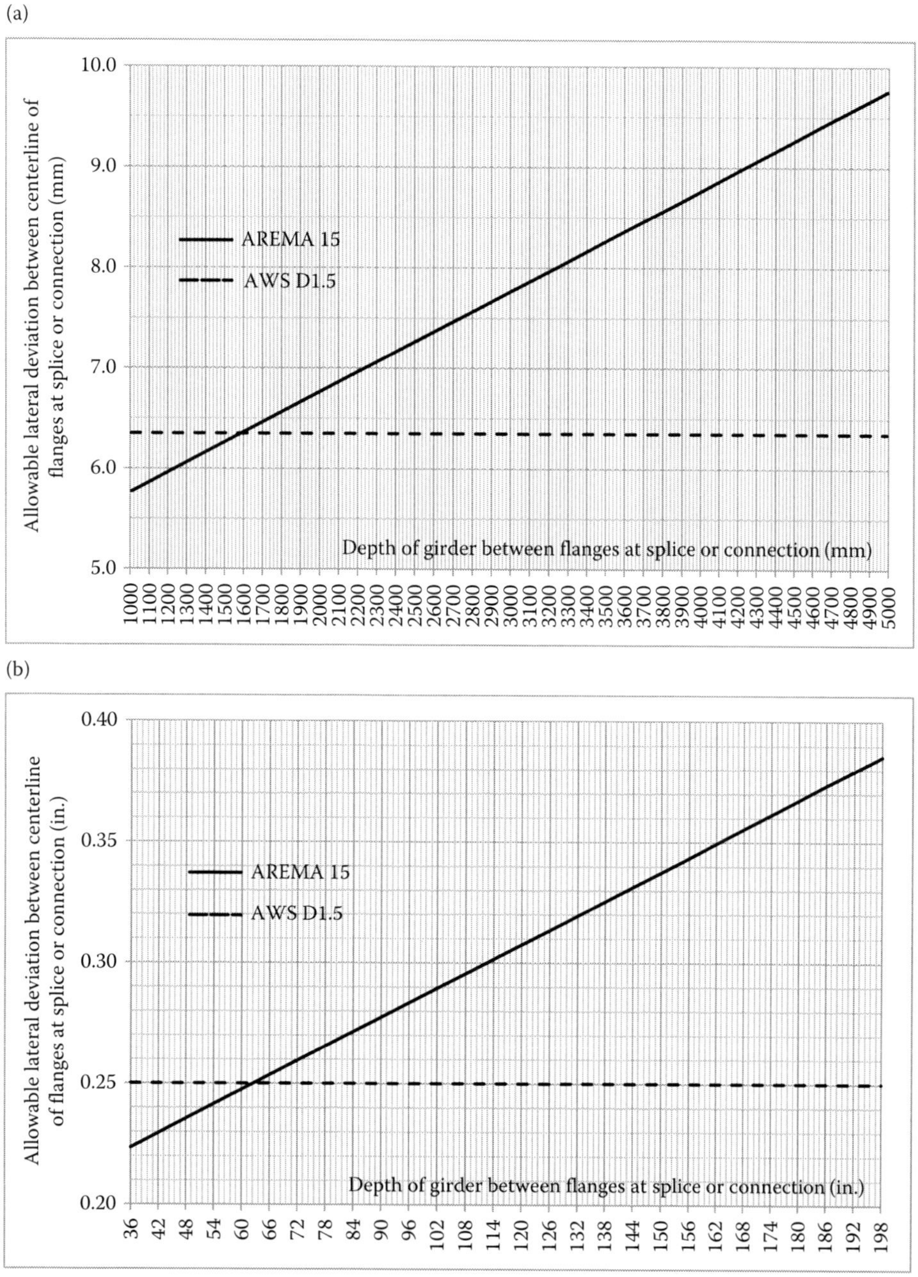

FIGURE 10.14 Allowable deviations of flanges at splices and connections (AREMA and AWS). (a) SI units (b) Imperial or US customary units.

Allowable tolerances for hole locations,* deck joints, and other member characteristics may also be specified by the design engineer in special circumstances. These can often be accommodated by fabricators, but typically at increased cost.

* Hole location deviations within a hole pattern or related to connection location. Hole location tolerances are often not specified, but good practice often limits hole location deviations to 1.5 mm (1/16 in.).

Mill and fabrication tolerances for members and connections may have accumulating effects for complex shop assemblies or erections. If required, shop assembly tolerances for entire assemblies* should be agreed upon by the design engineer, owner, and fabricator.

Specific requirements for erection tolerances typically relate to member deviations affecting position and/or alignment. Erection tolerances depend on the erection method to be employed and should be reviewed by the design engineer, owner, fabricator, and erector to ensure they are attainable for both fabrication and erection.

10.4. BOLTING OF PLATES AND SHAPES

Shop and almost all field connections require bolting. Bolting requirements vary for snug-tight (ST), pretensioned (PT), and slip critical joints. The type of joint required by the engineer's design depends on whether connection resistance to shear and/or tension is attained through the strength of the bolts or the friction on faying surfaces induced from bolt tension. Dynamic tensile loading requires PT or SC bolt installation. AREMA (2015) recommends the use of SC connections. There are additional specific faying surface requirements for SC joints. Good shop connection cleaning and coating practice should provide a slip coefficient of about 0.50 (see Chapter 9). All connections must be tightened progressively from the most rigid part of the joint to curtail relaxation of the previously tensioned bolts.

A ST connection is attained when all the connection plies are in firm contact and the bolt nuts cannot be removed without a wrench. PT and SC joints are brought to the ST condition before pretensioning of the bolts† in the connection.

PT joints are typically required if the connection is subjected to load reversals, fatigue loading without reversal, tensile fatigue loading on ASTM F-3125 Grade A325M (A325) bolts, or tension loading on Grade A490M (A490) bolts. SC joints are generally required for shear resistance with fatigue load reversal, where oversize or slotted holes are used in the connection, or where no slip of faying surfaces is tolerable in the superstructure. AREMA (2015) provides recommendations for the allowable shear in ASTM F3125 Grades A325M (A325) and A490M (A490) bolts in SC connections.

The minimum bolt tension required for PT and SC joints is provided in AREMA (2015) and RCSC (2014) (see also Chapter 9). The minimum required bolt pretension can be obtained by turn-of-nut, calibrated wrench, and tension control techniques. All these methods of bolt installation require that a tension calibrator be used to verify‡ the connection. Bolt installation by calibrated wrench also requires supplemental wrench calibrations using a tension calibrator.

Turn-of-nut pretension is attained at a specified nut rotation from the ST condition, depending on the bolt length and the angle of the pieces or plies to the bolt axis. The nut and end of the bolt should be marked before turning for subsequent inspection.

Twist-off-type tension-control (TC) bolts in accordance with ASTM F3125 Grades F1852M (F1852) or F2280M (F2280) may be tensioned from ST condition using a special TC bolt installation wrench that works in conjunction with the bolt spline mechanism. The spline at the end of the TC bolt shears off at the correct bolt tension using the special torqueing tool.

Direct tension indicators (DTIs) in accordance with ASTM F959M (F959) may be used to establish minimum required pretension from the ST condition through the compression of protrusions to a proper gap height.§

* Such as bents, frames, girder pairs, and decks.

† To 70% of minimum tensile strength (see Chapter 9).

‡ Typically, the verification pretension force is about 5% higher than installation pretension [e.g., 22 mm (7/8 in.) diameter Grade A325M (A325) bolts require a minimum verification pretension of about 180 kN (41 kips)].

§ Some proprietary DTIs provide evidence of proper installation through the release of liquid at the correct bolt tension.

Minimum required bolt pretension can also be achieved by the calibrated wrench method (typically using air impact wrenches) with daily calibration using a tension calibrator in order to adjust the wrench to the appropriate torque. Calibration should also occur due to any changes of conditions.* It is also feasible to use DTIs as tension calibrators for calibrated wrench installations to avoid the use of unreliable tables or equations relating wrench torque to pretension.

Bolt, nut, and washer assemblies for shop connections must be rotational capacity (RC) tested in tension calibrators to ensure proper strength and lubrication. RC testing is typically conducted by the fastener assembly manufacturer or distributor.

Each combination of production lots of bolts and nuts, that comply with the proof load tests of ASTM F606 (2009), and washers, of adequate hardness, are assembled and RC tested. A minimum of two assemblies from each combination of production lots† must be tested. AREMA (2015) specifies RC testing in accordance with ASTM F3125 (2015) and requires the use of a Skidmore-Wilhelm calibrator, or other device acceptable to the design engineer, for testing with a calibrated torque wrench. The torque measured, T_M, shall not exceed the following:

$$T_M \leq 0.25 P_M D \;(\text{m-N}), \qquad (10.1a)$$

where

P_M = is a measured bolt tension in the calibrator, kN,

D = the bolt diameter, mm,

or

$$T_M \leq 20.8 P_M D \;(\text{ft-lb}), \qquad (10.1b)$$

where

P_M = a measured bolt tension in the calibrator, kips·

D = the bolt diameter, in.

When the specified RC test nut rotation has been achieved, the bolt tension must be at least 15% greater than the required installation tension (see Chapter 9) as shown in Tables 10.5 and 10.6. If the bolt, nut, or washer in either of the two assemblies fails the RC test, the RC lot is rejected. The entire RC lot may be cleaned and relubricated for a single retest of the RC lot.

TABLE 10.5
Minimum Required RC Test Bolt Tension

	Minimum Required RC Test Tension (kN)	
Bolt Diameter, *D* (mm)	**Grade A325M Bolts**	**Grade A490M Bolts**
12	58	71
16	101	128
20	159	199
22	192	242
24	229	286
36	515	645

* For example, connection lubrication changes require recalibration of wrenches.

† Each production lot combination is referred to as an RC lot.

TABLE 10.6
Minimum Required RC Test Bolt Tension

Bolt Diameter, *D* (in.)	Minimum Required RC Test Tension (kips) Grade A325 Bolts	Grade A490 Bolts
1/2	14	17
5/8	22	28
3/4	32	40
7/8	45	56
1	59	74
1½	118	170

10.5 WELDING OF PLATES AND SHAPES

Shop and some field connections require welding. Welds must have at least the same strength,* ductility, fracture toughness, and corrosion resistance as the base metal being joined (see Chapter 2).

Welded field connections should be discouraged for dynamically loaded steel railway superstructures. AREMA (2015) recognizes this by restricting field welds to only minor connections not carrying live load stress and for joining deck plate sections that do not participate in carrying live load. Where field welding is permitted, it is preferable to prepare material for the proper fit and alignment of field groove welds under the more controlled conditions of the shop.

The welding processes used for steel railway superstructure shop fabrication are all variations of introducing an electric current through an electrode to create an electric arc to the base metal. The electric arc melts the welding rod or wire at one end of the arc and the base metal at other end of the arc. Pieces are then joined by solidification of the molten metal pool, which must be protected from atmospheric and localized contamination by shielding. Hydrogen in the atmosphere, oils, greases, and other fluids that may contact the welding area are particularly detrimental to weld quality. If inadequately shielded or shrouded, the solidified weld is subject to hydrogen embrittlement and cracking.

The amount of weld penetration and base metal melted, Heat affected zone properties, and weld metal grain structure are dependent on the size of the molten weld pool. The welding amperage, voltage, deposition speed, electrode size, weld geometry (angle and gap), and polarity control the size of the molten pool created during the welding process.

10.5.1 Shop Welding Processes

Shop welds for steel superstructures may be made by the submerged arc welding (SAW), shielded metal arc welding (SMAW), flux cored arc welding (FCAW), gas metal arc welding (GMAW), electroslag welding (ESW), or electrogas welding (EGW) processes. FCM must be fabricated using SMAW, SAW, FCAW, and/or, if approved by the design engineer, GMAW processes using metal cored electrodes. ESW and EGW processes are not permitted for FCM fabrication.

SAW process welds are made semi-automatically or automatically using bare welding wire submerged in granular flux shielding (blanketed). The molten flux adjacent to the weld deposit cools to a slag, which is removed. Unfused flux may be collected and recycled. SAW is particularly suited to welding low carbon and HSLA steel girders that require long flange-to-web fillet, partial penetration, or complete penetration welds (Figures 10.15 and 10.16). Long flange plates and large web plates (Figure 10.17) must be welded together with butt splices before making the flange-to-web

* These are matched-strength groove welds. Under-matched-strength welds may be specified by the design engineer provided that the minimum fracture toughness energy is 34 J (25 ft-lb) at −30°C (−20 F).

FIGURE 10.15 SAW flange-to-web welding with girder in vertical position. (Courtesy of Dr. K. Frank, Hirschfeld Industries, Austin, TX, USA. With permission.)

FIGURE 10.16 SAW flange-to-web welding with girder in horizontal position. (Courtesy of Y. Martin, Canam Bridges Inc., St. Foy, QC, Canada. With permission).

FIGURE 10.17 Web plate butt weld joint. (Courtesy of Y. Martin Canam Bridges Inc., St. Foy, QC, Canada. With permission.)

welds. The SAW process is also an effective means of creating girder flange and web plate butt splices. Heavy wires with high heat input may be required for thick elements to enable a reasonably fast butt weld deposition rate with adequate depth of penetration. Pre- and postweld treatments may be required to avoid weld cracking and residual stresses associated with SAW butt welding of relatively thick elements. The SAW process is also used for welded cover plates, for longitudinal stiffener connection to girders, and to fabricate truss members. Other elements are typically connected with SMAW welds.

SMAW process welds are made manually using welding rods (stick welding). SMAW welding is also referred to as manual metal arc welding. The welding rod electrode is filled or coated with flux material. The flux creates a shielding gas shroud and a molten flux layer to prevent contact with atmosphere and impurities. The molten flux solidifies into a slag that is removed following weld deposition. SMAW presents problems relating to weld continuity for the deposition of long welds, but is gravity independent and, therefore, may be used for any welding position. SMAW is comparatively less costly than other welding processes, and rather than moving pieces around the shop, DC and AC welding machines can be readily located at the pieces to be connected.

GMAW process welds are made by feeding bare welding wire, with the arc and molten weld metal shielded by an inert gas. GMAW welding is also referred to as metal inert gas (MIG) or metal arc gas shielded welding. The GMAW process is well suited to high alloy steel joining, but considerable welder skill is required.

FCAW process welds are made by feeding welding wire with flux in its core to create a gaseous shroud that protects the arc and molten weld pool (Figure 10.18). The flux cored shielding is sometimes supplemented by gas shielding similar to the GMAW process. A low-hydrogen electrode wire is required for critical welds to ensure weld metal ductility.* FCAW may be specified for vertical stiffeners and horizontal gusset plates with H4 electrodes.

ESW automatic process welds should be narrow gap (NG) welds made in the vertical position. ESW-NG process welds are made similar to SAW process welds with the electrode wire submerged in granular flux. An initial electric arc is formed to melt the flux, but the molten slag is electrically conductive and remains heated and in a molten condition.† Therefore, as the process commences, the molten slag pool stops the electric arc and current passes from electrode to substrate through the

FIGURE 10.18 FCAW welding. (Courtesy of Y. Martin, Canam Bridges Inc., St. Foy, QC, Canada. With permission.)

* Some specifications require low-hydrogen practice only.

† Due to electrical resistance of the molten slag between electrode and base metal.

molten slag rather than through an electric arc. Preheating is not required since much of the heat from welding is transferred to the base metal. The molten flux shields the molten weld pool. Material from about 25 mm (1 in.) thick to the thickest plates used in railway superstructures can be ESW process welded. ESW is not recommended for steel superstructures in FCM Zone 3, heat-treated (Q&T) low-alloy steel,* high-performance steel, and steel produced by TMCP (see Chapter 2). AWS D1.5M/D1.5 (2010) recommends ESW be used for welding only non-FCM high-strength low-alloy steels in accordance with ASTM A709M (A709) Grades 250 (36), 345 (50), 345S (50S), and 345W(50W). ESW process welds must be ultrasonic tested (UT) or radiograph tested (RT).

EGW automatic process welds are made with flux cored electrodes also in the vertical position. EGW process welds are made similar to ESW process welds except that the electric arc is maintained with shielding of the weld by external gases delivered to the welding area. The electrode flux provides deoxidization and slag to ensure weld cleanliness. The gases provide preheating and the molten slag temperature induces weld and base metal melting. Material from about 13 mm (1/2 in.) thick to about 50 mm (2 in.) thick can be EGW process welded. EGW is not recommended for dynamically loaded structures, welds in tension zones, welds in stress reversal, FCM, heat-treated (Q&T) low-alloy steel,† high-performance steel, and steel produced by TMCP (see Chapter 2).

AREMA (2015) does not permit the use of the ESW or EGW welding processes. However, continuing welding research into ESW process weld fatigue and fracture performance (particularly for connecting plates or shapes in tension or stress reversal) may precipitate future acceptance. EGW may be appropriate for groove welds in compression member butt splices, but premature weld pool solidification from weld deposit discontinuity issues must be overcome.

The welding process used must produce a high-quality weld at low cost. Therefore, the process used must be appropriate for fabrication shop equipment and productivity, welder qualification, and joint welding position conditions.‡ Hydrogen content, to eliminate the potential for weld cracking, is well controlled in the SAW and GMAW processes. Furthermore, the SAW and GMAW processes are appropriate for continuous welds and, because weld deposition depends on gravity, for horizontal or "downhand" welds. However, fabrication shops must be able to accurately layout, align, and fit-up joints in jigs§ or other positioning devices for the SAW or GMAW welding processes. Therefore, joints requiring vertical or overhead welds may be made with SAW or GMAW processes by rotating the joint or piece; otherwise, SMAW welding is generally used. SAW and GMAW welding processes are generally more productive and of lower cost than SMAW welding, but SMAW welding is often more versatile. In addition, qualified welder operators for SAW and GMAW welding may be less available than qualified SMAW welders in some areas.

10.5.2 Shop Welding Procedures

Design and shop drawings will typically specify the type of weld to be used in a joint as a fillet, Complete Joint Penetration groove, or partial joint penetration (PJP) groove weld. Fillet welds made in accordance with welding code provisions for size, orientation, and length do not generally require welding procedure qualification. Figure 10.19 shows a plate girder flange-to-web fillet weld (girder positioned horizontally for the SAW welding process). CJP and PJP groove welds require that the welding procedure be qualified by, or be in accordance with a prequalified, written Welding Procedure Specification.¶

---

* Unless subsequently heat treated.

† Unless subsequently heat treated.

‡ In general, it is good practice to avoid overhead welding and ensure adequate access at the joint for the welder and welding equipment.

§ Jigs and other positioning devices may also serve to prevent or minimize distortion during welding. Tack welds, if permitted, can be used to hold pieces in position for joint welding.

¶ The WPS document outlines detailed methods, practices, and procedures for weld production. AWS D1.5M/D1.5, the Canadian Welding Bureau, and other welding codes provide prequalified WPS for CJP and PJP groove welds.

FIGURE 10.19 Plate girder flange-to-web fillet weld. (Courtesy of Y. Martin, Canam Bridges Inc., St. Foy, QC, Canada. With permission.)

Qualification of WPS documentation is established by testing in accordance with the provisions of the applicable welding code. Prequalified WPS documents are based on use of proven materials, joint designs, and welding procedures. The WPS must indicate the type of groove weld, the type of joint (butt, corner, or "T"), the welding process (SMAW, FCAW, GMAW, or SAW), base metal thickness limitations, base metal preparation requirements,* applicable welding positions, detailed weld dimensions, and tolerances; and whether supplemental gas shielding is required for FCAW process welding. Shop drawings must reference the appropriate WPS to be used for welded connections. CJP groove welds must be made from one side with a backing plate (Figure 10.12) or from both sides after backgouging. PJP groove welds should not be used where tensile stresses are normal to the effective weld throat. In addition, PJP welds may not perform as well in fatigue as CJP welds. For railway superstructure fabrication, fillet, PJP, and CJP welds are typically specified depending on the joint load orientation and considering welded joint fatigue design requirements (see Chapter 9).

10.5.3 Effects of Welding on Plates and Shapes

Welded joints are efficient to fabricate and exhibit excellent performance if proper processes, procedures, and parameters are used to produce the connection. Nevertheless, detrimental effects from welded fabrication may occur that affect the subsequent performance of the welded connections. These damaging effects are typically weld flaws, cracking, distortion, residual stresses, and/or lamellar tearing.

10.5.3.1 Welding Flaws

Fabrication weld flaws are undercutting, porosity, pinholes, lack of weld fusion, and/or inclusions of weld metal impurities.

Weld toe undercuts and porosity are effective notches (localized stress concentrations) that can initiate fatigue cracking, particularly if the welded connection is subjected to cyclical or load reversal tensile stresses ranges. Undercutting may be controlled by reducing the electrical current. Porosity may be precluded by increasing current in the electrode, the use of low-hydrogen electrodes, and/or

* Plate edge preparation may be done by, for example, careful air-carbon arc gouging (Figure 10.20). However, a smoother prepared surface is generally attained through grinding.

increased shielding to reduce impurities in the weld metal. Undercutting and porosity defects may also be alleviated by reducing the weld deposition speed and the welding arc length.

Weld pinholes are stress concentrations, but not typically susceptible to cracking. Nevertheless, pinhole formation should be controlled by the use of dry electrodes and the removal of contaminants (e.g., rust, scale, coatings, and/or grease) from the surface of areas to be welded.

Lack of weld fusion or incomplete weld penetration detrimentally affects the strength of the connection, particularly if it occurs over a long length of weld.* Reduction of the weld deposition speed, increasing the welding current, and/or reducing the electrode size may be employed to preclude weld fusion defects.

Weld metal impurities can initiate fracture. The most common impurities are flux particles, commonly referred to as slag. Slag inclusions create discontinuities that are stress concentrations, which, if in a tensile stress range path, can initiate fatigue cracking. Slag cannot ascend out of the weld if the molten weld pool or puddle is too viscous and flux particles remain in the weld after cooling. Flux particle inclusions may be controlled by increasing the weld temperature and/or reducing the cooling rate (preheating also assists in diminishing the cooling rate). Slag inclusions may also be the result of inadequate weld cleaning between passes. The design of acute angle joints should be avoided in order to reduce the potential for slag inclusions from making difficult welds with insufficient cleaning.

10.5.3.2 Welding-Induced Cracking

Weld cracking can occur due to fabrication defects and/or through fatigue in service. Service-related weld cracking is precluded by the proper fatigue design of welded connection details (see Chapters 5 and 9). Welding-induced cracks are effective notches (localized stress concentrations) that can initiate fatigue cracking, particularly if subjected to tensile stress ranges. Fabrication weld solidification cracking may be avoided by using the appropriate welding procedure for the welding process, type and size of weld, material thickness, and joint geometry. Aspects of welding procedures that may typically affect the propensity for weld cracking are base metal edge preparation, electrode type, heat input, preheating, cooling rate, and weld deposition speed. Treatments to prevent weld cracking, such as mechanical peening or ultrasonic impact† between weld passes, are beneficial but relatively costly to include as part of a typical welding procedure.‡

Weld solidification cracking can be avoided at the outset by ensuring the proper geometry and smoothness of the base metal edge preparation,§ and through the use of the correct electrodes and fluxes for the welding process being employed (e.g., low-hydrogen electrode coverings are typically required for the SMAW process).

High heat input electrodes, which may precipitate weld cracking, should not be used for steel railway superstructure fabrication. In addition, welding preheat and interpass temperatures must be controlled in order to regulate the cooling rate and avoid weld cracking. Preheating is a powerful tool for weld cracking control as the application of preheat reduces the potential for brittle fracture by decreasing the temperature gradient between the weld and HAZ of the base metal. Preheating also promotes hydrogen dispersion from the weld to avoid cracking due to hydrogen embrittlement during cooling. Welding codes normally specify a maximum preheat temperature [typically about 230°C (450 F)] and minimum preheat and interpass temperatures based on the grade and thickness of the steel base material for the SMAW, SAW, GMAW, and FCAW processes. Nevertheless, the use of dry low-hydrogen electrodes may enable the elimination of preheating for welds made with the SAW process. Thicker plates (e.g., long girder flanges) require greater minimum preheat and interpass temperatures.

* This may occur with automatic and semi-automatic welding processes.

† Peening imparts compressive stresses to the weld surface. Ultrasonic impact treatment imparts compressive stresses into the weld via high-frequency impact and improves the surface weld profile.

‡ Weld treatment should not be used for the final weld pass due to surface damage and inspection needs.

§ For example, the difference between edges prepared by arc gouging and grinding.

Depositing large weld beads at slow speed can also cause weld cracking during automatic welding processes such as SAW. Weld cracking may also occur due to the stressing of joints during fabrication.

The base metal HAZ adjacent to the weld may be hard and brittle after welding and, therefore, also susceptible to cracking. HAZ cracking risk is increased with greater weld cooling rate and base metal carbon equivalence (see Chapter 2). Welding arc strikes on the base metal near welds may also create conditions for localized base metal embrittlement and cracking susceptibility.

10.5.3.3 Welding-Induced Distortion

Welding-induced distortion can occur during fabrication from weld shrinkage during cooling. Weld contraction is restrained by the cooler base material adjacent to the weld, thereby creating stresses in the weld causing plastic deformation and distortion of the shapes or plates being welded.

Weld shrinkage may be reduced through control of weld size, gap, edge preparation, heat input, number of passes, sequencing, and joint restraint. Treatments to prevent welding-induced distortions such as mechanical peening or ultrasonic impact between weld passes* are beneficial but relatively costly to include as part of a typical welding procedure. An effective shop procedure to resist welding-induced distortions is to preset or prebow (precamber) shapes and plates such that the welding-induced distortions are counterbalanced by the preset shape or plate. The angular distortion of girder flanges may be effectively avoided by precambering and through the use of sequenced weld deposition.

Larger welds will induce greater distortion and, therefore, weld size should be minimized. Fillet weld profiles are minimized by flat or concave surfaces (Figure 10.19) that do not affect throat size and weld strength. Groove welds should use a minimum root gap and have edge bevels of up to about 30 degrees. Welding-induced distortion caused by weld shrinkage can also be minimized by reducing electrode heat input and increasing the number of passes to complete the weld.

Weld sequencing and symmetric welding (Figure 10.13) are very effective in balancing forces and reducing welding-induced distortions. In some cases, using strong jigs and devices to hold the work in fixed position are necessary to prevent or minimize distortion during welding.

If welding-induced distortion occurs due to an inability to properly control weld shrinkage through procedures or external restraint, heat straightening may be employed in the shop. Heat straightening can be a relatively costly part of fabrication, but most fabricators have shop procedures for heat-straitening plates or shapes.† For example, flange plate distortion caused by butt welding must be heat straightened prior to welding the flange plate to the web plate. Heat straightening can be effectively used to level girder flange plates by the application of heat to the top surface of the flange along the web plate line.

In addition to distortion, weld shrinkage may create residual stresses caused by restraint to shrinkage from the base materials adjacent to the weld. Furthermore, if jigs and other devices are used to restrain the work, residual stresses are likely to be induced in the welded joint.

10.5.3.4 Welding-Induced Residual Stresses

Residual stresses exist in steel plates and shapes due to hot rolling (see Figures 10.1 and 10.2), plate cutting (see Figures 10.6 and 10.20), and welding. Residual stresses are generally not detrimental to the performance of structures subjected to only static load application and fabricated with modern ductile steel. However, since stability is not influenced by ductility, residual stresses must be considered for compression member design. Compressive load capacity determined in accordance with AREMA (2015) is established based on allowable compression stresses considering the potential for residual stresses (see Chapter 6).

* This may be necessary for very heavy butt joints.

† Many shop procedures for heat straightening are based on the guidelines in FHWA Report FHWA-IF-99-004: Heat-Straightening Repairs of Damaged Steel Bridges.

FIGURE 10.20 Plate edge preparation by air-carbon arc gouging. (Courtesy of Dr. K. Frank, Hirschfeld Industries, Austin, TX, USA. With permission.)

Nevertheless, for structures subjected to dynamic live load application, residual stresses may require relief after fabrication or more explicit consideration during design. Welding-induced residual stresses can occur during fabrication from restraint to weld shrinkage during cooling. Figures 10.21 and 10.22 illustrate the residual stress patterns due to welding of flat plates and plate girders, respectively. The cyclical application of live load may create localized inelastic stresses which, when superimposed on residual stresses, may precipitate localized yielding prior to yielding in the nominal stress regime. Modern ductile steels generally preclude localized yielding through stress redistribution.* However, high residual tensile stresses combined with high live load tensile stresses may increase the risk of brittle fracture at cold temperatures.

Relieving potentially large residual stresses induced from welding should be considered when fabricating steel railway superstructures for cyclical service load applications, erection in cold environments, and where concerns exist about stress corrosion in tensile zones. Consequently, locations

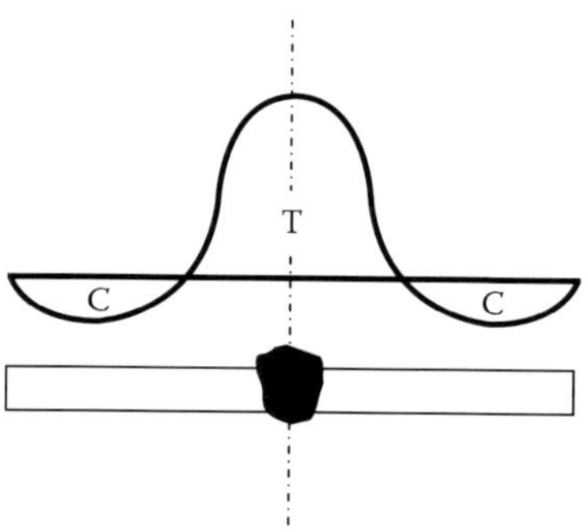

FIGURE 10.21 Residual stresses in flat plates due to welding.

* For example, if bending members experience slight inelastic deformation upon initial live load application, steel ductility will typically cause subsequent deformations from live load application to be within the elastic range.

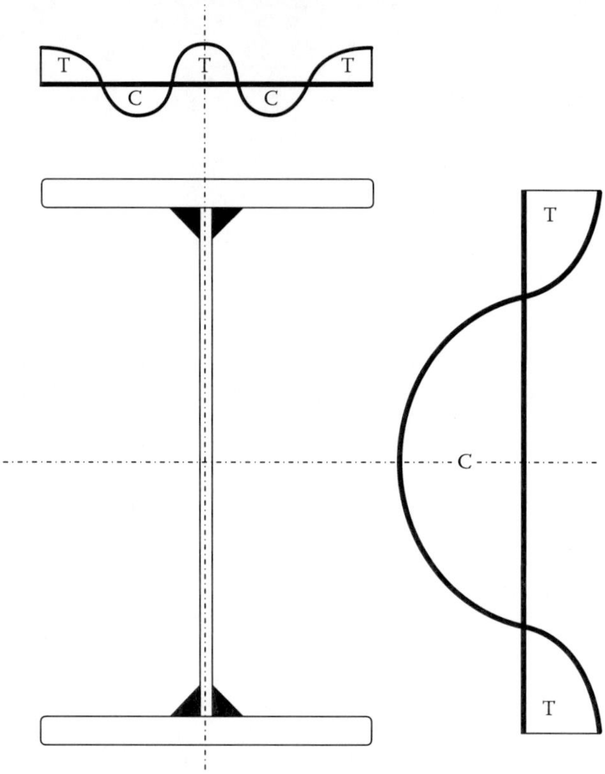

FIGURE 10.22 Residual stresses in plate girders due to welding.

with potentially high residual stress and subject to tensile live load stress* should be stress relieved by heat treatment between 600°C (1100 F) and 700°C (1300 F).

10.5.3.5 Welding-Induced Lamellar Tearing

Welding-induced lamellar tearing can also occur during fabrication from weld shrinkage during cooling. Weld contraction is restrained by the cooler base material adjacent to the weld creating stresses in the weld that exceed yield stress. Welding temperature induced lamellar tearing occurs in thick material due to the triaxial stresses created by restraint in the through thickness direction. The triaxial strains prevent ductile behavior and create conditions for fracture. Weld metal shrinkage may also create lamellar tearing at inclusions due to the temperature-induced strains instigated by through thickness restraint. Figure 10.23 schematically illustrates a typical stepped lamellar tear that may occur in thick flange plates due to flange-to-web welding (Figures 10.15 and 10.16). Flange plate thickness in excess of about 50 mm (2 in.) requires careful welding control (sequence, preheat, and weld size) to avoid lamellar tearing. In addition to joints with thick plates, lamellar tearing may also occur from the fabrication of highly constrained joints and other heavy welded fabrications.

The risk of lamellar tearing from welded fabrication may be reduced by using improved mill controls,† designing simple welded joint geometries, using bolted splices for heavy sections, and monitoring welding procedures.‡

---

* For example, butt splice welds in flange plates in flexural tension zones.

† Through thickness ductility is improved by sulfur reduction and inclusion control at the steel production mill.

‡ Welding sequence, preheat, and weld size control.

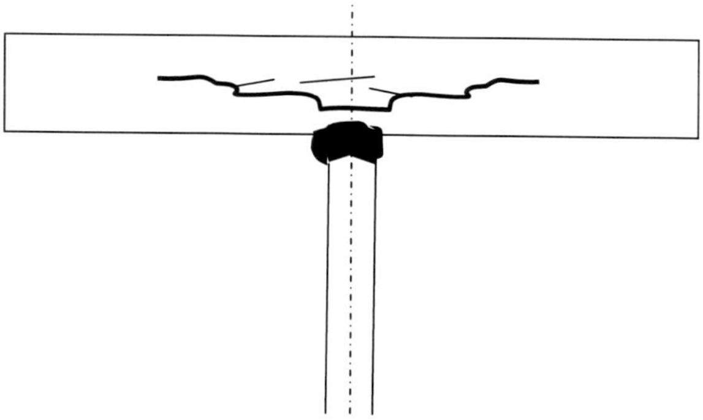

FIGURE 10.23 Lamellar tearing in thick flange plates.

10.6 COATING OF STEEL PLATES AND SHAPES FOR RAILWAY SUPERSTRUCTURES

Shop coatings for members, as outlined in the design documentation, should include surface preparation provisions, paint specifications, and the shop dry film thickness (DFT) requirements. In many cases, the use of atmospheric corrosion-resistant (weathering) steel* precludes the need for coatings, except at localized areas subject to accelerated corrosion.† Nevertheless, in some cases,‡ a corrosion protection coating system is specified. Corrosion protection is accomplished by coating the steel with sacrificial elements§ or by encapsulation of the steel (painting or waterproofing).

Commercial blast cleaning¶ (Figure 10.8) of atmospheric corrosion-resistant steel and near white blast cleaning** for zinc primer coatings are typically specified. Steel blast cleaning is usually performed in a blast cabinet or manually in a sealed room.

Zinc†† and aluminum are applied to the steel surface by hot dipping (galvanizing)‡‡ or thermal spraying (metalizing).

AREMA (2015) provides guidance to design engineers and owners regarding appropriate paint systems for steel railway superstructures. Three-coat paint systems are commonly used, consisting of an inorganic zinc primer, epoxy intermediate coat,§§ and urethane top coat¶¶ (Figure 10.24). Two-coat systems using an organic zinc primer have also been used and one-coat systems are being developed. Blast cleaned splice plate and connection faying surfaces should only receive the prime coat in order to provide adequate surface friction (slip coefficient) for the SC connection (see Chapter 9).

Waterproofing coatings are usually applied to cleaned steel deck surfaces.*** Many modern steel plate decks are waterproofed by a cold spray applied elastomer that binds to the steel or a synthetic rubber membrane††† glued to the steel substrate. The elastomer and synthetic rubber membrane may

* Such as ASTM A588M (A588), A709M (A709), and CSA G40.21 Grade 350A steels.

† Such as top surfaces of beams and girder bearing areas.

‡ Such as the use of atmospheric corrosion resistant steel in industrial or marine environments or where nonatmospheric corrosion resistant steel is used.

§ Such as zinc or aluminum.

¶ See SSPC SP6—Commercial Blast Cleaning.

** See SSPC SP10—Near White Blast Cleaning.

†† Zinc is superior to aluminum as a sacrificial element.

‡‡ Tank sizes available for hot-dip galvanizing may limit the length of members that can be coated to about 15 m (50 ft).

§§ The intermediate coat must bind to the primer.

¶¶ The top coat must be hard and bind to the intermediate coat.

*** Waterproofing system manufacturers should provide steel and concrete surface condition criteria and cleaning methods for the review of the design engineer and/or owner.

††† Typically ethylene propylene dieneterpolymer membrane, butyl rubber membrane, or combination sheet product.

FIGURE 10.24 Shop painted beam. (Courtesy of Y. Martin, Canam Bridges Inc., St. Foy, QC, Canada. With permission.)

need to be protected against puncture from stone ballast pressures with cushioning panels or mats.* AREMA (2015) provides recommendations for the application of waterproofing and protection systems.

10.7 QC AND QA OF FABRICATION

QC is performed and documented by the fabricator in accordance with an approved QC plan (QCP) and QA is executed and recorded on behalf of the design engineer or owner. QA inspections do not relieve the fabricator of QC responsibilities. QC and QA reviews and inspections are vitally important to ensure that the fabrication of plates and shapes into superstructure assemblies will perform as intended and permit effective erection. Following review of shop (or detail) drawings, QC and QA inspections should carry on through material procurement and the fabrication process.

10.7.1 QC Inspection of Fabrication

The QCP for QC inspections should include instructions regarding

- The inspection of raw materials and documentation (e.g., mill certificates and supplementary test results),
- Visual inspections for overall dimensions and geometry,
- Bolted connection inspections,
- The examination of WPS qualifications, welding equipment, and welder or welding operator qualifications, and
- The QC nondestructive testing (NDT) of fillet and PJP welds in main members by magnetic particle testing (MPT) (Figure 10.25), CJP groove welds by UT (Figure 10.26), butt joints by UT or RT (Figure 10.27), and "T" and corner joints by UT.†

* These may be reinforced asphalt-impregnated, elastomeric, or other panels and mats engineered for use in similar load and operation environments.

† AWS D1.5M/D1.5 outlines procedures for RT of groove welds in butt joints, UT of other groove welds, and the related QC acceptance criteria.

FIGURE 10.25 MPT of weld. (Courtesy of Y. Martin, Canam Bridges Inc., St. Foy, QC, Canada. With permission.)

FIGURE 10.26 Ultrasonic testing of weld. (Courtesy of Y. Martin, Canam Bridges Inc., St. Foy, QC, Canada. With permission.)

FIGURE 10.27 RT of weld. (Courtesy of Y. Martin, Canam Bridges Inc., St. Foy, QC, Canada. With permission.)

The QC NDT test frequency* depends primarily on the stress state and welding process. It typically involves the inspection of the following:

- 100% of the welds in joints in tension or stress reversal,
- 100% of a specified weld length† each side of the point of maximum tension in web plate butt welds and 25% of the remainder of the web weld length,
- 100% of the welds in 25% of the joints in compression or shear, or 25% of the welds in 100% of the joints in compression or shear, and
- 100% of a specified weld length for fillet and PJP welds in main members.

QC inspections of FCM must also include the supplemental inspections required by the specified FCP,‡ which generally involves the inspection of the following:

- Tension butt welds and approved butt joint weld repairs by both UT and RT,
- T- and corner joint welds in tension and repaired groove welds by UT, and
- Fillet weld repairs by MT (over a specified length).

10.7.2 QA Inspection of Fabrication

QA requirements specified by the design engineer or owner typically include a variety of reviews and inspections to ensure the fabrication is in accordance with the design intent.

10.7.2.1 Shop or Detail Drawing Review

Shop drawings are reviewed by the QA inspector and used during the course of the QA inspection. However, in many cases, the design engineer, and/or technical staff working under the direction of the design engineer, is also tasked with shop drawing review to ensure the shop drawings accurately represent the intent of the design drawings. Shop drawing reviews by QA inspectors and/or the owner's design representatives do not discharge the fabricator of the responsibility for the accuracy of the shop drawings.

10.7.2.2 Inspection of Raw Materials

This generally involves visual inspection for overall dimensions§ and geometry of plates and shapes, including the review of mill certificates outlining material mechanical and physical properties.¶ Visual inspection for laminations, plate flatness, dimensional tolerances, and/or edge damage before and, if necessary, after storage is required. Rolled shapes and steel castings are inspected for dimensional, straightness, and twist tolerances** as well as for the presence of fins, scabs, and rolling defects prior to use in fabrication.

10.7.2.3 Inspection of Fabricated Members

The QA inspector (QAI) should monitor, inspect, and verify that the work performed by the fabricator meets the requirements of the design drawings, specifications, shop drawings, and, in North America, AREMA (2015) and AWS D1.5M/D1.5 (2010) or CSA W59 (2013). The QAI must inspect fabrication work and tolerances, and indicate and record defects to be corrected

---

* The requirements apply to both shop and field welds. However, AREMA (2015) does not permit field welding of main members.

† Typically between 1/6 and 1/3 of web weld length.

‡ Typically AWS D1.5M/D1.5, Section 12 in the United States.

§ Including plate thickness.

¶ Some materials without, for example, mill certified toughness specification may require supplemental testing such as CVN testing in accordance with ASTM A673/A673M.

** Outlined for shapes in ASTM A6M/A6.

and pieces to be rejected. The QAI must advise the design engineer and fabricator of any deviation from the design drawings, shop drawings, and/or specifications. The fabricator will issue nonconformance reports (NCRs)* outlining the defects, and the proposed repairs for review and approval of the design engineer. QA inspection of fabrication work may include examination and monitoring of the following:

- Equipment conditions such as the size and quality of punches and dies, cutting apparatus, and the proper setup and securing of drilling or reaming templates,
- Member dimensions and geometry,
- Member straightness, and final sweep or camber. Also, the control and use of heat and/or forces to obtain camber† or correct sweep in accordance with the fabricator's QCP. Buckles, bends, twists, kinks, or other defects in fabricated members must be noted during inspection and repaired,
- Steel plate during cutting (for internal defects and other material or fabrication difficulties),
- Bolt hole locations, edge distances, diameter, geometry,‡ alignment, and reaming or drilling defects such as burrs, tears, and chips in the bolt holes,
- Stiffener and connection locations,
- Flatness of flanges at bearing areas, bearing plates, bearing assemblies, and shoes. Also, the contact condition of milled bearing surfaces and proper surface finish and protection of machined surfaces,
- Heat charts for normalizing, stress relieving, and other heat treatments,
- Fabrication of the correct number of members or pieces,
- Application of rust-preventive material when required, and coverings to prevent contamination of coated surfaces, and
- Legibility and position of erection and shipping marks. Also, loose pieces to be fastened in place for shipment.

10.7.2.4 Assembly Inspection

The QAI must inspect fabrication shop assemblies, and indicate and record defects to be corrected and pieces to be rejected. Inspection of fabrication shop assemblies may include examination and monitoring of the following:

- Assembly lay-downs, templates, jigs, and positioning devices. Also, the overall dimension and geometry control methods used for the assembly,
- Bolted connection clearances and fit-up accuracy,
- Welded assemblies for root gap, edge preparation, backing bars, alignment and cleanliness; and the size, quality, and location of tack welds,
- Girders or other members with reamed or drilled holes (inspection of positioning, securing, and match marking),
- Splice plate orientation,§ fill plates, and all plies in contact when assembled,
- Camber blocking, if required, is correct prior to drilling, and
- Match marking of the assembled members and the preparation of match-mark diagrams.

---

* NCRs may also result from the fabricator's own QC inspections.

† Rolled shapes may be cambered by heat application. Girder camber should be cut into the web plate. Truss camber is achieved by varying the length of chord members by an amount equal to the axial deformation from the specified camber load.

‡ Holes must be cylindrical and perpendicular to the surface being reamed or drilled.

§ Flange splice plates must have the rolling direction parallel to girder flanges.

10.7.2.5 Bolting Inspection

The QAI must examine and monitor the installation of fasteners in the shop to ensure compliance with the design and the appropriate standard or specification,* which may include inspection and observation of the following:

- Proper fastener segregation and storage, and
- The installation and verification testing procedures, including, if required, RC testing.†

Bolting inspections are generally more reliable if done during bolt installation rather than after the bolted joints are completed.

10.7.2.6 Welding Inspection

The QAI must inspect fabrication shop welding before, during, and after fabrication. Prior to welding, inspection of fabrication shop welding may include examination and monitoring of the following:

- Welding equipment and calibration (in accordance with the QCP), including electrode drying ovens,
- Condition and storage of welding consumables,
- WPS qualification for joints (electrode type, size, wire type, wire size, current, arc voltage, speed of weld deposition, number passes, preheat, and run-on/run-off pieces for SAW process welds) (Figure 10.28),
- Welder and welding operator qualifications and welding test pieces are qualified to the appropriate code,
- Qualification documentation of those performing QC NDT for the fabricator in accordance with the QCP,
- Parent material and welding consumables properties,
- Cleanliness of the surfaces to be welded, and
- Weld root face and opening, bevel angle, and alignment of parts are within tolerances.

The QAI must indicate and record defects to be corrected and welds to be rejected during and after completion of the welding. This will require inspection and monitoring of the following:

- Welder or welding operator compliance with approved WPS amperage, arc voltage, speed of travel (weld deposition), sequence, electrode extension, shielding gas flow rate, slag cleaning between passes, and preheat, interpass, and postheat temperatures,
- Cleaning and back-gouging of welds, including the removal of unsound metal and contaminations (e.g., copper and/or carbon),
- Weld starts and stops (stopping short of snipes or plate edges without craters),
- Weld deposit quality after cleaning‡ for appearance, profile, size, contour, and surface defects (craters, lack of fusion, cracks, porosity, and/or slag inclusions) of fillet and groove welds,
- QC NDT§ performed in accordance with the QCP,
- Shape and plate dimensional tolerances after distortion from weld cooling,

* North American railway superstructure bolting is performed in accordance with AREMA Chapter 15, Part 3 and the RCSC specifications for structural joints using A325M (A325) or A490M (A490) bolts.

† RC testing is typically performed by the manufacturer, distributor, or supplier of the bolt, nut, and washer assemblies in accordance with AREMA (2015) Part 3 and ASTM F3125.

‡ Typically by wire brush or grit blasting.

§ For QC RT verify edge blocks if necessary (when plate edge is final edge in structure) and that each radiograph film represents a unique section or piece.

FIGURE 10.28 Run-on/run-off pieces for SAW butt weld. (Courtesy of Dr. K. Frank, Hirschfeld Industries, Austin, TX, USA. With permission.)

- QA NDT for weld defects and discontinuities, such as dye-penetrant testing (DPT), MPT, UT, phased array ultrasonic testing (PAUT), and/or RT, and
- Repair procedures for fabrication errors.*

10.7.2.7 Coatings Inspection

The QAI must also inspect fabrication shop coatings, and indicate and record defects to be corrected. Inspection of fabrication shop coating may include examination and monitoring of the following:

- Coating containers and batch numbers,
- Batch straining, mixing, and sampling,
- Cleaning and surface preparation of base metal prior to coating,
- Application of coating to ensure in accordance with manufacturer and/or specification requirements,
- Coating of difficult to access areas and treatment of faying surfaces,
- Dry spray, runs, sags, and other paint application defects,
- Curing of each coat,
- Thickness of coating (wet or dry per specification), usually DFT is measured prior to the application of subsequent coats, and
- Sufficient drying of coating prior to loading for shipment.

The QAI typically submits weekly reports concerning review and inspection of raw materials (geometry, tolerances, and material documentation), as well as member, connection (bolting and welding), and assembly fabrication. A summary report following completion of fabrication is typically submitted to the design engineer and owner.

* For welding it is important that the QCI and the QAI have clear understanding of the interpretation of the QCP regarding the acceptance or rejection of welds.

10.7.2.8 Final Inspection for Shipment

The QAI must inspect fabricated members and assemblies prior to shipment through

- A final visual examination of the assemblies,
- Review of railway* or highway route clearances,
- Observation of assembly handling, loading, and security,†
- Review of QCI reports covering the materials to be shipped, and
- Review of piece quantities and match marking.

A major constituent of the QC and QA effort is expended on the critical task of welding inspection. Modern NDT techniques allow for high-quality and cost-effective welding inspection, and should be utilized for the inspection of welded steel railway superstructure fabrication.

10.7.3 NDT for QC and QA Inspection of Welded Fabrication

Appropriate NDT techniques depend on the welded joint stress state, geometry, and welding process used. QC NDT inspection requirements are specified in the fabricator's QCP, based on the appropriate welding code, standard, guidelines,‡ or specification.§ QA NDT inspection requirements are typically specified by the design engineer and/or owner. QA NDT inspection requirements are also usually modeled on applicable welding codes, standards, guidelines, or specifications. DPT, MPT, UT, and RT methods are commonly used NDT techniques for QC and QA inspection of welded fabrication.

10.7.3.1 Dye-Penetrant Testing

Dye-penetrant testing (DPT) works by capillary action and is appropriate for locating surface cracks, flaws, and porosity. DPT is most commonly used during field inspections and is usually not considered appropriate for shop QC or QA testing. Nevertheless, DPT is used by fabricators as a rapid means of locating cracks during fabrication.¶

10.7.3.2 Magnetic Particle Testing (Figure 10.25)

MPT uses a magnetic field and fine iron powder to locate surface cracks, porosity, slag inclusions, lack of fusion, undercutting, gas pockets, and inadequate weld penetration. Therefore, MPT is appropriate for fillet weld inspection and it is typically specified that 100% of girder flange-to-web fillet welds** be MPT inspected.

10.7.3.3 Ultrasonic Testing (Figure 10.26)

Ultrasonic testing (UT) can locate surface and subsurface flaws in the weld and the adjacent HAZ of the parent material using high-frequency sound wave reflection. The voltage impulse associated with sound wave reflection is observable on oscilloscope cathode ray tubes (CRT). However, UT does not provide a permanent defect record other than the operator's written interpretation of flaw characteristics. Therefore, operator's experience and aptitude is critical for the correct interpretation of the pulse-echo patterns on the oscilloscope CRT.

UT is particularly useful for complex fabrications, as access for UT is required from only one side of the piece or joint. UT is also required, in special circumstances, for base metal integrity investigations. UT is generally required for through (full penetration or CJP) and butt welds, and it is typically specified that 100% of girder flange-to-web CJP welds be UT inspected.

---

* Railway dimensional and weight clearances are generally available from the railway company Clearance Bureau.

† Railway loading is typically inspected by railway company personnel. QAI should periodically observe methods and supports used to prevent damage during shipping.

‡ Such as AASHTO/NSBA (2002).

§ Such as AWS D1.5M/D1.5 in the United Sates and CSA W59 in Canada.

¶ Particularly, useful for fillet welds.

** Often used in through plate girders. Deck plate girder flange-to-web welds are, in some cases, specified as CJP welds.

10.7.3.4 Phased Array Ultrasonic Testing

Phased array ultrasonic testing (PAUT) determines defects by trigonometric calculations from sound paths at various angles. Similar to RT, which provides a permanent film record, PAUT provides a permanent digital record of defects. PAUT is not currently used extensively in North America but is expected to gain acceptance once included as an NDT inspection method in AWS D1.5M/D1.5.

10.7.3.5 Radiographic Testing (Figure 10.27)

Radiographic testing (RT) is a volumetric test rather than a surface test such as MPT or UT. RT discharges X-rays or gamma rays through the weld to create an image on photosensitive film. The film indicates surface and subsurface cracks, inclusions, porosity, lack of fusion, insufficient penetration, and undercutting (radiation passes more readily through defects). RT is typically better at locating porosity and slag inclusions than cracks. RT flaw size allowances for tension and compression welds in terms of plate thickness* are specified in the appropriate bridge welding codes. RT provides a permanent film record but requires special skill to interpret defects on the film.

Access from both sides of the material is required and there is a need to shut down the area of the shop where RT is being performed. RT is generally required for through (full penetration or CJP) and butt welds in flat plates,† and it is typically specified that all flange and web CJP butt welds be RT inspected.‡ If required, butt joints with defects found on RT films may be subsequently UT to accurately locate the weld defects found on the films.

BIBLIOGRAPHY

AASHTO/NSBA, 2002, Steel bridge fabrication QC/QA guide specification, AASHTO/NSBA Steel Bridge Collaboration, Washington, DC.

AASHTO/NSBA, 2008, Steel bridge fabrication guide specifications, AASHTO/NSBA Steel Bridge Collaboration, Washington, DC.

AISC, 2010, Code of standard practice for steel buildings and bridges, AISC, Chicago, IL.

American Railway Engineering and Maintenance-of-Way Association (AREMA), 2015, Steel structures, Chapter 15, in Manual for Railway Engineering, Lanham, MD.

American Society for Testing and Materials (ASTM) F3125, 2015, Standard specification for high strength structural bolts, steel and alloy steel, heat treated, 120 ksi (830 MPa) and 150 ksi (1040 MPa) minimum tensile strength, inch and metric dimensions, ASTM Standards, West Conshohocken, PA.

ANSI/AISC, 2014, Code of standard practice for steel buildings and bridges, AISC, Chicago, IL.

ASTM F606, 2009, Standard test methods for determining the mechanical properties of externally and internally threaded fasteners, washers, direct tension indicators, and rivets, ASTM Standards, West Conshohocken, PA.

AWS D1.5M/D1.5, 2010, Bridge welding code, AASHTO/AWS, Miami, FL.

CSA W59, 2013, Welded steel construction (metal arc welding), Canadian Standards Association (CSA), Toronto, ON.

Durkee, J., 2014, Steel bridge construction, in *Bridge Engineering Handbook*, 2nd edition, Chen, W.F., and Duan, L., ed., CRC Press, Boca Raton, FL.

FHWA, 1998, *Heat-Straightening Repairs of Damaged Steel Bridges,* FHWA-IF-99-004, Washington, DC.

Ghosh, U.K., 2006, *Design and Construction of Steel Bridges*, Taylor & Francis, London.

Medlock, R., 2014, Steel bridge fabrication, Chapter 2, in *Bridge Engineering Handbook: Construction and Maintenance*, 2nd edition, Chen, W.F., and Duan, L., ed., CRC Press, Boca Raton, FL.

RCSC, 2014, Specifications for structural joints using A325M (A325) or A490M (A490) bolts, Research Council on Structural Connections (RCSC), AISC, Chicago, IL.

Schlafly, T., 2011, Fabrication and erection, Chapter 2, in *Structural Steel Designer's Handbook*, 5th edition, Brockenbrough, R.L., ed., McGraw-Hill, New York.

---

* The allowable flaw sizes are greater for thicker plates.

† So that film can be placed under the welded joint. RT is not appropriate for angular joint weld inspection, such as flange-to-web joint CJP welds.

‡ In some cases, it is specified by the design engineer that compression flange butt welds be treated as tension butt welds in regard to RT defect tolerances.

11 Construction of Steel Railway Bridges

Superstructure Erection

11.1 INTRODUCTION

Superstructure erection is the culmination of a process that commences with planning and proceeds through design and fabrication. Constructability must be considered in the planning, design, and fabrication stages of bridge projects in order to achieve cost- and schedule-effective construction.

For typical steel railway superstructures, attention to constructability involves an understanding of the railroad operating practices and the usual erection methods and procedures used by contractors. The methods and procedures should be simple and in accordance with usual and accepted practices. Methods and procedures for complex superstructure erection projects vary among contractors and consistent erection methods are infrequent. Therefore, if possible with respect to the project schedule and contractual arrangements, clear lines of communication between a designer and an erection contractor are beneficial.

The erection of steel railway superstructures may be performed by the steel fabricator, general contractor, specialty erection contractor, or railroad construction forces. Determination of the most cost- and schedule-effective means of erection depends on resource availability and erection complexity. In any case, the erection of new steel railway superstructures on new or existing substructures must be carefully planned considering safety, cost, schedule, and quality of construction.

Construction safety is paramount in the railroad industry that is keenly focused on industrial (operational and infrastructure), employee, and public safety. The erection methodology employed must ensure the safety of employees, the public and railroad operations, and comply with all applicable regulations regarding occupational health and safety. Good erection contractors are proud of their safety record and unapprehensive about presenting relevant safety data and information to railroad companies and bridge owners. Construction safety is a vitally important aspect of erection planning.

Cost and schedule are inextricably related in the erection of railway superstructures. There are substantial operating costs associated with the interruption of railway traffic* which must be considered in conjunction with erection methodology planning. It is not uncommon for railroad companies to impose strict time constraints† on construction activities on high-density lines. Therefore, careful attention to erection sequencing and scheduling is required in order to minimize any interruption to railway traffic. Erection contractors are challenged to safely erect quality superstructures within a rigorously mandated schedule. Erection planning is critical to ensure that safe and scheduled erection methods and procedures are established that may be implemented by experienced and qualified erectors with good crews and equipment.

---

* Interruptions to railway traffic can have detrimental effects on customers of the railroad and general social economic activity.

† For example, Class 1 railroad companies in North America often restrict daily normal construction activities to between 4 and 8 h time blocks, which may occur in any 24 h period. In addition, these normal construction activity blocks may only be available for a few days a week when traffic volumes are lower. Special construction activities, such as complex span installations, may require longer time blocks, which must be arranged with the railroad company many weeks or months in advance of the work.

Railway bridges are expected to have a long life and require minimal maintenance in a relatively punitive load and operating environment. These requirements are particularly relevant during design and fabrication, but erection methods may also influence the quality of the completed superstructure. Therefore, the quality of construction must support the designer's and owner's superstructure life and maintenance expectations.

The erection plan needs to consider site access,* site conditions, weight and dimensions of the superstructure segments being erected,† and railroad operational requirements. Modern planning tools such as physical models, computer models, and project scheduling software are often used to develop a cost- and schedule-effective methodology from which labor and equipment requirements for various procedures and phases of the erection can be established.

Superstructure erection may be uncomplicated by constructing temporary diversions or diverting railroad traffic by detour. However, in many cases, land ownership, regulation, permitting, and cost precludes the construction of diversions, and detouring railroad traffic is costly and may have detrimental effects on the railroads customers. Therefore, many steel railway superstructure erection projects are performed on the existing railway alignment using erection techniques that minimize the interruption of railroad traffic.

Superstructure erection typically involves the use of cranes, falsework, and techniques involving the use of stationary or moveable frames. Cranes may be supported on land or on barges in watercourses. Erection using cranes from a temporary land structure such as an adjacent bridge or berm with culverts is often effective,‡ but may be prohibitive due to land ownership, regulation, and cost issues. Falsework may be supported on foundations§ or on barges in watercourses.¶ Superstructure erection on the existing railway alignment at remote or difficult sites may be effectively accomplished using stationary and movable erection frames. The frames are typically constructed at track level for the longitudinal movement of superstructure assemblies during the erection procedure.

Complex and long-span superstructure erection usually requires specialized methods, procedures, and equipment such as launching, cantilever,** tower and cable, and catenary high-line erection techniques. For complex steel railway superstructures,†† it is beneficial to consult with experienced erectors or erection engineers through the planning, design, and fabrication stages of the project.

Recommendations relating to the erection of steel freight railway bridges are included in American Railway Engineering and Maintenance-of-Way Association (AREMA, 2015).‡‡

11.2 ERECTION PLANNING

Erection planning involves the use of engineering and scheduling principles to establish the methods and procedures for safe and efficient execution of the superstructure erection.

In some circumstances, an adjacent temporary diversion track (shoo-fly) may be constructed to allow uninterrupted rail traffic during the superstructure erection. Typically, the temporary

* Many railway bridges are in remote areas with poor access.

† This can range from relatively small lightweight components in "stick-build" construction using small derricks or cranes to maneuvering large superstructure segments using large cranes or other specialized equipment.

‡ Crane capacity requirements are greatly reduced if cranes can be used adjacent to the location of new superstructure erection.

§ Depending on the geometry of the falsework, soil/rock conditions, and weight supported, foundations may be shallow or deep (typically piles).

¶ The 1912 replacement of the St. Lawrence River bridge crossing at Montreal, Canada (see Chapter 1) involved the longitudinal sliding of 123 m (405 ft) double track trusses with the leading end supported on falsework on a barge.

** Cantilever or semicantilever construction precludes or reduces the need for falsework. It is a useful erection technique for long-span steel trusses (see Chapter 1).

†† For example, long-span girders, skewed spans, curved spans, and complex through spans with floor systems (e.g., multiple track spans).

‡‡ Recommended practice for the erection of steel railway bridges is outlined in Chapter 15 (Part 4) of AREMA (2015).

diversion consists of a berm or grade with culverts or temporary bridge (often a timber or steel trestle). However, the feasibility from land ownership, regulatory, and construction cost perspectives requires thorough investigation. Similarly, the new superstructure may be erected on new substructures on a permanent adjacent alignment if clearance, land ownership, regulatory, and cost* matters are not onerous.

However, unless the new superstructure is to be erected on an adjacent inactive or low traffic density track,† railroad operational scheduling constraints must be met by an erection plan that includes methods, procedures, schedules, and resources (labor and equipment) focused on accelerated bridge construction (ABC). ABC techniques that provide for the safe and rapid erection of superstructures have been pillars of railway superstructure erection for many decades. The principles that enable successful ABC for railway superstructure erection are:

- Repetition and standardization,‡
- The use of relatively light materials (steel),
- The erection of simple spans,
- The design of square spans, where possible, to eliminate skewed construction,
- Modular fabrication of the superstructure,§ and
- No-load shop fabrication of camber (see Chapter 10).

With the requisite ABC principles established, ABC for superstructure erection planning and execution involves:

- Consideration of material shipments from fabricators to the site,
- Development and assessment of erection methods and equipment (e.g., cranes, guys/derricks, launching equipment and structures, falsework, specialized transport vehicles, and/or barges),
- Establishing storage yards for verification and arrangement of assemblies, subassemblies, and members,
- Creating assembly yards¶ (near the erection site, if possible, to avoid excessive on-site handling and transportation),
- Erection monitoring (horizontal and vertical alignments during erection**) and final inspection, and
- If required, application of field coatings and inspection.

11.2.1 Erection Methods and Procedures Planning

The erection contractor must first establish an erection methodology. The erection methods and procedures will depend on erection scope, site characteristics (layout and topography), schedule, and resource requirements.

* Careful review of the track and grade construction requirements associated with a permanent bridge location adjacent to the existing bridge alignment is required.

† If the adjacent track is a temporary detour track or a new permanent alignment, track outages should be of little concern. However, on tracks that are existing busy mainline tracks, track outages for construction on adjacent tracks may be limited by the proximity of cross-over tracks and/or traffic density.

‡ Trestles and viaducts (see Chapter 1), in particular, are appropriate for the repetitive installation of standard span types.

§ Long ballasted through plate girder (BTPG) superstructures are often fabricated with floor system modules consisting of about four to six transverse floorbeams welded to the steel deck plates. These modules can be erected and field welded or bolted between the girders.

¶ For smaller projects with good site access, storage and assembly may occur in the same yard.

** Particularly important for field splices and connections which may be made with superstructure elements supported on falsework and/or with holding cranes.

The site characteristics and access can be assessed by modern internet mapping software and site visits. The erection method should consider relevant site characteristics such as:

- Property and right-of-way lines,
- Roads, tracks, utilities, channels, waterways, overhead wires, and obstructions,
- Physical and schedule constraints related to regulatory obligations,*
- Staging and material storage areas,
- Site and yard access,
- Material delivery locations and orientations,
- Work area boundaries and location of substructures, and
- Any other relevant site-specific information.

This information should be clearly indicated on the erection plans that outline the erection methods and procedures.

Fabricated materials and assemblies may be furnished by the owner and/or general contractor. These materials and assemblies will be shipped by road, rail, or barge to storage and/or assembly yards at, or nearby, the site with good access to the erection location (Figure 11.1). Typically, cranes are used to load and unload materials and assemblies from the fabricator that are stored at, or received from, the assembly yard. Mobile cranes are often used, but stationary crane derricks may be used in large storage and assembly yards. Members should be shipped within fabrication tolerances (see Chapter 10), but on some occasions may require straightening or other treatments (and related inspections) after approval of the owners representative.†

Minimizing interference with trains during erection is crucial and, therefore, schedule is of paramount importance in the planning of erection methods and procedures for steel railway superstructures. Railway superstructure erection typically occurs within prescribed track outages‡ to minimize

FIGURE 11.1 Stick-building of truss span on falsework at site assembly yard. (Courtesy of the Author, Canadian Pacific Engineering, Calgary, AB, Canada.)

* For example, site habitation, cultural, environmental, fish, and wildlife constraints related to the erection schedule.

† Depending on size and complexity of project, this could be the design engineer or project construction engineer.

‡ Typically less than about 4–8 h for only a few days per week on many Class 1 railroad main line tracks.

interruption to railway traffic and allow for transportation planners to efficiently resume railroad operations. Therefore, the erection methodology must consider the execution of many activities of the erection procedures within relatively short time durations (track blocks).* In addition, bridges over navigable waterways or high-density roads may impose additional erection scheduling constraints based on roadway and waterway (commercial and recreational) transportation needs.

The design and fabrication documentation must be accurate and complete in order that the erector can develop resource (labor and equipment) requirements for the various activities of the erection procedures, in particular, for establishing equipment requirements based on member and assembly weight, sizes, and final erected location. Equipment availability and/or accessibility will also affect erection methods and procedures. An assessment of the availability of conventional equipment, such as cranes with the capacity and reach required for the site, is essential. Also, depending on site conditions, the need for, and availability of, specialized erection equipment† may require consideration.

Based on site characteristics, schedule, and equipment requirements (and availability), erection methods and procedures may be established. Erection methods and procedures are typically outlined on project erection drawings that indicate the following:

- Location and elevation of superstructure bearings on substructures, anchor bolt voids, and shim heights at bearings,
- Substructure preparations required for the erection of the superstructure,
- Location and orientation of members and assemblies,
- Elevations of the top of girders and trusses at bearings and splices,
- Erection and wind loads‡ on members and assemblies,
- Expected deflections from loads on members and assemblies,
- Location, type, capacity, and wheel or outrigger loads of cranes and other equipment or devices used for lifting or moving loads,§
- Location and type of temporary supporting structures,
- Lifting¶ and temporary support locations and reactions** on members and assemblies,
- Required bracing to ensure girder or truss stability and/or reinforcing of members,††
- Temporary support structures or methods used for elevation and alignment of splices prior to bolting,
- Fit-up procedures for connections and bolt tightening sequences,
- Directions for release of temporary support systems, and
- Sequence of the erection procedures.

The steel superstructure erection procedures and drawings may require considerable erection engineering and must be in conformance with the engineering design drawings, specifications, special provisions, shop drawings, and all other available information‡‡ related to the design and fabrication of the superstructure to be erected. Compliance is ensured by the bridge owner's (typically a railroad company) representatives' review of erection engineering calculations, procedures, and drawings. However, review and approval of the erection planning and engineering documentation by the bridge owner's representatives' does not relieve the erection contractor of responsibilities relating to the safety, schedule, cost, and/or quality of construction.

---

* Including the time to safely clear equipment and personnel for train operations.

† For example, barges and heavy load moving equipment such as self-propelled modular transporter (SPMT) vehicles.

‡ Erection and wind loads should be indicated separately.

§ For example, hydraulic jack locations, strokes, and capacities.

¶ A critical consideration for ensuring stability during the lifting, splicing, and placing of beams and girders.

** For review of the effect on the erected member or assembly, and for temporary support design. In some cases, member strengthening or bracing may be required.

†† Typically done in consultation with the design engineer.

‡‡ For example, camber diagrams, match-marking diagrams, and fastener bills of material (BOM).

11.2.2 Erection Methods and Equipment Planning

The site conditions and/or time constraints typically associated with railway superstructure erection projects compel a focus on equipment used for erection. As such, erection methodology and equipment become closely associated. Erection methodologies for typical steel railway superstructures involve the use of cranes, falsework, lateral skidding,* barges, and/or stationary and movable frames.

Modern trends in erection procedures and equipment have involved the use of heavier cranes,† purpose-developed erection equipment (such as stationary and movable frame and gantry systems), and SPMT vehicles.

11.2.2.1 Erection with Cranes and Derricks

Superstructure erection by crane is the most commonly used method of construction. Mobile cranes (Figure 11.2a)‡ require access and bearing area supports (crane pads), but are used for superstructure erection more frequently than stationary derricks, which often require a site-specific design, fabrication, and erection. A stationary crane is shown in Figure 11.2b.

11.2.2.1.1 Derrick Cranes

Derrick cranes, even though relatively costly to design fabricate and erect, are often essential for the lifting and movement of heavy loads at long reaches§ not feasible with mobile cranes. When used, derrick cranes are typically variations of the stiff leg (Figure 11.3a), guyed (Figure 11.3b), or A-frame (Figure 11.3c) types, but other derrick arrangements¶ may also be suitable for specific sites. Typical guyed derricks have a mast of about 40 m (130 ft) and a slightly shorter boom of about 30 m (100 ft)** with a capacity of up to 180 tonnes (200 tons). Stiffleg derricks with booms up to 80 m

FIGURE 11.2 (a) Large latticed boom crawler crane and small telescopic boom truck-mounted crane. (Courtesy of the Author, Canadian Pacific Engineering, Calgary, AB, Canada.) and (b) Stationary crane (Courtesy of the Author, Calgary, AB, Canada.)

* Sliding or rolling using hydraulic jacks and winches.

† Modern infrastructure typically allows the transport and shipment of large crane components. These are typically erected using smaller cranes at or near the lifting site.

‡ Figure 11.2 shows a large latticed boom crawler crane and a relatively small telescopic boom truck-mounted crane.

§ Stationary derrick cranes are generally used for complex and long-span superstructure erection.

¶ For example, gin pole, breast, and Chicago boom types.

** The boom is shorter than the mast to facilitate boom movement under the guys for short radius lifts.

(260 ft) and capacities of over 450 tonnes (500 tons) have been used for the erection of long-span structures.

Stationary derrick cranes may also be mounted on barges (Figure 11.3d*), railroad cars (Figure 11.3e), and underframe travelers to achieve mobility.

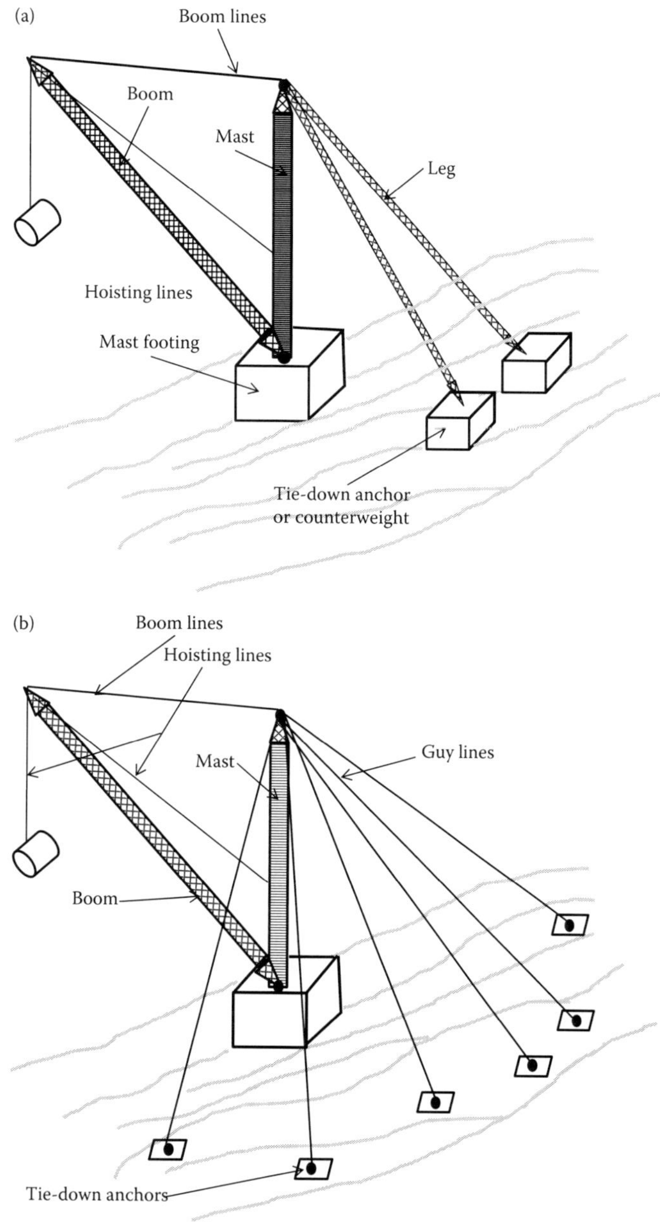

FIGURE 11.3 (a) Schematic of a stiff leg derrick crane. (b) Schematic of a guyed derrick crane. (c) Schematic of an A-frame derrick crane. (d) Schematic of a barge mounted derrick crane. (e) Schematic of a rail car mounted derrick crane.

(*Continued*)

* Shown in US Customary and Imperial units only.

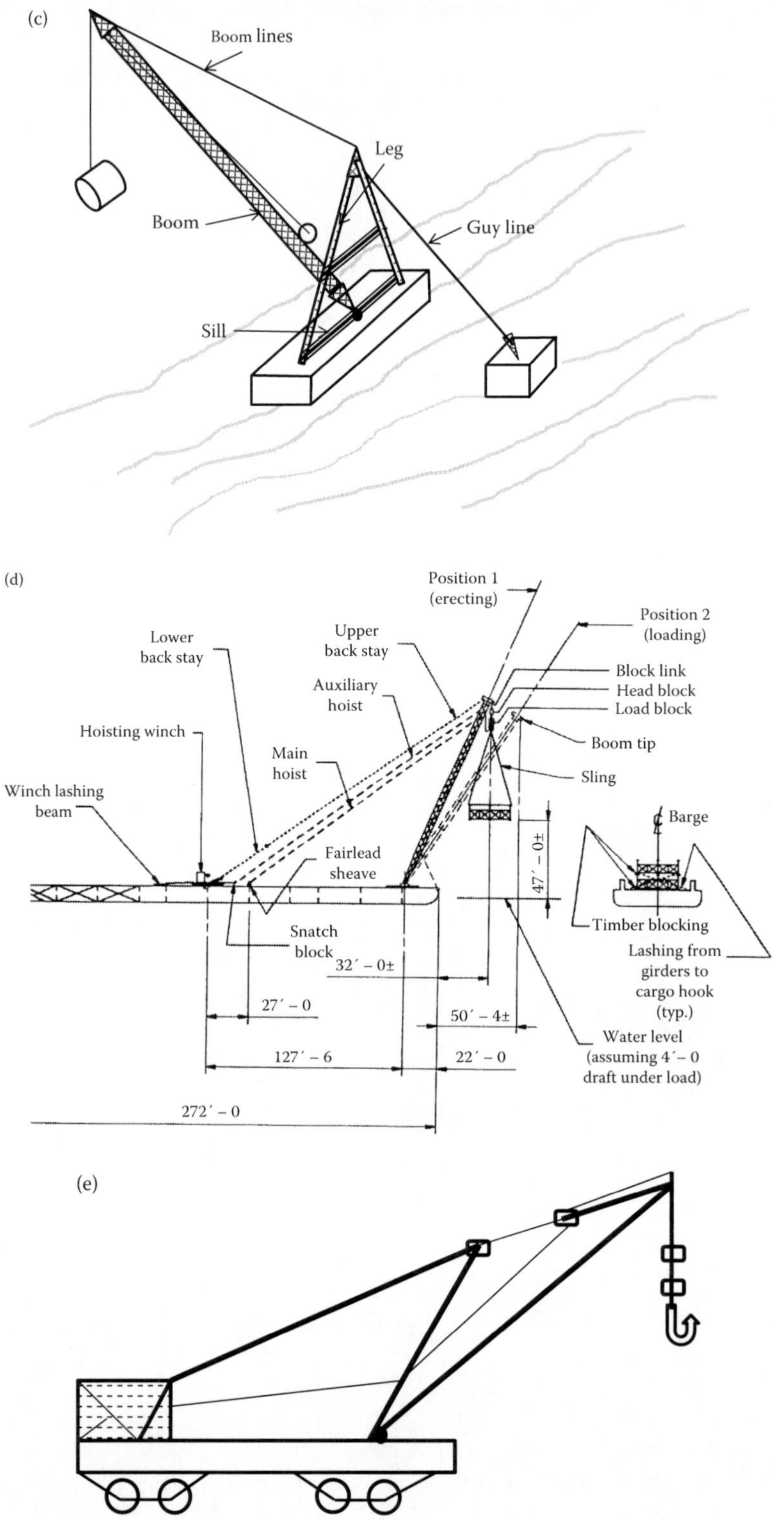

FIGURE 11.3 (CONTINUED) (a) Schematic of a stiff leg derrick crane. (b) Schematic of a guyed derrick crane. (c) Schematic of an A-frame derrick crane. (d) Schematic of a barge mounted derrick crane. (e) Schematic of a rail car mounted derrick crane.

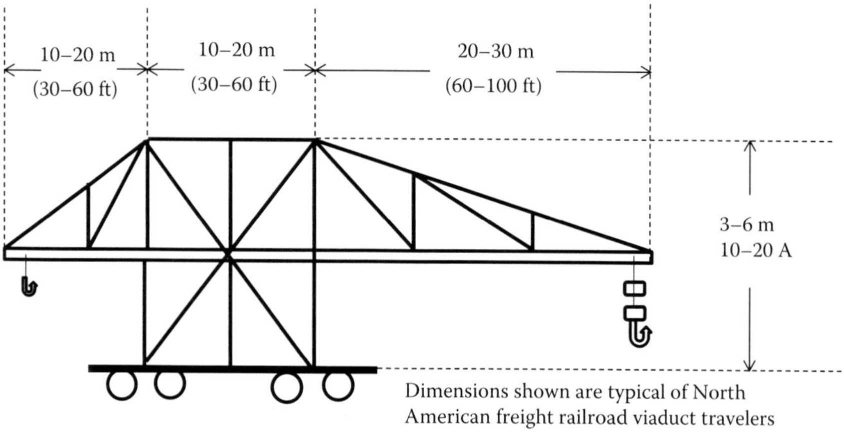

FIGURE 11.4 Schematic of a viaduct traveler used to sequentially erect short spans.

A variation of the underframe or railway car-mounted derrick crane is the viaduct traveler (Figure 11.4) used to sequentially erect the relatively short-span superstructures of steel railway viaducts.

11.2.2.1.2 Mobile Cranes

Regularly used mobile cranes include crawler, all-terrain, and truck-mounted cranes. Locomotives cranes may also be used for some railway superstructure erection projects. The appropriate mobile crane depends on the site (access and crane pads) and lifting (weight and radius) requirements of the superstructure erection procedures. Where site conditions allow mobile crane access and mobility, the sequential installation of multiple superstructures of similar weight and dimension using rail cars to transport new and existing spans to and from the site can be very effective from a cost and schedule perspective.

Mobile crane capacity is given as the lifting capacity at the shortest lifting radius. Mobile crane load ratings for short booms or close radii lifts may be governed by the structural strength of the boom or other components. Crawler crane load ratings for long boom lifts may be controlled by boom tip deflection restrictions. However, for the majority of typical loads at usual lifting radii, load ratings are governed by mobile crane stability (resistance to tipping or overturning moment).

11.2.2.1.2.1 Crawler Cranes (Figures 11.5 and 11.6) Crawler crane bases support the cabin, engine, winches, boom, and other mechanical devices, and provide a large area for load bearing and stability,* but need to be transported to the site by truck or rail. Crawler cranes use counterweights for stability and with latticed booms can have very high lifting capacity (rated load capacity). The rated load is generally taken as 75% of the stability load.

Crawler cranes may be used for heavy lifts of up to 1500 tonnes (1650 tons) with main booms at very small radius but are more typically used for lifts in the 150 tonnes (165 tons) to 400 tonnes (440 tons) range with booms at close radius. Typically, lifting capacity substantially diminishes for lifting radii greater than about 15 m (50 ft). Nevertheless, light loads can usually be lifted at radii greater than about 60 m (200 ft) and up to 120 m (400 ft), depending on the crawler crane capacity. Jibs may be added to the main booms for extended lifting radii, but load capacity is often severely reduced.

Crawler cranes are generally more readily available and are of lower cost than truck-mounted cranes.

* Primarily due to the large distance between treads, which can typically range from 5 m (15 ft) to over 12 m (40 ft).

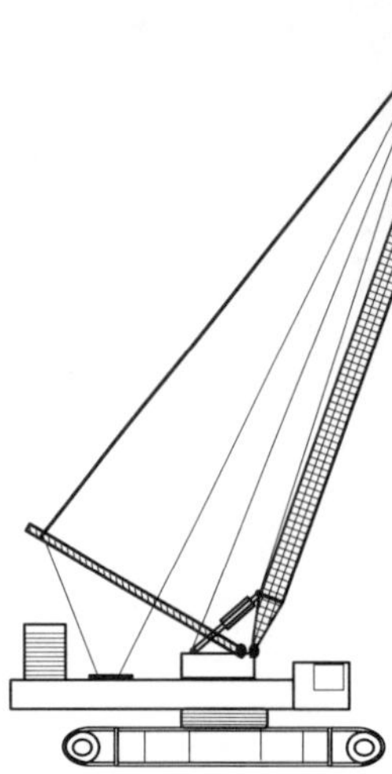

FIGURE 11.5 Schematic of a crawler crane.

FIGURE 11.6 Crawler crane lifting new DPG span onto existing rehabilitated substructures. (Courtesy of the Author, Canadian Pacific Engineering, Calgary, AB, Canada.)

11.2.2.1.2.2 Truck-Mounted Cranes (Figures 11.2 and 11.7)* Truck-mounted cranes may be very large and heavy, and need to be assembled at site or travel under special provisions of the highway authorities. These cranes use jacking outriggers and/or counterweights for stability and can have telescopic or latticed booms. Truck-mounted cranes with latticed booms can accommodate high capacity lifts. The rated load is generally taken as 85% of the stability load.

Truck-mounted cranes may be used for heavy lifts of up to 1000 tonnes (1100 tons) with main booms† at very close radius, but are more typically used for lifts in the 50 tonnes (55 tons) to 400 tonnes (440 tons) range with booms at close radius. Lifting capacity typically substantially

* Figure 11.2 shows a large latticed boom crawler crane and a relatively small telescopic boom truck-mounted crane.

† Truck-mounted cranes have telescopic and lattice booms. Lattice boom cranes generally have greater lifting capacity.

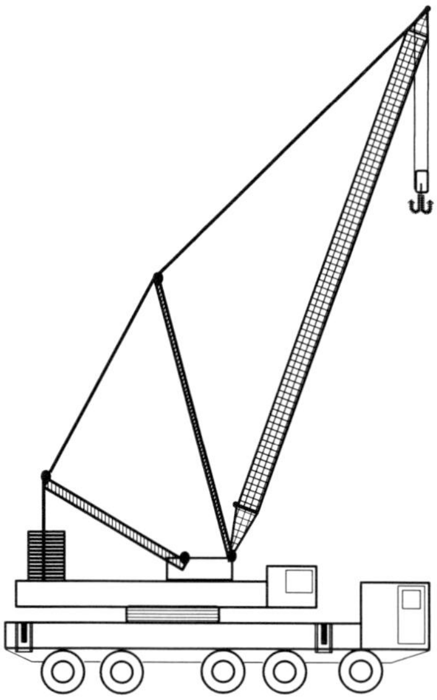

FIGURE 11.7 Schematic of a truck-mounted crane, Calgary, AB, Canada.

diminishes for lifting radii greater than about 15 m (50 ft), but light loads can typically be lifted at radii up to about 60 m (200 ft), and in some cases, up to about 100 m (300 ft). Jibs may be added to the main booms for extended lifting radii, but load capacity is often severely reduced.

Truck-mounted cranes are generally less readily available than crawler cranes, but are more versatile. Nevertheless, if required, even greater versatility can be attained through the use of all-terrain cranes.

11.2.2.1.2.3 All-Terrain Cranes (Figure 11.8) All-terrain cranes are self-propelled rubber tired cranes with smaller, but often substantial enough, rated lifting capacities. These cranes use outriggers for stability and have telescopic booms. Rough terrain cranes can travel more readily in very poor site conditions, but have limited rated lifting capacities.

All-terrain cranes are typically used for lifts up to about 200–300 tonnes (220–330 tons) with booms at close radius, but some all-terrain cranes have lifting capacities of up to about 500 tonnes (550 tons). Light loads can typically be lifted at radii up to about 60 m (200 ft), and in some cases up to 90 m (300 ft).

At some railroad sites, access by crawler, truck-mounted, and even all-terrain cranes may be prohibitive. In such cases, the use of locomotive cranes may be considered.*

11.2.2.1.2.4 Locomotive Cranes (Figures 11.9 and 11.10) Locomotive cranes are self-propelled railway equipment enabling circular swings of relatively short booms. The cranes use outriggers for stability but have a low capacity for lifts perpendicular to the track.

Locomotive cranes are typically used for lifts up to about 200–300 tonnes (220–330 tons) with booms at close radius, and some, although rare, have lifting capacities of up to about 600 tonnes (650 tons). Light loads can typically be lifted at radii up to about 25 m (75 ft).†

For crawler, truck-mounted, all-terrain, and locomotive cranes, technical data sheets outlining mobile crane working ranges and lifting capacities at various lifting radii must be consulted during erection planning and must be available for the equipment used at the erection site for lift

* Locomotive crane availability, track time, and lifting requirements may prohibit their use at some sites.

† A major disadvantage of locomotive cranes is their relatively small boom ranges.

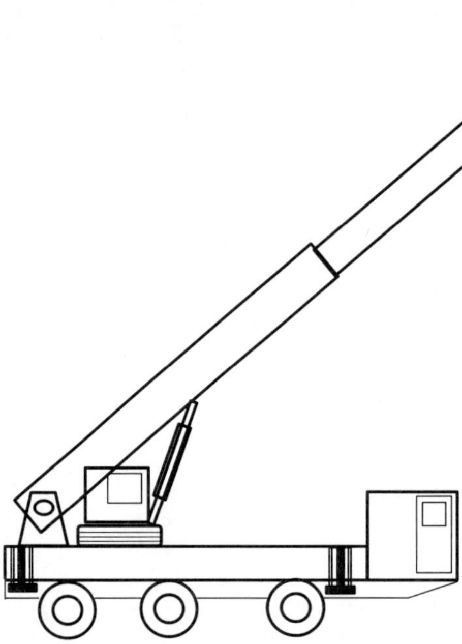

FIGURE 11.8 Schematic of an all-terrain crane.

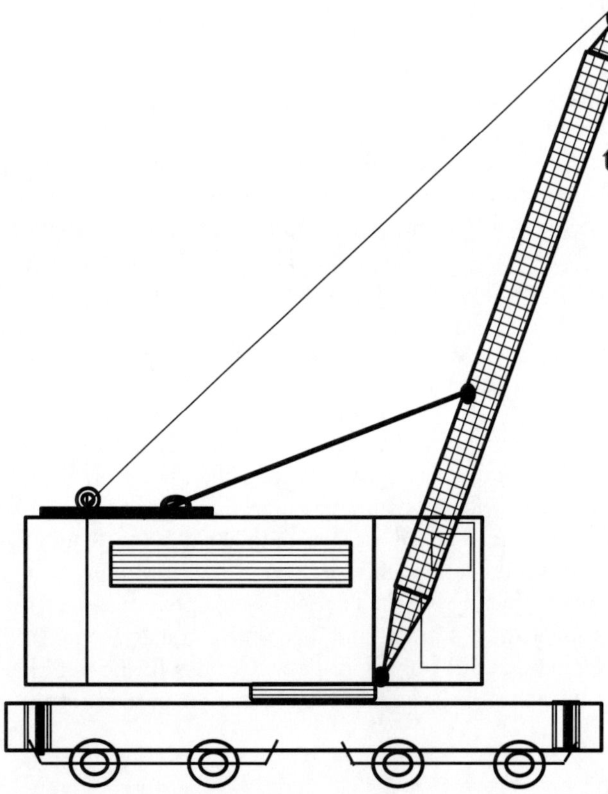

FIGURE 11.9 Schematic of a locomotive crane.

FIGURE 11.10 Locomotive crane lifting short span. (Courtesy of the Author, Canadian Pacific Engineering, Calgary, AB, Canada.)

verification. For truck-mounted, all-terrain, and locomotive cranes, lifting capacities at various lifting radii with and without outriggers must be considered.

11.2.2.1.2.5 Specialty Cranes Erection planners must have a good understanding of the type and capacity of cranes appropriate and available for the site. Nevertheless, in unusual circumstances, specialty cranes, such as the light traveling cranes used to stick-build trusses in cantilever construction, may be required for the erection of the superstructure. These cranes travel along the span being erected and have booms that are able to swing 180° to lift members arriving on trolleys or rail cars from behind. Other specialty cranes, usually for large bridges, may be designed and constructed to accommodate the particular site and erection methodology.

11.2.2.1.2.6 Holding Cranes For many railway simply supported span erections, temporary supports are not required* and crawler and truck-mounted cranes may be used to lift (from ground, structure, or barge) completely assembled (shop or field) superstructures directly onto suitably prepared substructures. This precludes potential individual girder or truss instability during lifting, and in the erected condition, that would require temporary falsework support, compression bracing, and/or holding cranes. Nevertheless, for procedures involving the erection of single girders or trusses, holding cranes to ensure stability may be required. Holding cranes may also be used in place of falsework. However, in some situations, falsework is necessary, such as at splice locations when in-air splicing with holding cranes is difficult.

11.2.2.2 Erection on Falsework and Lateral Skidding of Superstructures

Falsework is temporary support for the erection of superstructures. Large trusses and arches may be constructed on intrusive falsework (Figure 11.11). However, intrusive falsework may not be appropriate due to watercourse depth or flow, roadway, marine, and/or railroad use under the superstructure. Also, the extent† of falsework required may be cost prohibitive. Trestle-type falsework is useful

* With the exception of steel–concrete composite superstructures that may require shoring of the concrete deck (see Chapter 7).

† In particular, height of the falsework required.

for the support of through truss floor systems* from which cranes can stick-build (Figure 11.1) or move (Figure 11.12) the trusses.

Falsework is often used to support superstructures for lateral skidding (rolling or sliding) into position. Cylindrical rollers, chain-action rollers, low-friction sliding shoes, and greased rails or beams may be employed to facilitate the lateral movement of a superstructure. Movement occurs by pushing with horizontal hydraulic jacks fastened to the falsework cap or needle beam of the falsework, or by pulling with crane leads or winch and pulley systems.

The falsework is typically nonintrusive as it supports only the ends of the superstructures (Figures 11.13 through 11.15). Typically, the falsework may be installed with minimum interruption

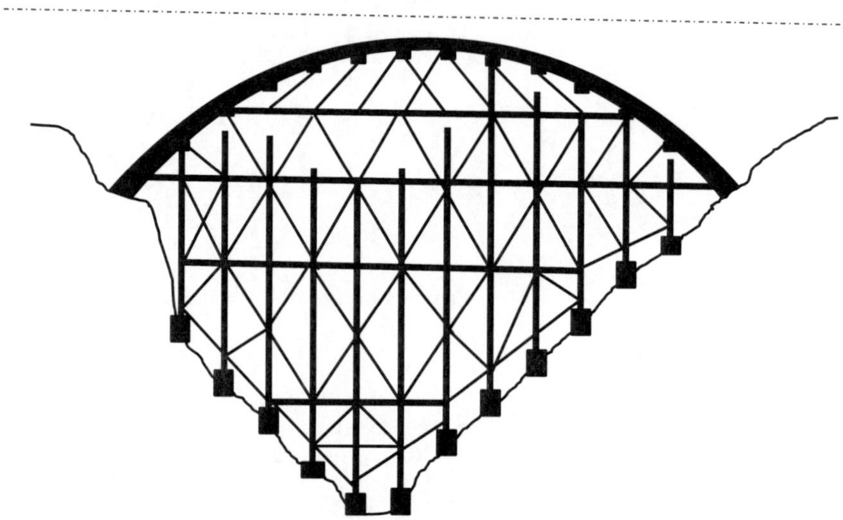

FIGURE 11.11 Schematic of intrusive falsework.

FIGURE 11.12 Nonintrusive emergency falsework. (Courtesy of the Author, Canadian Pacific Engineering, Calgary, AB, Canada.)

* Typically, the floorbeams are supported on transverse trestle bents.

FIGURE 11.13 Nonintrusive falsework for lateral rolling erection of new BTPG spans. (Courtesy of the author, Canadian Pacific Engineering, Calgary, AB, Canada.)

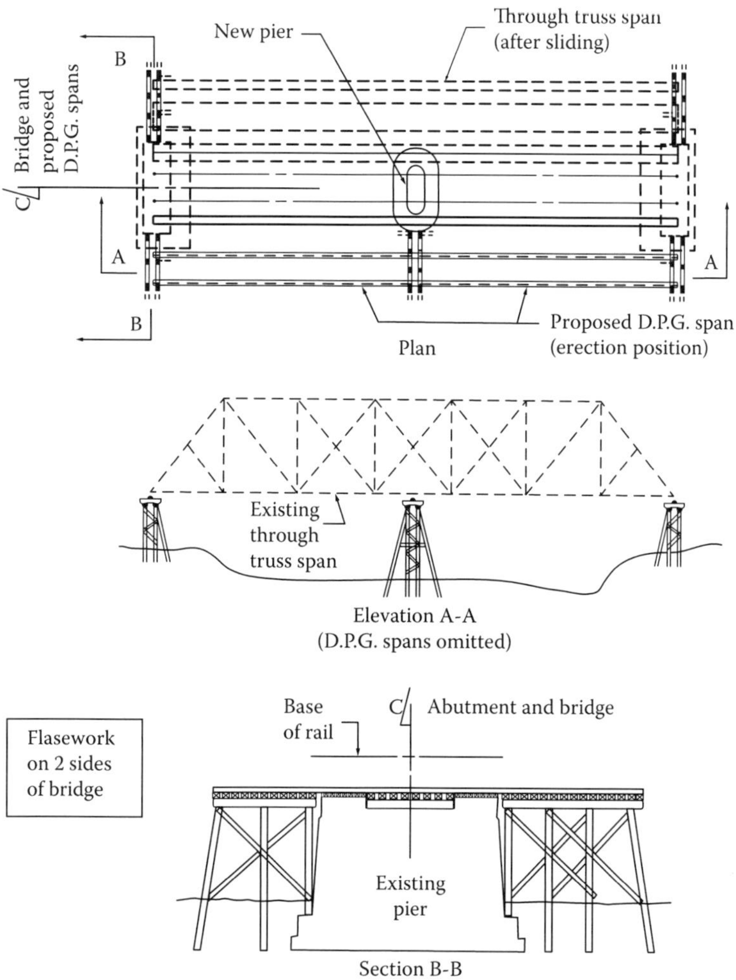

FIGURE 11.14 Nonintrusive falsework on both sides of substructures for lateral skidding erection of new spans.

to railroad traffic. In particular, this is the case for open deck superstructures (see Chapter 3), where pile-supported falsework may be constructed in short time blocks by temporarily shifting of bridge deck ties (sleepers) and installing caps or needle beams without affecting the existing superstructure and railroad traffic. The superstructures can be staged* and then skidded into position with minimum interruption to railway traffic.

Nonintrusive falsework may be constructed on both sides of the superstructure to be replaced. The new superstructure is erected on the falsework and the existing and new superstructures are successively jacked up onto rolling or sliding devices and moved laterally (Figure 11.14). Railroad traffic may be resumed while the old superstructure is demolished† and removed by crane (on a barge in watercourses). However, due to site conditions, proximity of existing adjacent structures, and/or cost, it is sometimes feasible to construct nonintrusive falsework on only one side of the existing superstructure. Figure 11.15 outlines the use of a railcar and blocking to install a new superstructure in the place of an existing superstructure laterally moved (jacked and slid or rolled) onto falsework. With railroad traffic is re-established, the old superstructure and railcar may be removed by crane (on a barge if in a watercourse).

Falsework used for composite steel–concrete construction must be designed and constructed in accordance with the superstructure design assumptions (i.e., whether shored or unshored curing of the concrete deck is specified, see Chapter 7).

In general, falsework must be engineered in accordance with the appropriate codes, standards, and best practices for the vertical loads of the supported superstructure during the various activities and procedures of erection. Loads from personnel and equipment, and lateral and longitudinal loads due to wind, ice, water, sliding,‡ and rolling resistance§ must also be considered for safe falsework design. The determination of the falsework design loads and their changes during various erection procedures requires careful deliberation by erection engineers, particularly for extensive falsework and/or heavy superstructures.¶

11.2.2.3 Erection by Flotation with Barges

Calm water, adequate depth, and the ability to, if necessary, monitor water elevations are typically necessary for erection by flotation. Barges may be used to support falsework on which the superstructure is supported** (Figure 11.16), stationary derrick cranes (Figure 11.3d), or mobile cranes (Figure 11.17) to erect the superstructure.

A consideration for erection by falsework on a barge is the proximity of assembly yards to the barge. The flotation operation typically involves the assembly on, or lifting of the span onto, the barge,†† towing of the barge to the erection site, and lowering the assembly into place.‡‡

---

* To further reduce interruption of rail traffic, the superstructure deck and track is staged with the superstructure.

† In some cases, demolished by cutting into manageable segments with torches, saws, and lances.

‡ Bearing pads of polytetrafluoroethylene (PTFE) with stainless steel surfaces may be used during the sliding process. The static and dynamic coefficients of friction depend on the roughness of the stainless steel surface, lubricants used on the contact surface, pressure applied from superstructure weight, and sliding speed. When the coefficient of sliding friction is unknown, erection engineers have typically used a conservative estimate of 15% for static (break-away) friction and 8%–10% for dynamic sliding friction.

§ Rolling resistance is considerably less than sliding resistance and it is common to use low-profile chain-action rollers to enable the lateral movement of superstructures. These rollers (especially when new) have very low coefficients of static and dynamic rolling friction.

¶ Steel, due to its relatively light weight, is beneficial for falsework design and economy.

** Typically, girder or/truss spans are supported on trestle-type falsework, bents, and/or towers mounted on barges. The 1912 construction of the St. Lawrence River Bridge (see Chapter 1) used falsework supported on a barge to support the end of the new trusses as they were longitudinally slid into position.

†† In some cases, superstructures are assembled on the falsework on the barge. However, in other cases, the assembly takes place nearby and the superstructure assembly is lifted onto the falsework on the barge. Barges are typically fixed by "spuds" during superstructure assembly or relocation on barges, making these operations free from barge stability considerations.

‡‡ This can be done by careful change to the barge ballast (i.e., adding water into barge compartments) and/or with hydraulic jacks mounted on the supporting falsework.

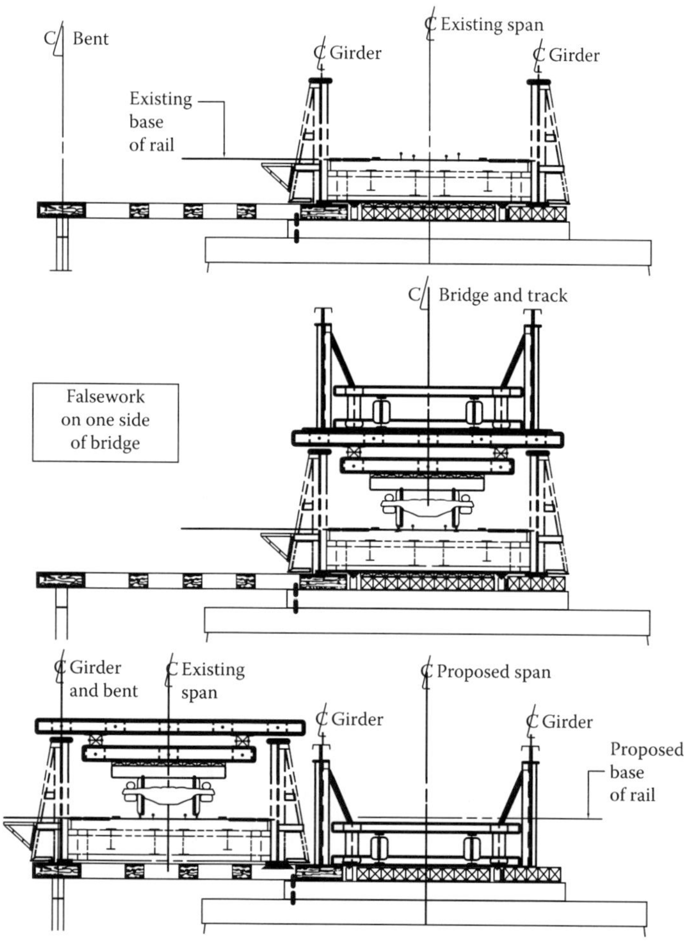

FIGURE 11.15 Nonintrusive falsework on one side of substructures for lateral skidding erection of new spans.

FIGURE 11.16 Superstructure supported on falsework on barge. (Courtesy of the Author, Canadian Pacific Engineering, Calgary, AB, Canada.)

FIGURE 11.17 Crawler crane on barge for pile driving. (Courtesy of the Author, Canadian Pacific Engineering, Calgary, AB, Canada.)

Mobile cranes are commonly supported on barges for superstructure erection over water. Mobile crane charts are typically only applicable up to a very small angle of tilt* from out-of-level crane pad or outrigger supports, or the tilting of barges.

Barge stability is of prime concern for erection by flotation methods.

11.2.2.4 Erection with Stationary and Movable Frames

Crane access by land or water, and/or the construction of falsework may not be feasible at some sites. In such cases, longitudinal erection techniques with stationary or movable frames may be considered. Site-specific stationary frame construction may be cost prohibitive for typical railway superstructure replacement projects, and erection methods that utilize reusable movable frames

* Typically, between 1% and 3%, depending on the mobile crane manufacturer's specifications.

(gantries) may be used to minimize the interruption of railway traffic during superstructure erection. A method and procedures for the replacement of an existing steel span using a stationary frame are shown in Figure 11.18.* A movable frame used for the erection of a two-span bridge is shown in Figure 11.19.

An erection methodology utilizing a movable frame (gantry) for the sequential installation of spans is shown in Figures 11.20 and 11.21. The erection frames may be moved by disconnecting

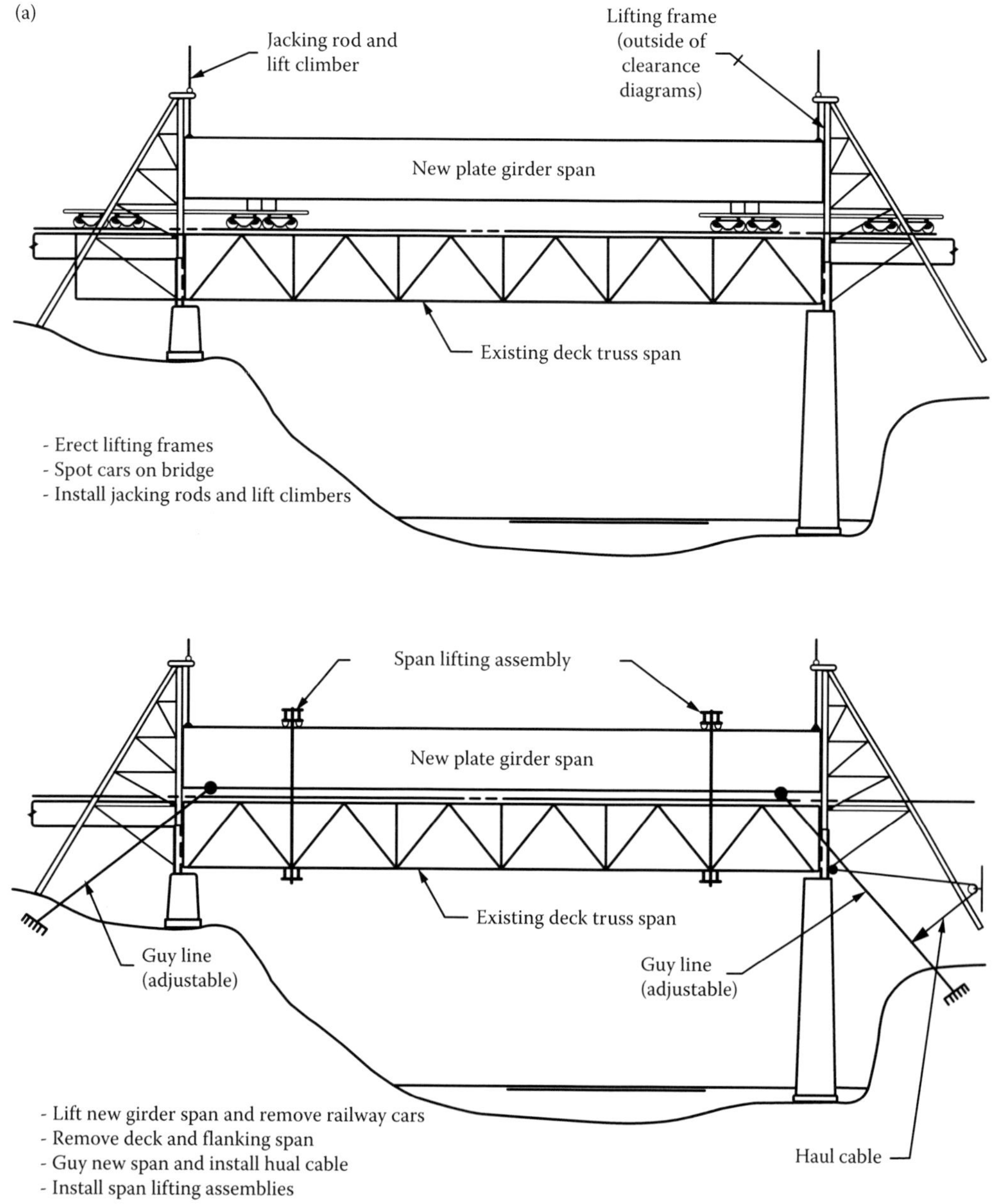

FIGURE 11.18 Replacement of steel span using a stationary frame. (a) Position new girder span, (b) remove existing truss span, and (c) install new girder span.

(Continued)

* A methodology and procedure for the replacement of a deck truss (DT) span with a deck plate girder (DPG) span is shown in Figure 11.18.

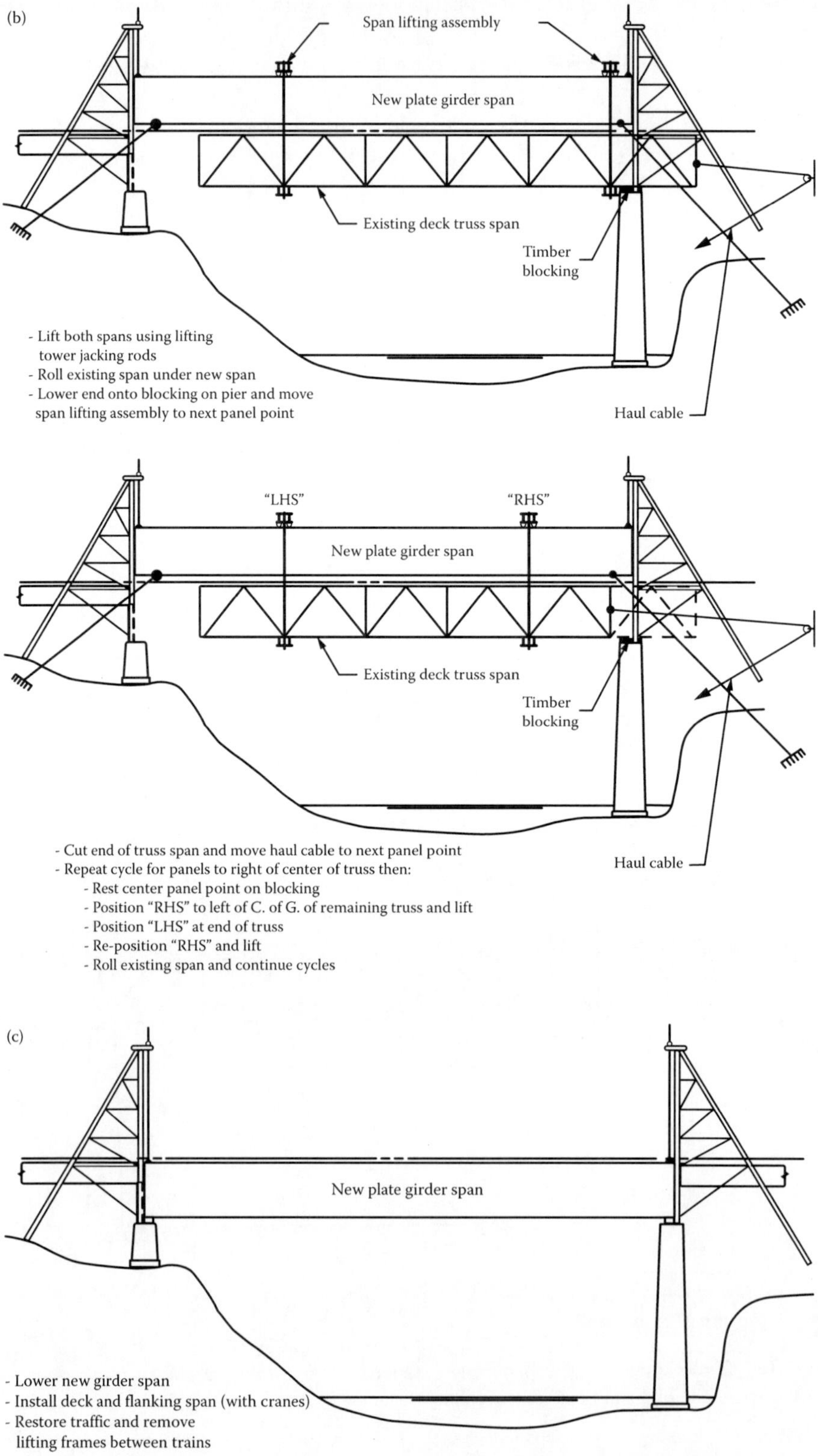

FIGURE 11.18 (CONTINUED) Replacement of steel span using a stationary frame. (a) Position new girder span, (b) remove existing truss span, and (c) install new girder span.

FIGURE 11.19 Sequential replacement of steel spans using a movable frame. (Courtesy of the Author, Canadian Pacific Engineering, Calgary, AB, Canada.)

from the supports* and jacking the frames onto railcars (Figure 11.20) or on trolleys with railroad car wheels for relocation and connection as shown in Figure 11.21.† A movable frame used for the erection of a multiple span bridge is shown in Figure 11.22.

11.2.2.5 Other Erection Methods

Specialized methods, procedures, and equipment are associated with launching, cantilever, tower and cable, catenary high-line, and SPMT erection techniques.

11.2.2.5.1 Erection by Launching

Superstructure assemblies may be longitudinally launched by pulling or pushing a span across the opening between substructures and/or temporary falsework supports. Superstructures are typically

* Typically, the substructures.

† Alternative methods of relocating movable frames such as using sliding or rolling devices on channel rails on the superstructure may also be appropriate at some bridge projects.

incrementally pulled* with derricks or wire rope winches with snatch blocks† to avoid sudden movements during horizontal movement of the superstructure. Superstructures are typically incrementally pushed with hydraulic jacking systems.

Launching may involve the use of moving falsework supports at the leading end, temporary protective noses,‡ and counterweights to launch between intermediate supports§ during the longi-

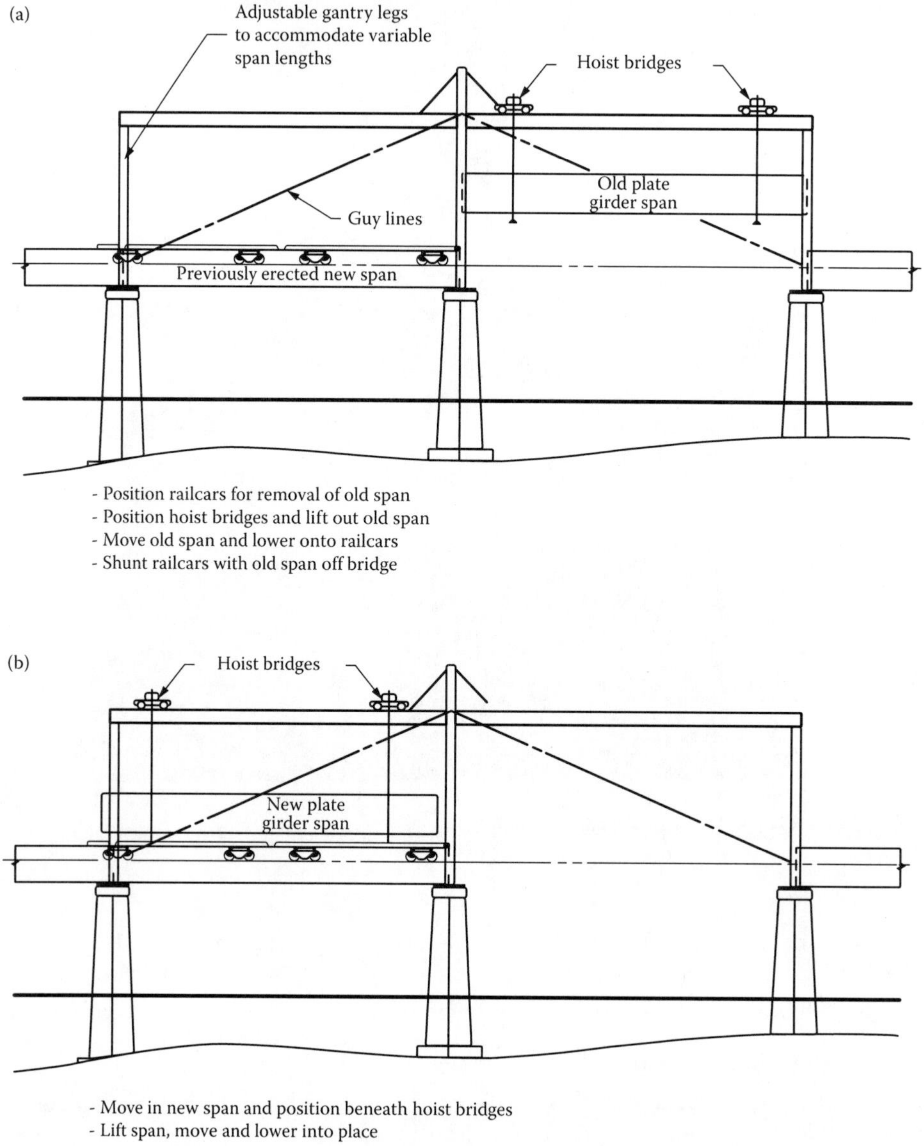

FIGURE 11.20 Procedure for sequential span erection with two-span movable frame. (a) Remove existing girder span. (b) Position and install new girder span. (c) Move gantry to new location.

(Continued)

* In some cases, intermediate falsework support may be provided.

† Pulley systems to assist with the winching of loads.

‡ The protective noses are often inclined to assist with the leading end "landing" on temporary falsework and/or permanent substructures.

§ Typically temporary falsework or permanent substructures.

(c)

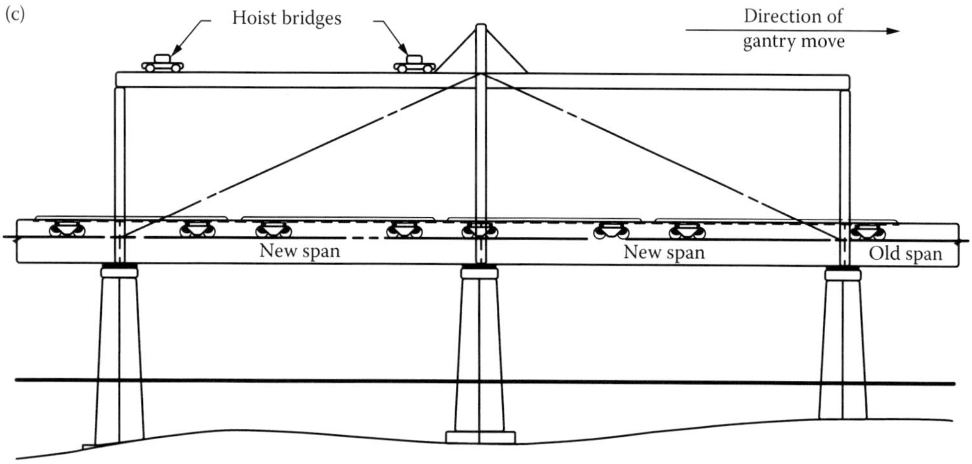

- Position railcars for gantry move
- Install cross beams and jacks

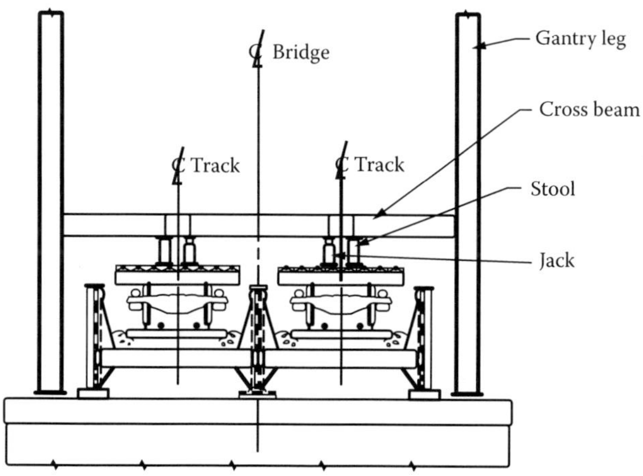

- Remove nuts from anchors of gantry legs
- Jack gantry and support on stools
- Move gantry to new lovation

FIGURE 11.20 (CONTINUED) Procedure for sequential span erection with two-span movable frame. (a) Remove existing girder span. (b) Position and install new girder span. (c) Move gantry to new location.

tudinal movement. Chain-action rollers fixed to the superstructure bearings and sliding plates* are typically used to reduce resistance to longitudinal movement during launching operations. In addition to a protective nose, means of load restraint and robust rolling mechanisms are required for control of long-span launches.

The longitudinal launching of superstructures requires careful consideration of schedule† and considerable erection engineering expertise and experience.

---

* Sliding plates are typically of PTFE (low coefficient of kinetic friction) material bonded to a stainless steel substrate, which may be on elastomers for cushioning.

† The launching method procedures must be able to be performed within the necessary time constraints imposed by modern railroad operations.

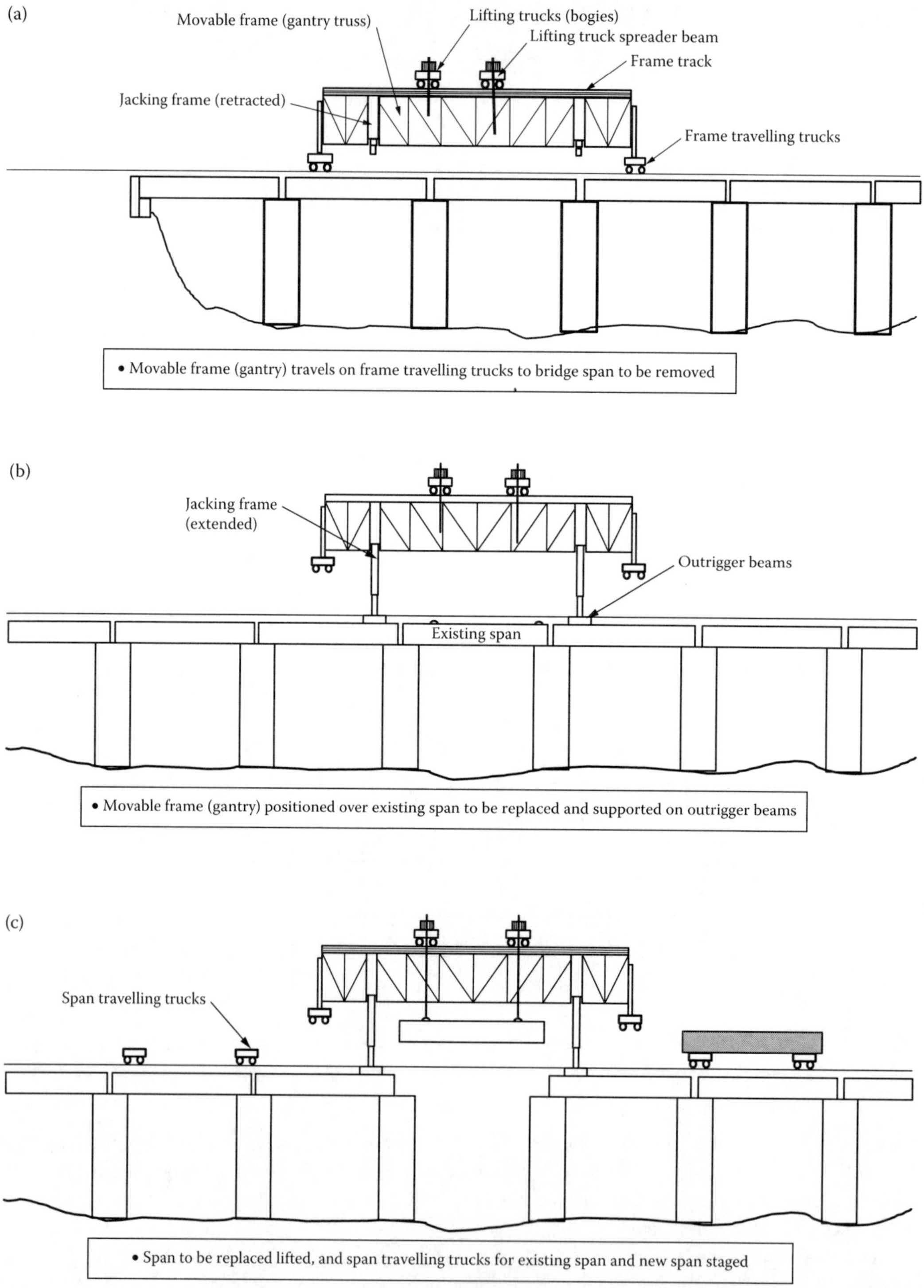

FIGURE 11.21 Procedure for sequential span erection with single-span movable frame. (a) Move and position erection gantry, (b) lift erection gantry, (c) Lift existing span, (d) position existing span for movement, (e) move existing span, (f) move new span, (g) position new span for installation, and (h) install new span and move erection gantry.

(Continued)

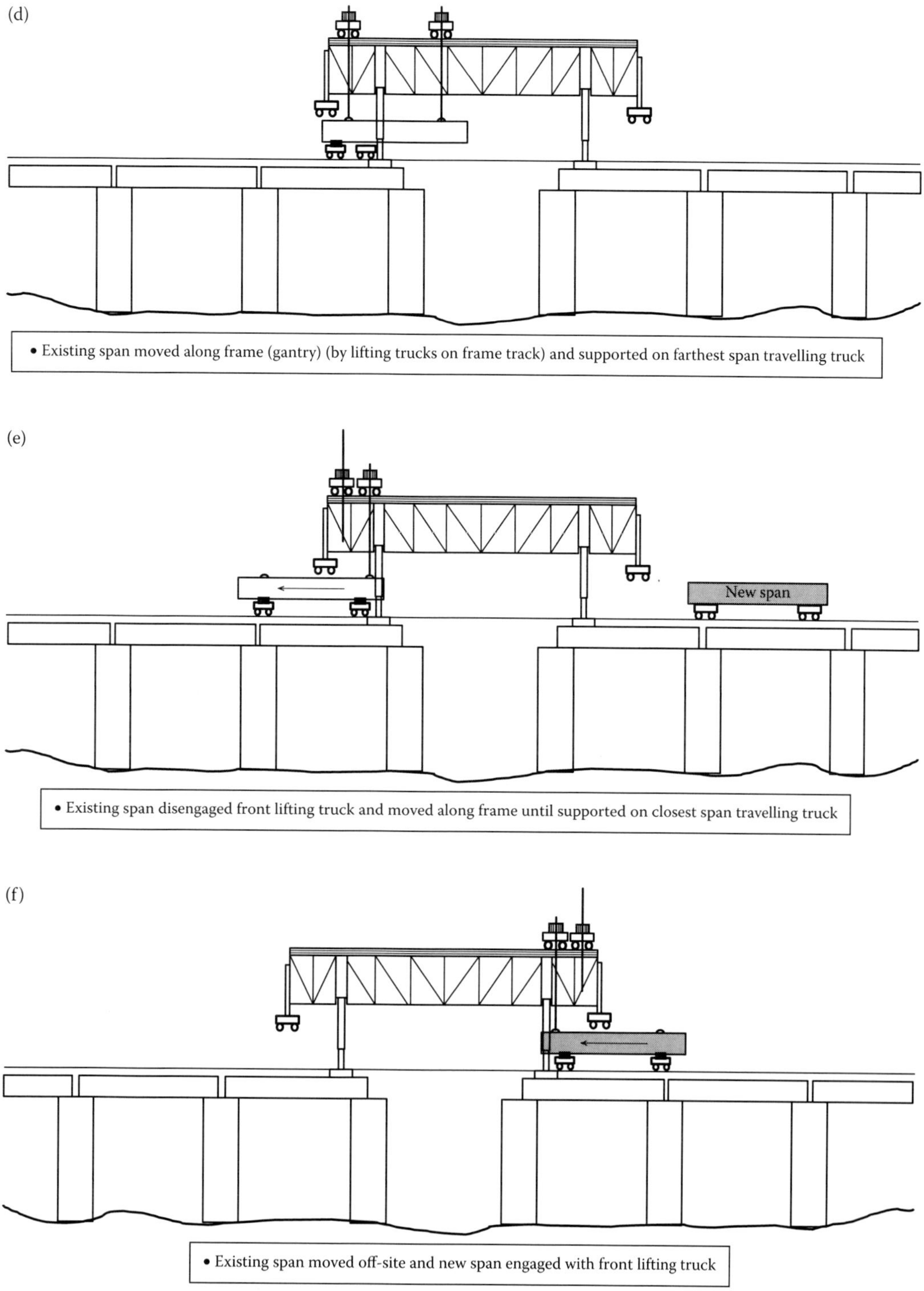

FIGURE 11.21 Procedure for sequential span erection with single-span movable frame. (a) Move and position erection gantry, (b) lift erection gantry, (c) Lift existing span, (d) position existing span for movement, (e) move existing span, (f) move new span, (g) position new span for installation, and (h) install new span and move erection gantry.

(Continued)

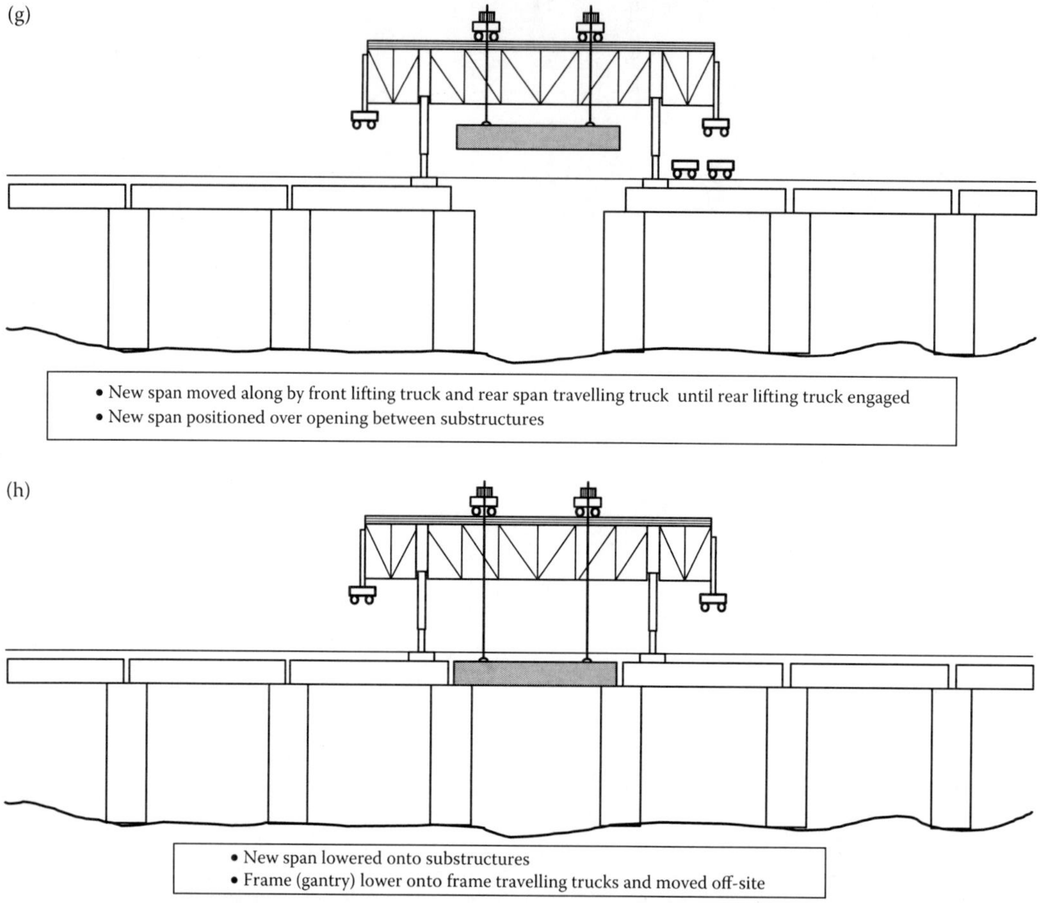

FIGURE 11.21 (CONTINUED) Procedure for sequential span erection with single-span movable frame. (a) Move and position erection gantry, (b) lift erection gantry, (c) Lift existing span, (d) position existing span for movement, (e) move existing span, (f) move new span, (g) position new span for installation, and (h) install new span and move erection gantry.

11.2.2.5.2 Erection by Cantilever Construction

Cantilever construction of superstructures precludes the need for falsework and has been used extensively for the erection of long-span steel superstructures* (see Chapter 1). The cantilever construction of steel railway truss superstructures can be performed by the erection of assemblies by crane† or the truss cantilevers may be stick-built using light traveler cranes.‡

The cantilever construction of superstructures requires considerable erection engineering expertise and experience. Cantilever construction of large superstructures is generally undertaken by specialty contractors with extensive erection engineering capabilities.

11.2.2.5.3 Erection by Tower and Cable

Guyed tower and cable erection procedures also require the engagement of specialized erection engineering and execution experience. Tower and cable construction is typically used in the construction of arches across deep canyons.

---

* Particularly trusses.

† Typically mounted on barges.

‡ Often, specifically designed and fabricated for each project.

FIGURE 11.22 Span erection with single-span movable frame. (Courtesy of the Author, Canadian Pacific Engineering, Calgary, AB, Canada.)

11.2.2.5.4 Erection by Catenary High-Line

High-line catenary cable systems used to longitudinally move and lower superstructure assemblies into position (typically on permanent substructures or falsework for field splicing) also require expert and specific erection engineering and execution experience. This method of superstructure erection is typically used for construction in difficult terrain or across deep canyons.

11.2.2.5.5 Erection by Heavy Load Transporters

The erection of entire spans is typically necessary for many railway superstructure replacement projects in order to minimize the interruption to railroad traffic. Superstructure erection using SPMTs may be another means of controlling project cost and schedule for some erection projects. Figure 11.23 shows SPMT installation of an entire multiple track steel railway span.

FIGURE 11.23 Span erection with SPMT. (Courtesy J. Richter. © Amtrak, Philadelphia, PA, USA.)

The SPMT consists of a modular platform* on an array of wheels used to move large and/or heavy superstructure assemblies. Wheel sets each side of a spine beam form the axle lines for bogies typically arranged between two and six axle lines. The bogies are typically about 3 m (10 ft) wide with length defined by the number of axle lines, which are typically spaced at about 1.5 m (5 ft). Four-line and six-line bogies are commonly used for SPMT erection. Typical four-axle bogies are about 3 m (10 ft) wide and 6 m (20 ft) long.† Several bogies may be coupled together in both the longitudinal and lateral directions to form the modular platform. Each bogie includes hydraulic suspension control fixtures, wheelset drive motors,‡ brake air lines, and steering assembly.§ Wheelsets¶ are connected to pendulum axles** supported by hydraulic cylinders mounted to turntables†† under the modular platform. The hydraulic cylinders for suspension control are interconnected with the hydraulic cylinders of adjacent wheelsets to form a hydraulic stability zone. The hydraulic stability zones ensure stability of the load (superstructure) on the modular platform. However, SPMTs move very slowly, which must be considered when planning erection procedures involving their use.

Many aspects of erection procedures and equipment planning involve the physical principles of applied science, which must be adeptly investigated by construction and/or erection engineers to ensure safe and effective superstructure erection.

11.3 ERECTION ENGINEERING

Erection engineering involves the use of scientific principles to plan and design the methods, procedures, and resources required for the safe and efficient execution of all stages of superstructure erection. Steel railway superstructure erection typically involves crane or derrick lifting and setting,‡‡ skidding (lateral sliding or rolling) on falsework, and/or longitudinal installation with stationary or movable erection frames.

Erection engineers must plan and design the procedures and structures necessary to realize safe and efficient erection while also considering the effects on the superstructure being erected. Erection with cranes, derricks, and falsework may be from land or barges in watercourses. Erection frames are typically supported on or near the superstructure being erected. Erection engineering of procedures may involve the investigation of crane and/or barge stability, and the techniques, structures, and devices required for moving superstructures on falsework§§ or with erection frames.¶¶

Erection procedure sequence is important for final geometry because fabrication and erection tolerances may affect member deflections, alignment, and geometry. Erection procedure sequence may also affect member stability. Therefore, procedures, particularly for staged construction, may require that the erection engineer determine methods of controlling stress and distortions in the superstructure during erection.

Experienced erection engineers are rare but critical for complex projects (Durkee, 2014). The level of erection engineering effort required depends on the methods and procedures to be used, which are usually related to the complexity of the superstructure erection project. A relatively

* The modular platforms are created by combining spine beams with wheel sets together to carry large superstructure assemblies.

† With axle spacing of 1.5 m. The lengths of SPMTs are measured from center to center of couplers so that the axle spacing remains constant when additional bogies are coupled.

‡ Driven wheelsets have hydraulic drive motors mounted in the hubs of the axles. However, typically, not all wheelsets are driven.

§ The steering assembly may be either mechanical or electronic.

¶ Each wheelset has two tires each side of the pendulum axle.

** Wheelsets are attached to the lower legs of pendulum axles to accommodate uneven travel surfaces.

†† Each wheelset is mounted to a turntable and may be steered by mechanical or electronic controls.

‡‡ Typically used for lateral installation of superstructures, but may also be used in longitudinal erection procedures (often using track mounted cranes).

§§ Typically used for lateral installation of superstructures.

¶¶ Typically used for longitudinal installation of superstructures.

common and straightforward erection engineering assessment involves crane equilibrium stability analysis* to verify the procedure for lifting members into place while ensuring the structural stability (buckling resistance) of the members until temporarily or permanently braced.

The design and fabrication documentation must be accurate and complete in order that the erector can develop appropriate methods, procedures, and resources (labor and equipment) required for the erection project scope and complexity. Then, proper erection drawings and documents can be prepared that clearly outline the methods, procedures, structures, and equipment required for the safe and effective† erection of the superstructure.

11.3.1 Erection Engineering for Member Strength and Stability

Control of stresses and deformations of girder or truss superstructures during erection by cranes, derricks, falsework, and/or stationary or movable erection frames is essential for safety. Components of the finished superstructure that may provide strength, stiffness, and/or stability to the superstructure assembly‡ may not be in place during erection. Therefore, the erection engineer must consider the strength and stability of partially constructed superstructures§ resisting wind and other construction loads during various procedures of the erection. This may require structural analyses of the superstructure for several events of the erection procedures.¶ Nevertheless, even where superstructure stability is critical, many erection contractors do not conduct adequate analytical evaluations of superstructure behavior during various procedures of the erection method. In lieu, these erection contractors may use rules of thumb to maintain member stability during erection, which may be unsafe practice for even some relatively simple procedures. However, while this approach has been used by many experienced erection contractors with success for typical girder erection procedures, it has also led to some girder stability failures during erection.

For girder erection, the lateral-torsional buckling of I-sections (see Chapter 7) during supporting,** lifting and handling is a critical safety consideration. Increasing the lateral-torsional strength of the unbraced girder is typically not economical and restraint against lateral-torsional buckling is usually employed. Restraint at girder supports and/or at intermediate sections may be required during erection procedures to prevent twisting (or lay-over) of the girder. The restraint is typically provided by temporary bracing and/or supports, or by holding cranes until adequate permanent bracing is installed.†† The erection procedure must ensure that enough cross frames are installed before the removal of holding cranes or falsework supports. It is often most economical to provide temporary bracing to prevent lateral and twisting deformations. The temporary bracing must be designed with the strength and stiffness to provide stability, and resist wind and other lateral loads during erection.

Where feasible, it is most effective to ship superstructures fully assembled or construct the complete span on the ground or falsework. This will avoid the need for strength and stability investigations, or the temporary bracing, required for the erection of individual girders or trusses.

Erection engineering stability calculations are also required when girder and system stability are of concern in other situations during an erection method. Procedures that create second-order amplification of lateral-torsional buckling effects and/or include cantilever erection situations that were not anticipated during design require careful evaluation. For example, in erection procedures where girders are cantilevered over substructures for field splicing, the erection engineer must

---

* Crane charts are typically used to review crane stability for the lifts required by the erection procedures.

† From both cost and schedule perspectives.

‡ Typically an individual girder or truss.

§ For example, cantilevered sections during erection procedures.

¶ For complex superstructures or erection procedures, the design engineer can be of considerable assistance to the erection engineer through sharing of digital models, calculations, drawings, and other design documentation.

** Procedures relating to both temporary girder support at staging or on falsework, and permanent support must be considered.

†† Typically in conjunction with erection of the adjacent girder.

carefully assess girder deflections and alignment. Erection procedures for girders involving long cantilevers are also of considerable concern with respect to lateral-torsional stability.

The erection contractor typically has the responsibility for the strength and stability of members during construction. However, the superstructure design engineer can be of considerable assistance to the erection engineer by providing strength and stability design data for the design of temporary bracing resisting erection, wind, stability, and concrete deck forces.*

Further, but simple, erection considerations for lifting of the girders of Examples 7.1a and b (see Chapter 7) are outlined in Examples 11.1a and b. Examples 11.2a and b also outline basic girder stability considerations for long girders that require temporary support during lifting. If required, more complex lifting arrangements may be investigated using linear and geometrically nonlinear finite element analysis (FEA).

Example 11.1a

During fabrication and erection of the 30 m long girder in Example 7.1a, the entire girder will be effectively laterally supported during fabrication, erection staging, and in final position on false-work. However, the erection procedure requires that the girder will be lifted at two locations at distance, *a*, from each end of the girder (see Figure E11.1) and that erection lifts will not be done coincident with windy conditions.

Girder self-weight ~ 1.15(820) ~ 950 kg/m (with 15% contingency load)

$w_{girder} = (950)(9.81)/1000 = 9.32$ kN/m.

For typical railway girder lifting arrangements, the maximum bending stress will occur between the lifting locations within the length, $b = L - 2a$.

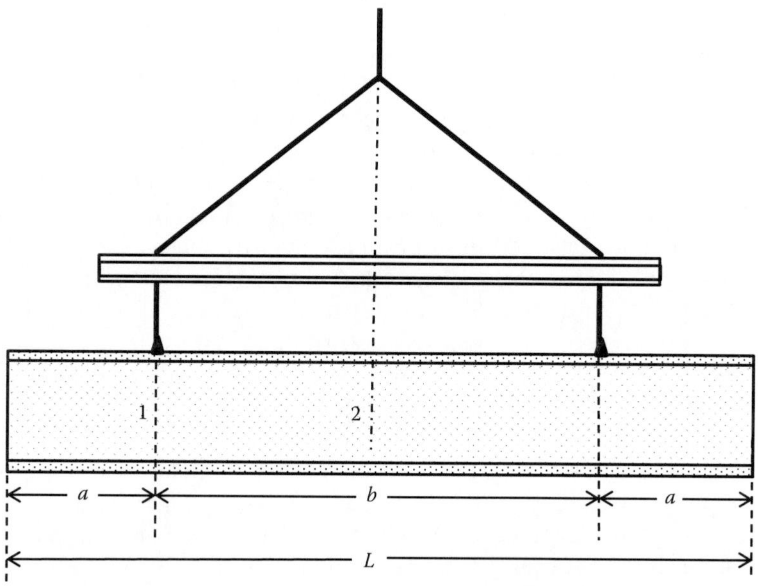

W = uniformly distributed load to be lifted

$M_1 = -wa^2/2$

$M_2 = ((wL(L-2a))/2b)[((L(L-2a))/4b) - a]$

$L_{u1} = a$

$L_{u2} = b$

FIGURE E11.1

* Concrete deck pours may cause deep girder rotations and lateral movements during the deck pours which may affect the installation of cross frames.

$$M_1 = \frac{-w_{girder}a^2}{2} = 4.66a^2 \text{ kN-m } (a \text{ in m}),$$

$$M_2 = \frac{w_{girder}L}{2}\left(\frac{L}{4} - a\right) = 139.8(7.50 - a) \text{ kN-m } (a \text{ in m}),$$

using an allowable bending stress of $1.25(0.55)F_y = 0.69F_y$ (see Chapter 4),

$$\left(\frac{L}{r_y}\right)_{1,2} = \sqrt{\frac{0.69(350) - (M_{1,2}/S)}{0.69(350)^2/6.3\pi^2(200000)}} = \sqrt{\frac{241.5 - (M_{1,2}/S)}{0.00680}} \quad (M/S \text{ in MPa}).$$

Figure E11.2 is a plot of L_{u1}/a and L_{u2}/b against length, a, to provide the lifting location parameters that ensure lateral-torsional stability of the girder during lifting onto the falsework. Figure E11.2 indicates that instability of the girder end cantilevers of length, a, will not occur. It also indicates that instability of the girder of length, b, between lifting points will not occur when $a \geq 6.4$ m. It is interesting to note that instability of the girder end cantilevers governs for $a \geq 10.0$ m. The girder should be lifted at a distance greater than 6.4 m from each end of the girder. The maximum unsupported length, L_u, is = 30 − 2(6.4) = 17.2 m, which is similar to the estimate made in Example 7.1a in Chapter 7.

The erection engineer may wish to increase the factor of safety (FS) related to the unsupported length of the girder lift.

Using an FS = 1.5, Figure E11.2 indicates that instability of the girder end cantilevers will not occur for $a \leq 11.5$ m and that instability of the girder of length, b, between lifting points will not occur when $a \geq 9.2$ m. The girder should be lifted at a distance greater than 9.2 m but not more than 11.5 m from each end of the girder.

Using an FS = 2.0, Figure E11.2 indicates that instability of the girder end cantilevers will not occur for $a \leq 8.5$ m and that instability of the girder of length, b, between lifting points will not occur when $a \geq 10.7$ m. Considering an FS of 2.0, it will not be feasible to lift the girder without supplementary support.

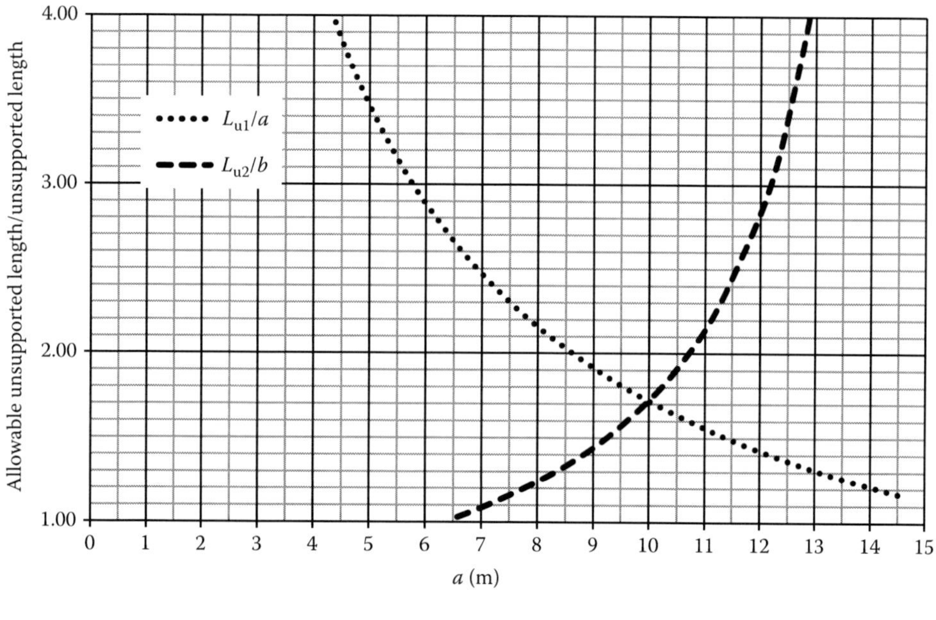

FIGURE E11.2

Example 11.1b

During fabrication and erection of the 90 ft long girder in Example 7.1b, the entire girder will be effectively laterally supported during fabrication, erection staging, and in final position on falsework. However, the erection procedure requires that the girder will be lifted at two locations a distance, a, from each end of the girder (see Figure E11.1) and that erection lifts will not be done coincident with windy conditions.

Girder self-weight ~ 1.15(521) ~ 600 lb/ft (with 15% contingency load).

For typical railway girder lifting arrangements, the maximum bending stress will occur between the lifting locations within the length, $b = L - 2a$,

$$M_1 = \frac{-w_{\text{girder}} a^2}{2} = 0.60a^2 \text{ kip-ft } (a \text{ in ft}),$$

$$M_2 = \frac{w_{\text{girder}} L}{2}\left(\frac{L}{4} - a\right) = 27.0(22.50 - a) \text{ kip-ft } (a \text{ in ft}),$$

using an allowable bending stress of $1.25(0.55)F_y = 0.69F_y$ (see Chapter 4), and

$$\left(\frac{L}{r_y}\right)_{1,2} = \sqrt{\frac{0.69(50000) - (M_{1,2}/S)}{0.69(50000)^2/6.3\pi^2(29\times 10^6)}} = \sqrt{\frac{34500 - (M_{1,2}/S)}{0.957}} \quad (M/S \text{ in psi}).$$

Figure E11.3 is a plot of L_{u1}/a and L_{u2}/b against length, a, to provide the lifting location parameters that ensure lateral-torsional stability of the girder during lifting onto the falsework. Figure E11.3 indicates that instability of the girder end cantilevers of length, a, will not occur. It also indicates that instability of the girder of length, b, between lifting points will not occur when $a \geq 8.5$ ft. It is interesting to note that instability of the girder end cantilevers governs for $a \geq 30$ ft. The girder should be lifted at a distance greater than 8.5 ft from each end of the girder. The maximum unsupported length, L_u, is = 90 − 2(8.5) = 73.0 ft, which is 81% of the girder length, similar to the estimate made in Example 7.1b in Chapter 7. The wide top flange benefits girder stability

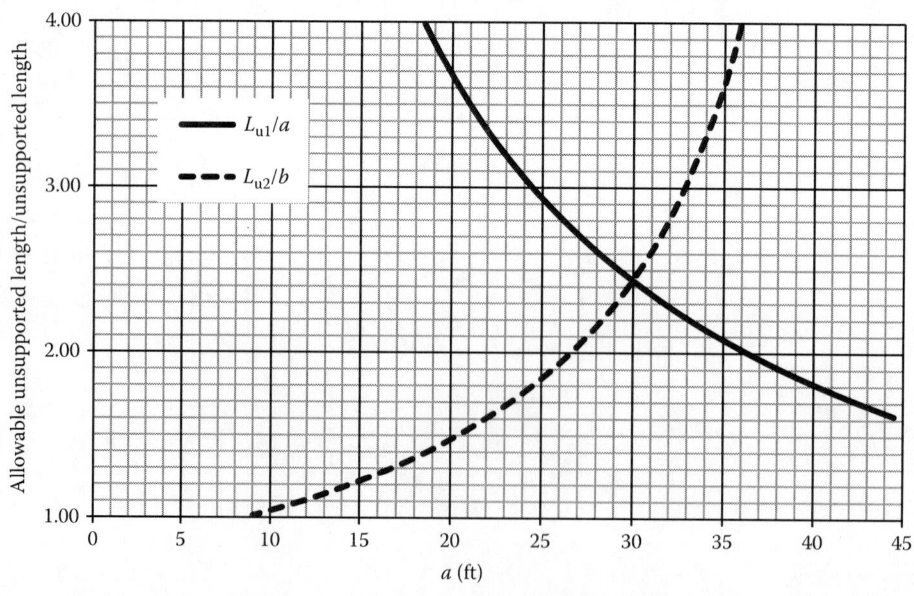

FIGURE E11.3

(see Example 11.1a for a girder with a narrow top flange and allowable unsupported length of only 57% of girder length).

The erection engineer may wish to increase the FS related to the unsupported length of the girder lift.

Using an FS = 1.5, Figure E11.3 indicates that instability of the girder end cantilevers will not occur and that instability of the girder of length, *b*, between lifting points will not occur when $a \geq 20.5$ ft. The girder should be lifted at a distance greater than 20.5 ft from each end of the girder.

Using an FS = 2.0, Figure E11.3 indicates that instability of the girder end cantilevers will not occur for $a \leq 36$ ft and that instability of the girder of length, *b*, between lifting points will not occur when $a \geq 26.5$ ft. The girder should be lifted at a distance greater than 26.5 ft but not more than 36 from each end of the girder.

Example 11.2a

An 80 m long continuous girder is to be lifted into final position and held until temporary bracing is installed. What are the lifting location parameters if the girder has the following properties:

$w_{girder} = 1500$ kg/m
$S = 110 \times 10^6$ mm^3
$r_y = 104$ mm
$F_y = 350$ MPa

Plots of L_{u1}/a and L_{u2}/b against length, *a*, are made (as shown in Example 11.1a) to provide lifting location parameters. Figure E11.4a indicates that instability of the girder end cantilevers of length, *a*, will occur for $a \geq 18.5$ m. However, it also indicates that instability of the girder of length, *b*, between lifting points will occur when $a \leq 31.5$ m. It will not be feasible to lift the girder without the use of holding cranes.

If a holding crane is used at midspan, Figure E11.4b illustrates that girder stability may be ensured for length, a ~20 m. For $a > 20$ m instability of the girder end cantilever will occur and for $a < 20$ m instability of the girder between lifting locations will occur. These lifting parameters are too contiguous.

If holding cranes are used at two locations in the length, *b*, between lifting locations, Figure E11.4c indicates that instability of the girder end cantilevers of length, *a*, will not occur for $a \leq 19$ m. Figure E11.4c also indicates that instability of the girder of length, *b*, between lifting

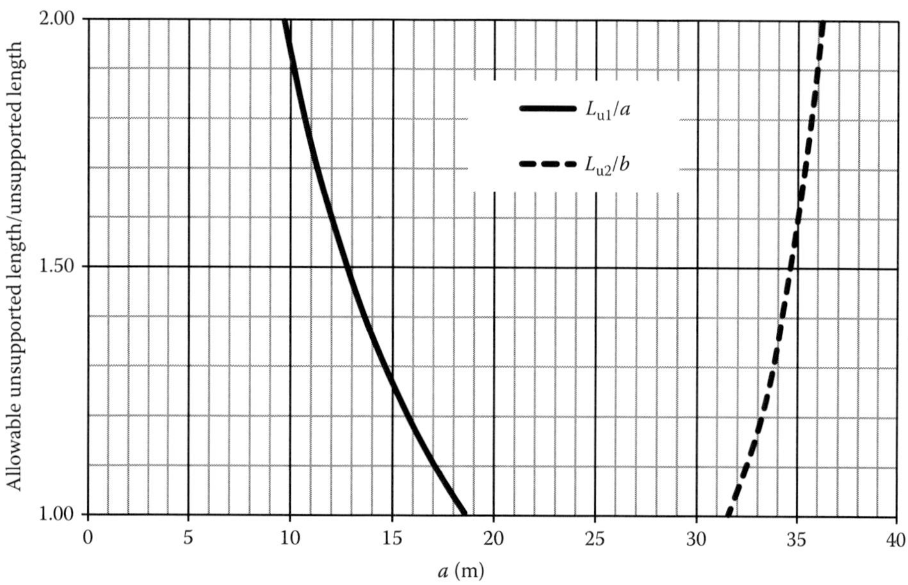

FIGURE E11.4a

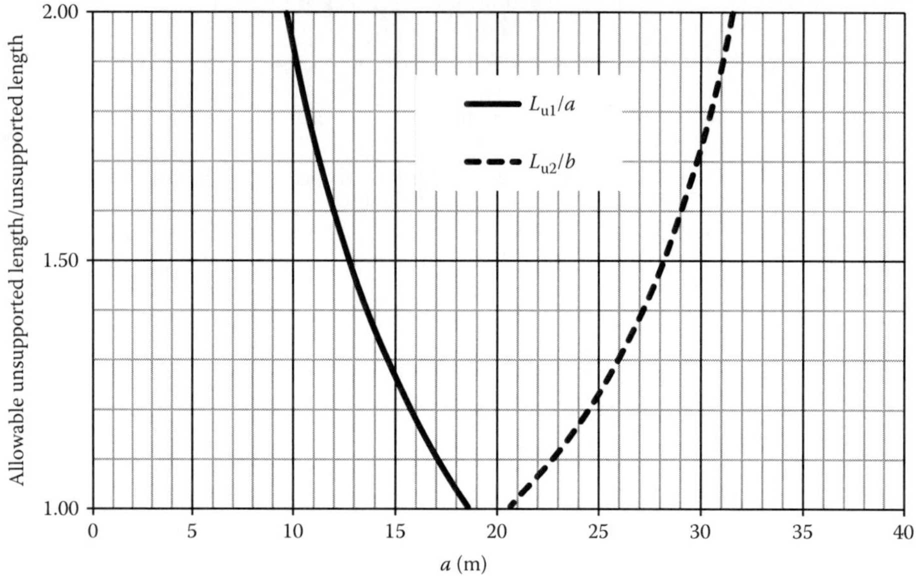

FIGURE E11.4b

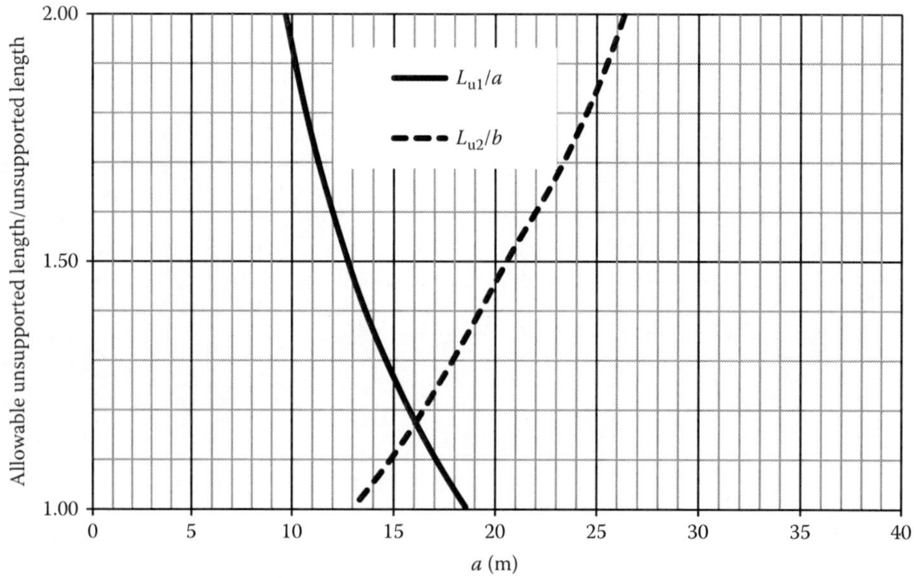

FIGURE E11.4c

points will not occur when $a \geq 13$ m. The girder should be lifted at a distance between 13 and 19 m from each end of the girder with holding cranes at the third points between the lifting locations.

If the erection engineer considers an FS related to the unsupported length of 1.5, Figure E11.4c indicates that instability of the girder end cantilevers of length, a, will not occur for $a \leq 12.5$ m. Figure E11.4c also indicates that instability of the girder of length, b, between lifting points will not occur when $a \geq 20.5$ m. Considering an FS of 1.5, it will not be feasible to lift the girder without holding cranes at the third points between the lifting locations.

Example 11.2b

A 250 ft long continuous girder is to be lifted into final position and held until temporary bracing is installed. What are the lifting location parameters if the girder has the following properties:

$w_{girder} = 2000$ lb/ft
$S = 6500$ in.^{3},
$r_y = 4.25$ in.,
$F_y = 50$ ksi

Plots of L_{u1}/a and L_{u2}/b against length, a, are made (as shown in Example 11.1b) to provide lifting location parameters. Figure E11.5a indicates that instability of the girder end cantilevers of length, a, will occur for $a \geq 60$. However, it also indicates that instability of the girder of length, b, between lifting points will occur when $a \leq 105$ ft. It will not be feasible to lift the girder without the use of holding cranes.

If a holding crane is used at midspan, Figure E11.5b illustrates that girder stability may be ensured for length, $a = 60$ ft. However for $a > 60$ ft instability of the girder end cantilever will occur and for $a < 60$ ft instability of the girder between lifting locations will occur. These lifting parameters are, too, contiguous.

If holding cranes are used at two locations in the length, b, between lifting locations, Figure E11.5c indicates that instability of the girder end cantilevers of length, a, will not occur for $a \leq 60$ ft. It also indicates that instability of the girder of length, b, between lifting points will not occur when $a \geq 40$ ft. The girder should be lifted at a distance between 40 and 60 ft. from each end of the girder with holding cranes at the third points between the lifting locations.

Figure E11.5c indicates that using an FS ≥ 1.25 renders it infeasible to lift the girder, even with holding cranes at the third points between the lifting locations.

Clear erection procedures must be established for the method used to erect truss superstructures. These procedures are often based on structural analyses of the truss under erection and wind loads during the various procedures of the erection method. These structural analyses may reveal that, during erection procedures, some truss members must resist forces opposite to the forces to be resisted in service.

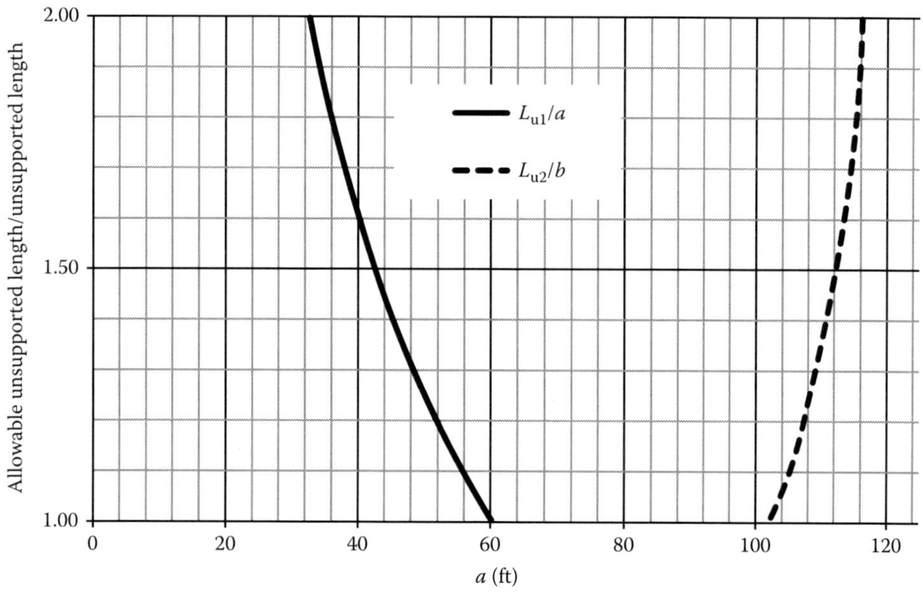

FIGURE E11.5a

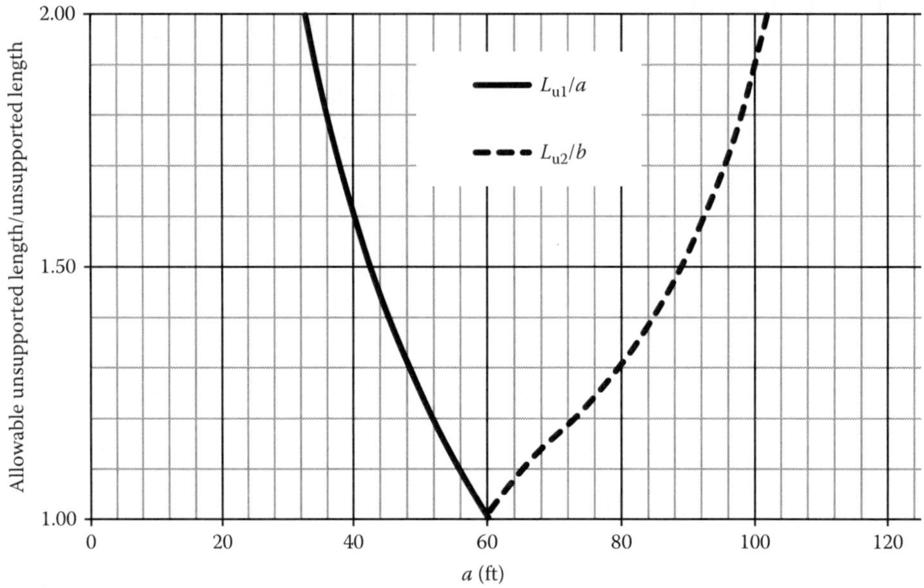

FIGURE E11.5b

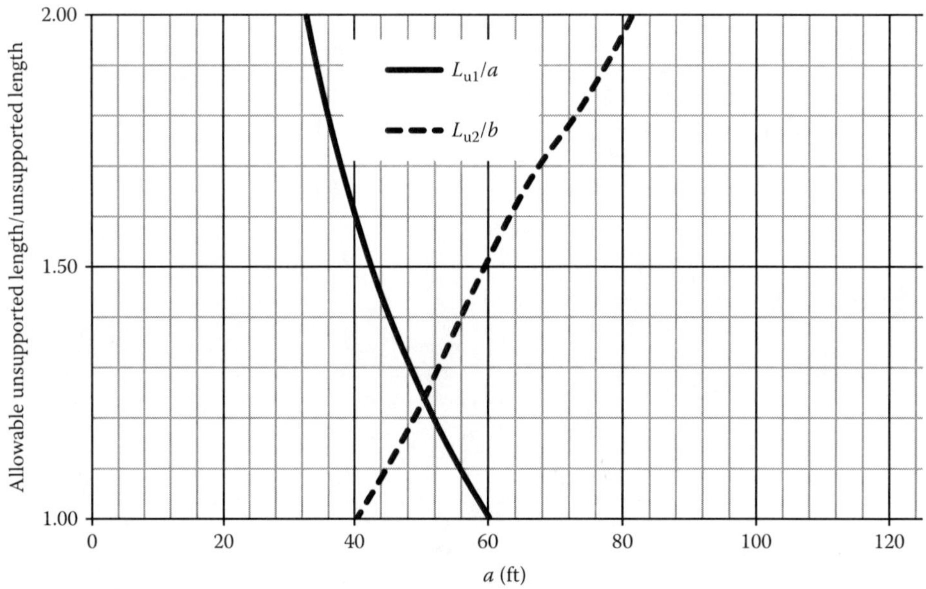

FIGURE E11.5c

The longitudinal launching of trusses also requires skilled erection engineering analysis. The longitudinal rolling and sliding of trusses entails care to avoid point load conditions on chord members. This may be precluded with temporary support beams between truss panel points, rolling on the floor system stringers and/or on roller nests fixed to truss panel points.

Field joints to connect members should be made without exceeding the calculated erection stresses until the complete connection is made. The allowable shear on drift pins [hardened steel with yield stress greater than 345 MPa (50 ksi)] is 138 MPa (20 ksi), and during erection they are considered to share the load with fully torqued bolts in the same plane. The erection drawings should indicate the number of bolts and/or pins required for the release of crane hold or other temporary supports at field splices, and in cross frames and diaphragms.

Erection plans must also show wind speed, and limitations on construction dead and live load for various erection method procedures. In addition, some erection methods and procedures require that the erection engineer considers uplift and cantilever end deflections* during erection.

The stresses due to erection loads in members and connections may exceed usual allowable stresses by 25% in steel freight railway bridges (AREMA, 2015). This may be increased to 33% greater than usual allowable stresses for load combinations including erection and wind loads (see Chapter 4).

If the erection procedures are established early enough, modifications to the permanent structure required by the erection method may be arranged with the fabricator by the erection contractor following the approval of the designer. However, it is often difficult for erection contractors to establish erection procedures early in the fabrication process. In such cases, the erection engineer should communicate with the design engineer concerning the reinforcement and strengthening of members, and/or bracing requirements for members during erection. Nevertheless, the erection contractor remains responsible for damage or detrimental overstress due to the erection method procedures.

11.3.2 Erection Engineering for Cranes and Derricks

Erection engineering for derricks and mobile cranes is similar. Strength and stability considerations govern, and for derricks, specialized engineering design† is usually required. Mobile crane strength and stability design is by the crane manufacturer and is summarized in crane working range and lifting capacity charts.‡

11.3.2.1 Stationary Derricks

Erection engineers must design guyed derricks with due consideration of the mast support loads and the design of the guys and their anchorages. The design of stiffleg derricks involves the consideration of mast and sill support loads. The structural analysis and design of the masts, booms, guys, sills, legs, foundations, anchorages, and other members of guyed and stiffleg stationary derricks using appropriate safety factors§ can be accomplished using usual structural engineering principles with the applicable codes and/or guidelines.¶

11.3.2.2 Mobile Cranes

The structural strength of the booms, jibs, and other members of mobile cranes is determined by the manufacturer and included in crane capacity charts for the mobile crane.** Crane capacity charts may also include limitations related to boom tip deflection.†† Nevertheless, for most lifting weights and distances, crane stability (resistance to tipping or overturning moment) governs mobile crane capacity and is the basis for the majority of the crane capacity chart data. The rated capacity is generally taken as 75% or 85% of the stability load.‡‡

Equilibrium of forces for the crawler crane shown in Figures 11.24a and b provides the overturning weight (or tipping load), W, as

$$W = \frac{W_f D_f + W_c D_c + W_m D_m - W_b D_b}{D}. \tag{11.1}$$

* For some erection methods, a critical aspect of erection engineering is the calculation of cantilever end deflections and monitoring of the deflections during erection.

† Sometimes site specific.

‡ Presented in terms of boom length and lifting radius.

§ For example, guys are often designed with a safety factor of 3.

¶ Shapiro et al. (2000) outline principles and methods for stationary derrick design.

** Typically, the structural strength of telescopic or latticed booms governs for only heavy lifts at small lifting radii.

†† Typically, boom tip deflections of telescopic or latticed booms govern for only long boom lifts.

‡‡ Typically, 75% for crawler cranes and 85% for truck cranes with outriggers.

Similarly, equilibrium of forces for the truck-mounted crane with outriggers shown in Figures 11.25a and b provides the overturning weight, W, as

$$W = \frac{W_{f1}(D_o - D_{f1}) + W_{f2}(D_o + D_{f2}) + W_c D_c + W_m D_m - W_b D_b - W_r D}{D}, \tag{11.2}$$

where

W_f, W_{f1}, W_{f2} = the weight of frames supporting cables
W_c = the weight of the counterweight
W_m = the weight of the crane (machinery, body, etc.)
W_b = the weight of the crane boom
W_r = the weight of rope sheaves, etc.

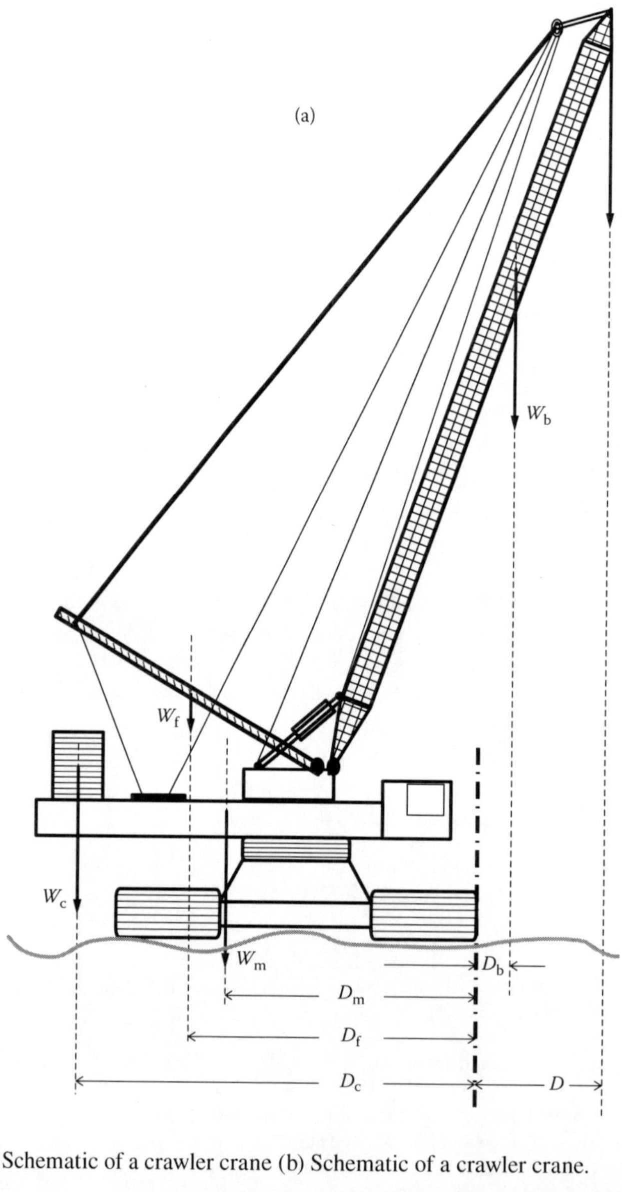

FIGURE 11.24 (a) Schematic of a crawler crane (b) Schematic of a crawler crane.

(Continued)

(b)

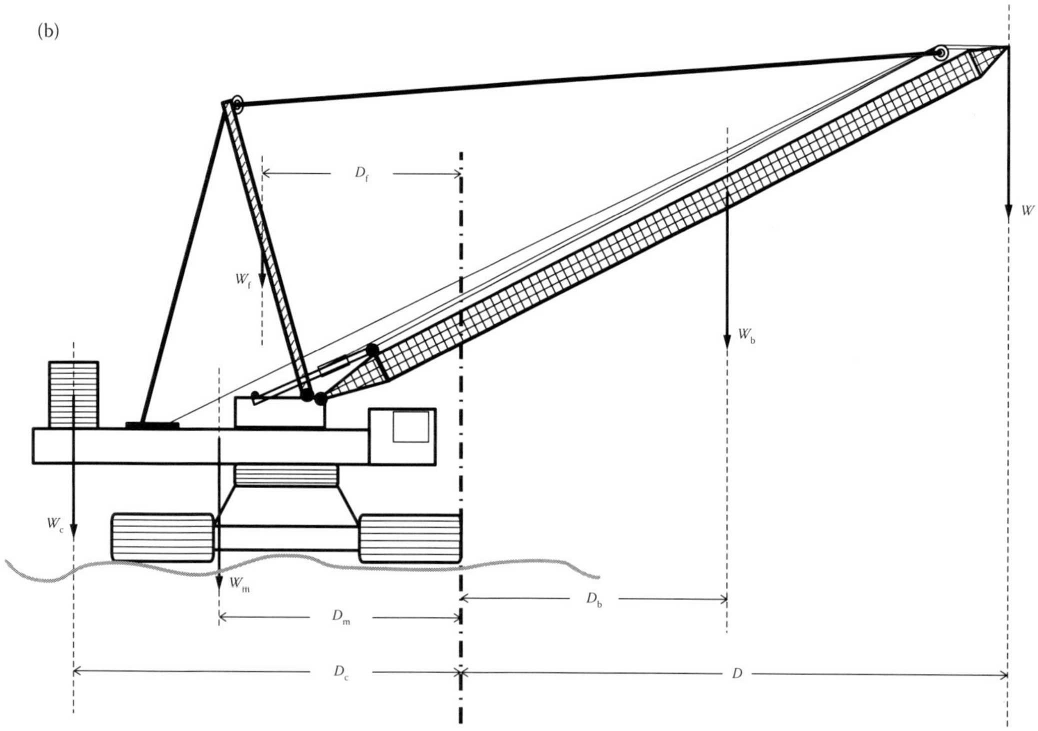

FIGURE 11.24 (CONTINUED) (a) Schematic of a crawler crane, (b) Schematic of a crawler crane.

D_f, D_c, D_m, D_b = the distance from center of gravity of frames, counterweight, crane body/machinery, and boom to tipping fulcrum, respectively

D_o = the distance between the crane center of rotation and the tipping fulcrum

D_{f1}, D_{f2} = the distance from the center of gravity of frames to the crane center of rotation

Since W_f, W_{f1}, W_{f2}, W_c, W_m, D_c, D_m, D_o, and D_{f2} are constants, W_cD_c, W_mD_m, $W_{f1}D_o$, $W_{f2}D_{f2}$, and $W_{f2}D_o$ are also constants. Therefore, only W_fD_r, W_bD_b, $W_{f1}D_{f1}$, and W_rD are variable. However, D_r, D_b, and D_{f1} can be expressed in terms of D, and crane charts and tables for W and D for various boom lengths can be readily developed. Such crane charts have been developed by crane manufacturers based on the geometry and weight of their cranes. Example 11.3 outlines some simple crane stability calculations for crawler cranes.

Example 11.3a (SI Units)

The girder of Example 11.1a weighs 950(30) = 28,500 kg and is to be erected with a crawler crane similar to that shown in Figure 11.24 with D = 30 m. The crawler crane has the following specifications:

W_f = 5000 kg
W_m = 60,000 kg
W_b = 90,000 kg
D_f = 5 m
D_c = 9.5 m
D_m = 7 m

The boom geometry is such that D_b = 13 m. What counterweight is required?

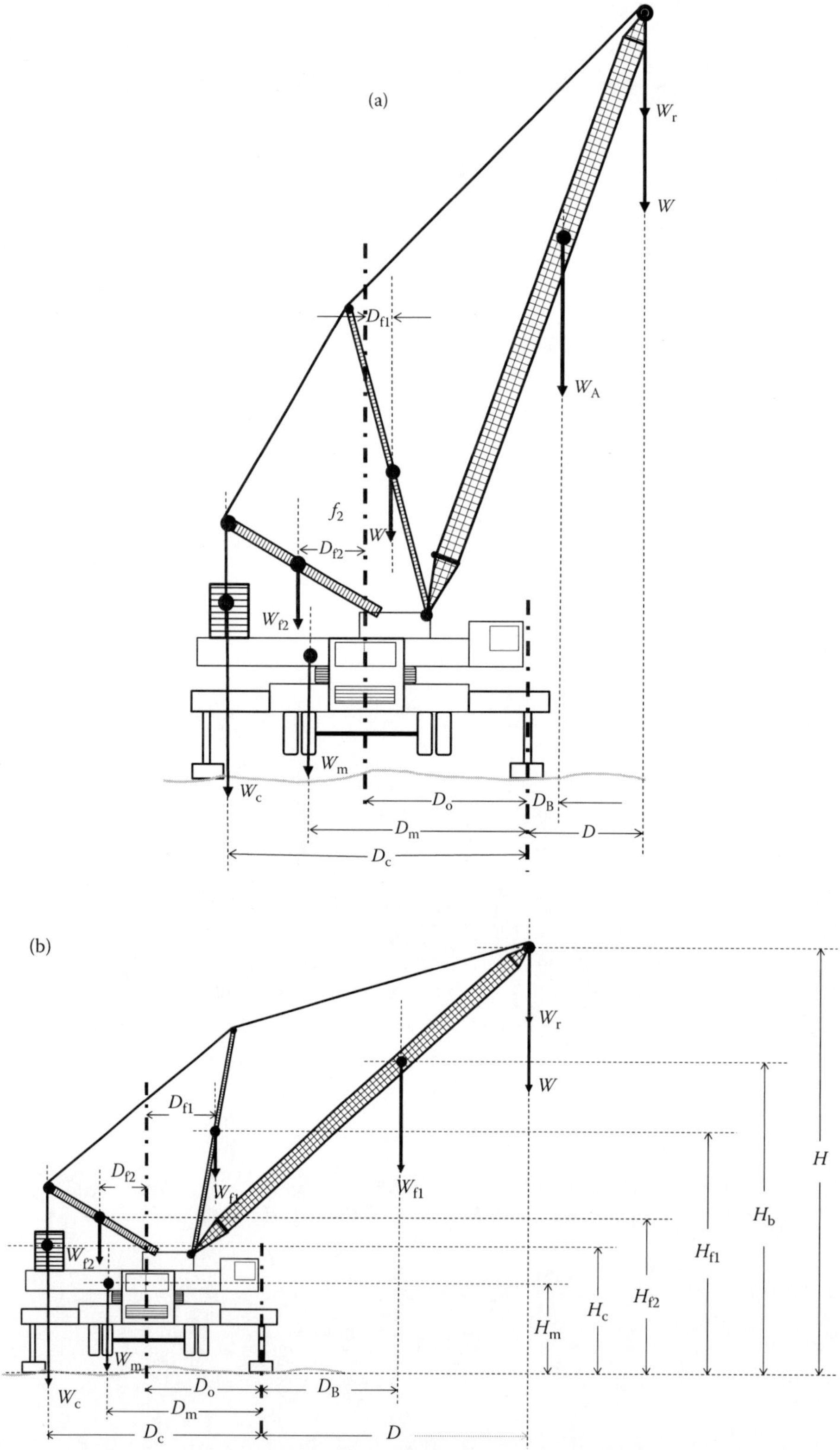

FIGURE 11.25 (a) Schematic of a truck-mounted crane. (b) Schematic of a truck-mounted crane.

$$W_c = \frac{WD + W_b D_b - W_m D_m - W_f D_f}{D_c},$$

$$= \frac{\left(28.5(30) + (90)(13) - 60(7) - (5)(5)\right)1000}{9.5},$$

$$= \frac{\left(855 + 1170 - 420 - 25\right)1000}{9.5} = 166,316 \text{ kg},$$

$$= 166.3 \text{ tonnes.}$$

With an FS = 1/0.75 = 1.33, the required counterweight is 1.33(166.3) = 221 tons. For this crane each counterweight plate weighs 7500 kg and 29 counterweight plates are required.

Example 11.3b (US Customary and Imperial Units)

The girder of Example 11.1b weighs 600(90) = 54,000 lb and is to be erected with a crawler crane similar to that shown in Figure 11.24 with D = 80 ft. The crawler crane has the following specifications:

W_f = 10,000 lb
W_m = 142,000 lb
W_b = 155,000 lb
D_f = 15 ft
D_c = 35 ft
D_m = 23 ft

The boom geometry is such that D_b = 32 ft. What counterweight is required?

$$W_c = \frac{WD + W_b D_b - W_m D_m - W_f D_f}{D_c},$$

$$= \frac{\left(54(80) + (155)(32) - 142(23) - (10)(15)\right)}{35},$$

$$= \frac{\left(4320 + 4960 - 3266 - 150\right)}{35} = 167.5 \text{ kips} = 83.77 \text{ tons.}$$

With an FS = 1/0.75 = 1.33, the required counterweight is 1.33(83.77) = 111 tons. For this crane each counterweight plate weighs 10,000 lb and 22 counterweight plates are required.

The use of manufacturer's crane working range and capacity charts precludes the need to perform similar, or even more complex,* crane stability calculations for multiple procedures of the erection method. Nevertheless, whether crane capacity is based on stability calculations, tables, or charts, the planned locations for all crane picks, crane type,† pick radius, support conditions (e.g., outriggers, mats, barges, cribs, and/or trestles), and rigging, clamp, lifting lug, spreader beam and lifting beam weights, and details must be made known to the erection engineer.

* For example, the installation of a jib at the end of the boom or conditions requiring outriggers to be used or not used.
† Make and model.

11.3.3 Erection Engineering for Falsework

Erection engineers must design temporary supports and falsework for the various erection method procedures from information obtained in the design,* fabrication,† and erection‡ documents.

Temporary supports and falsework must be designed for the vertical loads of the supported superstructure during the various erection procedures, and the associated personnel and equipment loads. In many situations, lateral and longitudinal loads due to wind, earth, ice, water pressure, and/or superstructure movement (sliding and rolling resistance§) also need to be considered. These loads are typically adjusted to reflect the duration of erection. The structural analysis and design of the temporary support or falsework members, bracing, connections, bearings,¶ and foundations using appropriate safety factors can be accomplished using usual structural engineering principles with the applicable codes, recommendations, and/or guidelines. These codes, recommendations, and/or guidelines** outline the minimum design requirements for typical construction method loads and load combinations for superstructures during erection. AREMA (2015)†† recommends falsework design loads and criteria specific to railway superstructure erection.

The AREMA minimum combined dead and live design load for falsework is 4.8 kPa (100 psf).‡‡ The recommended live loads are the actual weight of equipment to be supported by the falsework applied as concentrated loads at the points of contact and a uniform load of not less than 1.0 kPa (20 psf). A live load of 1.1 kN/m (75 lb/ft) applied at the outside edge of deck overhangs is also recommended. The minimum recommended horizontal load in any direction due to equipment, erection procedure, wind, and/or other horizontal forces applied to the falsework is 2% of the total dead load. AREMA (2015) also recommends that the falsework be designed with sufficient rigidity to resist the horizontal design load without considering the weight of the supported superstructure. The minimum recommended wind pressures for falsework design applied to the gross projected area of the falsework and any unrestrained portion of the permanent structure, excluding areas between falsework posts or towers where diagonal bracing is not used, are shown in Table 11.1, where $Q = (48 + 31.5W)$ but not greater than 480 Pa in which W is the width of the falsework system

TABLE 11.1
Wind Pressures for Railway Heavy-Duty Steel Shoring Design

	Wind Pressure, Pa (psf)	
Height of Falsework above Ground, m (ft)	**Members over, and Bents Adjacent to, Traffic Openings**	**Other Members**
0–9 (0–30)	$2.0Q$	$1.5Q$
9–16 (30–50)	$2.5Q$	$2.0Q$
16–30 (50–100)	$3.0Q$	$2.5Q$
Greater than 30 (100)	$3.5Q$	$3.0Q$

* Typically, design drawings and specifications.

† Typically, shop drawings and BOM.

‡ Typically erection procedure drawings and written procedures.

§ Rolling resistance is considerably less than sliding resistance so it is common to use low-profile chain-action rollers to move superstructures horizontally.

¶ On which the superstructure rests.

** For example, ASCE/SEI 37-14: *Design Loads on Structures during Construction*.

†† Chapter 8, Part 28.

‡‡ Regardless of slab thickness for falsework supporting concrete decks during steel–concrete composite superstructure construction. For steel–concrete composite superstructure construction, dead load includes the weight of concrete, reinforcing steel, forms, and falsework. The weight of concrete, reinforcing steel, and forms must not be taken as less than 2600 kg/m^3 (160 pounds per cubic ft) for normal strength concrete.

TABLE 11.2
Wind Pressures for Railway Falsework Design

Height of Falsework above Ground, m (ft)	Wind Pressure, Pa (psf) — Members over, and Bents Adjacent to, Traffic Openings	Other Members
0–9 (0–30)	960 (20)	720 (15)
9–16 (30–50)	1200 (25)	960 (20)
16–30 (50–100)	1440 (30)	1200 (25)
Greater than 30 (100)	1680 (35)	1440 (30)

in meters, or $Q = (1 + 0.2W)$ but not greater than 10 psf in which W is the width of the falsework system in feet. Furthermore, AREMA (2015) recommends the minimum wind pressures in Table 11.2 on 2.2 times the total projected area of all the elements in the tower face normal to the wind applied on each heavy-duty steel shore having a vertical load carrying capacity greater than 133 kN (30 kips) per leg.

AREMA (2015) recommends that falsework structural steel be designed in accordance with the AISC Manual of Steel Construction, with exception of the allowable flexural compression stress, which should be $\leq 0.60F_y$. Alternatively, the allowable stresses for structural steel falsework may be in accordance with Table 11.3. AREMA (2015) also recommends that if the steel to be used for falsework construction is in good condition but unidentifiable, allowable stresses should be based on the allowable stresses for ASTM A36 steel [F_y = 250 MPa (36,000 psi)], with exception of the allowable flexural compression stress, which should be in accordance with Table 11.3. The modulus of elasticity is assumed to be 207,000 MPa (30×10^6 psi).

Falsework for some superstructure erection projects may be constructed of wood. AREMA (2015) recommends the allowable stresses in Table 11.4 for wood falsework construction. The modulus of elasticity is assumed to be 11,000 MPa (1600×10^3 psi).

AREMA (2015) also recommends that for falsework supporting the decks of steel–concrete composite construction, the deflection due to concrete weight must not exceed the member span/240.*

TABLE 11.3
Allowable Stresses for Steel Railway Falsework Design

Stress	Allowable MPa	Allowable psi	Limitations
Axial tension	151.7	22,000	
Flexural tension	151.7	22,000	
Axial compression	$110.3-0.0026\left(\frac{L}{r}\right)^2$	$16{,}000-0.38\left(\frac{L}{r}\right)^2$	$\frac{L}{r} \leq 120$
Flexural compression	$82{,}000\left(\frac{Ld}{bt}\right) \leq 151.7$	$12{,}000\left(\frac{Ld}{bt}\right) \leq 22{,}000$	For unidentified and A36 steel
Shear	100	14,500	On gross section of web
Web crippling	186	27,000	For rolled shapes
Connections	In accordance with AISC Manual of Steel Construction		

* Not considering camber strips.

TABLE 11.4
Allowable Stresses for Wood Railway Falsework Design

Stress	Allowable MPa	Allowable psi
Compression perpendicular to grain	3.10	450
Compression parallel to grain	$33{,}000\left(\frac{L}{d}\right)^2 \leq 11.0$	$480{,}000\left(\frac{L}{d}\right)^2 \leq 1600$
Flexural stress for members with nominal depth >203 mm (8 in.)	12.4	1800
Flexural stress for members with nominal depth ≤203 mm (8 in.)	10.3	1500
Horizontal shear	965	140
Axial tension	8.3	1200
Connections	Per National Design Specification of Wood Construction by National Forest Products Association (NFPA) without reductions specified for wet conditions. NFPA allowable shear for bolts in single shear shall be reduced by 0.75.	
Piles	Maximum load on timber piles used for falsework = 400 kN (90 kips).	

11.3.4 Erection Engineering for Cranes, Derricks, and Falsework on Barges

Superstructure erection with cranes, derricks, and falsework mounted on barges is often utilized in watercourses. Once crane, derrick, and falsework strength and stability is established, barge stability (tilting) needs consideration during all erection procedures and barge loading sequences. Mobile cranes are commonly supported on barges for superstructure erection in watercourses, but mobile crane charts are only applicable up to a maximum angle of tilt. Therefore, barge stability is of prime concern for erection by flotation methods. If the barge tilts by angle, θ, to the horizontal (Figure 11.26), the crane overturning weight (or tipping load), *W*, is

$$W = \frac{W_{f1}(D_o - D_{f1} - \theta H_{f1}) + W_{f2}\left(D_o + D_{f2} - \theta H_{f2}\right) + W_c(D_c - \theta H_c) + W_m\left(D_m - \theta H_m\right) - W_b(D_b + \theta H_b)}{D + \theta H} - W_r. \tag{11.3a}$$

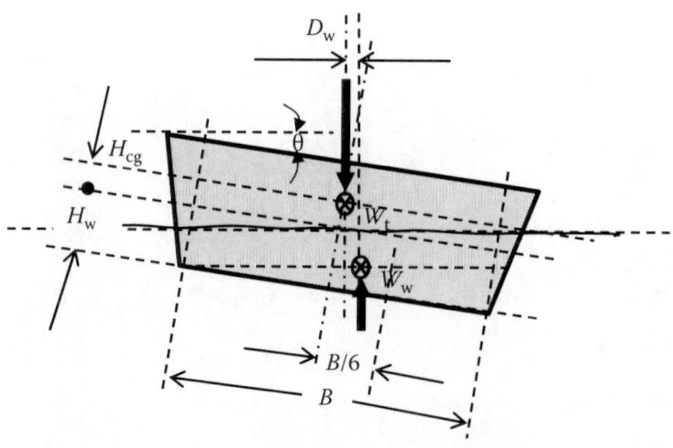

FIGURE 11.26 Barge tilt at angle, θ.

If the effects of the boom foot masts are conservatively neglected ($W_{f1} = W_{f2} = 0$), the overturning weight (or tipping load), W, is

$$W = \frac{W_c(D_c - \theta H_c) + W_m(D_m - \theta H_m) - W_b(D_b + \theta H_b)}{D + \theta H} - W_r. \tag{11.3b}$$

For a barge supporting self-weight, crane weight, and load, $W_t = W_{barge} + W_c + W_m + W_b + W_r + W$ with a barge tilting moment, $M_{barge} = M_a + W_t(H_{cg})\theta$,
where

H_{cg} = the distance between the centroid of the load, W_t, and the waterline

M_a = applied moment = $M_{crane} + M_{load} = W_m(D_m - \theta H_m) + W_c(D_c - \theta H_c) - W_b(D_b + \theta H_b) - (W + W_r)(D + \theta H)$, which for small θ is $M_a = W_m(D_m) + W_c(D_c) - W_b(D_b) - (W + W_r)(D)$

For barge tilt at an angle, θ, the centroid of the displaced water, D_w, shifts and is

$$D_w = \frac{\frac{B}{6}(B^2/2)\sin\theta}{H_w(B)} \sim \frac{B^2\theta}{12H_w}, \tag{11.4}$$

where

B = the width of the barge

H_w = the draft of barge (distance from waterline to bottom of barge)

W_w = the weight of the displaced water = $\rho_w(H_w)(B)(L)$

ρ_w = density of water = 1000 kg/m^3 (62.4 lb/ft^3)

L = the length of the barge

The moment arm, D_w, of the tilting resisting force of the displaced water is

$$D_w = \frac{M_{barge}}{W_t} = \frac{M_a + W_t(H_{cg})\theta}{W_t}$$

$$= \frac{W_m(D_m - \theta H_m) + W_c(D_c - \theta H_c) - W_b(D_b + \theta H_b) - (W + W_r)(D + \theta H) - W_t(H_{cg})\theta}{W_t}, \tag{11.5}$$

which when substituted into Equation 11.4 yields

$$\theta = \frac{12M_a}{B^3 L\rho_w - 12W_t H_w}. \tag{11.6}$$

The angle can be estimated by calculation and crane capacities adjusted, if necessary.* Example 11.4 outlines a simple barge stability calculation with a mobile crane.

Example 11.4a (SI Units)

The crawler crane in Example 11.3a is placed at the center of a barge weighing 75,000 kg with $L = 25$ m, $B = 15$ m, and $H_w = 2$ m. The crane is required to lift a 28,500 kg girder onto falsework at a distance of $D = 20$ m ($D_b = 13.5$ m) with a $W_c = 100{,}000$ kg counterweight. Estimate how much the barge will list,

$$W_t = W_{barge} + W_c + W_m + W_b + W_r + W = (75 + 100 + 60 + 90 + 5 + 28.5)1000 = 358{,}500 \text{ kg},$$

* For small tilt angles (typically about 1°) crane capacities are usually negligibly affected.

$$M_a = W_m(D_m) + W_c(D_c) - W_b(D_b) - (W + W_r)(D)$$
$$= (60(7) + 100(9.5) - 90(13.5) - (28.5 + 5)(20))1000 = -515{,}000 \text{ kg-m},$$

$$\theta = \frac{12(-515{,}000)}{15^3(25)(1000) - 12(358{,}500)(2)} = -0.083 \text{ rad} = -4.7°.$$

This lift will require redesign using a larger barge.

Using a barge weighing 115,000 kg with L = 25 m, B = 25 m (barges may have to be lashed together), and H_w = 2.5 m

$$W_t = W_{barge} + W_c + W_m + W_b + W_r + W = (115 + 100 + 60 + 90 + 5 + 28.5)1000 = 398{,}500 \text{ kg},$$

$$M_a = W_m(D_m) + W_c(D_c) - W_b(D_b) - (W + W_r)(D)$$
$$= (60(7) + 100(9.5) - 90(13.5) - (28.5 + 5)(20))1000 = -515{,}000 \text{ kg-m},$$

$$\theta = \frac{12(-515{,}000)}{25^3(25)(1000) - 12(398{,}500)(2.5)} = -0.016 \text{ rad} = -0.9°.$$

This lift is likely appropriate without any adjustments to crane capacity.

Example 11.4b (US Customary and Imperial Units)

The crawler crane in Example 11.3b is placed at the center of a barge weighing 120 tons with L = 70 ft, B = 40 ft, and H_w = 2.0 ft. The crane is required to lift a 54,000 lb girder onto the permanent substructures at a distance of D = 40 ft (D_b = 23.5 ft). Estimate how much the barge will list.

The counterweight for crane stability on the barge is

$$W_c = \frac{WD + W_b D_b - W_m D_m - W_f D_f}{D_c},$$
$$= \frac{54(40) + (155)(23.5) - 142(23) - (10)(15)}{35},$$
$$= \frac{2160 + 3643 - 3266 - 150}{35} = 68.2 \text{ kips}.$$

A 1.33(68.2) = 90.7 kip = 45 ton counterweight is required.

$$W_t = W_{barge} + W_c + W_m + W_b + W_r + W = (240 + 90.7 + 142 + 155 + 10 + 54)1000 = 691{,}700 \text{ lb},$$

$$M_a = W_m(D_m) + W_c(D_c) - W_b(D_b) - (W + W_r)(D)$$
$$= (142(23) + 90.7(35) - 155(23.5) - (54 + 10)(40))1000 = 238{,}000 \text{ lb-ft},$$

$$\theta = \frac{12(238{,}000)}{40^3(70)(62.4) - 12(691{,}700)(2)} = 0.011 \text{ rad} = 0.6°.$$

This lift is likely acceptable without any adjustments to crane capacity.

11.3.5 Erection Engineering for Stationary and Movable Frames

Stationary frames for the erection of superstructures are typically constructed of structural steel. Nevertheless, whether constructed of steel or wood, the erection engineering of stationary frames

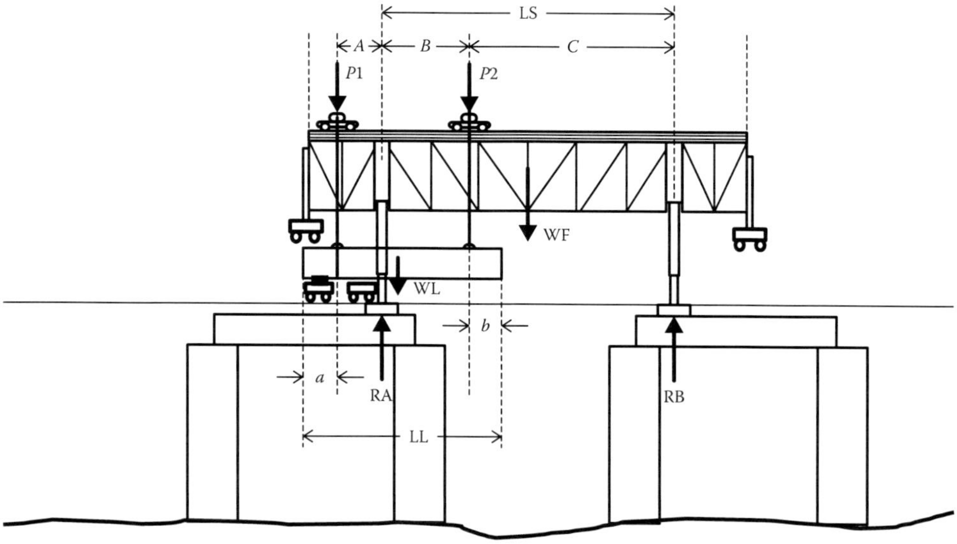

$P1$ = the weight of the lifting truck and spreader beam + ((LL/2 − b)/(LL − a − b)WL
$P2$ = the weight of the lifting truck and spreader beam + ((LL/2 − a)/(LL − a − b)WL
WL = the weight of the load (span + deck + railway track)
RA = 0.5 WF + $P1(1 + A/\text{LS}) + P2(C/\text{LS})$
RB = 0.5 WF + $P2(B/\text{LS}) - P1(A/\text{LS})$
WF = the weight of the movable frame (gantry truss) (span frame track + frame travelling trucks + miscellaneous equipment)

FIGURE 11.27 Loads on a movable frame.

is similar to the requirements for falsework. The stationary frames and their foundations must be designed for the vertical loads of the supported superstructure during the various erection procedures and, in some cases, personnel and equipment loads. In many situations, lateral and longitudinal loads due to wind, earth, ice and/or water pressure, and/or superstructure movement also require consideration. These loads are typically adjusted to reflect the duration of erection. The structural analysis and design of the stationary frame members, bracing, connections, and foundations using appropriate safety factors can be accomplished using usual structural engineering principles with the applicable codes, recommendations, and/or guidelines.

The erection engineering of movable frames used for superstructure erection may be more complex. For example, the procedures shown in Figure 11.21 would require erection engineers to investigate the strength of the frame (gantry truss), jacking frames, lifting devices (tracks, spreader beams), and other components for the worst case forces from the loads, WL, in various positions as illustrated in Figure 11.27 (shown for the removal of an existing span but also required for the installation of a new span, which is typically heavier). In addition to the structural aspects of the erection procedures, the erection engineer must design and/or verify various mechanical, hydraulic, and electrical control systems associated with the erection method and equipment.

11.3.6 Engineering for Other Erection Methods

11.3.6.1 Erection Engineering for Launching

Launching requires careful consideration of the horizontal forces on temporary falsework and/or substructures during the launching procedure. Erection engineers must also consider the need for stabilizing cables where alignment is difficult and/or during windy conditions. A critical erection engineering function is the calculation of cantilever end deflections under various conditions to enable monitoring during the erection procedures.

11.3.6.2 Erection Engineering for Cantilever Construction

Cantilever construction also involves some of the same erection engineering considerations as launching. Erection engineers must typically undertake a rigorous structural analysis* of the truss members under various conditions to determine any strengthening or stability† requirements for the superstructure being erected. In addition, erection engineers must design temporary tension links between the anchor and cantilever spans, compression buffers and vertical members between the tension link at the top chord, and the compression buffer at the bottom chord of through truss superstructures. DT superstructures erected by cantilever construction have similar tension tie and compression buffer requirements at intermediate piers.

11.3.6.3 Engineering for Tower and Cable, and Catenary High-Line Erection

Superstructure erection using towers and cables, or catenary high-lines may be used for construction across deep canyons or rapidly moving watercourses. These specialized erection methods require that erection engineers perform a rigorous structural analysis of both superstructure being erected and the erection structures.

11.3.6.4 Engineering for SPMT Erection

Superstructure erection using SPMTs is becoming more prevalent. Erection engineers must understand the operation, load,‡ and travel configuration in order to determine SPMT capacity. Erection engineers often assume external forces with out-of-level conditions on the SPMT to safely determine capacity.

SPMT capacity is determined by considering the weight of the superstructure and location of its center of gravity, superstructure dimensions, support locations, and allowable point loads on the superstructure, and if required, locations for securing with cables. SPMT capacity may be governed by the vehicle's structural, hydraulic, stability, or ground bearing capabilities.

The transporter platform and spine beam must resist the various loading conditions associated with the erection procedures. The structural capacity of the platform can be verified with SPMT manufacturer's platform loading diagrams. The structural capacity of the spine beam may be determined by a beam on an elastic foundation analysis considering the transporter configuration, number of axle lines, and superstructure support locations.§ SPMT manufactures generally prepare spine beam loading diagrams so that structural capacity may be readily established.

The hydraulic capacity of an SPMT is the sum of the capacities of the wheelset¶ hydraulic cylinders. The slow SPMT travel speed will not reduce hydraulic capacity via the inertial effects of the superstructure. However, hydraulic capacity may be affected by external loads such as inertial forces from acceleration and deceleration, wind on the superstructure, and dynamic forces.** In some cases, a ten to twenty percent reduction in the maximum hydraulic capacity is used to account for external load effects.

To ensure stability of the superstructure on the SPMT's modular platform, hydraulic stability zones are established by interconnection of the hydraulic cylinders of adjacent wheelsets. Hydraulic stability zones are typically triangular to establish three-point loading for resistance to rocking.†† Therefore, the transporter will be hydraulically stable provided the combined center of gravity of the superstructure and the transporter remains within the hydraulic stability triangle. The tilting

* For large span or complex cantilever construction projects, FEA is often used to evaluate superstructure stresses and deflections at various stages of the erection procedure.

† For example, member stiffness increase or member bracing requirements.

‡ Weight and location on the SPMT.

§ Typically at two or more locations along the spine beam.

¶ For all driving and nondriving axles.

** Forces related to the SPMT travel path.

†† Four-point hydraulic stability zones provide a greater stability area, but are prone to rocking.

limit for an SPMT is the angle of the transporter platform that results in the combined center of gravity reaching the edge of the hydraulic stability zone. SPMT manufacturers typically prepare charts to determine the tilting limits as a function of the combined center of gravity. During stable tilting the combined center of gravity moves within the stability area and two zones of the three-point hydraulic suspension will experience increases in load. SPMT manufacturers also typically prepare charts to determine the overload limit as a function of the combined center of gravity and superstructure weight.*

The pressure exerted on the ground by the SPMT must also be considered by the erection engineer. The tire contact pressure† and the pressure from the gross weight of the loaded SPMT over an area, determined as the distance from the front to rear axle and the transporter width, must not exceed the allowable ground bearing capacity.

11.4 ERECTION EXECUTION

Erection planning and engineering must be performed with a prevailing attention to safety. Erection execution must also consider safety as paramount. Railroads are a safety-intensive industry that demands contractor awareness of safety during the execution of superstructure erection.

Superstructure erection is preceded by site material storage with fabricated members and assemblies on blocking, and with additional measures to ensure that main members are stored upright and shored at supports. Interior site storage is typically required for fasteners in containers and machine-finished pieces.

Superstructures are commonly erected by cranes on temporary falsework,‡ existing substructures§ (Figure 11.28), or on new foundations and substructures designed, following geotechnical investigation, in accordance with Chapter 8 of AREMA (2015).

Construction surveys establish the longitudinal and transverse center lines of temporary falsework, and existing or new foundations, anchor bolt locations, and bridge seat elevations to ensure

FIGURE 11.28 Erection of complete span. (Courtesy of D. Ostby, Koppers Railroad Structures, Madison, WI, USA. With permission.)

* Often referred to as payload.
† This is the tire inflation pressure, which is a short-term load that will typically not govern regarding ground bearing capacity.
‡ Typically, for subsequent sliding or rolling into final position on existing or new substructures.
§ Often with bridge seat and other rehabilitation prior to erecting superstructures.

that the superstructure is erected to the correct lines and limits. In particular, it is important to confirm the accurate horizontal and vertical alignments of bearings and anchor bolts before erection. Blocking and/or tie-down devices for bearings will limit bearing movements during erection, especially for longer spans with multirotational bearings.* Erection execution must also consider the ambient temperature for proper setting of the bearings. Blocking and falsework elevations to accommodate fabricated camber (see Chapter 10) are also required.

Personnel safety must be considered whether erection is to be executed using cranes, falsework, lateral skidding, barges, stationary and movable frames, SPMT, or other methods. Safety provisions, such as personal protective equipment,† engineered horizontal tie-off lines, guard rails and cables, safety nets, scaffolding, welding safety equipment, and erection procedures that ensure that no persons are under loads, are all critical to the safe erection of steel superstructures.

11.4.1 Erection by Mobile Cranes

Superstructure erection by mobile crawler and truck-mounted cranes is often used where site and lifting conditions are appropriate. Large cranes are typically transported by truck to the erection site in pieces and assembled on site using smaller cranes. For all cranes, the appropriate winches (wire ropes and sheaves for heavy loads), snatch blocks (change direction of the load suspended from winches), spreader beams, and slings must be rigged to lift the load. Two-leg slings will decrease in capacity as the angle increases and four-leg slings are often used for lifting heavy loads. Longitudinal spreader beams are used to lift superstructure members at appropriate locations, but are not necessary when lifting superstructures with two cranes‡ (Figure 11.29). Where possible, the lifting points on superstructures should use existing bolt holes, designated or approved by the design engineer, for the installation lifting devices.

11.4.2 Falsework Construction

Falsework is typically used to support superstructures for skidding, stick-building,§ concrete deck support, support at splices,¶ control of stability,** control of differential deflections due to skew, and establishing elevations that account for tolerances and replicate shop blocking dimensions.††

Falsework should be constructed under superstructure stiffeners‡‡ for adjustment of elevations by jacking. Also, falsework must be founded on footings or piles capable of supporting the falsework loads by soil bearing, or pile skin friction and bearing, capacity. Falsework used to support concrete decks must be field monitored for the settlement of the falsework when concrete is being placed. Deck pour overhang brackets must be placed to not cause web plate distortion and rotation from equipment and concrete weight.

It is generally required to install all connections before dismantling falsework. The responsibility for falsework design,§§ construction, and dismantling generally rests with the erection contractor.

* Bearing fabricators should ship bearings as a unit with individual bearing components secured with tabs. If the tabs are temporarily tack welded to bearing components, the welds should be ground smooth after bearing installation.

† Includes equipment such as fall arrest harnesses and lines, and vision, hearing, and respiratory protection.

‡ Tandem crane lifts require skilled operators and vigilant on-the-ground instructions from an experienced crane supervisor. The use of multiple cranes may be limited by site constraints.

§ Trusses may be stick-built with cranes or derricks using falsework to support the truss floor system at panel points. The falsework is typically left in place until all field connections are completed.

¶ Holding cranes are typically not effective for field splicing.

** Typically until temporary or permanent cross frames are erected.

†† If available from the fabricator.

‡‡ Designers may need to know where falsework supports are likely in order to design bearing stiffeners at the falsework support locations. Generic bearing stiffeners are often designed.

§§ Often reviewed and approved by the design engineer, project manager, and/or owner without acquiring any responsibility for the design.

FIGURE 11.29 Erection with tandem cranes. (Courtesy of D. Ostby, Koppers Railroad Structures, Madison, WI, USA. With permission.)

11.4.3 Erection Fit-Up

The superstructure must be fabricated and assembled in accordance with the planned erection fit-up method. Fabrication assembly and fit-up may be full (without transverse members) or complete (with transverse members) in the no-load, steel dead load, or full dead load conditions (see Chapter 10).

No-load fit-up requires that members be fabricated to fit in the field as though no dead load including self-weight exists on the superstructure. Steel dead load fit-up necessitates that members be fabricated to fit in the field as though steel dead load exists during erection. Full dead load fit involves the fabrication of members to fit in the field as though full noncomposite dead load exists on the superstructure. Therefore, since girder webs and trusses should be vertical in the full dead load condition, cross frames, diaphragms, floorbeams, and other transverse members must be fabricated to reflect this requirement for no-load or steel dead load girder and truss erection procedures.* Steel railway superstructures are typically fabricated under the no-load condition† and often erected under no-load or steel dead load conditions. Splices and bolted connections in primary members

* Also the case for medium and long skewed spans that will rotate under dead load. External forces to make the girder out of plumb until dead load is added may be required as part of the erection procedure.

† With provision for camber, if necessary.

should be fully assembled in the shop before reaming or drilling to ensure they are within the tolerances required for field erection.

In general, erection tolerances should not exceed cumulative mill and fabrication tolerances (see Chapter 10). Therefore, shop assembly is an important precursor to successful field erection. Superstructure assembly blocked on the ground at site and then pinned and bolted in the no-load, steel dead load, or full dead load condition (depending on shop assembly method) may also ensure that tolerances are acceptable for erection. Erection tolerances for plate girder or rolled shape superstructures in the steel dead load condition are typically provided as deviations from the horizontal alignment in terms of the member length, web verticality (plumbness) as a function of the web depth, and vertical alignment in relation to member length.

11.4.4 Erection of Field Splices and Connections

Girder and truss field splices and connections are typically made by supporting the superstructure on blocking and/or falsework. In some cases, supporting the superstructure in the crane falls of holding cranes is used.* Field splices and connections may be made by welding or bolting. AREMA (2015) recommends that field welding of only minor connections and steel deck plates be permitted. Therefore, bolted field splices and connections are typically used for steel railway superstructure erection.

11.4.4.1 Welded Field Splices and Connections

Welds result in no section deductions, are typically more aesthetically amenable than bolted joints, and CJP welds are efficient from a strength perspective. However, field welding is often difficult from a weld location and position perspective. Field welding is typically costly,† requires alignment to within close tolerances,‡ and inclement weather susceptible.§ The AREMA (2015) restrictions on field welding¶ also recognize the inherent difficulties associated with field weld quality, fatigue strength,** and weld defect inspection and repair.

Bolted field splices and connections often require supplemental corrosion protection measures, but are simpler to align,†† install, and inspect. Field bolt installation requires less skill than welding, and is typically less costly and more reliable than field welding. Therefore, bolted connections are generally preferred in field erection procedures.

11.4.4.2 Bolted Field Splices and Connections

Connections and splices are made in the fit-up condition to allow for tolerances and geometry adjustments.‡‡ The connections and splices are fit-up using pins, temporary fit-up bolts, and/or permanent bolts.§§ Fit-up bolts and pins must secure connections to resist erection and wind forces until final permanent bolting. It is good practice during fit-up to use many full size pins.

Drift pins are used to line up (or fair) holes, but must not be driven such that the bolt holes are damaged.¶¶ Nevertheless, moderate reaming of some misaligned holes may be required. In such

---

* This is often difficult (particularly concerning wind) and not typically necessary unless girders or trusses are very long and deep.

† Field welding requires skilled SMAW welders and provision of an adequate working area.

‡ This is often difficult at the erection site.

§ Typically, enclosures are required to protect the welding operation, to preheat, and to foster weld quality.

¶ Prohibits field welds with exception of only minor connections and deck plate sections that do not carry live load stresses. Deck plate welds are usually made along the top flange of transverse floorbeams. They can often be made under well-controlled conditions in the down-hand position in shop prepared joints.

** Field welding may result in a lower Fatigue Category Detail (see Chapter 5), which could lead to a larger member at design or costly strengthening during erection.

†† With fit-up bolts and drift pins.

‡‡ Adjustments are usually made to the vertical and horizontal alignments.

§§ High strength or temporary fit-up bolts are used in combination with pins during fit-up. It is typically acceptable to use the final high-strength steel bolts for fit-up provided they are tightened only once, and if galvanized, tightened no more than ST.

¶¶ Split drift pins may be used for alignment of holes in field connections to avoid damage to the hole.

cases, the number, location, and extent of misalignment of holes requiring reaming should be reviewed by the erection and design engineers prior to reaming. To avoid fit-up problems during the erection of cross frames, diaphragms, and bracing members, it is often appropriate to field drill one side of the member.

The erection engineer will typically specify the required number of pins and bolts for fit-up and when bolts may be finally tightened in each connection.

It is good practice to leave temporary supports (blocking and/or falsework) or holding cranes in place until all bolts are tightened. However, in some cases, this may not be practical and the erection engineer must determine the number of bolts and pins required for the removal of temporary supports or release of holding cranes. It is unacceptable to allow railway live load on superstructures with connections in the fit-up condition using pins or temporary fit-up bolts.

Accurate pinning and bolting is critical for horizontal and vertical alignments of the superstructure and a survey of the erected superstructure geometry prior to final tightening of the bolts is recommended. Bolts are tightened after the proper superstructure geometry is attained.* Field bolts are commonly tightened by the turn-of-nut (TON) method or by using impact wrenches.†

11.4.4.2.1 Field Bolting Procedures

Field bolting requirements vary for snug-tight (ST), pretensioned (PT), and slip-critical (SC) joints. Field connections for primary members of railway superstructures require SC bolt installation. All bolted connections must be tightened progressively from the most rigid part of joint to curtail relaxation of previously PT bolts.

An ST connection is attained when all plies are in firm contact and nuts cannot be removed without wrench. PT and SC joints are brought to the ST condition before pretensioning of bolts in the connection. PT and SC joints require a minimum bolt tension, which can be obtained by TON, calibrated wrench, and tension control techniques (Table 11.5).

TABLE 11.5
Minimum Bolt Installation Tension

	Minimum Required Installed Bolt Tension (kN)	
Bolt Diameter (mm)	**Grade A325M Bolts**	**Grade A490M Bolts**
12	50	62
16	88	111
20	138	173
22	167	210
24	199	249
32	316	454
36	448	561
	Minimum Required Installed Bolt Tension (kips)	
Bolt Diameter (in.)	**Grade A325M Bolts**	**Grade A490M Bolts**
5/8	19	24
3/4	28	35
7/8	39	49
1	51	64
1¼	71	102
1½	103	148

* For example, by jacking and/or winching.

† Bolting with impact wrenches is fast but impact wrenches require, at a minimum, daily calibration and testing.

TON pretension is attained at a specified nut rotation from the ST condition, depending on bolt length and angle of pieces or plies to the bolt axis. The nut and end of bolt should be marked before turning for inspection.

Twist-off-type tension-control (TC) bolts may be PT from ST condition using a special twist-off-type TC bolt installation wrench that works in conjunction with the bolt spline mechanism. The spline at end of bolt shears off at the correct bolt tension using a special torqueing tool.

Direct tension indicators (DTIs) may be used to establish minimum required pretension from the ST condition through the compression of protrusions to a proper gap height.

Minimum required bolt pretension can also be achieved by the calibrated wrench method (typically using air impact wrenches) with daily calibration using a tension calibrator in order to adjust the wrench to the appropriate torque for the connection. Calibration should also occur due to any changes of conditions.* It is also feasible to use DTIs as tension calibrators for calibrated wrench installations. It is recommended to avoid the use of unreliable tables or formulas relating wrench torque to pretension.

11.4.4.2.2 Field Bolting Verification and Testing

Preinstallation verification tests may be required to ensure that field connections behave as intended by the design. Preinstallation verification of fastener assemblies and PT or SC installation procedures ensure that connections perform as required prior to field installation of fasteners. Tests on fastener assembly samples using tension calibration equipment may be used for preinstallation verification. Fastener assemblies, with exception of twist-off types, which do not meet the preinstallation verification criteria,† may be cleaned, relubricated, and retested.

To ensure that installed bolts achieve the required strength without excessive plastic deformation of the bolt or nut threads, rotational capacity (RC) testing may also be conducted in tension calibrators to evaluate the performance of the lubricant and compatibility of the bolted assemblies to those tested (see Chapter 10). For galvanized fasteners, RC testing must demonstrate that the lubricated galvanized nut may be rotated from the ST condition without plastic deformation‡ much greater than the rotation required for PT or SC installation.§ When required, RC testing of field bolted connections and splices may occur in the fabrication shop, but are typically conducted by erection contractors in the field.

11.4.5 Field Erection Completion

QC during erection is the responsibility of the erection contractor. Nevertheless, it is typical that the owner engages an independent construction project engineer to conduct and/or direct quality assurance (QA) reviews and tests during erection. The owner's construction project engineer must also lead communications between the erector, fabricator, designer, and owner.

In addition to incorrect geometry related to the erection procedures, thermal misalignments due to heat energy from the daily path of the sun may affect the horizontal and vertical alignments of superstructures during the erection. Therefore, a vigilant survey at the completion of the superstructure erection should be made in order to determine the extent and location of geometric inaccuracies so misalignments that must be corrected.¶ If required, alignment and leveling to within erection tolerances is typically achieved by applying forces with hydraulic jacks. Large movements may require the use of synchronized hydraulic jacks on traversing bases such as that shown in Figure 11.30. Safety blocking (timber or steel depending on load magnitude) must be properly installed for jacking operations.

---

* For example, connection lubrication changes require recalibration of wrenches.

† The pre-installation verification test must typically develop tension 5% greater than the design bolt tension (see Tables 11.5 and 11.6).

‡ Stripping or galling of nut on bolt.

§ Typically about twice the ST rotation.

¶ Including measurement of bearing rotations created during the erection procedures.

FIGURE 11.30 Frame for horizontal moving of truss. (Courtesy of the Author, Canadian Pacific Engineering, Calgary, AB, Canada.)

Blocking should be snug* against the superstructure at appropriate locations determined by the erection and design engineers. Following erection to within acceptable tolerances, if required, field coatings (painting† and/or metalizing) are applied, inspected, and tested.

Most erection projects executed by experienced contractors with the appropriate resources (manpower and equipment), and who engage in suitably thorough erection planning and engineering, will be well executed from a safety, cost, and schedule perspective.

BIBLIOGRAPHY

AASHTO/NSBA, 2003, Guidelines for design for constructability, G 12.1–2003, AASHTO/NSBA Steel Bridge Collaboration, Washington, DC.

AISC, 2010, Code of standard practice for steel buildings and bridges, AISC, Chicago, IL.

American Railway Engineering and Maintenance-of-Way Association (AREMA), 2015, Steel structures, in *Manual for Railway Engineering*, Chapter 15, Lanham, MD.

American Society of Testing and Materials (ASTM) F3125, 2015, Standard specification for high strength structural bolts, steel and alloy steel, heat treated, 120 ksi (830 MPa) and 150 ksi (1040 MPa) minimum tensile strength, inch and metric dimensions, ASTM Standards, West Conshohocken, PA.

ANSI/AISC, 2014, Code of standard practice for steel buildings and bridges, AISC, Chicago, IL.

AWS D1.5M/D1.5, 2010, Bridge welding code, AASHTO/AWS, Miami, FL.

Durkee, J., 2014, Steel bridge construction, in *Bridge Engineering Handbook*, 2nd edition, Chen, W.F. and Duan, L., Eds, CRC Press, Boca Raton, FL.

RCSC, 2014, Specifications for structural joints using A325M (A325) or A490M (A490) Bolts, Research Council on Structural Connections (RCSC), AISC, Chicago, IL.

Shapiro, H. I., Shapiro, J. P., and Shapiro, L. K., 2000, *Cranes and Derricks*, 3rd edition, McGraw-Hill, New York.

Unsworth, J.F., 1993, Reconstruction of railway bridges under traffic, Technical University of Nova Scotia, Halifax, NS.

* In some cases, a small gap of typically not more than about 25 mm (1 in.) is acceptable.

† Many railroads do not paint atmospheric corrosion resistant steel (see Chapter 2), but when required, commonly use a three-coat paint system consisting of a zinc-rich primer, epoxy intermediate coat, and polyurethane topcoat.

Appendix A: Design of a Ballasted through Plate Girder (BTPG) Superstructure

A1 SCOPE OF SUPERSTRUCTURE DESIGN

Replace an existing 107′ long pin-connected through truss (TT) with a 32.6 m long BTPG span. The span crosses a road with a minimum vertical clearance requirement of 5.0 m.

Existing substructures are founded on bedrock and have been salvaged through rehabilitation. The bridge seats are 7390 mm wide.

A2 GENERAL INFORMATION

The tangent track on the proposed BTPG span will consist of 180 mm deep × 2500 mm long track ties on 20 mm tie plates with 450 mm ballast from the base of rail to the top of the deck plate. A track lift of 70 mm is proposed in conjunction with span replacement.

To estimate flange width for clearance considerations, Equation 7.58 for local buckling of the top (compression) flange is

$$\frac{b}{2t_{\mathrm{f}}} \leq 0.35\sqrt{\frac{E}{F_{\mathrm{y}}}} \leq 8.4.$$

Assume flange thickness of between 40 and 65 mm (typical maximum thickness, see Chapter 10). For $40 \leq t_{\mathrm{f}} \leq 65$; $672 \leq b \leq 1092$:

Assume 880 mm (average) with 150 mm extension of the bearing plate beyond bottom flange and 200 mm edge distance. Therefore, the girder spacing ≤ 7390 − 880 − 2(150) − 2(200) = 5810 mm.

For this rail line, a 5.0 m lateral clearance is required as shown in Figure A.1. Therefore, maximum top flange width ≤ 5810 − 5000 ≤ 810 mm.

Use a 5.8 m girder spacing to accommodate lateral clearance requirements and bridge seat width.

A3 LOADING AND MATERIALS

Cooper's EM360 with alternate live load (445 kN axle load)
Steel with $F_{\mathrm{y}} = 350$ MPa
Bolts: 22 mm ASTM F3125 Grade A325M bolts with $F_{\mathrm{u}} = 830$ MPa
Welding: submerged arc welding or shielded metal arc welding to applicable bridge code using E70XX electrodes ($F_{\mathrm{u}} = 483$ MPa).

A4 LAYOUT OF THE PROPOSED SUPERSTRUCTURE

Plan, elevation, sections, and clearances of the proposed BTPG span are shown in Figure A.1.

Length of span between centers of bearings = 32,000 mm (assuming a 600 mm bearing plate).

The distance from the existing road surface to the existing base of rail = 6.50 m.

The construction depth (distance from the proposed base of rail to the underside of the steel girder), $d_{umax} = 6500 + 70 - 5000 = 1570$ mm. Allow 150 mm greater than minimum clearance. Construction depth $= d_u = 1570 - 150 + 20 = 1440$ mm (20 mm tie plate).

$$d_b = 20 + 180 + 450 = 650\,\text{mm}.$$

A5 FLOOR SYSTEM DESIGN

A5.1 General Information

Forty-one floor beams (2 end floor beams and 39 intermediate floor beams) are spaced at 800 mm c/c.

Minimum deck plate width $\geq 2[2900 - ((900 + 200 + 630)/2 + (810/2))] \geq 2(2900 - 1270) \geq 3260$ mm.

Use $\simeq 5000$ mm to fasten knee brace to the deck plate.

A5.2 Deck Plate

A5.2.1 Loads and Forces

A5.2.1.1 Dead Loads

Track = (300(9.81)/(2.5 + 2(0.45)))/(1000) = 0.87 kPa (distributed to a 3.4 m deck plate width assuming 2.5 m tie with 1:1 distribution through ballast with $d_b = 450$ mm).

$$\text{Ballast} = (1920)(9.81)(d_b)/(1000)^2 = 0.0188(d_b)\ \text{kPa}.$$

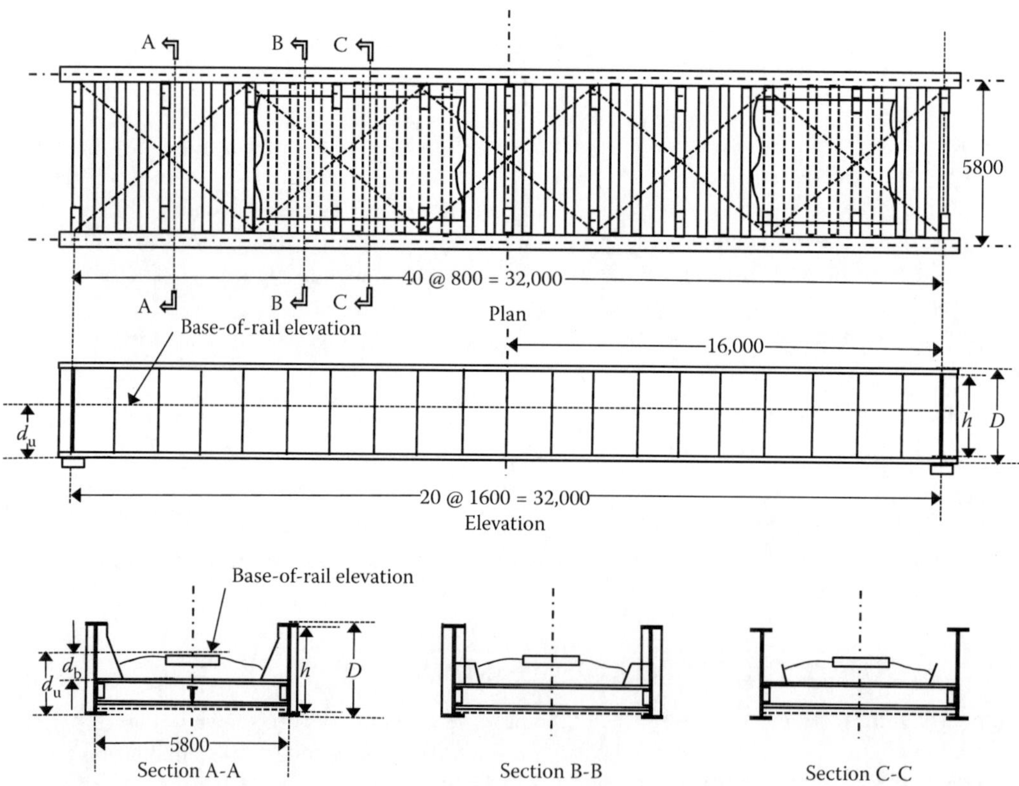

FIGURE A.1 Plan, elevation and cross-sections of span.

(Continued)

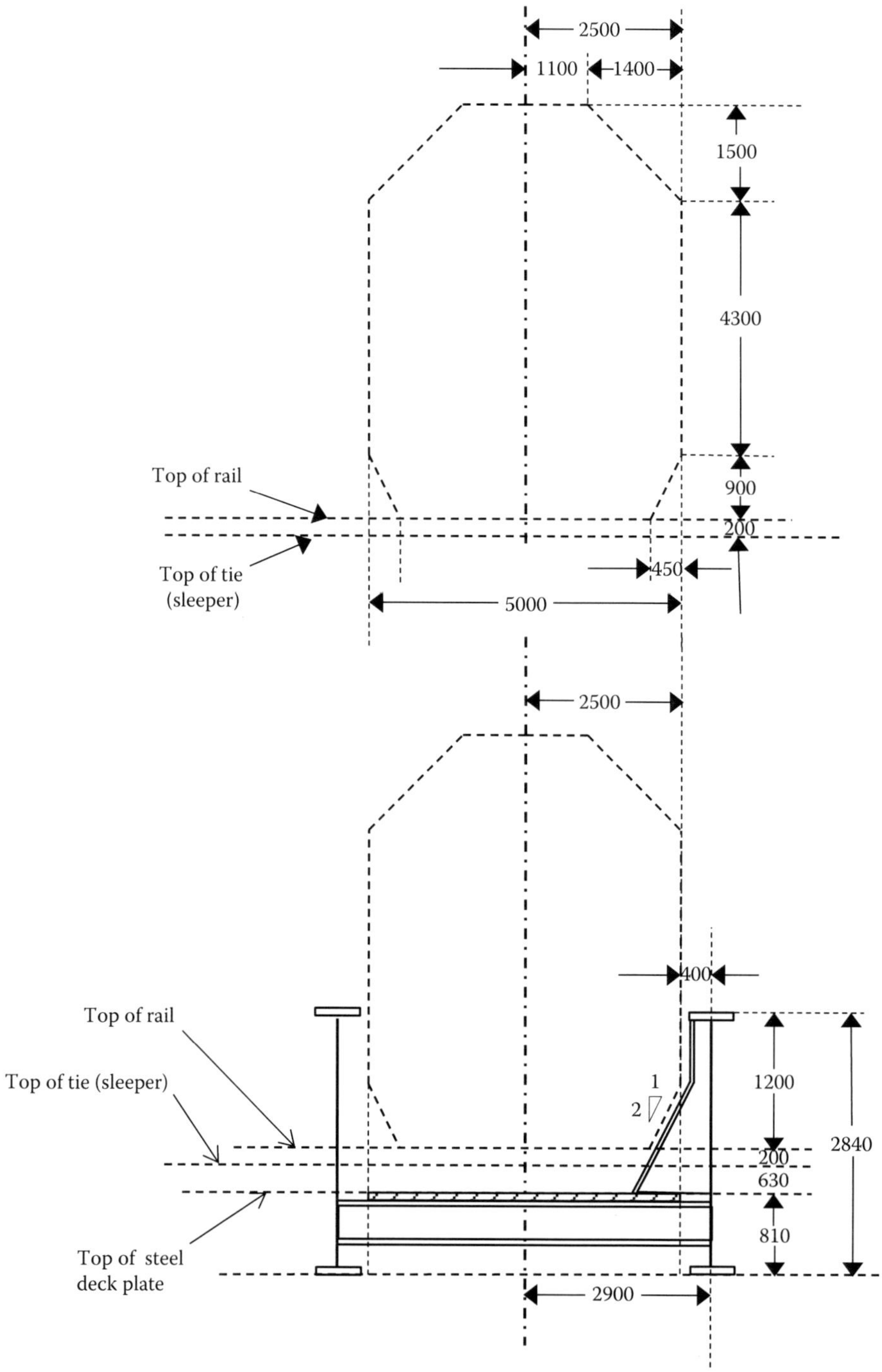

FIGURE A.1 (CONTINUED) Plan, elevation and cross-sections of span.

Assume a 600 mm ballast depth to account for up to 150 mm of future track raises.

Ballast = 11.3 KPa; use 12.0 kPa to account for contingencies related to ballast degradation and water retention.

Waterproofing = 50 (9.81)/1000 = 0.50 kPa.

Deck plate (assume a 24 mm plate) = (24)7850(9.81)/1000² = 1.85 kPa
w_{DL} = 0.87 + 12.0 + 0.50 + 1.85 = 15.2 kPa.

Model as continuous beam:

V_{DL} = 1.25(15.2)(0.80) = 15.2 kN/m (maximum from two span continuous at interior support)
M_{DL} = 0.125(15.2)(0.80)(0.80) = 1.22 kNm/m (maximum from two span continuous at interior support).

A5.2.1.2 Live Load

Longitudinal distribution = 915 + 450 + 24 − 180 − 12 = 1197 mm < 1525
Lateral distribution = 2500 + 450 + 24 − 180 − 12 = 2782 mm < 4250 mm
LL = 445/((1.197)(2.782)) = 133.6 kPa (alternate live load axle)
V_{LL} = 1.25(133.6)(0.8) = 133.6 kN/m (maximum from two span continuous at interior support)
M_{LL} = 0.125(133.6)(0.8)² = 10.69 kNm/m (maximum from two span continuous at interior support).

Vertical impact

I_{max} = 0.90(40.0%) = 36.0% (Figure 4.22)
I_{mean} = 0.65 (36.0) = 23.4% (Table 4.9)
$V_{LL+I(max)}$ = (1.36) 133.6 = 181.7 kN/m
$M_{LL+I(max)}$ = (1.36) 10.69 = 14.5 kNm/m
$M_{LL+I(mean)}$ = (1.23) 10.69 = 9.45 kNm/m.

A5.2.1.3 Wind Forces

Wind load on train = 4.38 kN/m at 2.4 m above the top of rail (including lower 1.2 m of train not exposed to wind due to girder height, but this height will typically be only about 20% of train height)

w_{WLL} = [4.38 (2.4)/1.5]/(0.90) = 7.7 kPa (wheel spacing = 1.5 m rail spacing distributed over 2(0.45) = 0.90 m width at deck plate elevation)

$V_{wLL} \simeq$ 1.25(7.7 (0.8)/2) = 3.9 kN/m (maximum from two span continuous at interior support)
$M_{wLL} \simeq$ 7.7 (0.8)²/8 = 0.62 kNm/m (maximum from two span continuous at interior support).

A5.2.1.4 Load Combinations for Deck Plate Design (Table 4.10)

Load combination D1-A: DL + LL + *I* at 100% allowable stress:

$$V_{max} = 15.2 + 181.7 = 196.9\,\text{kN/m}$$

$$M_{max} = 1.22 + 14.5 = 15.7\,\text{kNm/m}.$$

Load combination D1-B: LL + *I* at allowable fatigue stress:

$$M_{range} = 9.45\,\text{kNm/m}.$$

Load combination D2-A: DL + LL + *I* + W_L at 125% allowable stress:

$$V_{max} = 15.2 + 181.7 + 3.9 = 200.8\,\text{kN/m}$$

$$M_{max} = 1.22 + 14.5 + 0.62 = 16.3\,\text{kNm/m}.$$

Due to the 2% increase in shear and 4% increase in bending moment from load case D1-A, load combination D2-A with F_L = 1.25 will not be considered.

A5.2.2 Deck Plate Design

$$t_{dp} \geq 196.9/(0.35(350)) \geq 1.61\,\text{mm}$$

$$S_{max} \geq 15.7(1000)/(0.55(350)) = 81.6\,\text{mm}^3/\text{mm}$$

$$S_{mean} \geq 9.45(1000)/(165) = 57.3\ mm^3/mm$$

$$t_{dp} \geq \sqrt{6(81.6)} \geq 22.1\ \text{mm for bending.}$$

Use a 22 mm plate with 3 mm allowance for corrosion = 25 mm.

Weight of deck plate = 25(5.0)(32.8)(7850)/1000 = 32,185 kg.

A5.3 Intermediate Floor Beams

A5.3.1 Loads and Forces

A5.3.1.1 Dead Loads

Track = 300(0.8)/2.5 = 96 kg/m distributed over 3.4 m (assuming 2.5 m tie and 1:1 distribution through ballast with d_b = 450 mm)

Ballast = (1920)(d_b)(800)/1000 = 1536(d_b) kg/m distributed over deck plate width ≃ 2[2900 − ((900 + 200 + 630)/2 + (810/2))] ≥ 2(2900 − 1270) ≥ 3260 mm ≥ 3.260 m (Figure A.1).

Assume a 600 mm ballast depth to account for up to 150 mm of future track raises.

Ballast = 922 kg/m; use 1000 kg/m over 3.26 m length to account for contingencies related to ballast degradation and water retention
Waterproofing = 50(0.8) = 4 kg/m over 3.26 m length
Deck plate (25 mm plate) = 25(800)7850/1000^2 = 157 kg/m over 5.0 m length
Floor beam = 125 kg/m (assumed weight of the floor beam section) over 5.8 m length.

The intermediate floor beam dead loading is shown in Figure A.2a.

V_{DL} = [9.85(3.26) + 0.94(3.4) + 1.54(5.0) + 1.23(5.8)]/2 = (32.11 + 3.20 + 7.70 + 7.13)/2 = 25.1 kN
M_{DL} = 9.85(3.26)(5.8/4 − 3.26/8) + 0.94(3.40)(5.8/4 − 3.40/8) + 1.54(5.00)(5.8/4 − 5.00/8) + 1.23(5.80)(5.8/4 − 5.80/8) = 33.48 + 3.28 + 6.35 + 5.17 = 48.3 kNm.

A5.3.1.2 Live Load

Longitudinal distribution:

Floor beams spaced at 800 mm < axle spacing = 1500 mm

$$P = 1.15(445)\frac{(800)}{1500} = 272.9\ \text{kN.} \tag{4.60}$$

Lateral distribution:

No lateral distribution, $P/2$ applied at location of wheel loads

$$V_{LL} = 272.9/2 = 136.5\ \text{kN}$$

$$M_{LL} = 136.5(5.8 - 1.5)/2 = 293.5\ \text{kNm.}$$

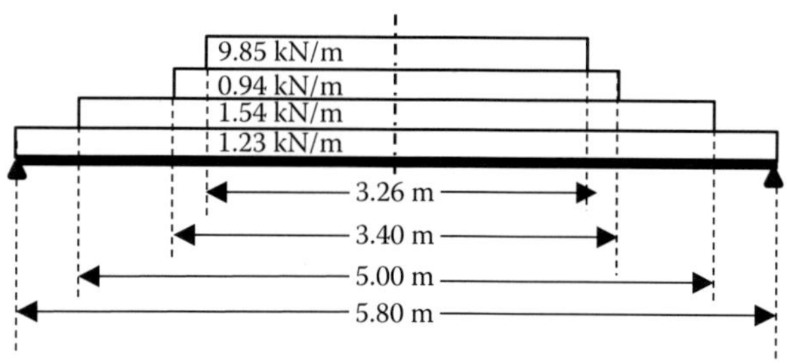

FIGURE A.2 (a) Dead loads on intermediate floorbeams.

Vertical impact:

$I_{max} = (0.90)39.3\% = 35.4\%$ (Figure 4.22)
$I_{mean} = 0.35(35.4) = 12.4\%$ (Table 4.9).

Rocking effect:

RA = 0.20 W (1525) = 305 W
RR = FR (5800)
FR = 0.053(W)
RE = FR (100)/W = 5.3%.

Total impact:

$I_{max} = 35.4\% + 5.3\% = 40.7\%$
$I_{mean} = 12.4\%$
$V_{LL+I(max)} = (1.407)\ 136.5 = 192.1\ \text{kN}$
$M_{LL+I(max)} = (1.407)\ 293.5 = 413.0\ \text{kNm}$
$M_{LL+I(mean)} = (1.124)\ 293.5 = 329.9\ \text{kNm}$.

A5.3.1.3 Wind Forces on Loaded Superstructure

4.38 kN/m at 2.4 m above the top of rail:

$w_{WLL} = 4.38\ (2.4)/1.5 = 7.0\ \text{kN/m}$
$V_{wLL} = (7.0\ (5.8)/2) = 20.3\ \text{kN}$
$M_{wLL} = 7.0\ (5.8)^2/8 = 29.4\ \text{kNm}$.

A5.3.1.4 Load Combinations for Intermediate Floor Beam Design (Table 4.10)

Load combination D1-A: DL + LL + *I* at 100% allowable stress:

$$V_{max} = 25.1 + 192.1 = 217.2\ \text{kN} = R_{max}$$

$$M_{max} = 48.3 + 413.0 = 461.3\ \text{kNm}.$$

Load combination D1-B): LL + *I* at allowable fatigue stress:

$$M_{range} = 329.9\ \text{kNm}.$$

Load combination D2-A: DL + LL + I + W_L at 125% allowable stress:

$$V_{max} = 25.1 + 192.1 + 20.3 = 237.5\ \text{kN} = R_{max}$$

$$M_{max} = 48.3 + 413.0 + 29.4 = 490.7\ \text{kNm}.$$

Due to the 9% increase in shear and 6% increase in bending moment, load combination D2-A with $F_L = 1.25$ will not be considered.

A5.3.2 Intermediate Floor Beam Design

$$A_{web} \geq 217.2(1000)/(0.35(350)) \geq 1773\ \text{mm}^2$$

$$S_{min} \geq 461.3(1000)^2/(0.55(350)) = 2396 \times 10^3\ \text{mm}^3$$

$$S_{mean} \geq 329.9(1000)^2/(165) = 1999 \times 10^3\ \text{mm}^3.$$

Effective width for a 25 mm deck plate

$$b_{\text{eff}} \le (2t_{\text{p}})(0.43)\sqrt{\frac{E}{F_{\text{y}}}} = (2)(25)(0.43)\sqrt{\frac{200{,}000}{350}} = 514\,\text{mm} \qquad (7.57)$$

$b = b_{\text{eff}} + b_{\text{beam}} = 514 + b_{\text{beam}} \le$ floor beam spacing $\le 800\,\text{mm}$.

Try W 610 × 113:

$$b_{\text{beam}} = 228\,\text{mm}$$

$$b = 514 + 228 = 742\,\text{mm} \le 800\,\text{mm, OK.}$$

Maximum and mean (fatigue range) stresses:

The intermediate floor beam section and properties are shown in Figure A.2b and Table A.1, respectively.

$$y_{\text{s}} = \frac{(4377.6 + 9260.2)(1000)}{14{,}400 + 14{,}960} = 464.5\,\text{mm}$$

$$I_{\text{g}} = (371.0 + 357.1 + 875 + 0.60)10^6 = 1603.7 \times 10^6\,\text{mm}^4.$$

The intermediate floor beams at the center of each bracing panel (Figure A.1) will be connected to the bottom lateral bracing at mid-span. Assuming 2–25 mm dia. holes:

$I_{\text{n}} = 1603.7 \times 10^6 - 2(25)(17)((17/2)\text{–}464.5)^2 = (1603.7 - 176.8) \times 10^6 = 1427.0 \times 10^6\,\text{mm}^4$
$S_{\text{t}} = 1603.7 \times 10^6/(608 + 22 - 464.5) = 9690.0 \times 10^3\,\text{mm}^3$
$S_{\text{b}} = 1427.0 \times 10^6/(464.5) = 3072.1 \times 10^3\,\text{mm}^3$
$A_{\text{web}} = (608 - 2(17))(11) = 6314\,\text{mm}$
$\sigma_{\text{max}} = 461.3 \times 10^6/3072.1 \times 10^3 = 150.2\,\text{MPa} \le 0.55(350) \le 192.5\,\text{MPa, OK.}$
$\sigma_{\text{range}} = 329.9 \times 10^6/3072.1 \times 10^3 = 107.4\,\text{MPa} < 165\,\text{MPa, OK.}$
$\tau_{\text{max}} = 217.2\,(1000)/(6314) = 34.4\,\text{MPa} \le 0.35(350) \le 122.5\,\text{MPa, OK.}$

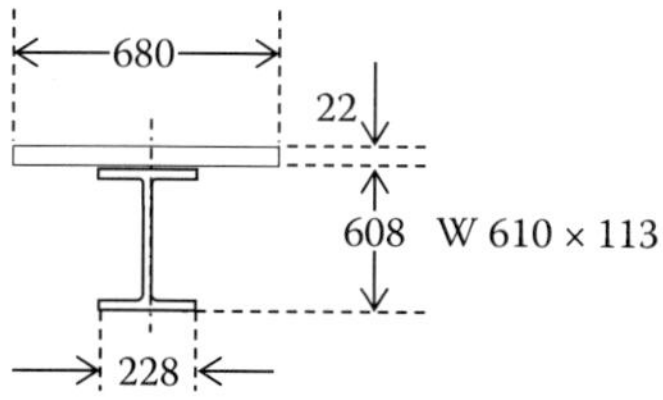

FIGURE A.2 (b) Intermediate floorbeam cross-section.

TABLE A.1
Intermediate Floorbeam Section Properties (22 mm Deck)

Section	A (mm²)	y (mm)	A_y (mm³)	$y_s - y$ (mm)	$A(y_s - y)^2$ (mm⁴)	I_o (mm⁴)
W 610 × 113	14,400	304.0	4377.6×10^3	160.5	371.0×10^6	875×10^6
22 mm deck plate	14,960	619.0	9260.2×10^3	−154.5	357.1×10^6	603.4×10^3

LL + *I* deflection:

P_{LL+I} = (1.407)1.15(445)(800)/1500 = 384.0 kN

$$\Delta_{max} = \frac{384.0(1000)(2150)\left(3(5800)^2 - 4(2150)^2\right)}{24(200{,}000)\left(1603.7 \times 10^6\right)} = 8.9 \text{ mm}$$

$$\Delta_{all} = \frac{\text{Span}}{640} = \frac{5800}{640} = 9.1 \text{ mm, OK.}$$

A5.3.3 Intermediate Floor Beam Connections

R_{max} = 217.2 kN
No. bolts single shear = 1.25(217.2)(1000)/(1)(120)(380) = 6.0
No. bolts double shear = 3.0
No. bolts bearing (on W 610 × 113, web = 11 mm) = (213.8)(1000)/((540)(22)(11)) = 1.6.

Use 6 bolts in girder web and 3 bolts in floor beam web with double angle connections (Section C-C in Figure B.1).

Use 3 bolts in angle and 3 bolts in intermediate stiffener connection to girder web; and 3 bolts in floor beam web. (Sections A-A and B-B in Figure B.1.)

Use W 610 × 113 with for intermediate floor beams.

A5.3.4 Intermediate Floor Beam Weight

Use W 610 × 113 intermediate floor beams. Weight of intermediate floor beams = 113(5.8)((32,000/800) − 1) = 25,560 kg.

A5.4 End Floor Beams

End floor beams are designed for dead load, live load, and wind load (on live load). End floor beams are also designed as jacking beams. The end floor beams as jacking beams will reverse stresses in the flanges and will typically govern design of the relatively small end floor beams lifting considerable dead load.

A5.4.1 End Floor Beams Carrying Live Load

A5.4.1.1 Loads and Forces

A5.4.1.1.1 Dead Loads (Table 4.1) Track = 300(400 + 300)/1000 = 210 kg distributed over 3.4 m (assuming 2.5 m tie with 1:1 distribution through ballast with d_b = 450 mm)

Ballast = (1920)(d_b)(700)/1000 = 806(d_b) kg/m over 3.26 m length.

To account for contingencies related to ballast depth (e.g., future track raises) and waterproofing materials, use 850 kg/m over length of floor beam.

Waterproofing = 50(0.4) = 2 kg/m over 3.26 m length
Deck plate (25 mm plate) = 25(700)7850/1000^2 = 137.4 kg/m over 5.0 m length
Floor beam = 125 kg/m (assumed weight of the floor beam section).

The end floor beam dead loading is shown in Figure A.3a.

V_{DL} = [8.36(3.26) + 0.61(3.4) + 1.35(5.0) + 1.23(5.8)]/2 = (27.25 + 2.07 + 6.75 + 7.13)/2 = 21.6 kN
M_{DL} = 8.36(3.26)(5.8/4 − 3.26/8) + 0.61(3.40)(5.8/4 − 3.40/8) + 1.35(5.00)(5.8/4 − 5.00/8) + 1.23(5.80)(5.8/4 − 5.80/8) = 28.42 + 2.13 + 5.57 + 5.17 = 41.3 kNm.

A5.4.1.1.2 Live Load

P = 272.9 kN
V_{LL} = 136.5 kN

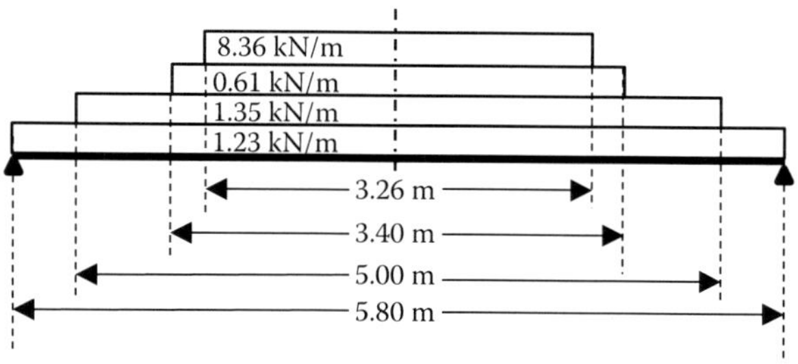

FIGURE A.3 (a) Dead loads on end floorbeams.

$$M_{LL} = 136.5(5.8 - 1.5)/2 = 293.5\,\text{kNm}$$

Vertical impact:
To account for end stiffness effects (transition from open track to superstructure) that may augment dynamics, impact will be increased by 25% for end floor beam design.

Total impact:

$$I_{max} = 1.25(40.7) = 50.9\%$$
$$I_{mean} = 1.25(12.4) = 15.5\%$$
$$V_{LL+I(max)} = (1.509)\ 136.5 = 206.0\,\text{kN}$$
$$M_{LL+I(max)} = (1.509)\ 293.5 = 442.9\,\text{kNm}$$
$$M_{LL+I(mean)} = (1.155)\ 293.5 = 339.0\,\text{kNm}.$$

A5.4.1.1.3 Wind Load on Loaded Superstructure

$$V_{wLL} = 20.3/2 = 10.2\,\text{kN}$$
$$M_{wLL} = 29.4/2 = 14.7\,\text{kNm}.$$

A5.4.1.1.4 Load Combinations for End Floor Beam Design (Table 4.10) Load combination D1-A: DL + LL + I at 100% allowable stress:

$$V_{max} = 21.6 + 206.0 = 227.6\,\text{kN} = R_{max}$$
$$M_{max} = 41.3 + 442.9 = 484.2\,\text{kNm}.$$

Load combination D1-B: LL + I at allowable fatigue stress:

$$M_{range} = 339.0\,\text{kNm}.$$

As for intermediate floor beams, load combination D2-A is neglected.

A5.4.1.2 End Floor Beam Design

Use W 610 × 113 (same section as intermediate floor beams).

Maximum and fatigue stresses:

The distance from girder bearing stiffeners to the end of girder will be about 300 mm.

$$b = 680/2 + 300 = 640\,\text{mm}.$$

The end floor beam section and properties are shown in Figure A.3b and Table A.2, respectively.

$$y_s = \frac{(4377.6 + 8715.5)(1000)}{14{,}400 + 14{,}080} = 459.7\ \text{mm}$$

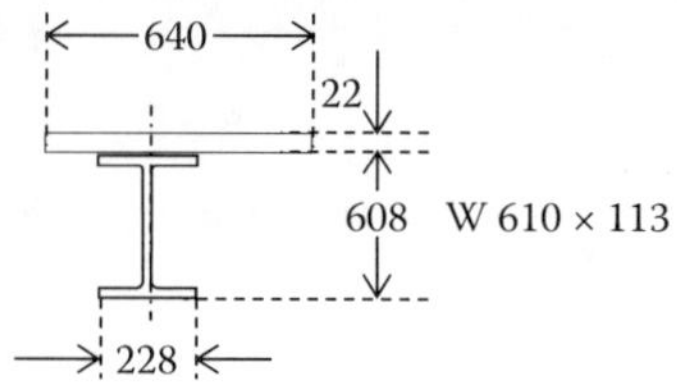

FIGURE A.3 (b) End floorbeam cross-section.

TABLE A.2
End Floorbeam Section Properties (22 mm Deck)

Section	A (mm^2)	y (mm)	A_y (mm^3)	$y_s - y$ (mm)	$A\,(y_s - y)^2$ (mm^4)	I_o (mm^4)
W 610 × 113	14,400	304.0	4377.6 × 10^3	155.7	349.1 × 10^6	875 × 10^6
22 mm deck plate	14,080	619.0	8715.5 × 10^3	−159.3	357.3 × 10^6	567.9 × 10^3

$I_g = (349.1 + 357.3 + 875 + 0.57) \times 10^6 = 1582.0 \times 10^6\,\text{mm}^4$
$S_t = 1582.0 \times 10^6 /(608 + 22 - 459.7)= 9289.3 \times 10^3\,\text{mm}^3$
$S_b = 1582.0 \times 10^6 /459.7 = 3441.3 \times 10^3\,\text{mm}^3$
$A_{web} = 6314\,\text{mm}$
$\sigma_{max} = 484.2 \times 10^6/3441.3 \times 10^3 = 140.7\,\text{MPa} \le 0.55(350) \le 192.5\,\text{MPa}$, OK.
$\sigma_{range} = 339.0 \times 10^6/3452.5 \times 10^3 = 98.2\,\text{MPa} < 165\,\text{MPa}$, OK.
$\tau_{max} = 227.6\,(1000)/(6314) = 36.1\,\text{MPa} \le 0.35(350) \le 122.5\,\text{MPa}$, OK.

LL + I deflection:

$$P_{LL+I} = (1.509)1.15(445)(800)/1500 = 411.9\,\text{kN}$$

$$\Delta_{max} = \frac{411.9(1000)(2150)\left(3(5800)^2 - 4(2150)^2\right)}{24(200{,}000)\left(1603.7 \times 10^6\right)} = 9.6\text{ mm}$$

$$\Delta_{all} = \frac{\text{Span}}{640} = \frac{5800}{640} = 9.1\text{ mm, 5\% over, OK.}$$

Also see Section A5.4.2 for end floor beam strengthening required for jacking span.

A5.4.1.3 End Floor Beam Connections

$R_{max} = 227.6\,\text{kN}$
No. bolts single shear = 1.25(227.6)(1000)/(1)(120)(380) = 6.2
No. bolts double shear = 3.1
No. bolts bearing (on W 610 × 113 web = 11 mm) = (227.6)(1000)/((540)(22)(11)) = 1.7.

For bolted bearing stiffeners, use 3 bolts in angle and 3 bolts in bearing stiffener connection to girder web; and 3 bolts in end floor beam web.

For welded bearing stiffeners, use 3 bolts in angle and bearing stiffener welded connection (to resist 227.6/2 = 113.8 kN) to girder web; and 3 bolts in end floor beam web.

The use of end floor beams for jacking the span will require more bolts. Also see Section A5.4.2 for end floor beam jacking span connection requirements.

A5.4.2 End Floor Beam as Jacking Beam

Use end floor beam to jack up span. Place jacks at 1650 mm from end of floor beam.

A5.4.2.1 Loads and Forces

A5.4.2.2 Load Combinations for Jacking Beam Design (Table 4.10)

Load combination C1: DL at 150% allowable stress:

$R_{DL} \simeq 822\ (32.6/32.0) = 837\,\text{kN}$ (see Section A.6.2.1 for span dead load reaction); use 850 kN for lifting force at each jack to account for weight contingencies

$$M_{DL} = 850(1.65) = 1403\,\text{kNm}.$$

A5.4.2.3 End Jacking Floor Beam Design

The W 610 × 113 end floor beam section and properties are shown in Figure A.3b and Table A.2, respectively.

$$y_s = \frac{(4377.6 + 8715.5)(1000)}{14,400 + 14,080} = 459.7\ \text{mm}$$

$I_g = (349.1 + 357.3 + 875 + 0.57) \times 10^6 = 1582.0 \times 10^6\,\text{mm}^4$
$S_t = 1582.0 \times 10^6/(608 + 22 - 459.7) = 9289.3 \times 10^3\,\text{mm}^3$
$S_b = 1582.0 \times 10^6\ /459.7 = 3441.3 \times 10^3\,\text{mm}^3$
$A_{web} = 6314\,\text{mm}$
$\tau_{DL} = 850\ (1000)/(6314) = 135\,\text{MPa} \le (1.5)0.35(350) \le 183.8\,\text{MPa}$, OK shear in end floor beam web for jacking.
$d = 630\,\text{mm}$
$A_{fc} = 228(17) = 3876\,\text{mm}^2$
$A_w = (11)(608 - 2(17)) = 6314\,\text{mm}^2$
$L_p = 5800 - 2(1650) = 2500\,\text{mm}$
$A_{yc} = (459.7 - 17)(11) + 228(17) = 8746\,\text{mm}^2$
$I_{yc} = (459.7 - 17)(11)^3/12 + 17(228)^3/12 = 17.658 \times 10^6\,\text{mm}^4$
$r_{yc} = (17.658 \times 10^6/8746)^{1/2} = 44.9\,\text{mm}.$

The allowable compressive stress is the larger of (Equations 7.44 and 7.45):

$$F_{call} = 1.5\left(\frac{0.13\pi E}{L_p d(\sqrt{1+\upsilon})/A_f}\right) = 1.5\left(\frac{0.13\pi(200,000)}{2500(630)(\sqrt{1+0.3})/(3876)}\right) = 264.5\text{MPa}$$

or

$$F_{call} = 1.5\left(0.55F_y - \frac{0.55F_y^2}{6.3\pi^2 E}\left(\frac{L_p}{r_{cy}}\right)^2\right) = 1.5\left(0.55(350) - \frac{0.55(350)^2}{6.3\pi^2(200,000)}\left(\frac{2500}{44.9}\right)^2\right)$$

$$= 1.5\,(192.5 - 16.8) = 263.6\ \text{MPa}.$$

$F_{call} = 264.5\,\text{MPa} \le 1.5(0.55)(350) \le 288.8\,\text{MPa}$, OK.

$\sigma_{DLc} = 1403 \times 10^6/3441.3 \times 10^3 = 408\,\text{MPa} \ge 264.5\,\text{MPa}$, fails at bottom flange in compression for jacking.

$\sigma_{DLt} = 1403 \times 10^6/9289.3 \times 10^3 = 151\,MPa \le (1.5)0.55(350) \le 288.8\,MPa$, OK at top flange in tension for jacking.

Strengthen end floor beams with bolted cover plate (welded cover plate poor fatigue detail for live load stress ranges) at bottom flange of end floor beam.

A5.4.2.3.1 End Jacking Floor Beam Bottom Cover Plate Design Try a 20 mm × 200 mm bolted cover plate.

The jacking end floor beam section and properties are shown in Figure A.4 and Table A.3, respectively.

$$y_s = \frac{(40.0+4665.6+8997.1)(1000)}{4000+14,400+14,080} = 421.9 \text{ mm}$$

$I_g = (678.6 + 138.0 + 663.7 + 875 + 0.69) \times 10^6 = 2356.0 \times 10^6\,mm^4$
$S_t = 2356.0 \times 10^6 /(608 + 22 - 421.9) = 11,321 \times 10^3\,mm^3$
$S_b = 2356.0 \times 10^6 /421.9 = 5584 \times 10^3\,mm^3$
$A_{web} = 6314\,mm$
$\tau_{DL} = 135\,MPa \le (1.5)0.35(350) \le 183.8\,MPa$, OK shear in end floor beam web for jacking.
$d = 650\,mm$.
$A_{fc} = 228(17) + 200(20) = 7876\,mm^2$
$A_w = (11)(608 - 2(17)) = 6314\,mm^2$
$L_p = 5800 - 2(1650) = 2500\,mm$
$A_{yc} = (459.7 - 17)(11)+228(17) + 20(200) = 12,746\,mm^2$
$I_{yc} = (459.7 - 17)(11)^3/12 + 17(228)^3/12 + 20(200)^3/12 = 30.991 \times 10^6\,mm^4$
$r_{yc} = (30.991 \times 10^6/12,746)^{1/2} = 49.3\,mm$.

The allowable compressive stress is the larger of (Equations 7.44 and 7.45):

$$F_{call} = 1.5\left(\frac{0.13\pi E}{L_p d(\sqrt{1+\upsilon})/A_f}\right) = 1.5\left(\frac{0.13\pi(200,000)}{2500(650)(\sqrt{1+0.3})/(7876)}\right) = 521 \text{ MPa}$$

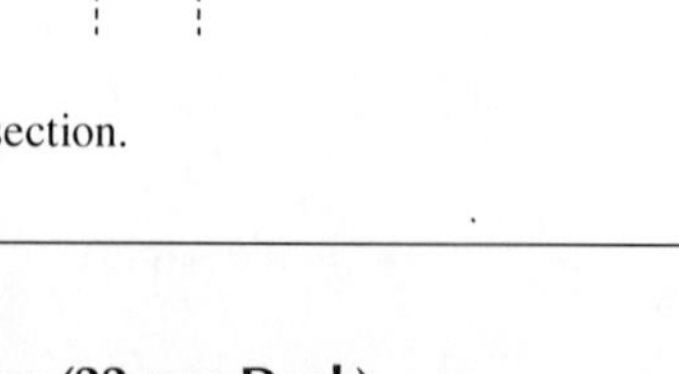

FIGURE A.4 Jacking beam cross-section.

TABLE A.3
Jacking Beam Section Properties (22 mm Deck)

Section	A (mm²)	y (mm)	A_y (mm³)	$y_s - y$ (mm)	$A(y_s - y)^2$ (mm⁴)	I_o (mm⁴)
20 mm × 200 mm cover plate	4000	10	40.0×10^3	411.9	678.6×10^6	0.13×10^6
W 610 × 113	14,400	324	4665.6×10^3	97.9	138.0×10^6	875×10^6
22 mm deck plate	14,080	639	8997.1×10^3	−217.1	663.7×10^6	0.56×10^6

or

$$F_{call} = 1.5\left(0.55F_y - \frac{0.55F_y^2}{6.3\pi^2 E}\left(\frac{L_p}{r_{cy}}\right)^2\right) = 1.5\left(0.55(350) - \frac{0.55(350)^2}{6.3\pi^2(200{,}000)}\left(\frac{2500}{49.3}\right)^2\right)$$

$$= 1.5(192.5 - 13.9) = 268 \text{ MPa.}$$

F_{call} = 1.5(0.55)(350) = 288.8 MPa.

σ_{DLc} = 1403 × 10^6/5584 × 10^3 = 251 MPa ≤ (1.5)0.55(350) ≤ 288.8 MPa, OK at bottom flange in compression for jacking.

σ_{DLt} = 1403 × 10^6/11,321 × 10^3 = 124 MPa ≤ (1.5)0.55(350) ≤ 288.8 MPa, OK at top flange in tension for jacking.

A5.4.2.3.2 End Jacking Floor Beam Bottom Cover Plate Connection

$$Q_{cp} = (20(200)(421.9 - 10))/2356.0 \times 10^6 = 699.3 \times 10^{-6}\,\text{mm}^{-1}$$

$$q_{max} = (850)(1000)(699.3 \times 10^{-6}) = 594 \text{ N/mm.}$$

Space bolts at maximum of 150 mm c/c:

$$v_h = 594(150)/1000 = 89.2\,\text{kN}$$

$$f_{vbolt} = 120(380)/1000 = 45.6\,\text{kN}$$

Number of bolts = 89.2/45.6 = 2.0; use 2 bolts = 1 bolt on each side of the end floor beam web plate in bottom flange.

A5.4.2.3.3 End Floor Beam to Deck Plate Weld Floor beam section properties are shown in Table A.3 for end floor beams.

Width of plate = 640 mm

$$y_s = \frac{(40.0 + 4665.6 + 8997.1)(1000)}{4000 + 14{,}400 + 14{,}080} = 421.9 \text{ mm}$$

$$I_g = (678.6 + 138.0 + 663.7 + 0.13 + 875 + 0.56) \times 10^6 = 2356.0 \times 10^6\,\text{mm}^4$$

$$\frac{Q}{I_g} = \frac{25(640)(620.5 - 421.9)}{2356.0 \times 10^6} = 1.35\ x\ 10^{-3}\ \text{mm}^{-1}$$

$$\vartheta = \frac{VQ}{I} = (850)(1000)(1.35 \times 10^{-3}) = 1148\ \text{N/mm}$$

$$\text{Weld size} \geq \frac{1148}{2\left(\frac{\sqrt{2}}{2}\right)(135)} = 6.0 \text{ mm on throat of fillet weld each side of flange}$$

$$\text{Weld size} \geq \frac{1148}{2(122.5)} = 4.7\ \text{mm on base metal each side of flange}$$

Use 6 mm fillet welds.

A5.4.2.3.4 Web Crippling for Jacking Concentrated Force

$R_{max} = 850$ kN (for $L_F = 1.0$)

The minimum web plate thickness is (Equation 7.80b)

$t_w \geq \dfrac{P}{0.75F_y(L_B+2k)} = \dfrac{850(1000)}{0.75(350)(100+2(36))} = 18.8$ mm (assuming a 100 mm × 100 mm jacking plate used)

W 610 × 113 web thickness = 11 mm; therefore, stiffener is required.

Maximum bearing stiffener width = (228 − 11)/2 = 108.5

Try a 100 × 16 mm plate on each side of end floor beam web at each jacking location (1650 mm from end of floor beam).

A5.4.2.3.4.1 End Floor Beam Bearing Stiffener Design as Compression Member The bearing stiffener is designed as a compression member in accordance with Equations 7.81a and b:

$$A_{ebs} = 2A_{bs} + 12(t_w)^2 = 2(100)(16) + 12(11)^2 = 4652 \text{ mm}^2$$

$$I_{ebs} = 2I_{bs} + 2A_{bs}\bar{y}^2 + t_w^4 = 2\left(\frac{16(100)^3}{12}\right) + 2(100)(16)(55.5)^2 + (11)^4 = 12{,}538 \times 10^3 \text{ mm}^4$$

$$r_{esb} = \sqrt{\frac{I_{esb}}{A_{esb}}} = \sqrt{12{,}538 \times \frac{10^3}{4652}} = 51.9 \text{ mm}$$

$$\frac{h}{r_{ebs}} = \frac{608 - 2(17)}{51.9} = 11.1.$$

The allowable compressive stress, F_{call}, from Equations 7.83, 7.84a, or 7.85 depends on h/r_{ebs}. For $\dfrac{h}{r_{ebs}} \leq 0.839\sqrt{\dfrac{200{,}00}{350}} \leq 20.0$

The allowable compressive stress, F_{call}, from Equation 7.83 is

$$F_{call} = 1.5(0.55F_y) = 288.8 \text{ MPa}.$$

The allowable force on the bearing stiffener = $P_{call} = F_{call}(A_{ebs}) = 288.8(4652)/1000 = 1344$ kN ≥ 850 kN, OK.

A5.4.2.3.4.2 End Floor Beam Bearing Stiffener Design for Bearing Stress The allowable bearing stress for milled stiffeners and parts in contact yields (Equation 7.87):

$$A'_{bs} \geq \frac{R}{0.83F_y} \geq \frac{(850)(1000)}{0.83(350)} \geq 2925 \text{ mm}^2 \text{ (with } L_F = 1.0)$$

$A'_{bs} = 2(100 - 8)(16) = 2944 \text{ mm}^2$, OK with 8 mm clearance for rolled beam flange-to-web fillet.

A5.4.2.3.4.3 End Floor Beam Bearing Stiffener Design for Local Buckling Stress For local buckling of the bearing stiffener plate with a free edge, Equation 7.88 indicates

$$t_{bs} \geq \frac{b_{bs}}{0.43}\sqrt{\frac{F_y}{E}} \geq \frac{b_{bs}}{0.43}\sqrt{\frac{350}{200{,}000}} \geq \frac{b_{bs}}{10.28} \geq \frac{100}{10.28} \geq 9.7 \text{ mm, 16 mm used, OK.}$$

Use a 100 × 16 mm plate on each side of the end floor beam web plate.

A5.4.2.3.4.4 End Floor Beam Bearing Stiffener Connection to Beam Web

$$t_{weld} \geq \frac{850(1000)}{(2)(574)135\sqrt{2}} \geq 3.9 \text{ mm } (\text{with } L_F = 1.0).$$

Use 6 mm fillet weld on each side of bearing stiffener.

A5.4.2.4 End Jacking Floor Beam Connection to Girder

$$R_{max} = 850\,\text{kN (with } L_F = 1.0).$$

End floor beam connected to bearing stiffener with bolts:

No. bolts single shear = (850)(1000)/((1)(120)(380)) = 18.6
No. bolts double shear = 9.3
No. bolt bearing (on W 610 × 113 web = 11 mm) = (850)(1000)/((540)(22)(11)) = 6.5.
For bolted bearing stiffeners, use 10 bolts in angle and 10 bolts in bearing stiffener connection to girder web; and 10 bolts in end floor beam web.
For welded bearing stiffeners, use 10 bolts in angle and bearing stiffener welded connection (to resist 850/2 = 425 kN) to girder web; and 10 bolts in end floor beam web.

Bearing stiffener welded to girder web:

$$\text{Weld size} \geq \frac{425(1000)}{2(\sqrt{2}/2)(135)(608)} = 3.7 \text{ mm on throat of fillet weld each side of flange}$$

$$\text{Weld size} \geq \frac{425(100)}{2(122.5)(608)} = 2.9 \text{ mm on base metal each side of flange}$$

Minimum fillet weld size = 6 mm
Maximum fillet weld size = 16 − 2 = 14 mm
6 mm fillet welds can transfer end jacking force combined with 10 bolts.

A5.4.3 End Jacking Floor Beam Weight

Use a W 610 × 113 with 20 mm × 200 mm bolted bottom cover plate (2 bolts in bottom flange at 150 mm max.)

Weight of end floor beams = $(113 + (20)(200)7850/(1000)^2)$ ((5.8)(2)) = 144.4(2)(5.8) = 1675 kg.

A5.5 Deck Plate to Floor Beam Connection

Assume full width of plate (800 mm) with uncorroded section (t_p = 25 mm). Floor beam section properties are shown in Table A.4 for intermediate floor beams.

$$y_s = \frac{(4377.6 + 12,410)(1000)}{14,400 + 20,000} = 488.0 \text{ mm}$$

TABLE A.4
Intermediate Floorbeam Section Properties (25 mm Deck)

Section	A (mm²)	y (mm)	A_y (mm³)	$y_s - y$ (mm)	$A(y_s - y)^2$ (mm⁴)	I_o (mm⁴)
W 610 × 113	14,400	304.0	4377.6×10^3	184.0	487.5×10^6	875×10^6
22 mm deck plate	20,000	620.5	$12,410.0 \times 10^3$	−132.5	351.1×10^6	1041.7×10^3

$$I_g = (487.5 + 351.1 + 875 + 1.04) \times 10^6 = 1714.6 \times 10^6 \text{ mm}^4$$

$$\frac{Q}{I_g} = \frac{25(800)(620.5 - 488.0)}{1714.6 \times 10^6} = 1.55 \times 10^{-3} \text{ mm}^{-1}$$

$$\vartheta = \frac{VQ}{I} = 227.6(1000)\left(1.55 \times 10^{-3}\right) = 352.8 \text{ N/mm}$$

$$\text{Weld size} \geq \frac{352.8}{2(\sqrt{2}/2)(135)} = 1.9 \text{ mm on throat of fillet weld each side of flange}$$

$$\text{Weld size} \geq \frac{352.8}{2(122.5)} = 1.4 \text{ mm on base metal each side of flange.}$$

Use 6 mm fillet welds (end jacking floor beam connection to the deck plate requires 6 mm for strength, and intermediate floor beam connection to the deck plate requires 6 mm as minimum size).

A6 GIRDER DESIGN

A6.1 General Information

Length of span = 32,600 mm = 32.6 m
Length between bearing stiffeners = 32,600 – 2(300) = 32,000 mm = 32.0 m
Distance between girders = 5.8 m
Knee brace spacing = 4(800) = 3200 mm < 3658 mm (maximum allowable spacing).

A6.2 Loads and Forces

A6.2.1 Dead Loads

Track = 300(9.81)/1000 = 2.94 kN/m
Ballast and waterproofing = (1920(0.6)+50)(5.0)(9.81)/1000 = 59.0 kN/m (for 600 mm ballast depth)
Deck plate = $25(5.0)7850(9.81)/(1000)^2 = 9.6$ kN/m
Floor beams = 42(113)(5.8)(9.81)/(32,600) = 8.3 kN/m
Girder = estimate as 40,000 kg = 40,000(9.81)/(32,600) = 12.0 kN/m
Girder dead load = (2.9 + 59.0 + 9.6 + 8.3)/2 + 12.0 = 39.9 + 12.0 = 51.9 kN/m
$V_{DL} = 51.9(32.0)/2 = 830.4$ kN
$M_{DL} = 51.9\,(32.0)^2/8 = 6643$ kNm.

A6.2.2 Live Load

$w_{LLV} = 88$ kN/m (Figure 5.19a)
$w_{LLM} = 75$ kN/m (Figure 5.22a)
$V_{LL} = 88(32.0)/2 = 1408$ kN
$M_{LL} = 75(32.0)^2/8 = 9600$ kNm.

Impact:
Vertical impact:

$I_{max} = (0.90)24\% = 21.6\%$ (Figure 4.22)
$I_{mean} = 0.35(21.6) = 7.6\%$ (Table 4.9).

Rocking effect:
RE = 5.3%.

Total impact:

$I_{max} = 21.6\% + 5.3\% = 26.9\%$
$I_{mean} = 7.6\%$
$V_{LL+I(max)} = (1.27)\ 1408 = 1787\,kN$
$M_{LL+I(max)} = (1.27)\ 9600 = 12{,}182\,kNm$
$M_{LL+I(mean)} = (1.08)\ 9600 = 10{,}330\,kNm.$

A6.2.3 Wind Load on Loaded Superstructure

4.38 kN/m at the rate of 2.4 m above the top of rail

$w_{WLL} = 4.38\ (2.4)/1.5 = 7.0\,kN/m$
$V_{WLL} = 7.0\ (32.0)/2 = 112.0\,kN$
$M_{WLL} = 7.0\ (32.0)^2/8 = 896.0\,kNm.$

A6.2.4 Load Combinations for Girder Design (Table 4.10)

Load combination D1-A: DL + LL + I at 100% allowable stress:

$V_{max} = 830.4 + 1787 = 2617\,kN$
$M_{max} = 6643 + 12{,}182 = 18{,}825\,kNm.$

Load combination D1-B: LL + I at allowable fatigue stress:

$M_{range} = 10{,}330\,kNm.$

Load combination D2-A: DL + LL + I + W_L at 125% allowable stress:

$V_{max} = 830.4 + 112.0 + 1787 = 2729\,kN$
$M_{max} = 6643 + 896 + 12{,}182 = 19{,}721\,kNm.$

Load combination D2-A may be neglected from girder design.

A6.3 Preliminary Design of Girders

A6.3.1 General

Depth of girder based on practical requirements that $L/15 \le D \le L/10$:

$$32(1000)/15 = 2130\,mm \le D \le 32(1000)/10 = 3200\,mm \text{ (average of 2665 mm).}$$

Assuming web dimensions based on stability (flexural buckling) requirements, for $F_y = 350\,MPa$, $h/t_w \le 135$ (7.60b) and

$$h_{opt} = \sqrt[3]{\frac{3M}{2f_b}\left(\frac{h}{t_w}\right)} = \sqrt[3]{\frac{3(18{,}825)(1000)^2}{2(0.55)(350)}(135)} = 2705 \text{ mm.} \qquad (7.39b)$$

Try $h = 2750\,mm$.

A6.3.2 Web Plate

Web thickness for strength = $t_w = 2617(1000)/(0.35(350)(2750)) = 7.8\,mm$.

Web thickness for flexural stability without longitudinal stiffeners = t_w =

$$\frac{h}{5.64}\sqrt{\frac{F_y}{E}} = \frac{2750}{5.64}\sqrt{\frac{350}{200{,}000}} = \frac{2750}{135} = 20.3 \text{ mm.} \qquad (7.60b)$$

Web thickness for shear stability without intermediate transverse stiffeners = t_w =

$$\frac{h}{2.12}\sqrt{\frac{F_y}{E}} = \frac{2750}{2.12}\sqrt{\frac{350}{200{,}000}} = \frac{2750}{50.6} = 54.3 \text{ mm.} \tag{7.66}$$

The maximum spacing, a, of intermediate stiffeners is Equation 7.103

$$a \le 1.95t_w\sqrt{\frac{E}{\tau}} \le 1.95t_w\sqrt{\frac{200{,}000}{(2617(1000)/2750t_w)}} \le 28.27t_w^{3/2}.$$

For a practical stiffener spacing of 2(800) = 1600 mm

$$t_w \ge \left(\frac{a}{28.27}\right)^{2/3} \ge \left(\frac{1600}{28.27}\right)^{2/3} \ge 14.7 \text{ mm}$$

Minimum plate thickness = 12 mm.

Try a 2750 mm × 16 mm web plate with longitudinal and transverse stiffeners.

A6.3.3 Flange Plates

$$A_f = \frac{M}{f_b h} - \frac{A_w}{6} = \frac{18{,}825(1000)^2}{0.55(350)(2750)} - \frac{2750(16)}{6}$$

$$= 35{,}560 - 7333 = 28{,}228 \text{ mm}^2 \text{ (for maximum moment).} \tag{7.35}$$

$A_f = \dfrac{10{,}330(1000)^2}{110(2750)} - \dfrac{2750(16)}{6} = 34{,}149 - 7333 = 26{,}816\ \text{mm}^2$ (for bolted stiffener moment range). f_b = 110 MPa = allowable stress for fatigue Category B with no welded attachments and transverse web stiffeners bolted to the web.

$A_f = \dfrac{10{,}330(1000)^2}{83(2750)} - \dfrac{2750(16)}{6} = 45{,}257 - 7333 = 37{,}924\ \text{mm}^2$ (for welded stiffener moment range). f_b = 83 MPa = allowable stress for fatigue Category C with transverse web stiffeners welded to the web.

Practical requirements that $D/4 \le b \le D/3$ for $D \simeq 2900$ mm yield a flange width range of between 725 mm and 967 mm. The flange thickness required for A_f ranges between 30 and 55 mm.

Equation 7.58 controls local buckling of the top (compression) flange as

$$\frac{b}{2t_f} \le 0.35\sqrt{\frac{E}{F_y}} = 8.4$$

$$t_f \ge \frac{b}{16.8}.$$

For b = 725 mm, $t_f \ge 43$ mm

For b = 967 mm, $t_f \ge 58$ mm

Try 45 × 725 top and bottom flanges, A_f = 725(45) = 32,625 mm^2.

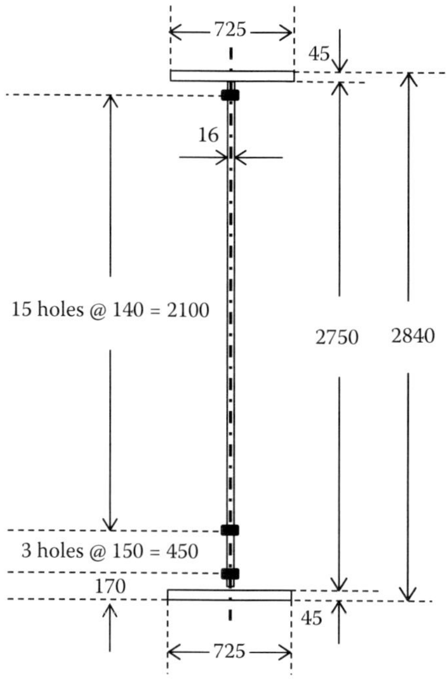

FIGURE A.5 Girder cross-section.

TABLE A.5
Girder Gross Section Properties

Element	Size	A (mm²)	y (mm)	A_y (mm³)	$y_s - y$ (mm)	$A\,(y_s - y)^2$ (mm⁴)	I_o (mm⁴)
Top flange	725 × 45	32,625	2817.5	91.92×10^6	−1398	$63{,}717 \times 10^6$	5.51×10^6
Web	2750 × 16	44,000	1420.0	62.48×10^6	0	0	$27{,}729 \times 10^6$
Bottom flange	725 × 45	32,625	22.5	0.73×10^6	1398	$63{,}717 \times 10^6$	5.51×10^6

A6.4 Detailed Design of Girders

A 2750 mm × 16 mm web plate with longitudinal and transverse stiffeners and 45 × 725 top and bottom flanges.

The girder gross section and properties are shown in Figure A.5 and Table A.5, respectively.

$$y_s = \frac{(91.92+62.48+0.73)(10^6)}{32{,}625+44{,}000+32{,}625} = 1420 \text{ mm}.$$

$I_g = (63{,}717 + 0 + 63{,}717 + 5.51 + 27{,}729 + 5.51)10^6 = 155{,}174 \times 10^6\,\text{mm}^4$
$S_{tf} = S_g = 155{,}174 \times 10^6/1420 = 109{,}277 \times 10^3\,\text{mm}^3$
$S_{bf} = S_g = 155{,}174 \times 10^6/1420 = 109{,}277 \times 10^3\,\text{mm}^3$.

The girder net section and properties are shown in Figure A.5 and Table A.6, respectively.

$$y_h = \frac{27680}{19} = 1457 \text{ mm}$$

$y_n = (91.92 + 62.48 + 0.73)10^6 - 400(27{,}680))/(32{,}625 + 44{,}000 + 32{,}625 - 7600) = 1417\,\text{mm}.$

TABLE A.6
Girder Net Section Properties

Hole	y (mm)	A (mm²)	I_o (mm⁴) × 10⁶
1	170	400	662.4
2	320	400	516.9
3	470	400	389.6
4	620	400	280.1
5	760	400	194.3
6	900	400	124.0
7	1,040	400	69.5
8	1,180	400	30.7
9	1,320	400	7.5
10	1,460	400	0
11	1,600	400	8.1
12	1,740	400	32.1
13	1,880	400	71.6
14	2,020	400	126.8
15	2,160	400	197.7
16	2,300	400	284.4
17	2,440	400	386.7
18	2,580	400	504.5
19	2,720	400	638.3
Total	27,680	7600	4525.2

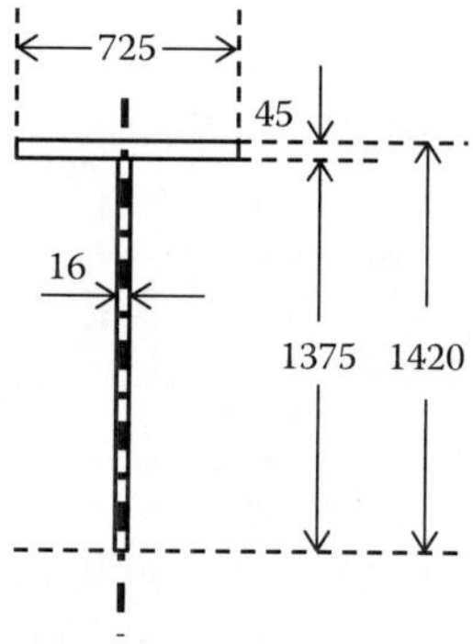

FIGURE A.6 Girder cross-section in compression.

AREMA recommends use of the gross section neutral axis.

$I_h = 7600(1457 - 1420)^2 + (4525.2 \times 10^6) = (7.80 + 4525.2) \times 10^6 = 4533.0 \times 10^6\,\text{mm}^4$
$I_n = I_g - I_h = (155{,}174 - 4533.0) \times 10^6 = 150{,}641 \times 10^6\,\text{mm}^4$
$S_n = 150{,}641 \times 10^6/1420 = 106{,}085 \times 10^3\,\text{mm}^3$ (due to the thin web plate only 3% reduction of S_g)
$S_{tf} = S_g = 109{,}277 \times 10^3\,\text{mm}^3$
$S_{bf} = S_n = 106{,}085 \times 10^3\,\text{mm}^3$
$A_w = 2750(16) = 44{,}000\,\text{mm}^2$.

A6.4.1 Design of Top Flange

Spacing of top flange lateral support = knee brace spacing = 4(800) = 3200 mm.

The section in compression (Figure A.6) has the properties:

$$A_{yt} = (725)(45) + (1375)(16) = 32{,}625 + 22{,}000 = 54{,}625\ \text{mm}^2$$

$$I_{yt} = 45(725)^3/12 + 1375(16)^3/12 = 1429.5 \times 10^6\ \text{mm}^4$$

$$r_{yt} = \sqrt{\frac{I_{yt}}{A_{yt}}} = \sqrt{\frac{1429.5 \times 10^6}{54625}} = 161.8\ \text{mm}$$

$L_p = 3200\ \text{mm}$

$$\frac{L_p}{r_{yt}} = \frac{3200}{161.8} = 19.8 \le 5.55\sqrt{\frac{E}{F_y}} \le 132.7. \qquad (7.43)$$

The allowable compressive stress is the larger of (Equations 7.44 and 7.45):

$$F_{call} = \left(\frac{0.13\pi E}{L_p d(\sqrt{1+\upsilon})/A_f}\right) = \left(\frac{0.13\pi(200{,}000)}{3200(2840)(\sqrt{1+0.3})/32{,}625}\right) = 257\ \text{MPa}$$

or

$$F_{call} = 0.55F_y - \frac{0.55F_y^2}{6.3\pi^2 E}\left(\frac{L_p}{r_{cy}}\right)^2 = 0.55(350) - \frac{0.55(350)^2}{6.3\pi^2(200{,}000)}\left(\frac{3200}{161.8}\right)^2 = 192.5 - 2.1 = 190.4\ \text{MPa}.$$

However, since 257 MPa > 192.5 MPa ($0.55F_y$), $F_{call} = 0.55F_y = 192.5$ MPa

$\sigma_{tf} = \sigma_c = 18{,}825 \times 10^6/109{,}277 \times 10^3 = 172.3\ \text{MPa} \le 192.5\ \text{MPa}$, OK.

Use a 725 × 45 plate for top flange.

A6.4.2 Design of Web Plate

$F_{vall} = 0.35(350) = 122.5$ MPa

$\tau_w = 2617(1000)/44{,}000 = 59.5\ \text{MPa} \le 122.5\ \text{MPa}$, OK.

Use a 2750 × 16 plate for web.

A6.4.3 Design of Bottom Flange

$F_{tall} = 0.55F_y = 0.55(350) = 192.5$ MPa

$F_{fat} = 110$ MPa Category B (no welded attachments)

$\sigma_{bfmax} = \sigma_t = 18{,}825 \times 10^6/106{,}085 \times 10^3 = 177.5\ \text{MPa} \le 192.5\ \text{MPa}$, OK.

$\sigma_{bfrange} = 10{,}330 \times 10^6/106{,}085 \times 10^3 = 97\ \text{MPa} \le 110\ \text{MPa}$, OK.

Use a 725 × 45 plate for bottom flange.

A6.4.4 Girder Deflection

$$\Delta_{all} = \frac{L}{640} = \frac{32{,}000}{640} = 50.0\ \text{mm}$$

From Equation 5.47

$$\Delta_{LL+I} = \frac{5w_{e\Delta}L^4}{384EI} = \frac{0.104M_{LL+I}L^2}{EI} = \frac{0.104(12{,}182\times10^6)(32{,}000)^2}{200{,}000(155{,}174\times10^6)} = 41.8\ \text{mm} \le 50.0\text{mm, OK.}$$

A6.4.5 Flange to Web Fillet Welds

$$V_{DL} = 830.4\,\text{kN}$$

$$V_{max} = 830.4 + 1787 = 2617\,\text{kN}.$$

A6.4.5.1 Top Flange to Web Weld

$$Q_{tf} = (45(725)(1375 + 45/2))/155{,}174 \times 10^6 = 293.8 \times 10^{-6}\,\text{mm}^{-1}$$

$$q_{max} = 2617(1000)(\ 293.8 \times 10^{-6}) = 768.9\ \text{N/mm}$$

$$t_{weld} \geq \frac{768.9}{135\sqrt{2}} \geq 4.0\ \text{mm on throat of fillet weld}$$

$$t_{weld} \geq \frac{768.9}{2(122.5)} \geq 3.1\ \text{mm.}$$

Use 6 mm fillet weld (minimum size).

A6.4.5.2 Bottom Flange to Web Weld

$Q_{bf} = (45(725)(1375 + 45/2))/155{,}174 \times 10^6 = 293.8 \times 10^{-6}\,\text{mm}^{-1}$
$q_{max} = 2617(1000)(293.8 \times 10^{-6}) = 768.9$ N/mm
$q_{range} = 1787(1000)(293.8 \times 10^{-6}) = 525$ N/mm

$$t_{weld} \geq \frac{768.9}{135\sqrt{2}} \geq 4.0\ \text{mm on throat of fillet weld}$$

$$t_{weld} \geq \frac{768.9}{2(122.5)} \geq 3.1\ \text{mm on base metal}$$

or

$$t_{weld} \geq \frac{525}{110\sqrt{2}} \geq 3.4\ \text{mm for Category B detail}$$

$$t_{weld} \geq \frac{525}{83\sqrt{2}} \geq 4.0\ \text{mm for Category B}'\ \text{detail.}$$

Use 6 mm fillet weld for top and bottom flange to web T-joints.

A6.4.6 Girder Longitudinal Stiffeners

Equation 7.60 indicates that longitudinal stiffeners are required where

$$\frac{h}{t_w} \geq 4.18\sqrt{\frac{E}{f_c}} \geq 142.4\ \left(\text{with } f_c = 172.3\ \text{MPa}\right)$$

$$\frac{h}{t_w} = \frac{2750}{16} = 171.9;\ \text{therefore, longitudinal stiffeners are required.}$$

No longitudinal stiffener is required for $t_w = 2750/142.4 = 19.3$ mm (also by Equation 7.89).

The required stiffener moment of inertia for a single longitudinal stiffener about the face of the web plate (Equation 7.91) is

TABLE A.7

Element	Size	A (mm²)	y (mm)	A_y (mm³)	$y_s - y$ (mm)	$A(y_s - y)^2$ (mm⁴)	I_o (mm⁴)
Top Flange	725 × 45	32,625	2817.5	91.92 × 10⁶	−1380	62,131 × 10⁶	5.51 × 10⁶
Longitudinal stiffener	200 × 12	2,400	2245.0	5.39 × 10⁶	−807	1563 × 10⁶	0.03 × 10⁶
Web	2750 × 16	44,000	1420.0	62.48 × 10⁶	18	14.0	27,729 × 10⁶
Bottom flange	725 × 45	32,625	22.5	0.73 × 10⁶	1415	65,357 × 10⁶	5.51 × 10⁶

$$I_{ls} = ht_w^3\left(2.4\left(\frac{a}{h}\right)^2 - 0.13\right) = 2750(16)^3\left(2.4\left(\frac{1600}{2750}\right)^2 - 0.13\right) = 7687\times10^3 \text{ mm}^4 \geq \frac{t_{ls}(b_{ls})^3}{12}.$$

The longitudinal stiffener will be welded to the web plate (in the flexural compression zone) at $h/5 = 2750/5 = 550$ mm from the underside of the top flange plate.

The girder gross section properties with longitudinal stiffener participation are shown in Table A.7.

$$y_s = \frac{(91.92+5.39+62.48+0.73)(10^6)}{32,625+2400+44,000+32,625} = 1438 \text{ mm}$$

$$I_g = (62,131 + 1563 + 14 + 65,357 + 5.51 + 0.03 + 27,729 + 5.51)10^6 = 156,805 \times 10^6 \text{ mm}^4$$

$$S_{ls} = 156,805 \times 10^6/(2245 - 1438) = 194,306 \times 10^3 \text{ mm}^3$$

$$\sigma_{ls} = 18,825 \times 10^6/194,306 \times 10^3 = 96.9 \text{ MPa} \leq 192.5 \text{ MPa}.$$

The maximum thickness of the longitudinal stiffener is Equation 7.95

$$t_{ls} \geq 2.39 b_{ls}\sqrt{\frac{f}{E}} = 2.39 b_{ls}\sqrt{\frac{96.6}{200,000}} = \frac{b_{ls}}{19.0}$$

$$b \leq \sqrt[4]{19.0(12)(7687\times10^3)} \leq 205 \text{ mm}$$

$$t_{ls} \geq \frac{205}{19.0} \geq 10.8 \text{ mm}.$$

Use a 2750 × 16 mm web plate with 200 × 12 longitudinal stiffener 550 mm below the underside of the top flange plate and welded with nominal 6 mm fillet welds to the web plate each side of the stiffener or use a 2750 × 20 mm web plate without longitudinal stiffeners.

A6.4.7 Girder Intermediate Stiffeners

Intermediate are stiffeners required when (Equation 7.66)

$$t_w \leq \frac{h}{2.12}\sqrt{\frac{F_y}{E}} = \frac{2750}{2.12}\sqrt{\frac{350}{200,000}} = 2750/50.6 = 54.3 \text{ mm}.$$

The maximum spacing, a, of intermediate stiffeners is (Equation 7.103)

$$a \leq 1.95 t_w\sqrt{\frac{E}{\tau}} \leq 1.95(16)\sqrt{\frac{200,000}{59.5}} \leq 1809 \text{ mm}.$$

The stiffener spacing used, a_o, must be

$$\leq h \leq 2750\,\text{mm},$$

or

$$\leq 2450\,\text{mm},$$

or

$$\leq a = 1809\,\text{mm}.$$

Stiffener spacing, a_o, is 2(800) = 1600 mm

$$a/h = 1809/2750 = 0.66.$$

The required stiffener moment of inertia (Equation 7.105) is

$$I_{ts} = 2.5 a_0 t_w^3 \left(\left(\frac{h}{a} \right)^2 - 0.7 \right) = 2.5(1600)(16)^3 \left((1.52)^2 - 0.7 \right) = 26,271 \times 10^3 \text{ mm}^4.$$

Stiffeners will be used on both sides of the web coincident with locations of every second transverse floor beam.

Try 150 × 100 × 13 angles on each side of the 16 mm web plate:

$$I_{ts} = \frac{12(316)^3}{12} + \frac{88(40)^3}{12} = 32,024 \times 10^3 \text{ mm}^4 \geq 26,271 \text{ mm}^4, \text{OK}.$$

Use 150 × 100 × 13 angles on each side of the 16 mm web plate at Sections B-B in Figure A.1. Floor beams and knee braces will connect to the outstanding leg of interior stiffener angles (6 bolts for intermediate floor beams and 18 bolts for end floor beams).

$$b_{ts} = 150 - 12 = 138\,\text{mm}$$

$$\frac{b_s}{t_s} \leq 0.50 \sqrt{\frac{E}{F_y}} = 12.0$$

$$t_s \geq \frac{138}{12.0} \geq 11.5 \text{ mm}.$$

The interior intermediate transverse stiffeners will be connected to the top flange with fillet welds. In addition, the first intermediate transverse stiffeners from the end of the girders will be ground to fit at the bottom flange.

A6.4.8 Girder Bearing Stiffeners

R_{max} = 830.4 + 1787 = 2617 kN.

The minimum web plate thickness is (Equation 7.80a)

$$t_w \geq \frac{R}{0.75 F_y (L_B + k)} = \frac{2617(1000)}{0.75(350)(600 + 45 + 6)} \geq 15.3 \text{ mm} \left(\text{for } L_B \geq 600 \text{ mm} \right).$$

Bearing stiffeners should be used as good practice and in case smaller bearing length, L_B, is used. Bearing stiffeners must be connected to both flanges and extend to near the edge of the flange on each side of the girder web plate.

A6.4.8.1 Girder Bearing Stiffener Design as Compression Member

Maximum bearing stiffener width ≃ (725 − 16)/2 = 355 mm.

300 × 25 mm plate on each side of the girder web plate:

The bearing stiffener is designed as a compression member in accordance with Equations 7.81a and b:

$$A_{ebs} = 2A_{bs} + 12(t_w)^2 = 2(300)(25) + 12(16)^2 = 18{,}072\ \text{mm}^2$$

$$I_{ebs} = 2I_{bs} + 2A_{bs}\bar{y}^2 + t_w^4 = 2\left(\frac{25(300)^3}{12}\right) + 2(300)(25)(158)^2 + (16)^4 = 487.0 \times 10^6\ \text{mm}^4$$

$$r_{esb} = \sqrt{\frac{I_{esb}}{A_{esb}}} = \sqrt{\frac{487.0 \times 10^6}{18072}} = 164\ \text{mm}$$

$$\frac{h}{r_{ebs}} = \frac{2750}{164} = 16.8.$$

The allowable compressive stress, F_{call}, from Equations 7.83, 7.84a, or 7.85 depends on h/r_{ebs}. For $h/r_{ebs}. \le 0.839\sqrt{200{,}000/350} \le 20.0$.

The allowable compressive stress, F_{call}, from Equation 7.83 is

$$F_{call} = 0.55F_y = 192.5\ \text{MPa}.$$

The allowable force on the bearing stiffener = $P_{call} = F_{call}(A_{ebs})$ = 192.5(18,072)/1000 = 3479 kN ≥ 2617 kN, OK.

A6.4.8.2 Girder Bearing Stiffener Design for Bearing Stress

The allowable bearing stress for milled stiffeners and parts in contact yields (Equation 7.87) is

$$A'_{bs} \ge \frac{R}{0.83F_y} \ge \frac{2617(1000)}{0.83(350)} \ge 9009\ \text{mm}^2$$

$A'_{bs} = 2(300-12)(25) = 14{,}400\ \text{mm}^2$, OK with 12 mm clearance for girder flange-to-web fillet welds.

A6.4.8.3 Girder Bearing Stiffener Design for Local Buckling Stress

For local buckling of the bearing stiffener plate with a free edge, Equation 7.88 indicates

$$t_{bs} \ge \frac{b_{bs}}{0.43}\sqrt{\frac{F_y}{E}} \ge \frac{b_{bs}}{0.43}\sqrt{\frac{350}{200{,}000}} \ge \frac{b_{bs}}{10.28}.$$

Width of bearing stiffener stressed to $0.55F_y$ (192.5 MPa) is

$$b_{bseff} = \frac{2617(1000)/192.5 - (12)(16)}{2(25)} = \frac{13{,}595 - 192}{2(25)} = 268.1\ \text{mm}$$

$$t_{bseff} \ge \frac{b_{bseff}}{10.28} \ge \frac{268.1}{10.28} \ge 26.1\ \text{mm}.$$

For b_{bseff} = 268 mm and t_{bseff} = 25 mm:

$f_c = 2617(1000)/268(25/26.1)(2)(25) + 192 = 200.9\ \text{MPa} \ge 192.5\,\text{MPa}$, 4% overstress at bearing stiffener considered OK.

Use a 300 × 25 mm plate on each side of the girder web plate.

A6.4.8.4 Girder Bearing Stiffener Connection to Web Plate

$$t_{\text{weld}} \geq \frac{2617(1000)}{(2)(2750)135\sqrt{2}} \geq 2.6 \text{ mm.}$$

Use 6 mm fillet weld on each side of bearing stiffener.

A6.4.9 Girder Camber

Girder dead load = 51.9 kN/m

$$\Delta_{\text{camber}} = \frac{5(51.9)(32{,}000)^4}{384(200{,}000)(155174 \times 10^6} = 22.8 \text{ mm.}$$

Camber 25 mm at the center of girder.

A6.5 BTPG Span Bracing

A6.5.1 Loads and Forces

A6.5.1.1 Girder Compression Flange Notional Bracing Force

Maximum notional transverse shear force for top flange buckling restraint = R_F = 0.025 $A_f f_c = 0.025(725(45))(177)/1000 = 144.4$ kN applied at top flange.

A6.5.1.2 Nosing Force

N = 360/4 = 90 kN applied at the top of rail.

A6.5.1.3 Wind Forces

w_{ul} = wind on unloaded superstructure = 2.39 kPa applied on 1.5 times the surface area
W_L = wind on loaded superstructure = 1.44 kPa applied on 1.5 times the surface area
W_t = wind on train = 4.38 kN/m at 2.4 m above the top of rail
V_B = 2.20 kN/m notional vibration load applied at top lateral bracing
V_B = 2.90 kN/m notional vibration load applied at bottom lateral bracing.

A6.5.1.4 Load Combinations for Bracing Design (Table 4.10)

Load combination D2-A: $W_L + N$ at 125% allowable stress
Load combination D4-A: W_L or LV at 100% allowable stress
Load combination D4-B: W_{uL} at 100% allowable stress
Load combination D5-A: BF + N + W_L at 125% allowable stress.

A6.5.2 Top Lateral Bracing (Knee Bracing)

A6.5.2.1 Top Lateral Bracing Loads and Forces

Top lateral bracing by knee braces resisting wind and girder compression flange forces.
Based on the geometry of the cross section in Figures A.1, A.7, and A.8:
Wind on unloaded span:

W_{TF} = (2.84/2)(2.39)(1.5) = 5.09 kN/m wind force at top flange.

Wind on loaded span:

W_{TF} = (2.84/2)(1.44)(1.5) = 3.07 kN/m wind force at top flange

$V_B = 2.20$ kN/m notional vibration load at top bracing.

Girder compression flange forces:

$$\mathrm{BF} = 0.025 A_f f_c = 0.025(725(45))(177)/1000 = 144.4 \text{ kN applied at } h_k = 990 + 200 + 460 - (45/2) = 1628\,\text{mm}.$$

Transverse wind shear on unloaded span = 5.09(3.2) = 16.3 kN on intermediate knee braces
Transverse wind shear on unloaded span = 5.09(32.6/2) = 83.0 kN on end knee braces
Transverse wind shear on loaded span = 3.07(3.2) = 9.8 kN on intermediate knee braces
Transverse wind shear on loaded span = 3.07 (32.6/2) = 50.0 kN on end knee braces
Transverse bracing force shear on loaded span = 144.4 kN maximum on intermediate knee braces.

Load combination D4-A: W_L at 100% allowable stress:

$P_i = 9.8$ kN on intermediate knee braces
$P_e = 50.0$ kN on end knee braces.

Load combination D4-B: W_{uL} at 100% allowable stress:

$P_i = 16.3$ kN on intermediate knee braces
$P_e = 83.0$ kN on end knee braces.

Load combination D5-A: BF + W_L at 125% allowable stress:

$P_i = 144.4 + 9.8 = 154.2$ kN on intermediate knee braces
$P_e = 0 + 50.0 = 50.0$ kN on end knee braces.

A6.5.2.2 Intermediate Knee Brace Design

Intermediate knee brace geometry is shown in Figure A.7

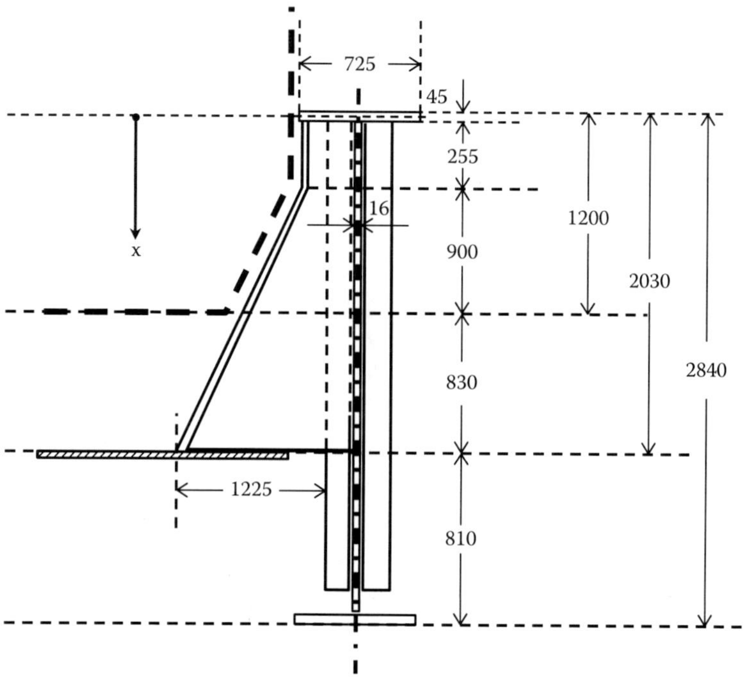

FIGURE A.7 Intermediate knee brace dimensions.

$$h_k = 2030 - (45/2) = 2008\,\text{mm}$$

$$w_k = ((900 + 830)/2) + (725/2) = 1228\,\text{mm; use } 1225\,\text{mm.}$$

$P = 154.2\,\text{kN}$ applied at 2008 mm above the deck plate at 125% allowable stress.

Preliminary design of knee brace flange as axial frame member (knee brace web plate provides lateral support):

Axial force in knee brace = (154.2)(2) = 308.4 kN
Bending moment induced force in knee brace flange = 154.2(2008/1225) = 252.8 kN
The axial stress in the knee brace flange = $(308.4 + 252.8)(1000)/(A_{kf})) \leq 192.5\,\text{MPa}$

$$A_{kf} \geq (308.4 + 252.8)(1000)/192.5 = 2915\,\text{mm}^2.$$

For local buckling of the knee brace flange plate, Equation 7.88 indicates

$$t_{kf} \geq \frac{b_{kf}}{0.43}\sqrt{\frac{F_y}{E}} \geq \frac{b_{kf}}{0.43}\sqrt{\frac{350}{200{,}000}} \geq \frac{b_{kf}}{10.28}$$

$$b_{kf} \geq \sqrt{10.28(2915)} = 173\ \text{mm}$$

$$t_{kf} \geq \frac{b_{kf}}{10.28} \geq \frac{173}{10.28} \geq 16.8\ \text{mm.}$$

Try 200 mm × 20 mm knee brace flange plate, $t_{kf} \geq \dfrac{b_{kf}}{10.28} \geq \dfrac{200}{10.28} \geq 19.5$ mm, OK.

$A_{kf} = 200(20) = 4000\,\text{mm}^2 \geq 2915\,\text{mm}^2$, OK.

Try knee brace with a 200 × 20 flange plate welded to a 12 mm web plate bolted to a 150 × 100 × 13 stiffener bolted through a girder web plate to a 150 × 100 × 13 stiffener.

$M = P(x) = 154.2(x)$ kNm, where $0 \leq x \leq 2.01$ m, at 125% allowable stress.

The section properties of the intermediate knee brace (Figure A.8) at $x = 277.5$, $x = 1200$ mm and $x = 2008$ mm are shown in Tables A8, A.9, and A.10, respectively.

Try a 12 mm knee brace web plate with

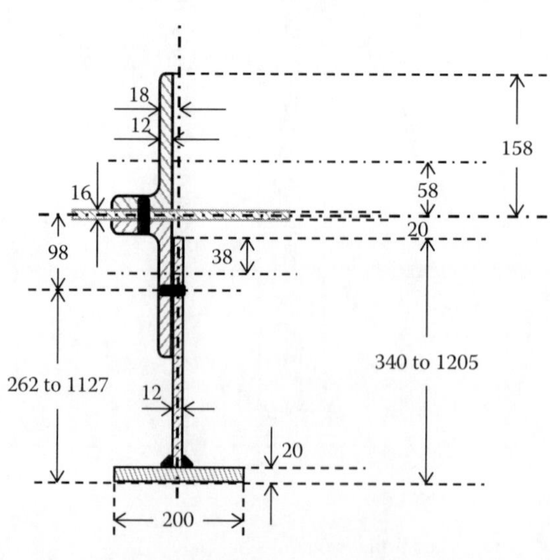

FIGURE A.8 Intermediate knee brace cross-section.

Width = 320 mm at $x = 277.5$ mm (Figures A.7 and A.8)

Width = 801 mm at $x = 1200$ mm (Figures A.7 and A.8)

Width = 1185 at $x = 2008$ mm (Figures A.7 and A.8).

Maximum width of girder web plate in knee brace section $= b_{kf} \leq 10.28 t_{kf} \leq 10.28(16) \leq 164$ mm.

The section properties and stresses of the intermediate knee brace at $x = 277.5$ mm (Table A.8) are

$$y_{s1} = \frac{(1287+944.6+930.2+691.2+40)(1000)}{3080+2624+3080+3840+4000} = \frac{3893(1000)}{16,624} = 234.2 \text{ mm}$$

$I_{kf1} = (104.1 + 41.5 + 14.2 + 11.3 + 201.0 + 7.03 + 0.06 + 7.03 + 32.8 + 0.13) \times 10^{6} = 419.0 \times 10^{6}\,\text{mm}^4$

$S_{kbflangel} = (419.0 \times 10^6)/(234.2) = 1789 \times 10^3\,\text{mm}^3$

$\sigma_{kbflasngel} = 154.2\,(1000)(277.5)/1789 \times 10^3 = 23.9$ MPa, OK.

$\tau_{kfwcbl} = 154.2(1000)/((320)(12)) = 40.2$ MPa in knee brace web plate, OK.

The section properties and stresses of the intermediate knee brace at $x = 1200$ mm (Table A.9) are

$$y_{s2} = \frac{(2707+2154+2350+4042+40)(1000)}{3080+2624+3080+9612+4000} = \frac{11,294(1000)}{22,396} = 504.3 \text{ mm}$$

TABLE A.8
Intermediate Knee Brace Section Properties (at $x = 277.5$ mm)

Element	Size	A (mm²)	y (mm)	A_y (mm³)	$y_s - y$ (mm)	$A(y_s - y)^2$ (mm⁴)	I_o (mm⁴)
Exterior stiffener	150 × 100 × 13	3080	418	1287×10^3	−183.8	104.1×10^6	7.03×10^6
Girder web	164 × 16	2624	360	944.6×10^3	−125.8	41.5×10^6	0.06×10^6
Interior stiffener	150 × 100 × 13	3080	302	930.2×10^3	−67.8	14.2×10^6	7.03×10^6
Knee brace web	12 × 320	3840	180	691.2×10^3	54.2	11.3×10^6	32.8×10^6
Knee brace flange	200 × 20	4000	10	40×10^3	224.2	201.0×10^6	0.13×10^6

TABLE A.9
Intermediate Knee Brace Section Properties (at $x = 1200$ mm)

Element	Size	A (mm²)	y (mm)	A_y (mm³)	$y_s - y$ (mm)	$A(y_s - y)^2$ (mm⁴)	I_o (mm⁴)
Exterior stiffener	150 × 100 × 13	3080	879	2707×10^3	−374.7	432.4×10^6	7.03×10^6
Girder web	164 × 16	2624	821	2154×10^3	−316.7	263.2×10^6	0.06×10^6
Interior stiffener	150 × 100 × 13	3080	763	2350×10^3	−258.7	206.2×10^6	7.03×10^6
Knee brace web	12 × 801	9612	420.5	4042×10^3	83.8	67.4×10^6	513.9×10^6
Knee brace flange	200 × 20	4000	10	40×10^3	494.3	977.2×10^6	0.13×10^6

$I_{kf2} = (432.4 + 263.2 + 206.2 + 67.4 + 977.2 + 7.03 + 0.06 + 7.03 + 513.9 + 0.13) \times 10^{6} = 2475 \times 10^6\,\text{mm}^4$
$S_{kbflange2} = (2475 \times 10^6)/(504.3) = 4907 \times 10^3\,\text{mm}^3$
$\sigma_{kbflasnge2} = 154.2\ (1000)(1200)/4907 \times 10^3 = 37.7\,\text{MPa}$, OK.
$\tau_{kfweb2} = 154.2(1000)/((801)(12)) = 16.0\,\text{MPa}$ in knee brace web plate, OK.

The section properties and stresses of the intermediate knee brace at $x = 2008$ mm (Table A.10) are

$$y_{s3} = \frac{(3952+3214+3594+8710+40)(1000)}{3080+2624+3080+14,220+4000} = \frac{19,510(1000)}{27,004} = 722.5\ \text{mm}$$

$I_{kf3} = (967.7 + 662.6 + 608.6 + 172.0 + 2030.5 + 7.0 + 0.06 + 7.0 + 1664 + 0.13) \times 10^6 = 6120 \times 10^6\,\text{mm}^4$

$S_{kfflange3} = (6120 \times 10^6)/722.5 = 8470 \times 10^3\,\text{mm}^3$
$\sigma_{kfflange3} = 154.2\ (1000)(2030)/(8470 \times 10^3) = 37.0\,\text{MPa}$, OK.
$\tau_{kfweb3} = 154.2(1000)/((1185)(12)) = 10.8\,\text{MPa}$ in knee brace web plate, OK.

A6.5.2.2.1 Intermediate Knee Brace Connection to Girder Intermediate Stiffeners The connection properties of the intermediate knee brace at $x = 22.5$ mm (Table A.8) are

$y_{s0} = 234.2$ mm
$I_{kf0} = 419.0 \times 10^6\,\text{mm}^4$
$Q = 200(20)(234.2 - 10) + 12(242)(234.2 - 141) = 897 + 271 = 1168 \times 10^3\,\text{mm}^3$
$Q/I = (1168 \times 10^3)/(419.0 \times 10^6) = 2.79 \times 10^{-3}\,\text{mm}^{-1}$
$q = VQ/I = 154.2\ (1000)\ (2.79 \times 10^{-3}) = 430$ N/mm.

Maximum bolt spacing = (380(120))/430 = 106 mm single shear; use maximum spacing of 100 mm from the top of knee brace to 400 mm below underside of top flange.

The connection properties of the intermediate knee brace at $x = 1200$ mm are $y_{s2} = 504.3$ mm
$I_{kf2} = 2475 \times 10^6\,\text{mm}^4$
$Q = 200(20)(504.3 - 10) + 12(723)(504.3 - 381.5) = 1977 \times 10^3 + 1065 \times 10^3 = 3043 \times 10^3\,\text{mm}^3$
$Q/I = (3043 \times 10^3)/(2475 \times 10^6) = 1.23 \times 10^{-3}\,\text{mm}^{-1}$
$q = VQ/I = 154.2\ (1000)\ (1.23 \times 10^{-3}) = 190$ N/mm.

Maximum bolt spacing = (380(120))/190 = 241 mm single shear; use maximum of 150 mm (sealing distance).

The connection properties of the intermediate knee brace at $x = 2008$ mm are $y_{s3} = 722.5$ mm
$I_{kf3} = 6120 \times 10^6\,\text{mm}^4$
$Q = 200(20)(722.5 - 10) + 12(1107)(722.5 - 573.5) = 4829 \times 10^3\,\text{mm}^3$
$Q/I = (4829 \times 10^3)/(6120 \times 10^6) = 0.79 \times 10^{-3}\,\text{mm}^{-1}$

TABLE A.10
Intermediate Knee Brace Section Properties (at x = 2008 mm)

Element	Size	A (mm²)	y (mm)	A_y (mm³)	$y_s - y$ (mm)	$A\ (y_s - y)^2$ (mm⁴)	I_o (mm⁴)
Exterior stiffener	150 × 100 × 13	3080	1283	3952 × 10³	−560.5	967.7 × 10⁶	7.03 × 10⁶
Girder web	164 × 16	2,624	1225	3214 × 10³	−502.5	662.6 × 10⁶	0.06 × 10⁶
Interior stiffener	150 × 100 × 13	3,080	1167	3594 × 10³	−444.5	608.6 × 10⁶	7.03 × 10⁶
Knee brace web	12 × 1185	14,220	612.5	8710 × 10³	110.0	172.0 × 10⁶	1664 × 10⁶
Knee brace flange	200 × 20	4,000	10	40 × 10³	712.5	2030.5 × 10⁶	0.13 × 10⁶

$q = VQ/I = 154.2\ (1000)\ (0.79 \times 10^{-3}) = 122$ N/mm.

Maximum bolt spacing = (380(120))/122 = 375 mm single shear; use maximum of 150 mm.

A6.5.2.2.2 Intermediate Knee Brace Connection to Girder Web The connection properties of the intermediate knee brace at $x = 22.5$ mm (Table A.8) are

$y_{s0} = 234.2$ mm
$I_{kf0} = 419.0 \times 10^6\ \text{mm}^4$
$Q_{int0} = 200(20)(234.2 - 10) + 12(320)(234.2 - 180) + 3080(302 - 234.2) = 1314 \times 10^3\ \text{mm}^3$
$Q_{int0}/I = (1314 \times 10^3)/(419.0 \times 10^6) = 3.14 \times 10^{-3}\ \text{mm}^{-1}$
$q_{int0} = VQ_{int0}/I = 154.2\ (1000)\ (3.14 \times 10^{-3}) = 483$ N/mm.

Maximum bolt spacing = (380(120))/483 = 94 mm; use maximum 90 mm spacing at the top of knee brace.

$Q_{ext0} = 3080(418 - 234.2) = 566.1 \times 10^3\ \text{mm}^3$
$Q_{ext0}/I = (566 \times 10^3)/(419.0 \times 10^6) = 1.35 \times 10^{-3}\ \text{mm}^{-1}$
$Q_{ext0} = VQ_{ext0}/I = 154.2\ (1000)\ (1.35 \times 10^{-3}) = 208$ N/mm.

Maximum bolt spacing = (380(120))/208 = 220 mm; use maximum of 150 mm.

The connection properties of the intermediate knee brace at $x = 1200$ mm are $y_{s2} = 504.3$ mm
$I_{kf2} = 2475 \times 10^6\ \text{mm}^4$

$Q_{int2} = 200(20)(504 - 10) + 12(781)(504 - 410.5) + 3080(763 - 504) = 3650 \times 10^3\ \text{mm}^3$
$Q_{int2}/I = (3650 \times 10^3)/(2475 \times 10^6) = 1.47 \times 10^{-3}\ \text{mm}^{-1}$
$q_{int2} = VQ_{int2}/I = 154.2\ (1000)\ (1.47 \times 10^{-3}) = 227$ N/mm.

Maximum bolt spacing = (380(120))/227 = 200 mm; use maximum of 150 mm.

$Q_{ext2} = 3080(879 - 504) = 1155 \times 10^3\ \text{mm}^3$
$Q_{ext2}/I = (1155 \times 10^3)/(2475 \times 10^6) = 0.47 \times 10^{-3}\ \text{mm}^{-1}$
$Q_{ext2} = VQ_{ext2}/I = 154.2\ (1000)\ (0.47 \times 10^{-3}) = 72$ N/mm.

Maximum bolt spacing = (380(120))/72 = 633 mm; use maximum of 150 mm.

The connection properties of the intermediate knee brace at $x = 2008$ mm are $y_{s3} = 722.5$ mm

$I_{kf3} = 6120 \times 10^6\ \text{mm}^4$
$Q_{int3} = 200(20)(722.5 - 10) + 12(1185)(722.5 - 612.5) + 3080(1167 - 722.5) = 5783 \times 10^3\ \text{mm}^3$
$Q_{int3}/I = (5783 \times 10^3)/(6120 \times 10^6) = 0.94 \times 10^{-3}\ \text{mm}^{-1}$
$q_{int3} = VQ_{int3}/I = 154.2\ (1000)\ (0.94 \times 10^{-3}) = 146$ N/mm.

Maximum bolt spacing = (380(120))/146 = 312 mm; use maximum of 150 mm.

$Q_{ext3} = 3080(1283 - 722.5) = 1726 \times 10^3\ \text{mm}^3$
$Q_{ext3}/I = (1726 \times 10^3)/(6120 \times 10^6) = 0.28 \times 10^{-3}\ \text{mm}^{-1}$
$Q_{ext3} = VQ_{ext3}/I = 154.2\ (1000)\ (0.28 \times 10^{-3}) = 43.5$ N/mm.

Maximum bolt spacing = 2(380(120))/43.5 = 1048 mm; use maximum of 150 mm.

A6.5.2.2.3 Intermediate Knee Brace Flange to Web Weld Connection The connection properties of the intermediate knee brace flange at $x = 2008$ mm are

$y_{s3} = 722.5$ mm
$I_{kf3} = 6120 \times 10^6\ \text{mm}^4$
$Q = 200(20)(722.5 - 10) = 2850 \times 10^3\ \text{mm}^3$
$Q/I = (2850 \times 10^3)/(6120 \times 10^6) = 0.47 \times 10^{-3}\ \text{mm}^{-1}$

$q = VQ/I = 154.2\ (1000)\ (0.47 \times 10^{-3}) = 72$ N/mm

$$t_{\text{weld}} \geq \frac{72}{135\sqrt{2}} \geq 0.4 \text{ mm on throat of fillet weld}$$

$$t_{\text{weld}} \geq \frac{72}{2(122.5)} \geq 0.3 \text{ mm.}$$

Use 6 mm fillet weld on each side.

The intermediate knee brace arrangement is shown in Figure A.9.

A6.5.2.3 End Knee Brace Design

End knee brace geometry is shown in Figure A.10.

$P = 83.0$ kN applied at 2008 mm above the deck plate at 100% allowable stress.

Intermediate knee braces designed for 154.2 kN at 125% allowable stress = 123.4 kN at 100% allowable stress. Use the same section as intermediate knee braces.

A6.5.2.3.1 End Knee Brace Connection to Girder Bearing Stiffeners The connection properties of the end knee brace at $x = 22.5$ mm (Figure A.11) (Table A.11) are

$$y_{s0} = \frac{(7076)(1000)}{25{,}464} = 278 \text{ mm}$$

$I_{kf0} = (817.1 + 145.6) \times 10^{6} = 962.7 \times 10^{6}$ mm^4

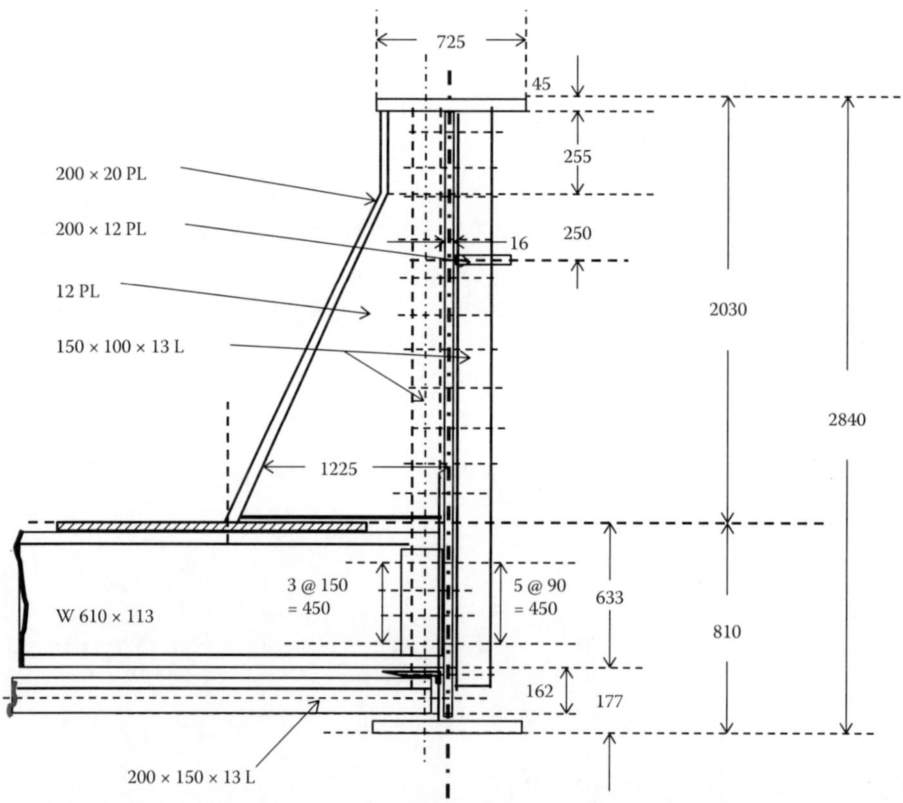

FIGURE A.9 Intermediate knee brace elevation.

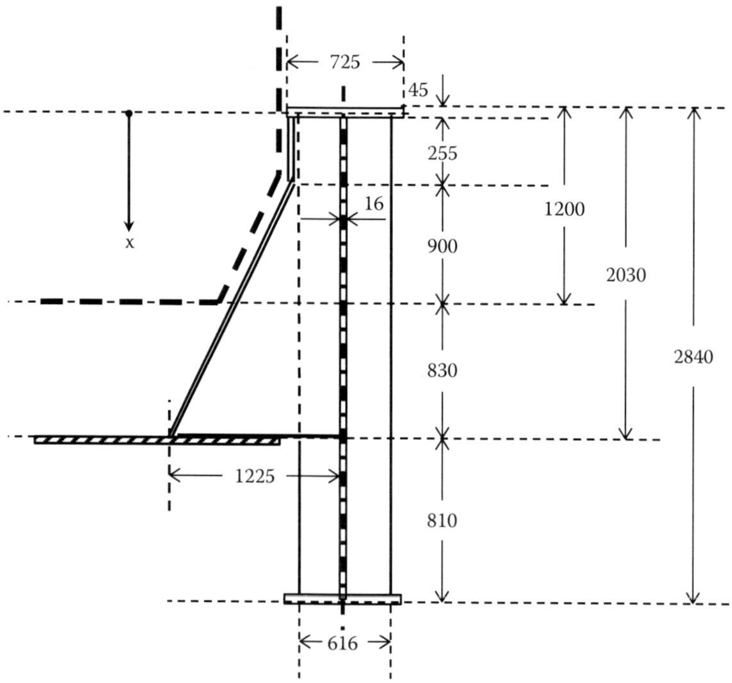

FIGURE A.10 End knee brace dimensions.

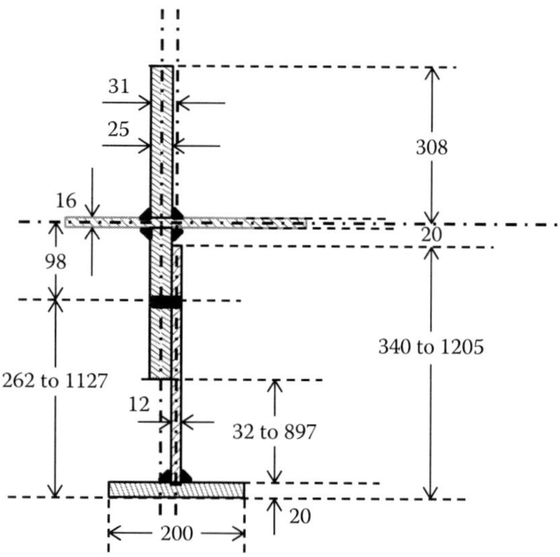

FIGURE A.11 End knee brace cross-section.

$Q = 200(20)(278 - 10) + 12(242)(278 - 141) = 1470 \times 10^3\,\text{mm}^3$
$Q/I = (1470 \times 10^3)/(962.7 \times 10^6) = 1.53 \times 10^{-3}\,\text{mm}^{-1}$
$q = VQ/I = 83.0\ (1000)\ (1.53 \times 10^{-3}) = 127\ \text{N/mm}.$

Maximum bolt spacing = (380(120))/127 = 360 mm single shear; use maximum of 150 mm. The connection properties of the end knee brace at $x = 2080$ mm (Table A.12) are

TABLE A.11
End Knee Brace Section Properties (at x = 22.5 mm)

Element	Size	A (mm²)	y (mm)	A_y (mm³)	$y_s - y$ (mm)	$A(y_s - y)^2$ (mm⁴)	I_o (mm⁴)
Exterior stiffener	300 × 25	7,500	518	3885 × 10³	−240.1	432.4 × 10⁶	56.3 × 10⁶
Girder web	164 × 16	2,624	360	945 × 10³	−82.1	17.7 × 10⁶	0.06 × 10⁶
Interior stiffener	300 × 25	7,500	202	1515 × 10³	75.9	43.2 × 10⁶	56.3 × 10⁶
Knee brace web	12 × 320	3,840	180	691 × 10³	87.9	36.8 × 10⁶	32.8 × 10⁶
Knee brace flange	200 × 20	4,000	10	40 × 10³	267.9	287.0 × 10⁶	0.13 × 10⁶
Σ		25,464		7076 × 10³		817.1 × 10⁶	145.6 × 10⁶

TABLE A.12
End Knee Brace Section Properties (at x = 2080 mm)

Element	Size	A (mm²)	y (mm)	A_y (mm³)	$y_s - y$ (mm)	$A(y_s - y)^2$ (mm⁴)	I_o (mm⁴)
Exterior stiffener	300 × 25	7,500	1383	10,373 × 10³	−536.6	2159 × 10⁶	56.3 × 10⁶
Girder web	164 × 16	2,624	1225	3,214 × 10³	−378.6	376.1 × 10⁶	0.06 × 10⁶
Interior stiffener	300 × 25	7,500	1067	8,003 × 10³	−220.6	364.9 × 10⁶	56.3 × 10⁶
Knee brace web	12 × 1185	14,220	612.5	8,710 × 10³	233.9	778.1 × 10⁶	1664 × 10⁶
Knee brace flange	200 × 20	4,000	10	40 × 10³	836.4	2798 × 10⁶	0.13 × 10⁶
Σ		35,844		30,339 × 10³		6477 × 10⁶	1777 × 10⁶

$$y_{s0} = \frac{(30,339)(1000)}{35,844} = 846.4 \text{ mm}$$

$I_{kf0} = (6477 + 1777) \times 10^{6} = 8254 \times 10^{6}\,\text{mm}^4$
$Q = 200(20)(846.4 - 10) + 12(1107)(846.4 - 573.5) = 6971 \times 10^3\,\text{mm}^3$
$Q/I = (6971 \times 10^3)/(8252 \times 10^6) = 0.84 \times 10^{-3}\,\text{mm}^{-1}$
$q = VQ/I = 83.0\,(1000)\,(0.84 \times 10^{-3}) = 70.1$ N/mm.

Maximum bolt spacing = (380(120))/70.1 = 650 mm single shear; use maximum of 150 mm.

A6.5.2.3.2 End Knee Brace Connection to Girder Web The connection properties of the end knee brace at x = 22.5 mm (Table A.11) are

$$y_{s0} = \frac{(7076)(1000)}{25,464} = 278 \text{ mm}$$

$I_{kf0} = (817.1 + 145.6) \times 10^{6} = 962.7 \times 10^{6}\,\text{mm}^4$
$Q_{int} = 200(20)(278 - 10) + 12(320)(278 - 180) + 25(300)(278 - 202) = 2018 \times 10^3\,\text{mm}^3$
$Q_{int}/I = (2018 \times 10^3)/(962.7 \times 10^6) = 2.10 \times 10^{-3}\,\text{mm}^{-1}$
$q_{int} = VQ_{int}/I = 83.0\,(1000)\,(2.10 \times 10^{-3}) = 174$ N/mm

$$t_{weld} \geq \frac{174}{135\sqrt{2}} \geq 0.9 \text{ mm on throat of fillet weld}$$

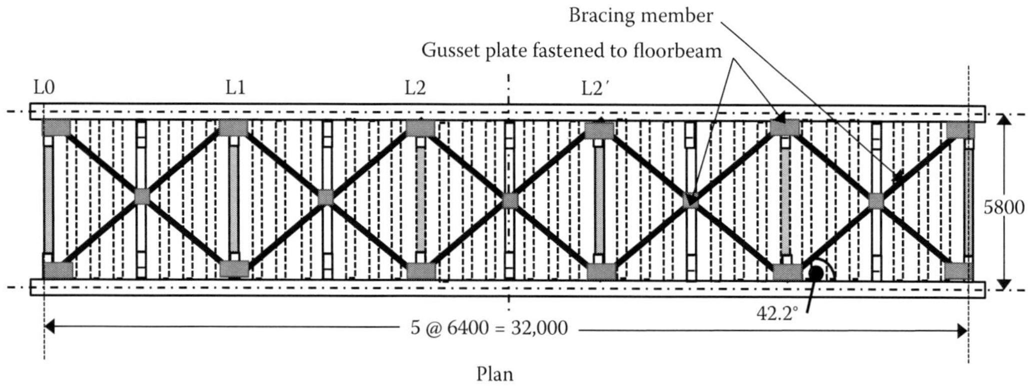

FIGURE A.12 Bottom lateral bracing plan.

$$t_{weld} \geq \frac{174}{2(122.5)} \geq 0.7 \text{ mm}.$$

Use 6 mm minimum fillet weld on each side.

The connection properties of the end knee brace at $x = 2080$ mm (Table A.12) are

$$y_{s0} = \frac{(30,339)(1000)}{35,844} = 846.4 \text{ mm}$$

$I_{kf0} = (6477 + 1777) \times 10^{6} = 8254 \times 10^{6}\text{mm}^4$
$Q_{int} = 200(20)(846.4 - 10) + 12(1185)(846.4 - 612.5) + 25(300)(1067 - 846.4) = 8326 \times 10^3\text{mm}^3$
$Q_{int}/I = (8326 \times 10^3)/(8254 \times 10^6) = 1.01 \times 10^{-3}\text{mm}^{-1}$
$q_{int} = VQ_{int}/I = 83.0\ (1000)\ (1.01 \times 10^{-3}) = 83.7$ N/mm.

Use 6 mm minimum fillet weld on each side.

A6.5.3 Bottom Lateral Bracing

Bottom lateral bracing by horizontal truss resisting wind and live load nosing forces. The lateral bracing members will be connected to every 8th floor beam (panel length = 8(800) = 6400 mm) as shown in Figures A.1 and A.12.

A6.5.3.1 Bottom Lateral Bracing Loads and Forces

Based on the geometry of cross sections in Figures A.1, A.7, A.8, and A10:

Wind on unloaded span:

$W_{BF} = (2.84/2)(2.39)(1.5) = 5.09$ kN/m wind force at bottom lateral bracing.

Wind on loaded span:

$W_{BF} = (2.84/2)(1.44)(1.5) = 3.07$ kN/m wind force at bottom lateral bracing

W_t = Wind on train = 4.38 kN/m at the top of rail (applied to bottom lateral bracing)

$V_B = 2.9$ kN/m notional vibration load at bottom lateral bracing (horizontal truss).

Nosing forces:

$N = 360/4 = 90$ kN at the top of rail (applied to bottom lateral bracing)

Transverse wind shear on unloaded span = 5.09(6.4) = 32.6 kN in intermediate panels

Transverse wind shear on unloaded span = 5.09(32.0/2) = 81.4 kN in end panels
Transverse wind shear on loaded span = (3.07 + 4.38)(6.4) = 7.45(6.4) = 47.7 kN in intermediate panels
Transverse wind shear on loaded span = 7.45 (32.0/2) = 119.2 kN in end panels
Transverse nosing force shear on loaded span = 90 kN.

Load combination D2-A and D5-A: W_L + N:

$$P_i = 47.7 + 90.0 = 137.7\,\text{kN at 125\% allowable stress}$$

$$P_e = 119.2 + 90.0 = 209.2\,\text{kN at 125\% allowable stress.}$$

Load combination D4-A: W_L or LV:

$$P_i = 47.7\,\text{kN at 100\% allowable stress} P_e = 119.2\,\text{kN at 100\% allowable stress.}$$

Load combination D4-B: W_{uL} at 100% allowable stress:

$$P_i = 32.6\,\text{kN at 100\% allowable stress}$$
$$P_e = 81.4\,\text{kN at 100\% allowable stress.}$$

A6.5.3.2 Design of Lateral Bracing Members

The forces in the bottom lateral members, assuming each member resists 1/2 of the panel shear force in both tension and compression, are shown in Table A.13.

$$L = \sqrt{6400^2 + 5800^2} = 8637 \text{ mm.}$$

Ends and center of bracing members supported at floor beams with knee braces (Figures A.10 and A.12). Lu = 0.90(8637/2) = 3887 mm (10% of length fastened to gusset plates).

A6.5.3.2.1 Compressive Design of Lateral Bracing Members

$$P = -155.7\,\text{kN.}$$

Considering elastic stability, inelastic stability, and compressive yielding:

$K = 0.75$ with bolted connections
$r_{min} \geq L/120 \geq 3887/120 \geq 32.4\,\text{mm}$.
Try 200 × 150 × 13
$r_x = 44.7\,\text{mm}$
$r_y = 64.0\,\text{mm}$
$r_z = 32.5\,\text{mm}$
$A = 4380\,\text{mm}^2$

$$\frac{L}{r_{min}} = \frac{3887}{32.5} = 119.6 \leq 120, \text{ OK.}$$

TABLE A.13
Bottom Lateral Bracing Forces

Panel	Wind Shear (kN)	Nosing Shear (kN)	Total Panel Shear (kN)	Force in Diagonals of Panel (kN)
L0–L1	47.7(5)/2 = 119.2	90	209.2	±(209.2/2)/sin (42.2) = ±155.7
L1–L2	119.2–47.7 = 71.5	90	161.5	± (161.5/2)/sin (42.2) = ±120.2
L2–L2′	71.5–47.7 = 23.8	90	113.8	± (113.8/2)/sin (42.2) = ±84.7

$$\frac{KL}{r} = \frac{0.75(4319)}{32.5} = 99.7 \geq 0.629\sqrt{\frac{E}{F_y}} \geq 15.0$$

$$\frac{KL}{r} = 99.7 \leq 5.034\sqrt{\frac{E}{F_y}} \leq 120$$

$$F_{call} = 0.60F_y - \left(635\frac{F_y}{E}\right)^{3/2}\left(\frac{KL}{r}\right) = 0.60(350) - \left(635\frac{(350)}{(200{,}000)}\right)^{yt}(99.7) = 210 - 116.8 = 93.2 \text{ MPa}$$

$F_c = -93.2(4380)/1000 = -408\,\text{kN} \geq -155.7\,\text{kN}$ (slenderness governs), OK.

A6.5.3.2.2 Design of Bracing Member to Lateral Gusset Plate Connections

P = lesser of
±(1.5) 155.7 = 233.6 kN or
−408 kN or
+0.55(350)(4380)/1000 = 843 kN

No. of bolts ≥ 233.6(1000)/((120)(380)) ≥ 233.6/45.60 ≥ 5.1 bolts single shear.
Use one row of 3 bolts and one row of 2 bolts staggered.

A6.5.3.2.3 Tensile Design of Lateral Bracing Members

$$P = +155.7\,\text{kN}.$$

Considering net effective area and tensile yielding:

$$A_n = 4380 - 2(25)(13) = 3730\,\text{mm}^2$$

$U_c = 0.60$ for short bolted connection on one leg of bracing member only

$$A_e = 0.60(3730) = 2238\,\text{mm}^2$$

$$F_{tall} = 0.55(350)(2238)/1000 = 431\,\text{kN}.$$

Use 200 × 150 × 13 (compression slenderness governs) with 5 bolts single shear each end of member.

A6.5.3.2.4 Design of Lateral Bracing Gusset Plate to Girder Web Connections

Tension on connection = 155.7 (sin 42.2) = 104.6 kN
Shear at connection = 155.7(cos42.2) = 115.3 kN.

Try 3 bolts.

$\phi_t = 104.6(1000)/(3(380)) = 91.8\,\text{MPa}$
$\phi_v = 115.3(1000)/3(380)) = 101.1\,\text{MPa}$

$$f_{vall} = 120\left(1 - \frac{91.8(380)}{270(1000)}\right) = 120(0.87) = 104.5 \text{ MPa} \geq 101.1 \text{ MPa, OK.}$$

Use minimum of 6 bolts (3 bolts in each adjacent panel) connecting the lateral bracing gusset plate to girder web. The bracing gusset plate arrangement is shown in Figure A.13.

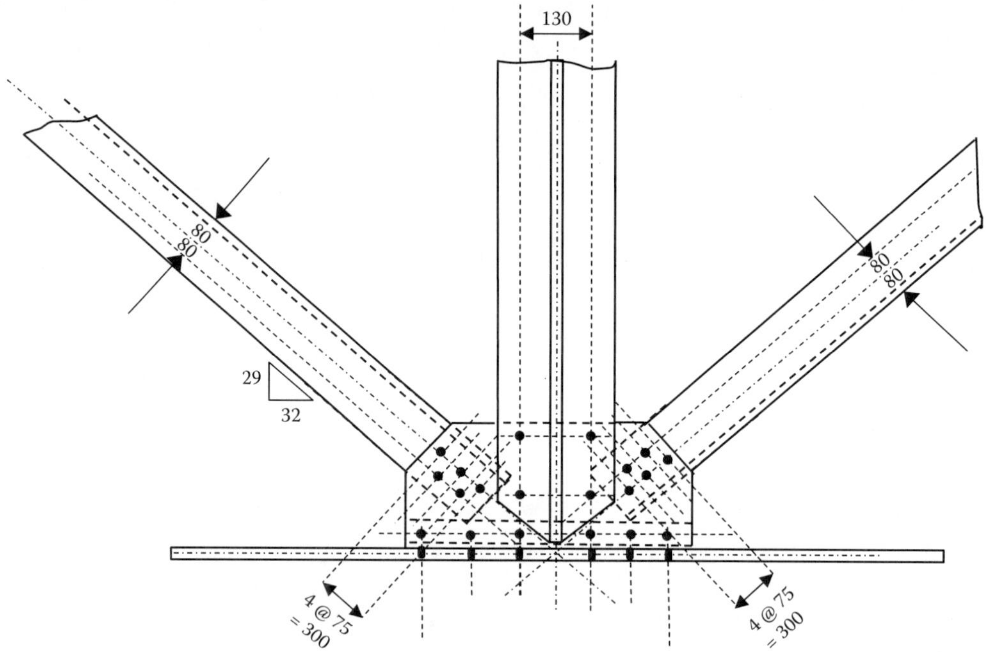

FIGURE A.13 Bottom lateral bracing gusset plate.

A7 SUMMARY OF BTPG DESIGN

A7.1 Deck and Floor System

A7.1.1 Deck: Grade 350 Weldable Steel Plate

8 – 5000 × 25 × 3200 PL, weight = 8 × 3140 kg = 25,120 kg
2 – 5000 × 25 × 3500 PL, weight = 2 × 3434 kg = 6,869 kg.

A7.1.2 Floor Beams: Grade 350 Rolled Sections

39 – W 610 × 113 section × 5800, weight = 39 × 113 × 5.8 = 25,561 kg

2 – W 610 × 113 section × 5800 with a 20 mm × 200 mm bolted bottom cover plate, weight = 2 × (113 + 31.4) × 5.8 = 1,675 kg.

A7.2 Girder and Stiffeners: Grade 350 Weldable Steel Plate

A7.2.1 Top Flanges

2 – 725 × 45 × 32,600 PL (shop CJP splices), weight = 2 × 8349 = 16,698 kg.

A7.2.2 Web Plates

2 – 2750 × 16 × 32,600 PL (shop CJP splices), weight = 2 × 11,260 = 22,520 kg.

A7.2.3 Bottom Flanges

2 – 725 × 45 × 32,600 PL (shop CJP splices), weight = 2 × 8349 = 16,698 kg.

A7.2.4 Girder Stiffeners

A7.2.4.1 Transverse Intermediate Stiffeners

76 – 150 × 100 × 13 L × 2750, weight = 76 × 24.2 × 2.75 = 5,058 kg.

TABLE A.14
Estimated Weight of Steel

Span Component	Item	Size (mm)	Unit Weight (kg)	No.	Total Weight (kg)
Deck	Intermediate deck plate	5,000 × 25 × 3,200	3,140	8	25,120
	End deck plate	5,000 × 25 × 3,500	3,434	2	6,896
	12 mm curb plates (600 mm above deck plate)	672 × 12 × 32,600	2,064	2	4,127
	12 mm curb brackets (600 mm above deck plate)	925 to 1,225 × 600	61	20	1,215
	Deck bolts, fills, etc.			5%	1,807
	Deck subtotal				39,165
Floor beams	Intermediate floor beams	W 610 × 113, 5,800 long	655	39	25,561
	End floor beams	W 610 × 113 w/cover plate, 5,800 long	838	2	1,675
	Connection angles	150 × 150 × 12 L, 450 long	11	108	1,188
	Bolts	22 M A325	0.30	≃900	270
	Floor beams subtotal				28,694
Girders	Girder top flange plates	725 × 45 × 32,600	8,349	2	16,698
	Girder web plates	2,750 × 16 × 32,600	11,260	2	22,520
	Girder bottom flange plates	725 × 45 × 32,600	8,349	2	16,698
	Girder intermediate web stiffeners	150 × 100 × 13 L, 2,750 long	66.5	76	5,058
	Stiffener bolts	22 M A325	0.30	≃ 800	240
	Girder end bearing stiffeners	300 × 25 × 2,750 PL	162	8	1,295
	Girder longitudinal stiffeners	200 × 12 × 32,600 PL	615	2	1,230
	Girder subtotal				63,739
Bracing	Intermediate knee braces	200 × 20 flange, 12 web	203	18	3,653
	End knee braces	200 × 20 flange, 12 web	203	4	812
	Knee brace bolts	22 M A325	0.30	≃ 650	195
	Intermediate panel bottom lateral bracing	200 × 150 × 13 L, 8,637 long	297	6	1,783
	End panel bottom lateral bracing	200 × 150 × 13 L, 8,637 long	297	4	1,188
	Bottom lateral bracing gusset plates	≃ 950 × 450 × 12 (to be sized by design drafter)	40	12 + 5 = 17	685
	Bottom lateral bracing bolts	22 M A325	0.30	≃ 320	96
	Bracing subtotal				8,412
Subtotal steel weight estimate					140,280
Contingency				5%	7,014
Total steel weight estimate					147,294say *147,500*

A7.2.4.2 Longitudinal Stiffeners

2 – 200 × 12 × 32,600 PL, weight = 2 × 615 = 1,230 kg.

A7.2.4.3 Bearing Stiffeners

8 – 300 × 25 × 2750 PL, weight = 8 × 162 = 1,295 kg.

A7.3 Span Bracing

A7.3.1 Top Bracing (Knee Bracing)

Flanges: 22 – 200 × 20 × (255 + $(1730^2 + 862.5^2)^{1/2}$)
PL = 22 – 200 × 20 × 2188 PL, weight = 22 × 69 = 1,511 kg
Webs: 22 – [(2030 × 334.5) +(1730 × (862.5/2))] × 12 PL = 22 – 1.43 mm^2 × 12 mm PL, weight = 22 × 134 = 2,953 kg
Total knee brace = 22 (69 + 134) = 22 × 195 = 4,465 kg.

A7.3.2 Bottom Bracing (Lateral Truss Bracing)

10 – 200 × 150 × 13 L × 8637, weight = 10 × 34.4 × 8.637 = 2971 kg.

A7.4 Superstructure Weight

Total estimated steel weights are summarized in Table A.14.
Estimated steel weight = 147,500 kg = 147,500/32.6 kg/m = 4,525 kg/m
Estimated floor system steel weight = 39,125 + 28,694 = 67,819 kg (48% of total estimated steel weight)
Estimated girders steel weight = 63,739 kg (45% of total estimated steel weight)
Estimated knee bracing weight = 4660 kg (7 % of girder estimated steel weight)
Estimated bottom lateral bracing weight = 3752 kg (< 3% of total estimated steel weight)
Total superstructure steel weight estimated in Table A.14 = 140,280 kg (not including 5% contingency).
Weight of steel estimated for girder design (see A6.2.1) = ((9.6 + 8.3)/2 + 12.0)32.6 = 683.0 kN = 69,620 kg per girder
Total superstructure steel weight estimated for design = 69,620(2) = 139,240 kg.
Estimated weight for design is 99% of that in Table A.14, which is very much acceptable.

A8 DESIGN, FABRICATION, AND ERECTION DRAWINGS

Once the design calculations (design brief) are peer reviewed and acceptable, a design drafting technician can then, in conjunction, and in consultation, with the design engineer, prepare design drawings. The design drawings are reviewed and approved by the design engineer.

The approved design drawings are forwarded to the fabricator for the production of shop (or detail) drawings (see Chapter 10). The shop drawings are reviewed by the design engineer and/or design drafting technician. This review does not constitute warranty that the shop drawings are accurate, which responsibility rests with the fabricator.

The design and shop drawings (depending on contract award schedule) may be used by the erector to design an erection methodology and procedure, and prepare the requisite erection drawings (see Chapter 11).

Appendix B: Design of a Ballasted Deck Plate Girder (BDPG) Superstructure

B1 SCOPE OF SUPERSTRUCTURE DESIGN

Replace an existing 107′ long pin-connected deck truss (DT) with a 107′ long BDPG span. The span crosses a waterway with adequate hydraulic clearances.

Existing substructures are founded on bedrock and have been salvaged through rehabilitation. The bridge seats are 16 ft wide.

B2 GENERAL INFORMATION

The tangent track on the proposed BDPG span will consist of 7″ deep × 8′ long track ties on ¾″ thick tie plates with 18″ of ballast from the base of the rail to the top of the reinforced concrete deck.

B3 LOADING AND MATERIALS

Cooper's E90 with alternate live load (100 kip axle loads)
Steel with $F_y = 50$ ksi
Bolts: 7/8″ ASTM F3125 Grade 325M bolts with $F_u = 120$ ksi
Welding: submerged arc welding or shielded metal arc welding to applicable bridge code using E70XX electrodes
Reinforced concrete with $f_c' = 5000$ psi
Reinforcement with $F_u = 60$ ksi.

B4 LAYOUT OF PROPOSED SUPERSTRUCTURE

Plan, elevation, and cross section of the proposed BDPG span are shown in Figure B.1.

Deck slab width of 14 ft was used (maximum width in case of future double tracking with track centers ≥ 14 ft).

Length of span between centers of bearings = 105 ft (assuming a 2 ft long bearing plate).

Depth of ballast, d_b = 18″ (18 − 7) = 11″ of ballast under the tie.

Depth of reinforced concrete slab = 16 in. (appropriate thickness for transverse reinforcement size and spacing).

B5 DECK SLAB

B5.1 General Information

Reinforced concrete deck slab

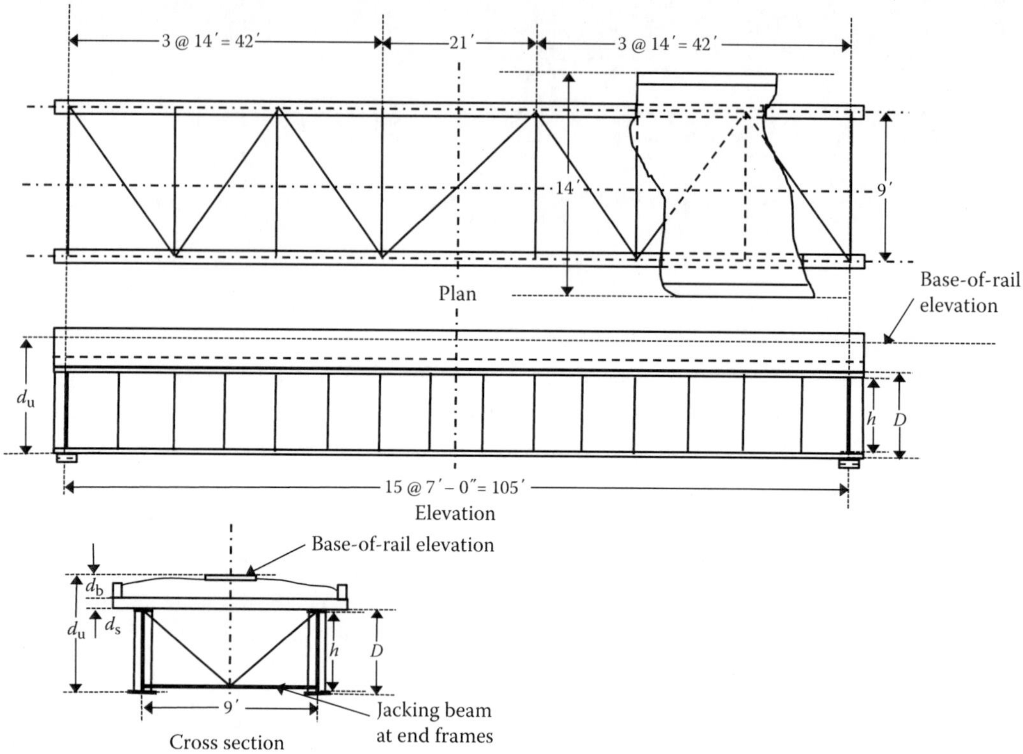

FIGURE B.1 Plan, elevation and cross section of span.

B5.2 Loads and Forces

B5.2.1 Dead Loads

Track = (200)/(8 + 2(18/12)))/(1000) = 18.2 psf (distributed to 11 ft deck slab width assuming 8 ft tie with 1:1 distribution through ballast with d_b = 18″).

Ballast = (120)(d_b/12) = 10.0(d_b) psf.

Assume a 24 in. ballast depth to account for up to 6 in. of future track raises.

Ballast = 240 psf; use 250 psf to account for contingencies related to ballast degradation and water retention.

Waterproofing = 10 psf.

Deck slab (assume 16 in. thick) = (16/12)150 = 200 psf

w_{DL} = 18.2 + 250 +10 + 200 = 478 psf.

Model as simply supported beam taking shear at d_s from support:

V_{DL} = (478(L)/2)(1 – (2d_s/L)) = (478(9)/2)(1 – (2(16/12)/9)) = 2151(1 – 0.30) = 1514 lb/ft length of slab

M_{DL} = 478(9²)/8 = 4840 lb-ft/ft length of slab.

B5.2.2 Live Load

Longitudinal distribution = 3 + (d_b + d_s – 7)/12 = 5.25 ft > 5 ft; use 5 ft

Lateral distribution = 8 + (d_b + d_s – 7)/12 = 10.25 ft

w_{LL} = 90(1000)/((5)(10.25)) = 1756 psf

Model as simply supported beam taking shear at d_s from support:

$V_{LL} = (1756(9)/2)(1 - (2(16/12)/9)) = 5561$ lb/ft length of slab
$M_{LL} = 1756(9)^2/8 = 17{,}780$ lb-ft/ft length of slab.

Vertical impact

$I = 60\%$ [American Railway Engineering and Maintenance-of-Way Association (AREMA) Chapter 8]
$V_{LL+I(max)} = (1.6)\ 5561 = 8898$ lb/ft length of slab
$M_{LL+I(max)} = (1.6)\ 17{,}780 = 28{,}447$ lb-ft/ft length of slab.

B5.2.3 Wind Forces

Wind load on train = 300 lb/ft at 8 ft above the top of rail

$w_{WLL} = [300(8)/5]/(3) = 160$ psf [wheel spacing = 5 ft rail spacing distributed over 2(18/12) = 3 ft width at deck slab elevation]

Model as simply supported beam taking shear at d_s from support:

$V_{wLL} \simeq (160(9)/2)(1 - (2(16/12)/9)) = 507$ lb/ft length of slab
$M_{wLL} \simeq 160(9)^2/8 = 1620$ lb-ft/ft length of slab.

B5.2.4 Load Combinations for Deck Slab Design (AREMA Chapter 8)

Group I: DL + LL + I at 100% service load allowable stresses
Group II: DL + W at 125% service load allowable stresses
Group III: DL + LL + $I + 0.5W + W_{LL}$ at 125% service load allowable stresses

Since W and W_{LL} effects are small for deck design, only Group I load combinations will be considered for the reinforced concrete slab design.

Model as simply supported beam:

$V_{max} = 1514 + 8898 = 10{,}412$ lb/ft length of slab
$M_{max} = 4840 + 28{,}447 = 33{,}287$ lb-ft/ft length of slab
$M_{LL+I} = 28{,}447$ lb-ft/ft length of slab.

B5.3 Deck Slab Design

Negative bending moments over the girders are typically of much smaller magnitude than positive bending moments at the center of the slab and, in most cases, minimum reinforcement is required. The designer can confirm this for each specific design. The design of the reinforced concrete slab for positive bending moment at the center of the slab will be considered in this example.

$$d = 16 - 2 = 14'' \text{ (depth at center of slab = 16'' with 2'' cover).}$$

Bending strength:

$$f_r = 7.5\sqrt{f'_E} = 7.5\sqrt{5000} = 530 \text{ psi (rupture strength)}$$

$I_c = 12(16^3)/12 = 4096$ in.^{4}
$M_{cr} = 530(4096)/(12(16/2)) = 22{,}613$ lb-ft/ft length of slab (cracking moment)
$A_{smin} = 1.2M_{cr}/f_s(jd) = 1.2(22.61)(12)/(24(0.875)(14)) = 1.11$ in.^{2}/ft length of slab
$A_{st} \geq M_{max}/(f_s jd)$
$f_s = 0.40(60) = 24$ ksi

$A_{st} \geq 33.29(12)/(24(0.875)(14)) \geq 1.36$ in.²/ ft length of slab
Try #8 bars @ 6″ c/c, $A_s = (12/6)(\pi(1^2)/4) = 1.57$ in.²/ ft length of slab ≥ 1.36 in.²/ ft length of slab $\geq A_{smin}$, OK.

Reinforcement fatigue strength:

$\Delta\sigma = 28.45(12)/(1.57(0.875)(14)) = 17.5$ ksi
$f_{min} = 4.84(12)/(1.57(0.875)(14)) = 3.0$ ksi
$\Delta\sigma_{all} = 21 - 0.33f_{min} + 8(r/h) = 21 - 0.33(3.0) + 8(0.3) = 22.4$ ksi ≥ 17.5 ksi, OK.

Shear strength:

$w_{max} = 478 + 1.6(1756) = 3288$ psf
M @ $V_s = M_s = 3288(14/2)(9(12)\text{-}14)/144 = 15,024$ lb-ft/ft length of slab

$$\tau_c = 0.9\sqrt{f_c'} + 1100\rho_w \frac{Vd}{M} \leq 1.6\sqrt{f_c'}$$

$V_s d/M_s = 10,412(14/12)/15,024 = 0.81 \leq 1.0$, OK
$\rho_w = A_s/(bd) = 1.57/(12(14)) = 0.0093$

$$\tau_c = 0.9\sqrt{5000} + 1100(0.0093)0.81 = 63.6 + 8.29 = 71.9 \text{ psi} \leq 1.6\sqrt{5000} \leq 113 \text{ psi}$$

$\tau = 10,412/((12)(14)) = 62.0$ psi ≤ 71.9 psi, OK
use reinforced concrete slab 16 in. thick
$E_c = 4030$ psi for $f_c' = 5000$ psi
$E_s = 29,000$ psi
$n = E_s/E_c = 7.2$; use 7
$b' = ((14)(12)/2)/7 = 12$ in.

B6 GIRDER DESIGN

B6.1 General Information

Length of span = 107 ft
Length between bearing stiffeners = 105 ft
Distance between girders = 9 ft.

B6.2 Loads and Forces

B6.2.1 Dead Loads

Track = 200 lb/ft = 0.2 kip/ft per span
Ballast and waterproofing = (120)(2)(13) + 10(13) = 3250 lb/ft = 3.25 kip/ft per span (for 24 in. ballast depth)
Deck slab = 150(14)(16/12) = 2800 lb/ft = 2.80 kip/ft per span
Deck curb and walkway ≃ 150(2)(2) + 70 + 85(3) = 600 + 70 + 255 = 925 lb/ft = 0.93 kip/ft per girder (walkway dead load of 70 lb/ft and live load of 85 psf on 3′ wide walkway taken as dead load for girder design)
Girder = estimate as 90,000 lb = 90,000/107 = 841 lb/ft = 0.84 kip/ft per girder.

B.6.2.1.1 Noncomposite Section Dead Load (Concrete Deck and Girder)

$w_{DLNC} = 2.80/2 + 0.84 = 2.24$ kip/ft per girder
$V_{DLNC} = 2.24(105)/2 = 117.6$ kips
$M_{DLNC} = 2.24(105)^2/8 = 3078$ kip-ft.

B.6.2.1.2 Composite Section Dead Load (Track, Ballast, Waterproofing, Curb, and Walkway)

$w_{DLC} = 0.2/2 + 3.25/2 + 0.93 = 2.66$ kip/ft per girder
$V_{DLC} = 2.66(105)/2 = 139.7$ kips
$M_{DLC} = 2.66(105)^2/8 = 3666$ kip-ft
(assuming curbs poured after deck is cured).

B.6.2.1.3 Total Dead Load

$w_{DL} = 2.24 + 2.66 = 4.50$ kip/ft
$V_{DL} = 117.6 + 139.7 = 257.3$ kips
$M_{DL} = 3078 + 3666 = 6744$ kip-ft.

B6.2.2 Wind Load on Loaded Superstructure

300 lb/ft @ 8 ft above the top of rail

$w_{WLL} = 300\,(8)/5 = 480$ lb/ft $= 0.48$ kip/ft per girder
$V_{WLL} = 0.48\,(105)/2 = 25.2$ kips
$M_{WLL} = 0.48\,(105)^2/8 = 661.5$ kip-ft.

B6.2.3 Cooper's E 90 Live Load

$w_{LLV} = (90/80)\,5900$ lb/ft $= 6638$ lb/ft (Figure 5.19b)

$w_{LLM} = (90/80)\,5150$ lb/ft $= 5794$ lb/ft (Figure 5.22b)
$V_{LL} = 6.638(105)/2 = 349$ kips
$M_{LL} = 5.794(105)^2/8 = 7985$ kip-ft.

Impact:

Vertical impact:
$I_{max} = (0.90)24\% = 21.6\%$ (Figure 4.22)
$I_{mean} = 0.35(21.6) = 7.6\%$ (Table 4.9).
Rocking effect:
RE = 11.1%.

Total impact:

$I_{max} = 21.6\% + 11.1\% = 32.7\%$
$I_{mean} = 7.6\%$
$V_{LL+I(max)} = (1.33)\,349 = 463$ kips
$M_{LL+I(max)} = (1.33)\,7985 = 10{,}597$ kip-ft
$M_{LL+I(mean)} = (1.08)\,7985 = 8592$ kip-ft.

B6.2.4 Load Combinations for Girder Design (Table 4.10)

Load combination D1-A: DL + LL + I at 100% allowable stress:

$V_{DLNC} = 2.24(105)/2 = 117.6\,kips$
$V_{DLC} = 2.66(105)/2 = 139.7\,kips$
$V_{LL+I(max)} = (1.33)\ 349 = 463\,kips$
$V_{NC} = 117.6\,kips$
$V_C = 139.7 + 463 = 603\,kips$
$M_{DLNC} = 2.24(105)^2/8 = 3078$ kip-ft
$M_{DLC} = 2.66(105)^2/8 = 3666$ kip-ft
$M_{LL+I(max)} = (1.33)\ 7985 = 10{,}597$ kip-ft
$M_{NC} = 3078$ kip-ft
$M_C = 3666 + 10597 = 14{,}263$ kip-ft.

Load combination D1-B: LL + I at allowable fatigue stress:

$M_{range} = 8592$ kip-ft.

Load combination D2-A: DL + LL + I + W_L at 125% allowable stress:

$V_{NC} = 117.6\,kips$
$V_C = 139.7 + 463 + 25.2 = 628\,kips$
$M_{NC} = 3078$ kip-ft
$M_C = 3666 + 10597 + 661.5 = 14{,}925$ kip-ft.

Load combination D2-A may be neglected from girder design.

B6.3 Preliminary Design of Girders

B6.3.1 General

Depth of girder based on practical requirements that $L/15 \le D \le L/10$:

$105(12)/15 = 84$ in. $\le D \le 105(12)/10 = 126$ in. (average of 105 in.).

Assuming web dimensions based on stability (flexural buckling) requirements, for $F_y = 50$ ksi, $h/tw \le$ 135 (Equation 7.60b), and $h_{opt} = \sqrt[3]{\frac{3M}{2f_b}\left(\frac{h}{t_w}\right)} = \sqrt[3]{\frac{3(14,263)(12)}{2(0.55)(50)}(135)} = 108$ in. (Equation 7.39b), try $h = 108$ in.

B6.3.2 Web Plate

Web thickness for strength = $t_w = (117.6 + 603)/(0.35(50)(108)) = 0.38$ in.

Web thickness for flexural stability without longitudinal stiffeners = $t_w = \frac{h}{5.64}\sqrt{\frac{F_y}{E}} = \frac{108}{5.64}\sqrt{\frac{50}{29000}} = 108/135 = 0.80$ in. (Equation 7.60b).

Web thickness for shear stability without intermediate transverse stiffeners = $t_w = \frac{h}{2.12}\sqrt{\frac{F_y}{E}} = \frac{108}{2.12}\sqrt{\frac{50}{29000}} = 108/50.6 = 2.1$ in. (Equation 7.66).

The maximum spacing, a, of intermediate stiffeners is (Equation 7.103)

$$a \le 1.95 t_w \sqrt{\frac{E}{\tau}} \le 1.95 t_w \sqrt{\frac{29,000}{(720.6/108 t_w)}}.$$

For a practical stiffener spacing of 21(12)/3 = 84 in.

$$t_w \geq \left(\frac{a}{128.5}\right)^{2/3} \geq \left(\frac{84}{128.5}\right)^{2/3} \geq 0.75 \text{ in.}$$

Minimum plate thickness = 0.50 in.

Try a 108 in. × 0.75 in. web plate with transverse stiffeners.

Clear spacing between stiffeners if assume 6″ stiffener angle leg = 108 − 4.5 = 103.5 in. and web thickness to resist flexural buckling = 103.5/135 = 0.77 in. Therefore, use 0.75 in. thick web without longitudinal stiffener.

B6.3.3 Flange Plates

$A_f = \dfrac{M}{f_b h} - \dfrac{A_w}{6} = \dfrac{14263(12)}{0.55(50)(108)} - \dfrac{108(0.75)}{6} = 57.63 - 13.50 = 44.13 \text{ in}^2$ (for composite section moment) (Equation 7.35).

$A_f = \dfrac{8592(12)}{16(108)} - \dfrac{108(0.75)}{6} = 59.67 - 13.50 = 46.17 \text{ in}^2$. (for bolted stiffener moment range).

f_b = 16 ksi = allowable stress for fatigue Category B with no welded attachments and transverse web stiffeners bolted to the web.

$A_f = \dfrac{8592(12)}{12\ (108)} - \dfrac{108(0.75)}{6} = 79.56 - 13.5 = 66.06 \text{ in}^2$. (for welded stiffener moment range).

f_b = 12 ksi = allowable stress for fatigue Category C with transverse web stiffeners welded to the web.

Practical requirements that $D/4 \leq b \leq D/3$ for $D \simeq 113$ in. yield a flange width range of between 28 in. and 38 in. The flange thickness required for A_f ranges between 1.22 and 1.65 in.

Equation 7.58 controls local buckling of the top (compression) flange as

$$\frac{b}{2t_f} \leq 0.35\sqrt{\frac{E}{F_y}} = 8.4$$

$$t_f \geq \frac{b}{16.8}.$$

For b = 28 in., $t_f \geq 1.67$ in.

For b = 38 in., $t_f \geq 2.26$ in.

Try 1.75 × 28 top and bottom flanges, A_f = 1.75(28) = 49.0 in.^{2}

B6.4 Detailed Design of Girders

A 108 in. × 0.75 in. web plate with transverse stiffeners and 1.75 in. × 28 in. top and bottom flanges supporting composite 16 in. thick by 168 in. wide concrete deck with $n = 7$.

B6.4.1 Girder Section Properties

B6.4.1.1 Noncomposite Gross Section Girder Properties

The noncomposite girder gross section and properties are shown in Figure B.2 and Table B.1, respectively.

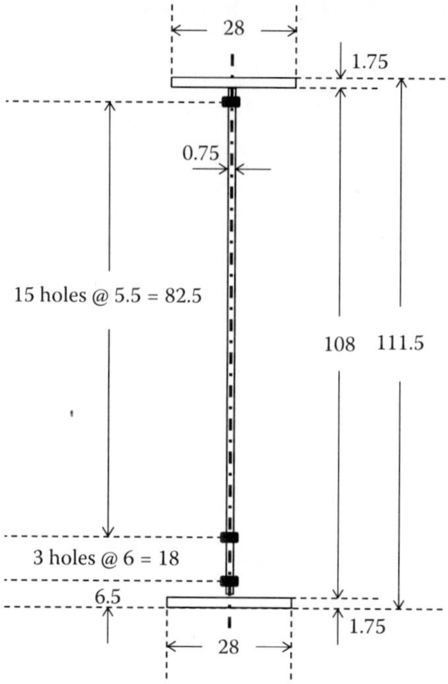

FIGURE B.2 Noncomposite girder cross section.

TABLE B.1
Noncomposite Girder Gross Section Properties

Element	Size	A (in.^{2})	y (in.)	A_y (in.^{3})	$y_s - y$ (in.)	$A(y_s - y)^2 \times 10^3$ (in.^{4})	$I_o \times 10^3$ (in.^{4})
Top flange	28 × 1.75	49.00	110.63	5420.6	−54.88	147.58	0.01
Web	108 × 0.75	81.00	55.75	4515.8	0	0	78.73
Bottom flange	28 × 1.75	49.00	0.875	42.88	54.88	147.58	0.01
Σ		179.00		9979.2		295.10	78.75

$y_g = 9979.2/179.0 = 55.75\,\text{in.}$
$I_g = (295.10 + 78.75)10^3 = 373.85 \times 10^3\ \text{in.}^4$
$S_{tf} = S_g = 373.85 \times 10^3/55.75 = 6706\ \text{in.}^3$

B6.4.1.2 Noncomposite Net Section Girder Properties

The noncomposite girder net section and properties are shown in Figure B.2 and Table B.2, respectively.

$$y_h = \frac{1089.5}{19} = 57.34\ \text{in.}$$

$y_n = (9979.2 - 0.75(1089.5))/(179.0 - 14.25) = 55.61\ \text{in.}$

AREMA recommends the use of the gross-sectional neutral axis.

$I_h = 14.25(57.34 - 55.75)^2 + 13140.2 = 36.0 + 13140.2 = 13{,}176\ \text{in.}^4 = 13.18 \times 10^3\ \text{in.}^4$
$I_n = I_g - I_h = (373.85 - 13.18) \times 10^3 = 360.67 \times 10^3\ \text{in.}^4$
$S_n = 360.67 \times 10^3/55.75 = 6470\ \text{in.}^3$ (3.5% reduction of S_g)

TABLE B.2
Noncomposite Girder Net Section Properties

Hole	*y* (in.)	*A* (in.²)	I_o (in.⁴)
1	6.5	0.75	1938.5
2	12.5	0.75	1508.0
3	18.5	0.75	1131.4
4	24.5	0.75	808.8
5	30	0.75	560.6
6	35.5	0.75	357.7
7	41	0.75	200.2
8	46.5	0.75	88.1
9	52	0.75	21.4
10	57.5	0.75	0
11	63	0.75	24.0
12	68.5	0.75	93.4
13	74	0.75	208.2
14	79.5	0.75	368.3
15	85	0.75	573.8
16	90.5	0.75	824.7
17	96	0.75	1120.9
18	101.5	0.75	1462.6
19	107	0.75	1849.6
Total	1089.5	14.25	13140.2

$S_{bf} = S_n = 6470$ in.^{3}
$A_w = 108(0.75) = 81.0$ in.^{2}.

B6.4.1.3 Composite Gross Section Girder Properties—Short-Term Loads, n = 7

The composite gross section and properties are shown in Figure B.3 and Table B.3, respectively.

Effective width of deck slab = $b_{eff} \leq 9/2 = 4.5$ ft = 54 in.
or $\leq 105/8 = 13.13$ ft = 157.5 in.
or $\leq 16(6) = 96$ in.
Deck slab overhang = $b_{over} = (168 - 108)/2 = 30$ in.
$b_{comp} = (b_{slab} + b_{over})/n = (54 + 30)/7 = 12$ in.
$y_c = 32923.3/371.0 = 88.74$ in.
$I_g = (671.6 + 82.85)10^3 = 754.46 \times 10^3$ in.^{4}
$S_{tc} = 754.46 \times 10^3/(127.5 - 88.74) = 19{,}465$ in.^{3} at top of slab
$S_{tf} = 754.46 \times 10^3/(127.5 - 16 - 88.74) = 33{,}149$ in.^{3} at top flange.

B6.4.1.4 Composite Net Section Girder Properties—Short-Term Loads, n = 7

The composite net section and properties are shown in Figure B.3 and Table B.4, respectively.

$$y_h = \frac{1089.5}{19} = 57.34 \text{ in.}$$

$I_h = 14.25(57.34 - 88.74)^2 + 27188.5 = 14050 + 27189 = 41{,}239$ in.^{4} = 41.24×10^3 in.^{4}
$I_n = I_g - I_h = (754.46 - 41.24) \times 10^3 = 713.22 \times 10^3$ in.^{4}
$S_{bf} = S_n = 713.22 \times 10^3/88.74 = 8037$ in.^{3}
$A_w = 108(0.75) = 81.0$ in.^{2}.

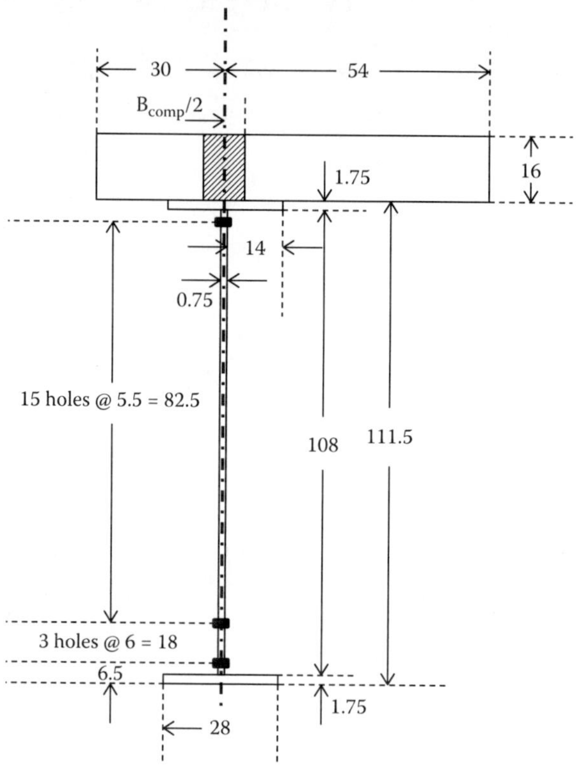

FIGURE B.3 Composite girder cross section.

TABLE B.3
Composite Girder Gross Section Properties (Short-Term Loads)

Element	Size	A (in.²)	y (in.)	A_y (in.³)	$y_s - y$ (in.)	$A(y_s - y)^2 \times 10^3$(in.⁴)	$I_o \times 10^3$ (in.⁴)
Deck slab ($n = 7$)	12 × 16	192.0	119.5	22944	−30.76	181.67	4.10
Top flange	28 × 1.75	49.0	110.63	5420.6	−21.89	23.48	0.01
Web	108 × 0.75	81.0	55.75	4515.8	32.99	88.16	78.73
Bottom flange	28 × 1.75	49.0	0.875	42.88	87.87	378.3	0.01
Σ		371.0		32923.3		671.6	82.85

B6.4.1.5 Composite Gross Section Girder Properties—Long-Term Loads, n = 21

The composite gross section and properties are shown in Figure B.3 and Table B.5, respectively.

$b_{comp} = (b_{slab} + b_{over})/n = (54 + 30)/21 = 4\,\text{in.}$
$y_c = 17627.3/243.0 = 72.54\,\text{in.}$
$I_g = (486.68 + 80.12)10^3 = 566.80 \times 10^3\ \text{in.}^4$
$S_{tc} = 566.80 \times 10^3/(127.5 - 72.54) = 10{,}313\ \text{in.}^3$ at top of slab
$S_{tf} = 566.80 \times 10^3/(127.5 - 16 - 72.54) = 14{,}548\ \text{in.}^3$ at top flange.

B6.4.1.6 Composite Net Section Girder Properties—Long-Term Loads, n = 21

The composite net section and properties are shown in Figure B.3 and Table B.6, respectively.

TABLE B.4
Composite Girder Net Section Properties (Short-Term Loads)

Hole	y (in.)	A (in.^{2})	I_o (in.^{4})
1	6.5	0.75	5072.6
2	12.5	0.75	4359.4
3	18.5	0.75	3700.2
4	24.5	0.75	3095.1
5	30	0.75	2587.8
6	35.5	0.75	2125.9
7	41	0.75	1709.3
8	46.5	0.75	1338.2
9	52	0.75	1012.4
10	57.5	0.75	732.0
11	63	0.75	496.9
12	68.5	0.75	307.2
13	74	0.75	163.0
14	79.5	0.75	64.0
15	85	0.75	10.5
16	90.5	0.75	2.3
17	96	0.75	39.5
18	101.5	0.75	122.1
19	107	0.75	250.1
Total	1089.5	14.25	27188.5

TABLE B.5
Composite Girder Gross Section Properties (Long-Term Loads)

Element	Size	A (in.^{2})	y (in.)	A_y (in.^{3})	$y_s - y$ (in.)	$A(y_s - y)^2 \times 10^3$(in.^{4})	$I_o \times 10^3$ (in.^{4})
Deck slab ($n = 21$)	4×16	64.0	119.5	7648	−46.96	141.1	1.37
Top flange	28×1.75	49.0	110.63	5420.6	−38.09	71.09	0.01
Web	108×0.75	81.0	55.75	4515.8	16.79	22.83	78.73
Bottom flange	28×1.75	49.0	0.875	42.88	71.67	251.66	0.01
Σ		243.0		17627.3		486.68	80.12

$$y_h = \frac{1089.5}{19} = 57.34 \text{ in.}$$

$I_h = 14.25(57.34 - 72.54)^2 + 16{,}432 = 3292 + 16{,}432 = 19{,}724 \text{ in.}^4 = 19.72 \times 10^3 \text{ in.}^4$
$I_n = I_g - I_h = (566.80 - 19.72) \times 10^3 = 547.1 \times 10^3 \text{ in.}^4$
$S_{bf} = S_n = 547.1 \times 10^3/72.54 = 7542 \text{ in.}^3$
$A_w = 108(0.75) = 81.0 \text{ in.}^2$.

B6.4.2 Design of Top Flange

B6.4.2.1 Design of Noncomposite Top Flange

Assume unshored construction because of waterway under superstructure.

TABLE B.6
Composite Girder Net Section Properties (Long-Term Loads)

Hole	y (in.)	A (in.^{2})	I_o (in.^{4})
1	6.5	0.75	3271.0
2	12.5	0.75	2703.6
3	18.5	0.75	2190.2
4	24.5	0.75	1730.9
5	30	0.75	1357.2
6	35.5	0.75	1029.0
7	41	0.75	746.1
8	46.5	0.75	508.6
9	52	0.75	316.4
10	57.5	0.75	169.7
11	63	0.75	68.3
12	68.5	0.75	12.2
13	74	0.75	1.6
14	79.5	0.75	36.3
15	85	0.75	116.4
16	90.5	0.75	241.9
17	96	0.75	412.8
18	101.5	0.75	629.0
19	107	0.75	890.6
Total	1089.5	14.25	16,431.8

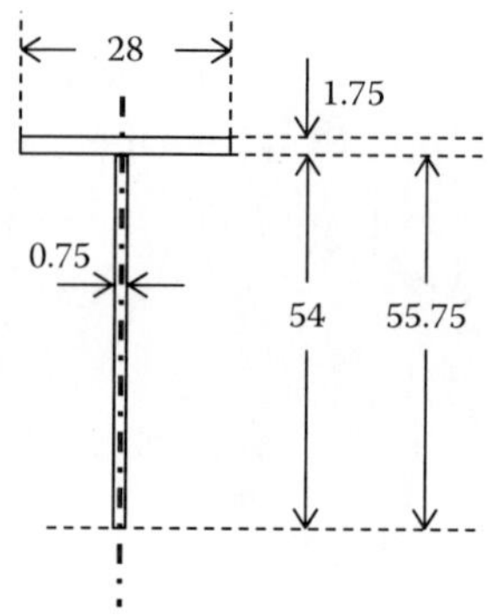

FIGURE B.4 Noncomposite girder cross section in compression.

Increase noncomposite DL on girder by 100 lb/ft to account for formwork and miscellaneous equipment).

$M_{DLNC} = 2.34(105)^2/8 = 3225$ kip-ft.

Assume vertical bracing between girders installed before deck pour with a maximum $L_p = 3(21)$ ft $= 63$ ft $= 756$ in. (maximum laterally unsupported length).

The section in compression during deck pouring (Figure B.4) has the properties:

$A_{yt} = (28)(1.75) + (54)(0.75) = 49.0 + 40.5 = 89.5$ in.^{2}

$I_{yt} = 1.75(28)^3/12 + 54(0.75)^3/12 = 3203$ in.^{4}

$$r_{yt} = \sqrt{\frac{I_{yt}}{A_{yt}}} = \sqrt{\frac{3203}{89.5}} = 5.98 \text{ in.}$$

$$\frac{L_p}{r_{yt}} = 756/5.98 = 126.4 \le 5.55\sqrt{\frac{E}{F_y}} \le 133.7. \qquad (7.43)$$

The allowable compressive stress is the larger of (Equations 7.44 and 7.45):

$$F_{call} = \left(\frac{0.13\pi E}{L_p d(\sqrt{1+\upsilon})/A_f}\right) = \left(\frac{0.13\pi(29{,}000)}{756(111.5)(\sqrt{1+0.3})/49.0}\right) = 6.0 \text{ ksi}$$

or

$$F_{call} = 0.55F_y - \frac{0.55F_y^2}{6.3\pi^2 E}\left(\frac{L_p}{r_{cy}}\right)^2 = 0.55(50) - \frac{0.55(50)^2(1000)}{6.3\pi^2(29{,}000)}\left(\frac{756}{5.98}\right)^2 = 27.5 - 12.2 = 15.3 \text{ ksi.}$$

$F_{call} = 15.3$ ksi.
$\sigma_{nctf} = 3225(12)/6706 = 5.8$ ksi, OK.

B6.4.2.2 Design of Composite Top Flange

B6.4.2.2.1 Design of Concrete Deck Top Flange $\sigma_{ctc} = 10597(1000)(12)/((7)19465)+(3225+3666)(1000)(12)/((21)10{,}313) = 933 + 382 = 1315$ psi $\le 0.4(5000) \le 2000$ psi, OK.

B6.4.2.2.2 Design of Steel Plate Top Flange Top flange plate continuously laterally supported for composite loads.

The allowable compressive bending stress = $F_{call} = 0.55F_y = 27.5$ ksi.

$\sigma_{ctf} = 5.8 + 3666(12)/14548 + 10597(12)/33149 = 5.8 + 3.0 + 3.8$ ksi $= 12.6$ ksi $\le 0.55(50) \le 27.5$ ksi, OK.

Use a 28 × 1.75 plate for top flange.

B6.4.3 Design of Web Plate

$F_{vall} = 0.35(50) = 17.5$ ksi
$\tau_w = 720.6/(0.75(108)) = 8.9$ ksi, OK
Use a 108 × 0.75 plate for web.

B6.4.4 Design of Bottom Flange

B6.4.4.1 Design of Noncomposite Bottom Flange

Increase noncomposite DL on girder by 100 lb/ft to account for formwork and miscellaneous equipment)

$M_{DLNC} = 2.34(105)^2/8 = 3225$ kip-ft
The allowable tensile stress = $F_{tall} = 0.55F_y = 27.5$ ksi
$\sigma_{ncbf} = 3225(12)/6470 = 6.0$ ksi, OK.

B6.4.4.2 Design of Composite Bottom Flange

$\sigma_{cbf(max)} = 6.0 + 3666(12)/7542 + 10597(12)/8037 = 6.0 + 5.8 + 15.8\,\text{ksi} = 27.6\,\text{ksi} \simeq 0.55(50) \leq 27.5\,\text{ksi}$, OK.

For $L > 100$ ft, the allowable fatigue stress is 18 ksi, Category B (no welded attachments)

$\sigma_{cbf(range)} = 8592(12)/8037 = 12.8\,\text{ksi} \leq 18\,\text{ksi}$, OK.

Use a 28 × 1.75 plate for bottom flange.

B6.4.5 Bending Stresses in the Composite Section

Top and bottom flange stresses are calculated for short-term dead loads on noncomposite ($n = 1$), long-term dead loads on composite ($n = 21$), and live loads on composite ($n = 7$) sections. The composite section resists long-term dead loads and live loads differently (due to concrete creep). The maximum flexural stresses are shown in Figure B.5.

B6.4.6 Design for Deck Maintenance

During deck maintenance under slow speed traffic, the noncomposite section should temporarily resist dead load, live load, and 50% of impact at no greater than 40% allowable stresses.

$w_{DL} = w_{NCDL} + w_{CDL} = 2.24 + 2.66 = 4.90$ k/ft
$M_{DL} = 4.90(105)^2/8 = 6753$ kip-ft
$M_{LL} = 7985$ kip-ft
$0.5M_I = 0.5(10597 - 7985) = 1306$ kip-ft
$M_{DM} = 6753 + 7985 + 1306 = 16{,}044$ kip-ft
$\sigma_{tf} = 16044(12)/6706 = 28.7\,\text{ksi}$.

During deck maintenance consider top flange supported by bracing at 21′

$$\frac{L_p}{+_{yt}} = \frac{21(12)}{5.98} = 42.1 \leq 5.55\sqrt{\frac{E}{F_y}} \leq 133.7. \quad (7.43)$$

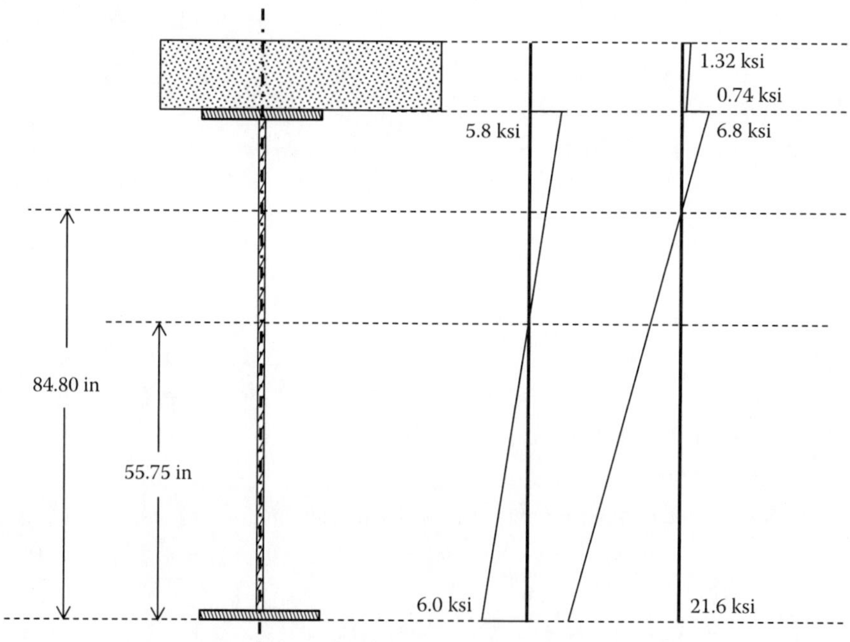

FIGURE B.5 Composite section flexural stresses.

The allowable compressive stress, increased by 40%, is the larger of (Equations 7.44 and 7.45):

$$F_{\text{call}} = 1.4\left(\frac{0.13\pi E}{L_{\text{p}}d(\sqrt{1+\upsilon})/A_{\text{f}}}\right) = 1.4\left(\frac{0.13\pi(29{,}000)}{252(111.5)(\sqrt{1+0.3})/49.0}\right) = 1.4(18.1) = 25.3\ \text{ksi}$$

or

$$F_{\text{call}} = 1.4\left(0.55F_{\text{y}} - \frac{0.55F_{\text{y}}^2}{6.3\pi^2 E}\left(\frac{L_{\text{p}}}{r_{\text{cy}}}\right)^2\right) = 1.4\left(0.55(50) - \frac{0.55(50)^2}{6.3\pi^2(29{,}000)}\left(\frac{252}{5.98}\right)^2\right)$$

$$= 1.4(27.5 - 1.4) = 1.4(26.2) = 36.6\text{ksi.}$$

$F_{\text{call}} = 36.6\,\text{ksi} \le 1.4(0.55(50)) \le 1.4(27.5) \le 38.5\,\text{ksi}$.
$\sigma_{\text{tf}} = 28.7\,\text{ksi} \le 36.6\,\text{ksi}$, OK.
$\sigma_{\text{bf}} = 16044(12)/6470 = 29.8\,\text{ksi} \le 1.4(0.55)(50) \le 38.5\,\text{ksi}$, OK.
$V_{\text{DM}} = 117.6 + 139.7 + 349 + 0.4(463 - 349) = 663\,\text{kips}$.
$\tau = 663/81.0 = 8.2\,\text{ksi} \le 1.4(0.35(50)) \le 1.4(17.5) \le 24.5\,\text{ksi}$, OK.

B6.4.7 Girder Deflection

B6.4.7.1 Deflection of Noncomposite Girder

Deflections may affect concrete pour and control final slab thickness.

$$\Delta_{\text{NC}} = \frac{5w_{\text{DLNC}}L^4}{384EI_{\text{nc}}} = \frac{5(2.34)(105(12))^4}{384(29000)(373{,}850)(12)} = 0.59\ \text{in.}$$

Unshored deflection of 0.59 in. should be mitigated with girder camber.

B6.4.7.2 Deflection of Composite Girder

$$\Delta_{\text{all}} = \frac{L}{640} = \frac{105(12)}{640} = 2.0\ \text{in.}$$

From Equation 5.47

$$\Delta_{\text{LL+I}} = \frac{5w_{e\Delta}L^4}{384EI} = \frac{0.104M_{\text{LL+I}}L^2}{EI} = \frac{0.104(10597)(12)(105(12))^2}{29000(754460\ )} = 1.0\ \text{in.} \le 2.0\ \text{in., OK.}$$

$\Delta_{\text{LL+I}} = \dfrac{L}{1260}$, which provides a stiff span that is appropriate for concrete deck performance.

B6.4.8 Flange to Web Fillet Welds

$V_{\text{DLNC}} = 117.6\,\text{kips}$
$V_{\text{DLC}} = 139.7\,\text{kips}$
$V_{\text{LL}+I} = 463\,\text{kips}$.

B6.4.8.1 Top Flange to Web Weld

The noncomposite girder gross section properties are:

$y_{\text{g}} = 9979.2/179.0 = 55.75\,\text{in}$.
$y_{\text{tfweld}} = 55.75\text{-}(1.75/2) = 54.88\,\text{in}$.

$I_g = (295.10 + 78.75)10^3 = 373.85 \times 10^3$ in.^{4}
$(Q/I)_{DLNC(n=1)} = (49.0)(54.88)/373.85 \times 10^3 = 7.19 \times 10^{-3}$ in.$^{-1}$
$q_{DLNC(n=1)} = 117.6(7.19 \times 10^{-3}{}_{)} = 0.85$ kips/in.

The composite ($n = 7$) gross section properties are:

$y_c = 32923.3/371.0 = 88.74$ in.
$y_{tfweldpl} = 127.5 - 88.74 - 16 - (1.75/2) = 21.9$ in. for the top flange plate
$y_{tfweldslab} = 127.5 - 88.74 - 8 = 30.76$ for the top flange slab
$I_g = (671.6 + 82.85)10^3 = 754.46 \times 10^3$ in.^{4}
$(Q/I)_{LL+I(n=7)} = (192.0(30.76)+49.0(21.9))/\ 754.46 \times 10^3 = 9.25 \times 10^{-3}$ in.$^{-1}$
$q_{LL+I(n=7)} = 463(9.25 \times 10^{-3}) = 4.28$ kips/in.

The composite ($n = 21$) gross section properties are:

$y_c = 17627.3/243.0 = 72.54$ in.
$y_{tfweldpl} = 127.5 - 72.54 - 16 - (1.75/2) = 38.1$ in. for the top flange plate
$y_{tfweldslab} = 127.5 - 72.54 - 8 = 47.0$ for top flange slab
$I_g = (486.68 + 80.12)10^3 = 566.80 \times 10^3$ in.^{4}
$(Q/I)_{DLC(n=21)} = (64.0(47.0) + 49.0(38.1))/566.80 \times 10^3 = 8.60 \times 10^{-3}$ in.$^{-1}$
$q_{DLC(n=21)} = 139.7(8.60 \times 10^{-3}) = 1.20$ kips/in.
$q_{tf} = 0.85 + 4.28 + 1.20 = 6.33$ kips/in.

A vertical shear force from live load is often considered for open and noncomposite deck construction. Nevertheless, the effect of this force will be considered for this composite deck design.

$q_w = 1.80(W)/5 = 1.80(100)/5 = 36.00$ kips/ft $= 3.00$ kips/in.

$$q_{tfw} = \sqrt{q_{tf} + q_w} = \sqrt{6.33^2 + 3.00^2} = 7.00 \text{ kips/in.}$$

Allowable weld stress = 19.0 ksi $\leq 0.35\ (50) \leq 17.5$ ksi.

$$t_{weld} \geq \frac{6.33}{19\sqrt{2}} \geq 0.24 \text{ in. on throat of fillet weld}$$

$$t_{weld} \geq \frac{6.33}{2(17.5)} \geq 0.18 \text{ on base metal.}$$

If vertical shear from direct live load is considered $t_{weld} \geq \dfrac{7.00}{19\sqrt{2}} \geq 0.26$ in.

B6.4.8.2 Bottom Flange to Web Weld

The noncomposite girder net section properties are:

$y_n = 55.75$ in.
$y_{bfweld} = 55.75\text{-}(1.75/2) = 54.9$ in.
$I_n = I_g - I_h = (373.85 - 13.18) \times 10^3 = 360.67 \times 10^3$ in.^{4}
$(Q/I)_{DLNC(n=1)} = 49.0(54.9)/360.67 \times 10^3 = 7.46 \times 10^{-3}$ in.$^{-1}$
$q_{DLNC(n=1)} = 117{,}6(7.46 \times 10^{-3}) = 0.88$ kips/in.

The composite ($n = 7$) net section properties are:

$y_c = 32923.3/371.0 = 88.74$ in.

$y_{\text{bfweldpl}} = 88.74 - (1.75/2) = 87.9$ in. for the bottom flange plate
$I_n = I_g - I_h = (754.46 - 41.24) \times 10^3 = 713.22 \times 10^3$ in.^{4}
$(Q/I)_{\text{LL}+I(n=7)} = 49.0(87.9)/713.22 \times 10^3 = 6.04 \times 10^{-3}$ in.$^{-1}$
$q_{\text{LL}+I(n=7)} = 463(6.04 \times 10^{-3}) = 2.79$ kips/in.

The composite ($n = 21$) net section properties are:

$y_c = 17627.3/243.0 = 72.54$ in.
$y_{\text{bfweldpl}} = 72.54 - (1.75/2) = 71.7$ in. for the bottom flange plate
$I_n = I_g - I_h = (566.80 - 19.72) \times 10^3 = 547.1 \times 10^3$ in.^{4}
$(Q/I)_{\text{DLC}(n=21)} = 49(71.7)/547.1 \times 10^3 = 6.42 \times 10^{-3}$ in.$^{-1}$
$q_{\text{DLC}(n=21)} = 139.7(6.42 \times 10^{-3}) = 0.90$ kips/in.
$q_{\text{bf(max)}} = 0.88 + 2.79 + 0.90 = 4.57$ kips/in.
$q_{\text{bf(range)}} = 2.79$ kips/in.

$$t_{\text{weld}} \geq \frac{4.57}{19.0\sqrt{2}} \geq 0.17 \text{ in. on throat of fillet weld}$$

$$t_{\text{weld}} \geq \frac{4.57}{2(17.5)} \geq 0.13 \text{ in. on base metal}$$

or

$$t_{\text{weld}} \geq \frac{2.79}{18\sqrt{2}} \geq 0.11 \text{ in. for Category B detail}$$

$$t_{\text{weld}} \geq \frac{2.79}{14.5\sqrt{2}} \geq 0.14 \text{ in. for Category B}' \text{ detail (if backing bar used).}$$

Use a ¼ in. fillet weld for top and bottom flanges to web T-joints. Some engineers specify CJP welds for deck plate girder spans.

B6.4.9 Concrete Deck to Steel Top Flange Plate Connection

$V_{\text{DL}} = 117.6 + 139.7 = 257.3$ kips
$w_{\text{DL}} = 2(257,3)/105 = 4.90$ kip/ft
$V_{\text{DL}}(x) = 4.90(52.5 - x)$
$V_{\text{LL}} = 349$ kips
$V_{\text{LL}+I} = 463$ kips.

From Equation 5.28:

$V_{\text{LL}} = w_{ev} b^2/2L$.

From Figure 5.19b for Cooper's E90 live load end shear on a 105 ft girder span:

$w_{ev} = 6650$ lb/ft for b = 105 ft.

From Figure 5.19b for Cooper's E90 live load quarter point shear on a 105 ft girder span:

$w_{ev} = 6975$ lb/ft for b = 78.75 ft
$V_{\text{LL}} = \text{w}_{ev}\text{b}^2/210$
$V_{\text{LL}+I} = (463/349)\ \text{w}_{ev}\text{b}^2/210 = \text{w}_{ev}\text{b}^2/158$.

Table B.7 uses Steinman's Chart (similar to Figure 5.25 and published by Transactions ASCE Vol. 86 for Cooper's E60 load) to obtain the equivalent uniform live load for shear, w_{ev}, at 10.5 ft intervals

TABLE B.7
Shear Force at Locations along Span

a (ft)	b (ft)	Wev (lb/ft)	Positive V_{LL+I} (kips)	Wheel 1 Shear (kips)	Corrected Positive V_{LL+I} (kips)	Negative V_{LL+I} (kips)	V_{range} (kips)	V_{DL} (kips)	V_{max} (kips)
0	105	6650	463	0	463	0	463	257	720
10.5	95.5	6795	392	−0.6	391	23.0	414	206	620
21.0	84.0	6940	310	−5.6	304	46.0	350	154	504
26.25	78.75	6975	274	−7.8	266	57.5	324	129	453
31.5	73.5	7090	242	−10.1	232	69.0	301	103	404
42.0	63.0	7260	182	−14.6	167	92.0	259	51.5	311
52.5	52.5	7700	134	−19.1	115	115	230	0	230

and at the quarter point (26.25 ft). In some cases, a correction must be made to the shear values from Steinman's Chart because the first wheel is neglected. The correction factor is $W(1 - ((b+8)/L))$. For $W = 45$ kips and $L = 105$ ft, the correction factor is $45(1 - ((b + 8)/105))$. The correction factor reduces the positive shear and applies when $b \leq 97$ ft. The effect of the correction is less than 10% up to the quarter point of the span and is a maximum of 14% at the center of the girder. The correction factor is often neglected as it is small and the calculated positive live load shear will be slightly conservative (see Chapter 5). The negative live load shear is estimated linearly from 0 to 115 kips.

The composite ($n = 7$) gross section properties are:

$y_c = 88.74$ in.

$y_{tfweldslab} = 127.5 - 88.74 - 8 = 30.76$ for top flange slab

$I_g = 754.46 \times 10^3$ in.^{4}

$(Q/I)_{slab} = 192(30.76)/754.46 \times 10^3 = 7.83 \times 10^{-3}$ in.$^{-1}$

$q_{range} = V_{range}(7.83 \times 10^{-3})$

$q_{max} = V_{max}(7.83 \times 10^{-3})$.

Use 7/8″ diameter × 6″ long shear studs at maximum 24″ spacing.

$A_{stud} = 0.60$ in.^{2}

$(L/d)_{stud} = 6/0.875 = 6.9 > 4$, OK.

Allowable maximum horizontal shear force $= S_m = 20{,}000(0.6) = 12{,}000$ lbs.

TABLE B.8
Shear Stud Spacing at Locations along Span

	Fatigue Stress		Maximum Stress			
a (ft)	V_{range} (kips)	s_{range} (in.)	V_{max} (kips)	s_{max} (in.)	s_{reqd} (in.)	s_{design} (in.)
0	463	6.6	720	7.1	6.6	6
10.5	414	7.4	620	8.2	7.4	6
21.0	350	8.8	504	10.1	8.8	8
26.25	324	9.5	453	11.3	9.5	8 or 10
31.5	301	10.2	404	12.6	10.2	10
42.0	259	11.8	311	16.4	11.8	12
52.5	230	13.3	230	22.2	13.3	12

Allowable fatigue horizontal shear force = S_r = 10,000(0.6) = 6,000 lbs (L > 100 ft).
Spacing of studs = $s_{range} = N(6.0)/q_{range} = 766.3N/V_{range}$
or = $N(10.0)/q_{max} = 1277N/V_{max}$.

Table B.8 shows the row spacing, s, required along the girder for four lines of 7/8 in. × 6 in. shear studs welded to each flange.

B6.4.10 Girder Intermediate Stiffeners

Intermediate stiffeners are required when (Equation 7.66)

$$t_w \le \frac{h}{2.12}\sqrt{\frac{F_y}{E}} = \frac{108}{2.12}\sqrt{\frac{50}{29000}} = 108/51.1 = 2.1 \text{ in.}$$

The maximum spacing, a, of intermediate stiffeners is (Equation 7.103)

$$a \le 1.95 t_w \sqrt{\frac{E}{\tau}} \le 1.95(0.75)\sqrt{\frac{29000}{8.9}} \le 83.5 \text{ in.}$$

The stiffener spacing used, a_o, must be

$\le h \le 108$ in., or
≤ 96 in., or
$\le a = 83.5$ in.
Stiffener spacing, a_o, is 21(12)/3 = 84 in.
a/h = 84/108 = 0.78.

The required stiffener moment of inertia (Equation 7.105) is

$$I_{ts} = 2.5 a_0 t_w^3 \left(\left(\frac{h}{a} \right)^2 - 0.7 \right) = 2.5(84)(0.75)^3 \left((1.29)^2 - 0.7 \right) = 84.44 \text{ in}^4.$$

Stiffeners will be used on both sides of the web at 84 in. spacing.
Try 6 × 4 × ½ angles on each side of the ¾ in. web plate.

$$I_{ts} = \frac{0.5(12.75)^3}{12} + \frac{3.5(1.75)^3}{12} = 87.92 \text{ in.}^4 \ge 84.44 \text{ in.}^4, \text{ OK.}$$

Use 6 × 4 × ½ angles on each side of the ¾ in. web plate.
Vertical bracing members will connect to the outstanding leg of interior stiffener angles.

$$b_{ts} = 6 - 0.5 = 5.5 \text{ in.}$$

$$\frac{b_s}{t_s} \le 0.50\sqrt{\frac{E}{F_y}} = 12.0$$

$$t_s \ge \frac{5.5}{12.0} \ge 0.46 \text{ in.}$$

The interior intermediate transverse stiffeners at vertical brace frames will be connected to the top flange with fillet welds.

B6.4.11 Girder Bearing Stiffeners

R_{DL}= 117.6 +139.7 = 257.3 kips
R_{LL} = 349 kips
R_{LL+I} = 463 kips
R_{max} = 257.3 + 463 = 720 kips.

The minimum web plate thickness is (Equation 7.80a):

$$t_w \geq \frac{R}{0.75F_y(L_B+k)} = \frac{720}{0.75(50)(24+1.75+0.25)} \geq 0.75 \text{ in. } \left(\text{for } L_B \geq 24 \text{ in.}\right)$$

Bearing stiffeners should be used as good practice and in case smaller bearing length, L_B, is used. Bearing stiffeners must be connected to both flanges and extend to near the edge of the flange on each side of the girder web plate.

B6.4.11.1 Girder Bearing Stiffener Design as Compression Member

Maximum bearing stiffener width ≃ (28 − 0.75)/2 = 13.6 in.

Use a 12 × 1 plate on each side of the girder web plate.

The bearing stiffener is designed as a compression member in accordance with Equations 7.81a and b:

$$A_{ebs} = 2A_{bs} + 12(t_w)^2 = 2(12)(1) + 12(0.75)^2 = 30.8 \text{ in.}^2$$

$$I_{ebs} = 2I_{bs} + 2A_{bs}\bar{y}^2 + t_w^4 = 2\left(\frac{1(12)^3}{12}\right) + 2(12)(1)(6.375)^2 + (0.75)^4 = 1120 \text{ in.}^4$$

$$r_{esb} = \sqrt{\frac{I_{esb}}{A_{esb}}} = \sqrt{\frac{1120}{30.8}} = 6.04$$

$$\frac{h}{r_{ebs}} = \frac{108}{6.04} = 17.9.$$

The allowable compressive stress, F_{call}, from Equations 7.83, 7.84a, or 7.85 depends on h/r_{ebs}. For $\frac{h}{r_{ebs}} \leq 0.839\sqrt{\frac{29{,}000}{50}} \leq 20.2.$

The allowable compressive stress, F_{call}, from Equation 7.83 is

$$F_{call} = 0.55F_y = 27.5 \text{ ksi.}$$

The allowable force on the bearing stiffener = $P_{call} = F_{call}(A_{ebs})$ = 27.5(33.0) = 908 kips ≥ 720 kips, OK.

B6.4.11.2 Girder Bearing Stiffener Design for Bearing Stress

The allowable bearing stress for milled stiffeners and parts in contact yields (Equation 7.87) $A'_{bs} \geq \frac{R}{0.83F_y} \geq \frac{720}{0.83(50)} \geq 17.35 \text{ in.}^2$

$A'_{bs} = 2(12 - 0.5)(1) = 23.0 \text{ in.}^2$, OK with ½ in. clearance for girder flange-to-web fillet welds.

B6.4.11.3 Girder Bearing Stiffener Design for Local Buckling Stress

For local buckling of the bearing stiffener plate with a free edge, Equation 7.88 indicates

$$t_{bs} \geq \frac{b_{bs}}{0.43}\sqrt{\frac{F_y}{E}} \geq \frac{b_{bs}}{0.43}\sqrt{\frac{50}{29{,}000}} \geq \frac{b_{bs}}{10.36}.$$

Width of bearing stiffener stressed to $0.55F_y$ (27.5 ksi) is

$$b_{bseff} = \frac{720/27.5-(12)(0.75)}{2(1)} = \frac{26.2-9.0}{2(1)} = 8.6 \text{ in.}$$

$$t_{bseff} \geq \frac{b_{bseff}}{10.36} \geq \frac{8.6}{10.36} \geq 0.83 \text{ in.}$$

For $b_{bseff} = 8.6$ in. and $t_{bseff} = 1$ in.:

$$f_c = \frac{720}{8.6(1/0.83)(2)(1)+9.0} = 24.2 \text{ ksi} \leq 27.5 \text{ ksi, OK.}$$

Use a 12 × 1 plate on each side of the girder web plate.

B6.4.11.4 Girder Bearing Stiffener Connection to Web Plate

$$t_{weld} \geq \frac{720}{(2)(108)19.0\sqrt{2}} \geq 0.12 \text{ in.}$$

Use ¼ in. fillet weld on each side of bearing stiffener.

B6.4.12 End Jacking Beam

Use end beam to jack up span. Place jacks at 2 ft from the centerline of girder.

B6.4.12.1 Loads and Forces

Load Combinations for Jacking Beam Design (Table 4.10)

Load combination C1: DL at 150% allowable stress:

R_{DL}= 117.6 +139.7 = 257.3 kips; use 280 kips for lifting force at each jack to account for weight contingencies

$M_{DL} = 280(2) = 560$ kip-ft.

B6.4.12.2 End Jacking Beam Design

Try W 24 × 104.

$d = 24$ in.
$A_f = 12.75(0.75) = 9.56$ in.^{2}
$A_w = (0.5)(24 - 2(0.75)) = 11.25$ in.^{2}
$r_y = 2.91$ in.
$L_p = 9 - 2(2) = 5$ ft $= 60$ in.
$S_x = 258$ in.^{3}
$\tau_w = 280/11.25 = 24.9$ ksi

$F_{vall} = 1.5(0.35)(50) = 26.3\,\text{ksi} \geq 24.9\,\text{ksi}$, OK
$\sigma_f = 560(12)/258 = 26.0\,\text{ksi}$.

The allowable compressive stress is the larger of

$$F_{call} = 1.5\left(\frac{0.13\pi E}{L_p d(\sqrt{1+\upsilon})/A_f}\right) = 1.5\left(\frac{0.13\pi(29{,}000)}{60(24)(\sqrt{1+0.3})/(9.56)}\right) = 103.4\text{ ksi}$$

or

$$F_{call} = 1.5\left(0.55F_y - \frac{0.55F_y^2}{6.3\pi^2 E}\left(\frac{L_p}{r_{cy}}\right)^2\right) = 1.5\left(0.55(50) - \frac{0.55(50)^2}{6.3\pi^2(29{,}000)}\left(\frac{60}{2.91}\right)^2\right)$$

$$= 1.5(27.5 - 0.8) = 40.1\text{ ksi}$$

$F_{call} = 1.5(0.55)(50) = 41.3\,\text{ksi}$ (maximum allowable compressive flexural stress) $\geq 26.0\,\text{ksi}$, OK.

B6.4.12.2.1 End Jacking Beam Bearing Stiffener Design Web crippling for jacking concentrated force:

$R_{max} = 280\,\text{kips}$ (for $L_F = 1.0$).

The minimum web plate thickness is (Equation 7.80b)

$t_w \geq \frac{P}{0.75F_y(L_B + 2k)} = \frac{280}{0.75(50)(4+2(1.5))} = 1.07$ in. (assuming a 4 in. × 4 in. steel jacking plate used), since $t_w = 0.5$ in. bearing stiffener required.

Maximum bearing stiffener width = (12.75 − 0.5)/2 = 6.13 in.

Try a 6 × ½ plate on each side of end jacking beam web at each jacking location (24 in. from the end of beam).

B6.4.12.2.1.1 *End Jacking Beam Bearing Stiffener Design as Compression Member* The bearing stiffener is designed as a compression member in accordance with Equations 7.81a and b:

$$A_{ebs} = 2A_{bs} + 12(t_w)^2 = 2(6)(0.5) + 12(0.5)^2 = 9.0\text{ in}^2.$$

$$I_{ebs} = 2I_{bs} + 2A_{bs}\bar{y}^2 + t_w^4 = 2\left(\frac{0.5(6)^3}{12}\right) + 2(6)(0.5)(3.25)^2 + (0.5)^4 = 72.44\text{ in}^4.$$

$$r_{esb} = \sqrt{\frac{I_{esb}}{A_{esb}}}\sqrt{\frac{72.44}{9.0}} = 2.84\text{ in.}$$

$$\frac{h}{r_{ebs}} = \frac{24 - 2(0.75)}{2.84} = 7.9.$$

The allowable compressive stress, F_{call}, from Equations 7.83, 7.84a, or 7.85 depends on h/r_{ebs}. For $\frac{h}{r_{ebs}} \leq 0.839\sqrt{\frac{29000}{50}} \leq 20.2$.

The allowable compressive stress, F_{call}, from Equation 7.83 is

$$F_{call} = 1.5(0.55F_y) = 41.3\text{ ksi}.$$

The allowable force on the bearing stiffener = $P_{call} = F_{call}(A_{ebs}) = 41.3(9) = 372\text{ kips} \geq 280\text{ kips}$, OK.

B6.4.12.2.1.2 *End Jacking Beam Bearing Stiffener Design for Bearing Stress* The allowable bearing stress for milled stiffeners and parts in contact yields (Equation 7.87) $A'_{bs} \geq \dfrac{R}{0.83F_y} \geq \dfrac{(280)}{0.83(50)} \geq 6.75\text{ in.}^2$ (with $L_F = 1.0$)

$A'_{bs} = 2(6-0.5)(0.5) = 5.50\text{ in.}^2$; a thicker stiffener is required with ½ in. clearance for rolled beam flange-to-web fillet.

Try a 6 × 5/8 in. plate on each side of the jacking beam web.

$$A'_{bs} = 2(6-0.5)(0.625) = 6.88\text{ in.}^2\text{, OK.}$$

B6.4.12.2.1.3 *End Jacking Beam Bearing Stiffener Design for Local Buckling Stress* For local buckling of the bearing stiffener plate with a free edge, Equation 7.88 indicates

$t_{bs} \geq \dfrac{b_{bs}}{0.43}\sqrt{\dfrac{F_y}{E}} \geq \dfrac{b_{bs}}{0.43}\sqrt{\dfrac{50}{29{,}000}} \geq \dfrac{b_{bs}}{10.36} \geq \dfrac{6}{10.36} \geq 0.58$ in., 0.625 in. thick bearing stiffener used, OK.

Use a 6 × 5/8 plate on each side of the end jacking beam web plate.

B6.4.12.2.1.4 *End Jacking Beam Bearing Stiffener Connection to Beam Web* $t_{weld} \geq \dfrac{280}{(2)(21)19.0\sqrt{2}} \geq 0.25$ in. (with $L_F = 1.0$).

Use ¼ in. fillet weld on each side of the bearing stiffener.

B6.4.12.2.2 End Jacking Beam Connection to Girder Bearing Stiffener $R_{max} = 280$ kips (for $L_F = 1.0$)

End jacking beam connected to bearing stiffener with bolts:
No. bolts single shear = 280/((1)(17.0)(0.60)) = 27.5
No. bolts double shear = 13.7
No. bolts bearing (on W 24 × 104 with $t_w = 0.50$ in.) = 280/((78)(0.875)(0.5)) = 10.3.
Use 14 bolts in girder web (with one connection angle) and bearing stiffener weld resisting 140 kips, and 14 bolts in jacking beam web.

Bearing stiffener welded to girder web:

$$\text{Weld size} \geq \frac{140}{2(\sqrt{2}/2)(19.0)(24)} = 0.22\text{ in. on throat of fillet weld each side of flange.}$$

$$\text{Weld size} \geq \frac{140}{2(17.5)(24)} = 0.17\text{ in. on base metal each side of flange.}$$

¼ in. fillet welds can transfer end jacking force combined with 14 bolts.
A5.4.3 End jacking floor beam weight
Weight of end floor beams = 2(104)(9) = 1872 lb.

B6.4.13 Girder Camber

$w_{DLNC} = 2.80/2 + 0.84 = 2.24$ kip/ft per girder
0.59 in. camber for noncomposite section deflection.
$w_{DLC} = 0.2/2 + 3.25/2 + 0.93 = 2.66$ kip/ft per girder

$$\Delta_{DLC} = \frac{5(2.66)(105(12))^4}{12(384)(29000)(754.46 \text{ x } 10^3)} = 0.33 \text{ in.}$$

Camber girder at center = 0.59 + 0.33 = 0.92 in., use camber of 1″ (cut into girder web).

B6.5 BDPG Span Bracing

B6.5.1 Loads and Forces

B6.5.1.1 Girder Compression Flange Notional Bracing Force

Top flange buckling fully restrained by composite concrete deck, bracing force, BF, = 0.

B6.5.1.2 Nosing Force

$N = 90/4 = 22.5$ kips applied at the top of rail.

B6.5.1.3 Wind Forces

w_{ul} = wind on unloaded superstructure = 50 psf applied on 1.5 times the surface area
w_l = wind on loaded superstructure = 30 psf applied on 1.5 times the surface area
W_t = wind on train = 300 lb/ft at 8 ft above the top of rail
V_B =150 lb/ft notional vibration load applied at bottom lateral bracing
V_B =200 lb/ft notional vibration load applied at top lateral bracing.

B6.5.1.4 Load Combinations for Bracing Design (Table 4.10)

Load combination D2-A: W_L + N at 125% allowable stress
Load combination D4-A: W_L or LV at 100% allowable stress
Load combination D4-B: W_{uL} at 100% allowable stress
Load combination D5-A: BF +N + W_L at 125% allowable stress.

B6.5.2 Top Lateral Bracing (Concrete Deck)

Top lateral bracing by concrete deck resisting wind and nosing forces.
Wind on unloaded span:

W_{TF} = ((111.5/2+16+18)/12)(50)(1.5) = 561 lb/ft wind force at top flange (additional wind load on curb).

Wind on loaded span:

W_{TF} = (89.8/12)(30)(1.5) = 337 lb/ft wind force at top flange (additional wind load on curb)
W_t = Wind on train = 300 lb/ft at the top of rail (applied to top lateral bracing)
V_B = 200 lb/ft notional vibration load at top bracing

Uniformly distributed transverse load on concrete deck = 337 + 300 = 637 lb/ft
Concentrated transverse load on concrete deck = 22.5 kips

V_{lat} = 637(105)/2 +22500 = 55,920 lbs =55.92 kips

$\tau_{lat} = 55920/16(168) = 21$ psi.

The 16 in. thick by 168 in. deep concrete deck is adequate for these loads.

B6.5.3 Bottom Lateral Bracing

Bottom lateral bracing by horizontal truss resisting wind forces. The lateral bracing panel length = 14 ft = 168 in. in the three end panels and 21 ft = 252 in. in the center panel as shown in Figure B.1.

B6.5.3.1 Bottom Lateral Bracing Loads and Forces

Wind on unloaded span:

$W_{BF} = (111.5/(12(2))(50)(1.5) = 348$ lb/ft wind force at bottom lateral bracing.

Wind on loaded span:

$W_{BF} = (111.5/(12(2))(30)(1.5) = 209$ lb/ft wind force at bottom lateral bracing
$V_B = 150$ lb/ft notional vibration load at bottom lateral bracing (horizontal truss)
Transverse wind shear = 348(14)/1000 = 4.9 kips in intermediate panels
Transverse wind shear = 348(21)/1000 = 7.3 kips in center panel
Transverse wind shear = 348(105/2)/1000 = 18.3 kips in end panels.

Load combination D2-A, D4-A, D4-B, and D5-A:

$P_i = 4.9$ kips at 100% allowable stress
$P_c = 7.3$ kips at 100% allowable stress
$P_e = 18.3$ at 100% allowable stress.

B6.5.3.2 Design of Lateral Bracing (Tension Only Members)

The forces in the bottom lateral members, assuming each member resists the panel shear force in tension, are shown in Table B.9.

$$L_{0-3} = \sqrt{9^2 + 14^2} = 16.6 \text{ ft} = 200 \text{ in.}$$

$$L_{3-3'} = \sqrt{9^2 + 21^2} = 22.9 \text{ ft} = 274 \text{ in.}$$

TABLE B.9
Bottom Lateral Bracing Forces

Panel	Wind Shear (kips)	Force in Diagonals of Panel (kips)
L0–L1	$\left(\frac{4.9(6)+7.3}{2}\right)$	$\frac{+(18.3)}{\sin(32.7)} = +33.8$
L1–L2	18.3 – 4.9 =13.4	$\frac{+(13.4)}{\sin(32.7)} = +24.8$
L2–L3	13.4 – 4.9 = 8.5	$\frac{+(8.5)}{\sin(32.7)} = +15.7$
L3–L3′	8.5 – 4.9 = 3.6	$\frac{+(3.6)}{\sin(23.2)} = +9.1$

$r_{min} \geq L/200$

$r_{min0-3} = 200/200 = 1.0$ in.

$r_{min3-3}' = 274/200 = 1.37$ in.

Try 8 × 8 × ½ angle.

$A = 7.80$ in.²

$r_{min} = 1.56$ in.

B6.5.3.2.1 Design of Bracing Member to Lateral Gusset Plate Connections P = lesser of ±(1.5) 33.8 = 50.7 kips or

0.55(50)(A_e)

No. of bolts ≥ 50.7/((17.0)(0.6)) ≥ 5.0 bolts single shear.

Use one row of 2 bolts and one row of 3 bolts.

B6.5.3.2.2 Tensile Design of Lateral Bracing Members $P = +50.7$ kips.

Considering net effective area and tensile yielding:

$A_n = 7.80 - 2(1)(0.5) = 6.80$ in.²

$U_c = 0.60$ for short bolted connection on one leg of bracing member only

$A_e = 0.60(6.80) = 4.08$ in.²

$F_{tall} = 0.55(50)(4.08) = 112.2$ kips > 50.7 kips, OK.

Use 8 × 8 × ½ with 5 bolts single shear each end of member.

B6.5.3.2.3 Design of Lateral Bracing Gusset Plate to Girder Web Connections Tension on connection = 50.7 (sin 32.7) = 27.4 kips

Shear at connection = 50.7(cos 32.7) = 42.6 kips.

Try 5 bolts.

$f_t = 27.4/(5(0.60)) = 9.13$ ksi

$f_v = 42.6/5(0.60)) = 14.20$ ksi

$$f_{vall} = 17.0\left(1 - \frac{9.13(0.60)}{39}\right) = 17.0(0.86) = 14.61 \text{ ksi} \geq 14.20 \text{ ksi, OK.}$$

Use minimum of 10 bolts (5 bolts in each adjacent panel) connecting lateral bracing gusset plate to girder web.

B6.5.4 End Vertical Brace Frames

B6.5.4.1 Loads and Forces

Uniformly distributed transverse load = 637 lb/ft

Concentrated transverse nosing load = 22.5 kips

End panel load = 22.5 + 637(105)/(2(1000)) = 55.9 kips

Consider 50% transferred to deck slab

$P_T = 0.5(55.9) = 28.0$ kips.

B6.5.4.2 Design of Diagonal Braces

Vertical projection of diagonal brace length ≃ 75 in.

$$F_{diag} \sim \frac{28.0\sqrt{(75)^2 + 54^2}}{54} \sim 28.0(1.71) = 47.9 \text{ kips } r_{min} > 0.75(92.4)/120 = 0.58 \text{ in.}$$

Try 4 × 4 × ½.
r_{min} = 0.767 in.
A = 3.77 in.²

$$\frac{kL}{r_{min}} = \frac{0.75(108)}{0.767} = 105.6 \le 5.034\sqrt{\frac{29000}{50}} \le 121.2$$

$$f_{call} = 0.60(50{,}000) - \left(17500\frac{50}{29000}\right)^{3/2}(90.4) = 30{,}000 - 165.74(90.4) = 30{,}000 - 14{,}975$$

$$= 15{,}025 \text{ psi} = 15.03 \text{ ksi}$$

F_{call} = 15.03(3.77) = 56.6 kips ≥ 47.9 kips, OK.
A_{net} = 0.60(3.77 − 1(0.5)) = 1.96 in.² (0.60 for shear lag at short bolted connection on one leg of bracing member only)
F_{tall} = 0.55(50)(1.96) = 54.0 kips ≥ 47.9 kips, OK.

B6.5.4.3 Diagonal Brace to Bearing Stiffener Connection

Connection to the bearing stiffener and gusset plate on end jacking beam:

Bolts bearing = 1.5(47.9)/(40(7/8)(0.5)) = 4.1 bolts
Bolts shear = 1.5(47.9)/(17.0(0.60)) = 7.0 bolts.

Use 7 bolts and a double shear connection with 4 bolts, and a larger angle will be used for the connection to the bearing stiffener and gusset plate.

Use double angle 6 × 4 × ½ diagonals at end frames for connection.

B6.5.4.4 Gusset Plate to Jacking Beam Connection

Gusset plate connection to jacking beam:

H_v = 55.9 kips
l_{weld} = 1.5(55.9)/((0.25)(17.5)(2)$^{1/2}$) = 13.6 in., use 14 in. minimum length of ¼ in. fillet weld.

Figure B.6 shows the end cross section of the BDPG span.

B6.5.5 Intermediate Vertical Brace Frames

B6.5.5.1 Loads and Forces

Intermediate (14 ft) panel load = 22.5 + 637(14)/1000 = 31.4 kips
Center panel (21 ft) panel load = 22.5 + 673(7+10.5)/1000 = 34.3 kips
Design braces for 34.3 kips and consider 50% transferred to deck slab

P_T = 0.5(34.3) = 17.1 kips.

B6.5.5.2 Design of Diagonal Braces

Vertical projection of diagonal brace length ≃ 97 in.

Diagonal braces:

$$F_{diag} \sim \frac{17.1\sqrt{(97)^2 + 54^2}}{54} \sim 17.1(2.06) = 35.2 \text{ kips}$$

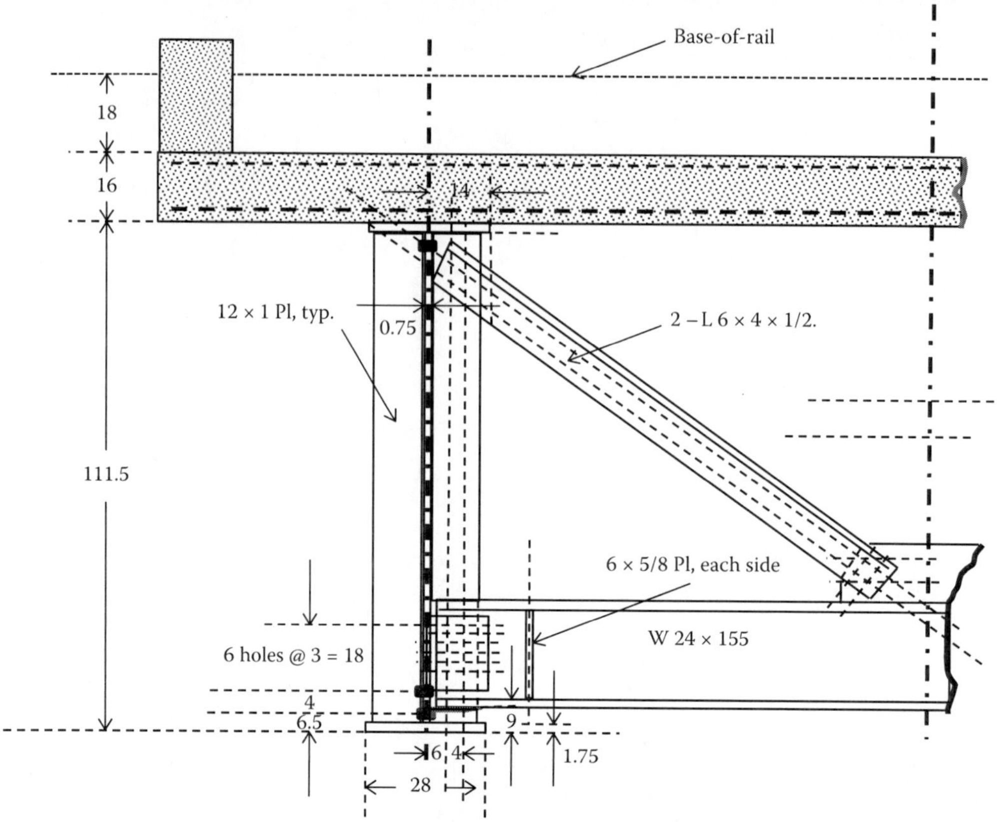

FIGURE B.6 Cross section of span at end brace frame.

$r_{min} > 0.75(111.0)/120 = 0.69$ in.
Try $4 \times 4 \times ½$.
$r_{min} = 0.767$ in.
$A = 3.77$ in.^{2}

$$\frac{kL}{r_{min}} = \frac{0.75(111.0)}{0.767} = 108.5 \le 5.034\sqrt{\frac{29000}{50}} \le 121.2$$

$$f_{call} = 0.60(50{,}000) - \left(17500\frac{50}{29000}\right)^{\frac{3}{2}}(108.5)$$

$$= 30{,}000 - 165.74(108.5) = 30{,}000 - 17{,}989 = 12011 \text{ psi} = 12.01 \text{ ksi}$$

$F_{call} = 12.01(3.77) = 45.3$ kips ≥ 35.2 kips, OK
$A_{net} = 0.6(3.77 - 1(0.5)) = 1.96$ in.^{2}
$F_{tall} = 0.55(50)(1.96) = 54.0$ kips ≥ 35.2 kips, OK.

B6.5.5.3 Diagonal Brace to Intermediate Stiffener Connection

Connection to intermediate stiffener:

Bolts bearing = 1.5(35.2)/(40(7/8)(0.5)) = 3.0 bolts

Bolts shear = 1.5(35.2)/(17.0(0.60)) = 5.2 bolts.

Use 6 bolts and a double shear connection with 3 bolts, and a larger angle will be used for the connection to the intermediate stiffener and gusset plate.

Use double angle 6 × 4 × ½ diagonals at intermediate frames for connection.

B6.5.5.4 Gusset Plate to Jacking Beam Connection

Gusset plate connection to jacking beam:

$l_{weld} = 1.5(34.3)/((0.25)(17.5)(2)^{1/2}) = 8.3$ in., use 9″ minimum length of ¼ in. fillet weld.

B6.5.5.5 Horizontal Transverse Brace

Horizontal transverse brace:

$F_{tran} = 17.1 + 1.5(50/1000)(111.5/(2(12))(7+10.5) = 17.1 + 6.1 = 23.2$ kips.
Try 4 × 4 × ½.
$r_{min} = 0.767$ in.
$A = 3.77$ in.^{2}

$$\frac{kL}{r_{min}} = \frac{0.75(92.4)}{0.767} = 90.4 \leq 5.034\sqrt{\frac{29000}{50}} \leq 121.2$$

$$f_{call} = 0.60(50{,}000) - \left(17500\frac{50}{29000}\right)^{3/2}(52.8)$$

$$= 30{,}000 - 165.74(105.6) = 30{,}000 - 17{,}503 = 12{,}497 \text{ psi} = 12.50 \text{ ksi}$$

$F_{call} = 12.50(3.77) = 47.1 \text{ kips} \geq 23.2 \text{ kips}$, OK
$A_{net} = 0.6(3.77 - 1(0.5)) = 1.96$ in.^{2}
$F_{tall} = 0.55(50)(1.96) = 54.0 \text{ kips} \geq 23.2 \text{ kips}$, OK.

B6.5.5.6 Horizontal Transverse Brace to Intermediate Stiffener Connection

Connection of horizontal transverse brace to intermediate stiffener:

Bolts bearing = 1.5(23.2)/(40(7/8)(0.5)) = 2.0 bolts
Bolts shear = 1.5(23.2)/(17.0(0.60)) = 3.4 bolts.

Use 4 bolts; a larger angle may be required for the connection to the intermediate stiffener.

B6.5.5.7 Gusset Plate to Jacking Beam Connection

Gusset plate connection to jacking beam:

$H_v = 34.3$ kips
Bolts bearing = 1.5(34.3)/(40(7/8)(0.5)) = 2.9 bolts
Bolts shear = 1.5(34.3)/(17.0(0.60)) = 5.0 bolts.

Use 5 bolts to connect the gusset plate to a horizontal transverse member of intermediate brace frames.

Figure B.7 shows the cross section of the BDPG span at an intermediate brace frame.

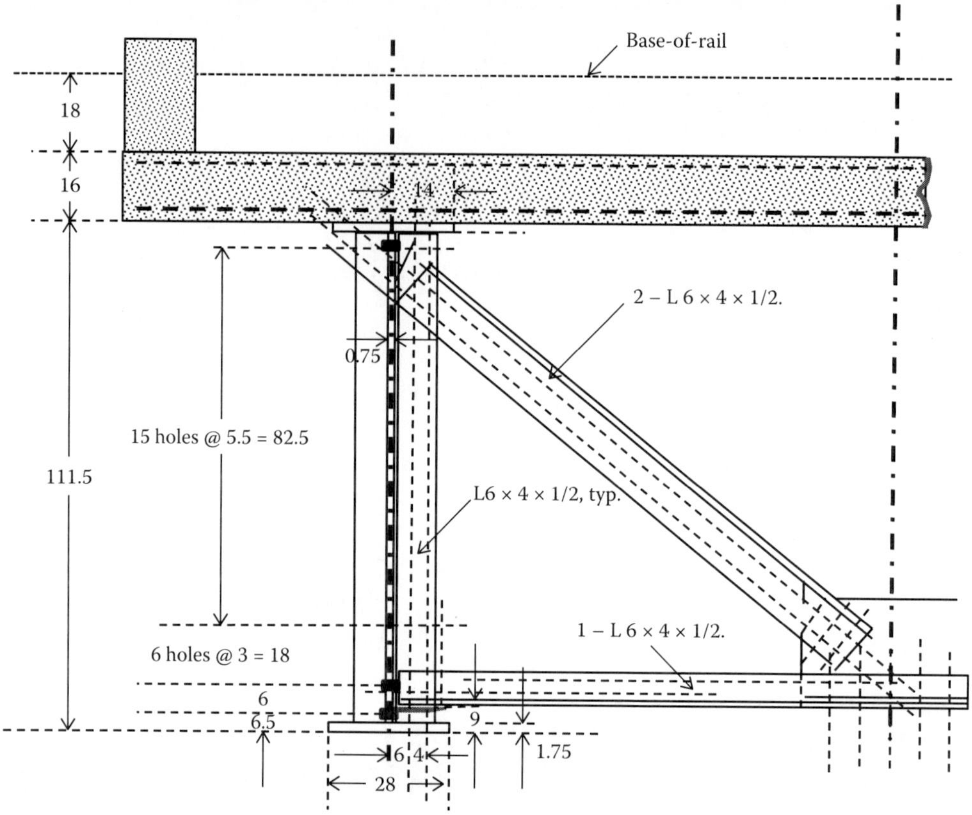

FIGURE B.7 Cross section of span at intermediate brace frame.

B7 SUMMARY OF BDPG DESIGN

B7.1 Deck

Deck: reinforced concrete with

1 – (((16 × 168) + 2(16 × 18))/144) 107 = 2425 ft^3
Weight of deck = 2425(150) = 363,750 lbs
Weight per girder = 181,875 lb
Load on girder = 1700 lb/ft/girder = 1.70 kip/ft/girder
Noncomposite load = 1.4 kip/ft/girder
Composite (curbs) = 0.3 kip/ft/girder
Bottom flexural reinforcement: #8 bars @ 6″ c/c.

B7.2 Girder and Stiffeners

Grade 350 weldable steel plate

B7.2.1 Top Flanges

2 – 28 × 1.75 × 107 PL (shop CJP splices), weight = 2(28)(1.75(107)(490)/144 = 35,682 lbs.

B7.2.2 Web Plates

2 – 108 × ¾ × 107 PL (shop CJP splices), weight = 2(108)(0.75)(107)(490)/144 = 58,984 lbs.

B7.2.3 Bottom Flanges

2 – 28 × 1.75 × 107 PL (shop CJP splices), weight = 2(28)(1.75(107)(490)/144 = 35,682 lbs.

B7.2.4 Girder Stiffeners

B7.2.4.1 Transverse Intermediate Stiffeners

56 – 6 × 4 × ½ L × 108, weight = 56(16.3)(9)= 8,215 lbs.

B7.2.4.2 Bearing Stiffeners

8 – 12 × 1 × 108 PL, weight = $8(12)(1)(108)(490)/12^3$ = 2940 lbs.

B7.2.5 End Jacking Beams

2 – W 24 × 104 × 108, weight = 2(104)(9) = 1,872 lbs.

B7.3 Span Bracing

B7.3.1 Bottom Bracing (Laterlal Truss Bracing)

6 – 8 × 8 × ½ *L* × 200, weight = 6(26.5)(16.7) = 2,655 lbs
1 – 8 × 8 × ½ *L* × 274, weight = 1(26.5)(22.8) = 604 lbs
6 – 6 × 4 × ½ × 108, weight = 6(16.3)(9) = 880 lbs
Bottom lateral bracing weight = 2655 + 604 + 880 = 4,139 lbs.

B7.3.2 Vertical Bracing

24 – 6 × 4 × ½ *L* × 111, weight = 24(16.3)(111/12) = 3,619 lbs
8 – 6 × 4 × ½ *L* × 75, weight = 8(16.3)(75/12) = 815 lbs.

B7.4 Superstructure Weight

Total estimated steel weights are summarized in Table B.10.

Estimated noncomposite concrete weight = 363,750 lbs = 3400 lbs/ft = 3.40 kips/ft/span = 1.70 kips/ft/girder.

Estimated composite concrete weight = 64,200 lbs = 600 lbs/ft = 0.60 kips/ft/span = 0.30 kips/ft/girder.

Estimated noncomposite steel weight = 155,248 lbs = 1451 lbs/ft = 1.45 kips/ft/span = 0.73 kips/ft/girder (0.84 kips/ft/girder estimated for design).

Estimated composite steel weight = 7,490 lbs = 70 lbs/ft/girder = 0.07 kips/ft/girder.

Estimated noncomposite total weight = 3.40 + 1.45 = 4.85 kip/ft/span = 2.43 kip/ft/girder (2.24 kips/ft/girder estimated for design = 8% low, OK for noncomposite design).

Estimated composite total weight (excluding track, ballast, and waterproofing) = 0.60 + 0.07 = 0.67 kip/ft/span = 0.34 kip/ft/girder.

Estimated composite total weight (including track, ballast, and waterproofing) = 0.67 + 3.25 + 0.2 = 4.12 kip/ft/span = 2.06 kip/ft/girder (2.66 kips/ft/girder estimated for design = 29% high, OK for composite dead load which is only about 25% of the total load on the girders).

Estimated total weight (including track, ballast and waterproofing) = 2.43 + 2.06 = 4.49 kips/ft/girder (4.90 kips/ft/girder estimated for design = 9% high, OK).

Estimated bracing systems weight = 11,147 lbs (< 7% of total estimated steel weight).

TABLE B.10
Estimated Weight of Steel

Span Component	Item	Size (in.)	Unit Weight (lbs)	No.	Total Weight (lbs)
Deck	Concrete slab and reinforcing	16 × 168 × 1284	363,750	1	363,750
	Concrete curbs and reinforcing	18 × 16 × 1284	32,100	2	64,200
	Walkway	36 × 1284	34,775	1	7,490
	Deck subtotal				462,725
Girders	Girder top flange plates	28 × 1.75 × 1284	17,841	2	35,682
	Girder web plates	108 × 0.75 × 1284	29,447	2	58,894
	Girder bottom flange plates	28 × 1.75 × 1284	17,841	2	35,682
	Girder intermediate web stiffeners	6 × 4 × 1/2*L*, 108 long	147	56	8,215
	Stiffener bolts	7/8 A325	0.67	≃500	336
	Girder end bearing stiffeners	12 × 1 × 108 PL	368	8	2,940
	Girder subtotal				141,749
Bracing and Jacking Beam	Bottom lateral members	8 × 8 × ½*L* 200	443	6	2,655
	Bottom lateral members	8 × 8 × ½*L* 274	604	1	604
	Bottom lateral members	6 × 4 × ½*L* 108	147	6	880
	Lateral bracing bolts	7/8 A325	0.67	≃150	100
	Intermediate vertical brace frame members	6 × 4 × ½*L*, 111 long	151	24	3,619
	End vertical brace frame members	6 × 4 × ½*L*, 75 long	102	8	815
	Vertical brace frame bolts	7/8 A325	0.67	≃200	134
	Bottom lateral gusset plates		≃180	8	1,440
	Vertical brace frame gusset plates		≃150	6	900
	Jacking beams	W 24 × 104 × 108 long	936	2	1,872
	Jacking beam connection angles		≃50	8	400
	Jacking beam bolts		0.67	≃120	80
	Bracing and jacking beam subtotal				13,499

(Continued)

TABLE B.10 (*Continued*)
Estimated Weight of Steel

Span Component	Item	Size (in.)	Unit Weight (lbs)	No.	Total Weight (lbs)
Subtotal noncomposite concrete weight estimate					363,750
Subtotal composite concrete weight estimate					64,200
Total concrete weight estimate					427,950
Noncomposite steel weight estimate	Assuming bracing and jacking beams installed prior to deck curing				155,248
Subtotal composite steel weight estimate	Walkway				7,490
Contingency	On noncomposite steel weight			5%	7,762
Total steel weight estimate					170,500

B8 DESIGN, FABRICATION, AND ERECTION DRAWINGS

Once the design calculations (design brief) are peer reviewed and acceptable, a design drafting technician can then, in conjunction with consultation with the design engineer, prepare design drawings. The design drawings are reviewed and approved by the design engineer.

The approved design drawings are forwarded to the fabricator for the production of shop (or detail) drawings (see Chapter 10). The shop drawings are reviewed by the design engineer and/or design drafting technician. This review does not constitute warranty that the shop drawings are accurate, which responsibility rests with the fabricator.

The design and shop drawings (depending on contract award schedule) may be used by the erector to design an erection methodology and procedure, and prepare the requisite erection drawings (see Chapter 11).

Appendix C: Units of Measurement

In most of the world, the Système Internationale (SI) units of measure are used. Few nations, with the notable exception of the Unites States, currently use Imperial (or US Customary) units of measure. Some countries, such as Canada, effectively practice steel superstructure design, fabrication, and erection in both SI and US Customary or Imperial units. Therefore, in this second edition, both SI and US Customary or Imperial units of measure are used. Some units are based on "soft" conversions available in the literature, but others are, by necessity, "hard" conversions based on the recognized conversion factors outlined in this appendix. All efforts at accuracy have been made, but conversion errors are not uncommon and may arise in any work, including superstructure design, fabrication, and erection.

The US Customary or Imperial systems of units rationally express engineering units in terms of length (foot, ft) and force (pound, lb). Mass (slug) is derived from Newton's second law of motion as slug = lb-s^2/ft. On the contrary, if engineering units in terms of length (foot, ft) and mass (pound-mass, lbm) are used, the derived unit, in accordance with Newton's second law of motion, is a poundal = 32.174 lb-s, which is awkward for engineering design, fabrication, and erection. The Imperial or US Customary ton = 2000 lb (force unit).

Nevertheless, the SI system of units expresses engineering units in terms of length (meter, m) and mass (kilogram, kg). Force (newton, N) is derived from Newton's second law of motion as newton = kg-m/s. The SI ton = 1000 kg (mass unit).

It is generally good practice that design, fabrication, and erection utilize a consistent system of units of measure. Nonetheless, in some cases, unit conversions must be made in the course of design and construction. Commonly used engineering units, based on the basic conversion factors of Tables C.1 and C.2, are indicated in Tables C.3 and C.4.

TABLE C.1

Basic SI Unit	Equivalent Imperial or US Customary Basic Unit
1 m	3.281 ft
1 N (force unit)	0.2248 lb (force unit)
1 kg (mass unit)	2.205 lbm (mass unit)

TABLE C.2

Imperial or US Customary Basic Unit	Equivalent SI Basic Unit
1 ft	0.3048 m
1 lb (force unit)	4.448 N
1 lbm (mass unit)	0.4536 kg

TABLE C.3
SI Units to Imperial (or US Customary) Units

Measure	SI Unit	Equivalent Imperial or US Customary Unit
Length	1 mm	0.0394 in.
	1 m	3.281 ft
Area	1 mm^2	0.00155 $in.^2$
	1 m^2	10.764 ft^2
Volume	1 m^3	61,024 $in.^3$
	1 m^3	35.315 ft^3
Inertia	1 mm^4	2.403×10^{-6} $in.^4$
	1 m^4	115.86 ft^4
Force	1 N	0.2248 lb (force)
	1 kN	0.2248 kip = 224.82 lb (force)
	1 N/m	0.0685 lb/ft
	1 kN/m	0.0685 kip/ft
Mass	1 kg	0.0685 slug (lb-s^2/ft)
	1 kg	2.205 lbm
	1 ton	1.102 ton (mass) = 2205 lbm
	1 kg/m	0.6720 lbm/ft
Speed	1 m/s	3.281 ft/s
	1 km/hour (kph)	0.6215 mile/hour (mph)
Acceleration	1 m/s^2	3.281 ft/s^2
Gravitational acceleration	9.807 m/s^2	32.174 ft/s^2
Stress (or pressure)	1 MPa (N/mm^2)	145.14 lb/$in.^2$ (psi)
	1 MPa (N/mm^2)	0.1451 kip/$in.^2$ (ksi)
	1 kPa (kN/m^2)	20.886 lb/ft^2
	1 kPa (kN/m^2)	0.0209 kip/ft^2
Bending moment (or torque)	1 N-mm	0.00885 lb-in.
	1 N-m	0.7377 lb-ft
	1 kN-m	0.7377 kip-ft
Density	1 kg/m^3	36.13×10^{-6} lbm/$in.^3$
	1 kg/m^3	0.0624 lbm/ft^3

TABLE C.4
Imperial (or US Customary) Units to SI Units

Measure	Imperial or US Customary Unit	Equivalent SI Unit
Length	1 in.	25.4 mm
	1 ft	304.8 mm
	1 ft	0.3048 m
Area	1 $in.^2$	645.16 mm^2
	1 ft^2	0.0929 m^2
Volume	1 $in.^3$	16,387 mm^3
	1 ft^3	0.0283 m^3
Inertia	1 $in.^4$	416,231 mm^4
	1 ft^4	0.00863 m^4
Force or weight	1 lb (force)	4.448 N
	1 kip = 1000 lb (force)	4.448 kN
	1 lb/ft (force)	14.594 N/m

(*Continued*)

TABLE C.4 (*Continued*)
Imperial (or US Customary) Units to SI Units

Measure	Imperial or US Customary Unit	Equivalent SI Unit
	1 kip/ft	14.594 kN/m
Mass	1 slug (lb-s^2/ft)	14.594 kg
	1 lbm	0.4536 kg
	1 ton (mass) = 2000 lbm	0.907 ton
	1 lbm/ft	1.488 kg/m
Speed	1 ft/s	0.3048 m/s
	1 mile/hour (mph)	1.609 km/hour (kph)
Acceleration	1 ft/s^2	0.3048 m/s^2
Gravitational acceleration	32.174 ft/s^2	9.807 m/s^2
Stress or pressure	1 lb/in.^{2} (psi)	0.00689 MPa (N/mm^2)
	1 kip/in.^{2} (ksi)	6.895 MPa (N/mm^2)
	1 lb/ft^2	0.04788 kPa (kN/m^2)
	1 kip/ft^2	47.880 kPa (kN/m^2)
Bending moment or torque	1 lb-in.	113.0 N-mm
	1 lb-ft	1.3556 N-m
	1 kip-ft	1.3556 kN-m
Density	1 lbm/in.^{3}	27,680 kg/m^3
	1 lbm/ft^3	16.0185 kg/m^3

Index

Note: Page numbers followed by *f*, *t* and *n* indicate figures, tables and notes, respectively.

A

AAR. *See* Association of American Railroads (AAR)
ABC. *See* Accelerated bridge construction (ABC)
Accelerated bridge construction (ABC), 541
Aerodynamic effects, 164
Air-carbon arc gouging, 527
AISC. *See* American Institute of Steel Construction (AISC)
Allowable stress design (ASD), 91, 92, 261
  allowable shear stress, 336
  factor of safety, 328
  in linear elastic analysis in, 376
Allowable weld stresses, 353*t*
Alloying elements, 40*t*
All-terrain cranes, 549, 550*f*
Aluminum, 40
American Institute of Steel Construction (AISC), 298, 506
American railway bridge
  all-wrought-iron bridge, 17
  engineering practice, 31
  steel railway bridges, 23
American Railway Engineering and Maintenance-of-Way Association (AREMA), 33–34, 35*f*, 99–100, 197*n*
  axial compression, 423
  bottom lateral bracing, 247
  bracing design force, 258
  bridge clearance, 69
  deck thickness, 86
  derailment load, 190
  design criteria, 91, 106, 144, 152, 168, 188
  dynamic forces, 125
  fracture toughness requirement, 50, 261*n*
  lateral force, 159, 187–188
  open bridge decks, 83, 143
  performance-based approach, 100
  recommendations, 50–51, 75, 91, 359–360, 423, 515–516, 540, 580
  scour design recommendations, 68–69
  seismic force, 187
  spacing, 90
  span types, 84
  static train live load, 106
  steel freight railway bridges, 100–101
  stress concentration factors for, 284*t*
  weathering steel, 53
  wind load, 168
American Railway Engineering Association (AREA), 33, 34, 35*f*, 115
American Society of Civil Engineers (ASCE), 24
  Structural Engineering Institute (SEI), 105
Amplitude cyclical loading, 119
Angle of twist (θ), 402
  in beams and girders, 330
  equations for, 402
  in torsion equation, 329
AREA. *See* American Railway Engineering Association (AREA)
AREMA. *See* American Railway Engineering and Maintenance-of-Way Association (AREMA)
ASCE. *See* American Society of Civil Engineers (ASCE)
ASD. *See* Allowable stress design (ASD)
Ashtabula Bridge, 20, 21*f*
  failure, 21
Assembly inspections, 533
Association of American Railroads (AAR), 160
Austenitic microstructure, 41
Automatic cutting, 503
Axial compression members, 291
  allowable compressive force, 300
  batten plates for, 309, 311
  bending and shear forces, 304
  and bending from concentrated transverse load, 417–418
  and bending from end moments, 419–420
  and bending from uniformly distributed transverse load, 416–417
  and biaxial bending, 423
  buckling coefficient, 300, 301, 302
  buckling strength, 306–308
  compression member design for steel railway superstructures, 302
  curves, 297, 299, 300, 301
  design, 262, 301, 308
  differential equation for, 415–420
  elastic buckling, 291–296
  elastic compression members, 291–296
  end conditions, 292, 293
  end restraint effect, 295
  and flexural loading, 420
  inelastic, 296
  inelastic buckling stress, 299
  inelastic compression members, 296–301
  interaction equations for, 420–423
  Johnson parabola, 299
  lacing bars for, 308
  perforated cover plates, 313
  serviceability of, 302–303
  slenderness ratio (C_{pc}), 313
  stay plates, 308
  in steel bridges, 302
  in steel railway superstructures, 304–314
    buckling strength of built-up compression members, 304–314
  strength of, 291
  and uniaxial bending, 414–423
  yielding of, 301
  yielding of compression members, 301

Axial forces
AREMA, 423
axial compression and uniaxial bending, 413, 414–423
bending, 413
combined, 413
influence lines for, 198*n*, 218–227, 234
loads on arch, 230–232
maximum, 231–232
moving loads, 230
Axial member splices, 482
beam and girder, 483
compression, 482
flange bolts, 483
tension, 482
web bolts, 483–495
Axial tension members, 277
bolted lacing bars, 290
design, 284, 288, 289–290, 302
effective net area (A_e), 279–280
fatigue strength of, 282–284
fatigue stress range, 283
lacing bar width, 289
net area (A_n) of, 278–280
serviceability, 284
shear deformation, 289
shear lag effects, 280
stay plates, 289
strength of, 277–292
stress concentration factors, 284
Axle spacing, 106

B

Baker, Benjamin, 16, 24
Ballasted deck plate girder (BDPG) span, 87, 140, 199, 201, 245, 380
deck slab, 635–638
general information, 635
girder design, 638–664
layout of proposed superstructure, 635
loading and materials, 635
scope of, 635
summary of, 664–667
Ballasted deck steel bridges
distribution of live load for, 160–162
Ballasted steel plate decks, 85–86
BDPG span, 87
BTPG span, 87
drainage consideration, 86
Ballasted through plate girder (BTPG) span, 87, 96, 199, 201, 360, 511*n*, 541*n*
floor system design, 596–610
general information, 595
girder design, 610–632
layout of proposed superstructure, 595–596
loading and materials, 595
scope of, 595
summary of, 632–634
Baltimore and Ohio (B&O), 1
Baltimore truss, 18, 19*f*
Barges
erection by flotation with, 554–556
stability, 566, 582
Batten plates, for compression members, 311–313
BDPG span. *See* Ballasted deck plate girder (BDPG) span
Beams and girders
beam bending, 328, 329
bending of laterally supported, 327–329
bending of laterally unsupported, 329–333
biaxial bending of, 336–337
box girder design, 360
equilibrium equation, 327, 329
girder flange proportioning, 337
lateral–torsional buckling curve, 332
lateral–torsional buckling moment, 330
lateral–torsional buckling stress, 331
parabolic transition equation, 332
plate girder design, 339–360
preliminary design of, 337–339
shear flow, 334
shearing of, 333–336, 378
shear stresses, 333–336
torsional warping constants, 330
Bearing components, expansion, 89
elastomeric bearings, 89
expansion hinged bearings, 89
flat steel plates, 89
linked bearings, 89
PTFE plates, 89
roller bearings, 89
Bearing stiffeners
bearing stresses, 355
compression member behavior, 354–355
cross section, 354
design, 355
girder, 353
local buckling, 356
web crippling, 353–354
Bearing stress, 453, 454
shear block failure, 453
shear block yield strength (P_y), 454
Bending
axial compression and biaxial, 423
axial compression and uniaxial, 414
axial tension and uniaxial, 413
Bending and shear, 346, 424
biaxial, 399
combined axial forces, 413–424
compression and biaxial bending, 423
compression and uniaxial bending, 414
differential equations, 415–420
general loading of, 415
interaction equations, 420–423
members, 413
plates, 424
plates shear and, 424
unsymmetrical, 401
unsymmetrical cross section, 400
Bending moment(s), 75, 199–212, 202, 419
change in, 201
on connections, 436, 471
Cooper's E80 load, 202
criteria for, 201
determination, 206–212, 216–218
equivalent uniform loads, 234, 238, 239, 240
influence line, 218, 227–229, 230, 232–233, 234
loads on arch, 230
maximum, 201–202, 214, 241–243

mid-span, 111–114, 120
moving load analysis, 198
simply supported spans, 244
supported spans, 237–240, 241–243
tangent track, 75
Bending of beam, 328
in buckled position, 329
laterally supported, 327–329
laterally unsupported, 329
lateral–torsional buckling curve, 332
lateral–torsional buckling stress, 331
out-of-plane bending, 329
torsional warping constants, 330
torsion equation, 329
Bernoulli–Euler equation. *See* Partial differential equation of motion
Bernoulli's equation, 164
Biaxial bending, 399–400
and axial compression, 423
beams and girders, 336–337
modulus of steel elasticity, 399
symmetrical axes, 400
unsymmetrical cross section, 400
Bills of materials (BOM), 497
Blackwell's Island (Queensboro) Bridge, New York, 28
Blocking, 592–593
Bollman truss, 10, 18*n*
bridge, 18
redundant nature of, 18*n*
Bolted connections, 446
allowable bolt stresses, 448
axially loaded members, 459
bolted joint design, 448–495
bolting processes, 446–447
bolt types, 447–448
joint types, 448
Bolted field splices and connections, erection of, 590–592
Bolting inspections, 534
Bolting, of plates and shapes, 518–520
Bolting, steel, 446
bolt pretension, (T_{bp}), 447
bolt types, 447
eccentrically loaded connections, 471
joint design, 448
joint types, 448
snug-tight bolt, 446
BOM. *See* Bills of materials (BOM)
Bottom lateral systems, 247–249
Box girder design, 339, 360
steel, 360
steel–concrete composite, 360
Bracing, 256*n*
cross, 159, 160
horizontal truss, 246
intermediate vertical, 255
knee, 256–258
lateral, 154–155, 160, 164, 190, 246, 247
portal, 250–255
sway, 255
Braking forces, 146
Bridge, 3–4*t*
arch, 21
axial force ($N(x)$) magnitude, 172
Bessemer steel, 25
crossing economics, 56
decks, 84–87
esthetics, 82–83
geometrics, 74–77
joint design, 101, 430, 448
stability, 90
Bridge crossing
economics, 56–57
hydraulics, 59
hydrology and hydraulics of, 58–69
public and technical requirements of, 58–70
scour at, 65–67, 65*f*
Bridge planning, 56
bridge crossing economics, 56
Bridge, skewed, 80–81
butt joints, 448
cantilever, 21–22, 24*n*, 26, 29
cast iron, 2, 8
CB, 56
centrifugal forces, 155–158
combined stress, 399
compressive design (σ_{ac}), 462
constricted discharge hydraulics, 59–62
contraction scour depth (d_c), 66
Cooper's specification, 33
corner joints, 448
critical sections review, 462
crossings, 65–67
CWR design, 170, 186–187
dead load, 105
deflection, 25
design, 32, 33, 57–58, 80, 186–187, 269
design impact load, 144–145
direct fixation decks, 86–87
ductility, 45
economical span length (l), 56
elastomeric bearings, 89
erection, 102
expression of function, 83
fabrication process, 100–101
first specification, 33
force consideration, 105
freight, 100–101
general design criteria, 33, 91, 102
geometry of track and bridge, 70–71
geotechnical investigation, 69–70
Gerber type, 24*n*
girder, 10–11
HSLA steels in, 47–48
hydraulic assessments, 58–69
iron, 2
local scour depth (d_l), 66, 67
material properties, 49–50
obstructed discharge hydraulics, 62–64
operating requirements, 57–58
parameters affecting, 142–143
pinned connections, 17
planning, 56
primary purpose, 55
regulatory requirements, 58
relationships, 173
river crossing profile, 59–60
scientific approach, 31
seismic dynamic analysis, 187

Bridge, skewed, (*cont.*)
site conditions, 58
spans, 80–81
square track support, 81
steel, 23
steel freight, 100–101
streambed velocity (V_s), 66
structural steel, 49–54
suspension, 16
timber, 10*n*, 19*n*
total scour depth (d_t), 67
tubular, 11
wind forces on steel, 164–170
Bridge, steel, 24, 29, 197
allowable bearing stress, 453, 454
allowable combined stress, 457–458
allowable shear fatigue stress, 458–459
allowable shear stress, 449, 452
allowable tensile fatigue stress, 458–459
allowable tension stress, 455
axial compression members in, 302
axial stress, 461
axial tensile design (σ_{at}), 462
axial tension members, 277
ballasted steel plate decks, 85–86
basic forms, 84
bearings, 88–90
bending moment, 198, 201, 202, 244
block shear, 460
cantilever bridges, 23, 25, 26–27, 28
compression member design, 301
design, 23, 197, 258, 262, 269
equivalent loads, 234
esthetics, 82–83
failure, 261–262
fatigue damage, 261, 268
fracture, 261
Fraser River Bridge, 24, 25
Hell Gate Bridge, 28, 30
influence, 213
lateral bracing systems, 246
lateral load analysis, 246
live load analysis, 197
moving loads, 200
pneumatic caisson method, 23
Quebec Bridge, 28, 29
serviceability design, 264
serviceability failure, 261
shear force, 199–201, 244
static analysis, 197
static and dynamic effects, 197
St. Lawrence Bridge, 30
St. Louis Bridge, 23
strength design, 262
strength failure, 261
structural analysis, 197, 245–246
tension members in, 289
vertical post end connections, 425
web members in trusses, 274
Bridge, steel, loaded members, 459
axial member end connection, 459
block shear strength, (P_{vs}), 460
bolt spacing (s_b), 459
edge distance, (e_b), 459
Bridge, steel, seismic forces on, 187
combined seismic design force, 187–188
equivalent static lateral distributed force ($p(x)$), 187
static analysis, 187
Bridge, steel, vertical effects on, 128–144
affecting parameters, 142–143
AREMA design, 169
AREMA design impact, 142
Bernoulli's equation, 164
C_D and *Re* relationship, 166
drag coefficient (C_D), 166–168
dynamic pressure (p), 164
empirical values, 139
equation of motion, 131
fatigue design, 144
FEA, 142
first vibration mode, 137–138
forced vibration solution, 139
Fourier transformation, 137
long-span bridge design, 140
mean impact loads, 145
moving concentrated load, 138
moving continuous load, 134, 136, 140
moving force equation, 138
moving train model, 125
Reynolds number (*Re*), 165–166
steady state solution, 140
undamped frequencies, 141
unloaded frequencies, 141, 142
wind flow, 165
Bridge superstructures, 87
Britannia Bridge, 11, 11*n*, 15, 15*f*, 31*n*
Brunel, 11*n*
BTPG span. *See* Ballasted through plate girder (BTPG) span
Buckling strength, 306–308
of built-up compression members, 304–314
elastic, 291, 295, 347
inelastic, 296, 299, 348
Butt and "T" joints, 426

C

CAD applications. *See* Computer aided design (CAD) applications
CAFL. *See* Constant-amplitude fatigue limit (CAFL)
Cambered superstructure assemblies, 509–511
Cantilever, 540
arm, 232
benefits, 22
Brunel, 11
deflection, 259–260
first bridge, 16, 23–24
method, 22, 24
span, 24*n*, 233–234
steel deck truss, 24
Telford, Thomas, 24*n*
trussed cantilever, 24
Cantilever construction, erection by, 564
engineering for, 586
Cantilever railway bridges, 26–27*t*
Carbon, 39
Carbon equivalence (CE), 46
Carbon equivalency equation, 46

Cast iron construction, 2
arch bridges, 2, 3–4*t*
Gaunless River Bridge, 2
trestles, 2, 5–7*t*
Catenary high-line, erection by, 565
engineering for, 586
CE. *See* Carbon equivalence (CE)
Centrifugal forces (CF), 155–158
calculation, 156
curved track, 155
determination of, 157
horizontal geometry, 71
lateral, 125
route geometrics, 71
CF. *See* Centrifugal forces (CF)
Charpy V-Notch (CVN), 46, 502*n*
CI. *See* Corrosion index (CI)
CJP. *See* Complete joint penetration (CJP)
Clapyron, B. P. E., 31
Clip angles, 476
CNC operations. *See* Computer numerically controlled (CNC) operations
Coatings
inspections, 535
of steel plates and shapes, 529–530
Column Research Council (CRC), 299
Combined forces, 187*n*
Commercial blast cleaning, 529
Common steel bolts, 447
Complete joint penetration (CJP), 101, 428, 514
Compliance, 543
Composite flexural members, 374
beams and girders shearing, 378
composite spans, 376
cross section, 375
flexural stress, 376
serviceability design of, 381–397
shear connection, 378
shear flow distribution, 379
shear resistance distribution, 381
slip and slip, 377
Compression member curve(s), 297
CRC and AREMA, 301
parabolic and linear, 300, 301
residual stress distributions, 299
strength curves, 302
weak and strong axis, 299
Compression members, built-up
AREMA minimum shear force, 308
batten plates design, 308, 311–313
buckling strength, 304, 306–308
comparison of designs, 320*t*
lacing bars design, 308
lacing bar thickness, 308
perforated plates design, 308, 313–320
Poisson's ratio (υ), 305
shear curvature (γ_v), 304
shear force ($V(x)$), 304
stay plates design, 308–309, 311
stay plate thickness, 311
Computer aided design (CAD) applications, 499
Computer numerically controlled (CNC) operations, 499
Connection assembly, 513
Constant-amplitude cycles, 119, 124, 134, 268–270
Constant-amplitude fatigue limit (CAFL), 269
Constricted discharge hydraulics, 59–62
constriction openings, 59
contraction coefficient, (C_c), 61
friction loss upstream (h_f), 61
Froude number (F), 61
minimum channel area (A), 60
USGS method, 61
Continuous span railway bridges, 17*t*
Continuous welded rail (CWR), 131
bridge thermal interaction, 170–171
forces, 170–186
Contraction at constrictions and obstructions, 64
contraction coefficient, (C_c), 64
Contraction scour, 66
Conwy Bridges, 15*f*
Cooper's (E80), 90, 106–124, 145, 150. *See also* Live load
bending moment, 202–212, 216–218, 268
design live load margins, 117
for projected railway equipment, 115–118
Cooper's E90
design live load, 115–118, 118*f*
Cooper's EM360, 106*n*, 107, 268
Corrosion index (CI), 47
Corrosion-resistant steel, 46–47, 48, 448, 529
load pattern, 203
maximum shear force, 234
wheel load, 202, 204–205
Cover plates, 266*n*, 289*n*
Crack growth behavior, 121
Cranes and derricks, erection with, 544–551
Crawler cranes, 547, 548*f*, 556, 576–577
CRC. *See* Column Research Council (CRC)
Critical buckling coefficients, 301*t*
Cugnot, Nicolas, 1*n*
Culmann, Karl, 19
Curve(s)
compression member, 297, 299, 300, 301, 302
geometry, 71
interaction, 421
moment–rotation, 438, 439, 475
S–N, 270, 272
stress–strain, 42
transition, 74
CVN. *See* Charpy V-Notch (CVN)
CWR. *See* Continuous welded rail (CWR)
Cyclical stress ranges, 110

D

d'Alembert's principle
of dynamic equilibrium, 135
d'Aubuisson equation, 64
differential equation, 171
failure criteria, 172
rail-to-deck-to-superstructure, 172, 175–186
safe rail separation criteria, 172–173
safe stress in CWR, 173
three-span bridge, 174
typical relationships, 174, 175
Dead loads, 105–106
anticipated future, 105
fit-up, 589
on steel railway bridges, 105, 106*t*

Deck plate girder (DPG), 86, 199, 201, 206, 245, 352
Deck trusses
  beam framing connections, 425
  camber, 274
Degree of freedom (DOF), 128, 133
Department of Transportation Federal Highway Administration (US), 505
Derailment load, 190–192
Derrick cranes, 544–547
Design, steel bridges. *See* Bridge planning; Bridge, steel
DFT. *See* Dry film thickness (DFT)
Diesel locomotive, 34
Differential equation, 415
  end moments, 419
  equivalent loads, 234
  failure, 261
  flexural loading, 420
  girder bearing, 353
  influence lines, 213
  instability criteria, 346
  interaction curves, 421
  lateral bracing systems, 246
  lateral load analysis of, 246
  live load analysis of, 197–198
  maximum shear force, 199, 245
  minimum thickness, 274
  nominal, 354
  other design criteria, 273–274
  secondary and bracing member connections, 425
  secondary stresses, 273–274
  serviceability design, 264
  strength design, 262
  web plate, 356–360
  web stiffeners, 357
Direct fixation deck steel bridges, distribution of live load for, 162–164
Direct tension indicators (DTIs), 518, 592
DPG. *See* Deck plate girder (DPG)
DPT. *See* Dye-penetrant test (DPT)
Drift pins, 590
Dry film thickness (DFT), 529
DTIs. *See* Direct tension indicators (DTIs)
Ductility, 41
  in bridges, 45
  higher-strength steels effect on, 48
  triaxial strains effect, 340*n*
Durability, 91
Dye-penetrant test (DPT), 505*n*, 536
Dynamic freight train, live load, 125–159
  centrifugal forces, 155–158
  design impact load, 144–145
  lateral forces from moving freight equipment, 159
  longitudinal forces due to traction and braking, 145–155
  rocking and vertical dynamic forces, 125–144

E

Eccentrically loaded connections, 471
  axially loaded connections, 480
  axial member splices, 482
  beam framing connections, 475–480
  shear and bending moments, 471
  shear and tension, 471
  shear forces, 471–477
  shear stress, 471
  tensile stress, 471–472
Elastic buckling, 291–296
  buckling force, 291
  centroidal axis load, 291
  critical force, 291
  eccentric axis load, 295
  eccentric to centroidal load, 295
  end conditions, 292, 293
  end restraint effect, 295
  equation, 332
  geometric imperfections, 296
  pure bending, 346–347
  rectangular plate stress, 343
Elastic critical buckling force
  for concentrically loaded members, 292*t*
Elasticity theory, 31
Electrogas welding (EGW), 520, 523
Electroslag welding (ESW), 520, 522–523
End vertical and portal bracing, 250–251
Engesser
  reduced modulus, 297
  tangent modulus, 297
Erection engineering, 566–587
  for cantilever construction, 586
  for cranes and derricks
    and falsework on barges, 582–584
    mobile cranes, 575–579
    stationary derricks, 575
  for falsework, 580–581
  for launching, 585
  for member strength and stability, 567–575
  for SPMT erection, 586–587
  for stationary and movable frames, 584–585
  for tower and cable, and catenary high-line erection, 586
Erection execution, 587–593
  erection fit-up, 589–590
  erection of field splices and connections, 590–592
    bolted field splices and connections, 590–592
    welded field splices and connections, 590
  falsework construction, 588
  field erection completion, 592–593
  by mobile cranes, 588
Erection planning
  erection by flotation with barges, 554–556
  erection on falsework and lateral skidding of superstructures, 551–554
  erection with cranes and derricks, 544–551
    derrick cranes, 544–547
    mobile cranes, 547–551
  erection with stationary and movable frames, 556–559
  for execution of superstructure erection, 540–566
  methods and procedures planning, 541–543
  other methods, 559–566
Euler–Bernoulli superstructure, 128
  moving load vehicle, 137–140
  moving mass vehicle, 134–137
Euler buckling. *See* Elastic buckling

F

Fabricated members, inspection of, 532–533
for shipment, 536
Fabrication
assembly, 589
nondestructive testing, 536–537
QA inspection of, 532–536
QC inspection of, 530–532
tolerances, 498, 514–515, 542
Fabrication planning
fabrication shop production scheduling and cost estimating, 499–500
material procurement for, 500–502
project cost estimating, 498
shop drawings for steel fabrication, 498–499
Factor of safety (FS), 262–263, 298
compression members, 262–263
tensile stresses, 263
Fairbairn, William, 11
Falsework, 540, 551–554
allowable stresses for railway design, 581–582*t*
construction, 588
erection engineering for, 580–581
wind pressures for railway design, 580–581*t*
Fastener holes, 508
Fatigue, 98, 99–100, 268–269, 283, 284
allowable fatigue stress range, 269
constant amplitude *S–N* curves for, 270, 272
constant amplitude stress cycles, 269, 271
design for railway equipment, 118–124, 144*f*
detail categories, 268*n*
limit state, 273
steel, 197
strength, 44*f*
stress range allowable, 283*t*
variable amplitude cycles, 269
FCAW. *See* Flux Cored Arc Welding (FCAW)
FEA software. *See* Finite element analysis (FEA) software
Federal Highway Administration (FHWA), 64
Field bolting
procedures, 591–592
verification and testing, 592
Field fasteners, 509
Field splices and connections, erection of, 590–592
Field welding, 590
Fillet welding, 524
Fillet welds, 429
Finite element analysis (FEA) software, 133, 142, 198, 244, 245–246, 586*n*
Fit-up, 514, 589
Fixed bearing component, 89
disc bearings, 89
elastomeric bearings, 89
fixed hinged bearings, 89
flat steel plates, 89
Flame cutting, 504
Flexural buckling
vertical, 342
of web plate, 343–345
Flexural members, 327
beams and girders, 327, 329, 333, 337, 378
box girder, 360
design of, 327–397
girder elements, 339, 356
plate girder, 339–360
serviceability design, 360–374, 381–397
shear connection, 378
steel and concrete spans, 376
strength design, 327–360, 374–381
web plate shear, 378–379
Floorbeam reaction
Cooper's live load, 244
equivalent loads, 242, 244
influence line, 215, 217
supported spans, 243–244
Flux Cored Arc Welding (FCAW), 101, 426, 427, 522
Force, 669
Forth Rail Bridge, 25*f*, 28
Four-axle locomotive superstructure, 129, 129*f*
Fourier integral transform, 137*n*
Fracture, 91
control plan (FCP), 46
critical members (FCM), 46, 88, 272, 499
limit state, 268*n*
toughness, 50*t*, 51*t*
Fraser River Bridge, 24, 25, 25*f*
Frequency distribution histogram
of stress ranges, 121
Froude number (*F*), 61

G

Garabit Viaduct, 18, 20, 20*f*
Gas Metal Arc Welding (GMAW), 522
Gaunless River Bridge, 2
Gerber type bridge, 24*n*
Girder(s), 509, 513, 590
bearing stiffeners, 353
box girder, 339, 360
bridge, 339
erection, 567–568
flange-to-web plate connection, 352
girder elements. *See* Stiffeners
hybrid, 339
modern plate girder, 339
plate elements, 339
plate girder, 339
preliminary design, 361–374
proportioning of, 337
steel box, 360
steel–concrete composite box, 360
tension flanges, 341
tension flange splices, 341–342
web plates, 346
web plates splices, 351–352
Girder compression flanges, 342
design requirements, 342
elastic buckling stress, 343, 346
lateral–torsional buckling, 342–343
plate buckling curve, 345
splices, 346
torsional buckling, 345–346
vertical flexural buckling, 343–345
width-to-thickness ratio, 345
Girder flange-to-web plate connection, 352
allowable weld stresses, 353

Girder flange-to-web plate connection, (*cont.*)
bottom flange-to-web connection, 353
forces transferred, 352
top flange-to-web connection, 352
Girder flange-to-web "T" joints, 445
allowable weld stresses, 446
horizontal shear flow, (q_f), 445
shear force (w), 446
Girder web plates, 346–351
circular interaction formula, 351
combined bending and shear, 350–351
elastic buckling, 347
flexural buckling, 347
inelastic buckling, 346
postbuckling shear strength of web, 349
splices, 351–352
stresses on, 347
ultimate shear buckling strength of web, 349–350
GMAW. *See* Gas Metal Arc Welding (GMAW)
Groove welds, 428–429, 514, 515
in butt welds, 433
CJP, 428–429, 445
Guardrail height, 91
Gusset plates, 461
PJP, 428

H

Harmonically varying equation, 140
Harmony, 83
Heat-affected zones (HAZ), 41, 45
Heat-treated low-alloy steels, 41, 48
Heavy load transporters, erection by, 565–566
Hell Gate bridge, 28, 30, 30*f*
High-performance steel (HPS), 39
hybrid applications, 49
hybrid girders, 339
plates, 48–49
welding, 49
High-strength low-alloy (HSLA), 39, 41
hybrid applications, 49
hybrid girders, 339
steel composition, 47
Holding cranes, 551
Hole location tolerances, 517*n*
Holes, 508
Hooke's Law, 327
differential equation of motion, 131*n*
in elastic buckling, 291, 295, 296
Horizontal geometry of bridge
bridge geometrics, 74–77
route (track) geometrics, 71–74
skewed bridges, 80–82
Horizontal truss bracing, 88, 246
Howe truss, 8–10
in bridge, in, 8
in collapsed railroad, 20
iron & wood, 10
wrought iron, 8
HPS. *See* High-performance steel (HPS)
HSLA. *See* High-strength low-alloy (HSLA)
Hydrogen diffusion, 508

I

Inclined chord truss. *See* Petit truss
Inelastic buckling, 346. *See also* Elastic buckling
under pure shear, 348
secant formula for, 296
Inertia, 135, 472
Influence lines, 213–234, 227–230
bending moment, 214, 215, 216–218
cantilever bridge span, 232–234
floorbeam reaction, 215, 217
maximum axial forces, 218–227
maximum effects, 227–229, 232
shear force, 213–215
use of, 244
Interaction curves, 421
Interaction dynamics
train (vehicle)–superstructure (bridge), 128–134
Interaction equations, 420
moment–curvature–axial compression, 420–423
relating to stability criterion, 423
transverse load, 416, 417
Interaction formula, 423
simply supported plate, 351
stability criterion, 423
Intermediate vertical and sway bracing, 255
Interruptions, to railway traffic, 539*n*
Intrusive falsework, 551, 552*f*
Iron railway bridges, 2–22
cast iron arch, 3–4*t*
cast iron trestles, 5–7*t*

J

Johnson parabola, 299
Joint design, bolt, 101, 448
Joint types, 447–448
lap joints, 448
live load impact, 105, 106–164
minimum bolt pretension, 448, 451
multirotational bearings, 89, 90
open deck bridges, 85
pedestrian walkways, 91–92
pretensioned bolt, 446
principal loads, 105
proportion and scale, 83
railway track on bridge decks, 84–85
redundancy, 88
rocking effects, 125–127
seismic forces, 187
snug-tight bolt, 446
span types, 83
stability, 90
steel bolts, 447–448
stringers, 87–88
substructures, 83
tensile strength, 448
tension forces and elongation, 447
"T" joints, 448
typical bolted joints, 449
typical connection, 464
vertical dynamic amplification, 125, 140
vertical effects, 128
Whitmore effective length, (w_e), 461

Whitmore stress block, 465
wind forces, 164–170

K

Kinzua Viaduct, 18, 20, 20*f*
Knee bracing in through spans, 256

L

Lacing bars
for compression members, 309–311
length, 289*n*
thickness, 311*t*
Lamellar tearing, 340*n*, 426, 528
in thick flange plates, 529
Lateral bracing systems, 246–256
bottom, 247–250
bracing design force, 258–259
end vertical and portal bracing, 250–255
horizontal truss bracing, 246
intermediate vertical brace, 255, 256
knee bracing, 256–258
sway bracing, 255
top lateral systems, 247
Lateral forces, 105, 159
bracing systems, 246
differential equation, 159
factors causing, 125, 159
freight equipment, 159
Lateral–torsional buckling, 342–343
effects, 567
flexural compression curve, 332
magnitude, 159
moment, 330
resisting flange deformations, 255
stress, 331
Lattice truss, 17
Launching, erection by, 559–564
engineering for, 585
Lehigh Valley Railroad, 17
Lenticular
spans, 2
truss, 11
Limiting stress, 92
Limit state design/load and resistance factor design (LSD/LRFD) calibration, 91
to allowable stress design, 96–98
using reliability methods, 98–99
Linville, J. H., 8*n*
Live load, 106, 116–117. *See also* Cooper's (E80)
bending moment, 198, 244
equivalent loads, 233
fatigue design, 118–124
freight train, 125
influence lines, 213
shear force, 198
simple, 83
static freight train, 106–124
structural analysis, 245
Local scour, 67
Locomotive cranes, 547, 549–551, 550–551*f*
Locomotive weights, 28*n*
Log–log linear relationship, 121*n*
Lognormal probability density, 95–96
Longitudinal beams, shop assembly of, 509
Longitudinal forces, 170–171
due to traction and braking, 145–155
Longitudinal restraint, 170
Long plate girder, 503
Long-Span Railway Bridges (book), 16

M

Machine bolts. *See* Common steel bolts
Magnetic particle test (MPT), 101, 505*n*, 530, 531*f*, 536
Manganese, 39
Manning's roughness coefficients, 62
Manual for Railway Engineering (MRE), 34
Mass, 669
Material failure, 44
Mild carbon steel composition, 47
Mobile cranes, 544, 556
all-terrain cranes, 549, 550*f*
on barges, 582
crawler cranes, 547, 548*f*
erection by, 589*f*
erection engineering for, 575–579
holding cranes, 551
locomotive cranes, 547, 549–551, 550–551*f*
specialty cranes, 551
truck-mounted cranes, 548–549, 549*f*
Modal superposition, 133
Modern railway bridges
bridge crossing economics, 56–57
bridge esthetics, 82–83
bridge stability, 90
detailed design of the superstructure, 102
erection considerations, 102
fabrication considerations, 100–101
general design criteria, 91–100
geometry of track and bridge, 70–82
pedestrian walkways, 90–91
planning of railway bridges, 56–82
preliminary design, 82–102
railroad operating requirements, 57–58
site conditions (public and technical requirements of bridge crossings), 58–70
steel railway bridge, 39–54, 83–90
carbon steels, 47
heat-treated low-alloy steels, 48
high-performance steels, 48–49
high-strength low-alloy steels, 47–48
material properties, 49–50
structural steel. *See* Structural steel
structural analysis for modern design, 91
structural design for modern fabrication, 92–100
Modern steel, 39
alloying elements, 39–40
corrosion resistance, 46–47
ductility, 45
fasteners in, 447
fracture toughness, 45–46, 50–51
making processes, 340
physical properties, 39, 40
railway bridges, 84. *See also* Railway superstructures

Modern steel, (*cont.*)
strength, 42–44
structural analysis for, 91
weldability, 46
Moment–rotation curves, 438
beam end connections, 439
bolted joint configurations, 475
Movable frames
erection engineering for, 584–585
erection with, 557, 559
Moving continuous load, 134, 136, 140
Moving load analysis, 246*n*
MPT. *See* Magnetic particle test (MPT)
MRE. *See* Manual for Railway Engineering (MRE)
Muller-Breslau principle, 245*n*

N

NCRs. *See* Nonconformance reports (NCRs)
NDT. *See* Nondestructive testing (NDT)
Newmark method, 133
Niagara Gorge suspension bridge, 16, 17, 18*f*
No-load fit-up, 589
Noncomposite flexural members, 327
beams and girders, 327, 329, 333, 337
box girders, 360
girder elements, 340, 356
plate girder, 337
serviceability design of, 360–374
Nonconformance reports (NCRs), 533
Nondestructive testing (NDT), 101, 530, 532
for QC/QA inspection of welded fabrication
dye-penetrant testing, 536
magnetic particle testing, 536
phased array ultrasonic testing, 537
radiographic testing, 537
ultrasonic testing, 536
Nonintrusive emergency falsework, 552, 554, 555
Normalizing, 41
Normal probability density, 93–95
North America
structural steel used in, 52*t*
"Nosing". *See* Wheel-to-rail interface
"Notional" forces, 258
NUCARS software, 133*n*

O

Obstructed discharge hydraulics, 62–64
minimum channel area (A), 64
normal depth Froude number (F_d), 64
river crossing profile with obstructions, 63
Yarnell equation, 64
Octahedral shear stress, 43
Open deck bridges, 85, 170
distribution of live load for, 159–160
DPG span, 86
TPG span, 86
Ordinary bolts. *See* Common steel bolts

P

Pacific railroad, 1*n*
Palmgren-Miner rule, 121
Paris-Erdogan power law, 121
Partial differential equation of motion, 131
Partial joint penetration (PJP), 101, 428, 523–524
Pedestrian loads, on steel railway bridges, 192–193
Pedestrian walkways, 90–91
Pennsylvania Railroad, 17
Perforated cover plates, for compression members, 313–314
Perry–Robertson formula, 296
Petit truss, 18*n*
Phased array ultrasonic testing, 537
Pins, 425*n*
PJP. *See* Partial joint penetration (PJP)
Plasma cutting, 502
Plate girder design, 339–360
bearing stiffeners, 353–356
bearing stresses, 355
compression member behavior of, 354–355
local plate buckling, 356
compression flanges and splices, 342–346
flange-to-web plate connection, 352–353
longitudinal web plate stiffeners, 356–357
main girder elements, 340
secondary girder elements, 356–360
tension flanges and splices, 341–342
transverse web plate stiffeners, 357–360
web plates and splices, 346–352
Plates and shapes
bolting of, 518–520
coating of steel, 529–530
cutting of, 502–504
heat treatment, 506–508
layout and marking of, 502
punching and drilling of, 508–509
QC and QA of fabrication, 530–537
shop assembly for fit-up of, 509–514
steel fabrication processes, 502–508
straightening, bending, curving, and cambering of, 505
surface preparation, 505–506
welding of, 520–528
Poisson's ratio (υ), 305
Polytetrafluoroethylene (PTFE), 89, 554*n*
Portageville Viaduct, 19, 19*f*
Pratt truss, 10, 17–18, 218–227
center counter, with, 9
examples of, 10
Pretensioned (PT) bolt
connections, 518, 519
installation, 446
Probabilistic structural design, 92–93
Probability distribution, 94
Progressive shop assembly, of longitudinal beams, girders, and trusses, 512
Propriety trusses, 8
Bollman truss, 10
Howe truss, 8
Pratt truss, 10
Warren trusses, 9
Whipple truss, 9
Pseudo-acceleration, 188
PTFE. *See* Polytetrafluoroethylene (PTFE)
Pure and warping effects, 333
Pure torsional moment resistance, 401

Q

QAI. *See* QA inspector (QAI)
QA inspection of fabrication
- assembly inspection, 533
- bolting inspection, 534
- coatings inspection, 535
- final inspection for shipment, 536
- inspection of fabricated members, 532–533
- inspection of raw materials, 532
- nondestructive testing (NDT) for, 536–537
- shop drawing reviews, 532
- welding inspection, 534–535

QA inspector (QAI), 532
QC inspection of fabrication, 530–532
- nondestructive testing (NDT) for, 536–537

Quebec Bridge, 29*f*
- cantilever-type bridge, 29
- collapse, 29
- completion, 29
- construction, 28

Quenching and tempering (Q&T) steels, 41

R

Radiographic testing (RT), 101, 530, 531*f*, 537
Railroad operating requirements, 57–58
Rail separation, 172–173
- bridge crossing discharge, 59

Railway bridge engineering
- design specifications, 32–34
- development of, 31–36
- modern steel design, 34–36
- strength of materials and structural mechanics, 31–32

Railway dead load(s), 106
- superimposed, 105
- superstructure, 105

Railway live load, 106–164
- analytical work, 150
- bearing forces, 151
- distribution of
  - for ballasted deck steel bridges, 160–162
  - for direct fixation deck steel bridges, 162–164
  - for open deck steel bridges, 159–160
- dynamic freight train
  - centrifugal forces, 155–158
  - design impact load, 144–145
  - lateral forces from moving freight equipment, 159
  - longitudinal forces due to traction and braking, 145–155
  - rocking and vertical dynamic forces, 125–144
- equilibrium equation, 148
- extensive testing, 150
- force equilibrium, 147
- horizontal effects, 146, 147
- horizontal reaction (*P*), 147
- lateral bracing, 154–155
- longitudinal forces, 145–146, 145–155
- rotational effects, 146, 147
- static freight train, 106–124
  - Cooper's design for projected railway equipment, 115–118
  - fatigue design for railway equipment, 118–124
- time history, 148
- traction bracing, 154–155
- vertical effects, 146–147

Railway live load distribution, 159–164
- ballasted deck bridges, 160–161
- beam spacing (*D*), 160
- direct fixation deck steel bridges, 162–164
- lateral distribution, 160, 161
- longitudinal distribution, 159, 160
- open deck bridges, 159–160

Railway superstructures. *See also* Superstructures, steel
- bolted connections, 446–495
- design, 34
- redundancy, 88
- welded connections, 425–446, 426

Raw materials, inspection of, 532
RCSC. *See* Research Council on Structural Connections (RCSC)
RE. *See* Rocking effect (RE)
Reliability index, 95
Requests for information (RFI), 497–498
Research Council on Structural Connections (RCSC), 505–506
Residual stresses, 341, 426, 501, 504, 526–528
- in flat plates, 527–528

RFI. *See* Requests for information (RFI)
River crossing profile, 59*f*, 60*f*
Rivets, 28*n*, 425*n*
RMC. *See* Root mean cube (RMC)
Rocking and vertical dynamic forces, 125–144
- dynamic load effect (LE_D), 125
- fundamental frequency (ω_1), 125
- impact factor (I_F), 125

Rocking effect (RE), 125–127
Roebling, John A., 16
Rolling mass, 146*n*
Root mean cube (RMC), 268
Rotational capacity (RC) testing, 592
Rough bolts. *See* Common steel bolts
Route (track) geometrics, 71–74
Royal Albert Bridge, 11, 11*f*
RT. *See* Radiographic testing (RT)
Runge–Kutta method, 133

S

SAW. *See* Submerged Arc Welding (SAW)
Secant formula, 296, 420
- axial compression and bending, 420
- eccentric load, 293, 296
- inelastic buckling, 296

Second axle, 200*n*
Section modulus, 472
Segregation, element, 40–41
Seismic forces, 187
- on steel railway bridges
  - equivalent static lateral force, 187–188
  - response spectrum analysis of steel railway superstructures, 188–190

Self-propelled modular transporter (SPMT) vehicles, 543*n*, 565–566
- erection engineering, 586–587

Self-weight deflections, 510*n*
Serviceability, 91, 99–100

Serviceability design
constant amplitude cycles, 268, 269–271
deflection criteria, 264–267
fatigue design criteria, 268–273
fatigue loading, 268
fatigue strength, 268–273
lateral deflections, 267
maximum flexural deflection, 264–266
truss deflections, 266
Shear force, 32, 75, 110, 199–212, 304, 382, 446
axial, 425
bolted connection, 471–472
compression member, 304, 308
on connections, 436, 471
criteria for, 200, 202
curvature effects, 75
eccentric, 425, 475
equivalent loads, 234, 235, 236
fillet welds, 437
influence lines, 213, 227–229, 233, 234
lag effects, 280
live load, 388
longitudinal, 289
moving load, 197
supported spans, 234–237, 241, 242, 245
transverse, 247, 258, 289–290
Shear force, maximum, 199–201, 241–243
absolute end shear, 199
change in, 200, 201
criteria for, 199, 201
equivalent loads, 234, 236, 237, 238
loads on arch, 230
supported spans, 234–237, 241–243
Shearing, 333–336, 378
of beams and girders, 333–336
design for shapes and plate girders, 336
of I-shaped sections, 335–336
postbuckling strength of web, 349
of rectangular beams, 333–334, 333–335
shear connection, 379–381
shear flow, 333
shear flow distribution, 379
shear resistance distribution, 381
shear stresses, 333–334
ultimate buckling strength of web, 349–350
web plate shear, 378
Shear stress, 449, 451, 452, 507
allowable tensile stress, (f_{bt}), 455
in bearing-type connections, 448–450
bending moment (M_b), 456
and elongation, 450
prying action, 455
prying tensile force (T_Q), 456
slip coefficients, 452
in slip-resistant connections, 450–453, 465–471
tension forces and elongation, 455
Shielded metal arc welding (SMAW), 426, 427, 522
Shop assembly, 590
Shop assembly, for fit-up of steel plates and shapes
of bolted splices and connections, 512–514
fabrication and erection tolerances, 514–518
fabrication of cambered superstructure assemblies, 509–511
fit-up for shop welded splices and connections, 514
of longitudinal beams, girders, and trusses, 511–512
Shop drawing reviews, by QA inspectors, 532
Shop welding
effects on plates and shapes
welding flaws, 524–525
welding-induced cracking, 525–526
welding-induced distortion, 526
welding-induced lamellar tearing, 528–529
welding-induced residual stresses, 526–528
procedures, 523–524
processes, 520–523
Short compression member buckling coefficient, 302*t*
Simple truss span railway bridges, 12–14*t*
Single-axle locomotive, 132*f*
Single-span movable frame, 562–564
Six-axle locomotive superstructure, 129, 129*f*
Skewed bridges, 80–82
Slip-critical (SC) fasteners, 499, 518
Slip-critical joints, 446–447
SMAW. *See* Shielded metal arc welding (SMAW)
S–N curves, 270, 272
Snug-tight bolt, 446, 518
Solidification cracking, 525
Solidity ratio (*f*), 166
for spans, 166
wind force, 167
Span railway bridges, 16–17
Spans, 374
strain distribution, 377
strain profile, 377
strength design, 374
stress profile, 379
web plate shear, 378
Span types, 154, 264
amplitude stress cycles, 124
Cooper's E80 load, 106–107
deflection criteria based on, 264
Specialty cranes, 551
Splice assembly, 513
Splices and bolted connections, 512–514
SPMT vehicles. *See* Self-propelled modular transporter (SPMT) vehicles
Sprung mass, 130, 133
SSPC specification. *See* Steel Structures Painting Council (SSPC) specification
Static bending, 110
Static freight train, live load, 106–124
Cooper's design for projected railway equipment, 115–118
fatigue design for railway equipment, 118–124
Stationary derricks, 575
Stationary frames
erection engineering for, 584–585
erection with, 556–559
Stay plates
for built-up compression members, 311
thickness, 311*t*
Steam locomotive geometry, 107
Steel bolts, types, 447–448
Steel box girders, 360
Steel bridge design. *See* Bridge, steel
Steel bridge failure, 268–269
fatigue damage, 261
fracture, 261

number of cycles, 261
serviceability failure, 261
strength failure, 261
yield failure stress, 261
Steel ductility, 45
Steel fabrication
material preparation, 502–508
punching and drilling of plates and shapes, 508–509
shop assembly for fit-up of steel plates and shapes, 509–514
shop drawings for, 498–499
Steel fracture toughness, 45–46
brittle fracture, 45
CVN test, 46
FCP, 46
requirements, 50–51
Steel members connections, 425
axially loaded members, 459
bolted connections, 446
bolted joint design, 449
bolting processes, 446
bolt stresses, 448
bolt types, 447
direct axial loads on connections, 432–434
eccentrically loaded connections, 436, 471
fillet welds, 429
flux cored welding, 427
groove welds, 428–429
high-strength steel bolts, 448
joint types, 429–430, 448
pretensioned bolt, 446
shielded metal welding, 426
snug-tight bolt, 446
steel bolts, 448
Steel mills, 500
tolerances, 501, 514
Steel plates and shapes. *See* Plates and shapes
Steel railway bridges
dead loads, 105–106, 106*t*
environmental and other design forces, 164–193
loads relating to overall stability of the superstructure, 190–192
pedestrian loads, 192–193
seismic forces on, 187–190
thermal forces from continuous welded rail on, 170–186
wind forces on, 164–170
history and development of, 1–36
loads and forces on, 105–194
other design criteria for
camber, 274
minimum thickness of material, 274
secondary stresses in truss members and girders, 273–274
web members in trusses, 274
railway live loads, 106–164
distribution of live load, 159–164
dynamic freight train live load, 125–159
static freight train live load, 106–124
structural analysis and design of, 197–274. *See also* Superstructures, steel
Steel railway bridge construction
superstructure erection, 539–593
erection engineering, 566–587
erection execution, 587–593
erection planning, 540–566
introduction, 539–540
superstructure fabrication
fabrication planning, 498–502
introduction, 497–498
plates and shapes, bolting and welding of, 518–537
steel fabrication processes, 502–518
Steel strength, 42
elastic yield, 42–43
fatigue, 44
octahedral shear stress, 43
tensile stress–strain behavior, 42
von Mises criterion, 43
yield stress, 43
Steel Structures Painting Council (SSPC) specification, 505, 506
Steel superstructures
design for, 91
fabrication for, 92–100
macroscopic fatigue strength of, 44
recommendations to, 99–100
Steel weldability, 46
carbon content and, 49
equivalency equation, 46
strength and, 44, 47
weld cracking, 46
Steinman's charts, 241, 244
Stephenson, Robert, 11*n*
Stiffeners, 339, 356, 359
bearing, 355
bending moment, 245
design, 261
longitudinal web plate, 356–357
transverse web plate, 357–360
Stiffleg derricks, 544–545
St. Lawrence Bridge, 30, 30*f*
St. Louis Bridge, 23, 23*f*
Stress, combined, 277, 399
on bridges, 399
critical sections review, 464
determination, 403, 404–413
members subjected to, 273–274
for steel superstructures, 273–274
truss end posts, 274
truss hangers, 274
truss members, 273
Stress, relieving, 41
Stress–strain curve(s), 42
compressive, 298
engineering tensile, 42
for structural steel, 291
Structural steel
bridges, 49
corrosion resistance, 46–47
disadvantage of HSLA, 48
ductility, 45
elastic yield strength of steel, 42–43
engineering properties of steel, 42–47
fatigue strength of steel, 43–44
fracture resistance, 45–46
heat-treated low-alloy steels, 48
HPS plates, 48–49
HSLA steels, 47–48

Structural steel (*cont.*)
manufacture of, 39–42
members, 399
mild carbon steel, 47
for modern North American railway superstructures, 50–54
for modern railway bridges, 53*t*
for railway superstructures, 41, 49–54
strength, 42–44
stress–strain curve, 291, 298
types of, 47–49
used in North America, 52*t*
weldability, 46
Stud welding, 427
St. Venant moment resistance. *See* Pure torsional moment resistance
Submerged Arc Welding (SAW), 49, 426, 427, 520–522
flange-to-web welding, 521
Superstructure eccentric loads, 436
beam framing connections, 438–439, 440–445
bending moment (M_{wa}), 439
bending stress, (f_{wa}), 439
connection angles deformation (Δ), 440
moment–rotation curves, 438, 439
rigid connection, 438
semirigid connection, 438
shear forces and bending moments, 436–437
shear forces and torsional moments, 437
tensile force, (T_P), 439
Superstructures, steel, 83–90, 116, 529–530. *See also* Bridge, steel
actual response spectrum, 188
AREMA design response spectrum, 188
ballasted bridge decks, 85–86
base metal weldability, 426
bridge bearings, 88–90
bridge decks for steel railway bridges, 84–86
bridge framing details, 87–88
CJP groove welds size, 428
derailment load, 190–192
design response spectrum, 188
direct axial loads, 432–434, 436
direct fixation decks, 86
dynamic analyses, 188
eccentric loads, 436
Euler–Bernoulli, 134–140
FCAW process, 427
fillet welds, 425–430
fillet welds size, 429
girder flange-to-web "T" joints, 445
groove welds, 428
joint types. *See* Superstructure steel, joints
lamellar tearing, 426
live load analysis of, 192–193, 197–246
bottom lateral systems, 247–249
end vertical and portal bracing, 250–251
horizontal truss bracing, 246
influence lines for effects, 213–234
intermediate vertical and sway bracing, 255
knee bracing in through spans, 256
lateral bracing systems, 246–256
lateral load analysis of, 246–256
maximum shear force and bending moment, 199–212, 234–245
modern structural analysis, 245–246
top lateral systems, 247
load combinations, 193–194
minimum effective length, 429
natural frequency of, 140–144
open-bridge decks, 85
pedestrian loads, 192–193
railway track on bridge decks, 84–85
residual stresses, 426
response spectrum analysis, 188–189
SAW process, 427
SMAW process, 426
structural analysis of railway, 91, 197–256
structural design, 260–274
failure modes, 261–262
for modern fabrication, 92–100
serviceability design, 264–273
strength design, 262–264
stud welding, 426, 427
Superstructure steel bolting, 446
axially loaded members, 459
bolting processes, 446
bolt types, 448
gusset plates, 461
joint design, 448
joint types, 448
Superstructure steel, joints, 429–430
allowable weld stresses, 430–431
butt joints, 429
corner joints, 429, 430
lap joints, 429, 430
stress concentrations, 431
"T" joints, 429, 430
weld line properties, 431–432, 433
Suspension systems, 128–130
Sway bracing
in deck spans, 255
intermediate vertical, 255
Système Internationale (SI) units of measure, 669–671

T

TAC. *See* Transportation Association of Canada (TAC)
Tandem cranes, 589*f*
Tay River Bridge, 21, 22*f*
after collapse, 22
failure, 22
TC bolts. *See* Tension-control (TC) bolts
Tensile yield stress, 42*f*, 277
Tension-control (TC) bolts, 518
Tension stress, 455
Tension zones, 268
Thermal cutting, 502–503
Thermal forces, 170–186
on steel railway bridges, 170–186
acceptable relative displacement between rail-to-deck and deck-to-span, 175–186
design for CWR, 186
safe rail separation criteria, 172–173
safe stress in CWR to preclude buckling, 173–175
Thermo-mechanical controlled processing (TMCP), 42, 49, 500
Through plate girder (TPG) span, 86, 200*n*, 360
Through truss (TT) spans, 200*n*

TMCP. *See* Thermo-mechanical controlled processing (TMCP)
Tolerances, 498, 501, 514–515, 516
TON method. *See* Turn-of-nut (TON) method
Top lateral systems, 247
Torsional buckling, 345–346
Torsional moment, 80, 264
 of inertia, 330
 pure torsion, 401
 torsional moment resistance (T), 401
 torsion differential equation, 401
 warping torsion, 401
Torsional warping constants, 330*t*
Total scour, 67
Tower and cable, erection by, 564
 engineering for, 586
TPG span. *See* Through plate girder (TPG) span
Track and bridge geometrics, 74–77
 bending moment, 75
 bridge geometrics, 74–77
 centrifugal force, 71
 consequences, 71
 horizontal eccentricities, 75
 horizontal geometry, 71–82
 horizontally curved bridges, 74
 route geometrics, 71–74
 shear force, 75
 shift effect eccentricity, (e_s), 75
 simple curve, 70, 71
 skewed bridges, 80–81
 superelevation (e), 71, 74–75
 track curvature, (e_c), 74, 75
 transition curves (L_s), 74
 vertical geometry, 82
Track eccentricity, 255*n*
Traction forces, 146
Transition curve(s), 74, 295
 in elastic buckling, 295
 length of, 74
 parabolic, 333
Transportation Association of Canada (TAC), 60
Transverse stiffeners, 349, 357–360, 514
Treatise on Bridge Building, A (book), 10*n*
Trestles, 5–7*t*
Trestle-type falsework, 551–552
"Truck hunting". *See* Wheel-to-rail interface
Truck-mounted cranes, 548–549, 549*f*, 578
Truck spacing, 106
Truss, 200*n*
 analysis development, 32
 Baltimore truss, 18
 composite, 8
 continuous trusses, 17
 field splices, 590
 gusset plate connection, 463
 hangers, 274
 Howe truss, 8–10
 lattice truss, 17
 lenticular truss, 11
 Petit truss, 18
 propriety, 8, 9
Truss members, 273, 277, 511
 on combined stresses, 273–274
 perforated cover plates, with, 266
 secondary stresses in, 273–274
 solutions for, 294
TT spans. *See* Through truss (TT) spans
Turn-of-nut (TON) method, 591
Twist-off-type tension-control (TC) bolts, 518, 592
Two-axle locomotive superstructure, 130, 130*f*, 132*f*
Two-span movable frame, 561

U

Ultrasonic testing (UT), 101, 530, 531*f*, 536–537
Unfinished bolts. *See* Common steel bolts
Uniaxial bending
 and axial compression, 414–423
 and axial tension, 413–414
Units of measurement, 669–671
Unsprung masses, 131, 133
Unsymmetrical bending, 336, 399, 400–413, 401
 angle of twist, 403
 equivalent static system, 403
 torsion, 401
US Customary or Imperial systems of units, 669–670
US Geological Survey (USGS), 61
UT. *See* Ultrasonic testing (UT)

V

Variable-amplitude cycles, 119, 269, 273
Vehicle (train)–bridge (superstructure) interaction (VBI), 128
Vehicle–track–deck–superstructure characteristics, 125
Vertical bracing, 88
Vertical flexural buckling, 343–345
Vertical flexural deflections (beams and girders), 264–266
Vertical geometry of bridge, 82
Vertical truss deflections, 266
Victoria Bridge, 15, 16, 16*f*, 31
von-Mises yield criteria, 261

W

Warren trusses, 9
 development of, 32
 horizontal, 246
 in railway bridge, 32
Waterproofing coatings, 529
Web plate stiffeners
 longitudinal, 356–357
 transverse, 357–360
Welded connections, 426
 allowable weld stresses, 430
 direct axial loads, 432–434
 eccentrically loaded, 436
 fillet welds, 429
 flux cored welding, 427
 groove welds, 428
 joint types, 429–430
 processes, 426
 shielded metal welding, 426
 stud welding, 427
 submerged welding, 427
 welded joint design, 430–446
 welding electrodes, 427
 welding processes for steel railway bridges, 427

Welded connections, (*cont.*)
weld line properties, 431–432
welds fatigue strength, 431
weld types, 428–429
Welded field splices and connections, erection of, 590
Welded joint design, 430
allowable weld stresses, 430
butt and "T" joints, 426
direct axial loads, 432
eccentrically loaded, 436
fatigue strength of welds, 431
lamellar tearing, 426
residual stress prevention, 426
stud welding, 427
submerged welding, 427
tension and uniaxial bending, 413–414
web plate, 351
welded connections, 426
welded joint design, 430
weld fatigue strength, 431
Welding
of plates and shapes, 520–528. *See also* Shop welding
welding electrodes, 427
welding inspection, 427, 534–535
welding processes, 427
weld line properties, 431–432
weld stresses, 430
weld types, 428
Welding procedure specifications (WPS), 499, 523*n*, 524
Welding processes, 427
fillet welds, 429
flux cored arc welding, 427
groove welds, 428
plug welds, 428
shielded metal arc welding, 427
slot welds, 428
stud welding, 427
submerged arc welding, 427
welding electrodes, 427
Weld shrinkage, 526
Wheel–rail contact forces, 131
Wheel-to-rail interface, 105
Whipple truss, 9
in railroad, in, 18
in steel railway bridge, 24
Wind
flow characteristics, 166*n*
forces, 164–170
gusts, 165
loads, 543
WPS. *See* Welding procedure specifications (WPS)
Wrought iron
acceptance in railways, 8*n*
ductility, 8
Fairbairn study, 15*n*
girders, 10–11
lenticular trusses, 11
Pratt truss bridges, 10
proponent, 8*n*
in railroad trusses, 8
tension members, 8, 10, 18*n*
tubular, 11
vertical members, 10
Wrought iron construction, 8
Ashtabula Bridge, 20
Baltimore Truss, 18
Britannia Bridge, 11, 15
cantilever construction method, 22
early, 3–4
end of, 22*n*
Garabit Viaduct, 18, 20
Howe truss, 8–10
Kinzua Viaduct, 18, 20
Niagara Gorge suspension bridge, 16, 18
Petit truss, 18*n*
pin-connected construction, 17
Portageville Viaduct, 19
Pratt truss, 17–18
railway suspension bridge, 16, 17
riveted connections, 8*n*
Royal Albert Bridge, 11
span railway bridges, 16–17
Tay River Bridge, 22
Tay River Bridge failure, 22
truss forms, 9
viaduct bridges, 18
Victoria Bridge, 15, 16
wrought iron truss railway bridges, 12–14
Wrought iron truss
lattice, 17
proponent of, 8*n*

Y

Yarnell equation, 64
Yarnell's pier shape coefficient, 64